STATISTICAL MECHANICS
AND THERMODYNAMICS

Statistical Mechanics and Thermodynamics

Claude Garrod
University of California, Davis

New York   Oxford
OXFORD UNIVERSITY PRESS
1995

Oxford University Press

Oxford   New York   Toronto
Delhi   Bombay   Calcutta   Madras   Karachi
Kuala Lumpur   Singapore   Hong Kong   Tokyo
Nairobi   Dar es Salaam   Cape Town
Melbourne   Auckland   Madrid

and associated companies in
Berlin   Ibadan

Library of Congress Cataloging-in-Publication Data
Garrod, Claude.
Statistical mechanics and thermodynamics / Claude Garrod.
p. cm.
Includes bibliographical references and index.
ISBN 0-19-508523-X (IBM software)
ISBN 0-19-509775-0 (Macintosh software)
1. Statistical mechanics. 2. Thermodynamics. I. Title.
QC174.8.G37 1995 530.1′3—dc20 94-2754

9 8 7 6 5 4 3 2 1

Printed in the United States of America
on acid-free paper

This book is dedicated to my sweetheart
Megan Kimberley Crawford (b. 10.27.93)

Preface

Thirty years ago statistical mechanics was becoming a somewhat peripheral topic in the physics curriculum. It was concerned mainly with the properties of gases and liquids—not very fashionable topics, then or now. Since that time there has been a striking change in the centrality and importance of the subject. The basic conceptual elements of statistical mechanics, such as phase transitions, spontaneous symmetry breaking, and renormalization theory, are now the working tools of both solid-state and high-energy physicists. The Monte Carlo method is used to explore the properties of both the quark-gluon plasma and the Hubbard model of high-T_c superconductors. This book is an attempt to bring some of the richness and variety of the field within the range of an introductory course in the subject.

The book treats statistical mechanics as an application of probability theory to systems of extremely large numbers of coordinates. This point of view is fairly traditional and noncontroversial. However, the concept of the thermodynamic limit is formulated more explicitly and used more frequently than is common in books at this relatively low level of mathematical sophistication. (Here at Davis the book has been used in manuscript form, during the last five years, for both the undergraduate and graduate introductory courses in statistical mechanics and thermodynamics.) I have tried to put a very consistent emphasis upon the fact that the driving principle determining the equilibrium properties of macroscopic systems is the progression of the system to the macrostate of maximum probability. This is particularly evident in the treatment of the foundations of thermodynamics, given in Chapter 5, which is based on the principle of maximum entropy at fixed energy, rather than the principle of minimum energy at fixed entropy. Although the two principles have equal mathematical validity, it is the author's opinion that the first leads to a clearer understanding of the underlying physics of thermodynamic phenomena.

The book has been structured for maximum flexibility. Although it contains more than ample material for a full year course, it can also conveniently be used for a single-semester course. Almost half of the book is devoted to extensive supplements to each of its ten chapters. The supplements serve two purposes. First, they provide a large library of worked-out applications of the theory presented in the regular chapters. Second, in the guise of exercises, they supply a variety of sometimes major extensions of the material presented in the chapters. For example, real-space renormalization is merely mentioned toward the end of Chapter 9, but in the Supplement to Chapter 9 it is presented in some detail in the form of a series of computational exercises. This structure allows the instuctor, either to include the subject of real-space renormalization, if he or she has the time and inclination to do so, without having to supply supplemental notes on the topic, or to leave it out without having to jump around within the chapter and worry that the omitted material will be essential for later topics.

The book comes with disks containing two types of computer programs. The first type is a set of graphics-oriented display programs that allow the user to see a Monte Carlo calculation as it develops. (Either IBM-pc or Macintosh versions are available.) These programs are intended to assist the student in developing physical intuition regarding the equilibrium states of systems. They also illustrate, quite graphically, some of the problems that Monte Carlo calculations encounter in trying to reach true equilibrium. The second type is a set of Fortran programs that apply the Monte Carlo method to a variety of physical systems. The Fortran programs are intended to be used, and often modified, by the student in a number of computationally oriented problems that are included in Chapters 8 and 9.

Many people have made informal cotributions to the book during personal conversations with the author, but two people have been particularly helpful. They are Rajiv R. P. Singh and James P. Hurley, both of UC, Davis, to whom I would like to express my sincere appreciation.

Table of Contents

STATISTICAL MECHANICS
AND THERMODYNAMICS

Chapter 1
The Theory of Probability

1.1 STATISTICAL MECHANICS AND PROBABILITY

Statistical mechanics is the theoretical bridge between the microscopic world and the macroscopic world. The microscopic world is a quantum mechanical world, in which electrons, protons, and neutrons organize themselves into atoms, molecules, crystals, and other structures on an atomic length scale. The macroscopic world is the tangible world of trees and rocks and water. The microscopic structure of these things is usually completely hidden from our view. It is only by a complex chain of reasoning that the fundamental laws of physics, which govern the structure and interactions of the unseen microscopic constituents of things, can be shown to give rise to the observed physical properties of ordinary objects. That chain of reasoning is the subject of statistical mechanics.

A typical macroscopic object, a glass of water, contains about 3×10^{25} molecules. The extreme compactness of exponential notation masks the vastness of that number. If that glass of water were mixed throughout all the oceans of the world, then a glass of water, drawn anywhere on earth, would contain many of the molecules that had been in it. Clearly, there is no possibility of ever knowing the exact microscopic state of a glass of water. In statistical mechanics we will have to work with very incomplete information about the systems we are analyzing. Typically, we can express our information about the state of the system in 3 or 4 numbers, such as the system's volume, mass, temperature, and pressure, while a complete specification would need about 10^{26} separate numbers. It is remarkable that, with this meager information, we can predict so many important things with high precision and reliability. For other things we can make only probabilistic predictions. But, even for those, the methods of statistical mechanics permit us to make accurate predictions of the probabilities.

The general mathematical discipline involved in making predictions with limited information is probability theory. In a real sense, statistical mechanics is a branch of probability theory. However, the systems under analysis in statistical mechanics have certain special characteristics that have allowed the subject to develop a subtlety and predictive power that are unmatched by any other subject in which statistical methods are used. This chapter will be devoted to an introduction to the general ideas of probability theory. All the following chapters will specialize the theory to the problem of calculating the physical properties of aggregate matter.

1.2   THE DEFINITION OF PROBABILITY

The reader certainly has some idea of what probability means and most likely knows how to make simple calculations of probabilities. In order to construct a mathematical theory of probability, we will first look at a problem in probability that is easy to solve and then abstract the rules we use to solve that problem and restate them in a more precise and general way.

The problem we will consider is to calculate the probability of throwing six at dice. It will help to assume that the two dice are of different colors, let us say red and blue. Each time the dice are thrown we get two numbers. The number on the red die we call R, and that on the blue we call B. The pair of numbers (R, B) will be denoted by the single symbol x. The value of x is always some element from the 36-element set

$$\Omega = \{(1,1), (1,2), \ldots, (6,5), (6,6)\} \tag{1.1}$$

If the dice are well made, it is reasonable to assume that each of the 36 possible values of x has the same probability of occurrence. The probability of occurrence of x is defined as the fraction of times that x occurs in a very long sequence of throws. We write it as $P(x)$. Regardless of whether the dice are well made or not, two things follow immediately from the definition of $P(x)$. They are that

$$0 \leq P(x) \leq 1 \tag{1.2}$$

and

$$\sum_{x \in \Omega} P(x) = 1 \tag{1.3}$$

For well-made dice, $P(x) = 1/36$ for all x. To calculate the probability of throwing six we must first see which combinations give $R + B = 6$. They are

$$x = (1,5), (2,4), (3,3), (4,2), \text{ and } (5,1) \tag{1.4}$$

We then sum $P(x)$ over those five values of x to get the probability that $R + B = 6$.

$$\mathbf{P}[R+B=6] = \frac{5}{36} \tag{1.5}$$

Now let us try to generalize the ideas used in this dice problem.

We assume that we are given a system (for example, the dice) whose possible configurations or states are described by a variable x that can take values within some set Ω. The elements of Ω we will call the possible *microstates* of the system and Ω will be called the *microstate space.** At this point it is assumed that the set Ω contains only a finite number of microstates. Later, infinite and continuous sets of microstates will be considered.

There exists some procedure for constructing a sampling sequence of arbitrary length, x_1, x_2, ..., where $x_n \in \Omega$. (For example, repeatedly shake the dice and

---

*This is a deviation from standard probability theory terminology where Ω is called the *sample space*. It will be far more natural for our major application which is to statistical mechanics.

throw them from a height of at least 2 feet.) The sampling sequence must have the property that the function

$$P(x) \equiv \lim_{N \to \infty} \left(\frac{\text{The number of times } x_n = x \text{ for } n = 1, \dots, N}{N} \right) \qquad (1.6)$$

exists for all $x \in \Omega$. This definition of the probability function, $P(x)$, in terms of a sampling sequence is known as the frequency interpretation of probability. The following things should be noted:

1. An example of a sequence for which the limit defining $P(x)$ does *not* exist is the following. Suppose Ω consists of the two elements A and B. Then the sampling sequence

$$\{x_n\} = A, B, A, B, A, A, B, B, A, A, A, A, \dots \qquad (1.7)$$

   does not define a function $P(x)$. Therefore the existence of the limit defining $P(x)$ is not a completely trivial restriction on acceptable sampling sequences.

2. We have not imposed any requirement of randomness or unpredictability on the sampling sequence. Although many of the cases we will consider lead to unpredictable sampling sequences, others do not. For example, a very important case in statistical mechanics is one in which the sampling sequence is the sequence of dynamical states of an isolated mechanical system at a sequence of time instants, t_1, t_2, t_3, ... For a deterministic dynamical system these are not at all unpredictable. They will not have to be. All of the results of probability theory that will be used are mathematical consequences of the assumed properties of $P(x)$ and no property of randomness will be necessary in deriving those consequences.

The function $P(x)$ will be called the *fundamental microstate probability*. In most cases $P(x)$ is determined either by actually carrying out measurements on a real sampling sequence or from some sort of plausability argument, as was used previously for well-made dice. Formal probability theory then shows us how to calculate the probability of statements that depend upon x (for example, $R + B = 6$) in terms of the fundamental microstate probability.

Consider now some statement or equation $q(x)$ whose truth is a function of x. The probability of $q(x)$ would naturally be defined as

$$\mathbf{P}[q] = \lim_{N \to \infty} \left(\frac{\text{the number of times } q(x_n) \text{ is true for } n = 1, \dots, N}{N} \right) \qquad (1.8)$$

Any such statement defines a subset $Q \subset \Omega$ by $x \in Q$ if and only if $q(x)$ is true. It is then easy to show that

$$\mathbf{P}[q] = \sum_{x \in Q} P(x) \qquad (1.9)$$

The relation between statements and subsets of Ω suggests that the definition of $P(x)$ be generalized to a probability $\mathbf{P}[Q]$ assigned to any subset $Q \subset \Omega$ by

$$\mathbf{P}[Q] = \lim_{N \to \infty} \left(\frac{\text{the number of } x_n \in Q \text{ for } n = 1, \dots, N}{N} \right) \qquad (1.10)$$

A probability function that is defined for the subsets of Ω is called a *probability measure*.

Now let us see why we restricted ourselves to a system with a finite number of microstates. For finite Ω the definition of $\mathbf{P}[Q]$ obviously implies that

$$\mathbf{P}[Q] = \sum_{x \in Q} P(x) \tag{1.11}$$

and, in particular, that

$$\sum_{x \in \Omega} P(x) = \mathbf{P}[\Omega] = 1 \tag{1.12}$$

But suppose Ω consists of all the positive integers and we consider the sampling sequence $x_1, x_2, \ldots = 1, 2, \ldots$. That is, $x_n = n$. For this sampling sequence the limit defining $P(x)$ converges to zero for each $x \in \Omega$. Thus $P(x)$ is well defined. Also, using Eq. (1.10), we get $\mathbf{P}[\Omega] = 1$. But then

$$\sum_{x \in \Omega} P(x) = 0 + 0 + \cdots = 0 \neq \mathbf{P}[\Omega] = 1 \tag{1.13}$$

We see that certain properties of the probability function that can be proven always to hold for finite Ω are not automatic for infinite Ω. They must therefore be incorporated into the definition of an acceptable sampling sequence. Before describing the properties that will be needed we will introduce some elementary ideas of set theory. It is assumed that all the sets discussed here are subsets of some microstate space, Ω.

1.3   THE BASIC NOTIONS OF SET THEORY

The *union* of two sets, A_1 and A_2, written $A_1 \cup A_2$ or $A_1 + A_2$, is a set containing any element of Ω that is in at least one of the two sets. For example,

$$\{1, 2, 3\} \cup \{2, 3, 4\} = \{1, 2, 3\} + \{2, 3, 4\} = \{1, 2, 3, 4\} \tag{1.14}$$

The *intersection* of two sets, A_1 and A_2, written $A_1 \cap A_2$ or $A_1 \cdot A_2$, is a set containing any element of Ω that is in both A_1 and A_2. Thus

$$\{1, 2, 3\} \cap \{2, 3, 4\} = \{1, 2, 3\} \cdot \{2, 3, 4\} = \{2, 3\} \tag{1.15}$$

The notions of union and intersection generalize in an obvious way to arbitrary collections of sets. For example, given a sequence of sets, $A_1, A_2, \cdots$, then

$$x \in A_1 \cup A_2 \cup \cdots \text{ if and only if } x \in A_k \text{ for some } k \tag{1.16}$$

and

$$x \in A_1 \cap A_2 \cap \cdots \text{ if and only if } x \in A_k \text{ for every } k \tag{1.17}$$

The statement that every element of A_1 is also an element of A_2 (i.e., that A_1 is a *subset* of A_2) is written $A_1 \subset A_2$ (or else $A_2 \supset A_1$). The *complement* of a set A is written $\bar{A}$ and is a set containing all elements of Ω that are not elements of A. The *difference* between two sets, written $A_1 - A_2$, is a set containing those elements of A_1 that are not elements of A_2. It is easy to see that

$$A_1 - A_2 = A_1 \cdot \bar{A}_2 \tag{1.18}$$

and that

$$\bar{A} = \Omega - A \tag{1.19}$$

The *empty set* is always denoted $\emptyset$ and is the set containing no elements

$$\emptyset = \{\} \tag{1.20}$$

As part of the definition of the word *subset*, $\emptyset$ is considered to be a subset of every set. Two sets are said to be *disjoint* if they contain no common elements. The sets in a sequence, $A_1, A_2, \ldots$, are called *mutually disjoint* if $A_i \cap A_j = \emptyset$ whenever $i \neq j$.

1.4   THE MEASURE THEORY INTERPRETATION

We now return to the question of what are the requirements of an acceptable probability measure. A probability function $\mathbf{P}[A]$ defined for all $A \subset \Omega$, will be deemed acceptable if and only if it satisfies the following conditions

1. $\mathbf{P}[A] \geq 0$ for all $A \subset \Omega$.
2. $\mathbf{P}[\Omega] = 1$.
3. For any (finite or infinite) sequence, $A_1, A_2, \cdots$, of mutually disjoint sets

$$\mathbf{P}[A_1 \cup A_2 \cup \ldots] = \mathbf{P}[A_1] + \mathbf{P}[A_2] + \ldots \tag{1.21}$$

From (3) it is easy to show that $\mathbf{P}[\emptyset] = 0$. (Let $A_1 = \emptyset$.) Properties (1), (2), and (3) can be proven to hold whenever Ω has a finite number of elements, but if Ω contains an infinity of elements, then, as we have seen, these properties, in particular 3, are not an automatic consequence of the convergence of the limit defining $\mathbf{P}[A]$. They will be postulated to hold for any probability distribution we consider henceforth.

It is possible to dispense with the notion of a sampling sequence entirely and to base the study of probability theory completely on Postulates 1, 2, and 3. This defines a measure theory interpretation of probability. One simply postulates a function, $\mathbf{P}[A]$, defined for all subsets of a microstate space, Ω, and satisfying conditions 1, 2, and 3.* For a strictly mathematical study of probability theory the measure theory interpretation is definitely the most convenient one. However, in applying probability theory to the analysis of physical systems, the idea of a sampling sequence supplies an interpretation that is more concrete and pictorial and therefore helps in developing an intuitive understanding of the subject.

1.5   SOME EXAMPLES

Before going any further in the general theory, let us consider some simple problems in probability theory. We begin with a few of what are called *combinatorial problems*.

Question   How many distinct sequences of n integers, $(I_1, I_2, \ldots, I_n)$, can be constructed if the integers are restricted to the range $1 \leq I_j \leq N$?

Answer   A sequence of n objects is called an n-tuple. Two n-tuples, $(x_1, \ldots, x_n)$, and $(y_1, y_2, \ldots, y_n)$, are considered identical only if $x_j = y_j$ for all j. Otherwise

---

* If Ω contains a continuous infinity of elements, such as all the points in a two-dimensional rectangle, then there are certain technical restrictions on the collection of sets for which $\mathbf{P}[A]$ is assumed to be defined, but it would take us too deeply into measure theory to discuss them.

they are considered to be distinct. This should be contrasted to sets of n objects. Permutation of the objects in a set does not give a distinct set. That is, the set $\{1, 3, 5\}$ is identical to the set $\{5, 1, 3\}$, but the 3-tuple $(1, 3, 5)$ is not equal to the 3-tuple $(5, 1, 3)$. Now let us answer the question. Clearly I_1 can have N possible values. For any choice of I_1, I_2 can have N possible values. Thus there are N^2 possible choices of I_1 and I_2. Proceeding in this way, it is easy to see that there are N^n possible choices or $I_1, I_2, \ldots, I_n$.

Question   How many distinct sequences of length n can we construct from the first N integers if we demand that no two integers in the sequence be equal? We assume that $N \geq n$.

Answer   Again, there are N choices for the first integer, I_1. Having chosen I_1, there are now only $N-1$ choices for the second integer, I_2. Thus there are $N(N-1)$ choices of I_1 and I_2. Having chosen those two, there are only $N-2$ choices for I_3 and so forth. Thus the number of distinct n-tuples containing n different integers drawn from the first N integers is

$$N(N-1)\ldots(N-n+1) = \frac{N!}{(N-n)!} \qquad (1.22)$$

For $n = N$, this formula says that there are $N!$ permutations of N distinct objects.

Question   How many distinct sets of n integers can be constructed from the first N integers?

Answer   We can answer this by grouping the distinct n-tuples of the last question into classes, such that any two n-tuples in a class contain the same set of integers and are therefore permutations of one another. Clearly, each class contains $n!$ n-tuples, which is the number of permutations there are of n distinct objects. Each class corresponds to one of the sets of this question. Thus the answer to this question is given by dividing the answer to the last question by $n!$. The number obtained is called the *binomial coefficient*.

$$\binom{N}{n} = \frac{N!}{n!(N-n)!} \qquad (1.23)$$

Question   How many ways can N numbered balls be grouped into three sets such that the first set contains n_1 balls, the second set n_2 balls, and the third set $n_3 = N - n_1 - n_2$ balls?

Answer   There are $\binom{N}{n_1}$ ways of choosing the first set. Having done that we would be left with $N - n_1$ balls from which we can draw the second set in $\binom{N-n_1}{n_2}$ ways. The third set is then determined. Thus the answer is

$$\binom{N}{n_1}\binom{N-n_1}{n_2} = \frac{N!}{n_1! n_2! n_3!} \qquad (1.24)$$

The generalization of this formula is obvious.

Question   What is the probability of getting n heads in a simultaneous toss of N fair coins?

Answer   A single toss is described by giving the state (H = heads or T = tails) of each of the N coins. Thus $x = (c_1, c_2, \ldots, c_N)$, where c_j is the state of coin j and $c_j = H$ or T. The total number of possible microstates is 2^N. Consider any

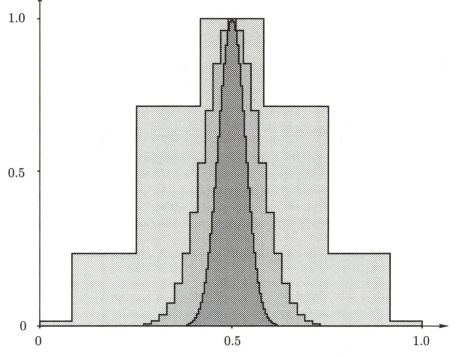

Fig. 1.1   The function $R(x)$ defined in Eq. (1.26), for the cases $N = 6$, 50, and 200. Since x must be an integer divided by N, the function is discontinuous. The wide curve is the case $N = 6$, the narrow one is the case $N = 200$. As N increases, the probability of getting significant deviations from 1/2 heads goes to zero.

particular microstate, $(c_1, c_2, \ldots, c_N)$. If we use the fact that coin 1 is a fair coin and we make the reasonable assumption that coin 1 is independent of the other $N - 1$ coins, then the probability of this microstate is the same as the probability of the microstate $(c_1^*, c_2, \ldots, c_N)$, where c_1^* is the "opposite" of c_1. Applying the same argument to different coins, we arrive at the conclusion that $P(x)$ has the same value for each of the 2^N possible microstates in Ω. Normalization then requires that $P(x) = 1/2^N$ for all $x \in \Omega$. Any n-head microstate defines a partition of the N coins into two sets, the n heads and the $N - n$ tails. Thus the number of n-head microstates is the same as the number of n-element sets that can be constructed from N distinct elements. Each of these microstates has a probability $1/2^N$. Therefore

$$\mathbf{P}[n \text{ heads}] = \frac{1}{2^N} \binom{N}{n} \tag{1.25}$$

For even N, the most probable number of heads is $N/2$. The ratio

$$R(x) = \frac{\mathbf{P}[xN \text{ heads}]}{\mathbf{P}[\frac{1}{2}N \text{ heads}]} \tag{1.26}$$

is plotted as a function of x for the three cases $N = 6$, 50, and 200 in Fig. 1.1. As the total number of coins increases the probability of obtaining a fraction of heads significantly different from $\frac{1}{2}$ goes to zero. This phenomenon of certain probability distributions becoming more and more sharply peaked as the number of elements in a system increases will be of central importance in our application of probability theory to systems with large numbers of particles. It is only this statistical effect that makes the macroscopic properties of large systems deterministic and predictable.

Question N weakly interacting particles move within a volume V. What is the probability of finding exactly n particles within some particular subvolume v within V?

Answer The probability that any particular particle will be found within v is $p = v/V$. Because the particle must be either inside of v or outside of v, the probability that it will be found outside of v is $1 - p$. If we ignore the interactions and therefore assume that the particles are statistically independent, then the probability that some particular set of n particles is within v while the remaining $N - n$ particles are outside v is equal to $p^n(1 - p)^{N-n}$. To get the probability of finding any set of n particles within v we must multiply this by the number of ways of choosing a set of n particles from a larger set of N particles. This gives

$$\mathbf{P}[n \text{ particles in } v] = P_n = \binom{N}{n} p^n (1 - p)^{N-n} \tag{1.27}$$

In the case that $v \ll V$ and $n \ll N$, this formula can be simplified by making the following approximations. (Note that $p = v/V \ll 1$.)

$$\frac{N!}{(N-n)!} = N(N-1)\cdots(N-n+1) \approx N^n \tag{1.28}$$

$$(1-p)^{N-n} \approx (1-p)^N = [(1-p)^{1/p}]^{pN} \approx e^{-pN} \tag{1.29}$$

With the definition, $\bar{n} \equiv Np$, P_n can be written as

$$P_n = \frac{\bar{n}^n}{n!} e^{-\bar{n}} \tag{1.30}$$

This is called the *Poisson distribution*. It is easy to confirm that $\bar{n}$ is actually the average value of n. That is, that

$$\bar{n} = \sum_{n=1}^{\infty} n P_n \tag{1.31}$$

If $v = V/2$, then, for obvious reasons, Eq. (1.27) for P_n becomes identical with Eq. (1.25) for the probability of obtaining n heads in a throw of N coins.

Question In a throw of three dice, what is the probability that at least one of the dice will show the number six?

Answer The probability that a given die will not show six is $\frac{5}{6}$. The dice are independent, and therefore the probability that the three will simultaneously not show the number six is $\left(\frac{5}{6}\right)^3$. The alternative is for at least one of the dice to show the number six. Therefore,

$$\mathbf{P}[\text{at least one six}] = 1 - \left(\frac{5}{6}\right)^3 = 0.42 \tag{1.32}$$

1.6 ENSEMBLES

Starting with a sampling sequence, one can construct yet another useful interpretation of the formulas of probability theory. Let us consider the case in which Ω

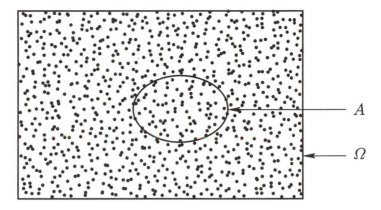

Fig. 1.2 An ensemble is a very large collection of points from the sampling sequence. The probability that the microstate will be found in the region A is the ratio of the number of ensemble points in A to the total number of points in the ensemble.

consists of all the points in some rectangle in the x-y plane. An associated sampling sequence would be a sequence of points, (x_1, y_1), $(x_2, y_2), \ldots$, all lying within the rectangle. If, for some very large value of N, we plot the first N points in the sampling sequence, then, to a high degree of accuracy, for any region $A \subset \Omega$,

$$\mathbf{P}[A] = \frac{\text{the number of points in } A}{\text{the number of points in } \Omega} \tag{1.33}$$

The collection of a large but finite number of points, taken from the sampling sequence, is called a *statistical ensemble* (see Fig. 1.2). Of course, Eq. (1.33) for $\mathbf{P}[A]$ only becomes exact in the limit that the number of points in the ensemble goes to infinity but, in that limit one would lose the simple picture of a discrete collection of points. Whenever the ensemble interpretation of probability is used, we will simply assume that N, the number of points in the ensemble, is sufficiently large to make any error due to finite N negligible.

1.7 PROBABILITY DENSITY

Using the ensemble picture, we can define a nonnegative function $P(x, y)$ called the *probability density*, as follows.

$$P(x, y) \, dx \, dy = \frac{\text{the number of points in } dx \, dy}{N} \tag{1.34}$$

In terms of the probability density, $\mathbf{P}[A]$ can be expressed as

$$\mathbf{P}[A] = \int_A P(x, y) \, dx \, dy \tag{1.35}$$

The phrase, *probability distribution*, will be used to refer to both a probability density function for a continuous variable, such as $P(x, y)$, or the set of discrete probabilities, $P(x)$, for a variable, x, that has only discrete values.

It will be left as an exercise for the reader (Problem 1.9), to derive, from the defining properties of a probability measure, the following facts, which are quite

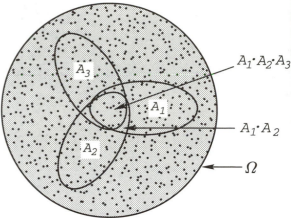

Fig. 1.3 The region $A_1 \cdot A_2$ is the overlap of the regions A_1 and A_2. It is clear that $\mathbf{P}[A_1]+\mathbf{P}[A_2]$ counts the ensemble points in $A_1 \cdot A_2$ twice, and therefore, $\mathbf{P}[A_1 + A_2] = \mathbf{P}[A_1] + \mathbf{P}[A_2] - \mathbf{P}[A_1 \cdot A_2]$. The other relations in Eq. (1.36) can be obtained in a similar way.

obvious from an ensemble picture (see Fig. 1.3).

$$
\begin{aligned}
&1. \quad \mathbf{P}[A_1 + A_2] = \mathbf{P}[A_1] + \mathbf{P}[A_2] - \mathbf{P}[A_1 \cdot A_2] \\
&2. \quad \mathbf{P}[A_1 - A_2] = \mathbf{P}[A_1] - \mathbf{P}[A_1 \cdot A_2] \\
&3. \quad \mathbf{P}[A_1 + A_2 + A_3] = \sum_i \mathbf{P}[A_i] - \sum_{i<j} \mathbf{P}[A_i \cdot A_j] + \mathbf{P}[A_1 \cdot A_2 \cdot A_3]
\end{aligned}
\tag{1.36}
$$

1.8   PROBABILITIES OF COMPOUND STATEMENTS

A compound statement is a statement composed of one or more component statements and the logical operators *and, or,* or *not.* Formulas for the probabilities of compound statements follow rather directly from the meaning of the logical operators. Given a system with a microstate space Ω and given two statements $q_1(x)$ and $q_2(x)$, then the probability that the statements are both true is given by

$$
\mathbf{P}[q_1 \text{ and } q_2] = \mathbf{P}[Q_1 \cdot Q_2] = \mathbf{P}[Q_1 \cap Q_2]
\tag{1.37}
$$

This is obvious, since q_1 is true if and only if $x \in Q_1$ and q_2 is true if and only if $x \in Q_2$. Thus "q_1 and q_2" is true if and only if $x \in Q_1 \cap Q_2$. The statement "q_1 or q_2" is true if either (or both) of the component statements is true. Therefore

$$
\mathbf{P}[q_1 \text{ or } q_2] = \mathbf{P}[Q_1 + Q_2] = \mathbf{P}[Q_1 \cup Q_2]
\tag{1.38}
$$

It is also easy to see that

$$
\mathbf{P}[\text{not } q] = \mathbf{P}[\bar{Q}] = 1 - \mathbf{P}[Q]
\tag{1.39}
$$

1.9   CONDITIONAL PROBABILITIES

"Given that the first two cards we are dealt in a game of poker are clubs, what is the probability that our hand will be a flush (all clubs)?" Questions of this type, involving *conditional probabilities*, are common. Their general form is: "Given statement q, what is the probability of statement q'?" Using the frequency interpretation of probability, the meaning of the question is: "If we only count those members of the sampling sequence that satisfy the relation '$q(x_n)$ is true,' what fraction of them will

Fig. 1.4   The fraction of the ensemble points in Q that are also in Q' is the ratio of the points in the overlap area to the points in the area Q. But that is just $\mathbf{P}[Q \cdot Q']/\mathbf{P}[Q]$.

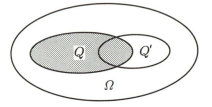

have $q'(x_n)$ true?" We will write the probability of q', conditioned on q as $\mathbf{P}[q'|q]$. In terms of the sampling sequence

$$\mathbf{P}[q'|q] = \lim_{N \to \infty} \left[\frac{\text{the number of times } q \text{ and } q' \text{ is true}}{\text{the number of times } q \text{ is true}} \right] \tag{1.40}$$

From this it is easy to see that

$$\mathbf{P}[q'|q] = \frac{\mathbf{P}[q' \text{ and } q]}{\mathbf{P}[q]} = \frac{\mathbf{P}[Q \cdot Q']}{\mathbf{P}[Q]} \tag{1.41}$$

This formula also follows easily from an ensemble picture. We are asking: "What fraction of the ensemble points in Q are also in Q'?" (See Fig. 1.4.)

The statement q' is said to be *statistically independent of* or *uncorrelated with* the statement q if the probability of q' being true, given that q is true, is equal to the probability of q' being true without that condition, that is, if

$$\mathbf{P}[q'|q] = \mathbf{P}[q'] \tag{1.42}$$

By using Eq. (1.41) we can see that statistical independence is a reciprocal relationship. That is, the statement "q' is statistically independent of q" implies that q is statistically independent of q'. Two statements are statistically independent if and only if

$$\mathbf{P}[q \text{ and } q'] = \mathbf{P}[q]\mathbf{P}[q'] \tag{1.43}$$

or equivalently

$$\mathbf{P}[Q \cdot Q'] = \mathbf{P}[Q]\mathbf{P}[Q'] \tag{1.44}$$

The probability of two statistically independent statements being simultaneously true is equal to the product of their separate probabilities. This is a fact that we have previously used in the examples.

Question   The shots of a certain marksman on a target are found to be distributed with a probability density $P(x, y) = Ce^{-r^2}$, where $r = \sqrt{x^2 + y^2}$ is the distance from the center of the bullseye, measured in centimeters. Given that a shot is at least 1 cm high, what is the probability that it is also at least 1 cm to the right?

Answer   The perforated target is an excellent representation of the statistical ensemble for the system. The question is: "Of those holes that lie above the line $y = 1$, what fraction lie to the right of the line $x = 1$?" (See Fig. 1.5.) The density of holes is of the form $D(x, y) = Ae^{-x^2 - y^2}$ and therefore the probability that $x > 1$, given that $y > 1$, is

$$\begin{aligned}
\mathbf{P}[x > 1 | y > 1] &= \frac{\int_1^\infty dx \int_1^\infty dy \, e^{-x^2 - y^2}}{\int_{-\infty}^\infty dx \int_1^\infty dy \, e^{-x^2 - y^2}} \\[2mm]
&= \frac{\int_1^\infty dx \, e^{-x^2}}{\int_{-\infty}^\infty dx \, |e^{-x^2}} \\[2mm]
&= \frac{\sqrt{\pi}}{2} \mathrm{erfc}(1) \approx 0.28
\end{aligned} \tag{1.45}$$

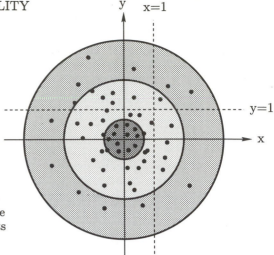

Fig. 1.5   We want the ratio of the 'ensemble points' (that is, bullet holes) in the upper right corner to the number of points above the dashed line.

where erfc(x) is the complementary error function, defined in the Mathematical Appendix. It is easy to see that $\mathbf{P}[x > 1|y > 1] = \mathbf{P}[x > 1]$, which shows that the two conditions are statistically independent. The statistical independence of the x and y distributions is a consequence of the fact that $P(x, y) = f(x)f(y)$. It can be shown (see Exercise 1.18) that the Gaussian function is the only rotationally symmetric probability distribution that yields statistically independent Cartesian coordinate distributions.

1.10   RANDOM VARIABLES

Any variable X with an associated probability distribution $P_X(x)$ is called a *random variable*.* The x and y coordinates of the shots in the above example are random variables. If X is discrete, then $P_X(x) = \mathbf{P}[X = x]$. For continuous X, $P_X(x)\,dx = \mathbf{P}[x < X < x + dx]$. Any function $Y(X)$ of the random variable X is itself a random variable with a probability distribution $P_Y(y)$. We now want to consider the problem of calculating $P_Y(y)$, given $P_X(x)$.

If X has only discrete values, then the possible values of Y must also be discrete, and the associated probability $P_Y(y)$ is given by

$$P_Y(y) = \mathbf{P}[Y = y] = \sum_{Y(x)=y} P_X(x) \tag{1.46}$$

where the sum is over all values of x that are mapped into the value y. If X and Y are continuous, then the problem of calculating the probability density function, $P_Y(y)$, associated with the variable $Y(X)$ is more challenging. We will first consider a simple example.

Question   Suppose that the two-dimensional random variable $\mathbf{X} = (X_1, X_2)$ has a probability density $P_\mathbf{X}(x_1, x_2)$. What is the probability density associated with the variable $R = \sqrt{X_1^2 + X_2^2}$?

---

* Why do we put the subscript X on $P_X(x)$? In this section we will be discussing the probability distributions of two or more different variables at once. If we simply write $P(x)$ and $P(y)$, rather than $P_X(x)$ and $P_Y(y)$, then we can be forced into absurd equations, such as $P(1) \neq P(1)$, where on the left we have $P(x)$ evaluated at $x = 1$ and on the right we have $P(y)$ at $y = 1$.

Answer   For any $r > 0$, the probability that $R < r$ is

$$\mathbf{P}[R < r] = \int \theta(r - R(\mathbf{x}))P_{\mathbf{X}}(\mathbf{x})\,d^2x \tag{1.47}$$

where the step function $\theta(x)$ is defined by

$$\theta(x) = \begin{cases} 0, & x < 0 \\ 1, & x > 0 \end{cases} \tag{1.48}$$

But

$$P_R(r)\,dr = \mathbf{P}[R < r + dr] - \mathbf{P}[R < r] = \frac{d\mathbf{P}[R < r]}{dr}\,dr \tag{1.49}$$

Using the fact that $d\theta(x)/\,dx = \delta(x)$, the Dirac delta function, we obtain

$$P_R(r) = \int \delta(r - R(\mathbf{x}))P_{\mathbf{X}}(\mathbf{x})\,d^2x \tag{1.50}$$

In general, if $\mathbf{X} = (X_1, X_2, \ldots, X_N)$ is an N-dimensional random variable with a probability density $P_{\mathbf{X}}(x_1, x_2, \ldots, x_N)$ and $Y(\mathbf{X})$ is any function of $\mathbf{X}$, then

$$P_Y(y) = \int \delta(y - Y(\mathbf{x}))P_{\mathbf{X}}(x_1, x_2, \ldots, x_N)\,d^Nx \tag{1.51}$$

1.11   AVERAGES AND UNCERTAINTIES

The *average* (or *mean* or *expectation*) *value* of a random variable x with a probability distribution $P(x)$ is written $\bar{x}$ or $\langle x \rangle$ and defined as

$$\bar{x} = \langle x \rangle = \int xP(x)\,dx \tag{1.52}$$

The *root-mean-square* (rms) *uncertainty* (or *fluctuation* or *dispersion*) in x is written Δx and defined by

$$(\Delta x)^2 = \langle (x - \bar{x})^2 \rangle = \int (x - \bar{x})^2 P(x)\,dx \tag{1.53}$$

Expanding the square and remembering that $\langle x \rangle \equiv \bar{x}$, one obtains

$$(\Delta x)^2 = \langle x^2 \rangle - 2\langle x \rangle \bar{x} + \bar{x}^2 = \langle x^2 \rangle - \langle x \rangle^2$$

1.12   CHEBYSHEV'S INEQUALITY

If a random variable x has an average value $\bar{x}$ with a very small uncertainty, then it is obvious that large deviations from $\bar{x}$ must be improbable. This relationship between the uncertainty in a random variable and the probability of deviations greater than any given magnitude is precisely expressed in an inequality due to Chebyshev. It states that, for any $a > 0$,

$$\mathbf{P}[(x - \bar{x})^2 > a^2] \leq \left(\frac{\Delta x}{a}\right)^2 \tag{1.54}$$

Before giving a proof of the inequality, let us consider some of its consequences. If we choose $a = \Delta x$, then the inequality is empty in that any probability must be less than or equal to one. However, for $a = K\Delta x$, with $K > 1$, Chebyshev's inequality shows that the probability of a deviation larger that $K\Delta x$ is less than or equal to $1/K^2$. Later we will use this inequality in proving a very fundamental theorem of statistical mechanics.

The proof of Chebyshev's inequality is straightforward. We let $y(x) \equiv (x - \bar{x})^2$. The range of the random variable y is $0 \le y < \infty$. Clearly

$$
\begin{aligned}
(\Delta x)^2 = \langle y \rangle &= \int_0^\infty P_y(y) y \, dy \\
&\ge \int_{a^2}^\infty P_y(y) y \, dy \\
&\ge a^2 \int_{a^2}^\infty P_y(y) \, dy \\
&= a^2 \mathbf{P}[y > a^2]
\end{aligned}
\tag{1.55}
$$

1.13   THE LAW OF LARGE NUMBERS

Two or more random variables are said to be *independent* if they are statistically uncorrelated. For example, if X, Y, and Z are random variables and $P(x, y, z) \, dx \, dy \, dz$ is the probability of simultaneously finding X between x and $x + dx$, finding Y between y and $y + dy$, and finding Z between z and $z + dz$, then X, Y, and Z are independent if and only if

$$
P(x, y, z) = P_X(x) P_Y(y) P_Z(z)
\tag{1.56}
$$

where P_X, P_Y, and P_Z are the probability densities associated with each variable separately.

Let $x_1, x_2, \ldots, x_N$ be N statistically independent random variables. The law of large numbers is a theorem concerning the uncertainty in the random variable

$$
x \equiv \frac{x_1 + \cdots + x_N}{N}
\tag{1.57}
$$

If, for example, the variables $x_1, x_2, \ldots, x_N$ are the x coordinates of N identical particles, then x is the x coordinate of their center of mass. If $x_1, x_2, \ldots, x_N$ represent the states (heads or tails) of N coins, with the provision that $x_j = 1$ if the jth coin is heads and $x_j = 0$ if it is tails, then x gives the fraction of coins that are heads. The content of the theorem is that, for large N, the uncertainty in x is much smaller than the uncertainties in the individual random variables, $x_1, x_2, \ldots, x_N$.

We assume that $x_1, x_2, \ldots, x_N$ have associated probability functions, $P_1(x_1)$, $\ldots, P_N(x_N)$, and uncertainties, $\Delta x_1, \ldots, \Delta x_N$. We define a *mean square uncertainty*, a, by

$$
a^2 = \frac{(\Delta x_1)^2 + \cdots + (\Delta x_N)^2}{N}
\tag{1.58}
$$

If the variables all have equal uncertainties, then $a = \Delta x_1 = \cdots = \Delta x_N$. In general, even for large N, a is of the same magnitude as the typical individual uncertainties.

Because the variables are statistically independent, the joint probability density for all N variables is the product of their individual probability densities.

$$P(x_1, x_2, \ldots, x_N) = P_1(x_1) \cdots P_N(x_N) \tag{1.59}$$

We will assume that the variables x_i are continuous. The modification of the proof for the case of discrete x_i consists simply in replacing integrals by sums. The average value and dispersion of the random variable x are given by

$$
\begin{aligned}
\bar{x} &= \int dx_1 \cdots \int dx_N \left(\frac{x_1 + \cdots + x_N}{N} \right) P_1(x_1) \cdots P_N(x_N) \\
&= \frac{1}{N} \int dx_1 \cdots \int dx_N \left(\sum_{i=1}^{N} x_i \right) P_1(x_1) \cdots P_N(x_N) \\
&= \frac{1}{N} \sum_{i=1}^{N} \int P_1(x_1) dx_1 \cdots \int x_i P_i(x_i) dx_i \cdots \int P_N(x_N) dx_N \\
&= \frac{\bar{x}_1 + \cdots + \bar{x}_N}{N}
\end{aligned}
\tag{1.60}
$$

and

$$
\begin{aligned}
(\Delta x)^2 &= \left\langle (x - \bar{x})^2 \right\rangle \\
&= \left\langle \left(\frac{(x_1 - \bar{x}_1) + \cdots + (x_N - \bar{x}_N)}{N} \right)^2 \right\rangle \\
&= \frac{1}{N^2} \int dx_1 \cdots \int dx_N \sum_{i,j}^{N} (x_i - \bar{x}_i)(x_j - \bar{x}_j) P_1(x_1) \cdots P_N(x_N)
\end{aligned}
\tag{1.61}
$$

Using the facts that

$$\int P_n(x_n) dx_n = 1 \tag{1.62}$$

$$\int (x_n - \bar{x}_n) P_n(x_n) dx_n = 0 \tag{1.63}$$

and

$$\int (x_n - \bar{x}_n)^2 P_n(x_n) dx_n = (\Delta x_n)^2 \tag{1.64}$$

one finds that

$$(\Delta x)^2 = \frac{\sum_n (\Delta x_n)^2}{N^2} = \frac{a^2}{N} \tag{1.65}$$

or

$$\Delta x = \frac{a}{\sqrt{N}} \tag{1.66}$$

Thus, for instance, the average of 100 independent variables has an uncertainty that is only one tenth as large as the uncertainties of the individual variables (assuming that the individual variables all have roughly equal uncertainties). The sharpening of the probability distributions illustrated in Fig. 1.1 is an example of the law of large numbers.

1.14   THE CENTRAL LIMIT THEOREM

The central limit theorem is the most powerful and important of those theorems of probability that are concerned with sums of independent random variables. It is a major extension of the law of large numbers. The central limit theorem shows that, not only does the distribution of $X = (x_1 + \cdots + x_N)/N$ become sharp, but the detailed form of $P_X(x)$ approaches a Gaussian function for large N regardless of what are the detailed forms of the separate probability densities, $P_1(x_1), \ldots, P_N(x_N)$. We will give the statement of the central limit theorem here but reserve the proof of the theorem for the Mathematical Appendix.

We consider an infinite sequence of independent random variables $x_1, x_2, \ldots$ with probability densities $P_1(x_1), P_2(x_2), \ldots$. It is no serious restriction to assume that the average value of each of the separate variables vanishes.

$$\int P_n(x_n) x_n \, dx_n = 0 \tag{1.67}$$

If X is the average of the first N variables and a is their mean square uncertainty (that is, $a^2 = (\Delta x_1^2 + \ldots + \Delta x_N^2)/N$), then, for very large N,

$$P_X(x) = (N/2\pi a^2)^{1/2} \, e^{-Nx^2/2a^2} \tag{1.68}$$

The uncertainty in X can be easily calculated from Eq. (1.68) and is

$$\Delta X = \frac{a}{\sqrt{N}} \tag{1.69}$$

This result was derived previously. What is new is the fact that, for large N, the detailed probability distribution for X is a Gaussian distribution even though no assumptions were made regarding the details of the probability distributions for the variables $x_1, x_2, \ldots, x_N$.

Fig. 1.6   The cylinder contains two moles of gas. The average value of the number of particles in either half is obviously one mole.

Question   A cylindrical container holds two moles of an ideal gas. (Fig. 1.6) What is the probability of finding more that $(1 + 10^{-8})$ mole in the right hand half of the cylinder?

Answer   We define the random variables, $\sigma_1, \sigma_2, \ldots, \sigma_N$, where $N = 2N_A$, by saying that $\sigma_n = +1$ if the nth particle is in the right-hand half of the container and $\sigma_n = -1$ otherwise. The variable σ_n takes its two possible values with equal probabilities. The average value of each σ_n is obviously zero. The square of the uncertainty in σ_n is

$$(\Delta\sigma_n)^2 = P_n(+1) \cdot (1)^2 + P_n(-1) \cdot (-1)^2 = 1 \tag{1.70}$$

The variable $X = (\sigma_1 + \cdots + \sigma_N)/N = (N_R - N_L)/N$, where N_R and N_L are the numbers of particles in the right and left sides, respectively. Equation (1.71) gives $a = 1$. The probability distribution associated with X is therefore

$$P_X(x) = (N_A/\pi)^{1/2} e^{-N_A x^2} \tag{1.71}$$

If $N_R > (1 + 10^{-8})N_A$, then $X > 10^{-8}$. Thus, letting $\epsilon = 10^{-8}$, the answer to the question is

$$p = \int_\epsilon^\infty P_X(x)\, dx = (N_A/\pi)^{1/2} \int_\epsilon^\infty e^{-N_A x^2}\, dx \tag{1.72}$$

Using the variable $u = x\sqrt{N_A}$, we can express this in terms of the complementary error function.

$$p = \frac{1}{\sqrt{\pi}} \int_{\epsilon\sqrt{N_A}}^\infty e^{-u^2}\, du = \frac{1}{2} \operatorname{erfc}(\epsilon\sqrt{N_A}) \tag{1.73}$$

Using the asymptotic approximation for the complementary error function that is valid for $z \gg 1$,

$$\operatorname{erfc}(z) \sim \frac{e^{-z^2}}{\sqrt{\pi} z} \tag{1.74}$$

we get

$$p \approx \frac{e^{-N_A \epsilon^2}}{2\epsilon\sqrt{\pi N_A}} = \frac{e^{-6 \times 10^7}}{2.7 \times 10^4} \tag{1.75}$$

This number is so small that it is physically indistinguishable from zero. Notice that the suppression of fluctuations in X is entirely a statistical effect. We have assumed that no interparticle forces exist, and therefore such things cannot be used to explain the uniformity of the density distribution.

PROBLEMS

1.1 Calculate the probability of getting the following things in one throw of dice.
(a) An odd number. (b) The number 7. (c) A multiple of 3.

1.2 5 spin variables, $s_1, s_2, \ldots, s_5$, in a magnetic field are statistically independent. Each can take the values $\pm \hbar/2$ with the probabilities

$$\mathbf{P}[s = \hbar/2] = p \text{ and } \mathbf{P}[s = -\hbar/2] = 1 - p$$

What is the probability that exactly 3 spins are up $(+\hbar/2)$?

1.3 10 computers are chosen at random from 100, of which 10 are defective. What is the probability that all 10 selected ones work?

1.4 Ignoring leap year, calculate the probability that, among 25 randomly chosen people, at least two have the same birthday.

1.5 12 books, containing a 4-volume series, are placed in random order on a shelf. What is the probability that the series is placed together and in order from left to right?

1.6 Let x be a pair of integers (m, n), each in the range from 1 to 10. The fundamental probability of x is $P(x) = (m + n)C$, where C is a constant. What is the probability that $m = n$?

1.7 How many different ways are there of seating 10 people at a round table with 12 chairs if two seatings that are related by a simple rotation are not considered as different?

1.8 Each of the 50 states supplies two U.S. senators. If N senators are chosen at random, what is the probability that they represent exactly K states? (Warning: this is a very difficult problem.)

1.9 From the assumptions listed in Section 1.4, derive Eq. (1.36).

1.10 Prove the following set identities, due to De Morgan. (a) $\overline{A \cup B} = \bar{A} \cap \bar{B}$; (b) $\overline{A \cap B} = \bar{A} \cup \bar{B}$ Feel free to use words or pictures, but do not use the set notations $+$ or $-$, because there is a danger of error, due to the fact that $(A + B) - B \neq A$.

1.11 Let A and B be two subsets of Ω and let $C_1 = A$, $C_2 = B - A$, and $C_3 = \overline{A \cup B}$. Show that $\sum_i \mathbf{P}[C_i] = 1$ and $\mathbf{P}[C_i \cap C_j] = 0$ for $i \neq j$.

1.12 A (finite or infinite) sequence of statements $q_1, q_2, \ldots$, is *complete and mutually exclusive* if the probability that one and only one of the statements is true is equal to one. For example, the set of statements "today is Monday", today is Tuesday", $\ldots$, today is Sunday" is complete and mutually exclusive. Assuming that $q_1, q_2, \ldots$ are complete and mutually exclusive, prove that, for any statement q,

$$\mathbf{P}[q] = \sum_i \mathbf{P}[q|q_i] \mathbf{P}[q_i]$$

1.13 For three well-made dice (red, blue, and green), given that $R + B + G = 12$, what is probability distribution for R?

1.14 x is a random variable with a probability distribution that is uniform within the interval $(0, 1)$ and zero outside that interval. What is the probability distribution of $y = \ln(x)$?

1.15 Show that it is impossible to have a point distributed in the x–y plane with the probability density $P(x, y) = C/(x^2 + y^2 + a^2)$, where C and a are constants.

1.16 Each of the 1 Ω resistors shown in Fig. 1.7 actually has a probability density for its resistance of $P(r) = \sqrt{\alpha/\pi} \exp[-\alpha(r - 1)^2]$. They are statistically independent. What is the probability distribution for the combined resistance r_{ab}? [Note: Assume that $\alpha \gg 1$ so that $P(r) \approx 0$ for the unphysical negative values of r. Then one can take all the integrals from $-\infty$ to ∞.]

Fig. 1.7

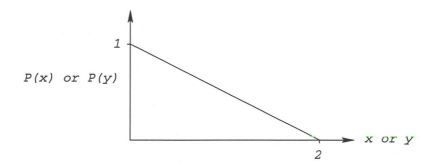

1.17 x and y are independent random variables, each with the range from 0 to 2 and the probability distribution shown in Fig 1.8. What is the probability function associated with the variable $s = x + y$? (Note that the range of s is 0 to 4.)

Fig. 1.8

1.18 Show that it is impossible for a point to be distributed over the infinite plane with a probability density of the form $P(x, y) = C/(x^2 + y^2 + 1)$.

1.19 A 1 inch needle is dropped on a sheet of paper on which parallel lines are drawn, 1 inch apart. What is the probability that it misses the lines?

1.20 A 1 inch needle is dropped on a sheet of paper on which 1 inch squares are drawn. What is the probability that it falls completely within one square?

1.21 $P(x, y)$ is uniform within the circle $x^2 + y^2 < a^2$. Calculate $P(x|y)$.

1.22 A harmonic oscillator oscillates with amplitude A. If the time t is chosen at random, what is the probability that $a \le x(t) \le a + da$? That is, find $P_x(a)$.

1.23 In a one-dimensional gas the particles have a velocity distribution $P(v) = (a/\pi)^{1/2} e^{-av^2}$. That is, the probability that a particle picked at random has velocity between v and $v + dv$ is $P(v)dv$. What is the probability that the next particle that passes the origin from left to right has velocity in the range v to $v + dv$?

1.24 A gas contains equal numbers of A particles and B particles. The A particles and B particles, respectively, have energy distributions $P_A(E)$ and $P_B(E)$, where

$0 \leq E < \infty$. Given that a particle has an energy larger than E_o, write an expression for the probability that it is an A particle.

1.25 Let x be a random variable that is restricted to the interval $[0, L]$. Assume that $\mathbf{P}[a < x < b] = f(b - a)$ for any $0 \leq a < b \leq L$. Show that $f(u) = u/L$.

1.26 Prove that, for the Poisson distribution, $\Delta n = \sqrt{\langle n \rangle}$.

The next three problems are concerned with *random walks*.

1.27 Each second a particle, which was initially at $x = 0$, jumps either left or right (each with a probability of $\frac{1}{2}$) a distance a. At time $t_n = n$ the particle is at location $x_k = ka$ with probability $P(n, k)$. Calculate $P(n, k)$ and show that, as n and k approach infinity, your result agrees with the central limit theorem.

1.28 Do Problem 1.27 for the case in which the particle jumps to the right with probability R and to the left with probability $L = 1 - R$.

1.29 Each second a particle, which was initially at $x = 0$, makes a jump. The probability distribution of the jump displacements is $p(a)$, where $-\infty < a < \infty$. Assume that $p(a) = p(-a)$ and that $\langle a^2 \rangle = \beta^2$. Use the central limit theorem to determine the probability distribution for the particle's position after a long time.

1.30 For a gas at STP (standard temperature and pressure), draw a plot of the probability that a cube of side a is completely empty as a function of a.

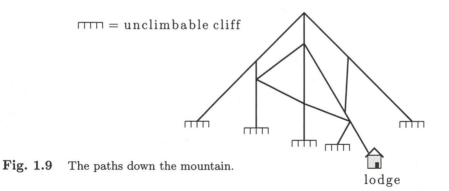

Fig. 1.9   The paths down the mountain.

1.31 Beginning at the peak and never going uphill, a hiker chooses the path randomly at each fork (see Fig. 1.9). What is the probability that she spends the night at the lodge?

Chapter 2
The Ideal Gas

2.1 THE CLASSICAL IDEAL GAS

As an introduction to the methods of statistical mechanics we will investigate the properties of a simple but very important system; the *ideal gas*. The ideal gas is defined as a collection of identical particles in which the interparticle forces are negligible. It is a reasonably accurate model of a real gas at low density.

Although it will be assumed that the interparticle forces are weak enough to be neglected, one should not picture the ideal gas as a system of particles with no interactions at all. Such a system would have very peculiar physical properties. If it began in some highly exceptional state, such as one in which all of the particles were moving with exactly the same speed but in various directions, then, because of the complete absence of interparticle collisions, the velocity distribution would never change and approach a more typical one in which the particles have a wide range of speeds. In short, such a system would never come to equilibrium. When we say that the interparticle interactions are negligible we mean that we can neglect the potential energy associated with them in comparison with the kinetic energy of the gas. However, at the same time, it is necessary to assume that the particles do have frequent collisions that transfer energy from one particle to another. In this and later chapters it will be demonstrated that these two assumptions are not mutually inconsistent by showing that both of them are satisfied for many real gases over a wide range of pressure and temperature.

Any gas in which the interaction potential is negligible is called an ideal gas. However, additional simplifying assumptions will be made in order to reduce the mathematical complexity of the analysis. It will be assumed that:

1. The particles have no internal degrees of freedom, such as rotation, vibration, or electronic excitation. We will therefore treat the particles as point particles. At room temperature this is an excellent approximation for all monatomic gases, particularly the noble gases (He, Ne, etc.), but not for diatomic or other molecular gases.

2. Quantum effects are negligible. This also is an excellent approximation at ordinary temperatures and pressures. Later in this chapter, we will carry out a quantum mechanical treatment of the ideal gas and determine the range of validity of the classical analysis.

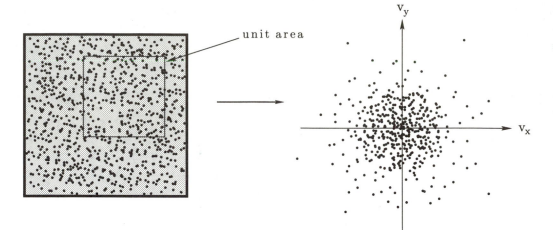

Fig. 2.1   The definition of the velocity distribution function for a two-dimensional gas. The velocity components of each particle in a unit area are plotted as a dot in the two-dimensional velocity space. When that is done, the density of dots at position (v_x, v_y) in the velocity space is defined as $f(v_x, v_y)$.

2.2   THE VELOCITY DISTRIBUTION

The focus of our study will be the velocity distribution function of an ideal gas (Fig. 2.1). It is defined by saying that, when the gas is at equilibrium, the number of particles per unit volume that have velocities within the velocity element $d^3\mathbf{v}$, centered at the velocity $\mathbf{v}$, is given by $f(\mathbf{v})\,d^3\mathbf{v}$. Due to the repeated scattering of the particles, the velocity distribution at equilibrium depends on the magnitude but not on the direction of $\mathbf{v}$.

In terms of the velocity distribution function it is a simple matter to calculate the pressure that would be exerted by the gas on the walls of its container. We look at a small area dA on one of the walls and assume that dA is perpendicular to the x axis with the inside of the container to the left. We consider only those particles that have velocities within some range $d^3\mathbf{v}$, centered at a velocity $\mathbf{v}$, and ask: "Of those particles, how many will strike the area dA within some short time interval dt?" Unless the particles are moving toward dA (that is, unless $v_x > 0$) the answer is obviously zero. If $v_x > 0$ then the answer to the question is (see Fig. 2.2.)

$$dN = f(\mathbf{v})\,d^3\mathbf{v}\,v_x\,dA\,dt \qquad (2.1)$$

It is assumed that each particle that strikes the wall rebounds elastically. The momentum delivered to the wall by a single rebounding particle is equal to the negative of the change in the momentum of the particle. That is, it is equal to $2mv_x$. Thus the total momentum delivered to the area dA during the time interval dt by those particles within the velocity range $d^3\mathbf{v}$ is

$$2mv_x^2 f(\mathbf{v})\,d^3\mathbf{v}\,dA\,dt \qquad (2.2)$$

Since the force that one object exerts on another is defined as the rate at which the first object transfers momentum to the second, we can calculate the force those particles exert on dA by dividing Eq. (2.2) by dt. The pressure is obtained by dividing the force by dA. Finally, the pressure due to particles of all possible velocities is obtained by integrating over all velocities that are aimed at the wall.

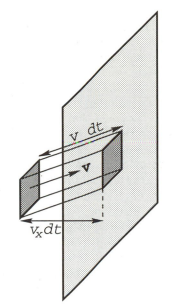

Fig. 2.2   Within the time interval dt, those particles with velocity **v** that lie in the cylinder hit the shaded area on the wall. The density of particles with velocities within the range $d^3\mathbf{v}$ is $f(\mathbf{v})\,d^3\mathbf{v}$. The volume of the cylinder is $v_x\,dt\,dA$.

A velocity is aimed at the wall only if v_x is positive.

$$p = \int_{v_x>0} 2mv_x^2 f(\mathbf{v})\,d^3\mathbf{v} \tag{2.3}$$

Because the integrand is an even function of v_x it will leave the value of p unchanged if we extend the integration to all v_x and drop the factor of 2. Also, since $f(\mathbf{v})$ depends only on the magnitude of **v**, v_x^2 can be replaced by $(v_x^2 + v_y^2 + v_z^2)/3$. This gives

$$p = \frac{1}{3}\int mv^2 f(\mathbf{v})\,d^3\mathbf{v} \tag{2.4}$$

Recalling the definition of $f(\mathbf{v})$, it is easy to see that the integral is equal to twice the kinetic energy per unit volume. Since it has been assumed that the potential energy is negligible, Eq. (2.4) implies that

$$p = \tfrac{2}{3}\frac{E}{V}, \tag{2.5}$$

a result the reader has probably seen before.

"... the trial already made sufficiently proves the main thing, for which I here allege it; since by it, it is evident that as common air, when reduced to half its wonted extent, obtained near about twice as forcible a spring (i.e. pressure) as it had before; so this thus compressed air being further thrust into half this narrow room, obtained thereby a spring about as strong again as that it last had, and consequently four times as strong as that of the common air. "
                —— Robert Boyle, *New Experiments Physico-Mechanical* (1662)
Boyle had deduced his famous law, that the pressure of a gas is inversely proportional to its volume, by capturing a quantity of air in the short closed end of a glass U tube, the other long open end of which was gradually filled with increasing amounts of mercury. The pressure was determined by the difference in the heights of the mercury columns, while the volume could be taken as proportional to the length of the entrapped air column.

2.3 THE MAXWELL–BOLTZMANN DISTRIBUTION

Equation (2.5) has been obtained without using any details of the velocity distribution function other than its angle independence. Therefore, the equation gives no information about $f(\mathbf{v})$. In order to determine $f(\mathbf{v})$ one must go beyond mechanics alone and use methods derived from probability theory. The derivation of the velocity distribution is a fairly intricate argument; so it will help if we first consider a simpler problem that can be solved with exactly the same method that will be needed for that derivation. We consider three students, A, B, and C, and three rooms, numbered 1, 2, and 3. Each of the students is told to pick one of the rooms at random and go into it. We assume that the rooms are arranged symmetrically, so that there is equal probability of a given student picking each of the three rooms. A *microstate* is defined by stating which room each student has chosen. For example, "A is in Room 2, B is in Room 3, and C is in Room 2" defines a microstate. Because each student has equal probability of entering each room, it is clear that every microstate is equally probable. We now define a *macrostate* by saying how many students are in each of the rooms. For example, "3 students are in Room 1, 0 students are in Room 2, and 0 students are in Room 3" defines a macrostate. The information needed to define a microstate is enough to allow one to calculate the corresponding macrostate but not vice versa. Thus, the microstates give more detailed information about the system. We now ask: 'What is the most probable macrostate?' Since each microstate has the same probability, the probability of any macrostate is proportional to the number of microstates that correspond to it. For example, the macrostate "$N_1 = 3$, $N_2 = 0$, and $N_3 = 0$" (N_n is the number of students in Room n.) has only one corresponding microstate, namely "A is in Room 1, B is in Room 1, and C is in Room 1". An easy calculation shows that the macrostate, "$N_1 = 0$, $N_2 = 1$, and $N_3 = 2$" has 3 corresponding microstates, while the macrostate, "$N_1 = N_2 = N_3 = 1$" has $3! = 6$ corresponding microstates and is thus the most probable distribution of students among rooms. The reader will soon see that, if we replace "students" by "particles" and "rooms" by "phase-space boxes" (soon to be defined), then the calculation to follow is essentially identical to the one that has just been done.

It will not greatly increase the difficulty of calculating the velocity distribution function if the problem is generalized to include an external potential field, $U(\mathbf{r})$, such as would be produced by a gravitational force. Thus, it is assumed that the energy of a particle of velocity $\mathbf{v}$ at location $\mathbf{r}$ is

$$E = \tfrac{1}{2}mv^2 + U(\mathbf{r}) \tag{2.6}$$

(It should be mentioned that it definitely would greatly increase the difficulty of the problem to include an interaction potential, that is, a potential function that depended on the distances *between* particles.) The distribution function, which has the same meaning as before, will now depend on both $\mathbf{r}$ and $\mathbf{v}$. Because of the existence of a force field, which defines a special direction at each location, it is not at all obvious that the distribution function $f(\mathbf{r}, \mathbf{v})$ will be independent of the direction of $\mathbf{v}$ (although the results of our calculation will show that it is so).

We are considering a system of N particles in a cubic box of volume V. The total energy of the system is E. By collisions, individual particles may change their energies, but the total energy remains fixed. The most essential assumption in our analysis is that N is an extremely large number. We conceptually decompose the volume into a very large number of little cubes, numbered 1 to K. The kth little cube is centered at $\mathbf{r}_k$ and has a volume we call $\Delta^3 r$. We also separate the infinite space of possible velocities into an infinite number of "velocity cubes", each

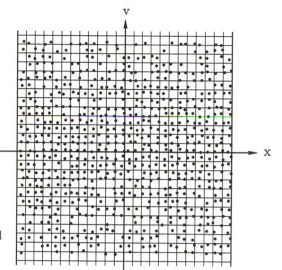

Fig. 2.3 The macrostate of the system is specified by giving the number of particles in each phase-space box. In reality, the phase-space is a six-dimensional space although it is shown here as two-dimensional.

of volume $\Delta^3 v$, labeled with an index $l = 1, 2, \ldots$. The detailed dynamical state of the system, which we will call the *microstate* of the system, is defined by specifying the exact position and velocity of each of the N particles. That is, the microstate is specified by the $2N$ vector variables, $\mathbf{r}_1, \mathbf{v}_1, \ldots, \mathbf{r}_N, \mathbf{v}_N$. If we permit a little bit of uncertainty in the definition of each particle's position and velocity, then the microstate of the system can be specified by giving the spatial and velocity box of each particle, that is, by giving the $2N$ integer variables $k_1, l_1, \ldots, k_N, l_N$. This step has the advantage of making the set of microstates discrete. The combination of a spatial box and a velocity box is called a *phase-space* box. It can be considered as a little box in a six-dimensional phase-space with coordinates (x, y, z, v_x, v_y, v_z). To calculate the distribution function associated with a given microstate it is quite unnecessary to know which particles are in which phase-space boxes. It is sufficient to know only how many particles are in each of the phase-space boxes. In fact, if, for a given microstate, the *occupation number* of the phase-space box specified by the integers k and l is called N_{kl}, then the distribution function at $(\mathbf{r}_k, \mathbf{v}_l)$ is

$$f(\mathbf{r}_k, \mathbf{v}_l) = N_{kl}/\Delta^3 r \Delta^3 v \tag{2.7}$$

We will say that the set of occupation numbers N_{kl} defines the *macrostate* of the system (see Fig. 2.3). Clearly, we are using the same definition of microstates and macrostates that we did previously in the "student and room" discussion.

We now ask the question: "How many different microstates correspond to a given macrostate?" This is just the question of how many ways there are of putting N distinguishable objects into boxes with a specified number of objects in each box. It was answered in the last chapter [see Eq. (1.24)]. The number of microstates corresponding to a given macrostate is

$$I = \frac{N!}{\prod_{(k,l)} N_{kl}!} \tag{2.8}$$

where the symbol $\prod_{(k,l)}$ means that a product is taken over all values of k and l. The product extends over all phase-space boxes. However, since $0! = 1$, and there are only a finite number of particles, only a finite number of terms in the product are not equal to unity. But a *factor* of one can be ignored. Thus the infinite product is actually a finite integer. Now that we know how many microstates correspond to any given macrostate we would like to determine which of the possible macrostates has the greatest probability of occurrence. From the principles of probability theory it is clearly impossible to do this without knowing something about the relative probabilities of the various microstates that contribute to a given macrostate. There are certain things that are known about the probability of finding the system in a given microstate.

1. A particle in the phase-space box (k, l) has an energy E_{kl}, given by

$$E_{kl} = \tfrac{1}{2}mv_l^2 + U(\mathbf{r}_k) \tag{2.9}$$

Therefore a macrostate defined by the occupation numbers N_{kl} has a total energy

$$E_{\text{tot}} = \sum_{(k,l)} N_{kl} E_{kl} \tag{2.10}$$

Unless E_{tot} is equal to the known system energy E, there is zero probability of finding the system in any microstate that yields that macrostate.

2. Also, unless N_{tot}, the sum of the occupation numbers for a given macrostate, is equal to the actual number of particles in the system N, there is no microstate that yields that macrostate.

Aside from these two restrictions, there is no obvious way of assigning probabilities to the vast number of possible microstates available to a large system.*

At this point we will make a completely unjustified assumption that those microstates that are not forbidden by the conditions on E_{tot} and N_{tot} are all equally probable. This is called the assumption of *equal a priori probabilities*. Much of Chapter 3 will be devoted to a careful analysis of this assumption. Here it will be used simply as a working hypothesis that will allow us to proceed in our calculation of the equilibrium distribution function $f(\mathbf{r}, \mathbf{v})$.

With the assumption of equal a priori probabilities for the allowed microstates, the probability of any allowed macrostate is proportional to I, the number of microstates corresponding to that macrostate. The *most probable macrostate* can be calculated by finding the maximum of I, considered as a function of the occupation numbers, N_{kl}, with the restrictions that $\sum N_{kl} E_{kl} = E$ and $\sum N_{kl} = N$. It turns out to be more convenient to calculate the maximum of the function

$$F = \log I = \log(N!) - \sum_{k,l} \log(N_{kl}!) \tag{2.11}$$

(Note: All logarithms in this book are natural logarithms—The base 10 logarithm is as much of a historical curiosity as a slide rule.) Because $\log I$ is a monotonically increasing function of I, the set of occupation numbers that maximizes $\log I$ also gives the maximum of I. In Eq. (2.11) one can use *Stirling's approximation* (derived in the Mathematical Appendix), which states that $\log K! \approx K(\log K - 1)$ for $K \gg 1$. Therefore, the function that will actually be maximized is

$$F = N(\log N - 1) - \sum_{k,l} N_{kl}(\log N_{kl} - 1) \tag{2.12}$$

---

* To get an idea of of the size of the number I in Eq. (2.8) when $N = N_A \approx 6 \times 10^{23}$, as it is for one mole of gas, we note that the dominant factor in Eq. (2.8) is the $N!$, which is of order N^N for large N. Thus $I \sim 6 \times 10^{23 \times 6 \times 10^{23}}$. That number has more than 10^{25} zeros!

Applying Lagrange's method (also derived in the Mathematical Appendix) to this problem, one ignores the constraints and maximizes the function

$$G = F - \alpha \sum_{k,l} N_{kl} - \beta \sum_{k,l} N_{kl} E_{kl} \qquad (2.13)$$

where α and β are two *Lagrange parameters* whose values will later be chosen so as to satisfy the constraints. Setting $\partial G / \partial N_{kl}$ equal to zero gives

$$\log N_{kl} = -\alpha - \beta E_{kl} \qquad (2.14)$$

or

$$N_{kl} = C e^{-\beta E_{kl}} \qquad (2.15)$$

where the constant $C = \exp(-\alpha)$. Using Eq. (2.9) to write E_{kl} in terms of $\mathbf{r}$ and $\mathbf{v}$ and using Eq. (2.7) to relate N_{kl} and $f(\mathbf{r}_k, \mathbf{v}_l)$, it is seen that the most probable distribution function (that is, the one with the greatest number of corresponding microstates) is of the form

$$f(\mathbf{r}, \mathbf{v}) = C \exp[-\beta(mv^2/2 + U(\mathbf{r}))] \qquad (2.16)$$

This is the *Maxwell–Boltzmann distribution*. The constants C and β must be chosen to give the correct values for the total number of particles and the total energy.

2.4   THE PARTICLE DENSITY

The density of particles, irrespective of velocity, is obtained by integrating $f(\mathbf{r}, \mathbf{v})$ over all velocities.

$$
\begin{aligned}
n(\mathbf{r}) &= \int f(\mathbf{r}, \mathbf{v}) \, d^3 v \\
&= C e^{-\beta U(\mathbf{r})} \int e^{-m\beta v^2/2} \, dv_x \, dv_y \, dv_z \\
&= C e^{-\beta U(\mathbf{r})} \left[\int_{-\infty}^{\infty} e^{-m\beta v_x^2/2} \, dv_x \right]^3 \\
&= C \left(\frac{2\pi}{m\beta} \right)^{3/2} e^{-\beta U(\mathbf{r})}
\end{aligned}
\qquad (2.17)
$$

The final line in Eq. (2.17) makes use of the Table of Integrals in the Mathematical Appendix. The potential function $U(\mathbf{r})$ always includes an arbitrary constant in its definition. If U is defined to be zero at some chosen location, $\mathbf{r}_o$, then

$$n(\mathbf{r}_o) = C \left(\frac{2\pi}{m\beta} \right)^{3/2} \qquad (2.18)$$

which allows us to write Eq. (2.17) in the form

$$n(\mathbf{r}) = n(\mathbf{r}_o) e^{-\beta U(\mathbf{r})} \qquad (2.19)$$

If we assume that this most probable distribution is the experimentally observed equilibrium distribution, then we see that, whenever the interaction potential energy can be neglected, the particle density at equilibrium varies with the negative exponential of the external potential.

2.5   THE EXPONENTIAL ATMOSPHERE

An important example of a system that exhibits a Maxwell–Boltzmann distribution is an ideal gas in a uniform gravitational field g. In that case $U(\mathbf{r}) = mgz$ and $n(z) = n(0)\exp[-\beta mgz]$. In the next section, we will show that the Lagrange parameter, β, is equal to $1/kT$, where k is Boltzmann's constant and T is the absolute temperature. Thus the density, as a function of z, has the form

$$n(z) = n(0)e^{-z/h} \tag{2.20}$$

with $h = kT/mg$. The *scale height*, h, is the altitude at which $n(z) = n(0)/e$. For nitrogen at room temperature (300K) the scale height is 9 kilometers, which is reasonably descriptive of the earth's atmosphere, although the earth's atmosphere is by no means at equilibrium. This number clearly illustrates the fact that the variation in density due to the gravitational field would be difficult to detect in ordinary laboratory experiments where the altitudes involved are about 1 meter and therefore $n(z)/n(0) \approx 1 - 1/9000$. There are, however, two ways of bringing h down to laboratory size. One method is by increasing the mass of the particles, m. This can be easily overdone by choosing macroscopic objects as the "particles". For $m = 1\,\mathrm{g}$ the scale height is much less than the size of an atom and is again unobservable. To get a scale height of about $10\,\mathrm{cm}$ at room temperature, we need $m \approx 4 \times 10^{-20}\,\mathrm{kg}$. This is the mass of a particle of 2.5×10^{7} atomic mass units, too large for a molecule but possible for the particles of a fine colloidal suspension. The second method is to increase g by means of a centrifuge. A modern ultracentrifuge can create an acceleration field of order $10^{5}\,\mathrm{m/s^2}$. In such a field a molecule of 10 atomic mass units has a scale height of $2.5\,\mathrm{cm}$.

2.6   THE IDENTIFICATION OF β

$f(\mathbf{r}, \mathbf{v})\, d^3r\, d^3v$ is the number of particles within d^3r that have velocities in the range d^3v. If we divide this by d^3r, multiply by $mv^2/2$, and integrate over $\mathbf{v}$, we obtain the density of kinetic energy at position $\mathbf{r}$. That quantity will be written as ε_K. Thus

$$
\begin{aligned}
\varepsilon_K &= Ce^{-\beta U(\mathbf{r})} \int e^{-m\beta v^2/2}(mv^2/2)\, dv_x\, dv_y\, dv_z \\
&= \tfrac{3}{2} Ce^{-\beta U(\mathbf{r})} \int e^{-\beta mv_x^2/2}(mv_x^2)\, dv_x \int e^{-\beta mv_y^2/2}\, dv_y \int e^{-\beta mv_z^2/2}\, dv_z \\
&= \tfrac{3}{2} Ce^{-\beta U(\mathbf{r})} \left(\sqrt{2\pi/m\beta^3}\right)\left(\sqrt{2\pi/m\beta}\right)^2
\end{aligned}
\tag{2.21}
$$

The second line makes use of the fact that each of the three terms in $v^2 = v_x^2 + v_y^2 + v_z^2$ gives an equal contribution to ε_K. The third line may require another look at the Table of Integrals. Equaation (2.17) now allows us to write ε_K in terms of the particle density.

$$\varepsilon_K = \tfrac{3}{2}n(\mathbf{r})/\beta \tag{2.22}$$

The first thing to notice is that the energy per particle, $\varepsilon_K/n(\mathbf{r})$, is independent of position. Although, at equilibrium, the density of particles varies from place to place as $e^{-\beta U}$, their distribution with respect to velocity is the same everywhere.

Equation (2.5), which relates the energy density to the pressure, and the ideal gas equation of state, give the relation $\varepsilon_K = \tfrac{3}{2}p = \tfrac{3}{2}n(\mathbf{r})kT$. A comparison of this relation with Eq. (2.22) allows us to identify the Lagrange parameter, β, with $1/kT$.

$$\beta = 1/kT \tag{2.23}$$

2.7   THE MAXWELL DISTRIBUTION

If no external force field exists, then $U = 0$ and the Maxwell–Boltzmann distribution becomes the simpler Maxwell distribution, which depends on velocity alone.

$$f(\mathbf{v}) = Ce^{-\frac{1}{2}mv^2/kT} \tag{2.24}$$

Using Eq. (2.17) with $U = 0$ and n equal to a constant, we can eliminate C and write the Maxwell distribution in terms of n and T.

$$f(\mathbf{v}) = \left(\frac{m}{2\pi kT}\right)^{3/2} ne^{-mv^2/2kT} \tag{2.25}$$

Because $v^2 = v_x^2 + v_y^2 + v_z^2$, the distribution function factors into three uncorrelated probability functions involving the three velocity components.

$$f(\mathbf{v}) = nP(v_x)P(v_y)P(v_z) \tag{2.26}$$

where $P(v_x)\,dv_x$ is the probability of finding a given particle with its x component of velocity in the range dv_x. Clearly, $P(v_x)$ is given explicitly as

$$P(v_x) = \left(\frac{m}{2\pi kT}\right)^{1/2} e^{-mv_x^2/2kT} \tag{2.27}$$

Although it may seem slightly surprising, Eq. (2.26) shows that the fact that a particular particle has a large x component of velocity in no way affects the probability of its having large y or z components.

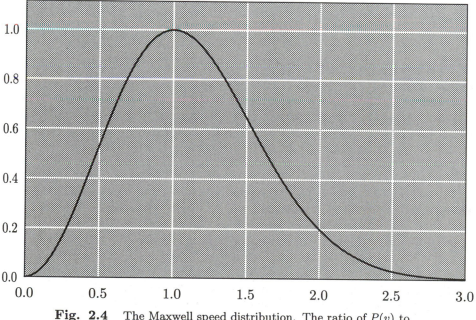

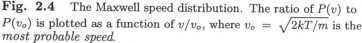

Fig. 2.4   The Maxwell speed distribution. The ratio of $P(v)$ to $P(v_o)$ is plotted as a function of v/v_o, where $v_o = \sqrt{2kT/m}$ is the *most probable speed.*

2.8   THE SPEED DISTRIBUTION

We define $F(v)\,dv$ as the density of particles with speeds within the range v to $v+dv$ (see Fig. 2.4). $F(v)\,dv$ is equal to $f(\mathbf{v})$ times $4\pi v^2 dv$, where the second factor is just the volume of the region in velocity space corresponding to the given speed range. Therefore,

$$F(v) = 4\pi v^2 n\left(\frac{m}{2\pi kT}\right)^{3/2} e^{-mv^2/2kT} \tag{2.28}$$

The average speed of the particles of a gas at temperature T is given by

$$\bar{v} = \frac{1}{n}\int_o^\infty F(v)v\,dv \tag{2.29}$$

The integration is straightforward and gives

$$\bar{v} = \left(\frac{8kT}{\pi m}\right)^{1/2} \tag{2.30}$$

In order to get an idea of the typical velocities involved in thermal motion, we evaluate $\bar{v}$ for helium and radon at $T = 300\,\mathrm{K}$.

$$\text{(He)} \ \bar{v} = 1.26 \times 10^3\,\mathrm{m/s}; \qquad \text{(Rn)} \ \bar{v} = 2.72 \times 10^2\,\mathrm{m/s} \tag{2.31}$$

The *root-mean-square* velocity is defined as $v_{\mathrm{rms}} = \left(\overline{v^2}\right)^{1/2}$. It can easily be calculated by making us of the fact that the average kinetic energy per particle is $\frac{3}{2}kT$. That is

$$\tfrac{1}{2}m\overline{v^2} = \tfrac{3}{2}kT \tag{2.32}$$

which gives

$$v_{rms} = (3kT/m)^{1/2} \tag{2.33}$$

2.9   THE VERY DILUTE GAS

In an ideal gas, there is a length ℓ, called the *mean free path*, that is defined as the average distance that a gas particle travels between collisions with the other gas particles. In Exercise 2.17 it is shown that, for particles of diameter D, the mean free path is given by the formula $\ell = 1/(\sqrt{2}\pi D^2 n)$, where n is the gas density. For argon, a typical noble gas, at standard temperature and pressure (STP), the mean free path is about $10^{-5}\,\mathrm{m}$. This is much larger than the typical distance between a particle and its nearest neighbor, which, for the same system, is about $10^{-9}\,\mathrm{m}$.

In the problems and exercises, there are a number of calculations involving the flow of particles from an ideal gas through a small hole in the container wall. These calculations involve a geometrical length a, namely, the size of the hole. Although the mean free path is much larger than the nearest neighbor distance, it is usually much smaller than the geometrical distance a. This means that a particle that is destined to pass through the hole will usually collide many times with other particles on its way out. When this is so (that is, when $a \gg \ell$), the calculation of flow through the hole is very complicated and will not be attempted in this book. Here we will restrict ourselves to making calculations in the opposite limit, in which $a \ll \ell$. Because ℓ is proportional to $1/n$, this case occurs only in a very dilute ideal gas. Thus, when it is stated that an ideal gas is *very dilute*, what is meant is that

"The modern atomists have therefore adopted a method which is, I believe, new in the department of mathematical physics, though it has long been in use in the section of Statistics. When the working members of Section F get hold of a report of the Census, or any other document containing the numerical data of Economic and Social Science, they begin by distributing the whole population into groups, according to age, income-tax, education, religious belief, or criminal convictions. The number of individuals is far too great to allow of their tracing the history of each separately, so that, in order to reduce their labour within human limits, they concentrate their attention on a small number of artificial groups. The varying number of individuals in each group, and not the varying state of each individual, is the primary datum from which they work."

—— J. C. Maxwell (in *Nature*, 1873)

the mean free path is much larger than any geometrical lengths that appear in the problem.

2.10 IDEAL QUANTUM SYSTEMS

The method that was used to derive the Maxwell–Boltzmann distribution for a classical system can be applied, with only minor modifications, to calculate the equilibrium properties of a quantum-mechanical system of noninteracting particles in an external potential. First, we will have to review the quantum mechanics of systems of many identical particles. The reader who is already familiar with this material or who is willing to accept the results of the analysis "on faith" can jump to Section 2.13.

A single particle in a potential field $U(\mathbf{r})$ is described by the Schrödinger equation

$$\left(-\frac{\hbar^2}{2m}\nabla^2 + U(\mathbf{r})\right)\psi(\mathbf{r}) = E\psi(\mathbf{r}) \tag{2.34}$$

Therefore, its Hamiltonian operator is

$$H = -\frac{\hbar^2}{2m}\nabla^2 + U(\mathbf{r}) \tag{2.35}$$

A system of two identical particles in a potential field $U(\mathbf{r})$ has a Schrödinger equation

$$\left(-\frac{\hbar^2}{2m}\nabla_1^2 + U(\mathbf{r}_1)\right)\psi(\mathbf{r}_1,\mathbf{r}_2) + \left(-\frac{\hbar^2}{2m}\nabla_2^2 + U(\mathbf{r}_2)\right)\psi(\mathbf{r}_1,\mathbf{r}_2) = E\psi(\mathbf{r}_1,\mathbf{r}_2) \tag{2.36}$$

where $\nabla_1^2 = \partial^2/\partial x_1^2 + \partial^2/\partial y_1^2 + \partial^2/\partial z_1^2$ with a corresponding equation for ∇_2^2. The Hamiltonian operator for that system is thus

$$H = -\frac{\hbar^2}{2m}\nabla_1^2 + U(\mathbf{r}_1) - \frac{\hbar^2}{2m}\nabla_2^2 + U(\mathbf{r}_2) \tag{2.37}$$

The generalization is obvious; a system of N identical particles in an external potential $U(\mathbf{r})$ is described by the Hamiltonian operator

$$H_N = \sum_{j=1}^{N}\left(-\frac{\hbar^2}{2m}\nabla_j^2 + U(\mathbf{r}_j)\right) \tag{2.38}$$

To calculate the N-particle energy eigenstates of this system it is only necessary to solve the following single-particle Schrödinger equation.

$$\left(-\frac{\hbar^2}{2m}\nabla^2 + U(\mathbf{r})\right)u_n(\mathbf{r}) = E_n u_n(\mathbf{r}) \tag{2.39}$$

The problem of actually solving this equation for particular potential functions will not be discussed here. It will simply be assumed that it has been done and that the complete set of single-particle eigenfunctions $u_n(\mathbf{r})$ and eigenvalues E_n is known.

A complete set of eigenfunctions for the N-particle problem can now be constructed simply by taking arbitrary products of the single-particle eigenfunctions. For example, for $N = 3$, the function

$$\psi(\mathbf{r}_1, \mathbf{r}_2, \mathbf{r}_3) = u_1(\mathbf{r}_1)u_1(\mathbf{r}_2)u_4(\mathbf{r}_3) \tag{2.40}$$

satisfies the Schrödinger equation

$$\sum_1^3\left(-\frac{\hbar^2}{2m}\nabla_j^2 + U(\mathbf{r}_j)\right)\psi = E\psi \tag{2.41}$$

with the eigenvalue $E = 2E_1 + E_4$.

2.11   WAVE FUNCTION SYMMETRIES—BOSONS

Even though the wave function $\psi(\mathbf{r}_1, \mathbf{r}_2, \mathbf{r}_3)$, given above, is a solution of the Schrödinger equation for a three-particle system, it is *not* an acceptable wave function for a real system of three identical particles. The wave function of any system of identical particles must satisfy a symmetry condition when the coordinates of any two particles are interchanged. The detailed form of the symmetry condition depends on the type of particle. There are two possibilities. Bose–Einstein particles are particles with integer values for their total angular momentum (measured in units of $\hbar$). Fermi–Dirac particles have half-integer ($\frac{1}{2}$, $\frac{3}{2}$, etc.) values for their total angular momentum. A system of Bose–Einstein particles (also called *bosons*) must have a wave function whose value is unchanged whenever the coordinates of any pair of particles are interchanged. That is, for a three-boson system,

$$\psi(\mathbf{r}_1, \mathbf{r}_2, \mathbf{r}_3) = \psi(\mathbf{r}_2, \mathbf{r}_1, \mathbf{r}_3) = \psi(\mathbf{r}_1, \mathbf{r}_3, \mathbf{r}_2), \text{ etc.} \tag{2.42}$$

The unsymmetric wave function of Eq. (2.40) may be converted to a properly symmetrized Bose–Einstein wave function by taking an equally weighted average over all permutations of the particle coordinates, $\mathbf{r}_1, \mathbf{r}_2$, and $\mathbf{r}_3$.

$$\psi(\mathbf{r}_1, \mathbf{r}_2, \mathbf{r}_3) = C\sum_{\text{perm}} u_1(\mathbf{r}_{\alpha_1})u_1(\mathbf{r}_{\alpha_2})u_4(\mathbf{r}_{\alpha_3}) \tag{2.43}$$

where $(\alpha_1, \alpha_2, \alpha_3)$ is a permutation of $(1, 2, 3)$, C is a normalization constant, and the sum is over all 3! permutations. The reader can easily verify that this wave function is properly symmetric and that it is still an eigenfunction of the Schrödinger equation with the same eigenvalue as before. This symmetrization procedure can be generalized as follows.

For a system of N bosons, we choose any finite sequence of nonnegative integers, $N_1, N_2, \cdots, N_K$ that satisfy the condition $N_1 + \ldots + N_K = N$. We call them *occupation numbers*. Some of the N_js may be zero. We then construct the function

$$\psi(\mathbf{r}_1, \ldots, \mathbf{r}_N) = C\sum_{\text{perm}} \underbrace{u_1(\mathbf{r}_{\alpha_1})\cdots u_1}_{N_1 \text{ factors}}\underbrace{u_2\cdots u_2}_{N_2 \text{ factors}}\cdots\underbrace{u_K\cdots u_K(\mathbf{r}_{\alpha_N})}_{N_K \text{ factors}} \tag{2.44}$$

where, in this case, $(\alpha_1, \ldots, \alpha_N)$ is a permutation of $(1, \ldots, N)$ and the sum is over all $N!$ permutations. It is obvious that the function ψ is symmetric in its arguments. It is also an eigenfunction of H_N with an energy eigenvalue equal to

$$E = N_1 E_1 + \cdots + N_K E_K \tag{2.45}$$

For an N-particle Bose–Einstein system, there is one such energy eigenfunction for each distinct set of occupation numbers satisfying the constraint

$$N_1 + \cdots + N_K = N \tag{2.46}$$

2.12 FERMIONS

For a system of Fermi–Dirac particles, the wave function must be completely anti-symmetric, rather than symmetric. That is, it must satisfy the condition (we give an example with $N = 3$)

$$\psi(\mathbf{r}_1, \mathbf{r}_2, \mathbf{r}_3) = -\psi(\mathbf{r}_2, \mathbf{r}_1, \mathbf{r}_3) = -\psi(\mathbf{r}_1, \mathbf{r}_3, \mathbf{r}_2) \tag{2.47}$$

In this case the N-particle wave functions are constructed somewhat differently. Since all fundamental fermions are spin-$\frac{1}{2}$ particles, we will only consider that case.* We have to introduce variables to describe the spin degrees of freedom of the particle. The state of a spin-$\frac{1}{2}$ particle can be defined by a combination of a position vector $\mathbf{r}$ that says where the particle is and a spin variable σ, whose only possible values are $+1$ or -1. The value of σ gives the z component of the spin angular momentum, divided by $\hbar/2$. We will use a single symbol, x_j, for the combination of the spatial variable and the spin variable of the jth particle.

$$x_j = (\mathbf{r}_j, \sigma_j) \tag{2.48}$$

As before, we first solve the single-particle Schrödinger equation for the sequence of single-particle eigenfunctions and eigenvalues

$$\left(-\frac{\hbar^2}{2m}\nabla^2 + U(x)\right)u_n(x) = E_n u_n(x) \tag{2.49}$$

We now choose any sequence of N single-particle eigenfunctions, $u_{K_1}, \ldots, u_{K_N}$, and construct the N-particle wave function

$$\psi(x_1, \ldots, x_N) = C \sum_{\text{perm}} (-1)^P u_{K_1}(x_{\alpha_1}) \cdots u_{K_N}(x_{\alpha_N}) \tag{2.50}$$

where, as for the boson case, the sum is taken over all $N!$ permutations of the variables $x_1, \ldots, x_N$ and where $(-1)^P$ is $+1$ for an even permutation and -1 for an odd permutation. A more compact way of writing Eq. (2.50) is to use the definition of a determinant and write ψ as (again we take $N = 3$)

$$\psi(x_1, x_2, x_3) = C \begin{vmatrix} u_{K_1}(x_1) & u_{K_1}(x_2) & u_{K_1}(x_3) \\ u_{K_2}(x_1) & u_{K_2}(x_2) & u_{K_2}(x_3) \\ u_{K_3}(x_1) & u_{K_3}(x_2) & u_{K_3}(x_3) \end{vmatrix} \tag{2.51}$$

---

*Nuclei of spin 3/2, 5/2, etc. may also be treated as single Fermi–Dirac particles by a trivial modification of the formalism given here.

Permuting the variables x_1 and x_2 in this equation merely results in a permutation of the first two columns of the determinant, which has the effect of changing its sign. Therefore, ψ has the necessary antisymmetry properties. If, in Eq. (2.51), any two of the indices $K_1, K_2,$ and K_3 are equal, then two rows of the determinant will be identical and therefore the determinant will vanish, yielding the unacceptable wave function $\psi = 0$. Thus, in choosing the functions $u_{K_1}, \ldots, u_{K_N}$, one must choose N *different* eigenfunctions. It is not difficult to verify that the function $\psi(x_1, \ldots, x_N)$ defined by Eq. (2.50) is a solution of the N-particle Schrödinger equation

$$\sum_{i=1}^{N} \left(-\frac{\hbar^2}{2m} \nabla_i^2 + U(x_i) \right) \psi = E\psi \tag{2.52}$$

with

$$E = \sum_{i=1}^{N} E_{K_i} \tag{2.53}$$

For Fermi–Dirac systems, one can also describe the N-particle eigenfunctions by a set of occupation numbers, $N_1, N_2, \ldots$ as follows. We define

$$N_K = \begin{cases} 1, & \text{if } K \in \{K_1, \ldots, K_N\} \\ 0, & \text{otherwise} \end{cases} \tag{2.54}$$

With this definition Eqs. (2.45) and (2.46) are still valid. However, whereas for boson systems there was one N-particle energy eigenstate for every distinct set of nonnegative integers summing to N, for fermions those integers are restricted to the values 0 and 1. This restriction is the *Pauli exclusion principle*.

2.13   MICROSTATES AND MACROSTATES

Having reviewed these aspects of many-particle wave functions, we are ready to apply the procedure used on the classical ideal gas to ideal quantum systems. The basic idea is to define a microstate and a macrostate and then to calculate the number of microstates corresponding to each macrostate. Assuming that the microstates have equal probabilities, one then calculates the most probable macrostate and identifies it with the macrostate of the system at equilibrium.

We have seen that, for a system of identical particles, the wave function is uniquely defined by specifying the set of quantum state occupation numbers, $N_1, N_2, N_3, \ldots$, where N_k gives the number of particles in the kth single-particle quantum state, which has energy E_k. If the particles are Bose–Einstein particles, then each N_k can be any nonnegative integer, but if the particles are Fermi–Dirac particles, then each N_k can only be either 0 or 1.

The identification of the microstates of the system seems fairly straightforward. The microstate of a system is the detailed dynamical state of the system. But one cannot specify more about a quantum-mechanical system than its wave function. Therefore we will identify the microstate of the system with the wave function of the system. If the system is known to have N particles and the system energy is known to be E, then any set of occupation numbers satisfying the conditions

$$\sum_{k} N_k = N \tag{2.55}$$

and

$$\sum_{k} N_k E_k = E \tag{2.56}$$

will define a possible microstate.

The definition of a macrostate is not quite so simple. Looking at the definition of a macrostate that was used in the classical analysis would strongly suggest that the macrostate also should be defined as the set of occupation numbers, N_k, since that gives the number of particles in each single-particle state. Such an identification of the system macrostates leads to very undesirable results. There is then exactly one microstate for each macrostate. Making an assumption of equal a priori probabilities for the microstates would trivially imply that the macrostates were also equally probable, eliminating any possibility of calculating a "most probable" macrostate.

Although defining the macrostate of the system as the set of occupation numbers has a great formal similarity to what was done before, it does not really stand up to close scrutiny. The macrostate is meant to be the realistically observable state of the system. However, for a macroscopic system, the spacing between adjacent discrete single-particle energy levels is so small that the detection of individual levels is many orders of magnitude beyond realistic measurement. For example, for particles of one atomic mass in a container of one cubic meter, the level spacing depends upon the energy and, at the energy value of $E = kT$ with $T = 300\,\mathrm{K}$, the spacing between adjacent levels is $3.7 \times 10^{-51}\,\mathrm{J}$ (see Problem 2.21). To get some idea of what this energy spacing means, we might note that, for a particle of typical thermal velocity ($\sim 10^3\,\mathrm{m/s}$), it would correspond to a change in velocity of $\Delta v = 2.2 \times 10^{-27}\,\mathrm{m/s}$. Such a change would be far beyond any practical experimental detection.

What we will do to define a meaningful macrostate of the system is to group the almost continuous energy spectrum into a large number of narrow energy ranges that we will call *energy bins*. In the case mentioned, if we choose 10^{12} bins within the range $0 < E < kT$, then each one would contain about 10^{19} discrete levels and have an energy range of about $10^{-33}\,\mathrm{J}$. The levels in the kth energy bin have an average energy of ε_k, which will be assigned to all the levels in that bin. The kth bin contains a large number K_k of discrete energy states. We now describe a macrostate of the system by specifying the number of particles, call it ν_k, within the states of the kth energy bin for all k.

2.14   QUANTUM DISTRIBUTION FUNCTIONS

We first want to calculate $I(K_k, \nu_k)$, the number of ways of distributing the ν_k indistinguishable particles within the K_k energy states of the kth bin. For the case of bosons, in which any number of particles can be put into a single state, that calculation is equivalent to determining how many ways there are of choosing K_k occupation numbers N_i satisfying the restriction, $\sum N_i = \nu_k$. This is left as a problem for the reader. The result is

$$I(K_k, \nu_k) = \frac{(K_k + \nu_k - 1)!}{(K_k - 1)!\nu_k!} \qquad \text{(bosons)} \qquad (2.57)$$

For fermions, where the N_i can only be 0 or 1, we can interpret the K_k quantum states as K_k coins, of which the filled ones are heads and the empty ones tails. The problem is then equivalent to the problem of determining the number of ways of arranging K_k coins to get ν_k heads. That was solved in the last chapter. The answer is just the binomial coefficient

$$I(K_k, \nu_k) = \frac{K_k!}{(K_k - \nu_k)!\nu_k!} \qquad \text{(fermions)} \qquad (2.58)$$

The total number of microstates corresponding to a given macrostate is obtained by multiplying the $I(K_k, \nu_k)$ for all k. That is,

$$I = \prod_k I(K_k, \nu_k) \qquad (2.59)$$

As before, the most probable macrostate will be determined by maximizing $\log I - \alpha N - \beta E$, where α and β are Lagrange parameters that have been introduced to take into account the restriction to fixed total particle number and fixed energy. The fermion case will be done explicitly, and the boson case left as an exercise. The function that must be maximized is

$$
\begin{aligned}
F = {}& \log \left(\prod_k \frac{K_k!}{(K_k - \nu_k)! \nu_k!} \right) - \alpha \sum_k \nu_k - \beta \sum_k \nu_k \varepsilon_k \\
= {}& \sum_k \Big[K_k (\log K_k - 1) - (K_k - \nu_k)(\log(K_k - \nu_k) - 1) \qquad (2.60) \\
& - \nu_k (\log \nu_k - 1) - \alpha \nu_k - \beta \varepsilon_k \nu_k \Big]
\end{aligned}
$$

The use of Stirling's approximation is valid when applied to the bin occupation numbers ν_k. Note, however, that it would be quite invalid if used for the individual state occupation numbers N_i. Setting $\partial F / \partial \nu_k = 0$, we get the following equation for the bin occupation numbers of the most probable macrostate.

$$\log[(K_k - \nu_k)/\nu_k] = \alpha + \beta \varepsilon_k \qquad (2.61)$$

which gives

$$\nu_k = \frac{K_k}{e^{\alpha + \beta \varepsilon_k} + 1} \qquad (2.62)$$

Since the energy differences of the states in the kth bin are negligible, Eq. (2.62) clearly implies that the average occupation of any individual state in that bin is $\nu_k / K_k = 1/(e^{\alpha + \beta \varepsilon_k} + 1)$. Thus, at equilibrium, the average occupation number of the nth single-particle state, of energy ε_n, is given by the well-known Fermi–Dirac distribution function

$$\bar{N}_n = f_{\mathrm{FD}}(\varepsilon_n) = \frac{1}{e^{\alpha + \beta \varepsilon_n} + 1} \qquad (2.63)$$

For bosons, the equivalent result is

$$\bar{N}_n = f_{\mathrm{BE}}(\varepsilon_n) = \frac{1}{e^{\alpha + \beta \varepsilon_n} - 1} \qquad (2.64)$$

The Lagrange parameter, β, has the same relationship to the absolute temperature as it had in the classical distribution. That is, $\beta = 1/kT$. The Lagrange parameter α, called the *affinity*, is essentially a normalization constant that must be adjusted so that the sum of $\bar{N}_n$ over all quantum states is equal to the number of particles in the system.

2.15   THE QUANTUM MECHANICAL IDEAL GAS

In order to understand the relationship between the quantum-mechanical distribution functions that we have just derived and the Maxwell distribution for a classical

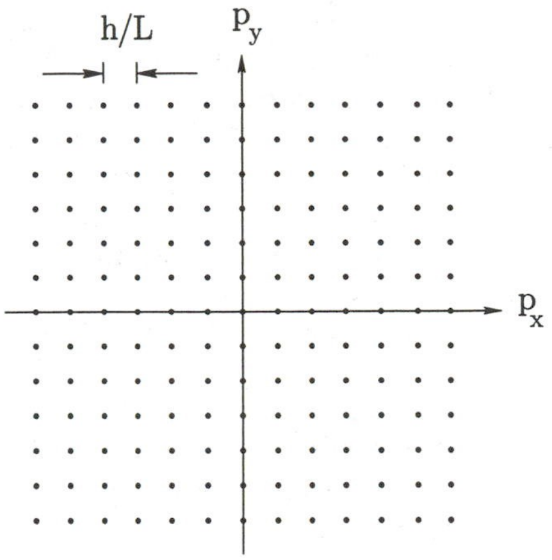

h/L P_y

Fig. 2.5 The possible values of the momentum of a particle in a periodic box form a lattice in momentum space. The picture shown would be appropriate for a particle in two dimensions. Along each axis, the spacing between allowed values is h/L. For a three-dimensional system the density of momentum eigenvalues is $(L/h)^3$.

P_x

ideal gas, we will consider a quantum mechanical ideal gas, that is, a system of non-interacting bosons or fermions with no external potential. If the external potential $U(\mathbf{r})$ is taken to be zero, then the single-particle Schrödinger equation is

$$-\frac{\hbar^2}{2m}\nabla^2 u(\mathbf{r}) = E u(\mathbf{r}) \tag{2.65}$$

If this equation is solved within a cube of volume L^3, using periodic boundary conditions,

$$u(x,y,z) = u(x+L,y,z) = u(x,y+L,z) = u(x,y,z+L) \tag{2.66}$$

then the single-particle eigenstates are plane waves

$$u(\mathbf{r}) = L^{-3/2}e^{i\mathbf{p}\cdot\mathbf{r}/\hbar} \tag{2.67}$$

of momentum

$$(p_x, p_y, p_z) = \frac{h}{L}(K_x, K_y, K_z) \tag{2.68}$$

where $h = 2\pi\hbar$ and K_x, K_y, and K_z are integers. If the momentum eigenvalues are plotted in a three-dimensional momentum space, they form a cubic lattice in which the distance between neighboring points (the lattice constant) is h/L (see Fig. 2.5). The corresponding energy eigenvalues are $E_p = p^2/2m$. The average occupation of one of the momentum eigenstates is

$$\bar{N}_p = \frac{1}{e^{\alpha + p^2/2mkT} \pm 1} \tag{2.69}$$

where the $+$ sign is taken for fermions and the $-$ sign for bosons. This must be compared with the Maxwell distribution, which, in momentum variables, is

$$f(\mathbf{p}) = Ce^{-p^2/2mkT} \tag{2.70}$$

2.16   THE CLASSICAL LIMIT

If $e^\alpha \gg 1$, then the second term in the denominator of Eq. (2.69) will be negligible, giving

$$\bar{N}_p \approx e^{-\alpha}e^{-p^2/2mkT} \tag{2.71}$$

and the quantum-mechanical distributions will reduce to the Maxwell distribution. One way of interpreting the criterion $e^\alpha \gg 1$ is to note that it implies that the average occupation of every momentum state is much less than one.

In order to express the condition $e^\alpha \gg 1$ in terms of the macroscopic parameters of the system, let us assume that it is valid and evaluate the normalization sum for the quantum-mechanical distribution.

$$\sum_{\mathbf{p}} \bar{N}_{\mathbf{p}} = \sum_{\mathbf{p}} e^{-\alpha - p^2/2mkT} = N \tag{2.72}$$

where N is the number of particles in the system and the sum is taken over all allowed momentum eigenvalues. Using the fact that the density of momentum eigenvalues in momentum space is $(L/h)^3 = V/h^3$, where V is the volume of the system, the sum over the discrete values of $\mathbf{p}$ may be converted to an integral over a continuous vector variable

$$\frac{Ve^{-\alpha}}{h^3} \int e^{-p^2/2mkT} d^3\mathbf{p} = N \tag{2.73}$$

The integral can be evaluated in terms of the standard Gaussian integral, given in the Table of Integrals.

$$\int e^{-p^2/2mkT} d^3\mathbf{p} = \left(\int_{-\infty}^{\infty} e^{-p_x^2/2mkT} dp_x \right)^3 = (2\pi mkT)^{3/2} \tag{2.74}$$

If we define a *thermal de Broglie wavelength* λ by

$$\lambda \equiv \frac{h}{\sqrt{2\pi mkT}} \tag{2.75}$$

then the normalization condition gives

$$e^{-\alpha} = (N/V)\lambda^3 \tag{2.76}$$

Therefore, $e^{-\alpha}$ is equal to the average number of particles in a volume λ^3. If that number is much less than one, then the classical approximation is justified. λ is the de Broglie wavelength of a particle with energy πkT, which is about twice the average energy of the particles in a gas at temperature T.

It is important to make clear what aspect of quantum mechanics is responsible for making the quantum distributions differ from the classical momentum distribution. It is not the discreteness of the quantum states, which is completely undetectable for a macroscopic system. Rather, it is the symmetry requirements on the many-particle wave functions. One way to see how the symmetry requirements affect the momentum distribution is to consider the mapping between classically described N-particle momentum states and quantum-mechanical ones. In order to simplify the analysis, we will artificially discretize the classical states so that the single-particle momentum states are mapped one-to-one. A classical N-particle momentum state is defined by giving the momentum values of the N distinguishable particles, $(\mathbf{p}_1, \mathbf{p}_2, \ldots, \mathbf{p}_N)$. If all N momentum values are different, then there are $N!$ classical states that map into a single Bose–Einstein or Fermi–Dirac N-particle quantum state. If we could ignore the probability of multiple occupancy of a single momentum state, then, averaging some quantity, such as the momentum distribution, over all classical N-particle states with total energy E, would be equivalent to averaging over $N!$ copies of the N-particle quantum states with the same energy and would therefore yield the same result. To see the effects of the quantum symmetry requirements, let us consider the mapping of a classical state in which $\mathbf{p}_1 = \mathbf{p}_2$. Then there are only $\frac{1}{2}N!$ classical states that are mapped into a single Bose–Einstein quantum state. Thus an average over the classical states would deemphasize this double-occupancy state by a factor of two in comparison with an average over Bose–Einstein N-particle states. That is, the Bose–Einstein symmetry requirement increases the probability of double occupancy in comparison with the classical system. In contrast, for a Fermi–Dirac system, there is no N-particle quantum state corresponding to the $\frac{1}{2}N!$ classical states. The Fermi–Dirac symmetry requirement prohibits double occupancy entirely. From this point of view, it is clear why the criterion for the validity of the classical approximation is that the average occupation of any single-particle quantum state be much less than one.

2.17   A CRITICAL EVALUATION

What are the deficiencies of the general calculational method used in this chapter? The major weaknesses of the technique are the following:

1. There is an obvious lack of logical precision and general principles in choosing our definitions of microstates and macrostates. It is not at all clear that our results do not depend on these somewhat arbitrary choices.

2. No justification, other than convenience, has been given for the assumption of equal a priori probabilities of all allowed microstates. Since that assumption is at the base of all the calculations, the whole logical structure is built upon sand.

3. There is no obvious way of applying the method to systems with interactions that cannot be neglected. For such systems, the energy cannot be written as a sum involving independent occupation numbers of any kind, nor is there any way of defining microstates and macrostates in terms of such occupation numbers.

4. In general, there is no logical "room" for adding extra assumptions, such as equal a priori probability. The time evolution of an actual system is determined by the laws of mechanics (or quantum mechanics). If the results of using any extra assumptions always agree with those of mechanics, then they are a logical consequence of the laws of mechanics, and it should be possible to show that fact. If they do not agree with the laws of mechanics, then the extra assumptions are wrong.

Although the method introduced in this chapter is very useful for calculating the properties of a wide variety of physical systems, it cannot really serve as a general logical foundation for statistical mechanics. In the next chapter we will sketch the outlines of such a general logical foundation, and, throughout the book, we will periodically return to the question of foundations to fill in more of the details.

PROBLEMS

2.1   (a) Make some reasonable estimate for the size of a cube containing 1000 grains of sand. (b) Using the above estimate, calculate the dimensions of a cube containing 10^{24} grains of sand.

2.2 For a relativistic particle, the relationships between velocity, momentum, and energy are $\mathbf{p} = m\mathbf{v}/\sqrt{1 - (v/c)^2}$ and $E = mc^2/\sqrt{1 - (v/c)^2}$. For an ideal gas of relativistic particles, determine whether the pressure is a function of the energy density, E/V, independent of the details of the velocity distribution? (Note that the relativistic energy defined here includes the rest energy, mc^2.)

2.3 For an extremely relativistic gas, $v \approx c$ for most of the particles. Show that, for such a system, the pressure is related to the energy density by $p = \frac{1}{3}E/V$.

2.4 If, from an ideal gas at temperature T, we choose two particles at random, then the velocities of the two particles are statistically independent. What are the probability distributions for their center-of-mass and relative velocities? Are those two random variables statistically correlated?

2.5 A container of volume V contains a very dilute ideal gas at density n and temperature T. The container is surrounded by vacuum and has a small hole of area A. At what rate do particles pass through the hole?

2.6 For the system described in Problem 2.5, calculate the rate at which the gas loses energy and calculate the temperature of the remaining gas as a function of time. Assume that the rate of leakage is small enough that the remaining gas is always uniform and at equilibrium.

2.7 For a classical ideal gas at temperature T, calculate the average value of $\sqrt{|\mathbf{p}|}$, where $\mathbf{p}$ is the momentum of a gas particle.

2.8 The law of partial pressures states that the pressure exerted by a mixture of ideal gases is equal to the sum of the pressures that would be exerted by each of the gases separately at the same temperature. Derive the law of partial pressures.

2.9 For the system described in Problem 2.5, assume that the wall is infinitely thin, so that a particle that enters the hole always exits the container without colliding with the side of the hole. Assume that, of the particles that leave during some time interval Δt, a fraction $\phi(v)\,dv$ have speed within the range dv. Calculate the speed distribution $\phi(v)$ and explain why it does not agree with Eq. (2.28).

2.10 The purpose of this problem is to estimate the average distance that a particle in a gas travels between collisions. That distance is called the *mean free path*. Take as a model of the gas a set of hard spheres of radius a. To simplify the calculation, assume that the spheres are frozen in their random locations. Assume that the average volume per particle is much larger than the hard sphere volume $[V/N \gg (4\pi/3)a^3]$. Consider a hard sphere moving through this set of fixed scatterers.   (a) Calculate the probability ΔP that it will be scattered within a short distance Δx. (b) Using that result, calculate the probability $P(x)$ that it will travel a distance x, without being scattered. (c) Calculate the average distance that the particle will move before being scattered. Compare your result with the exact formula given in Exercise 2.17. (d) Estimate the mean free path for neon at a temperature of $300\,\mathrm{K}$ and a pressure of 1 atmosphere, taking $a = 1.3\,\text{Å}$.

2.11 Given 30 students to distribute in 3 rooms, calculate the number of microstates that correspond to the macrostate "10 students are in each room."

2.12 A circular cylinder, with a radius of $10\,\mathrm{cm}$, contains a nitrogen gas at a pressure of one atmosphere and a temperature of $300\,\mathrm{K}$. If the cylinder is rotated about its axis, the centrifugal force increases the density at the periphery relative to that at the center. At what angular velocity would the ratio of the two densities be equal to 2?

2.13 The Maxwell distribution does not guarantee that the total energy is exactly E, but only that that is its average value. For one mole of ideal gas at 300 K, calculate the fractional uncertainty in the energy, $\Delta E/E$. (Hint: Assume that the velocities of the particles are N statistically independent random variables.)

Fig. 2.6

0 L

2.14 At $t = 0$, a one-dimensional classical ideal gas is contained in the region $0 < x < L$ (Fig. 2.6). The gas has an initial temperature T_o and a density n_o. The gas particles have mass m. The wall at $x = 0$ reflects all particles that hit it. The wall at $x = L$ allows any particle that hits it to pass through with a probability ϵ, that is much less than one and is independent of the particle's energy. Otherwise the particle is reflected with no change in its energy. (a) Calculate the rate at which the system loses particles and energy at $t = 0$. (b) Assuming that the particles remaining always have a Maxwell distribution, but with slowly varying density and temperature, derive and solve a differential equation for $T(t)$.

2.15 Generalize the derivation leading to the Maxwell–Boltzmann distribution to the case of an ideal gas containing two types of particles, A and B. Assume that there are N_A A particles and N_B B particles with masses m_A and m_B, respectively. Since they may interact differently with external force fields, they may, in general, experience different potentials, $U_A(\mathbf{r})$ and $U_B(\mathbf{r})$. The result should be that $f_i(\mathbf{r}, \mathbf{v}) = n_i(\mathbf{r}_{io})(m_i/2\pi kT)^{3/2} \exp[-(m_i v^2/2 + U_i(\mathbf{r}))/kT]$, where $i = A$ or B and $U_i(\mathbf{r}_{io}) = 0$.

2.16 Use the results of Problem 2.15 to show that, in a mixed ideal gas, the average kinetic energy per particle of the A particles is the same as that of the B particles. This is a restricted form of what is called the *equipartition theorem*. The general form will be derived later.

2.17 A system is composed of a large number N of elementary subsystems (one might picture the subsystems as atoms in a crystal lattice). Each subsystem can exist in one of three quantum states, with energies 0, 1, or 2 electron volts. Determine the number of microstates of the total system such that N_1 subsystems have energy $\varepsilon_1 = 0$ eV, N_2 subsystems have energy $\varepsilon_2 = 1$ eV, and N_3 subsystems have energy $\varepsilon_3 = 2$ eV. Assuming that each possible microstate has equal probability, determine the most probable values of N_1, N_2, and N_3, given that the total energy is $E = 0.5N$ eV.

2.18 Consider an ideal gas, composed of atoms that have K discrete possible internal energy states, of energy, ε_k $(k = 1, \ldots, K)$. Using an analysis similar to that which led to the Maxwell–Boltzmann distribution, derive the distribution function $f(k, \mathbf{v})$, where $f(k, \mathbf{v})\, d^3\mathbf{v}$ is the number of particles per unit volume in internal energy state k with velocity in $d^3\mathbf{v}$.

2.19 Using Lagrange's method, find the minimum of the function $x^2 + 2y^2$, subject to the constraint $x^3 + y^3 = 1$.

2.20 In 1876, J. Loschmidt argued that the Maxwell–Boltzmann distribution must be wrong. He said that, at equilibrium in a uniform gravitational field, an ideal gas should be warmer at the bottom than at the top because of the facts that every particle that moves from the bottom to the top slows down in the field and every particle that moves from the top to the bottom speeds up. Thus, in order to have the exchange of particles between bottom and top not change the distribution, it is necessary that the average speed of the particles at the bottom be larger than the average speed of the particles at the top. Explain why Loschmidt was wrong. Take the special case of a purely one-dimensional gas (in the z direction) of completely noninteracting particles in a constant gravitational field.

2.21 A globular cluster is a large collection of stars that is gravitationally held together in a more or less spherically symmetric distribution. Treat such a cluster as a gas of noninteracting, equal-mass stars in a gravitational potential $U(r)$. The gravitational potential must be determined *self-consistently*, using the Poisson equation, which can be written as $dU/dR = GM(R)/R^2$, where $M(R) = 4\pi m \int_o^R n(r)r^2\, dr$ is the mass of all the stars within a radius R. Choosing $U(0) = 0$ and using Eq. (2.19) for $n(r)$, show that no equilibrium solution exists for a single globular cluster in empty space. This shows that globular clusters are never truly at equilibrium. They very gradually "boil off" stars, causing the cluster to steadily, but very slowly, diminish in size. [Hint: For an isolated globular cluster, $n(r) \to 0$ as $r \to \infty$.]

2.22 At very high temperatures a hydrogen gas becomes completely ionized and may be viewed as a mixed ideal gas of protons and electrons. One can assume that relativistic effects become important when the rms velocity of the electrons is larger than $.1\,c$. At what temperature would that occur?

2.23 Consider a quantum-mechanical system of four particles in a one-dimensional harmonic oscillator potential. The one-particle energy eigenvalues are $(n + \frac{1}{2})\hbar\omega$ and are nondegenerate. For Bose–Einstein particles and Fermi–Dirac particles, determine the number of four-particle quantum states with an energy of $8\hbar\omega$. Assume that the Fermi–Dirac particles have zero spin, although that can be shown to be impossible.

2.24 Derive Eq. (2.57) for the number of ways of distributing ν bosons within K states. Hint: The diagrams in Fig. 2.7 show three ways of distributing five particles among three states. How many such diagrams are there?

Fig. 2.7   • | • • | • •   • • | • | • •   • • | • • | •

2.25 Consider a spinless particle of mass m, in a cubic periodic box of side L. (a) Calculate $N(E)$, the number of energy eigenfunctions with energies less than or equal to E. (b) $dN(E)/dE$ gives the number of eigenstates per unit energy interval. Its inverse is the average spacing between energy eigenvalues. Show that, for m equal to one atomic mass unit, $L = 1\,\mathrm{m}$, and $E/k = 300\,\mathrm{K}$, the estimate given in Section 2.13 for the energy spacing is correct.

2.26 Consider a system of spinless one-dimensional particles in a harmonic oscillator potential of angular frequency ω. The energy spectrum is then $\varepsilon_n = (n + \frac{1}{2})\hbar\omega$. Using Eqs. (2.63) and (2.64), evaluate the number of particles in the system as a function of the *affinity* α and the temperature T for the case $\alpha \gg 1$ for both fermions and bosons. (Hint: the infinite sum may be easily evaluated.)

Chapter 3
The Foundations of Statistical Mechanics

3.1 INTRODUCTION

This chapter will be devoted to a presentation of the general calculational principle of statistical mechanics and to its logical justification. The calculational principle will be a generalization, to systems that may include interparticle forces, of the assumption of equal a priori probability. The justification of that assumption will be based, in a fundamental way, on the fact that any macroscopic system contains a vast number of particles. The argument will be closely related to the Law of Large Numbers and the Central Limit Theorem, but it will require an extension of those ideas to variables that are statistically correlated.

The physical systems that are discussed in this chapter will be assumed to satisfy the laws of classical mechanics, rather than quantum mechanics. Actually, a certain improvement in logical elegance can be achieved by starting with a quantum-mechanical formulation and then deriving the corresponding classical mechanical rules by taking the limit $\hbar \to 0$. However, there is a danger that that logical elegance would be gained only at the cost of a clear picture of the meaning of the arguments. Therefore, the quantum-mechanical treatment will be postponed to a later chapter. Because it is convenient to use a Hamiltonian formulation of classical mechanics, the chapter will begin with a brief review of Hamilton's equations. It will then be necessary to introduce a certain amount of formalism, involving observables, statistical ensembles, and density functions, in preparation for treating the central topics of the chapter.

3.2 HAMILTON'S EQUATIONS

The dynamical state of a system of N point particles is completely determined by the values of $2N$ vector variables that give the positions of the particles and their momenta, $\mathbf{r}_1, \ldots, \mathbf{r}_N$ and $\mathbf{p}_1, \ldots, \mathbf{p}_N$. We define $2K$ scalar variables $(x_1, \ldots, x_K)$ and $(p_1, \ldots, p_K)$, where $K = 3N$, by the identifications

$$(x_1, \ldots, x_K) = (x_1, y_1, z_1, \ldots, x_N, y_N, z_N) \tag{3.1}$$

and
$$(p_1, \ldots, p_K) = (p_{1x}, p_{1y}, p_{1z}, p_{2x}, p_{2y}, p_{2z}, \ldots) \tag{3.2}$$

The energy of the system, written in terms of these variables, is called the *Hamiltonian function.*
$$E = H(x_1, \ldots, x_K, p_1, \ldots, p_K) = H(x, p) \tag{3.3}$$

where (x, p) is a compact symbol for the set of $2K$ variables. For a physical system composed of N identical point particles, each of mass m, in an external potential field $U(\mathbf{r})$, and with interparticle forces derivable from an interaction potential $\phi(|\mathbf{r}_i - \mathbf{r}_j|)$, the Hamiltonian function is

$$H(x, p) = \sum_{k=1}^{K} \frac{p_k^2}{2m} + \sum_{i=1}^{N} U(\mathbf{r}_i) + \sum_{i<j}^{N} \phi(|\mathbf{r}_i - \mathbf{r}_j|) \tag{3.4}$$

The time dependence of the $2K$ dynamical variables (x, p) is given by Hamilton's equations
$$\dot{x}_k(t) = \frac{\partial H(x, p)}{\partial p_k} \tag{3.5}$$

and

$$\dot{p}_k = -\frac{\partial H(x, p)}{\partial x_k} \tag{3.6}$$

where k runs from 1 to K.

3.3 PHASE SPACE

The $2K$ variables (x, p) may be interpreted as coordinates in a $2K$-dimensional space, called the *phase space* of the system. The $2K$-dimensional point (x, p) that represents the dynamical state of the system will be referred to as the *system point*. The time dependence of the complete set of $2K$ variables that describes the motions of all the particles in the system may then be pictured as the continuous motion of the single system point in the large-dimensional phase space.

The conservation of energy is a simple consequence of Hamilton's equations. If $x(t)$ and $p(t)$ is a solution of Hamilton's equations, then, by Eqs. (3.5) and (3.6),

$$\begin{aligned}
\frac{dE}{dt} &= \frac{dH(x(t), p(t))}{dt} \\
&= \sum_{k=1}^{K} \left(\frac{\partial H}{\partial x_k} \dot{x}_k + \frac{\partial H}{\partial p_k} \dot{p}_k \right) \\
&= \sum_{k=1}^{K} \left(\frac{\partial H}{\partial x_k} \frac{\partial H}{\partial p_k} - \frac{\partial H}{\partial p_k} \frac{\partial H}{\partial x_k} \right) = 0
\end{aligned} \tag{3.7}$$

3.4 THE ENERGY SURFACE

For any fixed value of E, the set of points in phase space that satisfy the equation $H(x, p) = E$ is called the *energy surface*, and will be denoted by the symbol $\omega(E)$. Since the phase space has $2K$ dimensions and the energy surface is defined by a single scalar equation, $\omega(E)$ is, except for special values of E, a $(2K-1)$-dimensional set. The reader should keep in mind the fact that, although our pictures will

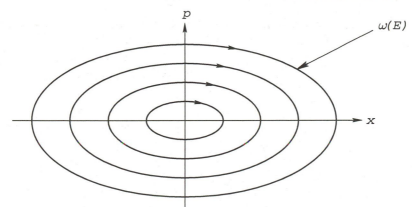

Fig. 3.1   For a simple harmonic oscillator, $H = p^2/2m + kx^2/2$. Therefore the set of points in phase space defined by $H = E$ is an ellipse. Because of energy conservation, the system point moves repeatedly around the ellipse defined by its initial energy. For a system with more than one coordinate, the phase space is more than two-dimensional and the set of points defined by $H = E$ is a many-dimensional surface.

show $\omega(E)$ as a one-dimensional surface in two-dimensional space, it is, for the macroscopic systems in which we are interested, usually an unimaginably convoluted surface in an unimaginably high dimensional space. Because of energy conservation, a system point that begins on the surface $\omega(E)$ is constrained to remain on that energy surface forever.

3.5   OBSERVABLES

A *physical observable* is any quantity that is determined by the dynamical state of the system, such as the system's kinetic energy or angular momentum. Therefore, a physical observable is simply any function $A(x, p)$ defined on the phase space. Two classes of observables are particularly important, in that they account for almost all observables that actually occur in physical measurements and calculations. They are *one-body* observables and *two-body* observables.

A one-body observable is a function $A(x, p)$ of the special form

$$A(x, p) = \sum_{i=1}^{N} a(\mathbf{r}_i, \mathbf{p}_i) \tag{3.8}$$

where $a(\mathbf{r}, \mathbf{p})$ is any function of $\mathbf{r}$ and $\mathbf{p}$. Two obvious cases of one-body observables are the kinetic energy, for which $a(\mathbf{r}, \mathbf{p}) = p^2/2m$, and the external potential energy, for which $a(\mathbf{r}, \mathbf{p}) = U(\mathbf{r})$. Another example is any component of the system's total angular momentum

$$\mathbf{L} = \sum_{i=1}^{N} \mathbf{r}_i \times \mathbf{p}_i \tag{3.9}$$

A two-body observable is a quantity that depends on the simultaneous state of a pair of particles, summed over all distinct pairs in the system. It is an observable of the general form

$$A(x, p) = \sum_{i \neq j} a(\mathbf{r}_i, \mathbf{p}_i, \mathbf{r}_j, \mathbf{p}_j) \tag{3.10}$$

where $a(\mathbf{r}, \mathbf{p}, \mathbf{r}', \mathbf{p}')$ is an arbitrary function of $\mathbf{r}$, $\mathbf{p}$, $\mathbf{r}'$, and $\mathbf{p}'$. The most important example of a two-body observable is the interaction potential energy of the system.

$$V_{\text{int}} = \sum_{i<j} \phi(|\mathbf{r}_i - \mathbf{r}_j|) = \frac{1}{2} \sum_{i \neq j} \phi(|\mathbf{r}_i - \mathbf{r}_j|) \tag{3.11}$$

in which case $a(\mathbf{r}, \mathbf{p}, \mathbf{r}', \mathbf{p}') = \frac{1}{2}\phi(|\mathbf{r} - \mathbf{r}'|)$. Notice that two-body observables are defined by a sum that excludes the terms with $i = j$.

3.6   THE MICROSTATE PROBABILITY

The central problem in the foundations of statistical mechanics is to devise some rational method of assigning a fundamental microstate probability function $P(x, p)$ to the set of possible dynamical states of a complex macroscopic system. Once we have decided upon the fundamental microstate probability, we can, by the general rules of probability theory, make statistical predictions about any observable of the system.

Since a Hamiltonian system is a completely deterministic system, an apparently simple and reasonable solution to this problem is to assign a probability one to the *actual* dynamical state of the system and a probability zero to any other dynamical state. For a system of a few particles, such as the solar system, this *is* a reasonable solution. However, for one mole of gas, the experimental determination of the dynamical state would require the precise and simultaneous measurement of 3.6×10^{24} scalar variables. Such an experimental operation is completely without practical meaning. Thus "determining the initial state of the system" is a string of words to which no real experimental operation corresponds. We must find some other solution. The fact that the known calculational methods of statistical mechanics can be used reliably to predict the equilibrium properties of large systems shows that a practical solution of the problem exists.

A *microstate probability density* will be taken to mean any function defined on the phase space that has the two properties that

$$P(x, p) \geq 0 \tag{3.12}$$

and

$$\int_\Omega P(x, p) \, d^K x \, d^K p = 1 \tag{3.13}$$

In Eq. (3.13) we have introduced the symbol Ω for the phase space of the system.

Before trying to justify the use of any definite probability function, we will look at the properties of two simple and important particular cases. Because of the possible interpretation of probability density functions in terms of statistical ensembles, it has become the practice to refer to the various important density functions as particular, named, *ensembles*.

3.7   THE UNIFORM ENSEMBLE

The uniform ensemble is the probability density function one obtains if a constant probability density is assigned to all states with energy less than or equal to E (Fig. 3.2). It is of the form

$$P(x, p) = C^{-1}\theta(E - H(x, p)) \tag{3.14}$$

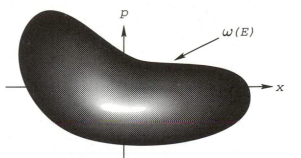

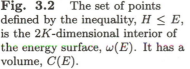

Fig. 3.2 The set of points defined by the inequality, $H \leq E$, is the $2K$-dimensional interior of the energy surface, $\omega(E)$. It has a volume, $C(E)$.

where the unit step function $\theta(t)$ is defined by

$$\theta(t) = \begin{cases} 0, & t < 0 \\ 1, & t \geq 0 \end{cases} \tag{3.15}$$

and $C(E)$ is a constant, determined by the normalization condition, Eq. (3.13). The value of $C(E)$, which is given by the integral

$$C(E) = \int_{\Omega} \theta(E - H(x,p)) \, d^K x \, d^K p, \tag{3.16}$$

can obviously be interpreted as the *volume* enclosed by the energy surface, $\omega(E)$.

3.8   C(E) FOR AN IDEAL GAS

In order to develop some feeling for these ideas, let us evaluate the normalization constant for a simple many-particle system, namely the ideal gas. We will later find many uses for the results of this calculation. For an ideal gas in a container of volume V,

$$H = \frac{1}{2m} \sum_{k=1}^{K} p_k^2 \tag{3.17}$$

where $K = 3N$, and the variables, $\mathbf{r}_1, \ldots, \mathbf{r}_N$ are each restricted to the inside of the container. The integral over the spatial variables gives a factor of V^N. Therefore,

$$C(E) = V^N \int \theta \left(E - \sum p_k^2 / 2m \right) d^K p \tag{3.18}$$

Defining K variables, $q_1, q_2, \ldots, q_K$, by $q_k = p_k / \sqrt{2m}$, we can calculate the integral for $C(E)$ as follows:

$$C(E) = V^N (2m)^{3N/2} \int \theta \left(R^2 - \sum_k q_k^2 \right) d^K q \tag{3.19}$$

where $R = \sqrt{E}$. The integral is equal to the volume of a K-dimensional sphere of radius R. It is evaluated in the Mathematical Appendix, with the result that

$$C(E) = \frac{V^N (2\pi m E)^{3N/2}}{(3N/2)!} \tag{3.20}$$

Since $3N/2$ is not necessarily an integer, $(3N/2)!$ must be defined in terms of the gamma function (again, see the Mathematical Appendix) by the relation

$$x! \equiv \Gamma(x+1) \tag{3.21}$$

The following calculation will reveal one of the bizarre geometrical properties of high-dimensional spaces. The uniform ensemble for an ideal gas describes a system that has a finite probability of being found with any energy between 0 and E. Let us now calculate the probability of finding the system with an energy less than xE, where $0 < x < 1$. This is easily seen to be the probability that a point, chosen at random within a $3N$-dimensional sphere of radius $R = \sqrt{E}$, will fall within the smaller sphere of radius $R = \sqrt{xE}$. It is equal to the ratio of the volume of a sphere of radius $\sqrt{x}R$ to that of a sphere of radius R.

$$\mathbf{P}[\text{energy} < xE] = x^{3N/2} \tag{3.22}$$

For $x = 0.999999$ and $N = N_A$ this is

$$\mathbf{P}[\text{energy} < 0.999999E] = (1 - 10^{-6})^{3N/2} \tag{3.23}$$
$$\approx e^{-9 \times 10^{17}}$$

The last line was obtained by noting that $\log \mathbf{P} = \frac{3}{2}N \log(1 - 10^{-6})$ and using the approximation, $\log(1 + x) \approx x$, that is valid for small x. Since we know that the energy must lie between 0 and E, this means that

$$\mathbf{P}[0.999999E \leq \text{energy} \leq E] = 1 - e^{-9 \times 10^{17}} \tag{3.24}$$

The uniform ensemble, although it appears to describe a system with a wide range of possible energies, actually describes a system with energy E with completely negligible uncertainty (see Fig. 3.3). Therefore, in practice, it is equivalent to a probability distribution that concentrates all the probability on the surface of the sphere. That probability distribution is called a *microcanonical ensemble*.

3.9   THE MICROCANONICAL ENSEMBLE

The density function that assigns equal probability to all microstates with energies in some narrow range from E to $E + \epsilon$ is of the form

$$P(x,p) = \frac{\theta(E + \epsilon - H) - \theta(E - H)}{C(E + \epsilon) - C(E)} \tag{3.25}$$

By integrating $P(x,p)$ over the phase space, it is easy to check that $C(E+\epsilon) - C(E)$ is the correct normalization constant. With this density function $P(x,p)$ is a positive constant if $E < H(x,p) < E+\epsilon$ and $P(x,p) = 0$ if $H(x,p)$ does not fall in that range. The limit of such a density function as ϵ goes to zero is called the *microcanonical probability density*. Using the fact that $d\theta(x)/dx = \delta(x)$, the Dirac delta function, and defining a function, $Q(E) = dC(E)/dE$, the microcanonical density can be written as

$$P(x,p) = \lim_{\epsilon \to 0} \frac{\theta(E + \epsilon - H) - \theta(E - H)}{C(E + \epsilon) - C(E)}$$
$$= \frac{\lim(\theta(E + \epsilon - H) - \theta(E - H))/\epsilon}{\lim(C(E + \epsilon) - C(E))/\epsilon} \tag{3.26}$$
$$= \frac{\delta(E - H(x,p))}{Q(E)}$$

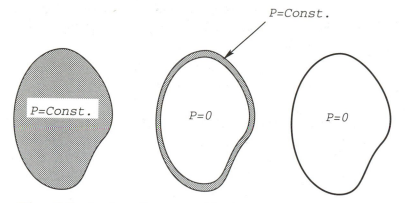

Fig. 3.3 In the uniform ensemble, the ensemble points are uniformly distributed within the volume enclosed by the energy surface. From a probability theory point of view, this means that the probability density function is constant throughout that volume and zero outside it. The intermediate ensemble (which was not given a name) has a uniform probability density between two energy surfaces. The extreme case of this is the microcanonical ensemble, in which all of the ensemble points are distributed over a single energy surface.

$Q(E)$ can also be defined directly in terms of the microcanonical density function by

$$Q(E) = \int_\Omega \delta(E - H(x,p))\, d^K x\, d^K p \tag{3.27}$$

Let us repeat that, in spite of the fact that the uniform and microcanonical density functions appear to be quite different, they always make experimentally indistinguishable predictions for large systems.

3.10 AVERAGE VALUES OF OBSERVABLES

Given any fundamental microstate density function $P(x,p)$, the average value (or expectation value) of any observable $A(x,p)$ is given by Eq. (1.52) of probability theory.

$$\bar{A} \equiv \langle A \rangle = \int_\Omega P(x,p) A(x,p)\, d^K x\, d^K p \tag{3.28}$$

For example, the average value of the external potential energy is given by

$$\langle U \rangle = \int_\Omega P(x,p) \sum_{i=1}^N U(\mathbf{r}_i)\, d^K x\, d^K p \tag{3.29}$$

But, if we define a particle density function $n(\mathbf{r})$ by saying that $n(\mathbf{r})\, d^3\mathbf{r}$ is equal to the probability of finding a particle in the volume element $d^3\mathbf{r}$, then $\langle U \rangle$ could also be given as

$$\langle U \rangle = \int_\mathbf{V} n(\mathbf{r}) U(\mathbf{r})\, d^3\mathbf{r} \tag{3.30}$$

where $\mathbf{V}$ is the three-dimensional region containing the N-particle system. (We will write the geometrical region in boldface to distinguish it from the numerical volume of the region, which will be written as V.)

Clearly, there must be some way of obtaining $n(\mathbf{r})$ from $P(x,p)$ that will make Eq. (3.29) identical with Eq. (3.30). It is easy to confirm that the following identification will do exactly that.

$$n(\mathbf{r}) = \int_\Omega P(x,p) \sum_{i=1}^N \delta(\mathbf{r} - \mathbf{r}_i)\, d^K x\, d^K p \tag{3.31}$$

where $\delta(\mathbf{r})$ is the three-dimensional Dirac delta function. Putting this into the right-hand side of Eq. (3.30), we get Eq. (3.29).

$$\begin{aligned}
\langle U \rangle &= \int_\Omega P(x,p) \sum_{i=1}^N \int_\mathbf{V} \delta(\mathbf{r} - \mathbf{r}_i)U(\mathbf{r})\, d^3\mathbf{r}\, d^K x\, d^K p \\
&= \int_\Omega P(x,p) \sum_{i=1}^N U(\mathbf{r}_i)\, d^K x\, d^K p
\end{aligned} \tag{3.32}$$

A more complicated example of a one-body operator is the angular momentum, whose expectation value is given by

$$\langle \mathbf{L} \rangle = \int_\Omega P(x,p) \sum_{i=1}^N (\mathbf{r}_i \times \mathbf{p}_i)\, d^K x\, d^K p \tag{3.33}$$

This also can be given in terms of a density function $F_1(\mathbf{r}, \mathbf{p})$, where $F_1(\mathbf{r}, \mathbf{p})\, d^3\mathbf{r}\, d^3\mathbf{p}$ is defined as the probability of finding a particle in the volume element $d^3\mathbf{r}$ with momentum in the range $d^3\mathbf{p}$. If, in analogy with Eq. (3.31), we define $F_1(\mathbf{r}, \mathbf{p})$ by

$$F_1(\mathbf{r}, \mathbf{p}) = \int_\Omega P(x,p) \sum_{i=1}^N \delta(\mathbf{r} - \mathbf{r}_i)\delta(\mathbf{p} - \mathbf{p}_i)\, d^K x\, d^K p \tag{3.34}$$

then $\langle \mathbf{L} \rangle$ is given by

$$\langle \mathbf{L} \rangle = \int F_1(\mathbf{r}, \mathbf{p})(\mathbf{r} \times \mathbf{p})\, d^3\mathbf{r}\, d^3\mathbf{p} \tag{3.35}$$

$F_1(\mathbf{r}, \mathbf{p})$ is called the one-particle phase-space density.* Given $F_1(\mathbf{r}, \mathbf{p})$, it is possible to calculate the average value of any one-body observable by a formula similar to Eq. (3.35).

The calculation of the average value of a two-body observable is slightly more complicated. If we define a two-particle phase-space density function $F_2(\mathbf{r}, \mathbf{p}, \mathbf{r}', \mathbf{p}')$ by

$$\begin{aligned}
F_2(\mathbf{r}, \mathbf{p}, \mathbf{r}', \mathbf{p}') = \int_\Omega P(x,p) \sum_{i \neq j}^N \delta(\mathbf{r} - \mathbf{r}_i)\, \delta(\mathbf{p} - \mathbf{p}_i) \\
\delta(\mathbf{r}' - \mathbf{r}_j)\, \delta(\mathbf{p}' - \mathbf{p}_j)\, d^K x\, d^K p
\end{aligned} \tag{3.36}$$

then the average value of the two-body observable $A = \sum a(\mathbf{r}_i, \mathbf{p}_i, \mathbf{r}_j, \mathbf{p}_j)$ is given in terms of F_2 by

$$\langle A \rangle = \int F_2(\mathbf{r}, \mathbf{p}, \mathbf{r}', \mathbf{p}')a(\mathbf{r}, \mathbf{p}, \mathbf{r}', \mathbf{p}')\, d^3\mathbf{r}\, d^3\mathbf{p}\, d^3\mathbf{r}'\, d^3\mathbf{p}' \tag{3.37}$$

---

* The *term phase space* is commonly used for both the $2K$-dimensional states of the complete system and the six-dimensional $(\mathbf{r}, \mathbf{p})$ states of a single particle.

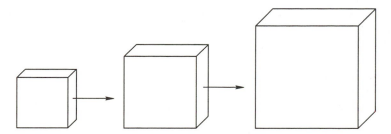

Fig. 3.4   The thermodynamic limit is defined by taking a sequence of systems whose volumes become larger and larger but whose particle density and energy density remain constant.

The physical interpretation of $F_2(\mathbf{r}, \mathbf{p}, \mathbf{r}', \mathbf{p}')$ is that $F_2(\mathbf{r}, \mathbf{p}, \mathbf{r}', \mathbf{p}') \, d^3\mathbf{r} \, d^3\mathbf{p} \, d^3\mathbf{r}' \, d^3\mathbf{p}'$ is the probability of finding a particle in $d^3\mathbf{r}$ with momentum in $d^3\mathbf{p}$ and simultaneously finding a different particle in $d^3\mathbf{r}'$ with momentum in $d^3\mathbf{p}'$. From the defining equations of F_1 and F_2, it is easy to see that they satisfy the following normalization conditions:

$$\int F_1(\mathbf{r}, \mathbf{p}) \, d^3\mathbf{r} \, d^3\mathbf{p} = N \tag{3.38}$$

and

$$\int F_2(\mathbf{r}, \mathbf{p}, \mathbf{r}', \mathbf{p}') \, d^3\mathbf{r} \, d^3\mathbf{p} \, d^3\mathbf{r}' \, d^3\mathbf{p}' = N(N-1) \tag{3.39}$$

3.11   THE THERMODYNAMIC LIMIT

The physical phenomena that are the chief topics of study in statistical mechanics; namely, the approach of a system to a time-independent equilibrium state and the physical properties of that state are characteristics that are peculiar to systems containing very large numbers of particles. Systems such as planetary systems with only a few degrees of freedom show no signs, after very long periods of time, of approaching a uniform, time-independent, "equilibrium" state. In contrast, a liquid, composed of a vast number of individual molecules, if isolated from outside influences, soon reaches a quiescent state and remains that way. Of course, there are systems such as star clusters with fifty or so "particles" that seem to be intermediate between a small system, for which any statistical analysis is inappropriate and of little value, and a system of Avagadro's number of particles, for which statistical methods offer the only hope of making progress. For these intermediate systems statistical methods give some useful but unreliable information. These systems do seem to settle down to fairly steady states for long periods of time, but will then exhibit large spontaneous fluctuations from the steady state. The exact times of the fluctuations are not predictable by statistical methods. The approach to a time-independent equilibrium state only becomes completely predictable and unambiguous in the limit of very large particle number. Also, the theoretical methods used become more powerful and reliable as N becomes larger. Clearly, the reasonable way to proceed is to study the physical characteristics of systems in the limit $N \to \infty$. Fortunately, the values of N encountered in real thermodynamic systems are so large that it is safe to assume that anything that would happen in the limit of infinite N has already happened for such systems to a high degree of approximation.

We cannot simply let $N \to \infty$, holding all other parameters fixed. If we are studying water, and we double the amount of water, then we have to double the size of the container in order to hold the density constant. We also have to double the system energy, in order to keep a constant energy per particle. We will see that, when these two corrections have been made, we then get reasonable limiting behavior as $N \to \infty$. This is the *thermodynamic limit*. It is defined more precisely in the following.

We begin with a system of N_o particles with a Hamiltonian function

$$H = \sum_{i=1}^{N_o} p_i^2/2m + \frac{1}{2} \sum_{i \neq j}^{N_o} \phi(r_{ij}) \qquad (3.40)$$

where $r_{ij} = |\mathbf{r}_i - \mathbf{r}_j|$. The variables $\mathbf{r}_i$ are restricted to a connected region $\mathbf{V_o}$, whose volume is V_o. We assume that the system has an energy E_o. Since our aim is to study the limiting properties of a system as N goes to infinity, we cannot consider only a single system, but must introduce a sequence of larger and larger systems. We do this as follows.

The system defined is considered as a *starting system* for the sequence. For each integer $N \geq N_o$, we consider an N-particle system with a Hamiltonian identical to that given in Eq. (3.40) but with N_o replaced by N. The particles of the N-particle system are restricted to a region $\mathbf{V}$ that is a uniformly expanded copy of the region $\mathbf{V_o}$. If we choose a *scale factor* s by $s^3 = N/N_o$ and scale up all dimensions of $\mathbf{V_o}$ by s, then N/V will be equal to N_o/V_o. $\mathbf{V}$ is then defined by

$$s\mathbf{r} \in \mathbf{V} \quad \text{if and only if} \quad \mathbf{r} \in \mathbf{V_o} \qquad (3.41)$$

The energy of the N-particle system is chosen to be $E = s^3 E_o$. This guarantees that the energy per particle is independent of s. We can express these relations in a uniform way by the three equations

$$\begin{aligned} N &\equiv s^3 N_o, \\ V &\equiv s^3 V_o, \\ \text{and} \qquad E &\equiv s^3 E_o \end{aligned} \qquad (3.42)$$

3.12 FLUCTUATIONS IN THE THERMODYNAMIC LIMIT

The logical foundation of statistical mechanics is based on a generalization of the Law of Large Numbers. In order to achieve the needed generalization, it is necessary to determine the behavior of the fluctuations (or statistical uncertainties) in macroscopic observables in the thermodynamic limit (Fig. 3.4). That will require that we precisely define the concept of a macroscopic observable, a task that will be postponed until later. At this point, two special classes of macroscopic observables will be introduced. The first class, called *uniform extensive* observables are of the form

$$A = \sum_{i=1}^{N} a(\mathbf{p}_i) \qquad (3.43)$$

The word *uniform* refers to the fact that they are independent of position. The word *extensive* means that $\langle A \rangle$ is proportional to the size of the system in the thermodynamic limit. The most obvious example of such an observable is the kinetic energy, for which $a(\mathbf{p}) = p^2/2m$. The fact that $\langle A \rangle$ goes to infinity along with N is inconvenient when studying the thermodynamic limit. That problem is eliminated in the definition of the second class of observables, called uniform *intensive* observables. A uniform intensive observable is either of the form

$$A = \frac{1}{N} \sum_{i=1}^{N} a(p_i) \tag{3.44}$$

or of the form

$$A = \frac{1}{V} \sum_{i=1}^{N} a(p_i) \tag{3.45}$$

Since V is exactly proportional to N in the thermodynamic sequence, the two forms are really equivalent. Again, the most obvious examples are the kinetic energy per particle and the kinetic energy per unit volume. In general, a uniform intensive observable is the value per particle or per unit volume of some function of momentum. The average value of any uniform intensive observable has a finite limit as N, V, and E go to infinity.

We now want to look at the expectation values and uncertainties of intensive observables in the thermodynamic limit. Although our aim will be to justify the use of the microcanonical ensemble, we will actually use the microcanonical ensemble density function in our analysis. Of course, it will then be necessary to guard against the use of circular reasoning. Let A be any uniform intensive observable. Because A is a one-body observable, the expectation value of A is given in terms of the one-particle phase-space density function.

$$\langle A \rangle = \frac{1}{N} \int d^3\mathbf{p} \int_{\mathbf{V}} d^3\mathbf{r}\, F_1(\mathbf{r}, \mathbf{p}) a(\mathbf{p}) \tag{3.46}$$

ΔA, the uncertainty in A, is defined by

$$(\Delta A)^2 = \langle (A - \langle A \rangle)^2 \rangle = \langle A^2 \rangle - \langle A \rangle^2 \tag{3.47}$$

The observable A^2 consists of a one-body part plus a two-body part.

$$A^2 = \frac{1}{N^2} \sum_{i=1}^{N} a^2(\mathbf{p}_i) + \frac{1}{N^2} \sum_{i \neq j}^{N} a(\mathbf{p}_i) a(\mathbf{p}_j) \tag{3.48}$$

Because the one-body part of A^2 has a factor of $1/N^2$, rather than $1/N$, preceding it, its expectation value will approach zero in the thermodynamic limit. It will therefore be dropped. $\langle A^2 \rangle$ can then be written in terms of the two-particle phase-space density.

$$\langle A^2 \rangle = \frac{1}{N^2} \int d^3\mathbf{p}\, d^3\mathbf{p}' \int_{\mathbf{V}} d^3\mathbf{r}\, d^3\mathbf{r}'\, F_2(\mathbf{r}, \mathbf{p}, \mathbf{r}', \mathbf{p}') a(\mathbf{p}) a(\mathbf{p}') \tag{3.49}$$

Combining Eqs. (3.46) and (3.49), we get

$$(\Delta A)^2 = \frac{1}{N^2} \int d^3\mathbf{p}\, d^3\mathbf{p}'\, a(\mathbf{p}) a(\mathbf{p}') \\ \times \int_{\mathbf{V}} d^3\mathbf{r}\, d^3\mathbf{r}'\, [F_2(\mathbf{r}, \mathbf{p}, \mathbf{r}', \mathbf{p}') - F_1(\mathbf{r}, \mathbf{p}) F_1(\mathbf{r}', \mathbf{p}')] \tag{3.50}$$

Recall now that $F_2(\mathbf{r}, \mathbf{p}, \mathbf{r}', \mathbf{p}')\, d^3\mathbf{r}\, d^3\mathbf{p}\, d^3\mathbf{r}'\, d^3\mathbf{p}'$ is the probability of simultaneously finding a particle in the phase-space element $d^3\mathbf{r}\, d^3\mathbf{p}$ and another particle in the phase-space element $d^3\mathbf{r}'\, d^3\mathbf{p}'$. In almost all cases that we will consider, the force that one particle exerts on another has an effective range of only a few angstroms. Beyond that distance it goes rapidly to zero. Therefore, if we choose two positions, $\mathbf{r}$ and $\mathbf{r}'$, that have a separation much greater than the range of interparticle interactions, one might expect very little statistical correlation between the distribution of particles at $\mathbf{r}$ and that at $\mathbf{r}'$. The statement that the particle distributions at $\mathbf{r}$ and $\mathbf{r}'$ are uncorrelated takes the mathematical form that $F_2(\mathbf{r}, \mathbf{p}, \mathbf{r}', \mathbf{p}') = F_1(\mathbf{r}, \mathbf{p})F_1(\mathbf{r}', \mathbf{p}')$. When this is true we say that the two-particle density *factors* into products of one-particle densities. For any finite value of $|\mathbf{r} - \mathbf{r}'|$ one would not expect the factorization to be exact. However, it certainly seems reasonable to assume that $F_2(\mathbf{r}, \mathbf{p}, \mathbf{r}', \mathbf{p}') \to F_1(\mathbf{r}, \mathbf{p})F_1(\mathbf{r}', \mathbf{p}')$ as $|\mathbf{r} - \mathbf{r}'| \to \infty$. This property is called *asymptotic factorization*. It will be seen that the existence or nonexistence of asymptotic factorization will crucially affect the validity and reliability of our calculational methods. Although the assumption of asymptotic factorization seems very reasonable for particles with short-range interactions, it is not always correct, even for particles with strictly finite-range interactions [that is, particles for which $\phi(r_{ij}) = 0$ when $r_{ij} > R$ for some finite distance R]. We will return to the question of the validity of this assumption later in this chapter.

There is a technical problem that appears at this point. For any finite value of N and any finite domain $\mathbf{V}$ we cannot let $|\mathbf{r} - \mathbf{r}'|$ approach infinity without having at least one of the two points move outside the system. It is only in the thermodynamic limit that infinite distances become available within the system. Therefore, the property of asymptotic factorization must be defined by coupling the limit $|\mathbf{r} - \mathbf{r}'| \to \infty$ and the thermodynamic limit (Fig. 3.5). We do so in the following way.

A thermodynamic limit sequence has been defined in terms of a size parameter, s. If we choose two points, $\mathbf{r}$ and $\mathbf{r}'$ ($\mathbf{r} \neq \mathbf{r}'$), both within the starting domain $\mathbf{V}_o$ then the points $s\mathbf{r}$ and $s\mathbf{r}'$ will both be in $\mathbf{V}$. We say that a system has the property of asymptotic factorization if the one-particle and two-particle phase-space density functions, calculated using the microcanonical ensemble, satisfy the condition that

$$\lim_{s \to \infty} \left[F_2(s\mathbf{r}, \mathbf{p}, s\mathbf{r}', \mathbf{p}') - F_1(s\mathbf{r}, \mathbf{p})F_1(s\mathbf{r}', \mathbf{p}') \right] = 0 \tag{3.51}$$

for all $\mathbf{r}, \mathbf{r}' \in \mathbf{V}_o$ with $\mathbf{r} \neq \mathbf{r}'$. The point of defining the property of asymptotic factorization is the following theorem, which will be called the *Zero Fluctuation Theorem*.

Theorem If the system has the property of asymptotic factorization, then, in the thermodynamic limit, $\Delta A \to 0$ for any uniform intensive observable.

Proof The proof is quite simple. In Eq. (3.50) we make the transformation of variables, $\mathbf{r} = s\mathbf{x}$ and $\mathbf{r}' = s\mathbf{x}'$. Then $d^3\mathbf{r}\, d^3\mathbf{r}' = s^6\, d^3\mathbf{x}\, d^3\mathbf{x}'$. The domain of integration of the variables $\mathbf{x}$ and $\mathbf{x}'$ is $\mathbf{V}_o$. Using the fact that $s^6/N^2 = 1/N_o^2$, we obtain the formula

$$(\Delta A)^2 = \frac{1}{N_o^2} \int d^3\mathbf{p}\, d^3\mathbf{p}'\, a(\mathbf{p})a(\mathbf{p}') \int_{\mathbf{V}_o} d^3\mathbf{x}\, d^3\mathbf{x}' \left[F_2(s\mathbf{x}, \mathbf{p}, s\mathbf{x}', \mathbf{p}') \right.$$
$$\left. - F_1(s\mathbf{x}, \mathbf{p})F_1(s\mathbf{x}', \mathbf{p}') \right] \tag{3.52}$$

In the limit $s \to \infty$ the right hand side clearly approaches zero.

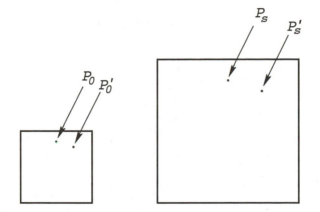

Fig. 3.5 As we go further and further in the thermodynamic sequence, the pair of "corresponding" points, P and P', become separated by larger and larger distances. The property of asymptotic factorization is that the physical properties at the two points become statistically uncorrelated.

A nice geometrical interpretation of this theorem can be obtained by using Chebyshev's inequality in the following way. For any $\epsilon > 0$ and any point (x,p) in the system phase-space Ω, we define the function $F_\epsilon(x,p)$ by

$$F_\epsilon(x,p) = \begin{cases} 1, & \text{if } |A(x,p) - \langle A \rangle| > \epsilon \\ 0, & \text{otherwise} \end{cases} \qquad (3.53)$$

That is, $F_\epsilon(x,p)$ is equal to one if and only if the value of observable A at the point (x,p) differs from its microcanonical average $\langle A \rangle$ by more than ϵ. Let us call the points at which this is true *exceptional points*, although, for very small ϵ, they may not seem at all exceptional. Given the microcanonical probability density $P(x,p)$ the probability that A differs from $\langle A \rangle$ by more than ϵ is

$$\begin{aligned} \mathbf{P}[|A - \langle A \rangle| > \epsilon] &= \int_\Omega F_\epsilon(x,p) P(x,p)\, d^K x\, d^K p \\ &= \frac{\int_\Omega F_\epsilon(x,p) \delta(E - H(x,p))\, d^K x\, d^K p}{\int_\Omega \delta(E - H(x,p))\, d^K x\, d^K p} \end{aligned} \qquad (3.54)$$

The second expression clearly defines the *fraction* of the points on the energy surface that are exceptional. But, by Chebyshev's inequality,

$$\mathbf{P}[|A - \langle A \rangle| > \epsilon] \leq \frac{(\Delta A)^2}{\epsilon^2} \to 0 \text{ as } s \to \infty \qquad (3.55)$$

No matter how small we make ϵ, in the thermodynamic limit, almost all the microstates on the energy surface $\omega(E)$ have values of the observable A that differ from the mean value by less than ϵ. In other words, the function $A(x,p)$ that we are averaging has the constant value $\langle A \rangle$ for almost all states on the energy surface. An obvious consequence of this fact is that any probability distribution that was restricted to the energy surface and did not, in the thermodynamic limit, become concentrated on the vanishing fraction of exceptional points would give the same average for any uniform intensive observable as is given by the microcanonical ensemble. If all probability distributions give the same result, then we are justified in using whichever one is the most convenient. That one is the microcanonical ensemble density.

Another way of justifying the use of the microcanonical ensemble is as follows. Imagine a macroscopic system that is initially in an exceptional state of energy E. As time progresses, the system point will move on the energy surface according to Hamilton's equations. Since only a very tiny fraction of the states on the energy surface are exceptional points, it is overwhelmingly probable that the system point will leave the small set of exceptional points and almost never return to them. In other words, it is overwhelmingly probable that the instantaneous value of any uniform intensive observable will smoothly move from its initial, exceptional, value to the microcanonical average value and then remain there. But that is exactly the behavior described as a movement to equilibrium in a macroscopic system. This will be our justification for identifying the microcanonical average value of an observable with the value of that observable in the thermodynamic equilibrium state.

Notice that we are not saying that the unexceptional points (the points that exhibit the average values of macroscopic observables) are, in any way, favored over the exceptional points. There are simply many many more of them. Thus a macroscopic system is driven to equilibrium, not by any dynamical principle that pushes the system into the unexceptional states, but by a purely statistical effect.

3.13   INFINITE-RANGE CORRELATIONS

Since it will form the logical foundation for almost everything that follows, it is worth our while to review carefully what has been proved in the last section and to display the faults and shortcomings of the analysis. Basically, we have shown that *if* the microcanonical two-particle density exhibits the property of asymptotic factorization, then the expectation values of *uniform intensive observables* become, in the thermodynamic limit, more or less independent of our choice of a fundamental microstate probability density. We have not proven that any particular system actually has the property of asymptotic factorization, nor have we proven anything at all about arbitrary physical observables.

The problem of extending the theorem to a sufficiently large class of physical observables will be taken up shortly. We will first consider the more important question of the general validity of asymptotic factorization.

Certainly it seems extremely reasonable that the local physical properties at two widely separated points in a system with strictly short-range interactions will be statistically uncorrelated. However, in spite of its apparent reasonableness, this property is not a universal characteristic of such systems. It is not at all impossible for systems with strictly short-range interactions to develop infinite-range correlations in the thermodynamic limit. For such systems, the values of macroscopic observables calculated with the microcanonical probability density are *unreliable* and, in general, disagree with the experimental values of the same observables.

In order to develop an understanding of the phenomenon of infinite-range correlation in systems with short-range forces, we will consider a simple system that exhibits it.*

The three-dimensional *Ising model* (cf. Fig. 3.6) is a cubic lattice containing

---

* This footnote is only for sophisticated readers. In the theory of critical phenomena, one says that a system has infinite-range correlations if the correlation functions do not approach zero exponentially fast at large distances. In the terminology being used here, infinite-range correlation means that the correlation function does not approach zero *at*

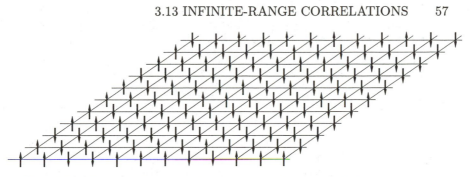

Fig. 3.6 The two-dimensional version of the Ising model, pictured here, is a very simple model of a two-dimensional magnetic solid. It is composed of a square lattice of particles. The state of each particle is completely specified by a spin variable that can have only two values, which are called *up* and *down*. Below a certain critical temperature it also exhibits infinite range correlations.

a particle at each lattice site. All mechanical degrees of freedom of the particles are ignored. However, each particle is assumed to have a spin degree of freedom that can have two possible values, which we will call *spin up* and *spin down*. We can interpret the two-valued spin variable as describing the state of a quantized magnetic moment that can point parallel or antiparallel to the z axis. Certain strongly anisotropic magnetic crystals are reasonably described by the model. We number the particles with an index i, running from 1 to N. We introduce N spin variables, $\sigma_1, \sigma_2, \ldots, \sigma_N$, that can take the values $\sigma_i = +1$ (for spin up) and $\sigma = -1$ (for spin down). Since there are no other degrees of freedom, the configuration of the complete system is described by the set of values of $\sigma_1, \sigma_2, \ldots, \sigma_N$. A given spin is assumed to interact only with its six *nearest neighbors*. The interaction energy of a pair of neighboring particles, with spin values σ_1 and σ_2, called $U(\sigma_1, \sigma_2)$, is assumed to have the form

$$U(+1, +1) = U(-1, -1) = -V \quad \text{and} \quad U(+1, -1) = U(-1, +1) = V \qquad (3.56)$$

This can be written as

$$U(\sigma_1, \sigma_2) = -V\sigma_1\sigma_2 \qquad (3.57)$$

The energy associated with any given configuration of the system $(\sigma_1, \sigma_2, \ldots, \sigma_n)$ is

$$E = -V \sum_{\text{NN}} \sigma_i \sigma_j \qquad (3.58)$$

where the sum contains one term for each nearest-neighbor (NN) pair of spins. The system has overall up–down symmetry. That is, the energy is unchanged if all the spins in a configuration are simultaneously flipped from up to down and vice versa.

We will assume that V is positive, so that a pair of nearest-neighbor spins has a lower energy if they are parallel than if they are antiparallel. Then the interaction would tend to align neighboring spins to be in the same direction. The lowest-energy configurations would be the two configurations in which all spins were up or all spins were down. In general, the more agreement there was between nearest-neighbor spins in a given configuration, the lower would be the energy for that configuration. An Ising model with positive V is called a *ferromagnetic* Ising

all for large distances. Thus a system at its critical point does *not* have infinite-range correlations in our terminology.

model. It is the simplest possible model of a real ferromagnetic material. For any configuration of the system we define the *mean magnetization* associated with that configuration as

$$m = \frac{1}{N} \sum_{i=1}^{N} \sigma_i \qquad (3.59)$$

Let us consider a three-dimensional Ising model that has the form of a very large cube. For example, we might take a cube of dimension 10^6 lattice spacings in each direction. Such a cube would contain $N = 10^{18}$ spins. The total number of possible configurations of the system would be the incredibly large number 2^N. For any given temperature T, there is a corresponding microcanonical ensemble with an energy $E(T)$. The calculation of the function $E(T)$ is a difficult problem that will be considered in later chapters. As the temperature ranges from zero to infinity, the corresponding energy ranges from $-3NV$ to 0 (see Problem 3.17). The reason why the Ising model has been introduced at this point is that it can be proved to have the following interesting characteristics.

1. For any energy, the microcanonical average value of the mean magnetization is zero. This is an obvious consequence of the up–down symmetry of the system. (For each configuration of mean magnetization m, there is another configuration, obtained by reversing all spins, that has the same energy but has mean magnetization equal to $-m$.)

2. If the temperature is greater than a certain critical value T_c, then, for very large N, almost all configurations in the corresponding microcanonical ensemble have mean magnetization $m \approx 0$.

3. If the temperature is less than T_c, then approximately half the configurations in the corresponding microcanonical ensemble have $m = +\mu$ while the other half have $m = -\mu$, where μ, called the *spontaneous magnetization*, is a function of T that is positive for $T < T_c$ and goes to zero at $T = T_c$ (Fig. 3.7). To be more precise, in the limit $N \to \infty$, the probability density associated with the random variable m approaches

$$P(m) = \begin{cases} \delta(m) & \text{if } T > T_c \\ \frac{1}{2}[\delta(m - \mu) + \delta(m + \mu)] & \text{if } T < T_c \end{cases} \qquad (3.60)$$

What is happening is that, as we lower the temperature, and therefore the energy, we are demanding a greater and greater degree of correlation between the spin states of neighboring particles. We are not demanding that the spins have any preferred orientation, either up or down, but only that neighboring spins should tend to have the *same* orientation. If σ_b is a neighbor of σ_a and σ_c is a neighbor of σ_b (but not of σ_a), then the correlation between σ_a and σ_b and that between σ_b and σ_c imply some degree of correlation between σ_a and σ_c, even though they do not interact directly. For small degrees of direct nearest-neighbor correlation, this induced correlation between more distant pairs of spins diminishes rapidly with distance. However, larger degrees of correlation between nearest neighbors can be achieved only by statistically coupling all the spins in the lattice. Those degrees of correlation can only be obtained with configurations in which more than half of the spins point in the same direction. The extreme case of this is the microcanonical ensemble for E equal to the lowest possible energy, namely $-3NV$. Then only two states of the system are possible, the state with all spins up or the state with all spins down. They have equal probability.

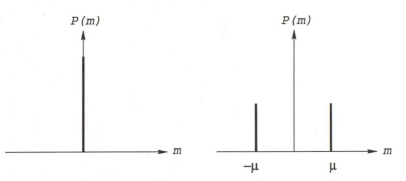

Fig. 3.7   If the temperature of the system is greater than T_c then the probability density for the average magnetization is strongly peaked at $m = 0$. At temperatures below T_c, there is almost zero probability of finding the system with $m = 0$, even though that is the ensemble average of m. Because of the nearest neighbor interactions, the spins tend to line up, either mostly up or mostly down, with equal probabilities for the two cases.

3.14   FLUCTUATIONS AND PHASE TRANSITIONS

Let us now relate this to our analysis of fluctuations in macroscopic observables. The mean magnetization, m, is clearly a macroscopic observable in the common sense of that phrase. That is, it is a large-scale observable feature of the system. Because of the up–down symmetry of the system, the microcanonical average of m is zero. But, according to the probability distribution illustrated in Fig. 3.7, if $T < T_c$, there is a negligible probability of ever actually observing the system with zero mean magnetization. It can also be shown that the zero microcanonical average does not, in any useful sense, represent the time average of the observed magnetization. Whether the magnetization has the value $+\mu$ or $-\mu$ is analogous to whether a coin is heads or tails. The coin has equal probability of having either face showing, but a real coin does not, every now and then, spontaneously flip over from one state to the other. What prevents any gradual drift in the state of a coin is that almost all of the intermediate states have much larger energy and are therefore unavailable. In the same way, almost all of the intermediate magnetization states of an Ising ferromagnet also have larger energy than the states with $m = \pm\mu$ and are therefore not observed if the system energy is too low.

Using the probability function given in Eq. (3.60), one can see that, for $T < T_c$, $\Delta m = \mu$. Thus, the fluctuation in the macroscopic observable, m, does not approach zero in the thermodynamic limit. Our theorem, relating fluctuations to infinite-range correlations, which could be easily modified to treat variables of the type m, therefore indicates that, for $T < T_c$, the corresponding microcanonical ensemble should exhibit infinite-range correlation. It is easy to see that it does. Suppose we choose an arbitrary configuration from the ensemble and look at the value of a particular spin, σ_i, in that configuration. If we find that $\sigma_i = +1$, then it is more likely that we have chosen one of the configurations with $m = +\mu$ than one of those with $m = -\mu$. But this information would then affect our estimate of the probability that some other spin in the configuration, say σ_j, is also $+1$, even if σ_j is at a great distance from σ_i. The phenomenon of spontaneous magnetization thus introduces statistical correlations between distant spins.

As we lower the temperature from above to below T_c, the change that occurs is called a *ferromagnetic phase transition*. Other types of phase transitions, such as liquefaction of a gas or solidification of a liquid, can also introduce infinite-range correlations and fluctuations in macroscopic observables into the microcanonical ensemble. For example, consider a system composed of water molecules. We assume that the water is in a container that is coated with hydrophobic material, such as Teflon, so that there is no tendency for the water to adhere to the surface of the container. We also assume that there is no gravitational field present. (The container is floating around in interstellar space.) Certainly there are values for the particle number, volume, and energy for which the state of the system would be part liquid and part vapor. Because of the effects of surface tension, it turns out that the equilibrium states of such systems are states in which the liquid component is in the form of one or a few large-sized pieces. We will call them blobs. However, the energy of the system is completely indifferent to the exact location of the blobs within the volume. (If we had not coated the container with Teflon, this would not be true. The liquid would then be attached to the walls.) Thus, the microcanonical ensemble will contain a collection of states in which the blobs are in all possible positions within the container. An average of the particle density at some fixed location, using that ensemble, will give a value that is intermediate between the liquid density and the vapor density. But, in any real system of this type, the blobs will exist in some particular places. As macroscopic objects, they will not make spontaneous transitions among all the locations they might conceivably have occupied. The local density of a real system will be quite nonuniform, in contrast with the uniform average density predicted by the microcanonical ensemble.

As might be expected, this fluctuation in the microcanonical average of the macroscopic density is associated with infinite-range correlation in the microcanonical ensemble. To see this, let us imagine that we choose some configuration in a microcanonical ensemble for this system. With the chosen configuration we see whether there is a particle in the volume element $d^3\mathbf{r}_1$. Remember that any configuration in the microcanonical ensemble will be composed of a few large blobs of liquid in some particular places. (Different configurations will have the blobs in different places.) Because the particle density in the liquid is much larger than that in the vapor, if we find a particle in $d^3\mathbf{r}_1$, then $d^3\mathbf{r}_1$ probably lies within one of the blobs. If we now consider a second volume element, $d^3\mathbf{r}_2$, at a macroscopic distance, such as one centimeter, from $d^3\mathbf{r}_1$, because the liquid blobs are of very large size, there will be a high probability of finding another particle in $d^3\mathbf{r}_2$. That this long-range correlation becomes of infinite range in the thermodynamic limit can only be proved by a more detailed analysis, but it is true.

This phenomenon, in which the expectation values of certain macroscopic variables obtained using the microcanonical ensemble are subject to large uncertainties, is associated with all known phase transitions. Since the study of phase transitions is one of the most important topics in statistical mechanics, the foregoing analysis seems to make any use of the microcanonical ensemble very suspect and unwise.

In fact, the problem of fluctuations in microcanonical averages of macroscopic observables has a very simple and general solution. Consider the liquid–vapor system. If, instead of assuming that the gravitational field is exactly zero, we assume that a very weak, but finite, gravitational field exists within the container, then the disagreement between the physically observed states and the predictions of the microcanonical ensemble disappears completely. The gravitational field will define a direction called *down*. In the physical system the liquid will all move as far as possible in that direction, which will uniquely determine its configuration. On the theoretical side, when the weak gravitational potential is added to the Hamiltonian, this redefines the energy surface, $\omega(E)$, and therefore the set of states in the microcanonical ensemble. It can be shown that, in the new microcanonical ensemble, virtually all the states exhibit liquid at the bottom of the container and vapor above it. The introduction of this weak *symmetry-breaking field* (so called because it breaks the exact translational symmetry of the original Hamiltonian) eliminates entirely the fluctuations in the microcanonical ensemble and, with them, the disagreement between the predictions of the microcanonical ensemble and physical observations. For the ferromagnetic Ising model the same thing could be accomplished by introducing an external magnetic field, no matter how weak, that would create an energy difference between the up-magnetized state and the down-magnetized state. With such a modification the microcanonical ensemble will predict only the lower-energy state with negligible uncertainty. For both systems, the introduction of a weak symmetry-breaking field eliminates the infinite-range correlations that lead to uncertainty in macroscopic observables.

3.15   THE STANDARD CALCULATIONAL PROCEDURE

One can therefore define the following standard calculational procedure.

1. Introduce a model Hamiltonian, H, to describe the physical system of interest. Using the microcanonical ensemble defined by the model Hamiltonian, calculate all the macroscopic properties of the system as a function of the total energy.
2. Check whether the correlation functions defined by the model Hamiltonian have the property of asymptotic factorization. If, for a particular range of E, they do have that property, then, within that energy range, almost all states of a given energy will exhibit the microcanonical average values for all macroscopic observables and we can safely expect any real system to exhibit those values of the same observables at equilibrium.
3. If, within a certain energy range, the microcanonical ensemble exhibits infinite-range correlations, then introduce a weak symmetry-breaking field to eliminate that phenomenon when making calculations within that energy range.

3.16   THE ERGODIC THEOREM

There is still another effect that can, and sometimes does, prevent the measured values of macroscopic observables from agreeing with their microcanonical averages. It is the existence of unexpected conservation laws. If, in addition to the energy, some other observable, $A(x, p)$, is exactly conserved, then the system point is restricted to move on the intersection of the energy surface and the surface defined by the equation, $A(x, p) = A_o$, where A_o is the initial value of the conserved quantity. This intersection constitutes a negligible fraction of the total energy surface, and thus there is nothing to guarantee that it is not composed primarily of exceptional points. In fact, if A is a macroscopic observable, then the average value of A on the intersection will be A_o, which might very well disagree with the microcanonical average of $A(x, p)$. In that case, the intersection would be composed entirely of exceptional points.

A simple, but not realistic, example of a system with extra conserved quantities (any conserved quantity in addition to the energy will be here called an *extra conserved quantity*) is a collection of interacting particles within a spherically symmetric three-dimensional harmonic oscillator potential. The Hamiltonian for the system is

$$H(x,p) = \sum p_n^2/2m + \sum_{i<j} v(r_{ij}) + \tfrac{1}{2}k \sum r_n^2 \tag{3.61}$$

In addition to the energy, the total angular momentum of the system is conserved. Therefore, if we happen to start the system with some nonzero angular momentum vector, then it will eventually approach a time-independent equilibrium state in which the mass of particles will rotate like a rigid body (see Problem 3.16). But the average over the energy surface of the angular momentum is zero, and, in fact, for a large system, almost all of the states on the energy surface have approximately zero angular momentum. Thus, for this system, the properties of the observed equilibrium state may not agree with those of the microcanonical average. This is an unrealistic model because the slightest deviation from spherical symmetry will introduce a coupling between the internal energy, associated with the complicated individual particle motions, and the rigid-body rotational energy that will act as a frictional force and eventually dissipate the macroscopic rotational motion.

There is a theorem, due to G. Birkhoff and called the *ergodic theorem*, that relates the time average of any observable to its microcanonical average. For an observable $A(x,p)$ and any initial point (x_o, p_o) the time average of A is defined as

$$\langle A \rangle_{x_o p_o} = \lim_{T \to \infty} \frac{1}{T} \int_o^T A(x(t), p(t)) \, dt$$

where $(x(t), p(t))$ is the solution of Hamilton's equations that started out, at $t = 0$, at (x_o, p_o). Birkhoff's theorem says that, in the absence of extra conserved quantities, the time average of any observable over the phase-space trajectory of a system point that starts anywhere on the energy surface will be equal to the microcanonical average of the same observable. (There are really some technical mathematical restrictions on the theorem, but they do not seriously weaken it.) Birkhoff's theorem is very nice. It guarantees that, in the absence of extra conservation laws, a system point that starts in an exceptional state will not just move around within a tiny island of exceptional states, but will eventually move to, and stay in, the much larger ocean of equilibrium states.

A system that does not have extra conservation laws is called an *ergodic* system. It seems to be almost impossible rigorously to prove that any realistic large system of interacting particles is ergodic. The only significant case of such a proof is a theorem, due to the Russian mathematician Ya. G. Sinai, that a system of billiard-ball particles is ergodic. While it is generally believed that most macroscopic classical systems do not have extra conservation laws, the possible existence of nonergodic systems is a serious problem in the foundations of statistical mechanics. The traditional conservation laws are associated with obvious and simple symmetries. The invariance of physical laws under translation in time leads to the conservation of energy. Invariance under spatial translation yields conservation of linear momentum, while rotational invariance implies the conservation of angular momentum. Such conserved quantities are relatively easy to predict from the nature of the system under study. However, during the 1960s it was discovered that systems that satisfy certain nonlinear wave equations conserve large numbers of other quantities that do not have such simple physical interpretations and whose definitions depend on the detailed nature of the wave equation. Also, all this discussion has been restricted to classical systems. When we consider macroscopic quantum-mechanical systems, the situation changes in an important way. It is not that the ergodic theorem is affected; there exists a quantum-mechanical version of the theorem that is essentially equivalent to the classical one. What changes is that there are many macroscopic quantum-mechanical systems that are known to have extra conserved quantities. The best example is a superconductor.

The magnetic flux encircled by a superconducting loop is strictly conserved as long as the loop is maintained in its superconducting state. This means that the total current flowing in the superconducting loop is also conserved. This is a robust conservation law that does not depend on maintaining some artificial perfect symmetry, as did the case of the system in a spherical potential. As in all cases of extra conservation laws, the observed equilibrium state of the system depends on its past history. In particular, the amount of magnetic flux that is trapped in the loop is equal to the amount of flux that was passing through the loop when it became superconducting. Identical systems at the same temperature can exhibit different time-independent equilibrium states. Only one of the possible equilibrium states is the microcanonical average equilibrium state. Therefore, the microcanonical average is not a reliable predictor of the observed equilibrium properties of such systems. This does not mean that we cannot use statistical mechanics to analyze superconductors. Actually, it is the only thing that we *can* use to understand them. It means only that we have to use it more carefully and intelligently. We must construct an ensemble of states with a given energy *and* a given value for the trapped flux. Once this extra conserved quantity is taken into account, the predictions of all other macroscopic observables becomes completely reliable. The difficulties involving superconductors and their resolution has obvious similarities to those involving spontaneous symmetry breakdown. It is an example of the fact that, in the real world, there is no cookbook procedure that can be reliably used without some thought.

3.17 GLASSES

If a system does not have infinite-range correlations and does not have extra conservation laws, then it should eventually approach an equilibrium state that exhibits the macroscopic features predicted by the microcanonical ensemble. However, "eventually" may be a very long time. Ordinary glass is the best example of a material that will remain essentially forever in a nonequilibrium state. In fact, pieces of ancient glass that have not been chemically weathered by rainwater or groundwater show no discernible signs of coming to equilibrium. Typical glasses are made by adding chemical impurities, particularly boron and oxygen, to molten silicon and then rapidly cooling the mixture. If, instead of rapid cooling, the material is subjected to very gradual cooling from the molten state, then, at a certain temperature that depends on the details of the mixture, crystals of silicon begin to form and grow. This behavior clearly indicates that the homogeneous, glassy state is not the true equilibrium state of the system at room temperature. What happens when the material is cooled rapidly is that the system gets caught in a state that is a local equilibrium state but is not the true global equilibrium state.

In Fig. 3.8 the probability of various values of some macroscopic variable x for two different imaginary systems is shown. The probability is proportional to the number of microstates that have that value of x, and therefore the graphs can be interpreted as the number of microstates of the two systems in terms of the parameter x. Graph A describes a normal system. If such a system is started in an improbable nonequilibrium state, its constant transitions from one microstate to another would soon cause its macroscopic features, such as the value of x, to move to their overwhelmingly more probable equilibrium values. Graph B describes what

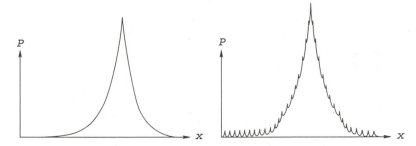

Fig. 3.8   For any given ensemble, the probability density for some macroscopic observable x is defined by saying that $P(x)\,dx$ is the fraction of ensemble members that have x within the range dx. A and B show the structure of $P(x)$ for some unspecified observable, such as the mass density or the potential energy, in the case of a *normal* system, in which $P(x)$ is a fairly smooth function with one extremely sharp maximum at the equilibrium value for x, and in case of a *complex* system, in which $P(x)$ still has a very large maximum at the equilibrium value but has a very rough structure containing many local maxima of various heights. It may take a very long time for such a system to make transitions from one local maximum to another. If the valleys between the local maxima are deep enough, it can take such a system an absurdly long time to come to true equilibrium.

is called a *complex system*. Glasses are the most common examples of complex systems. In a complex system, the function $P(x)$ has a large global maximum, corresponding to the equilibrium value of x, but it also has a very rough structure with many local maxima. Typically the graph has a fractal structure in which the roughness persists at all levels of magnification. Such a system, started in a nonequilibrium state, will most likely go to a nearby state of local equilibrium [a state in which $P(x)$ has a local maximum]. At any given temperature the thermal fluctuations in x will allow the system to jump from one local equilibrium state to another so long as the valleys between the two maxima are not too low. At higher temperatures larger jumps are possible. At very low temperature the system simply moves to the nearest local equilibrium state and stays there for an extremely long time until a very improbable fluctuation allows it to jump across the valley to a better local equilibrium state.

For a glass, the local equilibrium states describe configurations in which, on a very local scale, the atoms of the material are in energetically advantageous positions and have advantageous chemical bonding patterns. However, they are only local equilibrium states because, by breaking some of the bonds and doing some rearranging, one could find a still better configuration for the system. These rearrangements could be done at many different length scales but the larger the length scale the larger is the barrier to such a rearrangement and thus the more improbable is it that such a rearrangement will take place spontaneously. When the glass is heated to a liquid state the rough structure of the graph disappears entirely. For a normal liquid (in contrast to the supercooled glassy liquid) transitions from one macrostate to another are smooth and easy. Thus the molten glass would go to its true equilibrium state. If the glass is cooled very slowly, then, as the rough structure begins to develop, the system will always have enough energy to jump the valleys and remain in its globally most probable state. Thus, beginning with pure molten quartz (SiO_2), rapid cooling (called quenching) will produce silica glass, a microscopically amorphous material, but very slow cooling (called annealing) will produce a single quartz crystal.

The theoretical analysis of complex systems is a difficult problem that is still under development. It will not be treated any further in this book.

3.18 NONUNIFORM MACROSCOPIC OBSERVABLES*

3.18 NONUNIFORM MACROSCOPIC OBSERVABLES* The standard calculational procedure is based upon the zero-fluctuation theorem. The most glaring deficiency in that theorem is the limited set of observables to which it applies. In this and the next section, the theorem will be extended to nonuniform one-body and two-body macroscopic observables. The reader who is willing to accept, without detailed proof, that the no-fluctuation theorem can be extended to anything that, in common parlance, would be called a macroscopic variable, can skip to Section 3.20.

In trying to define nonuniform macroscopic observables, it is best to begin with a very specific case and then to generalize the procedure used there. As a starting system, we consider an ideal gas of N_o particles in a cubic box of volume $V_o = L_o^3$, subject to a uniform gravitational field, described by the potential $U_o(\mathbf{r}) = mgz$. The Hamiltonian for such a system is

$$H = \sum_{i=1}^{N_o} p_i^2/2m + \sum_{i=1}^{N_o} U_o(\mathbf{r}_i) \tag{3.62}$$

H is composed of two one-body observables. The kinetic energy is a uniform observable and is therefore of no use to us in our task of extending the set of macroscopic observables. The gravitational potential is not a uniform observable. The question to be considered is: "When we scale the system from (N_o, V_o, E_o) to (N, V, E), what potential energy function should be used in the larger system"? There are two natural possibilities. They are:

1. Keep the gravitational field, $\mathbf{g} = -g\hat{\mathbf{z}}$, fixed and uniform for the sequence of larger and larger systems. That is, use the same potential function for the scaled system as for the starting system. In that case, $U_s(\mathbf{r}) = U_o(\mathbf{r})$.

2. Scale the gravitational potential function so that the potential at a point in the scaled system is equal to the potential in the starting system at the "corresponding" point. To do so, one must modify the potential function by the transformation $U_s(\mathbf{r}) = U_o(\mathbf{r}/s)$. The gravitational field strength will then get weaker as the system size increases.

In Problem 3.18 the reader is asked to show that the first alternative, keeping a fixed gravitational field, leads to unphysical results in the thermodynamic limit. As the system size increases, the density and pressure at the bottom of the container approach infinity. In contrast, the second alternative, scaling the potential function, produces a density profile that is independent of the scale size. It is the expected exponential density function, derived in Section 2.6. This is not intended to show that, in real systems, the gravitational field is infinitely weak; of course, it is not. The thermodynamic limit is a mathematical device that is useful in studying the properties of systems that are actually finite but very large. This analysis shows how to apply that mathematical device to nonuniform systems.

The appropriate Hamiltonian function for the scaled system is

$$H = \sum_{i=1}^{N} p_i^2/2m + \sum_{i=1}^{N} U_s(\mathbf{r}_i) \tag{3.63}$$

$\star$ A star indicates an optional section that may be skipped without a serious loss of continuity.

where $U_s(\mathbf{r}) = U_o(\mathbf{r}/s)$. In general, we will define an one-body *extensive* observable by choosing an arbitrary one-body observable, $A = \sum a_o(\mathbf{r}_i, \mathbf{p}_i)$, in the starting system and extending it to the larger systems in the thermodynamic limit sequence by the rule

$$A = \sum_{i=1}^{N} a_s(\mathbf{r}_i, \mathbf{p}_i) \tag{3.64}$$

where $a_s(\mathbf{r}, \mathbf{p}) = a_o(\mathbf{r}/s, \mathbf{p})$.

A general way of defining corresponding one-body *intensive* observables is to begin with some arbitrary one-body observable in the starting system but extend it to the scaled system by the rule

$$A = \frac{1}{N} \sum_{i=1}^{N} a_s(\mathbf{r}_i, \mathbf{p}_i) \tag{3.65}$$

The factor of $1/N$ guarantees that $\langle A \rangle$ approaches a finite limit as N, V, and E go to infinity. It is left as a problem to prove that the zero-fluctuation theorem is valid for all one-body intensive observables. Notice that uniform intensive observables are now just a special case of one-body intensive observables in which the function a_o does not depend on $\mathbf{r}$.

Although the fluctuation in an *extensive* observable does not go to zero in the thermodynamic limit, it can be shown to become negligible in comparison with the average value of the observable. (See Problem 3.20.)

3.19   TWO-BODY MACROSCOPIC OBSERVABLES*

Two-body macroscopic observables can be defined in a way that closely parallels that used for one-body observables. A uniform extensive two-body observable is an observable of the form

$$A = \sum_{i \neq j}^{N} a(\mathbf{r}_{ij}, \mathbf{p}_i, \mathbf{p}_j) \tag{3.66}$$

where $\mathbf{r}_{ij} = \mathbf{r}_i - \mathbf{r}_j$. The most important example is the interaction potential energy, for which $a(\mathbf{r}, \mathbf{p}, \mathbf{p}') = \frac{1}{2}\phi(|\mathbf{r}|)$. Again, the word *uniform* refers to the fact that a depends only on relative coordinates and is therefore unchanged in a simultaneous translation of the coordinates of all the particles. A corresponding uniform intensive observable can be obtained by dividing the extensive observable by any extensive parameter, such as N or V. For example, the density of pairs of particles with relative position vector $\mathbf{q}$ is the average value of the observable

$$A = \frac{1}{V} \sum_{i \neq j}^{N} \delta(\mathbf{r}_{ij} - \mathbf{q}) \tag{3.67}$$

A nonuniform two-body observable is defined by choosing an arbitrary two-body observable in the starting system, written in terms of center-of-mass and relative coordinates.

$$A_o = \sum_{i \neq j}^{N_o} a_o(\mathbf{R}_{ij}, \mathbf{r}_{ij}, \mathbf{p}_i, \mathbf{p}_j) \tag{3.68}$$

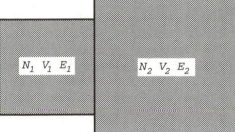

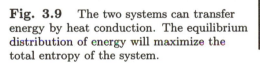

Fig. 3.9   The two systems can transfer energy by heat conduction. The equilibrium distribution of energy will maximize the total entropy of the system.

where $\mathbf{R}_{ij} = \frac{1}{2}(\mathbf{r}_i + \mathbf{r}_j)$ and $\mathbf{r}_{ij} = \mathbf{r}_i - \mathbf{r}_j$. In transporting the observable to further members of the thermodynamic limit sequence, only the center-of-mass coordinate is scaled. For the scaled system, A_s is defined as

$$A_s = \sum_{i \neq j}^{N} a_s(\mathbf{R}_{ij}, \mathbf{r}_{ij}, \mathbf{p}_i, \mathbf{p}_j) \tag{3.69}$$

with $a_s(\mathbf{R}, \mathbf{r}, \mathbf{p}, \mathbf{p}') = a_o(\mathbf{R}/s, \mathbf{r}, \mathbf{p}, \mathbf{p}')$. Higher order (three-body, etc.) macroscopic observables may also be defined, but the analysis becomes rather technical and is inappropriate for an introductory text. A zero fluctuation theorem can be proven for arbitrary intensive two-body macroscopic observables. It requires that the four-body phase-space density function, $F_4(1, 2, 3, 4)$ [where $1 \equiv (\mathbf{r}_1, \mathbf{p}_1)$, etc.] has an asymptotic factorization property of the form $F_4(1, 2, 3, 4) \rightarrow F_2(1, 2)F_2(3, 4)$ when the distance between the pair of variables $(\mathbf{r}_1, \mathbf{r}_2)$ and the pair of variables $(\mathbf{r}_3, \mathbf{r}_4)$ approaches infinity.

3.20   THERMODYNAMIC OBSERVABLES

The standard calculational procedure now seems to be broadly satisfactory. It permits one to calculate the equilibrium value of any macroscopic observable with a reasonable expectation that the value of the corresponding observable in a real system will approach the calculated value. However, it still has the following major deficiency. There is an important class of clearly macroscopic observables about which the standard calculational procedure gives no information whatever. They are what will be called *thermodynamic* observables, as opposed to mechanical observables. A mechanical observable is what we have been calling simply an observable. That is, it is any quantity for which there is a corresponding phase-space function, $A(x, p)$. The class of thermodynamic observables may be illustrated by considering its most important member, namely, the *entropy*. For a system of point particles, what function of their positions and momenta is their entropy? A system of a few particles simply does not have an entropy. Entropy is a concept that is only defined for macroscopic systems. This can be seen by considering the role played in thermodynamics by the entropy. We will now do so for a system composed of two components. The subject will be taken up again in a later chapter on thermodynamics, where it will be treated in much more detail.

The total entropy of a system of two components, such as that shown in Fig. 3.9, is the sum of the entropies of its parts. The entropy of each component is a function of that component's particle number, volume, and energy. Thus the total entropy $S_T = S_1(N_1, V_1, E_1) + S_2(N_2, V_2, E_2)$. If the systems are allowed to interact (for example, if they are put into thermal contact so that they may exchange energy), and the interaction is maintained for a long period of time, the system will come to an equilibrium state. The equilibrium state will be that state which maximizes the total entropy, taking into account the fundamental conservation laws of energy and particles and the constraints of the allowed interaction (that is, thermal contact allows only energy exchange but a channel between the two systems would allow both energy and particles to be exchanged). For example, if the two components shown in Fig. 3.9 are brought into thermal contact, then the condition that S_T be maximized, with the constraint that $E_1 + E_2 = E_T$, where E_T is the fixed total system energy, leads to the equilibrium condition

$$\frac{\partial S_1(N_1, V_1, E_1)}{\partial E_1} = \frac{\partial S_2(N_2, V_2, E_2)}{\partial E_2} \tag{3.70}$$

This Entropy Maximization Principle is the defining characteristic of the entropy. But, since its definition involves the concept of thermal equilibrium, the entropy can have no meaning for small systems, which never approach an equilibrium state. Entropy is an observable that emerges, along with the idea of equilibrium states, only in the thermodynamic limit, that is, only for systems of sufficient complexity. All other thermodynamic observables, such as free energy or enthalpy, involve the entropy or its derivatives in their definition and could thus be calculated if one had a statistical mechanical rule for calculating the entropy.

The fact that no phase-space function exists that represents the entropy will not mean that it cannot be calculated by statistical mechanics. The meaning of the entropy in statistical mechanics can be discovered by solving physical problems, using the rules of statistical mechanics, that can also be solved by thermodynamics, and then comparing the solutions obtained by the two methods. The first problem to be considered is the thermal equilibrium problem that, by thermodynamics, led to Eq. (3.70).

The collection of phase-space variables that describe subsystem 1 will be denoted as (x_1, p_1) and those for subsystem 2 as (x_2, p_2). The total system has energy E_T. The Hamiltonian is an expression that gives the energy in terms of the positions and momenta of the particles. If the interaction between the subsystems is very weak, then the total energy is just the energy of subsystem 1, which depends on its particles, plus the energy of subsystem 2, which depends only on its particles. In that case, which we assume is true, the Hamiltonian function for the complete system can be approximated as

$$H_T = H_1(x_1, p_1) + H_2(x_2, p_2) \tag{3.71}$$

The microcanonical probability density for the complete system is

$$P(x_1, p_1, x_2, p_2) = \frac{\delta(E_T - H_1 - H_2)}{Q_T(E_T)} \tag{3.72}$$

The question to be considered is: "What is the energy of subsystem 1 at equilibrium"? This is equivalent to asking for the equilibrium value of $\langle H_1 \rangle$. H_1 is a random variable in the sense of Section 1.10. The general rule for calculating the probability that a random variable $A(x)$ will have a value a, given the microstate probability density $P(x)$, is that

$$P_A(a) = \int_\Omega \delta(a - A(x)) P(x) \, d^N x \tag{3.73}$$

Therefore, the probability that $H_1(x_1, p_1) = E_1$ is given by

$$
\begin{aligned}
P_1(E_1) &= [Q_T(E_T)]^{-1} \int \delta(E_1 - H_1)\delta(E_T - H_1 - H_2)\, d^{K_1}x_1\, d^{K_1}p_1\, d^{K_2}x_2\, d^{K_2}p_2 \\
&= [Q_T(E_T)]^{-1} \int \delta(E_1 - H_1)\delta(E_T - E_1 - H_2)\, d^{K_1}x_1\, d^{K_1}p_1\, d^{K_2}x_2\, d^{K_2}p_2 \\
&= \frac{Q_1(E_1)Q_2(E_2)}{Q_T(E_T)}
\end{aligned}
\tag{3.74}
$$

where $E_2 = E_T - E_1$. Certainly H_1 is a macroscopic observable, and thus its measured value at equilibrium will almost surely be its most probable value. That value is determined by setting $\partial P_1(E_1)/\partial E_1 = 0$, which gives

$$
Q_1'(E_1)Q_2(E_2) - Q_1(E_1)Q_2'(E_2) = 0
\tag{3.75}
$$

or

$$
\frac{Q_1'(E_1)}{Q_1(E_1)} = \frac{Q_2'(E_2)}{Q_2(E_2)}
\tag{3.76}
$$

Taking into account that Q_1 depends also upon N_1 and V_1 and that Q_2 depends on N_2 and V_2, Eq. (3.76) can be written as

$$
\frac{\partial \log Q_1(N_1, V_1, E_1)}{\partial E_1} = \frac{\partial \log Q_2(N_2, V_2, E_2)}{\partial E_2}
\tag{3.77}
$$

This equilibrium condition will agree with Eq. (3.70), the thermodynamic condition for equilibrium, if we identify $S(N, E, V)$ for any system as

$$
S(N, E, V) = k \log Q(N, E, V) + K
\tag{3.78}
$$

where k and K are two constants that may, however, depend on N and V.

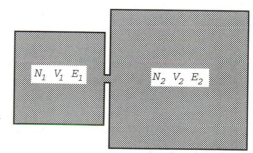

Fig. 3.10   The two systems can transfer both energy and particles. The final distribution of both quantities will be such as to maximize the entropy of the combined system.

In order to obtain information about the N dependence of S, it is necessary to consider the equilibrium of two subsystems that can exchange particles as well as energy. Such a situation is depicted in Fig. 3.10. The system contains a total of N_T particles and has an energy E_T. The problem is now to determine the values of E_1 and N_1 at equilibrium. Let the left domain be called $\mathbf{V}_L$. The characteristic function of $\mathbf{V}_L$ is defined as

$$
C_L(\mathbf{r}) = \begin{cases} 1, & \mathbf{r} \in \mathbf{V}_L \\ 0, & \text{otherwise} \end{cases}
\tag{3.79}
$$

As is shown in Exercise 3.4, the observable described by the phrase "the number of particles on the left" is

$$N_L(\mathbf{r}_1, \ldots, \mathbf{r}_N) = \sum_{i=1}^{N_T} C_L(\mathbf{r}_i) \tag{3.80}$$

For any particular configuration of the system, we can write $H_T = H_L + H_R$, where H_L is the energy of all the particles on the left and H_R is the energy of those on the right. The probability of finding exactly N_1 particles on the left with an energy E_1 is given by

$$P_L(N_1, E_1) = \frac{\int \delta(N_1 - N_L)\delta(E_1 - H_L)\delta(E_T - H_L - H_R)\, d^K x\, d^K p}{Q_T(N_T, V_T, E_T)} \tag{3.81}$$

The discrete Kronecker delta function, $\delta(N_1 - N_L(x))$, is zero except for configurations in which exactly N_1 particles are on the left. There are $\binom{N_T}{N_1}$ ways of choosing those particles. When that is taken into account, one finds that

$$P_L(N_1, E_1) = \frac{N_T!}{N_1! N_2!} \frac{Q_L(N_1, V_1, E_1) Q_R(N_2, V_2, E_2)}{Q_T(N_T, V_T, E_T)} \tag{3.82}$$

The most probable state is determined by setting $\partial P_L / \partial E_1 = 0$ and $\partial P_L / \partial N_1 = 0$. The first equation gives the same result obtained in Eq. (3.77). The second condition will give

$$\frac{\partial \log(Q_L/N_1!)}{\partial N_1} = \frac{\partial \log(Q_R/N_2!)}{\partial N_2} \tag{3.83}$$

which shows that $\log Q$ should be replaced by $\log(Q/N!)$ in Eq. (3.78). This change will have no effect on the solution of the energy equilibrium equation. There is one further modification that must be made. In Chapter 4, which treats quantum-mechanical systems, we will define the entropy as the logarithm of the number of discrete quantum states that have energy less than E. In order to prevent having the classical and quantum-mechanical definitions of the entropy differ by a constant (which would not have any effect on the equilibrium states that the entropy functions would predict), we must divide the phase-space integral that defines Q by the factor h^K, where $h = 2\pi\hbar$ is Planck's constant and $2K$ is the dimensionality of the phase-space. The appearance of Planck's constant clearly reveals the quantum-mechanical origin of this term. One might say that Planck's constant supplies a natural or fundamental unit of phase-space volume, namely h^K, just as the velocity of light supplies a fundamental unit of velocity. The net result is that, for any system that can be described by classical mechanics, the thermodynamic entropy is given by the formula

$$\begin{aligned} S &= k \log\left[\frac{1}{h^K N!} \int \delta(E - H(x, p))\, d^K x\, d^K p\right] \\ &= k \log[Q/h^K N!] \end{aligned} \tag{3.84}$$

The absolute temperature is defined in terms of the entropy by

$$\frac{1}{T} = \frac{\partial S}{\partial E} \tag{3.85}$$

Thus, the choice of a temperature unit fixes the arbitrary constant k. For the Kelvin temperature scale, k must be chosen to be Boltzmann's constant.

Because of the geometrical properties of high-dimensional spaces that were discussed in Section 3.7, it would make no difference if the microcanonical normalization integral Q were replaced in Eq. (3.84) by the normalization integral for the uniform ensemble C defined in Eq. (3.16).

3.21   THERMODYNAMIC CALCULATIONS

This formula, giving the entropy as a phase-space integral, is the basis for another method of computing many important macroscopic observables that is much more efficient than the standard calculational procedure. In Chapter 5 it will be shown that, once the entropy function is known, the equilibrium values of most macroscopic observables may be computed using thermodynamic identities, directly from that function. As an example, we will derive the equations of state of an ideal gas from Eq. (3.20), which gives the value of $C(N, E, V)$. From that equation, we get that

$$
\begin{aligned}
S &= k \log[C/h^K N!] \\
&= Nk(\log V + \tfrac{3}{2} \log E + \phi(N))
\end{aligned}
\tag{3.86}
$$

where ϕ is a function of N that could easily be calculated but will not be needed here. Using the thermodynamic formulas* $1/T = \partial S/\partial E$ and $p/T = \partial S/\partial V$, we immediately obtain the desired relations.

$$
\frac{1}{T} = \frac{3kN}{2E}
\tag{3.87}
$$

and

$$
\frac{p}{T} = \frac{kN}{V}
\tag{3.88}
$$

3.22   THE PRESSURE FORMULA*

The temperature is fundamentally a statistical variable. From the equation $1/T = \partial S/\partial E$, it is clear that it conveys information on how fast the number of microstates available to the system increases with energy. But the pressure, being a force over an area, is a mechanical variable. For a system of N particles in a cylinder closed by a piston, the force on the piston is a well-defined function of the positions and momenta of the particles. That is exactly what defines a mechanical variable. Therefore, the formula $p = T \, \partial S/\partial V$, which equates a mechanical variable to a statistical variable, is particularly remarkable. In Chapter 5, it will be derived by a thermodynamic argument. However, the thermodynamic argument, because it makes no reference to microscopic dynamics, somewhat hides the detailed mechanical origin of the relation. Therefore, in this section, we will give an alternative derivation of the formula, using a microcanonical ensemble.

We consider a system of N particles in a cylinder whose left wall is a moveable but fixed piston (Fig. 3.11). In other words, the piston could be located at any point, but while the particles of the system move, it remains fixed and is not a dynamical variable in the system. We model the interaction between the ith particle and the piston by a "wall potential" $U(x_i - X)$, which smoothly becomes infinite as $x_i - X \to 0$ but is equal to zero for $x_i - X$ larger than some fixed microscopic distance a. Thus the Hamiltonian of the system, when the piston is at position X, is

$$
H = H_o(\mathbf{r}_1, \ldots, \mathbf{r}_N, \mathbf{p}_1, \ldots, \mathbf{p}_N) + \sum_1^N U(x_i - X)
\tag{3.89}
$$

---

*The second relation is derived in the next section and also in Chapter 5.

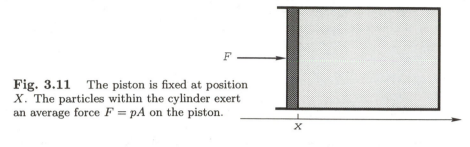

Fig. 3.11   The piston is fixed at position X. The particles within the cylinder exert an average force $F = pA$ on the piston.

where H_o contains the kinetic energy and interaction potentials for the N particles. For any finite energy, the wall potential guarantees that no particle's x coordinate becomes less than X. For any state of the system, the instantaneous force that the piston is exerting on the particles is

$$F = -\sum_{1}^{N} \frac{\partial U(x_i - X)}{\partial x_i} \tag{3.90}$$

The pressure on the wall is defined as F/A, where A is the area of the piston.

$$p = -A^{-1} \sum_{1}^{N} \frac{\partial U(x_i - X)}{\partial x_i} \tag{3.91}$$

The pressure resembles a one-body macroscopic observable, in that it is a one-body function, summed over all the particles in the system. But, in the thermodynamic limit, it would not be appropriate to scale the wall potential $U(x_i - X)$. Even for a macroscopic system, a particle only interacts with the wall when it is microscopically close to it. Therefore, it is really necessary to repeat the thermodynamic limit analysis separately for this case. The details have been left as an exercise (see Problem 3.25). One must assume that, for x and x' in the range $X < x, x' < X + a$,

$$F_2(x, sy, sz, \mathbf{p}, x', sy', sz, \mathbf{p}') \to F_1(x, sy, sz, \mathbf{p})\, F_1(x', sy', sz', \mathbf{p}') \tag{3.92}$$

as $s \to \infty$. That is, particles at large distances from one another on the surface of the piston are assumed not to be statistically correlated. With that assumption, the observed pressure is almost certain to be the average pressure for a microcanonical ensemble at the system energy. Thus, we can assume that

$$p = -A^{-1} \int \sum \frac{\partial U}{\partial x_i} \delta(E - H)\, d^{3N}r\, d^{3N}p \bigg/ \int \delta(E - H) d^{3N}r\, d^{3N}p \tag{3.93}$$

But, by the definition of H, it is easy to see that

$$-\sum \frac{\partial U}{\partial x_i} \delta(E - H) = -\frac{\partial}{\partial X} \theta(E - H) \tag{3.94}$$

which implies that

$$p = -A^{-1} \frac{\partial}{\partial X} \int \theta(E - H) d^{3N}r\, d^{3N}p \bigg/ \int \delta(E - H)\, d^{3N}r\, d^{3N}p$$

$$= -A^{-1} \frac{\partial C(E, X)}{\partial X} \bigg/ Q(E, X) \tag{3.95}$$

Using the facts that $Q(E, X) = \partial C(E, X)/\partial E$ and that $dV = -A\,dX$, where V is the container volume, the pressure can be written as

$$
\begin{aligned}
p &= \frac{\partial C/\partial V}{\partial C/\partial E} \\
&= \frac{\partial \log C/\partial V}{\partial \log C/\partial E} \\
&= \frac{\partial S/\partial V}{\partial S/\partial E} \\
&= T\frac{\partial S}{\partial V}
\end{aligned}
\qquad (3.96)
$$

PROBLEMS

3.1   For a system with a Hamiltonian function of the form of Eq. (3.4), verify that Hamilton's equations (Eqs. 3.5 and 3.6) are equivalent to Newton's equation $\mathbf{F}_i = m\,d^2\mathbf{r}_i/dt^2$.

3.2   For a particle, constrained to the z axis, in a uniform gravitational field, draw a few energy surfaces in the two-dimensional phase-space and indicate the trajectory of a system point in the phase-space.

3.3   Choose a potential function $U(x)$ so that the energy surfaces in the two-dimensional phase-space of the Hamiltonian, $H(x, p) = p^2/2m + U(x)$, are each composed of two disconnected parts.

3.4   (a) If a point is chosen at random in an N-dimensional unit sphere, what is the probability of falling inside the sphere of radius 0.99999999? (b) Evaluate your answer for $N = 3$ and $N = N_A$.

3.5   The unit cube in N dimensions is the set of points defined by the inequalities $0 \leq x_i \leq 1$ for $i = 1, \ldots, N$. The *inscribed sphere* is the largest N-dimensional sphere that can fit within the unit cube. It has a unit diameter. (a) If a point is chosen at random within the unit cube, what is the probability that it will fall within the inscribed sphere? (b) The *circumscribed sphere* is the smallest sphere that contains the unit cube. If a point is chosen at random in the circumscribed sphere, what is the probability that it will fall within the unit cube? (c) For very large N, give simplified, approximate forms for your answers to (a) and (b). (Hint: Simplify the logarithm of the answer and then exponentiate it.)

3.6   The transformation from four-dimensional Cartesian coordinates (x, y, z, w) to spherical coordinates (r, ψ, θ, ϕ) is $x = r \sin\psi \sin\theta \cos\phi$, $y = r \sin\psi \sin\theta \sin\phi$, $z = r \sin\psi \cos\theta$, and $w = r \cos\psi$. The ranges of the angles are $0 \leq \psi \leq \pi$, $0 \leq \theta \leq \pi$, and $0 \leq \phi \leq 2\pi$. Determine the volume element in spherical coordinates and, by direct integration, the volume of a four-dimensional sphere of radius R. Check your result with the general formula given in the Mathematical Appendix.

3.7 Consider a system composed of two weakly interacting subsystems, so that the Hamiltonian function of the system can be approximated by $H(x,p) = H_1(x_1, p_1) + H_2(x_2, p_2)$. Show that

$$Q(E) = \int Q_1(E_1) Q_2(E - E_1)\, dE_1 \qquad (3.97)$$

where Q_1 and Q_2 are the microcanonical normalization constants (defined in Eq. (3.27) for the subsystems separately. $(x_1, p_1) = (x_1, \ldots, x_K, p_1, \ldots, p_K)$ are the coordinates and momenta of subsystem 1 and $(x_2, p_2) = (x'_1, \ldots, x'_L, p'_1, \ldots, p'_L)$ are those of subsystem 2.

3.8 A system of N noninteracting one-dimensional particles are constrained to an interval of length L and have a total energy $E = N\varepsilon$. (a) Write an explicit formula for the uniform ensemble probability density $P(x_1, \ldots, x_N, p_1, \ldots, p_N)$. (b) By integrating over all coordinates and momenta except p_N, determine the probability distribution for the momentum of the Nth particle. (c) Take the logarithm of the result obtained in (b), assuming that $N \gg 1$, use Stirling's approximation to rewrite it, and then exponentiate the result to show that the momentum distribution associated with the uniform ensemble is just the Maxwell distribution.

3.9 (a) For N one-dimensional particles in a harmonic oscillator potential, using the uniform ensemble, write an expression for the probability density $P(x_1, \ldots, x_N, p_1, \ldots, p_N)$ and, by integrating over all coordinates and moments except x_N, calculate the probability density for the position of the Nth particle. (b) Assume that $N \gg 1$ and write a simpler, approximate, expression for your answer. (Hint: Take the logarithm of your answer, simplify it, and then exponentiate it.)

3.10 For a system of identical particles, $H = \sum p_i^2/2m + V(x)$. Verify the identity, $\langle \sum p_i^2/2m \rangle = m(\partial C/\partial m)/(\partial C/\partial E)$, where C is the normalization constant for the uniform ensemble. Check the identity for an ideal gas, using Eq. (3.20) for C.

3.11 For an ideal gas, calculate $Q(E)$ as an explicit function of E, m, N, and V.

3.12 Define the function $a(\mathbf{r}, \mathbf{p})$ for the one-body observable described by the phrase "the fraction of particles with kinetic energy greater than K".

3.13 Write a two-body observable whose average value is equal to the average number of pairs of particles with relative separation less than d. (Hint: Do not count $(1, 2)$ and $(2, 1)$ as *two* pairs.)

3.14 Define the function $a(\mathbf{r}, \mathbf{p}, \mathbf{r}', \mathbf{p}')$ for the two-body observable described by the phrase "the number of pairs of particles that are separated by a distance less than d and are approaching one another". (Hint: "approaching one another" means that the distance between them is getting smaller.)

3.15 Prove that the normalization integrals of F_1 and F_2, given in Eqs. (3.38) and

(3.39), are correct.

3.16 Consider a system of N noninteracting particles, confined to move within a smooth circular cylinder of radius R and length ℓ. The axis of the cylinder is the z axis. Suppose that, besides fixing the number of particles in the system, we fix the total energy to be E and the z component of angular momentum to be L_z. (a) By the method used to derive the Maxwell-Boltzmann distribution in the previous chapter, show that the equilibrium distribution function is of the form, $f(\mathbf{r}, \mathbf{v}) = C \exp[-\beta(mv^2/2 - m\Omega(xv_y - yv_x))]$. (b) Calculate the average velocity of particles at position (x, y, z) and show that it is identical to what would be obtained for rigid body rotation about the z axis with angular velocity Ω. (c) Show that the particle density $n(\mathbf{r})$ has the form $C \exp[-\phi(\mathbf{r})/kT]$, where $\phi = -\frac{1}{2}m\Omega^2(x^2 + y^2)$ is the centrifugal potential.

3.17 (a) Show that the minimum value of the energy per particle for a three-dimensional Ising lattice is $-3V$. (b) At very high temperatures, the spins in an Ising lattice become statistically independent random variables. Show that, in that case, $\langle E \rangle/N = 0$ and that $\Delta E/N \approx 0$.

3.18 William Cleghorn, one of the early proponents of the caloric theory of heat, according to which heat is a fundamental conserved substance, called caloric, whose density in a body determines the body's temperature, gave the following objection to the competing theory that temperature was a measure of the vibrational or translational motion of the minute particles that comprised the body. If heat and temperature were measures of the vibrational motions of the particles of solid bodies, then they would propagate through solid bodies at very high speeds, as vibrations are known to do. The temperature differences in an ordinary sized object would be expected to vanish in a time so small as to be difficult to measure. This is clearly contradicted by the observed slow diffusion of heat through objects. With your much more detailed knowledge of the internal structure of bodies, and the motions of the particles comprising them, can you give a clear refutation of Cleghorn's objection to the kinetic theory of heat?

3.19* [Starred problems are those that are associated with optional sections.] Consider a system of N noninteracting particles in a box of volume $V = L^3$ subject to a uniform gravitational field g. According to Eq. (2.20), the density at height z is given by $n(z) = n_o \exp(-z/h)$, where $h = kT/mg$ and n_o is determined by the normalization condition $\int_o^L dx\, dy\, dz\, n(z) = N$. (a) Take g as fixed and consider a thermodynamic sequence $N = s^3 N_o$ and $L = sL_o$. Show that, as $s \to \infty$, for any fixed z, $n(z) \to \infty$, but $n(sz) \to 0$. Thus the density in fixed or scaled units behaves badly in the thermodynamic limit. (b) Take $g = g_o/s$ [this is the prescription of Eq. (3.63)] and show that $n(z) = n_o$ and $n(sz) = n_o e^{-z/h_o}$, where h_o and n_o are independent of s. Thus, with this prescription, the density in scaled units gives the expected exponential distribution.

3.20* Show that any intensive one-body observable [an observable defined by Eq. (3.65)] satisfies the zero-fluctuation theorem.

3.21* Show that, if A is an extensive observable, as defined in Eq. (3.64), and $\langle A \rangle \neq 0$, then, in the thermodynamic limit, $\langle A \rangle$ is proportional to N but $\Delta A/\langle A \rangle \to 0$.

*3.22** For a system of particles in a box described by the inequalities $0 < x, y, z < L$, the particle density $n(x, y, z)$ can be written in terms of its Fourier coefficients n_{klm} as

$$n(x, y, z) = \sum_{k,l,m} n_{klm} \sin\left(\frac{k\pi x}{L}\right) \sin\left(\frac{l\pi y}{L}\right) \sin\left(\frac{m\pi z}{L}\right)$$

Show that $n_{klm} = \langle A_{klm} \rangle$, where A_{klm} is an intensive one-body observable as defined by Eq. (3.65).

*3.23** Consider a starting system composed of N_o particles with a total energy of E_o, within a symmetric three-dimensional harmonic oscillator potential $U_o = \frac{1}{2}k_o(x^2 + y^2 + z^2)$. Calculate the parameters C and β in a Maxwell–Boltzmann distribution [Eq. (2.16)] in terms of N_o, E_o, and k_o. Let $N = s^3 N_o$ and $E = s^3 E_o$ and determine how the force constant k_o must be scaled in order for C and β to be independent of s. Show that the required scaling satisfies Eq. (3.63).

3.24 Calculate the function $\phi(N)$ defined in Eq. (3.86).

3.25 Using the uniform ensemble, evaluate the entropy $S(N, E)$ for a system of N distinguishable one-dimensional harmonic oscillators. When the particles are distinguishable, the factor of $1/N!$ is left out of the definition of S given in Eq. (3.84).

*3.26** Using the formula for the pressure given in Eq. (3.91), the assumption about F_2 given in Eq. (3.92), and the fact that the area of the piston goes as $s^2 A_o$ in the thermodynamic limit, show that the pressure satisfies the no-fluctuation theorem.

Chapter 4
The Canonical Ensemble

4.1 QUANTUM ENSEMBLES

A statistical ensemble for a dynamical system is a collection of system points in the system phase space. Each system point represents a full system in a particular dynamical state. Therefore, a statistical ensemble may be pictured as a vast collection of exact replicas of a particular system. If the system under study is a bottle of beer, then the ensemble can be thought of as a huge warehouse filled with individual, noninteracting, bottles of beer. It cannot be pictured as a very large barrel of beer, which would be a single, different system. Each system in the ensemble has exactly the same Hamiltonian function, but, in general, different systems in the ensemble are in different dynamical states. Since the systems all have the same Hamiltonian function, it makes sense to represent the dynamical states of all the systems by points in a single phase space. Thus, the state of the complete ensemble is represented by a "dust cloud" in the phase space, containing a vast number of discrete points. The danger inherent in using this picture of an ensemble is that it is very easy to confuse it with the similar picture of a gas containing a vast number of point particles. This is unfortunate because the two things behave in quite different ways. The motion of a single gas particle is affected by the positions and velocities of its neighbors, because the gas particles interact. In contrast, each point in the ensemble representation is a system point for an isolated system that moves through the phase space according to Hamilton's equations with no regard for the states of the other isolated systems in the ensemble. For the uniform ensemble the dust cloud is of uniform density inside the energy surface and of zero density outside it. For the microcanonical ensemble the system points of the ensemble are confined to the energy surface itself.

A quantum system is defined by a Hermitian Hamiltonian operator rather than a Hamiltonian function. The possible energy states of the system are the solutions of Schrödinger's equation

$$H\psi_n(\mathbf{r}_1, \ldots, \mathbf{r}_N) = E_n\psi_n(\mathbf{r}_1, \ldots, \mathbf{r}_N) \tag{4.1}$$

A statistical ensemble for a quantum system will be pictured as a vast collection of replicas of the system. Each will have the same Hamiltonian operator, but, in general, different members of the ensemble will be in different quantum states. The quantum states are assumed to be energy eigenstates with energies within some particular energy interval. This is not the only reasonable way to choose the set of allowed microstates. A more general way is to allow any microstate that is a normalized linear combination of the energy eigenstates within the above-mentioned energy interval. That would include states that were not themselves exact energy eigenstates. In the problems, the reader will be asked to show that this more general procedure actually leads to exactly the same results as the assumption made here.

4.2   THE UNIFORM QUANTUM ENSEMBLE

For the uniform quantum ensemble the probability of finding a system, chosen at random from the ensemble, in eigenstate ψ_n is assumed to be

$$P_n = \theta(E - E_n)/\Omega(E) \tag{4.2}$$

This means that all states of energy less than E are equally likely but all states of energy greater than E have zero probability. The normalization constant Ω is equal to the number of quantum states of energy less than E.

$$\Omega(E) = \sum_{n=1}^{\infty} \theta(E - E_n) = \left\{ \begin{array}{c} \text{number of} \\ \text{eigenstates} \\ \text{with } E_n < E \end{array} \right\} \tag{4.3}$$

For a system of N particles in a volume V, Ω is a function of N and V also. In the Mathematical Appendix it is shown that, as $\hbar \to 0$

$$\Omega(N, E, V) \to \frac{1}{h^K N!} \int \theta(E - H(x, p)) \, d^K x \, d^K p \tag{4.4}$$

where $H(x, p)$ is the classical Hamiltonian function corresponding to the operator H. This identifies the thermodynamic entropy function as, essentially, the logarithm of the number of quantum states of the system with energy less than E.

$$S(N, E, V) = k \log \Omega(N, E, V) \tag{4.5}$$

It has been mentioned before that the value of the constant k depends on our unit of temperature. In relating thermodynamics to statistical mechanics, it is often convenient to use what will be called a rational system of units, in which the entropy is exactly equal to $\log \Omega$. The temperature is then measured in joules (that is, $\tau = kT$ is called the temperature). The entropy in rational units will be written as S^o. The relationship between S^o and S is very simple.

$$S^o = \log \Omega \tag{4.6}$$

and therefore,

$$S^o = S/k \tag{4.7}$$

In rational units,

$$\frac{\partial S^o}{\partial E} = \frac{1}{\tau} \quad \text{and} \quad \frac{\partial S^o}{\partial V} = \frac{p}{\tau} \tag{4.8}$$

Fig. 4.1 If the "spread" in δ_ϵ is much larger than the average spacing between energy levels, then the function $P_n(E) = C\delta_\epsilon(E - E_n)$ is a smooth function of E.

4.3   THE MICROCANONICAL QUANTUM ENSEMBLE

In defining the microcanonical quantum ensemble a small technical problem arises. It would be natural simply to replace $\theta(E - E_n)$ in Eq. (4.2) by $\delta(E - E_n)$. However, because the energy spectrum of a bound system is discrete, the probability function $P_n = \text{const.} \times \delta(E - E_n)$ would be a very irregular function of E. In fact, unless E were equal to one of the energy eigenvalues, all terms P_n would be zero and thus the probability could not be normalized. In reality, although the energy eigenvalues are discrete, the spacing between adjacent eigenvalues, for a macroscopic system, is extremely small. In Problem 4.2 the reader is asked to show that the energy spacing between neighboring levels in one mole of gas is of the order $\Delta E \sim e^{-N_A}$. If we replace the Dirac delta function by a function $\delta_\epsilon(E - E_n)$, which has a "spread" ϵ that is much larger than the energy spacing but much smaller than the uncertainty in any realistic energy measurement, then, for any value of E, the probability becomes a smooth function of n (Fig. 4.1). Thus the quantum microcanonical probability will be defined as

$$P_n = \delta_\epsilon(E - E_n)/\Omega'(E) \tag{4.9}$$

where $\Omega' = d\Omega/dE$ is the density of eigenstates as a function of energy. That is, $\Omega' dE$ is the number of energy eigenstates with eigenvalues in the range dE.

Differentiating Eq. (4.6), in the form $\Omega = e^{S^o}$, and using the thermodynamic relation $\partial S^o/\partial E = 1/\tau$, gives $\Omega' = e^{S^o}/\tau$. Taking the logarithm of this relation, we get

$$\log \Omega' = S^o - \log \tau \tag{4.10}$$

To estimate the relative sizes of the two terms, we note that, for one mole of helium at 0°C and atmospheric pressure S^o (the entropy in rational units) is a pure number equal to $15\,N_A$ (see Problem 4.3). In general, for any macroscopic system, the entropy function has the general form $S^o(N, E, V) = N s^o(\varepsilon, v)$, where s^o is the entropy per particle, ε is the energy per particle, and v is the volume per particle. The second term in Eq. (4.10) is not proportional to N, and, therefore, for a large system it is completely negligible in comparison to the first. Thus, an equally valid formula for the entropy is

$$S^o(N, E, V) = \log \Omega'(N, E, V)$$
$$= \log \left[\sum_n \delta_\epsilon(E - E_n) \right] \tag{4.11}$$

In quantum mechanical systems, observables are represented by Hermitian operators. It is clear from our interpretation of a quantum ensemble, that the expectation value of an observable A is given by

$$\langle A \rangle = \sum_n P_n (\psi_n | A | \psi_n) \tag{4.12}$$

where the sum extends over all energy eigenstates of the system and P_n is the probability of eigenstate ψ_n.

4.4   VIBRATIONAL SPECIFIC HEATS OF SOLIDS

As an example of the use of the quantum microcanonical ensemble, we will analyze the contribution to the heat capacity of a crystalline solid that is due to its quantized lattice vibrations. Such quantized lattice vibrations are called *phonons*, and thus the calculation will yield the phonon contribution to the specific heat of the solid.

We will first do the calculation using an approximation introduced by Albert Einstein. We consider a crystal composed of N atoms arranged in a cubic lattice. If all the atoms but one are held at their equilibrium positions and that atom is displaced slightly in the x direction, it will execute simple harmonic motion, with some angular frequency ω, parallel to the x axis. It could also oscillate parallel to the y or z axes. Thus, the single atom is equivalent to a set of three independent harmonic oscillators of equal frequencies. In the Einstein model, the complete crystal is assumed to be equivalent to a system of $K = 3N$ independent harmonic oscillators, all of angular frequency ω. Thus, to calculate the vibrational energy of the crystal, we must first determine the energy spectrum of a system composed of K identical harmonic oscillators. We number the oscillators with an index k, running from 1 to K. The kth oscillator has energy levels

$$E_{n_k} = \hbar\omega(n_k + \tfrac{1}{2}) \tag{4.13}$$

where $n_k = 0, 1, \ldots$. The *system* has one energy eigenstate for each set of quantum numbers $n_1, n_2, \ldots, n_K$.

$$E(n_1, n_2, \ldots, n_K) = E_o + \hbar\omega(n_1 + \cdots + n_K) \tag{4.14}$$

where $E_o = K\hbar\omega/2$. E_o may be dropped if we measure all energies relative to the ground-state energy. The energy eigenvalues of the system have a uniform spacing $\hbar\omega$, but, for large K, they are highly degenerate. The quantity $\Omega'(E)$ is the density of eigen*states*, not the density of eigen*values*. We can calculate $\Omega'(E)$ at the energy $E = \hbar\omega M$ by taking the number of eigenstates in the energy interval from $E - \hbar\omega/2$ to $E + \hbar\omega/2$ and dividing by $\hbar\omega$. Therefore

$$
\begin{aligned}
\Omega'(\hbar\omega M) &= \frac{1}{\hbar\omega} \left\{ \begin{array}{l} \text{the number of ways of choosing } K \\ \text{integers } n_1, n_2, \ldots, n_K \text{ adding up to } M \end{array} \right\} \\
&\equiv \frac{W(K, M)}{\hbar\omega}
\end{aligned}
\tag{4.15}
$$

The problem of computing $W(K, M)$ can be solved with the following trick. We interpret the integers as Bose–Einstein occupation numbers for K quantum states. Then $W(K, M)$ is the quantity given in Eq. (2.57). That is, $W(K, M) = (K + M - 1)!/(K - 1)!M!$ and

$$\Omega'(\hbar\omega M) = \frac{(K + M - 1)!}{\hbar\omega(K - 1)!M!} \quad (4.16)$$

Using Stirling's approximation, writing $E = \hbar\omega M$, and dropping terms that are negligible in comparison with K or M, which are assumed to be very large numbers, one obtains (see Problem 4.6)

$$S^o(E) = \log \Omega'(E) = \left(K + \frac{E}{\hbar\omega}\right) \log\left(K + \frac{E}{\hbar\omega}\right) - K \log K - \left(\frac{E}{\hbar\omega}\right) \log\left(\frac{E}{\hbar\omega}\right) \quad (4.17)$$

The thermodynamic identity $\partial S^o(E)/\partial E = 1/kT$ gives

$$\frac{1}{\hbar\omega}\left[\log\left(K + \frac{E}{\hbar\omega}\right) - \log\left(\frac{E}{\hbar\omega}\right)\right] = \frac{1}{kT} \quad (4.18)$$

which can be solved for E as a function of T.

$$\frac{E(T)}{K} = \frac{\hbar\omega}{e^{\hbar\omega/kT} - 1} \quad (4.19)$$

There are K identical oscillators; therefore $E(T)/K$ can be interpreted as the average energy per oscillator at temperature T.

The *specific heat* of a system is defined as the amount of energy required to raise the temperature of the system by one unit. Therefore, the vibrational specific heat of the crystal C_v, in the Einstein approximation, is given by

$$C_v = \frac{\partial E}{\partial T} = Kk(\hbar\omega/kT)^2 \frac{e^{\hbar\omega/kT}}{(e^{\hbar\omega/kT} - 1)^2} \quad (4.20)$$

At high temperatures (that is, for $kT \gg \hbar\omega$), $(e^{\hbar\omega/kT} - 1)^{-2} \approx (kT/\hbar\omega)^2$ and

$$C_v(T) \approx Kk \quad (4.21)$$

This result, that the specific heat of a solid is equal to Boltzmann's constant times the number of normal modes, was known as the *Dulong–Petit law*. It is the result one gets by using classical mechanics, rather than quantum mechanics, in the calculation. It was exactly the deviations from the Dulong–Petit law that occur at lower temperatures that Einstein was trying to explain. If $kT \ll \hbar\omega$, then $(e^{\hbar\omega/kT} - 1)^{-2} \approx e^{-2\hbar\omega/kT}$ and

$$C_v(T) \approx Kk(\hbar\omega/kT)^2 e^{-\hbar\omega/kT} \quad (4.22)$$

As T goes to zero the factor $(\hbar\omega/kT)^2$ goes to infinity but the exponential factor goes to zero much more rapidly and therefore dominates. (The best way to see this is to look at what happens to $\log C_v$ as $T \to 0$.) $C_v(T)$ goes to zero, as T goes to zero, essentially exponentially. The Einstein result for the vibrational specific heat is plotted in Fig. 4.2. The explanation for the vanishing of the specific heat at low T, as a consequence of quantum mechanics, was a great triumph of the Einstein theory. However, the detailed quantitative result disagreed with

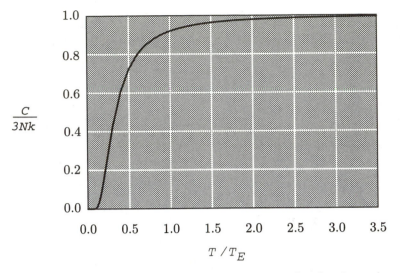

Fig. 4.2   The Einstein specific heat function. The plot shows the ratio of the Einstein prediction for the vibrational specific heat of a crystal to the Dulong—Petit, or classical prediction. The *Einstein temperature* is defined by the relation $kT_E = \hbar\omega$.

experimental measurements, which clearly indicated that $C_v(T) \to \text{const.} \times T^3$ as $T \to 0$. A more quantitatively accurate theory was first given by Peter Debye.

4.5   THE FREQUENCY DISTRIBUTION FUNCTION

In the Mathematical Appendix it is proved that, for small vibrations, the Hamiltonian function of a crystal of N atoms is exactly the same as the Hamiltonian function of a system of $K = 3N$ harmonic oscillators with different angular frequencies $\omega_1, \omega_2, \ldots, \omega_K$. Each angular frequency ω_k is associated with a possible normal-mode motion of the crystal lattice. A normal mode is a solution of the classical equations of motion in which all of the coordinates oscillate with the same frequency. That is, $x_i(t) = A_i \sin(\omega t + \phi_i)$, where the amplitude and phase of the oscillation are generally different for different coordinates, but the angular frequency is the same. (This sounds a bit confusing—when the particles are vibrating with any particular normal mode motion they all move with the same angular frequency but that common frequency will be different for different normal modes.) Using Eq. (4.19), for the average thermal energy of an oscillator of angular frequency ω, at temperature T, we can write the thermal energy of the crystal as a sum over all the normal modes of the crystal lattice.

$$E(T) = \sum_{k=1}^{K} \frac{\hbar\omega_k}{e^{\hbar\omega_k/kT} - 1} \tag{4.23}$$

We now define a function, called the *frequency distribution function* $D(\omega)$, by saying that $D(\omega)\,d\omega$ is equal to the number of normal modes with angular frequencies in the range $d\omega$ (see Fig. 4.3). With this definition, $D(\omega)$ satisfies the normalization condition

$$\int_o^\infty D(\omega)\,d\omega = K = 3N \tag{4.24}$$

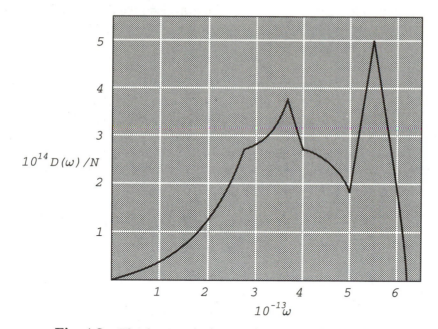

Fig. 4.3   The density of vibrational normal modes for aluminum. As can be seen in this typical example, $D(\omega)$ is usually a quite irregular function of ω that is only crudely approximated by the Debye theory given in Section 4.6.

In terms of the frequency distribution function, Eq. (4.23) can be written

$$E(T) = \int_{o}^{\infty} \frac{\hbar\omega}{e^{\hbar\omega/kT} - 1} D(\omega)\, d\omega \tag{4.25}$$

At high temperature $e^{\hbar\omega/kT} \approx 1 + \hbar\omega/kT$ and

$$E(T) \approx \int_{o}^{\infty} kTD(\omega)\, d\omega = 3NkT \tag{4.26}$$

which shows that the thermal energy at high temperatures is independent of the details of the frequency distribution function and is given by the Dulong–Petit law.

At very low temperature, the function $(e^{\hbar\omega/kT} - 1)^{-1}$ goes rapidly to zero with increasing ω. Therefore, at very low T, only the lowest-frequency normal modes contribute to the vibrational thermal energy of the crystal. But the low-frequency modes are long-wavelength vibrational waves. That is, they are simply sound waves. In a gas, sound waves are always longitudinally polarized waves. The vibratory motion of the particles is parallel to the direction of travel of the wave. The same is not true for sound waves in a solid. In a solid, two types of sound waves are possible: longitudinal waves and transverse waves. For a given direction of travel and wavelength (that is, for a given wave vector $\mathbf{k}$), there is one longitudinal sound wave and two perpendicularly polarized transverse sound waves. Both types of sound waves satisfy the relation $\lambda = v/\nu$, where v is the sound speed, but the value of v is different for the two types. For any type of wave motion, the relationship between the wave vector and the angular frequency is called the *dispersion relation* for the wave. For both types of sound waves the dispersion relation can be written in the form $\omega(k) = vk$, where $k = 2\pi/\lambda$. Using this, one can show (see Problem 4.9) that $D(\omega)$ is of the form

$$D(\omega) = \frac{V}{2\pi^2}\left(\frac{1}{v_L^3} + \frac{2}{v_T^3}\right)\omega^2 \equiv VA\omega^2 \tag{4.27}$$

where v_L and v_T are the longitudinal and transverse sound speeds and V is the volume of the crystal. Of course, this form is valid only for small values of ω. However, for small T, it is only small values of ω that contribute to the integral in Eq. (4.25). Therefore, at sufficiently low temperatures, we expect that the vibrational energy of a crystal is given by

$$\begin{aligned} E(T) &= VA \int_o^\infty \frac{\hbar\omega^3 d\omega}{e^{\hbar\omega/kT} - 1} \\ &= VA\frac{(kT)^4}{\hbar^3}\int_o^\infty \frac{x^3 dx}{e^x - 1} \\ &= VA\frac{(kT)^4}{\hbar^3}\frac{\pi^4}{15} \end{aligned} \tag{4.28}$$

where the integral has been evaluated using the Table of Integrals in the Mathematical Appendix. The vibrational specific heat, defined by $C_v = dE(T)/dT$, is thus expected to be proportional to T^3 at low temperatures. As mentioned before, this expectation is borne out by experiment. One class of crystals that do not satisfy this $C_v \approx T^3$ law at low temperatures is strongly anisotropic crystals, such as graphite. For those crystals the sound velocity is strongly directional, and therefore they do not satisfy the simple dispersion relation used previously.

4.6   THE DEBYE APPROXIMATION

An approximation, introduced by Peter Debye, is to use the frequency distribution function $D(\omega) = VA\omega^2$ for all ω up to a value ω_D, which is chosen to give a total of $3N$ normal modes. Above ω_D, $D(\omega)$ is assumed to vanish.

$$D(\omega) = \begin{cases} VA\omega^2, & \omega < \omega_D \\ 0, & \omega > \omega_D \end{cases} \tag{4.29}$$

This form will yield the correct behavior of C_v at low T, because it is correct for long wavelengths. It will also give the correct behavior at high T, because the high-T behavior is independent of the form of $D(\omega)$ except for its total normalization. In using the Debye approximation, it is best to treat the constant A as an adjustable parameter, chosen so as to obtain the best overall fit to the specific heat curve at intermediate temperatures. A is related to ω_D by the normalization condition

$$VA\int_o^{\omega_D}\omega^2 d\omega = 3N \tag{4.30}$$

which gives $VA = 9N/\omega_D^3$. In the Debye approximation

$$\frac{E}{N} = \frac{9\hbar}{\omega_D^3}\int_o^{\omega_D}\frac{\omega^3 d\omega}{e^{\hbar\omega/kT} - 1} \tag{4.31}$$

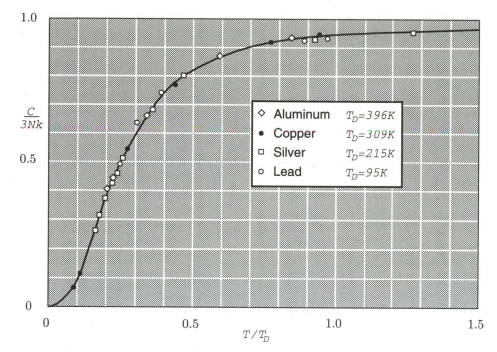

Fig. 4.4 A comparison of the Debye prediction (curve) with the experimental values of the vibrational specific heat for four metals. The data are plotted as a function of T/T_D. Values of T_D are chosen for each substance so as to give the best fit to the experimental data. (Data from G. Wannier, *Statistical Mechanics*, Dover Publ.)

or

$$\frac{E}{N\hbar\omega_D} = 9\lambda^{-4} \int_o^\lambda \frac{x^3\,dx}{e^x - 1} \tag{4.32}$$

where $\lambda = T_D/T$ and the *Debye temperature* T_D is defined by

$$kT_D = \hbar\omega_D$$

The integral cannot be evaluated in closed form, but expansions of it in a power series in λ and numerical tables of the integral can be found in standard handbooks.*

The vibrational specific heat is defined by $C_v = \partial E/\partial T$. At low temperatures it approaches the limit [see Eq. (4.28) and Fig. 4.4]

$$\frac{C_v}{Nk} \approx \frac{12\pi^4}{5}\left(\frac{T}{T_D}\right)^3 \tag{4.33}$$

4.7   THE CANONICAL ENSEMBLE

Both the uniform ensemble and the microcanonical ensemble describe isolated systems of known energy. By "isolated" we mean that the system does not exchange

---

*For example, *Handbook of Mathematical Functions*, M. Abromowitz and I.A. Stegun, National Bureau of Standards, p. 998. Also see Exercise 4.11.

energy with any external system. However, most experiments are carried out on systems that are not isolated. For example, when the equilibrium states of a chemical reaction are under study, the reaction vessel is maintained at constant temperature by being kept in contact with a large thermal reservoir, such as a water bath, that can absorb any energy evolved in the reaction. The statistical ensemble that describes a system of known temperature, rather than one of known energy, is the canonical ensemble. The canonical ensemble probability function can be derived by considering a system composed of two weakly interacting parts. The first part is a relatively small but still macroscopic system that will be called the *sample*, while the second part is a very large system, called the *reservoir*. The total system is assumed to be isolated and to have a total energy E. We will describe the total system by a microcanonical ensemble and ask the question: "What is the probability of finding the sample in an eigenstate of energy E_n"?

When we say that the two subsystems interact weakly, we mean that the energy eigenvalues of the combined system can be approximated by the sum $E_n + E_N$, where E_n is any energy eigenvalue of the sample and E_N is any energy eigenvalue of the reservoir. In the intermediate stages of the calculation, it is most efficient to lump all multiplicative factors that are independent of E_n into a single normalization constant that can be easily evaluated at the end.

If the total system has an energy E, then the probability of finding the sample in its nth eigenstate and the reservoir in its Nth eigenstate is const $\times \delta(E - E_n - E_N)$. If we sum this over all the eigenstates of the reservoir, we obtain the probability of finding the sample in its nth eigenstate, independent of the state of the reservoir.

$$P_n = \text{const.} \times \sum_N \delta(E - E_n - E_N) \tag{4.34}$$

But, by Eq. (4.11), the entropy function of the reservoir alone, at an energy $E - E_n$, would be given by

$$e^{S_R^o(E - E_n)} = \sum_N \delta(E - E_n - E_N) \tag{4.35}$$

The assumption that the sample is small relative to the reservoir allows S_R^o to be expanded to first order in a power series in E_n. Using the fact that $\partial S^o / \partial E = 1/kT$, we obtain

$$S_R^o(E - E_n) = S_R^o(E) - \beta E_n \tag{4.36}$$

where $\beta = 1/kT$ and T is the temperature of the reservoir at energy E. This allows the sum in Eq. (4.34) to be evaluated in the form

$$P_n = \text{const.} \times \sum_N \delta(E - E_n - E_N) = \text{const.} \times e^{-\beta E_n} \tag{4.37}$$

The final normalization constant is calculated by demanding that the sum over all n of P_n be unity.

$$(\text{const.})^{-1} = \sum_n e^{-\beta E_n} \equiv Z(\beta) \tag{4.38}$$

$Z(\beta)$ is called the *partition function* of the sample. It will, in general, depend on other parameters, such as the volume and number of particles in the sample, that determine the sample energy spectrum. In terms of the partition function, the probability that the system will be found in its nth energy eigenstate is

$$P_n = \frac{e^{-\beta E_n}}{Z} \tag{4.39}$$

It will now be shown that, just as the logarithm of the normalization constant for the uniform and microcanonical ensembles give the entropy function of the system, $\log Z(N, \beta, V)$ is equal to another important thermodynamic function.

4.8   THE CANONICAL POTENTIAL

Using the fact that the number of eigenstates in the energy interval dE is given by $\Omega'(E)dE = e^{S^o(E)}dE$, $Z(N, \beta, V)$ can be written as

$$Z(N, \beta, V) = \int e^{S^o(N,E,V) - \beta E} dE \tag{4.40}$$

Changing the variable of integration to $\varepsilon = E/N$, the energy per particle, and recalling that $S^o(N, E, V) = Ns^o(\varepsilon, v)$, where s^o is the entropy per particle and $v = V/N$, one obtains

$$Z(N, \beta, V) = N \int^{\infty} e^{N(s^o - \beta \varepsilon)} d\varepsilon \tag{4.41}$$

In the Mathematical Appendix, it is shown that, for any function $f(x)$ that has a single maximum at $\bar{x}$, as N goes to infinity,

$$\int e^{Nf(x)} dx \rightarrow \left(\frac{2\pi}{N|f''(\bar{x})|} \right)^{1/2} e^{Nf(\bar{x})} \tag{4.42}$$

This gives, for $Z(N, \beta, V)$,

$$Z(N, \beta, V) = \left(\frac{2\pi N}{|\partial^2 s^o / \partial \varepsilon^2|_{\bar{\varepsilon}}} \right)^{1/2} e^{N[s^o(\bar{\varepsilon}, v) - \beta \bar{\varepsilon}]} \tag{4.43}$$

where $\bar{\varepsilon}$ is given by $(\partial s^o(\varepsilon, v)/\partial \varepsilon)_{\bar{\varepsilon}} = \beta$. This is equivalent to the condition that

$$\left. \frac{\partial S^o(N, E, V)}{\partial E} \right|_{\bar{E}} = \beta = \frac{1}{\tau} \tag{4.44}$$

But this is exactly the thermodynamic relation between the energy and the temperature. This shows that, for a given reservoir temperature, the only sample energy that contributes significantly to the partition function integral is the expected thermodynamic energy for that temperature. Taking the logarithm of Z and dropping terms that are negligible in comparison with N, we get

$$\log Z(N, \beta, V) = S^o(N, \bar{E}, V) - \beta \bar{E} \equiv \phi(N, \beta, V) \tag{4.45}$$

where $\bar{E}$ is a function of N, β, and V defined by Eq. (4.44). In the future we will write simply $E(N, \beta, V)$ for $\bar{E}(N, \beta, V)$.

$\phi(N, \beta, V)$ is called the *canonical potential*. It plays the same role for a system at known temperature, as is played by the entropy function for a system of known energy. What those roles are will be fully developed in the chapter on thermodynamics. The thermodynamic relations, $\partial S^o / \partial E = \beta$ and $\partial S^o / \partial V = \beta p$, may be used to derive equivalent relations giving the energy and pressure in terms of the canonical potential.

$$\frac{\partial \phi}{\partial \beta} = \left(\frac{\partial S^o}{\partial E} - \beta \right) \frac{\partial E}{\partial \beta} - E = -E(N, \beta, V) \qquad (4.46)$$

and

$$\frac{\partial \phi}{\partial V} = \frac{\partial S^o}{\partial V} + \left(\frac{\partial S^o}{\partial E} - \beta \right) \frac{\partial E}{\partial V} = \beta p \qquad (4.47)$$

4.9   ENERGY FLUCTUATIONS IN THE CANONICAL ENSEMBLE

The canonical ensemble describes a system in contact with a reservoir. For such a system, the energy is not exactly fixed—the system can exchange energy with the reservoir. We will now evaluate the magnitude of the energy fluctuations. In the canonical ensemble, the expectation values of the energy and of the square of the energy are given by

$$\langle E \rangle = \frac{\sum E_n e^{-\beta E_n}}{\sum e^{-\beta E_n}} = \frac{-\partial Z / \partial \beta}{Z} \qquad (4.48)$$

and

$$\langle E^2 \rangle = \frac{\sum E_n^2 e^{-\beta E_n}}{\sum e^{-\beta E_n}} = \frac{\partial^2 Z / \partial \beta^2}{Z} \qquad (4.49)$$

This gives the following simple formula for the square of the energy fluctuation.

$$
\begin{aligned}
(\Delta E)^2 &= \langle E^2 \rangle - \langle E \rangle^2 \\
&= \frac{\partial^2 Z / \partial \beta^2}{Z} - \frac{(\partial Z / \partial \beta)^2}{Z^2} \\
&= \frac{\partial}{\partial \beta} \left(\frac{\partial Z / \partial \beta}{Z} \right) \\
&= \frac{\partial}{\partial \beta} \left(\frac{\partial \log Z}{\partial \beta} \right) \\
&= \frac{\partial^2 \phi}{\partial \beta^2} \\
&= -\frac{\partial E}{\partial \beta} \\
&= \tau^2 \frac{\partial E}{\partial \tau} \\
&= \tau^2 C
\end{aligned}
\qquad (4.50)
$$

where $C = \partial E / \partial \tau$ is the specific heat of the system, a quantity that is proportional to the size of the system. Written in terms of the specific heat per particle, $c = C/N$, and the energy per particle, $\varepsilon = E/N$, the ratio of the energy fluctuation to the energy is

$$\frac{\Delta E}{E} = \frac{\tau c}{\varepsilon \sqrt{N}} \qquad (4.51)$$

Clearly, the $1/\sqrt{N}$ dependence of the energy fluctuation is another manifestation of the law of large numbers. Thus, for a system of one mole, $\Delta E/E$ is of the order of 10^{-12}, and is therefore completely negligible. Because the energy fluctuation in the canonical ensemble is negligible for a macroscopic system, the predictions of the physical properties of such a system, made using the canonical ensemble, are identical with those made using the microcanonical ensemble, which has zero energy fluctuation. Although the canonical ensemble was introduced to describe a system in contact with a reservoir, it can perfectly well be used to describe an isolated system. The choice of which ensemble to use is purely one of convenience for a particular calculation; in fact, they give identical results.

4.10 THE CLASSICAL CANONICAL ENSEMBLE

An analysis, equivalent to that which led to Eq. (4.39) for the probability function of a quantum canonical ensemble (see Problem 4.18), would show that a classical system, in interaction with a reservoir at temperature T, has a probability $P(x,p)$ of being found in the dynamical state (x,p) given by

$$P(x,p) = I^{-1}e^{-\beta H(x,p)} \tag{4.52}$$

The classical normalization integral I can be related to the partition function Z by using the identity

$$I = \int e^{-\beta H}d^K x\, d^K p = \int_0^\infty dE \int \delta(E - H)e^{-\beta E}\, d^K x\, d^K p \tag{4.53}$$

and using Eq. 3.84, which gives the integral $\int \delta(E - H)\, d^K x\, d^K p$ in terms of S^o. One obtains

$$I = h^K N! \int_0^\infty e^{S^o - \beta E}\, dE \tag{4.54}$$

Equation (4.40) then gives, for a classical system,

$$Z(N,\beta,V) = \frac{I}{h^K N!} = \frac{1}{h^K N!}\int e^{-\beta H}\, d^K x\, d^K p \tag{4.55}$$

Written in terms of Z, rather than I, the canonical probability density function is

$$P(x,p) = e^{-\beta H(x,p)}/h^K N! Z \tag{4.56}$$

4.11 MOLECULAR GASES

The canonical ensemble is a convenient tool for studying the contributions of internal degrees of freedom, such as rotation and vibration, to the thermal properties of an ideal gas. It will be assumed that the rotational motion can be adequately described by classical mechanics. In Problem 4.19, the correct quantum mechanical treatment is described, and the range of validity of the classical treatment is determined. In order to have a definite focus, we will consider a specific system, namely, a gas of diatomic molecules with only rotational degrees of freedom. Such a molecule can be pictured as two point masses, m_A and m_B, at fixed distances, a and b, from

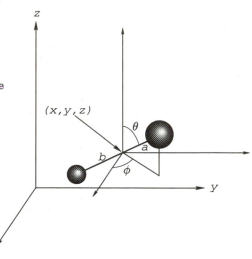

Fig. 4.5 The five coordinates that define the configuration of a diatomic molecule with a fixed internuclear distance. x, y, and z are the coordinates of the center of mass, while θ and ϕ are spherical angles that give the orientation of the axis in a local coordinate system.

their center of mass. By the definition of a center of mass it is necessary that $m_A a = m_B b$.

Diatomic molecules come in two classes, *homonuclear* molecules, in which the nuclei of the two atoms are identical, and *heteronuclear* molecules, in which they differ in some way. For a homonuclear molecule, the quantum mechanical wave function must be either symmetric or antisymmetric under exchange of the two nuclei. It must be symmetric if the number of fermions (neutrons and protons) in each nucleus is even, and it must be antisymmetric if that number is odd. Such symmetry requirements add a certain amount of extra complication to the analysis that we will avoid by restricting ourselves to heteronuclear molecules. (However, see Problem 4.20.) That a molecule is heteronuclear does not require that its atoms be of different chemical species. It is enough that they be different isotopes of the same element. It is even sufficient, if they are the same isotope, that the value of the z component of their nuclear spin be different. That is, if the two nuclei differ in any characteristic at all, then the analysis to be given is applicable.

The configuration of the molecule is described by five coordinates (x, y, z, θ, ϕ) (Fig. 4.5), where the first three give the location of the center of mass and the last two are spherical angles defining the direction of the line from B to A. For each of the five coordinates there is a corresponding canonical momentum. Thus the phase space of a single molecule is ten-dimensional, and the phase space of the N-particle system has $2K$ dimensions with $K = 5N$.

Our first task is to determine the Hamiltonian function for the rotational motion. It is easy to see that the rotational kinetic energy T_R is given by

$$T_R = \tfrac{1}{2}(m_A a^2 + m_B b^2)(\dot\theta^2 + \sin^2\theta\,\dot\phi^2) \equiv \tfrac{1}{2}I(\dot\theta^2 + \sin^2\theta\,\dot\phi^2) \tag{4.57}$$

where I is the moment of inertia of the molecule. The canonical angular momenta are

$$p_\theta = \frac{\partial T_R}{\partial\dot\theta} = I\dot\theta \qquad \text{and} \qquad p_\phi = \frac{\partial T_R}{\partial\dot\phi} = I\sin^2\theta\,\dot\phi \tag{4.58}$$

Writing the rotational kinetic energy in terms of canonical coordinates and momenta gives

$$T_R = \frac{1}{2I}\left(p_\theta^2 + \frac{p_\phi^2}{\sin^2\theta}\right) \tag{4.59}$$

The Hamiltonian function for a single molecule is therefore

$$H = \frac{1}{2m}(p_x^2 + p_y^2 + p_z^2) + \frac{1}{2I}\left(p_\theta^2 + \frac{p_\phi^2}{\sin^2\theta}\right) \tag{4.60}$$

In general, if a single molecule has a $2k$-dimensional phase space with a Hamiltonian $H(x,p)$, then the partition function for the N-particle system is, by Eq. (4.55),

$$Z = \frac{1}{N!}\left(\frac{1}{h^k}\int e^{-\beta H(x,p)}d^k x\, d^k p\right)^N \equiv \frac{z^N}{N!} \tag{4.61}$$

where z is the partition function for a single molecule. The canonical potential is given by

$$\phi = N\log z - \log N! \tag{4.62}$$

For a diatomic molecule, the integration over the center of mass coordinates and momenta gives

$$\frac{1}{h^3}\int_V d^3 x \int e^{-\beta p^2/2m}\, d^3 p = \frac{V}{\lambda^3} \tag{4.63}$$

where $\lambda = \sqrt{\beta h^2/2\pi m}$ is the thermal de Broglie wavelength of the molecule. The integration over the internal coordinates and momenta gives what we will call the rotational partition function.

$$\begin{aligned}
z(\text{rot}) &= \frac{1}{h^2}\int_o^{2\pi} d\phi \int_o^\pi d\theta \int_{-\infty}^\infty e^{-\beta p_\theta^2/2I}\, dp_\theta \int_{-\infty}^\infty e^{-\beta p_\phi^2/2I\sin^2\theta}\, dp_\phi \\
&= \frac{2I}{\beta\hbar^2}
\end{aligned} \tag{4.64}$$

When this is all put together, and Stirling's approximation is used for $N!$, one gets

$$\begin{aligned}
\phi(N,\beta,V) &= N\left(\log(V/N) + \frac{3}{2}\log(2\pi m/\beta h^2) + \log(2I/\beta\hbar^2) + 1\right) \\
&= N\left(\log(V/N) - \tfrac{5}{2}\log\beta + \log\left(2I(m/2\pi)^{3/2}k^{5/2}/\hbar^5\right) + 1\right)
\end{aligned} \tag{4.65}$$

The pressure and energy equations of state are given by the thermodynamic relations

$$\frac{p}{kT} = \frac{\partial\phi}{\partial V} = \frac{N}{V} \tag{4.66}$$

or

$$pV = NkT \tag{4.67}$$

and

$$E = -\frac{\partial\phi}{\partial\beta} = \frac{5}{2}kT \tag{4.68}$$

The internal degrees of freedom have no effect on the pressure equation for an ideal gas. This is natural, since the pressure is due to the momentum transferred to the wall by the rebounding particles and only center of mass momentum is involved in that transfer. The two rotational degrees of freedom each add a term $kT/2$ to the thermal energy per particle. This is an example of a general theorem, called the equipartition theorem, that will be given later in this chapter.

4.12   DIPOLAR MOLECULES IN AN ELECTRIC FIELD

If atom A is not identical to atom B, then the molecule will almost surely have a net electric dipole moment because electrons will be drawn somewhat from one atom to the other. We will assume that the molecule has a dipole moment μ pointing in the direction of atom A and that the gas is placed in a uniform electric field, $\mathbf{E} = \mathcal{E}\hat{\mathbf{z}}$. There will then be a potential energy that will depend on the orientation of the molecule.

$$V = -\mu\mathcal{E}\cos\theta \qquad (4.69)$$

With this modification, the Hamiltonian function for a single molecule is

$$H = \frac{1}{2m}(p_x^2 + p_y^2 + p_z^2) + \frac{1}{2I}\left(p_\theta^2 + \frac{p_\phi^2}{\sin^2\theta}\right) - \mu\mathcal{E}\cos\theta \qquad (4.70)$$

A recalculation of the rotational partition function (see Problem 4.23) gives

$$\phi(N,\beta,V) = N\left(\log(V/N) - \frac{5}{2}\log\beta + \log\left(\frac{\sinh(\beta\mu\mathcal{E})}{\beta\mu\mathcal{E}}\right) + \text{const.}\right) \qquad (4.71)$$

The pressure equation of state is unaffected by the change in the internal Hamiltonian. However, the energy equation of state now becomes

$$E = -\frac{\partial\phi}{\partial\beta} = \tfrac{5}{2}NkT + N\left[kT - \mu\mathcal{E}\coth(\mu\mathcal{E}/kT)\right] \qquad (4.72)$$

The function $E(T) - \frac{5}{2}NkT$, which is the dipolar contribution to the thermal energy, is plotted in Fig. 4.6.

For low temperature or large electric field strength (that is, for $kT \ll \mu\mathcal{E}$), the axes of the molecules only undergo small fluctuations about the z direction. In that case the two internal degrees of freedom act like a two-dimensional harmonic oscillator with an energy of $-\mu\mathcal{E}$ at the zero point and contribute an amount $-\mu\mathcal{E} + 2kT$ to the energy per particle (see Problem 4.24).

Using the first two terms of the expansion, $\coth x = 1/x + x/3 - x^3/45 + \cdots$, valid for small x, we find that the potential energy contribution to the thermal energy goes to zero as $1/T$ for large T. That is, for $kT \gg \mu\mathcal{E}$,

$$kT - \mu\mathcal{E}\coth(\mu\mathcal{E}/kT) \approx -\frac{\mu^2\mathcal{E}^2}{3kT} \qquad (4.73)$$

4.13   THE DIELECTRIC CONSTANT OF A GAS

The dielectric constant of a substance is defined by $\kappa = 1 + \chi_e$, where χ_e is the electric susceptibility and $\epsilon_o\chi_e$ is the ratio of the polarization of the medium to the electric field strength. The polarization of a molecular gas is simply the density of molecules times their average dipole moment. There are two contributions to the average dipole moment. One contribution, called the orientational dipole moment, is due to the lining up of the preexisting electric dipoles in the electric field. This

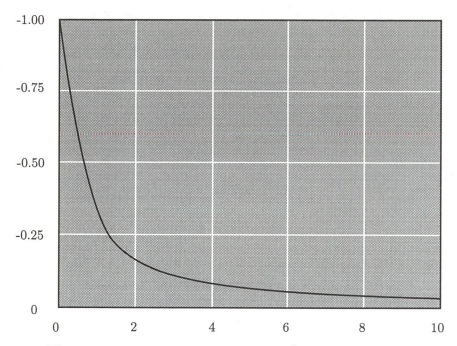

Fig. 4.6   A graph of the function $(E/N - \frac{5}{2}kT)/\mu\mathcal{E}$ in terms of the variable $x = kT/\mu\mathcal{E}$ for a dipolar gas in an electric field. This gives the part of the energy due to the interaction of the dipoles with the external field (divided by $\mu\mathcal{E}$). Note that the values on the vertical axis are negative.

is what will be calculated here. The other contribution, called the induced dipole moment, is due to a change in the internal state of the molecule in response to the electric field (the electrons and nuclei shift their positions relative to one another). Calculating the induced dipole moment is a problem in pure quantum mechanics and therefore it will be ignored in this book, although, even in polar molecules, it may be larger than the orientational dipole moment.

Clearly, the average dipole moment will be in the z direction, since that is the direction of the electric field. The average dipole moment is therefore

$$\bar{\mu}_z = \frac{1}{N}\langle\sum_i^N \mu\cos\theta_i\rangle \tag{4.74}$$

$\bar{\mu}_z$ will be evaluated by first deriving a thermodynamic relation that is valid in a much wider context. In particular, the derivation will not assume that the system is composed of noninteracting particles. We will only assume that the Hamiltonian function, in the presence of an electric field, is of the form

$$H = H_o - \sum_{i=1}^N \mu\mathcal{E}\cos\theta_i \tag{4.75}$$

where H_o is the system Hamiltonian when $\mathcal{E} = 0$. If the system is described by K momentum variables and K coordinates, some of which are the angles $\theta_1, \ldots, \theta_N$, then the canonical potential is

$$\phi = \log\left[\int \exp\left(-\beta H_o + \beta\mathcal{E}\sum\mu\cos\theta_i\right)d^Kx\, d^Kp\right] \tag{4.76}$$

The derivative of ϕ, with respect to the electric field strength $\mathcal{E}$ is

$$\frac{\partial \phi}{\partial \mathcal{E}} = \beta \frac{\int (\sum \mu \cos \theta_j) \exp[-\beta H_o + \beta \mathcal{E} \sum \mu \cos \theta_i] \, d^K x \, d^K p}{\int \exp[-\beta H_o + \beta \mathcal{E} \sum \mu \cos \theta_i] \, d^K x \, d^K p} \tag{4.77}$$

Comparing this with Eq. (4.74), we see that

$$N \bar{\mu}_z = kT \frac{\partial \phi}{\partial \mathcal{E}} \tag{4.78}$$

Equation (4.71), for ϕ, then gives the average dipole moment as

$$\bar{\mu}_z = \mu \coth(\beta \mu \mathcal{E}) - \frac{1}{\beta \mathcal{E}} \tag{4.79}$$

To calculate the dielectric constant we need only the term that is linear in $\mathcal{E}$. We may therefore use the expansion of $\coth x$, valid for small x, to get

$$\bar{\mu}_z \approx \frac{\mu^2}{3kT} \mathcal{E} \tag{4.80}$$

The electric susceptibility is obtained by multiplying by the particle density n and dividing by the field strength.

$$\epsilon_o \chi_e = \frac{\mu^2 n}{3kT} \tag{4.81}$$

As was stated before, this calculation ignores the contribution of the induced dipole moment to the polarization of the gas. Therefore, a single measurement of the dielectric constant of the gas could not be used, in combination with Eq. (4.81), to determine the permanent dipole moment of the molecule (Fig. 4.7). However, the induced dipole moment is independent of temperature, while the orientational dipole moment has the characteristic $1/T$ behavior shown in Eq. (4.81). Thus, a measurement of χ_e at a number of different temperatures does allow a separation of the two effects and, therefore, a determination of the molecular dipole moment μ. Before the introduction of microwave spectroscopy, this was the standard technique for determining molecular dipole moments. (See Problems 4.21 and 4.22.)

4.14   THE VIBRATIONAL ENERGY OF A GAS

The eigenvalues of the square of the angular momentum of a quantized rotating diatomic molecule are $\hbar^2 \ell(\ell + 1)$, where $\ell = 0, 1, \ldots$. The corresponding energy eigenvalues are $E_\ell = \hbar^2 \ell(\ell + 1)/2I$. Therefore, quantum mechanical effects will become noticeable when kT is of the order of $\hbar^2/2I$. For diatomic molecules like O_2, Cl_2, or KCl the quantity $\hbar^2/2kI$ is about one Kelvin; a temperature at which the substance does not exist as a gas. It is only for the isotopes of hydrogen that quantum mechanical effects are significant in the rotational thermal energy.

In contrast, the "stretching mode" of most diatomics demands a quantum mechanical treatment. For this vibrational mode, the significant temperature would be $T = \hbar \omega / k$, where ω is the vibrational angular frequency. For typical diatomic molecules, such as the three mentioned, $\hbar \omega / k$ is a few thousand Kelvin, far above room temperature. In fact, at temperatures near room temperature, the stretching

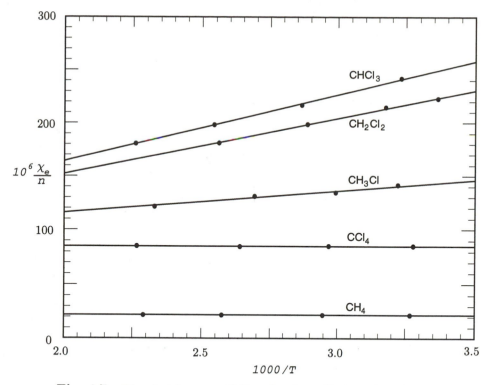

Fig. 4.7   The electric susceptibility of various chlorinated methanes. n is in moles/m^3 and T in K.

vibrational motion is completely unexcited. For those temperatures, the approximation of taking the interatomic distance as fixed is an accurate one.

The potential associated with the interaction between the atoms in a diatomic molecule is of the general form shown in Fig. 4.8. Near the equilibrium distance the potential function may be approximated as a quadratic, leading to the energy spectrum of a one-dimensional harmonic oscillator. The contribution of the vibrational modes of all N diatomic molecules is just the thermal energy of N identical harmonic oscillators, given in Eq. (4.19).

$$E_{\text{vib}}(T) = \frac{N\hbar\omega}{e^{\hbar\omega/kT} - 1} \qquad (4.82)$$

This should be added to the translational and rotational energies given in Eq.

4.15   THE SCHOTTKY SPECIFIC HEAT

As another example of the use of the canonical ensemble, we will consider a solid composed of N atoms (or molecules) and assume that each atom may be in one of two possible quantum states with energies 0 and ε. We want to calculate the specific heat per particle, $C = \partial(E/N)/\partial T$, as a function of temperature.

The atoms in a solid, being localized in space, may be treated as distinguishable particles.* If, in addition to being distinguishable, the particles do not interact, then

---

* Solid helium, in which there is a significant amplitude for particle exchange, is the only exception to this rule.

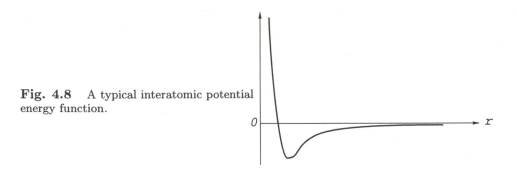

Fig. 4.8   A typical interatomic potential energy function.

the energy eigenvalues of the system are of the form

$$E = E_{\alpha_1} + \cdots + E_{\alpha_N} \tag{4.83}$$

where the index α_i numbers the quantum states of the ith particle. The partition function then becomes the Nth power of a partition function for a single atom. 4.68.

$$Z = \sum_{\{\alpha\}} \exp[-\beta(E_{\alpha_1} + \cdots + E_{\alpha_N})] = \left(\sum_\alpha e^{-\beta E_\alpha} \right)^N \equiv (Z_{\text{at}})^N \tag{4.84}$$

The canonical potential of the system is then N times the canonical potential of a single atom.

$$\phi = N \log Z_{\text{at}} \tag{4.85}$$

The partition function of a two-level atom has only two terms.

$$Z_{\text{at}} = 1 + e^{-\beta \varepsilon} \tag{4.86}$$

Thus

$$\phi(\beta) = N \log(1 + e^{-\beta \varepsilon}) \tag{4.87}$$

The thermal energy as a function of T is

$$E(T) = -\frac{\partial \phi}{\partial \beta} = N\varepsilon \frac{e^{-\beta \varepsilon}}{1 + e^{-\beta \varepsilon}} = \frac{N\varepsilon}{e^{\varepsilon/kT} + 1} \tag{4.88}$$

The specific heat per particle is

$$C = \frac{\varepsilon^2 e^{\varepsilon/kT}}{kT^2 (e^{\varepsilon/kT} + 1)^2} \tag{4.89}$$

A graph of $C(T)$ shows that the specific heat has a single large maximum at a temperature $T \approx 0.42\,\varepsilon/k$. This characteristic pattern is known as a Schottky anomaly (see Fig. 4.9).

4.16   SEPARABLE SYSTEMS
The system considered, composed of N noninteracting identical atoms, is a special case of a system whose energy spectrum is of the form of a sum of terms, each of

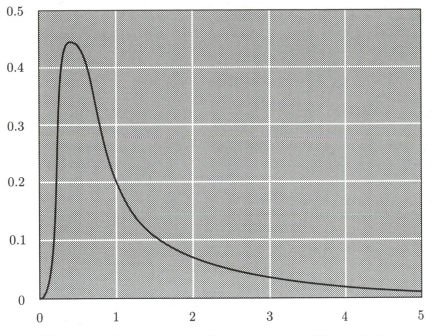

Fig. 4.9   The Schottky specific heat anomaly. C/k is plotted as a function of $x = kT/\varepsilon$.

which depends on an independent set of quantum numbers. For example, consider a system with energy levels

$$E = E_a + E_b + E_c \tag{4.90}$$

where a, b, and c are independent sets of quantum numbers. That the quantum numbers are "independent" means, for instance, that the allowed values of c do not depend on the values of a or b. This condition of independence is not satisfied by the quantum numbers (n, l, m) that index the hydrogenic wave functions. A system whose energy eigenvalues satisfy Eq. (4.90), with independent quantum numbers, has a partition function of the form

$$
\begin{aligned}
Z &= \sum_{a,b,c} e^{-\beta(E_a + E_b + E_c)} \\
&= \left(\sum_a e^{-\beta E_a} \right) \left(\sum_b e^{-\beta E_b} \right) \left(\sum_c e^{-\beta E_c} \right) \\
&= Z_A Z_B Z_C
\end{aligned}
\tag{4.91}
$$

and a canonical potential

$$\phi = \phi_A + \phi_B + \phi_C \tag{4.92}$$

A separable system may be composed of two or more physically separate systems that are in weak thermal contact, or it may simply be a single system with different weakly interacting degrees of freedom, such as the collection of translational, rotational, and vibrational degrees of freedom for a diatomic molecule.

4.17   THE EQUIPARTITION THEOREM

The equipartition theorem is a general formula for calculating the contribution to the thermal energy of terms that appear quadratically in the Hamiltonian function of a classical system. Let q be one of the coordinates in phase space (either an x_i or a p_i) and let Q be the set of $(2K-1)$ other coordinates. Suppose that the Hamiltonian function is of the form

$$H = A(Q)q^2 + H'(Q) \tag{4.93}$$

and that q has the range $-\infty < q < \infty$. Certainly, this is the form of the translational kinetic energy terms, where q is any component of the momentum of any particle and $A = 1/2m$. But it is also the form of the rotational kinetic energy terms for a diatomic molecule, where q is either p_θ or p_ϕ. In the canonical ensemble

$$E = \langle Aq^2 \rangle + \langle H' \rangle \tag{4.94}$$

and

$$\langle Aq^2 \rangle = \frac{\int Aq^2 e^{-\beta(Aq^2 + H')} \, dq \, dQ}{\int e^{-\beta(Aq^2 + H')} \, dq \, dQ} \tag{4.95}$$

Taking the q integral first, we get

$$\int Aq^2 e^{-\beta(Aq^2 + H')} \, dq \, dQ = \frac{\sqrt{\pi}}{2\beta^{3/2}} \int A^{-1/2} e^{-\beta H'} \, dQ \tag{4.96}$$

and

$$\int e^{-\beta(Aq^2 + H')} \, dq \, dQ = \frac{\sqrt{\pi}}{2\beta^{1/2}} \int A^{-1/2} e^{-\beta H'} \, dQ \tag{4.97}$$

which shows that

$$\langle Aq^2 \rangle = \frac{1}{2\beta} = \tfrac{1}{2}kT \tag{4.98}$$

Thus, each purely quadratic term in a classical Hamiltonian function contributes an amount $kT/2$ to the thermal energy.

This makes the task of calculating the thermal energy of a classical system of K one-dimensional harmonic oscillators completely trivial. Both the momentum variables and the coordinate variables appear quadratically in the Hamiltonian with the result that

$$\frac{E(T)}{K} = kT \tag{4.99}$$

Comparing this with the equivalent equation for quantum mechanical harmonic oscillators [Eq. (4.19)], we see that the high-temperature limit of the quantum mechanical result is equivalent to the classical limit.

PROBLEMS

4.1   A more general form of the uniform quantum ensemble could be constructed as follows. Let $|E_1\rangle, \ldots, |E_K\rangle$ be the set of all energy eigenstates with eigenvalues less than some given energy E. Instead of assuming that each member of the ensemble is in a definite energy eigenstate, we simply assume that each member of the ensemble has a wave function that is a linear combination of the first K eigenstates.

$$|\psi\rangle = \sum_{k}^{K} c_k |E_k\rangle = \sum_{k}^{K} (a_k + ib_k)|E_k\rangle$$

where c_k is a complex number and a_k and b_k are real. This is equivalent to assuming that there is zero probability of finding the system with energy larger than E. Assume that the probability distribution for the expansion coefficients is constant, except for the normalization condition $\langle\psi|\psi\rangle = \sum(a_k^2 + b_k^2) = 1$. Show that the ensemble average $\langle A\rangle$ of any operator, taken with this "more general" ensemble agrees exactly with $\langle A\rangle$ taken with the ordinary uniform ensemble. (Warning: this is a very hard problem.)

4.2 $\Omega(E)$ is the number of eigenstates with energy less than E. Therefore, $d\Omega/dE$ is the number of eigenstates per unit energy interval (the *eigenstate density*) and $(d\Omega/dE)^{-1}$ is the average spacing between energy eigenvalues (counting degenerate eigenvalues as having zero spacing). Evaluate the average spacing, using the classical approximation [Eq. (4.4)], for one mole of neon at STP, neglecting interparticle interactions.

4.3   Show that, for one mole of helium at $0°$ C and atmospheric pressure, using the classical approximation, $S^o = 15 N_A$. For the same system, evaluate $\log \tau$ and compare the two terms in Eq. (4.10).

4.4   The normal modes of a violin string with a fundamental angular frequency ω_o may be considered as a set of noninteracting harmonic oscillators. (a) Express the average vibrational energy of the system at temperature T, measured from the ground state, as an infinite series. Remember that there are two normal modes for each allowed wavelength. (b) For $kT \gg \hbar\omega_o$, the infinite sum may be approximated by an integral that can be found in the Table of Integrals. Calculate the thermal energy of the system in that limit. (c) For $kT \ll \hbar\omega_o$, the sum may be evaluated by using   the   fact   that $\exp(n\hbar\omega_o/kT) \gg 1$ for any positive integer n. Calculate the thermal energy of the system in that limit.

4.5   A quantum system is composed of K distinguishable subsystems. The kth subsystem has an energy spectrum $E_{n_k} = \varepsilon n_k^2$, where $n_k = 0, 1, 2, \ldots$. Calculate $\Omega(E)$ for the system, when $E \gg \varepsilon$.

4.6   Derive Eq. (4.17) from Eq. (4.16).

4.7   Shown in Fig. 4.10 is a simple model of a one-dimensional crystal. The springs all have spring constant k and, at equilibrium, the particles are all separated by length ℓ. Let x_n be the deviation from equilibrium of the nth particle. (a) Show that the equation of motion of the nth particle is $m\ddot{x}_n = k(x_{n+1} + x_{n-1} - 2x_n)$, where $n = 1, \ldots, N$ and x_o and x_{N+1} are defined to be zero. (b) Show that there

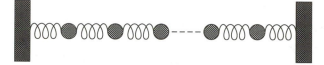

Fig. 4.10

are N solutions of the equations of motion of the form

$$x_n(t) = A\cos(\omega_K t)\sin(\pi\frac{K}{N+1}n)$$

where $K = 1,\ldots,N$ and, putting those solutions into the equations of motion, determine the normal mode angular frequencies ω_K. (c) Treating the system quantum mechanically, write the thermal energy per particle at temperature T as a sum over the variable K. Rewrite it as an integral, using the integration variable $k = \pi K/(N+1)$ which, for large N, has the range $0 < k < \pi$. The *thermal energy* is the energy at temperature T relative to the ground state energy. (d) Expand the integral obtained in (c) as a power series in $\hbar$, retaining only the first two nonvanishing terms. Verify that the first term gives the classical, Dulong–Petit value, which for a one-dimensional system is $E(T) = NkT$. (e) Take $\omega_o = \sqrt{k/m} = 10^{13}$ rad/sec (a reasonable value for real crystals) and $T = 300\,\mathrm{K}$. Determine the ratio of the second (quantum) term in the expansion to the first (classical) term. Is it negligible?

4.8   Equation (4.33) gives the specific heat for a crystal at low temperature, according to the Debye theory. Use this, along with the facts that $S^o = 0$ at $T = 0$ and $dS^o/dE = 1/kT$, to calculate $S^o(T)$ when $T \ll T_D$.

4.9   Consider an elastic material in the form of a cube of side L. If $\mathbf{u}(\mathbf{r},t)$ is the displacement from equilibrium of the material at location $\mathbf{r}$ at time t, then, for sound waves, $\mathbf{u}(\mathbf{r},t)$ satisfies the wave equation $\nabla^2\mathbf{u} - v^{-2}\partial^2\mathbf{u}/\partial t^2 = 0$. If we impose the condition that the material at the surface does not move, then $\mathbf{u}(x,y,z,t)$ satisfies *zero boundary conditions*,

$$\mathbf{u}(0,y,z,t) = \mathbf{u}(x,0,z,t) = \mathbf{u}(x,y,0,t) = 0$$

and

$$\mathbf{u}(L,y,z,t) = \mathbf{u}(x,L,z,t) = \mathbf{u}(x,y,L,t) = 0$$

The solutions of the wave equation, for vibrational waves, with zero boundary conditions at the surface, are $\mathbf{A}\sin(k_x x)\sin(k_y y)\sin(k_z z)$, where $\mathbf{A}$ is any vector and $k_i = \pi K_i/L$ $(i = x,y,z$ with $K_i = 1,2,3,\ldots$. Thus the possible values of the wave vector form a cubic lattice in k space with lattice spacing π/L. However, the possible wave vectors are limited to the positive octant (the region in k space with all components positive). (a) Show that the number of solutions with $|\mathbf{k}| < K$ is $K^3 L^3/6\pi^2$. (b) If $\mathbf{A}$ is parallel to $\mathbf{k}$ then the normal mode is said to be longitudinal. If the two vectors are perpindicular then it is said to be transverse. There is one longitudinal and two transverse modes for each wave vector. From the result of (a), using the dispersion relations $\omega = v_L k$ and $v_T k$ for longitudinal and transverse sound waves, derive Eq. (4.27) for the frequency distribution function.

4.10   In the Debye approximation, one demands that the total number of normal modes be equal to the number of coordinates, namely $3N$. A slightly better approximation would be to treat the longitudinal and transverse sound waves separately by demanding that there be N longitudinal vibrations (of velocity v_L) and $2N$ transverse vibrations (of velocity v_T). Using such an approximation, redo the Debye theory to the point of deriving a modified form of Eq. (4.32).

4.11 Show that the probability distribution in the energy, for a classical system, in contact with a reservoir at temperature T, is given by $P(E) = Z(\beta)^{-1} \exp[S^o(E) - \beta E]$.

4.12 A classical system has a Hamiltonian function of the form $H(x,p) = H_o(x,p) + \alpha h(x,p)$. The value of the canonical partition function is then a function of the parameter α. Show that $\langle h \rangle$, the canonical average of the observable $h(x,p)$, is given by

$$\langle h \rangle = -kT \frac{\partial \phi}{\partial \alpha}$$

4.13 Show that the formula in Problem 4.12 also holds for a quantum mechanical system, where H_o and h are Hermitian operators. (Warning: This problem requires a knowledge of first-order quantum mechanical perturbation theory.)

4.14 Calculate the partition function $Z(\beta)$ for a system of K noninteracting, quantum harmonic oscillators of angular frequency ω, and use your result to calculate the specific heat of such a system.

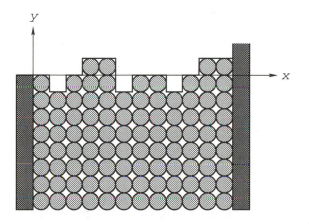

Fig. 4.11

4.15 In Fig. 4.11 is shown the "surface" or upper boundary of a two-dimensional square crystal. In this *solid-on-solid* model, we assume that there are no overhangs (filled lattice sites that lie above empty ones). The configuration of the surface can then be defined by N integer height variables $y_1, y_2, \ldots, y_N$. (In the figure, $y_1 = 0$ and $y_2 = -1$.) The surface energy is assumed to be proportional to the length of the surface. If we assign zero energy to the straight surface, then $E = \varepsilon \sum_{n=1}^{N} |y_n - y_{n-1}|$, where $y_o = 0$. The canonical partition function for the system is

$$Z(N, \beta) = \sum_{y_1} \cdots \sum_{y_N} \exp\left(-\beta \varepsilon \sum |y_n - y_{n-1}|\right)$$

(a) By a transformation of variables, $u_n = y_n - y_{n-1}$, evaluate $Z(N, \beta)$. (b) Determine the probability distribution for the "step height" u_n. It should be independent of n. (c) Using the central limit theorem, determine the probability distribution for the coordinate of the last step, y_N.

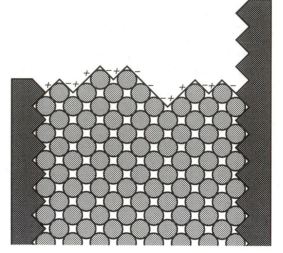

Fig. 4.12

4.16 Another model of the surface of a two-dimensional crystal is shown in Fig. 4.12. A configuration of the surface can be described by a sequence of N variables $\sigma_1, \ldots, \sigma_N$ that take the values $\sigma = \pm 1$ according to the scheme shown in the figure. The y coordinate of the right-hand end (in appropriate units) is $y = \sum \sigma_n$. If there is an interaction between the crystal particles and the right-hand wall, then it is reasonable to assume that the energy is given by $E = \varepsilon y$, where $\varepsilon < 0$ for an attractive interaction and $\varepsilon > 0$ for a repulsive one. (a) Calculate the canonical potential, ϕ, as a function of N, β, and ε. (b) Show that $\langle y \rangle = \partial\phi/\partial\varepsilon$ and calculate $\langle y \rangle$ as a function of N, β, and ε.

4.17 The Riemann zeta function is defined as the sum of the infinite series

$$\zeta(K) = \sum_{n=1}^{\infty} n^{-K}$$

Using the identities $e^x - 1 = e^x(1 - e^{-x})$ and $(1-x)^{-1} = \sum x^n$ derive the formula, given in the Table of Integrals, for $\int x^n \, dx/(e^x - 1)$.

4.18 Consider an isolated classical system composed of a large reservoir and a small but macroscopic sample. The Hamiltonian for the total system is $H_T(x, p, X, P) = H(x, p) + H_R(X, P)$. Using a microcanonical ensemble to describe the system, derive Eq. (4.52) for the probability distribution of the sample.

4.19 The rotational partition function of a diatomic molecule calculated by classical mechanics is $z(\text{rot}) = 2I/\beta\hbar^2$. Calculated quantum mechanically, $z(\text{rot})$ is given by the following sum over rotational states

$$z(\text{rot}) = \sum_{\ell} g_\ell \exp(-\beta\epsilon_\ell)$$

where $g_\ell = (2\ell + 1)$ is the degeneracy of the rotational state of angular momentum $\sqrt{\ell(\ell+1)}\hbar$ and $\epsilon_\ell = \ell(\ell+1)\hbar^2/2I$. (a) Show that, if the quantum mechanical sum is approximated by an integral over a continuous ℓ variable, the classical value is obtained. This procedure would be valid for small values of $\beta\hbar^2/2I$ ($kT \gg \hbar^2/2I$). (b) Evaluate the sum numerically for $\beta\hbar^2/2I = 1$, 0.5, and 0.1 and compare your result with the classical approximation.

4.20 For a homonuclear diatomic molecule, the rotational wave function must be even or odd under interchange of the nuclei; that is, $\psi(\mathbf{R}_A, \mathbf{R}_B) = \pm\psi(\mathbf{R}_B, \mathbf{R}_A)$. If nucleus A (which is identical with nucleus B) is composed of an even number of fermions (protons and neutrons), then the nucleus is a Bose–Einstein particle and the + sign holds; in the other case, the nucleus is a Fermi–Dirac particle and the − sign holds. For the case of a BE homonuclear diatomic molecule, because of the symmetry of the wave function, the angular momentum quantum number ℓ must be even. Therefore, the sum in Problem 4.19 is over $\ell = 0, 2, 4, \ldots$. (Actually, for homonuclear molecules, even the nuclear spin gets into the game. Exchanging the coordinates of the nuclei means also exchanging the nuclear spins. This adds another complication. The reader can now see why we avoided this case. For this problem, assume that the nucleus has a spin of zero, in which case the nuclei can be treated like point particles.) Do Problem 4.19(a) for the case of a BE homonuclear diatomic molecule and show that, when $kT \gg \hbar^2/2I$, $z(\text{rot}) \approx I/\beta\hbar^2$, which is exactly half the result for heteronuclear molecules.

4.21 Using the experimental data shown in Fig. 4.7, determine the permanent dipole moments of methane (CH_4) and methyl chloride (CH_3Cl). Note that the density in the figure is given in moles/m³.

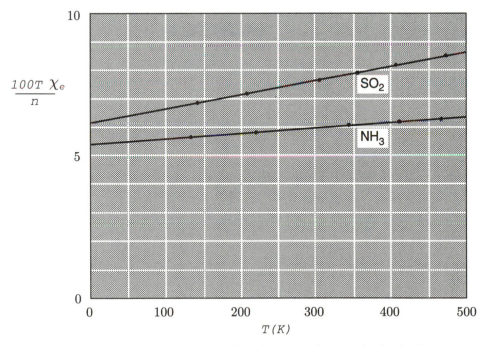

Fig. 4.13 Values of $100T\chi_e/n$, where n is the particle density in moles/m³ for NH_3 and SO_2.

4.22 Using the experimental data shown in Fig. 4.13, determine the permanent dipole moments of ammonia (NH_3) and sulfur dioxide (SO_2).

4.23 Using the Hamiltonian given in Eq. (4.70), calculate $z(\text{rot})$ for a dipolar molecule in an electric field.

4.24 Expand the energy for a dipolar gas in an electric field, given in Eq. (4.72), to first order in T to obtain a low-temperature approximation for the specific heat per particle.

4.25 A long polymer under tension τ is made up of N monomers that can each be in a compact or an elongated state. Thus the polymer has 2^N possible states. Assume that the length of the polymer is $L = N_c \ell_c + N_e \ell_e$, where $\ell_e > \ell_c$ and N_c and N_e are the numbers of compact and elongated monomers, respectively. Take the energy as $E = \tau(N\ell_e - L)$ and calculate the average length as a function of T and τ.

4.26 A weight W hangs on a chain of N links, each of length ℓ and negligible weight (see Fig. 4.14). Each link can rotate freely from $-\pi$ to π, but only in the plane of the paper. The energy of the system is given in terms of its configuration by $E = -Wg\ell \sum^N \cos\theta_n$. Ignore any kinetic energy and calculate the canonical potential of the system by integrating $e^{-\beta E}$ over all the possible configurations. You will need the integral, $\int_{-\pi}^{\pi} e^{z\cos\theta}\, d\theta = 2\pi I_o(z)$, where $I_o(z)$ is a modified Bessel function. Using the asymptotic form of the Bessel function $I_o(z) \sim e^z/\sqrt{2\pi z}$ valid for $z \gg 1$, determine the average energy of the system when $kT \ll gW\ell$ and show that it agrees with the prediction of the equipartition theorem.

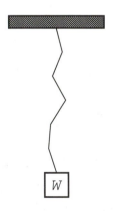

Fig. 4.14

Chapter 5
The Foundations of Thermodynamics

5.1 THERMODYNAMICS VERSUS STATISTICAL MECHANICS

The aim of statistical mechanics is to predict the equilibrium properties of a macroscopic system, starting from an adequate model of the microscopic Hamiltonian. However, long before any reliable information about the microscopic structure of matter existed, there was a well-developed science for predicting the equilibrium properties of macroscopic systems; namely, thermodynamics.

Historically, thermodynamics arose out of an attempt to improve the efficiency of heat engines. The mechanical engineering origin of the subject has left a lasting imprint upon it. For example, the most important law of thermodynamics is usually given in the form of a postulate about the impossibility of constructing certain kinds of engines. It is the author's opinion that it would be illogical at this point to make a digression into the study of heat engines, Carnot cycles, etc. We already have a clear idea of the basic mechanism that drives a macroscopic system toward a time-independent equilibrium state. It has little to do with the details of the microscopic dynamics. It is essentially a statistical effect. Almost all of the microstates available to the system have exactly the same macroscopic features, namely, those that define the equilibrium macrostate. Thus, virtually any trajectory in the microstate space of the system will cause the system's macrostate to move to the equilibrium state. This idea, stated in a way that makes no detailed reference to the microscopic structure of the system, should be the central principle of our thermodynamics. Macroscopic systems do not behave as they do in order to frustrate our attempts to build an engine that takes heat from a colder to a hotter body. Therefore, our basic thermodynamic laws should not be formulated in that way. Instead, our basic thermodynamic laws should reflect, simply and accurately, the true mechanism that is driving the system's macroscopic behavior. For example, when one drops a sugar cube into coffee, it dissolves. It is reasonable to ask *why* it dissolves. We would answer that the sugar molecules do not try, preferentially, to dissolve. There are simply so many more *dissolved* states than *undissolved* ones that, in the course of its fairly random motion, the sugar moves from the small set of undissolved states to the vastly larger set of dissolved ones. The traditional theory would answer the same question by showing that, if the sugar did not dissolve, then it would be possible to transfer heat from a colder to a hotter body. As an answer to the question: "Why does the sugar cube dissolve?", that explanation is practically worthless.

With this starting point, our formulation of the laws of thermodynamics will be rather unconventional.* Our objective in this chapter is to construct a thermodynamic theory specifically designed to complement statistical mechanics. Therefore, certain decisions, such as choices of variables and units, will be made with this in mind. In the next chapter, the theory will be recast in a form that is both more conventional and more convenient for applications to engineering processes. Before describing the principles of the subject we should discuss in more detail the relationship of thermodynamics to statistical mechanics and the distinction between the two subjects. A very important point is that thermodynamics is a self-contained subject. Its laws should be expressed entirely in terms of the observable macroscopic properties of substances, with no reference to their microscopic structure. The basic question considered by thermodynamics is the following: Given the initial macroscopic state of an isolated system, not necessarily at equilibrium, what will be the system's macroscopic state when it comes to equilibrium? Of course, this does not mean that the laws of thermodynamics cannot be used in answering a large variety of other questions, but only that this is the central question in the subject.

Statistical mechanics is related to thermodynamics in two fundamental ways. First, statistical mechanical arguments can be used to derive the basic axioms of thermodynamics from the laws of quantum or classical mechanics. Second, particular functions needed in thermodynamic calculations, such as the entropy function, can be computed from a model of the microscopic structure of a substance, using statistical mechanical rules. However, since thermodynamics is an independent science, there must be rules within thermodynamics that allow a determination of all the required functions without any recourse to statistical mechanics or any knowledge of the microstructure of the substances studied. Generally, the rules of thermodynamics show how these functions can be determined by experimental measurements, while the rules of statistical mechanics show how they can be computed by mathematical analysis. This may make it appear that statistical mechanics is inherently more powerful than thermodynamics. Why should we bother measuring functions that can be obtained by thought alone? However, when the system under consideration is something like a bucket of tar (which has a very complex microscopic structure) the elegant formulas of statistical mechanics are of no practical value but the simple operational rules of thermodynamics encounter no serious obstacles. In one sense, statistical mechanics *is* more powerful than thermodynamics in that thermodynamics can be derived from statistical mechanics but not vice versa.

The general tone of this chapter will be more formal and axiomatic than that of the previous chapters. This is necessary in order to construct a logically self-contained subject. We will have to be careful that we do not allow information, based on the microscopic models of particular substances, to slip into our fundamental thermodynamic laws. In choosing a system of fundamental axioms for

---

*Although unconventional, our formulation is not new. A similar treatment of thermodynamics was given in *Thermodynamics* by H. B. Callen (John Wiley & Sons, Publ.). Like much of thermodynamics, it could be traced back to J. W. Gibbs.

thermodynamics, we will use the following criterion: Anything that follows in a straightforward way from the laws of mechanics, quantum mechanics, or electromagnetic theory, such as the conservation of energy, angular momentum, or electric charge, will be considered as a part of those disciplines and will therefore not be included in the axioms of thermodynamics. Only those universal characteristics of macroscopic systems that do not follow in any simple and direct way from the laws of mechanics, quantum mechanics, or electrodynamics will be taken as postulates of thermodynamics. Generally, they will be properties of macroscopic systems that have no counterpart in systems with only a few degrees of freedom. In other words, the laws of thermodynamics are taken as those new characteristics that appear only in the thermodynamic limit. For example, it will be taken as an axiom that isolated systems approach time-independent equilibrium states, composed of a small number of uniform phases (gas, liquid, solid, etc.). In the traditional formulations of thermodynamics such characteristics are considered to be so obvious that they are unworthy of specific mention. In fact, they are the most exceptional characteristics of the macroscopic world. Nonequilibrium states produce grasshoppers, trees, and snowflakes. No such things are possible at equilibrium. Equilibrium states consist only of dull uniform phases. Certainly this fact is no trivial consequence of the underlying laws of nature, which are the same for equilibrium and nonequilibrium states.

5.2   EQUILIBRIUM STATES

Our first axiom of thermodynamics simply postulates the existence of equilibrium states.

Axiom 1. Any isolated macroscopic system will eventually come to a state in which all of its macroscopic properties remain constant. Such states are called *equilibrium* states.

By the word "isolated" we mean that the system does not exchange energy with anything external to it. The word does not preclude the existence of external force fields, such as gravitational or magnetic fields, as long as they are constant in time and are conservative fields. Such fields would then simply contribute to the definition of the conserved energy of the system.

The next few axioms describe general characteristics of the equilibrium states.

Axiom 2. The values of all other macroscopic observables in the equilibrium state are determined by a finite number of conserved variables.

In all the cases we will consider in this chapter the conserved variables are simply the energy of the system E, the volume of the system V, and the number of particles for each type of conserved particle. A *simple substance* is one that is composed of only one type of particle, such as neon, water, or copper. Initially we will restrict ourselves to simple substances, but the extension of our analysis to complex substances will be very simple. It would only complicate the notation to try to treat the most general case at the start.

Axiom 3. For sufficiently weak external fields, the equilibrium state consists of a finite number of homogeneous *components* or *phases*.

If we formulate this axiom only for zero external fields, then we eliminate any possibility of confirming the theory on the surface of the earth. Actually it is preferable to have a weak gravitational field present for a system composed of two phases, such as liquid and gas, because it pulls the denser phase together at the bottom of the vessel. Strong fields create nonuniform states in which the properties vary continuously with position. An atmosphere of continuously varying density is a good example. A minor problem is that our restriction to weak fields will require that we ignore the easily observed variation of pressure with depth in dense fluids. This restriction to weak fields is only a temporary device to simplify the introduction of thermodynamics. The generalization to continuously varying systems requires no new fundamental physics; it is simply more complicated mathematically. Axioms 2 and 3 are best considered as general descriptions of the sort of systems we will work with in this chapter.

5.3 THE SECOND LAW

The next axiom is the heart of thermodynamics. In the usual formulation of the subject, it appears as the Second Law (in a rather different form). It leads to the calculational scheme for determining the equilibrium state of a system. It is best to motivate and interpret it with some analysis from statistical mechanics. However, this introductory material should not be considered as part of the body of thermodynamics. The axioms of thermodynamics must be powerful enough and clear enough to be usable without external assistance.

To be specific, let us consider a macroscopic quantum mechanical system. We know that the macrostate of the system approaches the equilibrium state because that is the most probable macrostate, where the probability of any macrostate is taken as proportional to the number of quantum states exhibiting that combination of macroscopic features. We also know that, if the system is composed of two spatially separated macroscopic parts, then the number of quantum states corresponding to a given macrostate of the whole system is given by a product of the numbers of quantum states of each part corresponding to the macrostate of that part. [That is, for a system with separated parts, any quantum state of part 1 may be combined with any quantum state of part 2 to give a quantum state of the whole system. If K is the total number of such quantum states, then $K(\text{whole}) = K(\text{part 1}) \times K(\text{part 2})$.] Thus, for a system composed of two parts, the logarithm of the number of microstates of the whole is the sum of the logarithms of the numbers of microstates of the two parts. Also, the equilibrium state is the state that maximizes the number of microstates, which means that it is also the state that maximizes the logarithm of that number, which we call the entropy.

Axiom 4. (The Second Law)   There is a function of the macroscopic state of a system, called the *entropy*, with the following characteristics:

a. If a system is composed of a number of spatially separate parts, then the entropy of the whole is the sum of the entropies of the parts.

b. The equilibrium state of the system is that state that maximizes the entropy within the constraints of the system and the conservation laws of nature.

Notice that, except for properties (a) and (b), no rules are given about how to calculate the entropy functions of specific systems. Since such rules would depend on the detailed microscopic structure of the systems involved, they lie outside the realm of thermodynamics, which recognizes only the macroscopic features of systems. Somehow, we will have to use the given axioms to devise a scheme for determining the entropy functions needed.

5.4 EXTENSIVITY OF THE ENTROPY

Since, according to Axiom 3, general equilibrium states are composed of a few homogeneous phases, the obvious place to start our analysis is to consider samples composed of a single homogeneous phase. What does the Second Law tell us about the entropy of, let us say, a uniform sample of liquid with particle number N, energy

"It is an inference naturally suggested by the general increase of entropy which accompanies the changes occurring in any isolated material system that, when the entropy of the system has reached a maximum, the system will be in a state of equilibrium. Although this principle has, by no means, escaped the attention of physicists, its importance does not appear to have been duly appreciated. Little has been done to develop the principle as a foundation for the general theory of thermodynamic equilibrium."

"The general criterion for equilibrium can be stated simply and precisely: for the equilibrium of any isolated system, it is necessary and sufficient that, in all possible variations of the state of the system which do not alter its energy, the variation of its entropy shall either vanish or be negative."

—— J. Willard Gibbs

E, and volume V? Let us call the entropy function that we seek $S(N, E, V)$. (The entropy is a macroscopic observable. By Axiom 2, at equilibrium, all macroscopic observables are functions of N, E, and V. Therefore, S is a function of N, E, and V.) If we conceptually split the sample into K equal parts then, at equilibrium, each part will have a particle number N/K, an energy E/K, and a volume V/K. Thus we can use part (a) of Axiom 4 to state that

$$S(N, E, V) = KS\left(\frac{N}{K}, \frac{E}{K}, \frac{V}{K}\right) \tag{5.1}$$

or

$$S(\lambda N, \lambda E, \lambda V) = \lambda S(N, E, V) \tag{5.2}$$

where $\lambda = 1/K$. It is not difficult to construct a similar argument that will give Eq. (5.2) for any positive rational number L/K, and assuming continuity, we then get Eq. (5.2) for any positive real number. Thus we now know that the entropy function of any single-phase substance is a first-degree homogeneous function of its arguments.* Such a function is called an *extensive* function. The extensive property of the single-phase entropy functions significantly simplifies the task of determining them.

There is a technical point that should be clarified here. The number of particles may be measured either in individual particles or in moles (Avagadro's number of particles). An individual particle is both an impractical and theoretically inappropriate unit for a strictly macroscopic theory. However, in applying our formulas to statistical mechanics, it is an extremely convenient unit. Almost nothing depends on the unit of particle number, and so we will only indicate it when it is a relevant variable. When using an individual particle as our unit, the quantity $S(1, E, V)$ should not be interpreted as the entropy of a one-particle system, which is a meaningless phrase, but as $S(N, NE, NV)/N$ for very large N. For analyzing thermodynamic systems, it is more convenient to measure N in moles and so that is what we will do throughout most of this chapter. However, except where otherwise noted, the formulas are valid when interpreted in individual particles. Writing $s(E, V)$ for $S(1, E, V)$, the extensive property of $S(N, E, V)$ implies that

$$S(N, E, V) = Ns(\varepsilon, v) \tag{5.3}$$

where $\varepsilon = E/N$ and $v = V/N$.

*A homogeneous function of degree K is a function with the property that $f(\lambda x, \lambda y, \ldots) = \lambda^K f(x, y, \ldots)$ for any $\lambda > 0$.

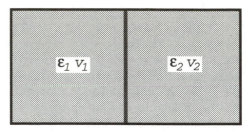

Fig. 5.1 A system composed of two moles of gas, separated by a movable, heat conducting barrier.

5.5 CONVEXITY OF THE ENTROPY

Part (b) of Axiom 4 also has important implications regarding the properties of the function $s(\varepsilon, v)$. We picture a two-mole sample of liquid, with an energy 2ε, at equilibrium within a cylinder of volume $2v$. We split the sample into two equal parts, of one mole each, by means of a movable, heat-conducting partition (see Fig. 5.1). Since the initial uniform state was an equilibrium state, the entropy must be a maximum when the samples on both sides of the partition are in the same state. In that case, each sample has energy ε and volume v. We now picture the system in a different state, but one with the same total energy and volume. Since this new state is not the equilibrium state but is not forbidden by any constraint (the partition only prevents particle transfer between the two halves) or conservation law, then it must have smaller entropy than the equilibrium state. This implies that

$$2s(\varepsilon, v) > s(\varepsilon_1, v_1) + s(\varepsilon_2, v_2) \tag{5.4}$$

whenever $\varepsilon_1 + \varepsilon_2 = 2\varepsilon$ and $v_1 + v_2 = 2v$. It is useful to write Eq. (5.4) in the form

$$s(\frac{\varepsilon_1 + \varepsilon_2}{2}, \frac{v_1 + v_2}{2}) > \frac{1}{2}(s(\varepsilon_1, v_1) + s(\varepsilon_2, v_2)) \tag{5.5}$$

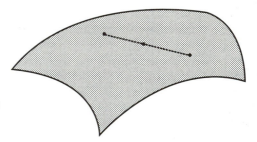

Fig. 5.2 If a convex function of two variables is plotted as a surface, it gives a convex surface. The midpoint of the line between any two points on a convex surface falls below the surface.

If we plot the function $s(\varepsilon, v)$ as a surface over the (ε, v) plane, Eq. (5.5) says that the surface always lies above any chord connecting two points on it. (See Fig. 5.2.) That is, the surface is *convex*. Any function satisfying Eq. (5.5) is also said to be convex. To be more precise, if we have a strict inequality, as in Eq. (5.5), then the function is said to be strictly convex. A function is simply convex if the inequality sign is replaced by $\geq$. A function whose graph is an upward-pointing cone is just convex (the surface of a cone contains straight lines), while a function whose graph is a paraboloid that opens downward is strictly convex.

5.6 THE THERMODYNAMIC TEMPERATURE

This section will be devoted to a study of the partial derivatives of the entropy function and their identification with measurable properties of the substance. We will first give them names and symbols. The variables α, β, and γ are defined by

$$\alpha = \frac{\partial S}{\partial N}, \qquad \beta = \frac{\partial S}{\partial E}, \qquad \text{and} \qquad \gamma = \frac{\partial S}{\partial V} \qquad (5.6)$$

where $S(N, E, V)$ is the entropy function of any simple phase. α is called the *affinity*, β the *inverse temperature* or *coldness*, and γ the *free expansion coefficient*. Their physical meanings will become clear as we proceed.

An *empirical temperature* is any macroscopic parameter with the property that, when two bodies are brought into contact, heat energy will flow spontaneously from the body with the higher value of the temperature to the body with the lower. It is easy to see that there is no unique empirical temperature, since any monotonically increasing function of one empirical temperature is itself an empirical temperature. A common practical empirical temperature is the length of the mercury column on a thermometer in contact with the body, but the length of an uncalibrated ten-penny nail would do just as well, although it might be a bit difficult to read.

Theorem $T = \beta^{-1}$ is an empirical temperature.

Proof Consider the system shown in Fig. 5.3, in which two substances are separated by a thermally conducting partition. The substances have initial energies E_1^o and E_2^o. Their entropy functions are $S_1(E_1)$ and $S_2(E_2)$. Since the constraints of the system prevent particle or volume exchange, we will not bother to indicate those quantities. As a function of E_1 the total entropy of the system is

$$S(E_1) = S_1(E_1) + S_2(E_T - E_1) \qquad (5.7)$$

where $E_T = E_1^o + E_2^o$. The condition of equilibrium is that $S(E_1)$ be a maximum.

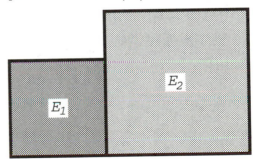

Fig. 5.3 In the equilibrium state, the energies are shared between the two components so as to maximize the total entropy.

This implies that the equilibrium value of E_1 satisfies $S'(E_1) = 0$, which gives $S_1'(E_1) = S_2'(E_2)$. Therefore, at equilibrium, $T_1(E_1) = T_2(E_2)$. However, this is not enough to show that T is an empirical temperature. We must show that at no intermediate stage on the movement towards equilibrium is energy flowing from lower T to higher T.

Graphs of the functions S_1 and S_2 are shown in Fig. 5.4. Of course, we do not know the actual form of the entropy functions of these unspecified substances. The values of E_1^o, E_1, E_2^o, and E_2 are indicated. We have assumed that substance 2 loses energy to substance 1, but the argument we will give would work just as well with the reverse assumption. At the final energies the two entropy graphs must have the same slope. But, by our previous argument, we know that both graphs must be convex, as they have been drawn. Thus at all points on the movement toward equilibrium

$$S_1'(E_1) > S_2'(E_2) \qquad \text{or} \qquad T_1(E_1) < T_2(E_2) \qquad (5.8)$$

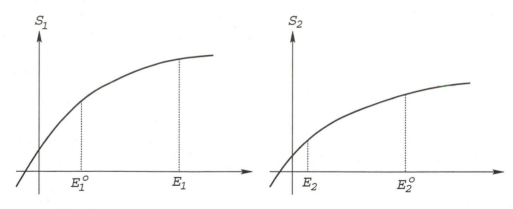

Fig. 5.4   The entropy functions of the two components.

Notice that, because of the convexity property of the entropy functions, the slope of each entropy function is a monotonically decreasing function of E. Thus the equation $S_1'(E_1) = S_2'(E_T - E_1)$ cannot have more than one solution. That it *must* have one solution with nonzero values of E_1 and E_2 actually depends on another axiom that will be presented later.

5.7   THE PRESSURE

Theorem $p = \gamma/\beta$ is the mechanical pressure in the substance.
Proof In Fig. 5.5 is shown a sample of fluid in a cylinder under a frictionless movable piston topped by a weight W. The system is surrounded by vacuum. If we let E_o denote the fixed total energy of the system, including the potential energy of the weight, and call the energy of the fluid, not including the gravitational potential energy of the weight, E, then the entropy of the system, as a function of the piston height h is

$$S(h) = S(N, E, V) = S(N, E_o - Wh, Ah) \tag{5.9}$$

where A is the area of the piston. The equilibrium value of h is given by setting $S'(h) = 0$. This gives

$$-W\frac{\partial S}{\partial E} + A\frac{\partial S}{\partial V} = 0 \tag{5.10}$$

or

$$p = \frac{\gamma}{\beta} = \frac{W}{A} \tag{5.11}$$

W/A is clearly the mechanical pressure in the fluid. Since p is the mechanical pressure in this situation and all macroscopic features are unique functions of N, E, and V at equilibrium, then p is the mechanical pressure at equilibrium in all situations.

5.8   THE FREE EXPANSION COEFFICIENT

The parameter γ is therefore related, in a simple way, to the pressure. In order to see the physical meaning of γ directly, consider the system shown in Fig. 5.6. If the plug is pulled, so that the gas can expand into the initially empty volume ΔV, then during the expansion the energy is constant, so that the change in entropy is given by

$$\Delta S = \frac{\partial S}{\partial V}\Delta V = \gamma\Delta V \tag{5.12}$$

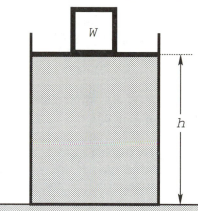

Fig. 5.5   The equilibrium value of h maximizes the system entropy.

Such an expansion into a vacuum is called a *free expansion*, which shows why we have called the parameter γ the *free expansion coefficient*; it is the ratio of the entropy change to the volume change in a free expansion.

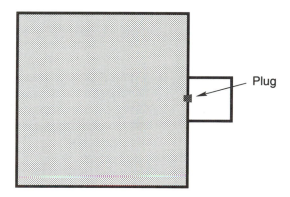

Fig. 5.6   When the plug is pulled, the gas expands into the vacuum, causing an entropy increase of $\gamma \Delta V$.

5.9   THE AFFINITY

Now let us consider the conditions of equilibrium between two phases of the same simple substance (for example, water and water vapor). We will assume that the two phases are liquid and gas and therefore call their entropy functions S_L and S_G. The total entropy is the sum of the entropies of the parts

$$S_T = S_L(N_L, E_L, V_L) + S_G(N_G, E_G, V_G) \tag{5.13}$$

Demanding that S_T be a maximum with respect to changes in N_L, E_L, and V_L, taking into account the constraints that $N_G = N_T - N_L$, $E_G = E_T - E_L$, and $V_G = V_T - V_L$, gives the three equilibrium conditions

$$\alpha_L = \alpha_G, \quad \beta_L = \beta_G, \quad \text{and} \quad p_L = p_G \tag{5.14}$$

Because α is defined as the derivative of S with respect to N, it is clear that, in the system considered, if α_L is greater than α_G, then a transfer of particles from the gas to the liquid would create a net increase in entropy of the system and would therefore occur spontaneously. Thus, we can say that, when possible, particles will spontaneously move from the substance with the smaller affinity to the substance with the larger affinity. The affinity plays the same role with respect to particle transfer as the coldness plays with respect to energy transfer. For a nonsimple substance (one with more than one type of particle) the entropy is a function of the particle numbers of all types, $S(N_A, N_B, \ldots, E, V)$. In that case there is a separate affinity for each type of particle. $\alpha_A = \partial S / \partial N_A$, $\alpha_B = \partial S / \partial N_B$, etc.. The physical meaning is unchanged. For example, in a system composed of air over water, water will evaporate into the air if $\alpha_{H_2O}(\text{air})$ is greater than $\alpha_{H_2O}(\text{liquid})$.

5.10 INTENSIVE VARIABLES

An *intensive* variable is any zero-order homogeneous function of N, E, and V, that is, a function with the property that $f(\lambda N, \lambda E, \lambda V) = f(N, E, V)$ for any $\lambda > 0$.

Theorem α, β, and γ are intensive variables.
Proof We know that

$$S(\lambda N, \lambda E, \lambda V) = \lambda S(N, E, V) \tag{5.15}$$

Differentiating both sides with respect to N, we get

$$\lambda \alpha(\lambda N, \lambda E, \lambda V) = \lambda \alpha(N, E, V) \tag{5.16}$$

which gives the desired result for α. The same could obviously be done for β and γ.

From the equation $\alpha(N, E, V) = \alpha(\lambda N, \lambda E, \lambda V)$, we can, by choosing $\lambda = 1/N$, see that $\alpha(N, E, V) = \alpha(1, \varepsilon, v)$, where ε and v are the energy and volume per mole. The function $\alpha(1, \varepsilon, v)$ will be written simply as $\alpha(\varepsilon, v)$. Clearly, we could also show that β and γ (or p) are also functions of ε and v.

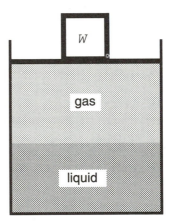

Fig. 5.7 Two phases in equilibrium at pressure p.

Theorem For any two-phase equilibrium state of a simple substance, α and β are functions of p.
Proof Consider two phases in equilibrium at an externally fixed pressure p (see Fig. 5.7). According to Eq. (5.14), the conditions for equilibrium are

$$
\begin{aligned}
p_G(\varepsilon_G, v_G) &= p \\
p_L(\varepsilon_L, v_L) &= p \\
\alpha_L(\varepsilon_L, v_L) &= \alpha_G(\varepsilon_G, v_G) \\
\beta_L(\varepsilon_L, v_L) &= \beta_G(\varepsilon_G, v_G)
\end{aligned}
\tag{5.17}
$$

These are four equations in the four unknowns ε_L, v_L, ε_G, and v_G. The only free parameter in the equations is p. For a given value of p two things are possible.

1. The equations have no solution, showing that a two-phase equilibrium state is not possible at that pressure. In this case, if the system initially contained two phases, one of them would disappear as the system approached equilibrium. Thus, this case is irrelevant to the situation we are considering.

2. The equations have a solution, $\varepsilon_L(p)$, $v_L(p)$, $\varepsilon_G(p)$, and $v_G(p)$. That the solution must be unique can be shown from the convexity property of the functions $s_L(\varepsilon, v)$ and $s_G(\varepsilon, v)$. Therefore, in this case, $\alpha(p) = \alpha_L(\varepsilon_L(p), v_L(p)) = \alpha_G(\varepsilon_G(p), v_G(p))$ and $\beta(p) = \beta_L(\varepsilon_L(p), v_L(p)) = \beta_G(\varepsilon_G(p), v_G(p))$ are functions of p.

This is important in that it allows us to keep a substance at a constant (although unknown) value of β during its expansion or compression (see Fig. 5.8).

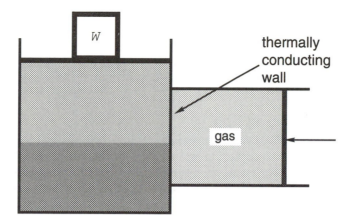

Fig. 5.8   If the piston on the right is moved very slowly, the gas passes through a sequence of states that have the same temperature but different volumes and energies.

5.11   QUASISTATIC PROCESSES

The gradual compression or expansion shown in Fig. 5.8 is our first example of a very important class of processes, called *quasistatic* processes. In a quasistatic process the change in the system brought about by any external agent is assumed to take place so slowly that the system has ample time to maintain internal equilibrium. In Fig. 5.8 the external agent is the force exerted on the piston in the cylinder containing the gas. That force is increased so gradually that only a tiny temperature difference develops between the gas and the two-phase system. It is obvious that some such assumption is necessary if we are to conclude that the gas and the two-phase system are at the same temperature.

Theorem   The quasistatic compression or expansion of a substance in an insulating cylinder does not change its entropy.
Proof   The force needed to maintain the piston of Fig. 5.9 in a fixed position is $F = pA$. If the piston moves an amount dx the work done on the gas, and therefore its change in energy, will be

$$dE = F\,dx = -p\,dV \tag{5.18}$$

But, by the definitions of β and γ and by the relation $\gamma = \beta p$, we get

$$
\begin{aligned}
dS &= \beta\, dE + \gamma\, dV \\
&= \beta(dE + p\, dV) \\
&= 0
\end{aligned}
\tag{5.19}
$$

We now have methods of changing the state of a substance that maintain either its temperature or its entropy constant. These processes will be important in devising a technique for measuring the entropy.

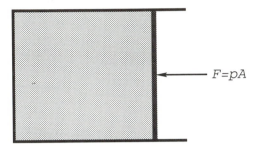

$$F = pA$$

Fig. 5.9   If the piston is slowly moved, the entropy is not changed.

5.12   HEAT TRANSFER

When the volume of a substance enclosed in a cylinder with a movable piston is quasistatically increased an amount dV by moving the piston, then the substance does an amount of work $p\, dV$ against the piston and its energy changes by an amount $dE = -p\, dV$. This is an important consideration when heat energy is added to a substance being held at constant pressure. If the amount of heat transferred to the substance is dQ, then the change in the internal energy of the substance will be equal to the heat added minus the work done on the piston as the substance expands due to its increase in temperature.

$$
dE = dQ - p\, dV
\tag{5.20}
$$

If heat is added to a substance in a closed vessel with a movable piston, then, using the facts that $dN = 0$, $\beta = 1/T$, and $\gamma = p/T$, one can see that the change in entropy is directly proportional to the heat added.

$$
dS = \frac{dE + p\, dV}{T} = \frac{dQ}{T}
\tag{5.21}
$$

5.13   AMBIGUITY IN THE ENTROPY FUNCTION

In Axiom 4 we postulated the existence of an additive entropy function, but we have not yet shown how to determine the entropy function of any substance. Before we do so, we have to investigate how rigidly these axioms actually define the entropy functions. That is, how many arbitrary parameters are there in the set of entropy functions, one for each phase of a substance, that satisfy Axiom 4. That question is answered by the following theorem. In this analysis it is important that we treat the case of nonsimple substances, and therefore we will consider a substance composed of two different types of particles. The generalization to an arbitrary number of particle types will be completely obvious.

Let $S_K(N_A, N_B, E, V)$ be the entropy function of the Kth phase of a substance composed of independently conserved particles of types A and B. We point out the fact that $S_K(N_A, 0, E, V)$ and $S_K(0, N_B, E, V)$ are then the entropy functions of the Kth phase of the two related simple substances.

Theorem The set of entropy functions

$$S_K^*(N_A, N_B, E, V) = \lambda S_K(N_A, N_B, E, V) + a_A N_A + a_B N_B \qquad (5.22)$$

predict, in all cases, the same equilibrium states as the set of entropy functions $S_K(N_A, N_B, E, V)$, for any $\lambda > 0$ and arbitrary values of a_A and a_B. The functions S_K^* are also extensive if the functions S_K are extensive.

Proof Using the entropy functions S_K^*, the equilibrium state of a system would be determined by maximizing the quantity

$$
\begin{aligned}
S_{\text{total}}^* &= \sum S_K^* \\
&= \lambda \sum S_K + a_A \sum N_{AK} + a_B \sum N_{BK} \qquad (5.23) \\
&= \lambda S_{\text{total}} + a_A N_A + a_B N_B
\end{aligned}
$$

over all possible distributions of energy, volume, and particle numbers among the various phases in the system, taking into account the constraints of the system and the fundamental conservation laws. Since both types of particles are assumed to be conserved, the term $a_A N_A + a_B N_B$ on the third line is effectively a constant and would have no effect on the maximizing distribution. Since $\lambda > 0$, the maximum of S_{total}^* is given by the same distribution that maximizes S_{total}.

One might be tempted to add another term of the form $bE + cV$ to the definition of S_K^*. This is forbidden by the existence of situations, such as that depicted in Fig. 5.5, in which E and V are not separately conserved (see Problem 5.2).

5.14 MEASUREMENT OF $S(N, E, V)$

We now have all the tools we need to attack the fundamental problem of determining the entropy function of a given substance. We proceed as follows.

1. We first choose some convenient simple substance and call it the *standard thermometric substance*. In practice, the standard substance is pure water. We place one mole of the standard substance in an insulating cylinder with a movable piston. Within the cylinder is a small electrical resistor. Well-controlled amounts of energy can be added to the substance by running a current through the resistor. Therefore, one can easily determine the energy difference between two states of the same volume. (See Fig. 5.10.)

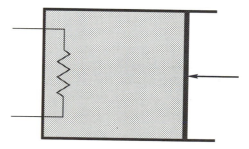

Fig. 5.10 The system whose entropy is to be measured.

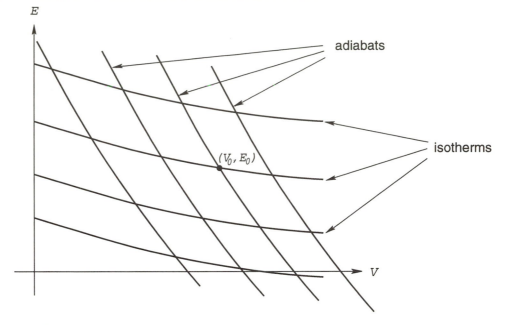

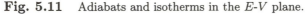

Fig. 5.11   Adiabats and isotherms in the E-V plane.

By quasistatically moving the piston and carefully keeping track of the work done in the process, we can determine the energy difference between two arbitrary states. The energy, being partly potential energy, always has some arbitrary constant associated with it. Therefore, we can choose some *standard state*, let us say a single-phase state of volume V_o and pressure p_o, and declare it to have energy E_o. Having done that, we can locate any physical state on the E-V plane by purely mechanical measurements.

2. Starting from any state, we can, by quasistatic expansion or compression of the standard substance within the insulated cylinder, determine a sequence of states that have equal (but unknown) entropy. A sequence of states with the same value of S is called an *adiabatic curve*. Thus, we can construct the set of adiabatic curves shown in Fig. 5.11.

3. We now return the system to the standard state (E_o, V_o). Making use of the arbitrary constant λ in the entropy functions, we are free to declare that the inverse temperature of the standard state of the standard substance is some arbitrarily chosen positive number β_o. Making use of the arbitrary constant a_A in the entropy functions, where A is the standard substance, we are free to declare that the entropy of the standard state of the standard substance is some arbitrarily chosen number S_o.

4. Starting from the standard state, and using a device like that shown in Fig. 5.8, we can determine a sequence of states (an *isothermal curve*) that all have the inverse temperature β_o (see Fig. 5.11). Since, during isothermal compression, the standard substance gives up energy to the two-phase system, the energy rises less rapidly with decreasing volume on an isothermal curve than on an adiabatic one.

5. Starting from the standard state, we can determine the entropy of any state (E_1, V_1) that lies on the isothermal curve by using the relation $dS = \beta_o(dE + p\,dV)$. (Remember that the pressure is directly measurable by mechanical means.)

$$S(E_1, V_1) = S_o + \beta_o \int_{(E_o, V_o)}^{(E_1, V_1)} (dE + pdV) \tag{5.24}$$

6. Using the fact that all the states on any adiabatic curve have the same entropy, we can then determine the entropy of one mole of the standard substance at any point in the (E, V) plane.

Since we have been dealing with one mole of the standard substance, the function we have actually determined is the entropy per mole and should more properly have been written as $s(\varepsilon, v)$. The entropy of an arbitrary amount of the standard substance would then be given by

$$S(N, E, V) = Ns(E/N, V/N) \qquad (5.25)$$

Having determined the entropy function of the standard substance, one can differentiate it to obtain the numerical value of the inverse temperature $\beta(\varepsilon, v) = \partial s / \partial \varepsilon$ for any state of the standard substance. By putting it in thermal contact with other substances, one can use the standard substance as a thermometer to determine the temperature of anything else.

Suppose we have one mole of another simple substance, which we will call B (A is the standard substance). We can easily determine $\beta(\varepsilon_B, v_B)$ by using the standard substance as a thermometer. Note that we cannot arbitrarily define β for some standard state of substance B. The single free multiplicative parameter λ has been used up in defining β for the standard state of the standard substance. But we do still have one free parameter, namely a_B. This is an additive parameter. We are therefore free to assign an arbitrary value to $s_B(\varepsilon_o, v_o)$, where (ε_o, v_o) is some chosen standard state of substance B. Since we know $\beta(\varepsilon_B, v_B)$ and we can measure $p(\varepsilon_B, v_B)$ mechanically, we can now integrate the differential $dS = \beta(dE + p\,dV)$ along any curve we like from the standard state to any other state in order to determine the entropy function of substance B.

All this has been done without any mention of a perfect gas. An essentially unique definition of thermodynamic temperature, namely $1/T = \partial S / \partial E$, emerges from the fundamental principles of thermodynamics. The value of T does not depend upon the specific properties of any particular substance. If we use substance B as our standard substance, the only change that it can make in the entropy functions we obtain is a different set of arbitrary constants, λ, a_A, and a_B. The new temperature scale would be exactly proportional to the old. Thus, the change would simply constitute choosing a different unit of temperature.

5.15   UNITS, RATIONAL AND PRACTICAL

As we mentioned before, the entropy is proportional to $\log K$, where K is the number of quantum states associated with a given macrostate. In a *rational* system of units, S is set exactly equal to $\log K$. However, this choice can only be made after a fully developed quantum theory exists that allows one to determine the number K for some substance. In reality, the development of thermodynamics preceded the development of quantum theory by about a century. Therefore, the early thermodynamicists were obliged to construct a somewhat arbitrary system of thermodynamic units that we will call the *practical* system. (These names should not be construed to imply that the practical system is irrational nor that the rational system is impractical. The word *practical* simply means "what is used in practice". The term *rational* is commonly used to describe any system of units in which the number of physical constants is reduced by using fundamental theoretical relationships.)

The two systems of units differ only by the values of the multiplicative constant λ. The modern version of the practical system is defined by choosing water as

"In the present state of physical science, therefore, a question of extreme interest arises: *Is there any principle on which an absolute thermometric scale can be founded?* It appears to me that Carnot's theory of the motive power of heat enables us to give an affirmative answer."

—— William Thompson (Lord Kelvin)

(Lord Kelvin was the first person to give, by an analysis equivalent to that given in this chapter, the definition of the absolute scale of temperature.)

the standard substance. The standard state is chosen as that unique value of pressure and temperature at which the three phases, solid, liquid, and gas, are simultaneously in equilibrium (the *triple point*), and it is declared that, at the triple point, $T = (\partial S/\partial E)^{-1} = 273.16$ exactly. In this system, the unit of temperature is the *Kelvin*, written K. The relationship between the entropy in the practical system and the number of quantum states is

$$S = k \log K \tag{5.26}$$

where k is a physical constant, called *Boltzmann's constant*, with the value and units of

$$k = 1.3807 \times 10^{-23}\,\text{J/K} \tag{5.27}$$

Equations 5.26 and 5.27 may be interpreted in two different ways. In the first way, one declares that the Kelvin is a dimensionless quantity. The entropy then has the physically bizarre units of joules. They are called bizarre because the entropy has no natural interpretation as an energy of any kind. In the second way, the Kelvin is taken to be another unit of energy. Boltzmann's constant is then merely a conversion factor between the two independently defined energy units. In this interpretation S is dimensionless. Interpreting the Kelvin as an energy unit is in accord with the common practice in low-temperature physics of specifying all energy values, such as quantum mechanical energy levels, in Kelvins. It is the second interpretation that will be used in this book.

In order to avoid confusion, whenever rational units are being specifically used, the entropy will be written as S^o and the temperature as τ. In rational units, the temperature τ is measured in Joules. The affinity, being the derivative of S with respect to N, depends on the units used for both S and N. There are four possibilities, but we will carefully restrict ourselves to two of them. When using rational units, we will always express N in particles and, when using practical units, in moles. The conversion from practical to rational units is given by

$$S^o = S/k, \qquad \tau = kT, \qquad \text{and} \qquad \alpha(\text{rat.}) = \alpha(\text{prac.})/R \tag{5.28}$$

Another point regarding notation is that in this chapter we have been using the symbol β to mean the inverse of the temperature in whatever system of units is being used. The well-established custom is to use β to mean only $1/\tau$. After this chapter we will strictly abide by that custom.

5.16   THERMODYNAMIC STATE SPACE

We will now consider a nonsimple substance with two types of conserved particles, A and B. According to Axiom 2, the set of independent thermodynamic variables for this system is the set (N_A, N_B, E, V). We view this as a point in a four-dimensional *thermodynamic state space* $\mathcal{S}_T$.

$$(N_A, N_B, E, V) \in \mathcal{S}_T \tag{5.29}$$

All properties of the equilibrium states of this system are unique functions of position in $\mathcal{S}_T$. For examples, the pressure on the containing walls and the volume occupied by the liquid phase (which might very well be zero) are functions $p(N_A, N_B, E, V)$ and $V_L(N_A, N_B, E, V)$ defined throughout $\mathcal{S}_T$. For a different system the thermodynamic state space would be different, but the general results given in this section would still hold with obvious adjustments.

The first question we want to consider is: "What are the boundaries of $\mathcal{S}_T$?" If one or both of the particles have absolutely incompressible hard cores, then there will be an inequality relating N_A, N_B, and V that defines the maximum packing density. It will be of the form $V_{\min}(N_A, N_B) < V$, where $V_{\min}$ is an extensive function of N_A and N_B. Even if the particles do not have absolutely hard cores, because the interaction potential becomes larger and larger as the particles are pushed closer and closer together, for any fixed value of E, there is a certain volume, $V_{\min}$, at which the quantum mechanical ground state energy becomes equal to E. For V less than $V_{\min}$, there are no physical states with the specified energy. This gives a more general inequality of the form

$$V_{\min}(N_A, N_B, E) < V < \infty \tag{5.30}$$

For given N_A, N_B, and V, the energy must be larger than some minimum energy that is just the ground-state energy of a system composed of N_A particles of type A and N_B particles of type B in a volume V. There is no natural upper bound on the energy for a system of particles. Thus

$$E_{\min}(N_A, N_B, V) < E < \infty \tag{5.31}$$

Actually, this is just the inequality given in Eq. (5.30), solved for E as a function of V.

There are two axioms of thermodynamics remaining to be presented. They concern the behavior of the entropy at the boundaries of $\mathcal{S}_T$.

A state is called *thermodynamically inaccessible* if it can never be brought about by bringing the system into contact with other systems. For example, a state with zero volume but nonzero particle number is thermodynamically inaccessible, because it would require infinite pressure to achieve it.

Axiom 5. The boundary of $\mathcal{S}_T$ consists of limit states that are thermodynamically inaccessible because, at all finite boundaries, the normal derivative of S approaches minus infinity. That is, $\partial S / \partial \mathbf{n} \to -\infty$. This implies that

$$
\begin{array}{llll}
\alpha_i \to \infty & \text{as} & N_i \to 0, & i = A, B \\
\beta \to \infty & \text{as} & E \to E_{\min}(N_A, N_B, V) & \\
\gamma \to \infty & \text{as} & V \to V_{\min}(N_A, N_B) &
\end{array}
\tag{5.32}
$$

To see how these limiting properties make the border states thermodynamically inaccessible, consider what would happen if we put the system we are considering in thermal contact with another system. Since the other system is also a thermodynamic system, it would naturally also satisfy Axiom 5. Thus, for each system, the

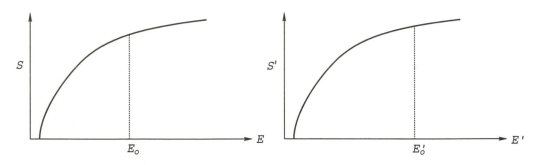

Fig. 5.12   Both curves have infinite slope at their left ends.

entropy, as a function of the energy, is given by a strictly convex curve that is vertical at its left end. The equilibrium condition is that the slopes of the two curves be equal ($\beta_1 = \beta_2$). Regardless of the initial total energy, this would obviously occur at some interior point on both curves. (See Fig. 5.12.)

Let us try to communicate, in a casual way, the physical content of the three conditions. The first says, for example, that as air gets drier and drier, it will eventually steal water from anything with which it comes into contact; the second says that the temperature approaches zero at the lowest possible energy. Since $\gamma = p/T$, the third line says that as the system is brought to its minimum volume, either the temperature goes goes to zero or the pressure goes to infinity. For quantum mechanical systems, it is the first possibility that always occurs. For classical systems, the first possibility occurs for interaction potentials that go smoothly to infinity; the second occurs for an interaction potential with a discontinuous hard core or for a perfectly ideal gas.

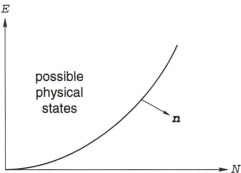

Fig. 5.13   For a fixed value of V, the lower boundary of the state space in the N–E plane is given by a curve that can be described either by the function $E_{\min}(N, V)$ or, equivalently, by the function $N_{\max}(E, V)$. $\partial S / \partial \mathbf{n} \to -\infty$ (where $\mathbf{n}$ is always the outward-pointing normal) as one approaches the boundary.

possible physical states

5.17   NEGATIVE TEMPERATURE

Axiom 5 has interesting consequences for any system in which either N or E is restricted to a finite range. For example, a system composed of hard-core particles (that is, impenetrable particles, like billiard balls) has an upper bound on the number of particles that can be fit within a given volume. Thus $0 \leq N \leq N_{\max}(V)$. For particles without absolutely hard cores, the potential energy rises to very high values as more and more particles are added at fixed volume. Thus at fixed energy and volume there is a maximum packing density. $0 \leq N \leq N_{\max}(E, V)$ (see

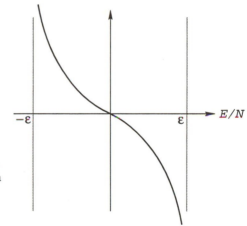

Fig. 5.14 The inverse temperature, as a function of $E/N\varepsilon$, for the system of spins in a magnetic field. At positive energies, β is negative.

Fig. 5.13). This is just another way of saying that the ground-state energy of the system increases with increasing density. That $\partial S/\partial \mathbf{n} \to -\infty$ at the boundary shown in Fig. 5.13 implies the two conditions that $T \to 0$ as $E \to E_{\min}(N,V)$ and that $\alpha \to -\infty$ as $N \to N_{\max}(E,V)$. The ideal Fermi–Dirac gas, to be considered in Chapter 7, is an excellent example of this phenomenon.

A more interesting case is that of a system whose range of possible energies is finite. The simplest example is a collection of N noninteracting half-integer magnetic moments in an external magnetic field. There are a number of real physical systems that are fairly well approximated by this simplified model. If the magnetic field strength is B and the projection of the magnetic moment in the field direction is $\pm m$, then a single magnetic moment has two possible energy states $\pm\varepsilon$, where $\varepsilon = mB$. The entropy function for such a system has been worked out in Exercise 3.14. It gives the following relationship between energy and inverse temperature (in rational units).

$$\beta = \frac{1}{2\varepsilon} \log\left(\frac{1 - E/N\varepsilon}{1 + E/N\varepsilon}\right) \tag{5.33}$$

The range of possible energies for the system is obviously $-N\varepsilon \le E \le N\varepsilon$. The plot of β in that range is shown in Fig. 5.14. In the positive half of the energy range $(0 < E < N\varepsilon)$ $\beta = \partial S^o/\partial E$ is negative and goes to $-\infty$ as E approaches its maximum, in agreement with Axiom 5. Naturally, a state with negative coldness is less cold (that is, hotter) than a state with positive coldness. If a system in such a state is put into contact with any system at positive β, energy will spontaneously flow from the hotter (negative coldness) system to the colder (positive coldness) one. This has the physically reasonable implication that, if two identical systems of magnetic moments, one in a positive energy state and one in a negative energy state, are put into thermal contact, then heat energy will flow from the higher-energy system to the lower-energy one, finally bringing the two systems to the same intermediate energy (which might be a negative or a positive coldness state).

The temperature, related to the coldness by $\tau = 1/\beta$, is an inconvenient variable to use in describing these systems. As the energy passes smoothly through zero, from left to right, τ goes to $+\infty$, then jumps to $-\infty$ and moves toward zero from the left. One must keep in mind that the negative temperature states are hotter than the positive temperature states. They are even hotter that the state at $\tau = \infty$.

Most systems contain kinetic energy terms in their Hamiltonian functions. Since the kinetic energy alone can range up to $+\infty$, such systems cannot have any upper bound to their energy. Also, we know that $T \to \infty$ (that is, $\beta \to 0$) as $E \to \infty$. Thus these *normal* systems always have positive temperatures. Any system in a negative temperature state is always unstable with respect to thermal contact with any normal system. No matter what the relative sizes of the two systems are, energy will continue to flow from the negative temperature (hotter) system to the normal system until both systems are left in a common *positive* temperature state. In this sense, the negative temperature states are somewhat exotic. However, it is not difficult to find systems of nuclear magnetic moments that are so well isolated from energy exchange with the other normal degrees of freedom of the substance that they can be maintained in negative temperature states for appreciable lengths of time. The trick used to move a system into such a negative temperature state is to first put it into a low positive temperature (large positive β) state in which a majority of the magnetic moments are parallel to the magnetic field. In such a state the system has a negative energy. The field direction is then reversed. Because the system of nuclear spins is isolated, it has no way of shedding its excess energy by flipping its magnetic moment to point in the new field direction. The system of nuclear spins is therefore left in a positive energy, negative β state.

5.18   NERNST'S LAW

Our sixth and last axiom is usually called Nernst's Law or the Third Law of Thermodynamics. We seem to have been somewhat profligate in our axiom creation.*
Axiom 6. (Nernst's Law)   All minimum energy states have zero entropy. That is,

$$\lim_{E \to E_{\min}} S(N_A, N_B, E, V) = 0 \tag{5.34}$$

regardless of the values of N_A, N_B, and V.

Before we discuss the consequences of Nernst's law, we should give some idea of the quantum mechanical origin of the law, for the explanation of Nernst's law must be given in quantum mechanical terms. As we will see, systems described by purely classical models do not satisfy Nernst's law. For classical systems, as T approaches zero, the entropy goes to minus infinity.

Using the fact that $S^o = \log K$, one can give a facile, but erroneous, explanation of the law by noting that, at the ground state energy, the system must be in its ground-state (which we will assume to be nondegenerate) and therefore K, the number of quantum states available to the system, must become unity. $\log K$ would then become zero. Such an effect would come into play when the probability of finding the system in its ground state was close to one. That happens when the total energy of the system is less that the energy separation between the ground state and the first excited state. For a truly macroscopic system, an energy of that magnitude would give a temperature on the order of $10^{-19}\,\mathrm{K}$, a temperature that is quite unattainable. In reality, the approach of the entropy to its limiting value becomes definitely noticeable at temperatures below $10\,\mathrm{K}$. This is a temperature at which the average energy *per particle* becomes comparable to the distance between the ground- and excited-state energy levels. The two criteria differ by a factor of

---

*But much less so than was David Hilbert who, in reformulating Euclidean geometry, increased the number of axioms from five to twenty three.

"The present discoveries in science are such as lie immediately beneath the surface of common notions. It is necessary, however, to penetrate the more secret and remote parts of nature, in order to absract both notions and axioms from things, by a more certain and guarded method"

—— Francis Bacon (*Novum Organum*)

10^{20}. The true origin of Nernst's law is the Bose–Einstein or Fermi–Dirac symmetry requirements on allowed wave functions and not merely the discreteness of the energy levels. At small values of the energy per particle, the density of energy eigenstates (that is, the number of different eigenstates of the system per unit of energy) is much less than it would be if there were no symmetry requirements imposed upon allowed quantum states. (For more details, see Problem 5.19.)

One great benefit of Nernst's law is that it provides a natural way of choosing all of the arbitrary *additive* constants, a_A, a_B, etc., in the entropy functions. The requirement that the entropy of every simple substance approach zero as T goes to zero fixes all of the additive constants. No new arbitrary constants appear in the entropy functions of nonsimple substances. Thus, the laws of thermodynamics leave only a single multiplicative arbitrary constant in the set of all the entropy functions for all substances.

Nernst's law has some unexpected and surprising consequences. Two examples are given in the following.

The Pressure Paradox: If a sample of any real substance is kept in a sufficiently small volume, then its pressure will remain finite as E approaches E_{min}. But

$$\frac{\partial S}{\partial V} = \beta p \qquad \text{and} \qquad \beta \to \infty \quad \text{as} \quad E \to E_{min} \tag{5.35}$$

Therefore

$$\lim_{E \to E_{min}} \left(\frac{\partial S}{\partial V} \right) = \infty \tag{5.36}$$

However, according to Nernst's law

$$\frac{\partial}{\partial V} \left(\lim_{E \to E_{min}} S \right) = 0 \tag{5.37}$$

which suggests that this is a rather nasty limit. The simplest system for which one can check that both of the above equations are true is the ideal Fermi gas, and therefore, the verification of this odd result will have to be postponed until we treat that system in detail.

The Mixing Paradox: Consider a container of volume $2V$, filled with a mixed ideal gas composed of one mole each of two types of particles. At very low temperatures the entropy of this system is essentially zero. Now consider two containers, each of volume V and containing one mole of the two gases separated. Again, at low temperature the entropy is negligible. Thus it should be possible reversibly to go from one system to the other. There is no "entropy of mixing" at very low temperature. (See Exercise 5.11 for a discussion of the entropy of mixing.) We will again have to postpone the verification of this until we treat low-temperature quantum gases.

5.19   THE GIBBS–DUHEM EQUATION

For a simple substance, the extensivity of the entropy is expressed by the equation

$$\lambda S(N, E, V) = S(\lambda N, \lambda E, \lambda V) \tag{5.38}$$

If we differentiate this equation with respect to λ and then set λ equal to one, we get

$$S = \alpha N + \beta E + \gamma V \tag{5.39}$$

This fundamental relation will be very useful in the future. Taking a differential of this relation gives

$$dS = \alpha\, dN + \beta\, dE + \gamma\, dV + N\, d\alpha + E\, d\beta + V\, d\gamma \tag{5.40}$$

But by the definitions of α, β, and γ we know that

$$dS = \alpha\, dN + \beta\, dE + \gamma\, dV \tag{5.41}$$

Subtracting this equation from the previous one gives the Gibbs–Duhem equation

$$N\, d\alpha + E\, d\beta + V\, d\gamma = 0 \tag{5.42}$$

It is useful to eliminate γ in favor of p to obtain

$$N\, d\alpha + (E + pV)\, d\beta + \beta V\, dp = 0 \tag{5.43}$$

This relates the changes in the intensive variables α, β, and p as the system goes from one equilibrium state to a neighboring one. The major use of the Gibbs–Duhem relation is to calculate α, the affinity of a substance, in terms of the more easily measured variables, p and T.

An *isobar* or isobaric curve is a sequence of equilibrium states of the same pressure. Along an isobar, $dp = 0$, and hence

$$N\, d\alpha + (E + pV)\, d\beta = 0 \tag{5.44}$$

or

$$\left(\frac{\partial \alpha}{\partial \beta}\right)_p = -(\varepsilon + pv) \tag{5.45}$$

Along an isotherm $d\beta = 0$. This gives

$$\left(\frac{\partial \alpha}{\partial p}\right)_\beta = -\beta v \tag{5.46}$$

The fact that $\partial^2 \alpha / \partial p\, \partial\beta = \partial^2 \alpha / \partial\beta\, \partial p$ leads to the following consistency equation, which will be useful in the next section.

$$\left(\frac{\partial \varepsilon}{\partial p}\right)_\beta = -p\left(\frac{\partial v}{\partial p}\right)_\beta + \beta\left(\frac{\partial v}{\partial \beta}\right)_p \tag{5.47}$$

5.20   THE ENTROPY OF AN IDEAL GAS

Although we have avoided the use of the ideal gas in formulating the fundamental laws of thermodynamics, there is no question about the fact that it occupies a uniquely important place in thermodynamic theory and practice. In this section the phrase *ideal gas* will be used in the restricted sense of a substance that satisfies the classical ideal gas equation of state. This excludes quantum ideal gases, which are systems of noninteracting particles, but do not satisfy that equation. In rational units the ideal gas equation takes the form $pV = N\tau$, or, in terms of the volume per particle,

$$v = \tau/p = 1/\beta p \tag{5.48}$$

Equation (5.47) then gives

$$\left(\frac{\partial \varepsilon}{\partial p}\right)_\beta = -p\left(\frac{-1}{\beta p^2}\right) + \beta\left(\frac{-1}{\beta^2 p}\right) = 0 \tag{5.49}$$

which shows that ε is not a function of pressure. Thus ε is a function of β alone. $\varepsilon = f(\beta)$. This function can be inverted to give $\beta(\varepsilon)$. Integrating the relation $(\partial s^o/\partial \varepsilon)_v = \beta$ gives

$$s^o(\varepsilon, v) = \int^\varepsilon \beta(\varepsilon)\, d\varepsilon + F(v) \tag{5.50}$$

where F is an unknown function of v. Equation (5.48), combined with the relation $\gamma = \partial s^o/\partial v$, implies that

$$\left(\frac{\partial s^o}{\partial v}\right)_\varepsilon = \beta p = \frac{1}{v} \tag{5.51}$$

which, upon integration, gives

$$s^o(\varepsilon, v) = \log v + G(\varepsilon) \tag{5.52}$$

with G an unknown function of ε. These two formulas for $s^o(\varepsilon, v)$ give the entropy per particle within an arbitrary constant.

$$s^o(\varepsilon, v) = \log v + \int^\varepsilon \beta\, d\varepsilon + C$$
$$\equiv \log v + g(\varepsilon) + C \tag{5.53}$$

From the extensivity relation for $S^o(N, E, V)$ we get

$$S^o(N, E, V) = N\left[\log(V/N) + g(E/N) + C\right] \tag{5.54}$$

The detailed form of the energy–temperature relationship, which defines $g(\varepsilon)$, depends on the internal structure of the particles and cannot be derived from the pressure equation of state. For point particles (a good approximation to the noble gases, He, Ne, etc.), we have the energy equation $\varepsilon = \frac{3}{2}\tau$, giving $\beta = 3/2\varepsilon$ and

$$S^o = N(\log(V/N) + \tfrac{3}{2}\log(E/N) + C) \tag{5.55}$$

Clearly, for this system, the minimum energy is zero ($E_{\min}(N, V) = 0$). A simple calculation will verify that all three parts of Axiom 5 are satisfied, which guarantees that the boundary states of $\mathcal{S}_T$ are thermodynamically inaccessible. However, it is also easy to see that

$$\lim_{E \to E_{\min}} S^o(N, E, V) = -\infty \tag{5.56}$$

which is a clear violation of Nernst's law. This violation of Nernst's law is not surprising, since the ideal gas equation of state is valid only when quantum mechanical effects can be neglected, and, therefore, it is not valid close to zero temperature.

5.21   THERMODYNAMIC POTENTIALS

As was noted in Section 4.7, most experiments are actually carried out at fixed temperature, not at fixed energy. Usually it is also much easier to maintain a fixed pressure on an experimental sample than to keep the sample's volume fixed. It would therefore be helpful if the conditions for thermodynamic equilibrium could be expressed in terms of T and p, rather than E and V. The convexity of S guarantees that such transformations of independent variables are possible. For example, at fixed N and V,

$$\frac{\partial \beta}{\partial E} = \frac{\partial^2 S}{\partial E^2} < 0 \tag{5.57}$$

Therefore, the transformation from E to β is one-to-one, which is what is needed in order to replace E by β as an independent variable (see Fig. 5.15).

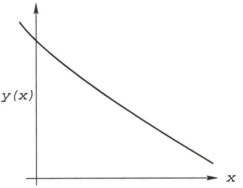

Fig. 5.15   If the slope of the curve $y(x)$ is always negative, then there is only one value of x that gives a particular value of y so that the function $y(x)$ can be inverted to give the function $x(y)$.

The transformation from E to β is most easily carried out by a method due to Legendre. We define a new thermodynamic function, which will be immediately recognized as the canonical potential, by

$$\phi = S - \beta E \tag{5.58}$$

The difference in ϕ between two neighboring thermodynamic states is

$$\begin{aligned} d\phi &= \alpha\, dN + \beta\, dE + \gamma\, dV - \beta\, dE - E\, d\beta \\ &= \alpha\, dN - E\, d\beta + \gamma\, dV \end{aligned} \tag{5.59}$$

It is clear from Eq. (5.59) that, if ϕ is expressed in terms of the variables N, β, and V, then

$$\left(\frac{\partial \phi}{\partial N}\right)_{\beta V} = \alpha$$

$$\left(\frac{\partial \phi}{\partial \beta}\right)_{NV} = -E \tag{5.60}$$

$$\text{and} \quad \left(\frac{\partial \phi}{\partial V}\right)_{N\beta} = \gamma = \beta p$$

The first and last of these equations give the affinity and the free expansion coefficient (or pressure) in terms of N, β, and V. The second gives the old variable E in terms of the new variable β. These formulas are identical with Eqs. (4.46) and (4.47), confirming the identification of ϕ as the canonical potential.

In Chapter 7, another ensemble will be defined that describes a system that can exchange both particles and energy with a large reservoir. It is called the *grand canonical ensemble*. The independent variables defining a grand canonical ensemble are α, β, and V. The logarithm of its normalization constant will be another thermodynamic potential, namely, the grand potential. From a thermodynamic viewpoint, the grand potential is the potential obtained in a Legendre transformation to the variables α, β, and V.

5.22 THE GRAND POTENTIAL

If ψ is defined everywhere in the thermodynamic state space $\mathcal{S}_T$ by

$$\psi = S - \alpha N - \beta E \tag{5.61}$$

then the change in ψ, in going from one point in $\mathcal{S}_T$ to a neighboring point, is

$$d\psi = -N\,d\alpha - E\,d\beta + \gamma\,dV \tag{5.62}$$

which shows that, if ψ is written in terms of the variables α, β, and V, then

$$\left(\frac{\partial \psi}{\partial \alpha}\right)_{\beta V} = -N$$

$$\left(\frac{\partial \psi}{\partial \beta}\right)_{\alpha V} = -E \tag{5.63}$$

$$\text{and} \qquad \left(\frac{\partial \psi}{\partial V}\right)_{\alpha \beta} = \gamma = \beta p$$

A particular choice of independent variables is called a *representation*. We have considered the (N, E, V), the (N, β, V), and the (α, β, V) representations. It is clear that a Legendre transformation could be used to eliminate any one or two of the extensive variables, N, E, and V, in favor of their thermodynamically *conjugate* variables, α, β, and γ. However, it is not possible to eliminate all extensive variables. There are two ways of seeing this. First, because of the Gibbs–Duhem relation, it is not possible to make arbitrary changes in the three variables, α, β, and γ. Thus, they cannot form a set of three independent variables. Second, if one attempted to make such a triple transformation by means of a Legendre transformation, the appropriate thermodynamic potential would be

$$\omega = S - \alpha N - \beta E - \gamma V \tag{5.64}$$

But, because of the fundamental relation, Eq. (5.39), the function ω would be identically zero in $\mathcal{S}_T$ and thus could not fill the role of a thermodynamic potential at all.

5.23 THERMODYNAMIC VARIATIONAL PRINCIPLES

In Fig. 5.16 is shown a system, containing three phases, in contact with a large thermal reservoir. Certainly, the condition that the system be in equilibrium with respect to energy transfer between the phases and between the system and the reservoir is that the temperature of every phase be equal to that of the reservoir. We now want to show that, in determining the equilibrium conditions with respect to all other variables (volumes and particle numbers of all phases), the total canonical

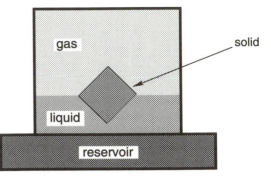

Fig. 5.16 A system containing three phases in contact with a thermal reservoir.

potential of the system, not including the reservoir, plays the same role as the total entropy for an isolated system. We number the phases, in any order, 1, 2, and 3. We assume that each phase contains two types of particles, such as sugar and water. The numbers of particles of the two types in phase i we call N_i' and N_i''. The total canonical potential is $\phi_T = \sum_i \phi_i(N_i', N_i'', \beta, V_i)$.

Theorem At equilibrium, the total canonical potential is a maximum with respect to all possible changes in particle numbers and volumes.

Proof The conditions for ϕ_T to be stationary with respect to a transfer of particles between phases i and j is that $\partial\phi_i/\partial N_i' = \partial\phi_j/\partial N_j'$ and $\partial\phi_i/\partial N_i'' = \partial\phi_j/\partial N_j''$. But these are just the already known equilibrium conditions that $\alpha_i' = \alpha_j'$ and $\alpha_i'' = \alpha_j''$. Similarly, the condition that ϕ_T be stationary with respect to changes in the volumes occupied by the three phases leads to another known equilibrium condition, namely, that $p_1 = p_2 = p_3$. That the stationary point is a maximum relies on the facts that, for any substance,

$$\frac{\partial^2 \phi}{\partial N^2} = \left(\frac{\partial \alpha}{\partial N}\right)_{\beta V} < 0 \tag{5.65}$$

and

$$\frac{\partial^2 \phi}{\partial V^2} = \beta\left(\frac{\partial p}{\partial V}\right)_{N\beta} < 0 \tag{5.66}$$

The first of these conditions says that adding particles at constant temperature and volume reduces the affinity for particles. The second says that compressing the system at constant temperature increases the pressure. Both conditions seem intuitively obvious, but they turn out to be surprisingly tricky to actually prove from the fundamental principles of thermodynamics (see Problem 5.6).

The statistical mechanism underlying this variational principle is that the number of quantum states of the system plus the reservoir, as a function of the system parameters, is equal to $\exp(\phi_T + \text{const.})$. Therefore, the statement that ϕ_T goes to a maximum is equivalent to the statement that the combined system goes to its most probable macrostate. It can also be shown that, for a system that can exchange energy and particles with a reservoir, the total grand potential of the system is a maximum at equilibrium.

5.24   THERMODYNAMIC FLUCTUATIONS

Thermodynamics has been defined as the collection of those universal properties of macroscopic systems that emerge only in the thermodynamic limit. From that point of view, it can be seen that a major item of thermodynamic theory has been omitted.

All our thermodynamic laws describe properties of the equilibrium state. In statistical mechanics, that state can be identified with the most probable macrostate. That most probable macrostate is the state that is actually observed at equilibrium because, in the thermodynamic limit, the fluctuations in macroscopic observables about their most probable values approach zero. The reason for this is that a macroscopic observable has been defined as an average of some type of "local" observable over a large region, and the values of the local observable in different portions of the large region are statistically independent. The statistical independence brings the Law of Large Numbers into play, suppressing the fluctuations.

However, for averages of large numbers of statistically independent variables, the Law of Large Numbers and the Central Limit Theorem give much more detailed information about the fluctuation probability than the simple fact that the fluctuations approach zero. The Law of Large Numbers gives a detailed estimate of the statistical uncertainty, while the Central Limit Theorem gives an explicit formula for the probability distribution of the small fluctuations that remain at any finite stage in the limiting procedure. Corresponding formulas can be given for the fluctuations in large but finite thermodynamic systems. They do not require a knowledge of the Hamiltonian function of the system, but rather are written in terms of thermodynamic potentials, which can be obtained by purely macroscopic measurements. Therefore, these formulas logically form part of the body of thermodynamic theory. Two special cases of the thermodynamic fluctuation formulas will be given, which should make the general principle clear.

We first consider energy fluctuations. If a thermodynamic system is in contact with a reservoir at temperature T, then its energy is not perfectly predictable because only the total energy of the system plus reservoir is exactly constant. We saw in this chapter, using thermodynamic methods, and in Chapter 4, using the canonical ensemble, that the average energy is given in terms of the canonical potential by

$$\bar{E} = -\frac{\partial \phi}{\partial \beta} \tag{5.67}$$

where $\beta = 1/kT$. We also saw in Chapter 4 that the mean-square fluctuation in the energy, $(\Delta E)^2$, could be written in terms of the canonical potential by a similar formula.

$$(\Delta E)^2 = \langle (E - \bar{E})^2 \rangle = \frac{\partial^2 \phi}{\partial \beta^2} \tag{5.68}$$

In thermodynamic fluctuation theory, this formula must be taken as a fundamental postulate. It is the equivalent, for a system containing a large number of particles, of the Law of Large Numbers for the fluctuations in the average value of a large number of independent random variables. In fact, for an ideal gas, it is exactly the same as the Law of Large Numbers (see Problem 5.17), but, for systems of interacting particles, this formula includes effects due to statistical correlations between the particles. Such effects fall outside the range of the Law of Large Numbers. Actually these formulas for $\bar{E}$ and $(\Delta E)^2$ can both be derived from a more general postulate that gives the detailed probability, $P(E)$, that the system will be found with energy E. This formula is the thermodynamic equivalent of the Central Limit Theorem in probability theory. In Problem 4.11, the reader was asked to show, using the canonical ensemble, that

$$P(E) = e^{S^o(E) - \beta E}/Z(\beta) \tag{5.69}$$

where $Z(\beta)$ is the partition function.

For a system that can exchange both energy and particles with a reservoir at affinity α and inverse temperature β, the average values of particle number and energy are given, in terms of the grand potential, $\psi(\alpha, \beta, V)$ by Eq. (5.63).

$$\bar{N} = -\frac{\partial \psi}{\partial \alpha} \quad \text{and} \quad \bar{E} = -\frac{\partial \psi}{\partial \beta} \tag{5.70}$$

The fluctuations in those quantities are given by formulas that will be derived in Chapter 7.

$$(\Delta N)^2 = \langle (N - \bar{N})^2 \rangle = \frac{\partial^2 \psi}{\partial \alpha^2} \tag{5.71}$$

$$(\Delta E)^2 = \langle (E - \bar{E})^2 \rangle = \frac{\partial^2 \psi}{\partial \beta^2} \tag{5.72}$$

The pattern should now be obvious; when certain variables can fluctuate, the first derivatives of the appropriate thermodynamic potential, *which must be written in rational units*, give the average values of the variables while the second derivatives give the mean-square fluctuations. The probability of finding the system with exactly N particles and energy E is given by a formula that will also be derived in Chapter 7 and is obviously analogous with Eq. (5.69).

$$P(N, E) = e^{S^o(N,E) - \alpha N - \beta E} / \Lambda(\alpha, \beta) \tag{5.73}$$

The normalization constant, Λ, for the probability distribution, called the grand partition function, is related to the thermodynamic potential, ψ, in the same way that the cononical partition function is related to the canonical potential. That is, $\psi = \log \Lambda$.

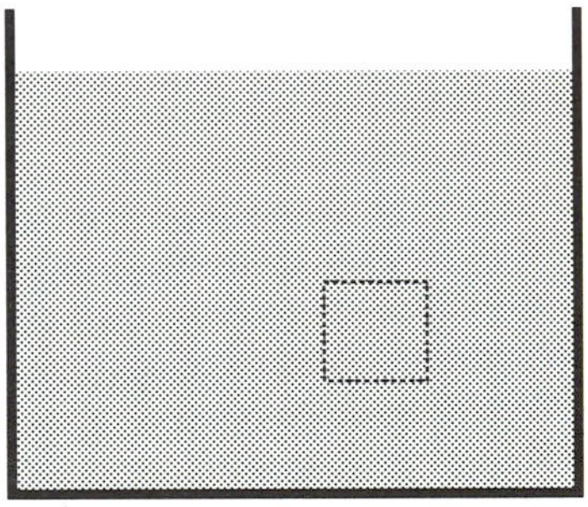

Fig. 5.17 The "sample" is a small subvolume within a larger solution of glucose in water.

5.25 THE AFFINITY OF A DILUTE SOLUTION

From Eq. (5.71), a formula for the affinity of a dilute solution can be derived that has a wide range of useful applications. In order to focus on a specific problem, we consider a large vessel containing a dilute solution of glucose in water (Fig. 5.17). The "sample" is taken as a small subvolume ΔV within the vessel. The rest of the system acts as a reservoir. Since the sample can exchange both particles and energy with the reservoir, the appropriate thermodynamic potential is $\psi(\alpha_g, \alpha_w, \beta, \Delta V)$, where α_g and α_w are the affinities for glucose and water, respectively. The fluctuation in the number of glucose molecules in the sample is given by

$$(\Delta N_g)^2 = \frac{\partial^2 \psi}{\partial \alpha_g^2} = -\frac{\partial N_g}{\partial \alpha_g} \tag{5.74}$$

But, if the solution is very dilute, then the glucose molecules drift in and out of the volume independently and the Poisson distribution is valid [see Eq. (1.29)]. This gives

$$(\Delta N_g)^2 = N_g \tag{5.75}$$

Therefore

$$\frac{\partial \alpha_g}{\partial N_g} = -\frac{1}{N_g} \tag{5.76}$$

This can be integrated to give

$$\alpha_g = -\log N_g + \text{const.} \tag{5.77}$$

The constant may depend on any variables other than N_g. Because α_g is an intensive quantity, it should be possible to express this relation in terms of the density of glucose molecules, $n_g = N_g/\Delta V$. The way to do so is to write the arbitrary constant as $\log \Delta V + f_g^o$, where f_g^o is another arbitrary constant. Then Eq. (5.77) becomes

$$\alpha_g = -\log n_g + f_g^o \tag{5.78}$$

We have been using rational units and, therefore, the affinity given here is defined as $\partial S^o/\partial N_g$ where N_g is the number of glucose molecules. In practical units α_g is defined as $\partial S/\partial N_g$ where $S = kS^o$ and N_g is the number of moles of glucose. The conversion from one system of units to the other adds a factor of the gas constant to the equation. Therefore, in practical units, the equivalent relation is

$$\alpha_g(\text{prac.}) = -R\log n_g + f_g \tag{5.79}$$

where $f_g = Rf_g^o$ and n_g is the molar density.

5.26 CONDITIONAL THERMODYNAMIC POTENTIALS

All the entropy functions that have been used so far have been equilibrium entropy functions. As such, they are functions only of the independently conserved variables N, E, and V. Such equilibrium entropy functions are adequate for doing minimal, bare, thermodynamics. That is, they allow the prediction of the equilibrium distributions of those same conserved quantities under conditions in which transfer of the quantities is allowed between subsystems in a larger isolated system. However, by extending the definition of an entropy function to include functions of other macroscopic variables, a much richer set of physical problems can be attacked with thermodynamic methods. The same extension can also be made for other thermodynamic potentials, such as the canonical and grand potentials.

The equilibrium entropy, $S^o(N, E, V)$ is defined as the logarithm of the number of quantum states of the system for fixed values of N, E, and V. Consider some macroscopic observable, $A(x, p)$, that is not a conserved quantity, such as the electric polarization or the kinetic energy. We could then define a *conditional entropy function* by selecting, from a microcanonical ensemble, only those states that have a specific value, say a, of the observable, $A(x, p)$. The logarithm of the normalization constant for that restricted ensemble would be an entropy that was a function of N, E, V, and a.

$$e^{S(N,E,V,a)} = \frac{1}{N!h^{3N}} \int \delta(E - H)\delta(a - A)\,d^{3N}x\,d^{3N}p \tag{5.80}$$

The conditional entropy function, $S(N, E, V, a)$, is equal to the logarithm of the number of quantum states available to the system, conditioned on the facts that the number of particles is N, the energy is E, the volume is V, and the value of the observable A is a. It might be compared with the conditional probability, $P(y|x)$, which is the probability of y, conditioned on the value of x.

Our first application of these conditional thermodynamic potentials will be to generalize the thermodynamic fluctuation formulas presented before. It is clear from Eq. (1.51) that, if we are using a microcanonical ensemble, then the probability of finding a system in the ensemble with the value a for the observable $A(x, p)$ is equal to

$$
\begin{aligned}
P_A(a) &= \int \delta(E - H)\delta(a - A)\, d^{3N}x\, d^{3N}p \bigg/ \int \delta(E - H)\, d^{3N}x\, d^{3N}p \\
&= e^{S(a)}/\Omega'(E)
\end{aligned}
\tag{5.81}
$$

For other ensembles, $S(a)$ would be replaced by the conditional thermodynamic potential appropriate to the ensemble. Expanding $S(a)$ to second order about $\bar{a}$, one obtains a Gaussian approximation to $P_A(a)$, which can be used to calculate the mean-square fluctuation in A.

$$
(\Delta A)^2 = \left| \frac{\partial^2 S(a)}{\partial a^2} \right|_{\bar{a}}^{-1}
\tag{5.82}
$$

where $\bar{a}$ is the value of a for which $\partial S(a)/\partial a = 0$.

PROBLEMS

5.1   Suppose $S(N, E, V)$ is an entropy function that gives the correct equilibrium states of a particular substance. Let $S^* = S + AE + BV$, where A and B are constants. Show that S^* will not give the correct equilibrium states for the situation shown in Fig. 5.5.

5.2   One mole of a substance satisfies the equations of state

$$T = \lambda \varepsilon^{2/3} v^{-1/2} \quad \text{and} \quad pv = \frac{3}{2} \varepsilon \qquad (5.83)$$

where λ is a constant. Determine $s(\varepsilon, v)$ within an arbitrary constant.

5.3   Consider a substance containing K different types of particles. For example, $K = 3$ for a water, sugar, and salt solution. Let the spatial densities of the particles be called $n_1, \ldots, n_K$. For a single phase of the substance, we can independently vary the K densities and the temperature, but all other physical quantities are then determined by those $K+1$ parameters. Thus, for a single phase, we have $K+1$ *free variables*. For an equilibrium state of the substance containing two phases in contact (a two-phase state), one has the $2K + 1$ variables $n_1', \ldots, n_K', n_1'', \ldots, n_K''$, and β, but these are restricted by $K + 1$ equilibrium conditions, $\alpha_1' = \alpha_1''$, $\ldots$, $\alpha_K' = \alpha_K''$, and $p' = p''$. This leaves only K free variables. Show that, for a state containing P phases in equilibrium, the number of free variables is $F = K - P + 2$. This is the *Gibbs phase rule*. Notice that it says that the triple point of a simple substance has no free variables—there is a unique temperature and pressure at which three phases of a simple substance can be in equilibrium.

5.4   The entropy per unit volume of a system composed of particles of types A and B is

$$s = \tfrac{3}{2}(n_A + n_B) \log\left(\frac{u}{n_A + n_B}\right) - n_A \log(n_A/C_A) - n_B \log(n_B/C_B) \qquad (5.84)$$

where n_A and n_B are the particle densities, u is the energy density, and C_A and C_B are constants. What is the entropy function $S(N_A, N_B, E, V)$?

5.5   Let $s(\varepsilon, v)$ be the molar entropy of a simple substance. In Exercise 5.6 it is shown that, due to the convexity properties of s, the 2×2 matrix

$$M = \begin{bmatrix} \partial^2 s/\partial \varepsilon^2 & \partial^2 s/\partial \varepsilon \partial v \\ \partial^2 s/\partial \varepsilon \partial v & \partial^2 s/\partial v^2 \end{bmatrix} \qquad (5.85)$$

has only negative eigenvalues. For an ideal monatomic gas, compute the eigenvalues explicitly and verify that they are negative.

5.6   Let $s(n, \varepsilon)$ be the entropy per unit volume (the entropy density), expressed as a function of the particle density and the energy density. Consider 2 moles of a simple substance, with an energy $2E$, in a cylinder of volume $2V$. The cylinder is divided into two equal volumes by an immovable partition that has a small hole in it. (a) Assuming that the uniform state is the equilibrium state, derive a convexity condition on the entropy function $s(n, \varepsilon)$ by an argument similar to that used in Section 5.5. (b) Following the method used in Exercises 5.6 and 5.7, show that $\partial \alpha(n, \beta)/\partial n < 0$.

5.7 Show that the entropy function given in Eq. (5.55) satisfies Axiom 5 but does not satisfy Nernst's Law.

5.8 Consider a monatomic ideal gas that can exchange energy and particles with a reservoir. (a) Using Eq. (5.73), show that the energy and particle number fluctuations are not statistically independent by showing that $\langle (N - \bar{N})(E - \bar{E}) \rangle \neq 0$. (b) Show that, if we define another energy variable, Q, by $Q(E, N) = E - N\bar{E}/\bar{N}$, then the fluctuations in N and Q are statistically independent. Give a physical explanation of this result.

5.9 Many ideal gases satisfy equations of state of the form $pV = NkT$ and $E = KNkT$. (For monatomic gases, $K = \frac{3}{2}$, for diatomic gases with only rotational kinetic energy, $K = \frac{5}{2}$, and, for diatomic gases in which rotational and vibrational degrees of freedom can be treated classically, $K = \frac{7}{2}$.) (a) For such an ideal gas, show that, along an adiabatic curve,

$$pV^{(K+1)/K} = \text{const.} \quad \text{and} \quad VE^K = \text{const.} \tag{5.86}$$

(b) Also show that, in the E–V plane, the adiabatic curves are steeper than the isotherms.

5.10 The interaction potential between most molecules has the following characteristics. (1) It is negligible at very large distances; therefore, very dilute gases satisfy the ideal gas equation, in which the interaction potential energy is neglected completely. (2) It is attractive (negative) at moderate distances, typically in the range of a few Ångstroms to a few tens of Ångstroms. Thus, as the density of the gas is increased, the deviations from the ideal gas law are initially in the direction of lower pressure (the particles are pulled toward one another). The attractive part of the interaction is also responsible for the eventual condensation of the gas into a liquid. (3) At small distances, the potential becomes strongly repulsive (large and positive). This is why most liquids are relatively incompressible. Taking these properties into account, van der Waals proposed the following approximate pressure equation of state:

$$p = \frac{RT}{v - v_o} - \frac{a}{v^2} \tag{5.87}$$

where v is the volume per mole and v_o and a are positive constants, different for each substance. A monatomic van der Waals gas is one that satisfies the additional condition that the molar energy ε approaches $\frac{3}{2}RT$ as $v \to \infty$. (a) Determine the molar entropy function $s(\varepsilon, v)$ of a monatomic van der Waals gas. (b) Show that $\varepsilon = \frac{3}{2}RT - a/v$ for a monatomic van der Waals gas. (c) Show that, in terms of the dimensionless variables $\tilde{p} = pv_o^2/a$, $t = RTv_o/a$, $\tilde{v} = v/v_o$, and $\tilde{\varepsilon} = \varepsilon v_o/a$, all monatomic van der Waals gases have the same equation of state, $\tilde{p} = t/(\tilde{v}-1) - 1/\tilde{v}^2$ and $\tilde{\varepsilon} = \frac{3}{2}t - 1/\tilde{v}$. (d) Plot the isotherms for the pressure equation of state in the $\tilde{p}$–$\tilde{v}$ plane over the range $1.3 < \tilde{v} < 10$ for $t = 7/27$, $8/27$, and $9/27$, and show that the first curve has unphysical characteristics. (To understand where the peculiar constant $8/27$ came from, see Exercise 5.20.)

5.11 A highly relativistic classical ideal gas of point particles satisfies the equations of state

$$pV = NkT = \tfrac{1}{3}E \tag{5.88}$$

For such a system, derive the entropy function $S(N, E, V)$.

5.12 In the nineteenth century, using Maxwell's equations, it was possible to show that the pressure exerted on the walls of an evacuated container by the electromagnetic radiation inside was related to the energy density of the radiation by the formula $p = \frac{1}{3}E/V$. The entropy function of that "system" of electromagnetic radiation would be a function $S(E, V)$. Assuming that the energy density is related to the temperature by an equation of the form $E/V = AT^\lambda$, where A and λ are constants, show that λ must be equal to 4. This result was first derived by Ludwig Boltzmann after being guessed at, from an analysis of experimental data, by Josef Stefan.

5.13 In Section 4.12 the canonical potential for a gas of dipolar molecules in an electric field was given as

$$\phi(N, \beta, V) = N\left[\log\left(\frac{V}{N}\right) - \frac{5}{2}\log\beta + \log\left(\frac{\sinh(\beta\mu\mathcal{E})}{\beta\mu\mathcal{E}}\right) + \text{const.}\right] \qquad (5.89)$$

Use this result to calculate the molar entropy of that system as a function of density and temperature.

5.14 For an ideal gas of diatomic molecules with the energy equation $E = \frac{5}{2}NkT$, calculate the grand potential, $\psi(\alpha, \beta, V)$, and verify Eq. (5.63).

5.15 For a system of N quantum mechanical harmonic oscillators, the entropy function $S^o(N, E)$ was shown to be

$$S^o = \left(N + \frac{E}{\hbar\omega}\right)\log\left(N + \frac{E}{\hbar\omega}\right) - N\log N - \frac{E}{\hbar\omega}\log\left(\frac{E}{\hbar\omega}\right) \qquad (5.90)$$

Show that this entropy function satisfies Axiom 5 and Nernst's law. (Note: there is no relevant "volume" for this system.)

5.16 Calculate the canonical potential $\phi(N, \beta, V)$ for the system of Problem 5.14 and verify the inequalities given in Eqs. (5.65) and (5.66).

5.17 For a monatomic ideal gas in the canonical ensemble the total energy of the system is the sum of the kinetic energies of all the particles, which are N statistically independent variables. Thus the Law of Large Numbers can be applied. Show that the Law of Large Numbers predicts the same value for $(\Delta E)^2$ as the thermodynamic fluctuation formula, $(\Delta E)^2 = \partial^2\phi/\partial\beta^2$.

5.18 For one mole of an ideal monatomic gas in equilibrium with a reservoir at STP, use Eq. (5.68) to calculate the ratio of ΔE to E.

5.19 In Section 5.18, it was stated that the source of Nernst's law is the Bose–Einstein or Fermi–Dirac statistics of the particles and not simply the discreteness of quantum mechanical energy levels. To show that this is so, let us consider an imaginary system of noninteracting *boltzons*. They are quantum mechanical point particles that have the usual Hamiltonian operator for an ideal gas, but they differ from real particles in that their quantum states do not have to satisfy any symmetry or antisymmetry conditions. Show that, in the thermodynamic limit, the system will satisfy the equations of state of a classical monatomic ideal gas at any temperature and pressure. From this one can conclude that the entropy function of the system satisfies Axiom 5 but not Nernst's law. (Hint: Look carefully at the section entitled The Classical Limit in the Mathematical Appendix.)

5.20 Most diatomic gases satisfy the equations of state, $pV = N\tau$ and $E = \frac{5}{2}N\tau$, over some appreciable temperature interval. (a) For such a gas, use Equations 5.45 and 5.46 to calculate the affinity, α, (within an arbitrary constant) as a function of p and τ. (b) Now combine the result in (a) with Eq. (5.39) to calculate the entropy function, $S^o(N, E, V)$.

5.21 Consider a container separated into two parts by means of a rigid partition that can pass water molecules but not glucose molecules. On one side is pure water but on the other is a dilute solution of glucose. Using the Gibbs–Duhem equation and Eq. (5.78) for α_g, show that, at equilibrium, the pressure in the solution will exceed that in the pure water by an amount (called the *osmotic pressure*) $\Delta p = n_g kT$. Notice that this is exactly what would be obtained by treating the glucose molecules as an ideal gas, although a derivation of the result using that picture would be very questionable.

5.22 In the canonical ensemble, the probability of finding a system in a state described by the phase-space point (x, p) is

$$P(x, p) = e^{-\beta H(x, p)} / h^K N! Z(N, \beta, V) \tag{5.91}$$

The uncertainty in the energy is defined by

$$(\Delta E)^2 = \int \left[H(x, p) - \bar{E} \right]^2 P(x, p)\, d^K x\, d^K p \tag{5.92}$$

Show that $(\Delta E)^2 = \partial^2 \phi / \partial \beta^2$, where ϕ is the canonical potential.

5.23 In Eq. (5.81), assume that the function $S(a)$ can be expanded about its maximum value, $S(\bar{a})$, as a Taylor series through second order and derive Eq. (5.82). Hint: Use the fact that $\Omega'(E) = \int \exp S(a)\, da$ and make the same expansion of S in both numerator and denominator.

5.24 A wire is held under tension by a force F. Assume that the entropy of the wire is a function $S(E, L)$ of its energy and length. Show that the tension force is related to the entropy function of the wire by $F = -T\, \partial S / \partial L$.

Chapter 6
Applications of Thermodynamics

6.1 INTRODUCTION

In the last chapter, a formulation of thermodynamics was presented that is specifically designed to be convenient for use in statistical mechanics. In this chapter, the subject of thermodynamics will be recast in its more customary form and a variety of practical applications of the theory will be exhibited. But first, the theory of chemical reaction equilibrium will be discussed using a mixture of thermodynamics and statistical mechanics.

6.2 CHEMICAL REACTIONS

The theory of chemical reaction equilibrium is one of the most successful applications of thermodynamics. As an example, we consider a dilute aqueous solution of three chemicals, A_2, B, and AB, which participate in the reaction

$$A_2 + 2B \leftrightarrow 2AB \tag{6.1}$$

The question we will consider is: At equilibrium, how are the concentrations of A_2, B, and AB related? We will assume that the system is isolated and work in a rational system of units. The entropy is some function of N_{A_2}, N_B, N_{AB}, N_w, E, and V, where N_w is the number of molecules of water.

$$S^o = S^o(N_{A_2}, N_B, N_{AB}, N_w, E, V) \tag{6.2}$$

According to the reaction equation, if $2k$ molecules of AB are produced by the reaction, then $dN_{A_2} = -k$ and $dN_B = -2k$. The change in entropy would be

$$
\begin{aligned}
dS^o &= \alpha_{A_2} dN_{A_2} + \alpha_B dN_B + \alpha_{AB} dN_{AB} \\
&= k(2\alpha_{AB} - \alpha_{A_2} - 2\alpha_B)
\end{aligned}
\tag{6.3}
$$

At equilibrium, S^o is a maximum with respect to all allowed changes, and thus dS^o should be zero for all very small values of k. The equilibrium condition is, therefore,

$$\alpha_{A_2} + 2\alpha_B = 2\alpha_{AB} \tag{6.4}$$

Comparing this with Eq. (6.1), the reaction equation, it is quite obvious how to write the equilibrium equation for any other chemical reaction. Notice that this is an exact result, that does not depend, as most of our future analysis will, upon any low density approximation.

For dilute solutions, Eq. (5.78) gives the affinities in terms of the densities.

$$\alpha_i = -\log n_i + f_i^o, \qquad i = A_2, B, \text{ or } AB \tag{6.5}$$

The constant $f_{A_2}^o$ is a function of all intensive variables other than n_{A_2}. But, if all chemical concentrations are very small, then we can, in evaluating f_{A_2}, set n_B and n_{AB} equal to zero. Thus, $f_{A_2}^o$, f_B^o, and f_{AB}^o can be assumed to be functions of pressure and temperature only. Using this relation for α_i in Eq. (6.4) gives

$$\log\left(\frac{n_{AB}^2}{n_{A_2}n_B^2}\right) = 2f_{AB}^o - f_{A_2}^o - 2f_B^o \equiv \log K \tag{6.6}$$

where K is called the *equilibrium constant* for this reaction. The equilibrium condition can thus be written as

$$\frac{n_{AB}^2}{n_{A_2}n_B^2} = K \tag{6.7}$$

This relation is called the *law of mass action* for the reaction $A_2 + 2B \leftrightarrow 2AB$. In general, the reaction

$$aA + bB + \cdots \leftrightarrow rR + sS + \cdots \tag{6.8}$$

will be in equilibrium when

$$\frac{n_R^r n_S^s \cdots}{n_A^a n_B^b \cdots} = K \tag{6.9}$$

where the integers, $a, b, \ldots, r, s, \ldots$ are called the *stoichiometric coefficients* of the reaction and

$$\log K = rf_R^o + sf_S^o + \cdots - af_A^o - bf_B^o - \cdots \tag{6.10}$$

6.3 VAN'T HOFF'S LAW

For a reaction of the form $aA + bB \leftrightarrow rR$, the changes in N_A, N_B, and N_R caused by the reaction are related by $\Delta N_A/a = \Delta N_B/b = -\Delta N_R/r$. The quantity $k = \Delta N_R/r$ measures the number of individual forward reactions minus the number of back reactions that have taken place during any time interval. For a given k, the *heat of reaction*, W, is the amount of energy that would have to be removed from the system in order to maintain it at constant temperature divided by k. It is the energy given off by a single reaction. Therefore, at constant temperature and volume,

$$E(N_A, N_B, N_R) = E(N_A + \Delta N_A, N_B + \Delta N_B, N_R + \Delta N_R) + kW \tag{6.11}$$

This simply states that the initial energy of the system is equal to its final energy plus the energy removed from the system.

Using the facts that $\Delta N_A = -ak$, $\Delta N_B = -bk$, and $\Delta N_R = rk$, Eq. (6.11) can be written, for small k, as

$$a\left(\frac{\partial E}{\partial N_A}\right) + b\left(\frac{\partial E}{\partial N_B}\right) - r\left(\frac{\partial E}{\partial N_R}\right) = W \tag{6.12}$$

But

$$\left(\frac{\partial E}{\partial N_A}\right) = -\frac{\partial^2 \phi}{\partial N_A\, \partial \beta} = -\frac{\partial \alpha_A}{\partial \beta} \tag{6.13}$$

Thus

$$\frac{\partial}{\partial \beta}(r\alpha_R - a\alpha_A - b\alpha_B) = W \tag{6.14}$$

Since $\alpha_i = -\log n_i + f_i^o$ and $(\partial n_i/\partial \beta)_{N,V} = 0$, we see that the temperature dependence of the equilibrium constant is related to the heat of reaction by *Van't Hoff's law*.

$$\frac{\partial \log K}{\partial \beta} = W \tag{6.15}$$

This yields the reasonable result that, if energy is released in a reaction $A + B \rightarrow C + D$, then increasing the temperature will decrease the equilibrium constant, which will increase the proportion of particles in the high-energy forms A and B.

In practical units the equilibrium constant K is defined by Eq. (6.9), with all densities expressed in moles/m^3. Then Van't Hoff's law has the form

$$R\frac{\partial \log K}{\partial(1/T)} = W \tag{6.16}$$

where W is given by Eq. (6.11) with all numbers expressed in moles. That is, W is the molar heat of the reaction.

6.4   IDEAL GAS REACTIONS

An excellent example of the powerful interaction of statistical mechanics and thermodynamics is provided by the study of reaction equilibrium among chemical components of a dilute gas. For an aqueous solution, the system is so complex and strongly interacting that there is little hope of actually carrying out the multidimensional integrations that define the thermodynamic potentials in statistical mechanics. However, for reactions among ideal gas constituents, it is often possible to make a purely theoretical calculation of the equilibrium constant. It is really quite remarkable that the properties of something as intrinsically complicated as a chemical reaction within a system containing a vast number of particles can be successfully predicted by purely mathematical analysis. We will describe two typical cases of ideal gas reaction calculations, although they use the concepts of statistical mechanics and therefore are somewhat out of place in this chapter.

The equilibrium condition in the presence of a chemical reaction

$$aA + bB + \cdots \leftrightarrow rR + sS + \cdots \tag{6.17}$$

is that

$$a\alpha_A + b\alpha_B + \cdots = r\alpha_R + s\alpha_S + \cdots \tag{6.18}$$

Therefore, one can predict the equilibrium condition for reactions among ideal gas components if one can calculate the affinity of each component as a function of its density and the temperature. That can be done by considering the various gases separately.

For an ideal gas of N identical molecules, of any type, the partition function of the system, Z, can be written in terms of the partition function of a single molecule, z. [See Eq. (4.61).]

$$Z = \frac{z^N}{N!} \tag{6.19}$$

The partition function of a single molecule can be further separated into a translational and an internal factor.

$$z = z(\text{trans})z(\text{int}) \tag{6.20}$$

For molecular gases, the translational partition function may always be calculated using classical mechanics.

$$z(\text{trans}) = V/\lambda^3(\beta) \tag{6.21}$$

where λ is the thermal de Broglie wavelength. The internal partition function is given by a sum over the internal energy levels of the molecule or atom, taking into account the degeneracy g_n of each energy eigenvalue.

$$z(\text{int}) = \sum_n g_n e^{-\beta\varepsilon_n} \tag{6.22}$$

The canonical potential of the gas is the logarithm of Z.

$$\begin{aligned} \phi &= \log(z^N/N!) \\ &= \log[z(\text{trans})z(\text{int})/N!] \\ &= N\left[\log(V/N) + 1 + \log\left(z(\text{int})/\lambda^3\right)\right] \end{aligned} \tag{6.23}$$

According to this equation, the affinity of an ideal gas, at any concentration, has the same form as the affinity of a dilute solution.

$$\alpha = \frac{\partial\phi}{\partial N} = -\log n + \log\left[z(\text{int})/\lambda^3\right] \tag{6.24}$$

Equation (6.10), for the equilibrium constant of a chemical reaction within a dilute solution, is valid at all concentrations for reactions among ideal gas constituents, since it was derived using only the above form for the affinity. The term f_i^o in the formula

$$\alpha_i = -\log n_i + f_i^o \tag{6.25}$$

is called the *chemical constant* of the ith component of the gas (or dilute solution). From Eq. (6.24), it is clear that a theoretical calculation of the chemical constant for any ideal gas component requires only a knowledge of the internal partition function for that component. That is,

$$f_i^o(T) = \log\left[z_i(\text{int})/\lambda_i^3\right] \tag{6.26}$$

One can see that, for an ideal gas (but not for a dilute solution), the chemical constant is a function of the temperature alone. It does not depend on the pressure.

6.5   MOLECULAR DISSOCIATION

At any fixed temperature and density, a gas of diatomic molecules reaches dissociation–recombination equilibrium with its atomic constituents. As the temperature is increased at constant density or the density is decreased at constant temperature, the equilibrium shifts in the direction of dissociation. If the dissociation of the diatomic molecule is taken to be the forward reaction, then the chemical reaction equation is of the form

$$AB \leftrightarrow A + B \tag{6.27}$$

For such a reaction, according to Eqs. (6.10) and (6.26), the equilibrium constant is given by

$$\log K = \log\left[z_A(\text{int})/\lambda_A^3\right] + \log\left[z_B(\text{int})/\lambda_B^3\right] - \log\left[z_{AB}(\text{int})/\lambda_{AB}^3\right] \tag{6.28}$$

From the fact that $m_{AB} = m_A + m_B$ and the definition of the thermal de Broglie wavelength, one can easily show that $\lambda_A \lambda_B / \lambda_{AB} = \lambda_\mu$, where μ is the reduced mass, $\mu = m_A m_B/(m_A + m_B)$. Thus, Eq. (6.28) can be written as

$$K = \frac{z_A(\text{int})z_B(\text{int})}{z_{AB}(\text{int})\lambda_\mu^3} \tag{6.29}$$

In evaluating the equilibrium constant, the following approximations will be made:

1. The calculation will initially be restricted to heteronuclear diatomic molecules, and the necessary modification for homonuclear molecules will be given at the end. (See Section 4.11 for the definition of homonuclear and heteronuclear molecules.)

2. Electronic excited states will be ignored in the molecule. Since electronic excitation levels are typically of the order of electron volts and one eV corresponds to a temperature of about 10,000 K, this is generally a good approximation at the dissociation temperatures of diatomic molecules. If the temperature is high enough to require the consideration of electronic excited states, then it is usually high enough to require the consideration of ionization phenomena also.

3. Hyperfine structure of the atomic ground states will be neglected. Hyperfine structure causes a splitting of otherwise degenerate energy levels due to the orientation of the nuclear spin $\mathbf{S}_{\text{nuc}}$ relative to the total electronic spin, $\mathbf{S}$, and the total orbital angular momentum $\mathbf{L}$. The hyperfine levels are very closely spaced, and thus, at any temperature at which a molecule is likely to dissociate, they can be considered as degenerate.

4. The molecular rotational levels will be treated classically. This is an excellent approximation at all temperatures relevant to molecular dissociation.

5. The vibrational levels of the molecule will be approximated by a pure harmonic oscillator spectrum. Because of the rapid convergence of the partition function sum over vibrational energy levels, this is generally a good approximation.

If the zero-energy state is taken to be the state in which all atoms are infinitely separated and at rest, then the ground-state energy values of the isolated atoms are zero and the atomic partition function is given by

$$z_A(\text{int}) = g_o + g_1 e^{-\beta \varepsilon_1} + \cdots \tag{6.30}$$

where ε_n and g_n are the energy and degeneracy of the nth level of atom A. Only those levels need to be taken into account for which ε_n is not much larger than kT. Often this includes only the ground state (see Table 6.2). A similar equation holds for $z_B(\text{int})$.

The internal energy levels of the AB molecule can be approximated by the formula

$$\varepsilon = -\varepsilon_B + \frac{l(l+1)\hbar^2}{2I_{AB}} + n\hbar\omega \tag{6.31}$$

where ε_B is the molecular binding energy (the absolute value of the difference between the ground-state energy of a molecule and the ground-state energies of its constituent atoms), I_{AB} is the moment of inertia of the molecule, $l(l+1)\hbar^2/2I_{AB}$ is the $(2l+1)$-degenerate rotational energy state, and $n\hbar\omega$ is the vibrational excitation energy (the zero-point vibrational energy is included in $-\varepsilon_B$). Since the energy levels are a sum of three independent terms, the internal partition function splits into factors.

$$z_{AB}(\text{int}) = e^{\beta\varepsilon_B} z_{AB}(\text{rot}) z_{AB}(\text{vib}) \tag{6.32}$$

Using the classical approximation, $z_{AB}(\text{rot})$ was calculated in Section 4.11.

$$z_{AB}(\text{rot}) = 2I_{AB}/\beta\hbar^2 \tag{6.33}$$

The vibrational partition function is easily calculated by using the sum for a geometrical series.

$$z_{AB}(\text{vib}) = \sum_{n=0}^{\infty} e^{-n\beta\hbar\omega} = \frac{1}{1 - e^{-\beta\hbar\omega}} \tag{6.34}$$

Combining the two factors gives

$$z_{AB}(\text{int}) = \frac{(2I_{AB}/\beta\hbar^2)e^{\beta\varepsilon_B}}{1 - e^{-\beta\hbar\omega}} \tag{6.35}$$

For homonuclear molecules whose nuclei have zero spin, the internal wave function of the molecule must have the symmetry property,

$$\psi(\mathbf{R}_1, \mathbf{R}_2) = \pm\psi(\mathbf{R}_2, \mathbf{R}_1) \tag{6.36}$$

where the $+$ sign holds if the nuclei (which are always identical for a homonuclear molecule) are Bose–Einstein particles and the $-$ sign holds if they are Fermi–Dirac particles. We will not consider the case of homonuclear molecules with nonzero spin because then the wave function also contains spin variables and the situation is considerably more complicated. It can be shown that the symmetry of a wave function depends only upon the value of ℓ, the angular momentum quantum number. If ℓ is even, then the wave function is unchanged when $\mathbf{R}_1 \rightleftharpoons \mathbf{R}_2$ and if ℓ is odd the wave function changes sign. Thus, for boson nuclei one must sum only over even values of ℓ in calculating $z_{A_2}(\text{rot})$ and for fermion nuclei one must sum only over odd values. In both cases, when the classical approximation is valid the net effect is simply to reduce $z_{A_2}(\text{rot})$ by a factor of $\frac{1}{2}$. Thus, for homonuclear molecules,

$$z_{A_2}(\text{int}) = \frac{(I/\beta\hbar^2)e^{\beta\varepsilon_B}}{1 - e^{-\beta\hbar\omega}} \tag{6.37}$$

	E_B	$\hbar\omega$	$\hbar^2/2I$
O_2	6.0×10^4	2250	2.07
S_2	5.1×10^4	702	0.42
Ca_2	1.4×10^3	65	0.066
Mg_2	3.6×10^3	51	0.13
CaO	4.7×10^4	1056	0.50
CaS	3.7×10^4	824	0.26
MgO	4.8×10^4	902	0.70
MgS	2.8×10^4	761	0.39

TABLE 6.1 Molecular parameters of O, S, Ca, and Mg diatomics, which are composed of atoms with zero nuclear spin; all energies are in Kelvins.

The values of the molecular parameters ε_B, ω, and I for a selection of heteronuclear and homonuclear diatomic molecules are given in Table 6.1.

6.6 IONIZATION–THE SAHA EQUATION

At standard temperature and pressure, hydrogen is a gas of diatomic molecules. If the pressure is kept constant, but the temperature is increased, then, at about 4000 K, the diatomic molecules dissociate and, by 6000 K, the gas is composed almost entirely of individual atoms. If the temperature is increased further, at about 10,000 K, the atoms begin to decompose into their constituent electrons and protons. That ionization process can be analyzed by the same formulas as were used for the dissociation of a diatomic molecule. The chemical reaction is (p and e stand for proton and electron)

$$H \leftrightarrow p + e \tag{6.38}$$

Both the proton and the electron are spin-$\frac{1}{2}$ particles and therefore have two possible spin states. In the absence of a magnetic field, the two spin states have equal energy, which we can take to be zero. Thus

$$z_p = 2 \quad \text{and} \quad z_e = 2 \tag{6.39}$$

The internal partition function of the hydrogen atom is

$$z_H = \sum_n g_n e^{-\beta\varepsilon_n} \tag{6.40}$$

where the nth hydrogenic energy eigenvalue is $\varepsilon_n = -\varepsilon_o/n^2$ with $\varepsilon_o = 2.18 \times 10^{-18}$ J. The degeneracy of the nth level is $g_n = 4n^2$, where the factor of 4 comes from the two possible spin states of the proton and electron in the hydrogen atom. We will show that, at the temperatures at which ionization occurs, it is a good approximation to neglect all states in the sum except the ground state. Then

$$z_H = 4e^{\beta\varepsilon_o} \tag{6.41}$$

Because of the large ratio of the proton mass to the electron mass, the reduced mass is negligibly different from the electron mass. Therefore, using Eq. (6.29), one gets an equation first derived by Saha in 1921.

$$\frac{n_p n_e}{n_H} = K(T) = \frac{z_p z_e}{z_H \lambda_e^3} = \frac{e^{-\beta\varepsilon_o}}{\lambda_e^3} \tag{6.42}$$

atom	levels
H	$^2S_{1/2}(0)$, $^2P_{1/2}(118349)$, $^2S_{1/2}(118349)$, $^2P_{3/2}(118349)$
D	$^2S_{1/2}(0)$, $^2P_{1/2}(118381)$, $^2S_{1/2}(118381)$, $^2P_{3/2}(118382)$
He	$^1S_0(0)$, $^2S_1(229982)$, $^1S_0(239221)$
Li	$^2S_{1/2}(0)$, $^2P_{1/2}(21443)$, $^2P_{3/2}(21443)$
Be	$^1S_0(0)$, $^3P_0(31622)$, $^3P_1(31623)$, $^3P_2(31626)$
B	$^2P_{1/2}(0)$, $^2P_{3/2}(23.02)$, $^4P_{1/2}(41443)$, $^4P_{3/2}(41450)$, $^4P_{5/2}(41459)$
C	$^3P_0(0)$, $^3P_1(23.60)$, $^3P_2(62.59)$, $^1D_2(14666)$
N	$^4S_{3/2}(0)$, $^2D_{5/2}(27657)$, $^2D_{3/2}(27668)$
O	$^3P_2(0)$, $^3P_1(228.04)$, $^3P_0(325.87)$, $^1D_2(22830)$
F	$^3P_{3/2}(0)$, $^3P_{1/2}(581.25)$, $^4P_{5/2}(147337)$
Ne	$^1S_0(0)$, $^XX_2(192854)$
Na	$^2S_{1/2}(0)$, $^2P_{1/2}(24395)$, $^2P_{3/2}(24420)$
Mg	$^1S_0(0)$, $^3P_0(31436)$, $^3P_1(31465)$, $^3P_2(31524)$
Al	$^2P_{1/2}(0)$, $^2P_{3/2}(161.20)$, $^2S_{1/2}(36469)$
Si	$^3P_0(0)$, $^3P_1(111.00)$, $^3P_2(321.27)$, $^1D_2(9063)$
P	$^4S_{3/2}(0)$, $^2D_{3/2}(16347)$, $^2D_{5/2}(16368)$
S	$^3P_2(0)$, $^3P_1(571.0)$, $^3P_0(825.3)$, $^1D_2(13292)$
Cl	$^2P_{3/2}(0)$, $^2P_{1/2}(1268)$, $^4P_{5/2}(103523)$
A	$^1S_0(0)$, $^XX_2(134010)$
K	$^2S_{1/2}(0)$, $^2P_{1/2}(18682)$, $^2P_{3/2}(18765)$
Ca	$^1S_0(0)$, $^3P_0(21808)$, $^3P_1(21883)$, $^3P_2(22036)$

TABLE 6.2 A table of the low-lying quantum states of the first 20 elements (including deuterium). The states are denoted with the usual LS-coupling symbol, which gives $^{2S+1}L_J$, where S is the total spin, L is the total orbital angular momentum (in spectroscopist's peculiar notation), and J is the combined angular momentum of the state. The symbol ^XX_J means that LS-coupling does not work for this state. The number in parentheses is the energy of the state in Kelvins. The degeneracy of each level is $2J + 1$

We must still show that the neglect of excited atomic states is justified. At 20,000 K, a temperature at which the gas is almost totally ionized, the ratio of the probability of finding an atom in the first excited state to that of finding it in the ground state is

$$\frac{\mathbf{P}[n = 2]}{\mathbf{P}[n = 1]} = 4e^{-(\varepsilon_2 - \varepsilon_1)/kT} \approx 0.01 \tag{6.43}$$

The factor of four in this formula is due to the fourfold degeneracy of the $n = 2$ state.

6.7   THE ENERGY REPRESENTATION

Among the variables N, E, V, and S, the entropy stands out as being of a fundamentally different character. N, E, and V are perfectly well defined for a system of three particles in a hard-walled box. They are mechanical or geometrical variables that have meanings outside of the domain of statistical mechanics or thermodynamics. In contrast, S is a statistical or probabilistic concept that has no obvious extension to small deterministic systems. Therefore, it was natural to base thermodynamics on a study of the function $S(N, E, V)$, which gives the new variable in terms of the old, familiar variables. However, for historical reasons, this is not the usual approach to thermodynamics. Thermodynamic information is usually presented in what is called the energy representation, in which E is given as a function of S, V, and N.

$$E = E(S, V, N) \tag{6.44}$$

(According to the definition of a representation that was given in Section 5.22, the energy representation should more properly be called the (S, V, N) representation.) The functions $S(N, E, V)$ and $E(S, V, N)$ contain the same physical information expressed in different forms. Given the values of three of the variables N, E, V, S, either function would allow one to determine the value of the fourth.

The partial derivatives of the function $E(S, V, N)$ can be expressed in terms of measurable properties of the system in the following way. We first write Eq. (5.39) for dS in the form

$$dS = \alpha dN + \frac{dE + p\,dV}{T} \tag{6.45}$$

Solving this for dE gives

$$dE = T\,dS - p\,dV - T\alpha dN \tag{6.46}$$

But it is a mathematical identity that

$$dE(S, V, N) = \left(\frac{\partial E}{\partial S}\right)_{VN} dS + \left(\frac{\partial E}{\partial V}\right)_{SN} dV + \left(\frac{\partial E}{\partial N}\right)_{SV} dN \tag{6.47}$$

which shows immediately that

$$\left(\frac{\partial E}{\partial S}\right)_{NV} = T, \qquad \left(\frac{\partial E}{\partial V}\right)_{SN} = -p \tag{6.48}$$

and

$$\left(\frac{\partial E}{\partial N}\right)_{SV} = -T\alpha \equiv \mu \tag{6.49}$$

where μ is called the *chemical potential* of the substance. In terms of the chemical potential, dE can be expressed as

$$dE = T\,dS - p\,dV + \mu\,dN \tag{6.50}$$

6.8   *F*, *H*, AND *G* IN THE ENERGY REPRESENTATION

S, V, and N are all extensive variables. By means of Legendre transformations, any one or two of them may be replaced by corresponding intensive variables. In considering such transformations we will restrict ourselves to the case of a simple substance and assume that N is fixed and equal to one mole. Therefore, the function $E(S, V)$ is the molar energy as a function of the molar entropy and the molar volume. Then, with $dN = 0$,

$$dE(S, V) = T\,dS - p\,dV \tag{6.51}$$

In order to eliminate S in favor of T, we define a function, called the *Helmholtz free energy*, by

$$F(T, V) \equiv E - TS \tag{6.52}$$

Then Eq. (6.51) for dE gives

$$dF(T, V) = -S\,dT - p\,dV \tag{6.53}$$

To transform from (S, V) to (S, p) we define the *enthalpy* function

$$H(S, p) \equiv E + pV \tag{6.54}$$

which satisfies the differential relation

$$dH(S, p) = T\,dS + V\,dp \tag{6.55}$$

and lastly, to transform from (S, V) to (T, p) we define the *Gibbs free energy*

$$G(T, p) \equiv E - TS + pV \tag{6.56}$$

which satisfies the differential relation

$$dG(T, p) = -S\,dT + V\,dp \tag{6.57}$$

6.9   THE MAXWELL RELATIONS

The equation $dE = T\,dS - p\,dV$ implies that $(\partial E/\partial S)_V = T$ and $(\partial E/\partial V)_S = -p$. But, these relations allow the mixed partial derivative to be written in two equivalent forms, $\partial^2 E/\partial S\partial V = \partial^2 E/\partial V\partial S$, giving the *Maxwell relation*

$$\left(\frac{\partial T}{\partial V}\right)_S = -\left(\frac{\partial p}{\partial S}\right)_V \tag{6.58}$$

In the same way, the facts that $(\partial F/\partial T)_V = -S$ and $(\partial F/\partial V)_T = -p$ lead to the Maxwell relation

$$\left(\frac{\partial S}{\partial V}\right)_T = \left(\frac{\partial p}{\partial T}\right)_V \tag{6.59}$$

and corresponding relations, involving H and G, give two further Maxwell relations

$$\left(\frac{\partial T}{\partial p}\right)_S = \left(\frac{\partial V}{\partial S}\right)_p \tag{6.60}$$

and

$$\left(\frac{\partial S}{\partial p}\right)_T = -\left(\frac{\partial V}{\partial T}\right)_p \tag{6.61}$$

6.10 THE USE OF THERMODYNAMIC TABLES

In making any practical use of thermodynamics one must refer to standard compilations of physical properties, such as the *Handbook of Chemistry and Physics*, to obtain the essential thermodynamic functions for any particular substance. In most cases, the physical properties are tabulated as functions of temperature and pressure, for the simple reason that, experimentally, those are usually the variables that are easiest to accurately control. Not every conceivable physical property is tabulated. For a given substance, one can typically find the molar entropy and the molar volume as functions of p and T. Also commonly available are the constant pressure molar heat capacity,

$$C_p = \left(\frac{\partial Q}{\partial T}\right)_p = T\left(\frac{\partial S}{\partial T}\right)_p \qquad (6.62)$$

the temperature coefficient of expansion,

$$\beta_p = V^{-1}\left(\frac{\partial V}{\partial T}\right)_p \qquad (6.63)$$

and the isothermal compressibility,

$$\kappa_T = -V^{-1}\left(\frac{\partial V}{\partial p}\right)_T \qquad (6.64)$$

The five quantities S, V, C_p, β_p, and κ_T are equivalent to the set of first and second derivatives of the molar Gibbs free energy. In particular, $S = -(\partial G/\partial T)$, $V = (\partial G/\partial p)$, $C_p = -T(\partial^2 G/\partial T^2)$, $\beta_p = (\partial^2 G/\partial T\,\partial p)/V$, and $\kappa_T = -(\partial^2 G/\partial p^2)/V$, as can be verified from Eq. (6.57). Therefore, in terms of these five quantities, it should be possible to express any quantity that involves first and second derivatives of any thermodynamic potentials as a function of p and T. This is enough to evaluate almost any quantity that is likely to appear in a normal thermodynamic analysis.

6.11 TRANSFORMATION OF VARIABLES

The thermodynamic space of a simple substance is three dimensional. However, when we fix the particle number at one mole, as we have, the set of possible equilibrium states becomes two dimensional. That means that fixing the values of any two independent thermodynamic parameters is enough to define the state of the substance and hence to determine the values of all other thermodynamic variables. It is not uncommon, in practical calculations, for very strange combinations of variables to appear naturally. For example, in certain processes one can show that the enthalpy H remains constant (see Problem 6.12). Determining how the temperature varies with the pressure in such a process requires that one calculate the partial derivative $(\partial T/\partial p)_H$, which is just the partial derivative of the function $T(p, H)$. It is therefore important that one be able to convert almost any conceivable combination of partial derivatives of thermodynamic variables to some standard form in which they may be evaluated. In this section a systematic procedure will be given to convert any expression involving partial derivatives of the thermodynamic quantities p, T, S, V, E, F, H, G, and μ into an expression involving only p and T and the five "handbook variables," S, V, C_p, β_p, and κ_T. This is equivalent to evaluating an arbitrary partial derivative in the p-T representation. The procedure makes use of the three types of information listed here.

1. Differential relations

$$\begin{aligned}
dE &= T\,dS - p\,dV \\
dF &= -S\,dT - p\,dV \\
dH &= T\,dS + V\,dp \\
dG &= -S\,dT + V\,dp
\end{aligned}$$

(6.65)

Of these, only the first needs to be memorized; the others are obtained by Legendre transformations, which give predictable sign changes in the terms.

2. Definitions of C_p, β_p, and κ_T

$$C_p = T\left(\frac{\partial S}{\partial T}\right)_p$$

$$\beta_p = V^{-1}\left(\frac{\partial V}{\partial T}\right)_p = -V^{-1}\left(\frac{\partial S}{\partial p}\right)_T$$

$$\kappa_T = -V^{-1}\left(\frac{\partial V}{\partial p}\right)_T$$

(6.66)

In the formula for β_p one of the Maxwell relations has been used.

3. Partial derivative identities Given any set of variables x, y, z, and w that have the property that the values of any two of the variables determine those of the others, then

$$\left(\frac{\partial x}{\partial y}\right)_z = \frac{1}{\left(\dfrac{\partial y}{\partial x}\right)_z}$$

(6.67)

$$\left(\frac{\partial x}{\partial y}\right)_z = -\frac{\left(\dfrac{\partial z}{\partial y}\right)_x}{\left(\dfrac{\partial z}{\partial x}\right)_y}$$

(6.68)

and

$$\left(\frac{\partial x}{\partial y}\right)_z = \frac{\left(\dfrac{\partial x}{\partial w}\right)_z}{\left(\dfrac{\partial y}{\partial w}\right)_z}$$

(6.69)

The first identity follows from the fact that, once the value of z has been fixed, x becomes a function of y and vice versa. But, for functions of one variable, $dx/dy = (dy/dx)^{-1}$.

The second identity can be derived from the differential relation

$$dz = \left(\frac{\partial z}{\partial x}\right)_y dx + \left(\frac{\partial z}{\partial y}\right)_x dy$$

(6.70)

If we set dz equal to zero, then the ratio of dx to dy is just $(\partial x/\partial y)_z$. This gives

$$\left(\frac{\partial z}{\partial x}\right)_y \left(\frac{\partial x}{\partial y}\right)_z + \left(\frac{\partial z}{\partial y}\right)_x = 0$$

(6.71)

which is equivalent to the desired identity.

The third identity can be derived by considering two neighboring points that have the same values of z. The changes in x, y, and w are related by

$$dx = \left(\frac{\partial x}{\partial w}\right)_z dw, \quad dy = \left(\frac{\partial y}{\partial w}\right)_z dw, \quad \text{and} \quad \left(\frac{\partial x}{\partial y}\right)_z = \frac{dx}{dy} \tag{6.72}$$

from which the identity follows immediately.

Our task is now to convert any partial derivative expression, such as $(\partial \mu/\partial p)_H$ or $(\partial S/\partial V)_p$, into an expression involving p, T, S, V, and derivatives of the form

$$\left(\frac{\partial(S \text{ or } V)}{\partial(p \text{ or } T)}\right)_{(T \text{ or } p)} \tag{6.73}$$

which can all be written in terms of handbook quantities. We will do this in two stages. Calling p, T, S, and V the *good* variables and E, F, H, G, and μ the *bad* variables, we will first eliminate all bad variables in favor of good variables. Then we will transform any resulting partial derivatives, which will only involve good variables, to the specific form shown in Eq. (6.73).

In eliminating the bad variables, the chemical potential μ is a special case. It is eliminated by using the fact that (see Problem 6.10) the chemical potential of any substance is always equal to the molar Gibbs free energy. Thus μ is simply replaced by G. We now introduce the convention that a and b denote good variables, α denotes a bad variable, and x, y, and z may be either good or bad.

To transform a term of the form $(\partial \alpha/\partial x)_y$, one uses the corresponding differential relation. For example, the relation, $dH = T\,dS + V\,dp$ gives

$$\left(\frac{\partial H}{\partial x}\right)_y = T\left(\frac{\partial S}{\partial x}\right)_y + V\left(\frac{\partial p}{\partial x}\right)_y \tag{6.74}$$

Notice that this will never introduce any new bad variables.

To eliminate a term of the form $(\partial a/\partial \alpha)_x$, one uses the first partial derivative identity.

$$\left(\frac{\partial a}{\partial \alpha}\right)_x = \frac{1}{\left(\frac{\partial \alpha}{\partial a}\right)_x} \tag{6.75}$$

followed by the appropriate differential relation to eliminate α. These two procedures will eliminate bad variables as either numerators or denominators in partial derivative expressions. The only remaining possibilities are expressions of the form $(\partial a/\partial b)_\alpha$. These are converted by the second partial derivative identity.

$$\left(\frac{\partial a}{\partial b}\right)_\alpha = -\frac{\left(\frac{\partial \alpha}{\partial b}\right)_a}{\left(\frac{\partial \alpha}{\partial a}\right)_b} \tag{6.76}$$

followed by the use of a differential relation to eliminate α entirely.

With these procedures we can reduce any partial derivative to ones containing only p, T, S, and V. If these are not in the desired form, shown in Eq. (6.73), then they can be converted to that form as follows.

For an expression of the form

$$\left(\frac{\partial(p \text{ or } T)}{\partial(S \text{ or } V)}\right)_{(T \text{ or } p)} \tag{6.77}$$

we use the first partial derivative identity.

For an expression of the form

$$\left(\frac{\partial(p \text{ or } T}{\partial(T \text{ or } p)}\right)_{(S \text{ or } V)} \tag{6.78}$$

we use the second partial derivative identity. The remaining possibilities are expressions of the form

$$\left(\frac{\partial(S \text{ or } V)}{\partial(p \text{ or } T)}\right)_{(V \text{ or } s)} \quad \text{or} \quad \left(\frac{\partial(S \text{ or } V)}{\partial(V \text{ or } S)}\right)_{(p \text{ or } T)} \tag{6.79}$$

The first of these is converted to the standard form by the procedure illustrated.

$$\left(\frac{\partial S}{\partial p}\right)_V \equiv \frac{\partial S(p, V)}{\partial p} = \frac{\partial S(p, T(p, V))}{\partial p} = \left(\frac{\partial S}{\partial p}\right)_T + \left(\frac{\partial S}{\partial T}\right)_p \left(\frac{\partial T}{\partial p}\right)_V \tag{6.80}$$

which reduces it to the immediately previous case. The second case requires the third partial derivative identity.

$$\left(\frac{\partial S}{\partial V}\right)_T = \frac{\left(\dfrac{\partial S}{\partial p}\right)_T}{\left(\dfrac{\partial V}{\partial p}\right)_T} \tag{6.81}$$

6.12   A FORMULA FOR $C_p - C_v$

An important example of the foregoing type of analysis is the following derivation of a formula for the difference between the constant pressure and constant volume molar specific heats. The definition of C_v is

$$C_v = \left(\frac{\partial Q}{\partial T}\right)_V = T\left(\frac{\partial S}{\partial T}\right)_V \tag{6.82}$$

But

$$\begin{aligned}
\left(\frac{\partial S}{\partial T}\right)_V &= \frac{\partial S(T, V)}{\partial T} = \frac{\partial S(T, p(V, T))}{\partial T} \\
&= \left(\frac{\partial S}{\partial T}\right)_p + \left(\frac{\partial S}{\partial p}\right)_T \left(\frac{\partial p}{\partial T}\right)_V
\end{aligned} \tag{6.83}$$

Using the facts that

$$T\left(\frac{\partial S}{\partial T}\right)_p = C_p, \tag{6.84}$$

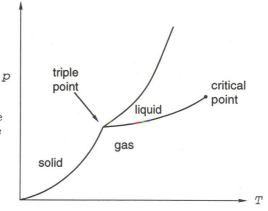

Fig. 6.1   A typical phase diagram. The gas–liquid transition line extends from the triple point to the critical point, at which the properties of the gas and the liquid become indistinguishable and thus the phase transition disappears.

$$\left(\frac{\partial S}{\partial p}\right)_T = -V\beta_p, \tag{6.85}$$

and

$$\left(\frac{\partial p}{\partial T}\right)_V = -\frac{(\partial V/\partial T)_p}{(\partial V/\partial p)_T} = \frac{\beta_p}{\kappa_T} \tag{6.86}$$

we obtain the desired formula for $C_p - C_v$.

$$C_p - C_v = TV\beta_p^2/\kappa_T \tag{6.87}$$

6.13   THE CLAUSIUS–CLAPEYRON EQUATION

The phase diagram of an imaginary substance is shown in Fig. 6.1. At most combinations of temperature and pressure, the equilibrium state of the substance is a single uniform phase: solid, liquid, or gas. These regions of single-phase states are separated by two-phase *coexistence curves*. It is impossible to compute the phase diagram of any substance using the laws of thermodynamics. The equilibrium phase at given p and T depends upon the details of the microscopic structure of the substance. However, the laws of thermodynamics do provide an important relation between the slope of any coexistence curve and the changes in S and V associated with the corresponding phase transition. In deriving this relation, called the Clausius–Clapeyron equation, we will consider the gas–liquid condensation curve. At any point (p, T) on that curve, a two-phase system composed of gas and liquid could remain at equilibrium within a single container. Besides the obvious requirements that the temperatures and pressures in the two phases be equal, the third condition for equilibrium between two phases is that the affinities of gas and liquid be equal. Since the temperatures are equal, this equality of the affinities implies the equality of the chemical potentials. Thus, all along the condensation curve,

$$\mu_L(p, T) = \mu_G(p, T) \tag{6.88}$$

But, as noted earlier, the chemical potential of any substance is equal to its molar Gibbs free energy, $G(p, T)$. If we consider two neighboring points on the condensation curve, then the condition that $d\mu_L = d\mu_G$, and Eq. (6.65), which states that $d\mu = -S\,dT + V\,dp$, where S and V are the molar entropy and volume, implies that

$$-S_L\,dT + V_L\,dp = -S_G\,dT + V_G\,dp \tag{6.89}$$

Solving for dp/dT gives the slope of the condensation curve in the p-T plane.

$$\frac{dp}{dT} = \frac{S_G - S_L}{V_G - V_L} \tag{6.90}$$

In converting one mole of liquid to one mole of gas at the same temperature, an amount of heat energy, called the *latent heat of evaporation L*, must be added to the substance. Since the process is carried out at fixed temperature, the equation $dS = dQ/T$ implies that $S_G - S_L = L/T$. Thus the Clausius–Clapeyron equation may be expressed in the alternative form

$$\frac{dp}{dT} = \frac{L}{T(V_G - V_L)} \tag{6.91}$$

A useful set of approximations for many substances is: (1) to consider L as a constant, (2) to ignore V_L in comparison with V_G, and (3) to use the ideal gas law for V_G, namely $pV_G = RT$, where R is the molar gas constant. Then, we obtain the equation

$$\frac{dp}{dT} = \frac{Lp}{RT^2} \tag{6.92}$$

which can be easily integrated by first rewriting it in the form

$$d(\log p) = \frac{dp}{p} = \frac{L}{R}\frac{dT}{T^2} = -\frac{L}{R}d\left(\frac{1}{T}\right) \tag{6.93}$$

yielding

$$\log p(T) = \text{const.} - \frac{L}{RT} \tag{6.94}$$

Figure 6.2 compares this approximate result with the experimental condensation curves for a number of substances.

6.14   THE CLASSIFICATION OF PHASE TRANSITIONS

The entropy function has a different form for each bulk phase of a substance. All other thermodynamic potentials, such as the canonical potential or the Helmholtz free energy, are defined in terms of the entropy, and therefore the same statement can be made for them; they have different forms for each phase of a substance. In fact, this can be used as the defining characteristic of a phase transition. For example, if we look at one mole of a simple substance, in the p-T representation, then the natural thermodynamic potential is the Gibbs free energy, $G(p,T)$. In Fig. 6.2, on any phase transition line, the values of $G(p,T)$ of the two adjacent phases must be equal. That is, $G(p,T)$ must be a continuous function over the whole plane. Away from the phase transition lines, the derivatives of G give the molar entropy and the molar volume, and they must therefore also be continuous. If we run along any phase transition line, then the rates of change of G on the two sides of the line must be equal. That is, the directional derivative of G in the direction along a phase transition line must also be continuous. This is exactly the condition that is expressed in the Clausius–Clapeyron equation. However, the derivatives of G in the direction normal to the line do not have to be continuous. A long time ago, it was suggested by P. Ehrenfest that phase transitions be classified according to

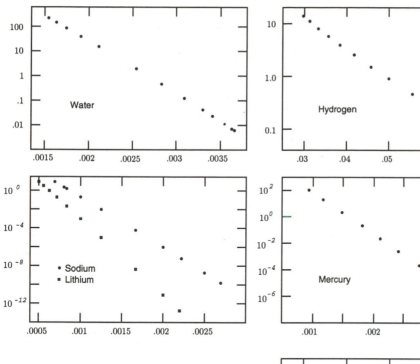

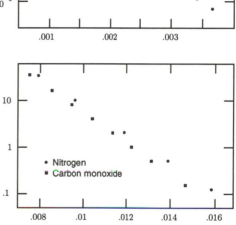

FIGURE 6.2 The condensation curves for a variety of different substances. The vertical axis shows the *vapor pressure* (in bars), on a log scale, in a two-phase equilibrium state. The horizontal axis gives $1/T$, the inverse temperature. According to Eq. (6.94), the points should lie on a line whose slope gives the latent heat of the transition. Notice that the pressure ranges over which the linear approximation is accurate are typically very large.

the behavior of the normal derivatives of G as we cross the phase transition line. If the normal derivative of G is discontinuous across the phase transition line then the phase transition was called a *first-order transition*. If the first derivative of G is continuous, but the second derivative has a discontinuity at the phase transition line, then the phase transition was called a *second-order transition*, and so on for third and higher derivatives. Unfortunately, nature has not cooperated with the Ehrenfest scheme. Except for the important case of first-order phase transitions, most of the phase transitions that actually occur do not fit into it.

It is now common practice to separate phase transitions into two classes. First-order phase transitions are those in which the microscopic structure of a substance undergoes a discontinuous change across the phase transition line. This class happens to agree with the old Ehrenfest class of first-order transitions. Second-order transitions are all those in which all microscopic structural properties of the substance vary continuously (but perhaps not differentiably) across the phase transition line. The most important subset of second-order phase transitions are the *critical-point* transitions. A critical-point transition separates two phases in which

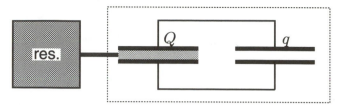

Fig. 6.3 The system, enclosed in the dashed-line box, consists of a dielectric capacitor and a vacuum capacitor in parallel, in contact with a thermal reservoir.

there are different numbers of thermodynamic variables. A ferromagnetic solid gives an excellent example. For a ferromagnet, above a certain temperature, called the *Curie temperature*, the pressure and temperature uniquely define all properties of the equilibrium state. If T is less than the Curie temperature, then the substance spontaneously magnetizes itself. The direction of the magnetization cannot be predicted from the values of p and T and therefore must be introduced as a new independent thermodynamic variable. However, the amount of the magnetization goes continuously to zero as T approaches the Curie temperature—thus the microscopic properties vary continuously across the transition and so it is a second-order phase transition. A more detailed discussion of this topic will be given in Chapter 9.

6.15 THERMODYNAMICS WITH E AND B FIELDS

If a substance is placed in an electric field, it becomes polarized. That is, the atoms or molecules that make up the substance develop electric dipole moments in the direction of the electric field, or, if they have permanent dipole moments, these tend to line up with the field. The dipole moment per unit volume is called the *electric polarization* and is denoted **P**. In a similar way, if a substance is placed in a magnetic field, it develops a magnetization **M** (a magnetic dipole moment per unit volume). The thermodynamic functions of the substance, such as the entropy or the free energy, will be affected by the degree of electric or magnetic polarization. In this section, the thermodynamic conditions that determine the equilibrium values of **P** and **M** will be derived. Thermodynamics in the presence of fields is a fairly subtle subject. Therefore, in order to deepen the reader's understanding of it, the calculation for the electric field case will be carried out twice, at different levels of sophistication. Naturally, we will obtain the same equilibrium condition by the two methods. Another reason for repeating the calculation is that the magnetic field case requires the more sophisticated method, because the simple method makes use of conserved electric charge and no equivalent magnetic charge exists.

We consider the system shown in Fig. 6.3. A slab of dielectric, of area A and thickness ℓ, lies between two capacitor plates. It is in thermal contact with a reservoir at temperature T. The capacitor plates are connected to a vacuum capacitor of capacity C_o. The system, shown by the dashed line, is composed of the two capacitors but not the reservoir. The Helmholtz free energy of a system is related to the canonical potential by the simple relation $F = -kT\phi$. Since the canonical potential is a maximum at equilibrium, it is clear that, in the equilibrium state, the value of F must be a minimum. Therefore, the equilibrium condition for the system is that the total free energy, $F = U - TS$ (where we have changed the symbol denoting the energy from E to U in order to prevent confusion with the electric field $\mathbf{E}$) must be a minimum in comparison with any other state to which the system might go without violating fundamental conservation laws. The free energy of the dielectric capacitor may be calculated by integrating f, the free energy per unit volume of the dielectric material (which includes the electric field energy density), over the volume of the dielectric. The free energy density of a piece of the dielectric depends on its density n and the temperature T, but it also depends on the strength of the electric field passing through it, since the existence of the field will add a field energy term and will also affect the microscopic distribution of the particles in the material. We could describe the electric field dependence by making f a function of the local electric field E, but we could also use other, equivalent, variables, such as the polarization P or the displacement field D to describe the field dependence. The most convenient formulation is obtained if f is taken as a function of n, T, and D. To calculate the total free energy of the system we must add the energy of the vacuum capacitor, which has no entropy. Thus

$$F = \ell A f(n, T, D) + \frac{q^2}{2C_o} \tag{6.95}$$

If the system is at equilibrium with the indicated charges on the plates, then F must be unchanged, to first order, by any small shift of charge dq from the dielectric capacitor to the vacuum capacitor. Thus

$$dF = \ell A \frac{\partial f}{\partial D} dD + \frac{q\, dq}{C_o} = 0 \tag{6.96}$$

But, by the definition of D, $D = Q/A$, and thus, $dD = dQ/A = -dq/A$, by charge conservation. This implies that

$$dF = \left(-\ell \frac{\partial f}{\partial D} + V\right) dq = 0 \tag{6.97}$$

for any infinitesimal dq, where $V = q/C_o$ is the voltage on the vacuum capacitor. We note that the voltage on the vacuum capacitor is the same as the voltage on the dielectric capacitor, which is given by $V = \ell E$, where E is the electric field within the dielectric material. We therefore obtain, as our condition for equilibrium, that the field within the dielectric is related to its free energy density by

$$E = \frac{\partial f(n, T, D)}{\partial D} \tag{6.98}$$

This equation shows that $\mathbf{D}$ and $\mathbf{E}$ are thermodynamically conjugate variables. Shortly it will be shown that $\mathbf{H}$ and $\mathbf{B}$ are also conjugate variables. These facts are really quite remarkable when one recalls that the definitions of these fields were devised by Maxwell with no consideration of thermodynamics at all.

For most materials, the displacement vector is proportional to the local electric field, the proportionality constant being ϵ, the dielectric constant of the substance. This would say that

$$\frac{\partial f(n, T, \mathbf{D})}{\partial \mathbf{D}} = \frac{\mathbf{D}}{\epsilon} \tag{6.99}$$

an equation that can be easily integrated to give

$$f(n, T, D) = f_o(n, T) + \frac{D^2}{2\epsilon}$$
$$= f_o(n, T) + \frac{1}{2}ED \tag{6.100}$$

where $f_o(n, T)$ is the Helmholtz free energy density of the material with no electric field.

We will now consider a more general situation, in which the field varies from point to point. One aim of this analysis is that it will illustrate the techniques required for doing thermodynamic calculations on spatially nonuniform systems. We consider a sample occupying a volume V_S in the vicinity of some fixed charges that give rise to an electric field (see Fig. 6.4). The Helmholtz free energy of the system, $F = U - TS$, will include the electric field energy, and thus F could be written in the form (E_3 is three-dimensional Euclidean space)

$$F = \frac{1}{2}\epsilon_o \int_{E_3 - V_S} E^2 d^3\mathbf{r} + \int_{V_S} f\, d^3\mathbf{r} \tag{6.101}$$

where f, the Helmholtz free energy per unit volume, will vary from point to point because of the variation in the electric field strength. The restriction of the electric field integral to the region outside the dielectric material is inconvenient for the analysis. We can eliminate it by defining another function $\tilde{f}$, which it is best to consider as a function of n, T, and $\mathbf{P}$ ($\mathbf{P}$ is the polarization of the dielectric), by

$$\tilde{f}(n, T, \mathbf{P}) \equiv f + \frac{\epsilon_o}{2}E^2 \tag{6.102}$$

With this definition,

$$F = \frac{\epsilon_o}{2} \int_{E_3} E^2 d^3\mathbf{r} + \int_{V_S} \tilde{f}\, d^3\mathbf{r} \tag{6.103}$$

For isotropic substances, $\tilde{f}$ depends only on the magnitude of $\mathbf{P}$, but for many crystalline solids it depends also on the orientation of the polarization vector with respect to the crystal axes. The electric field, in general, varies from point to point, and therefore the polarization at equilibrium would be expected to do so also. Thus the equilibrium value of the polarization is given by the vector function $\mathbf{P}(\mathbf{r})$ that minimizes the free energy defined in Eq. (6.103). This is obviously a more difficult minimization problem than any we have encountered before. It is solved by the following device.

Suppose $\mathbf{P}(\mathbf{r})$ is the function that minimizes F. Consider the change in F when we change the polarization from $\mathbf{P}(\mathbf{r})$ to $\mathbf{P}(\mathbf{r}) + \Delta\mathbf{P}(\mathbf{r})$, where $\Delta\mathbf{P}(\mathbf{r})$ is any extremely small vector function of $\mathbf{r}$. Call the change ΔF. Of course, if we change the polarization, then the electric field will also undergo a change from the equilibrium electric field $\mathbf{E}(\mathbf{r})$ to some new electric field, $\mathbf{E}(\mathbf{r}) + \Delta\mathbf{E}(\mathbf{r})$. Thus

$$\Delta F = \frac{\epsilon_o}{2} \int_{E_3} [(\mathbf{E} + \Delta\mathbf{E})^2 - E^2]\, d^3\mathbf{r} + \int_{V_S} [\tilde{f}(\mathbf{P} + \Delta\mathbf{P}) - \tilde{f}(\mathbf{P})]\, d^3\mathbf{r}$$
$$= \epsilon_o \int_{E_3} \mathbf{E} \cdot \Delta\mathbf{E}\, d^3\mathbf{r} + \int_{V_S} \frac{\partial\tilde{f}}{\partial\mathbf{P}} \cdot \Delta\mathbf{P}\, d^3\mathbf{r} \tag{6.104}$$

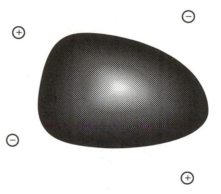

Fig. 6.4   A piece of dielectric material is in the vicinity of some fixed electric charges.

where we have used the fact that $\Delta\mathbf{P}$ is very small to neglect all higher-order terms in the small quantities $\Delta\mathbf{P}$ and $\Delta\mathbf{E}$, and where

$$\frac{\partial\tilde{f}}{\partial\mathbf{P}}\cdot\Delta\mathbf{P} \equiv \frac{\partial\tilde{f}}{\partial P_x}\Delta P_x + \frac{\partial\tilde{f}}{\partial P_y}\Delta P_y + \frac{\partial\tilde{f}}{\partial P_z}\Delta P_z \tag{6.105}$$

In the Mathematical Appendix it is shown, using the equations of electrostatics, that

$$\epsilon_o\int_{E_3}\mathbf{E}\cdot\Delta\mathbf{E}\,d^3\mathbf{r} = -\int_{V_S}\mathbf{E}\cdot\Delta\mathbf{P}\,d^3\mathbf{r} \tag{6.106}$$

Therefore,

$$\Delta F = \int_{V_S}\left(\frac{\partial\tilde{f}}{\partial\mathbf{P}} - \mathbf{E}\right)\cdot\Delta\mathbf{P}\,d^3\mathbf{r} \tag{6.107}$$

It is clear that ΔF cannot be negative, because $\mathbf{P}$ was defined as that polarization that minimized F. But, if for some $\Delta\mathbf{P}(\mathbf{r})$, ΔF were positive, then for the change, $-\Delta\mathbf{P}(\mathbf{r})$, ΔF would be negative, which is impossible. Thus ΔF cannot be negative nor positive (and thus must be zero) for any vector function $\Delta\mathbf{P}(\mathbf{r})$. This is possible only if

$$\frac{\partial\tilde{f}(n,T,\mathbf{P})}{\partial\mathbf{P}} = \mathbf{E}(\mathbf{r}) \tag{6.108}$$

which is thus the desired equilibrium condition for the polarization of a material in an electric field.

That the two equations that have been obtained for the equilibrium electric field are equivalent is a consequence of the electrostatic identity defining the polarization, $\mathbf{P} = \mathbf{D} - \epsilon_o\mathbf{E}$. (See Problem 6.20.)

6.16   THE MAGNETIZATION

The derivation of the equilibrium condition for the magnetization of a substance in an external magnetic field follows closely the analysis for the electric polarization. We consider a system that is composed of two parts, a ferromagnet substance that occupies the region $\mathbf{V}_F$ and another substance that occupies the region $\mathbf{V}_S$. (A ferromagnetic substance is one that has a net magnetization at equilibrium.) The combined region, $\mathbf{V}_F \cup \mathbf{V}_S$, is called $\mathbf{V}_C$. This extra complication is necessary because there is no magnetic equivalent to the fixed electric charges that were used in the last section.

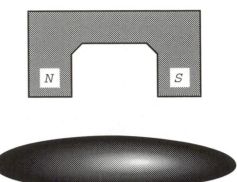

Fig. 6.5   A piece of material is in the vicinity of a ferromagnet.

As before, the Helmholtz free energy contains a term describing the field energy and a term describing the thermodynamic free energy of the specimen.

$$
\begin{aligned}
F &= \frac{1}{2\mu_o} \int_{E_3 - \mathbf{V}_C} B^2(\mathbf{r})\, d^3\mathbf{r} + \int_{\mathbf{V}_C} f\, d^3\mathbf{r} \\
&= \frac{1}{2\mu_o} \int_{E_3} B^2(\mathbf{r})\, d^3\mathbf{r} + \int_{\mathbf{V}_C} \tilde{f}(n, T, \mathbf{M})\, d^3\mathbf{r}
\end{aligned}
\tag{6.109}
$$

We follow the analysis of the last case exactly, making use of the following theorem, derived in the Mathematical Appendix.

$$
\frac{1}{\mu_o} \int_{E_3} \mathbf{B}(\mathbf{r}) \cdot \Delta \mathbf{B}(\mathbf{r})\, d^3\mathbf{r} = \int_{\mathbf{V}_C} \mathbf{B}(\mathbf{r}) \cdot \Delta \mathbf{M}(\mathbf{r})\, d^3\mathbf{r}
\tag{6.110}
$$

Notice the difference in sign between this equation and Eq. (6.106). Taking into account this sign difference, we obtain, by methods analogous to those used in the electrostatic case, the relation that

$$
\Delta F = \int_{\mathbf{V}_C} \left(\frac{\partial \tilde{f}}{\partial \mathbf{M}} + \mathbf{B} \right) \cdot \Delta \mathbf{M}\, d^3\mathbf{r} = 0
\tag{6.111}
$$

for all possible changes in the magnetization. Thus the equation for the magnetic field $\mathbf{B}$ at equilibrium is

$$
\mathbf{B}(\mathbf{r}) = -\frac{\partial \tilde{f}}{\partial \mathbf{M}}
\tag{6.112}
$$

This equation is valid for both ferromagnetic and nonferromagnetic substances. It is the magnetic analog of Eq. (6.108). To obtain the magnetic analog of Eq. (6.98), we recall that the relationship connecting the magnetization $\mathbf{M}$, the magnetic field $\mathbf{B}$, and the magnetic displacement $\mathbf{H}$ is

$$
\mathbf{H} = \frac{1}{\mu_o} \mathbf{B} - \mathbf{M}
\tag{6.113}
$$

From this, and the fact that $f = \tilde{f} + B^2/2\mu_o$, we obtain

$$
df = d\tilde{f} + \frac{1}{\mu_o} \mathbf{B} \cdot d\mathbf{B} = -\mathbf{B} \cdot d\mathbf{M} + \frac{1}{\mu_o} \mathbf{B} \cdot d\mathbf{B} = \mathbf{B} \cdot d\mathbf{H}
\tag{6.114}
$$

which implies that, if we write the free energy per unit volume (including the magnetic field energy) as a function of n, T, and $\mathbf{H}$, then the equilibrium magnetic field is given by

$$
\mathbf{B}(\mathbf{r}) = \frac{\partial f(n, T, \mathbf{H})}{\partial \mathbf{H}}
\tag{6.115}
$$

PROBLEMS

6.1   In a reaction vessel of volume $0.1\,\mathrm{m}^3$ at a temperature of $700\,\mathrm{K}$, 4×10^{-4} moles of H_2, 2×10^{-4} moles of I_2, and 2.09×10^{-3} moles of HI (hydrogen iodide) are in equilibrium with respect to the reaction $H_2 + I_2 \leftrightarrow 2HI$. If an additional 10^{-3} moles of HI are added to the vessel, which is kept at constant temperature, what will be the amounts of the three chemical species when equilibrium is reestablished?

6.2   Suppose the equilibrium constant for the reaction $A \leftrightarrow 2B$ is $2 \times 10^{-4}\,\mathrm{m}^{-3}$ and that for the reaction $B \leftrightarrow C$ is 0.3. What is the equilibrium constant for the reaction $A \leftrightarrow 2C$?

6.3   When 0.1 mole of HI is placed in a reaction vessel kept at $45°\,\mathrm{C}$, 22% of it decomposes by the reaction, $2HI \leftrightarrow H_2 + I_2$. If another 0.1 mole of H_2 is added to the vessel, how many moles of HI will there be at equilibrium?

6.4   At $25°$ C, with equal molar densities of of H_2O (gas) and H_2, the reaction $CO_2 + H_2 \leftrightarrow CO + H_2O$ is at equilibrium when the densities of CO and CO_2 are in the ratio, $n_{CO}/n_{CO_2} = 2.57 \times 10^{-6}$. Going from left to right, the reaction absorbs energy at the rate of 28.5 kilojoules/mole. What would be the equilibrium constant for the reaction at $500°$ C?

6.5   An ideal gas molecule has two possible internal energy states, with energies ε_a and ε_b. (a) Consider the transition between the two states as a chemical reaction, $a \leftrightarrow b$, and calculate the equilibrium constant as a function of T. (b) For this 'chemical reaction', verify Van't Hoff's law.

6.6   A pure dissociation reaction, $AB \leftrightarrow A + B$, takes place at fixed volume in a gas that initially contains only AB molecules at a density of n_o. The degree of dissociation is defined as the ratio of dissociated A atoms to the total density of A atoms. That is, $r \equiv n_A/(n_A + n_{AB})$. Show that $r = (\sqrt{\gamma + 1/4} - 1/2)/\gamma$, where $\gamma = n_o/K$ and K is the equilibrium constant of the reaction.

6.7   Show that, in a partially ionized hydrogen gas, the ratio of the electron density to the total particle density, $n_{tot} = n_{e^-} + n_{H^+} + n_H$, is given by $n_{e^-}/n_{tot} = \sqrt{\lambda(\lambda + 1)} - \lambda$, where $\lambda = e^{-\beta\varepsilon_o}/n_{tot}\lambda_e^3$.

6.8   Equation (6.15) gives the change in the equilibrium constant with respect to temperature at fixed volume. For gas reactions in a reaction vessel of fixed volume, this is useful, but for reactions in aqueous solution, what is desired is $(\partial K/\partial\beta)_p$. Show that, at fixed pressure, Van't Hoff's law has the same form, but the heat of reaction W must be defined by [compare with Eq. (6.11)]
$E(N_A, N_B, N_R, \beta, p) = E(N_A + \Delta N_A, N_B + \Delta N_B, N_R + \Delta N_R, \beta, p) + kW + p\Delta V$
where ΔV is the change in the volume of the system when k reactions take place at fixed pressure.

6.9   The molar volume of water is 18.02 cm^3 and that of ice is 19.63 cm^3 at $0°$ C and one atmosphere. The heat of fusion of the transition is 6000 Joules per mole. Taking these quantities as constants, determine the melting temperature at a pressure of 100 atmospheres.

6.10 Prove that the molar Gibbs free energy of any simple substance is equal to its chemical potential.

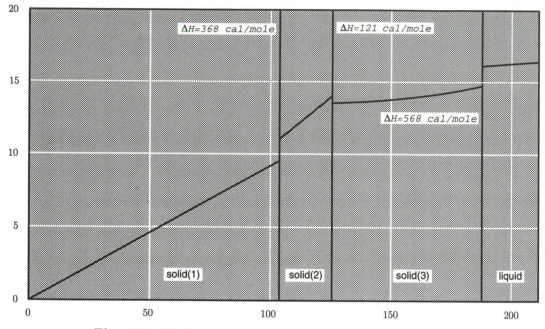

Fig. 6.6 C_p for one mole of HS (in calories) as a function of temperature (in K). The enthalpy changes at the phase transitions are indicated.

6.11 Figure 6.6 shows the constant pressure specific heat of hydrogen sulfide, maintained at a pressure of one atmosphere, from 0 K to its boiling point at 212.8 K. In that temperature interval, it passes through three different solid phases and a liquid phase. The enthalpy changes (in calories per mole) at the phase transitions are shown in the figure. Reasonable approximations to $C_p(T)$ in the four temperature intervals are

$$C_p = 0.0936\,T, \qquad 0 < T < 103.5\,\text{K}$$
$$C_p = 11.05 + 0.135(T - 103.5), \qquad 103.5 < T < 126.2\,\text{K}$$
$$C_p = 13.23 + 0.0066(T - 126.2) + 0.00022(T - 126.2)^2, \quad 126.2 < T < 187.6\,\text{K}$$
$$C_p = 16.2 + 0.005(T - 187.6), \qquad 187.6 < T < 212.8\,\text{K}$$

Draw a graph of the entropy per mole at one atmosphere, as a function of T in the temperature range, $0 < T < 212.8\,\text{K}$.

6.12 Figure 6.7 shows a fluid flowing through a tube of varying cross-sectional area. Assume that the flow is viscous and that the tube contains obstructions, such as the porous plug shown, and that heat is being added and leaking away. Assume also that everything, including the velocity at each point in space, is *constant in time*. Let $h(x)$ be the enthalpy per mole of the fluid at point x, and $v(x)$ be the flow velocity at x. Show that

$$h(x_2) + \tfrac{1}{2}mv^2(x_2) = h(x_1) + \tfrac{1}{2}mv^2(x_1) + Q$$

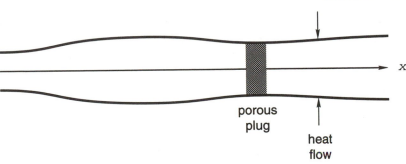

Fig. 6.7   A fluid flows in the x-direction through a pipe of varying cross-section.

where Q is the net rate at which heat is being added to the fluid *per mole* between points x_1 and x_2 and m is the molar mass of the fluid.

6.13   The flow of a gas through a porous plug is called a Joule-Thompson process. It is of great importance in practical refrigeration methods and is usually used at some point in the process of liquifying gases, such as nitrogen and helium. Consider the process shown in Fig. 6.8. Assume that the kinetic energy associated with the macroscopic flow velocity of the gas is negligible in comparison with the internal energy of the gas. Then, by the result of Problem 6.12, the enthalpy per mole is unchanged in flowing through the plug. Assume that the pressure and temperature differentials, Δp and ΔT, are small. Show that cooling occurs only if the initial temperature of the gas satisfies the relation $\beta_p T > 1$. [β_p is defined in Eq. (6.63).] The curve $p(T)$ that limits the region in the p-T plane in which this inequality is satisfied (defined by $\beta_p T = 1$) is called the Joule–Thompson *inversion curve*.

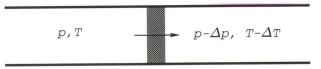

Fig. 6.8   In flowing through the porous plug the fluid undergoes a small drop in pressure and a small change in temperature.

6.14   The adiabatic coefficient of thermal expansion is defined by $V\beta_S = (\partial V/\partial T)_S$. Show that $\beta_S - \beta_p = -\kappa_T C_p/TV\beta_p$.

6.15   Let $L(T)$ be the latent heat per mole at a transition from phase A to phase B. Show that $T\,d(L/T)/dT = C_p(B) - C_p(A)$, where $C_p(A)$ and $C_p(B)$ are the constant pressure molar specific heats of the two phases close to the transition line.

6.16   A volume V_l of liquid water and a volume V_g of steam are in equilibrium at pressure p in a thermally insulated cylinder, topped by a piston. The pressure is increased a small amount, dp. Obtain an expression for the change in volume, dV. Ignore the compressibility of the liquid (but not that of the gas) and the molar volume of the liquid in comparison with that of the gas.

6.17   Express the Gibbs–Duhem relation in terms of the variables μ, T, and p.

6.18 In an isotropic substance the vector variables $\mathbf{P}$ and $\mathbf{E}$ may be replaced by their scalar magnitudes, P and $\mathcal{E}$. (The symbol, $\mathcal{E}$, is used to avoid confusion with the energy.) The molar entropy and the polarization can be written as functions of the particle density, the temperature, and $\mathcal{E}$. Show that $\partial s(n, T, \mathcal{E})/\partial \mathcal{E} = \partial P(n, T, \mathcal{E})/\partial T$.

6.19 A parallel plate capacitor filled with some dielectric is connected to a very large external capacitor at voltage V. The heat capacity of the parallel plate capacitor in this situation is found to be C_V, where in this context the subscript V stands for constant voltage, not volume, although one should assume throughout this problem that any volume change is negligible. The plate capacitor is now disconnected from the external capacitor. Its heat capacity is measured again and found to be C_Q, where the subscript Q means constant charge. Show that $C_V > C_Q$.

6.20 Equation (6.98) implies that, at constant n and T, $df = \mathbf{E} \cdot d\mathbf{D}$. Use this, and the definitions, $\mathbf{P} = \mathbf{D} - \epsilon_o \mathbf{E}$ and $\tilde{f} = f - \epsilon_o E^2/2$, to show that, at constant n and T, $d\tilde{f} = \mathbf{E} \cdot d\mathbf{P}$, which is equivalent to Eq. (6.108).

6.21 Equation (4.78) is equivalent to the equation $P = -\partial f'(n, T, E)/\partial E$, where $f' = -kT\phi/V$. Equation (6.108) states that $E = \partial \tilde{f}(n, T, P)/\partial P$. Show that $\tilde{f}$ and f' are related by a Legendre transformation from the variable P to the variable E.

Chapter 7
Quantum Gases

7.1 THE GRAND CANONICAL ENSEMBLE

The calculations done in Chapter 4 should make it clear that the canonical ensemble, in which the energy of the system is not fixed exactly, is much more convenient to use than the microcanonical ensemble. The canonical ensemble describes a system that is in thermal contact with a reservoir and can therefore exchange energy with it. The average energy of the system is determined by the reservoir temperature. Although the system described by the canonical ensemble does not have an exactly predictable energy, it does have an exact and definite number of particles. We will now introduce another ensemble that is designed to describe a system that can exchange both energy and particles with a much larger reservoir. It is called the *grand canonical ensemble*. In the grand canonical ensemble, both the number of particles and the energy are allowed to fluctuate, their average values being determined by the given values of the affinity and temperature.

In order to derive the probability distribution appropriate for such a situation, we will consider a system, like that shown in Fig. 7.1, composed of a small but macroscopic sample volume and a much larger particle reservoir. Both can exchange energy with a still larger heat reservoir at inverse temperature β. The system, composed of the sample and the particle reservoir (but not the heat reservoir),

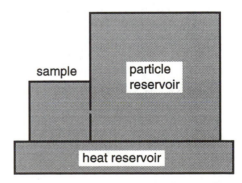

Fig. 7.1 A sample and a much larger reservoir that can exchange energy and particles. The two parts of 'the system' can exchange energy (but not particles) with a heat reservoir.

can be described by a canonical ensemble. It contains a total of N_T particles. We introduce a more compact symbol, z_i, for the six-dimensional phase-space coordinates $(\mathbf{r}_i, \mathbf{p}_i)$ of the ith particle. We assume that the particles in the sample volume do not interact with the particles in the reservoir volume. Thus, if particles 1 through N are in the sample volume while particles $N+1$ through N_T are in the reservoir volume, then the energy can be written as

$$E = H_S(z_1, \ldots, z_N) + H_R(z_{N+1}, \ldots, z_{N_T}) \qquad (7.1)$$

where H_S is the Hamiltonian function for the sample particles and H_R is that for the reservoir. Now, using a canonical probability density for the system, we can easily calculate the probability of finding particles numbered 1 through N in the sample volume at phase-space points, $(z_1, \ldots, z_N)$, with the other $N_T - N$ particles being somewhere in the reservoir. It is [see Eq. (4.56)]

$$\tilde{P}_N(z_1, \ldots, z_N) = e^{-\beta H_S(z_1 \ldots z_N)} \int_R e^{-\beta H_R} d^6 z_{N+1} \cdots d^6 z_{N_T} \Big/ h^{3N_T} N_T! Z_T \qquad (7.2)$$

where Z_T is the partition function of the system. The probability of finding *any* N particles at phase space points z_1 to z_N within the sample and all other particles in the particle reservoir is obtained by multiplying this by the number of ways of choosing N particles from N_T particles, that is, by the binomial coefficient.

$$P_N(z_1, \ldots, z_N) = \binom{N_T}{N} e^{-\beta H_S(z_1 \ldots z_N)} \int_R e^{-\beta H_R} d^6 z_{N+1} \cdots d^6 z_{N_T} \Big/ h^{3N_T} N_T! Z_T$$

$$= e^{-\beta H_S(z_1 \ldots z_N)} \int_R e^{-\beta H_R} d^6 z_{N+1} \cdots d^6 z_{N_T} \Big/ h^{3N_T} N!(N_T - N)! Z_T \qquad (7.3)$$

The integral over the reservoir phase space can be written in terms of the canonical potential of the reservoir with $N_T - N$ particles, which can then be expanded to first order in N. According to Eq. (4.55),

$$\int_R e^{-\beta H_R} d^{6(N_T - N)} z = h^{3(N_T - N)}(N_T - N)! \exp[\phi_R(N_T - N, \beta, V_R)]$$

$$\approx h^{3(N_T - N)}(N_T - N)! \exp[\phi_R(N_T, \beta, V_R)] \exp(-\alpha N) \qquad (7.4)$$

Using this in the equation for $P_N(z_1, \ldots, z_N)$, and lumping together all factors that are independent of N into one normalization constant that we will call Λ^{-1}, we get the desired *grand canonical probability distribution*

$$P_N(z_1, \ldots, z_N) = \frac{e^{-\alpha N - \beta H_S(z_1, \ldots, z_N)}}{\Lambda h^{3N} N!} \qquad (7.5)$$

Thus, if a system can exchange particles and energy with a reservoir at affinity α and coldness β, then, at equilibrium, the probability of finding the system to contain exactly N particles with positions $(\mathbf{r}_1, \ldots, \mathbf{r}_N)$ and momenta $(\mathbf{p}_1, \ldots, \mathbf{p}_N)$ is given by

$$P_N(\mathbf{r}_1, \mathbf{p}_1, \ldots, \mathbf{r}_N, \mathbf{p}_N) = \frac{e^{-\alpha N - \beta H(\mathbf{r}_1, \mathbf{p}_1, \ldots, \mathbf{r}_N, \mathbf{p}_N)}}{\Lambda h^{3N} N!} \qquad (7.6)$$

The normalization constant Λ, called the *grand partition function*, must be chosen so that the integration and sum of P_N over all possible states of the system is one. This gives

$$\Lambda(\alpha, \beta, V) = \sum_{N=0}^{\infty} \frac{e^{-\alpha N}}{h^{3N} N!} \int e^{-\beta H(1, \ldots, N)} d^{3N} x \, d^{3N} p \qquad (7.7)$$

The sum on N has been extended from what would be its natural limit, N_T, to infinity by using the fact that, for very large N, the $N!$ in the denominator makes the terms diminish very rapidly. We will see that the probability of obtaining a value of N that is much larger than the average value of N is negligibly small.

An equivalent calculation for a quantum mechanical system shows that the probability of finding the system with exactly N particles and in the N-particle quantum state $\psi_K(\mathbf{r}_1, \ldots, \mathbf{r}_N)$ of energy E_{NK} is

$$P_{NK} = \Lambda^{-1} e^{-\alpha N - \beta E_{NK}} \tag{7.8}$$

The normalization constant Λ must then be chosen so that the sum of P_{NK} over all N and K is unity.

$$\Lambda = \sum_N \sum_K e^{-\alpha N - \beta E_{NK}} \tag{7.9}$$

It can be shown that, as $\hbar \to 0$, the quantum mechanical expression for Λ reduces to the classical formula given in Eq. (7.7).

As expected (see Problem 7.7), the logarithm of the grand partition function gives the grand potential in rational units.

$$\log \Lambda(\alpha, \beta, V) = \psi(\alpha, \beta, V) = S^o - \alpha N - \beta E \tag{7.10}$$

Using the thermodynamic identity $S^o = \alpha N + \beta E + \gamma V$, ψ can be written in the simpler form

$$\log \Lambda = \psi(\alpha, \beta, V) = \beta p V \tag{7.11}$$

The probability of finding the system with exactly N particles, irrespective of the quantum state, is

$$P_N = \sum_K e^{-\alpha N - \beta E_{NK}} / \Lambda \tag{7.12}$$

This shows that the average number of particles is given by the expected thermodynamic relation.

$$
\begin{aligned}
\langle N \rangle &= \Lambda^{-1} \sum_N \sum_K N e^{-\alpha N - \beta E_{NK}} \\
&= -\Lambda^{-1} \frac{\partial}{\partial \alpha} \sum_N \sum_K e^{-\alpha N - \beta E_{NK}} \\
&= -\frac{\partial \Lambda / \partial \alpha}{\Lambda} \\
&= -\frac{\partial \log \Lambda}{\partial \alpha} \\
&= -\frac{\partial \psi}{\partial \alpha}
\end{aligned}
\tag{7.13}
$$

In an almost identical way, one can show that

$$\langle E \rangle = -\frac{\partial \psi}{\partial \beta} \tag{7.14}$$

7.2 SURFACE ADSORPTION

The surfaces of real vessels containing gases do not have the convenient but artificial properties of the hard-wall or periodic boxes we have been using in our calculations.

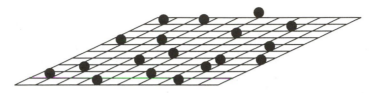

Fig. 7.2   A surface containing adsorption sites that can each accept a single particle.

Even when cleaned and polished, they are usually covered with adsorption sites, at which particles of the gas may temporarily attach themselves, due to interatomic forces between the gas molecules and the surface atoms (see Fig. 7.2). The disappearance of particles onto the surface or their reappearance in the gas as the temperature is raised causes a change in the apparent properties of the gas, which must be corrected for in order to interpret experimental measurements in terms of the true properties of the gas alone.

When investigating the phenomenon of adsorption, it is convenient to use a grand canonical ensemble, since the number of adsorbed particles is not fixed. The model that will be used is a surface containing K independent adsorption sites. Each site may be unoccupied or it may be occupied by a single gas particle, which may be in any of a finite number of discrete quantum states of energies $\varepsilon_1, \varepsilon_2, \ldots$. In this calculation, the system is only the set of adsorbed particles. The particles in the gas act as the reservoir and do not enter explicitly into the calculation. In summing over all states of the surface, we need only sum over the occupation number (0 or 1) at each site, and, for the occupied sites, over the set of energy states.

As usual, the most efficient procedure is to calculate the normalization constant Λ, which will then yield the thermodynamic potential ψ, from which other physical properties can be calculated. The easiest way of deriving the correct expression for Λ is to consider first the trivial cases of a single adsorption site and two adsorption sites. If $K = 1$, then N can be either 0 or 1, and the sum over quantum states only appears if $N = 1$.

$$\Lambda = \sum_{N=0}^{1} e^{-\alpha N} \sum_{k} e^{-\beta \varepsilon_k} = 1 + e^{-\alpha} \sum_{k} e^{-\beta \varepsilon_k} \tag{7.15}$$

For $K = 2$ there are two adsorption sites and therefore two occupation numbers, N_1 and N_2. Again, the sum over quantum states at a particular site only occurs when that occupation number is not zero.

$$\begin{aligned} \Lambda &= \sum_{N_1=0}^{1} \sum_{N_2=0}^{1} e^{-\alpha N_1 - \alpha N_2} \sum_{k_1} \sum_{k_2} e^{-\beta \varepsilon_{k_1} - \beta \varepsilon_{k_2}} \\ &= \left(\sum_{N_1=0}^{1} e^{-\alpha N_1} \sum_{k_1} e^{-\beta \varepsilon_{k_1}} \right) \left(\sum_{N_2=0}^{1} e^{-\alpha N_2} \sum_{k_2} e^{-\beta \varepsilon_{k_2}} \right) \\ &= \left(1 + e^{-\alpha} \sum_{k} e^{-\beta \varepsilon_k} \right)^2 \end{aligned} \tag{7.16}$$

For general K,

$$\Lambda = \left(1 + e^{-\alpha} \sum_{k} e^{-\beta \varepsilon_k} \right)^{K} \tag{7.17}$$

The sum $z_a = \sum \exp(-\beta \varepsilon_k)$ is just the canonical partition function for a single occupied adsorption site. Using the grand potential,

$$\psi(\alpha, \beta) = K \log(1 + e^{-\alpha} z_a) \qquad (7.18)$$

it is easy to calculate the fraction of occupied adsorption sites, $f_a = N_a / K$.

$$f_a = -\frac{\partial}{\partial \alpha} \log(1 + e^{-\alpha} z_a) = \frac{e^{-\alpha} z_a}{1 + e^{-\alpha} z_a} \qquad (7.19)$$

This can be solved for α as a function of f_a.

$$\alpha = \log\left(\frac{1 - f_a}{f_a}\right) + \log z_a \qquad (7.20)$$

It will be assumed that the adsorbing surface is in equilibrium with an ideal gas at pressure p. For any ideal gas, not necessarily monatomic, Eq. 5.46 says that

$$\frac{\partial \alpha(p, T)}{\partial p} = -\frac{1}{p} \qquad (7.21)$$

which can be integrated to give

$$\alpha = -\log p + \log g(T) \qquad (7.22)$$

where the function $g(T)$ cannot be determined by the ideal gas equation alone and is different for different ideal gases. Eliminating α between Eqs. (7.20) and (7.22) gives a relation between the gas pressure and the adsorption fraction.

$$\log\left(\frac{g}{p}\right) = \log\left(\frac{1 - f_a}{f_a}\right) + \log z_a \qquad (7.23)$$

which can easily be solved for f_a as a function of p and T.

$$f_a = \frac{p}{p + g/z_a} \qquad (7.24)$$

It is clear that, for fixed T, the adsorption site occupation probability goes from 0 to 1 as the pressure goes from 0 to infinity.

In order to proceed further, one must determine the functions $z_a(T)$ and $g(T)$. Let us consider the simplest case by assuming that the gas is a monatomic gas and that there is only one possible quantum state at each adsorption site. We assume that the state has a negative energy, $\varepsilon = -kT_o$, indicating a bound state. Then

$$z_a(T) = e^{T_o/T} \qquad (7.25)$$

The affinity of the gas can be calculated as a function of the pressure and temperature by first computing the partition function.

$$Z = \frac{1}{N! h^{3N}} \int e^{-\beta \sum p_n^2 / 2m} \, d^{3N} \mathbf{r} \, d^{3N} \mathbf{p}$$

$$= \frac{1}{N!} \left(\frac{V}{\lambda^3}\right)^N \qquad (7.26)$$

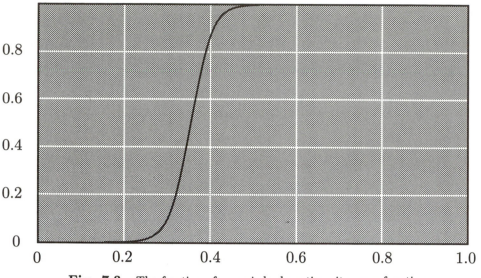

Fig. 7.3   The fraction of occupied adsorption sites as a function of the binding energy (in K), T_o, for the case of argon at STP. The vertical axis gives f_a and the horizontal axis is T_o/T_{STP}.

where λ is the thermal de Broglie wavelength. The canonical potential of the gas is

$$\phi = N\left(\log V - \log N + 1 - \log \lambda^3\right) \tag{7.27}$$

and the affinity is equal to $\partial\phi/\partial N$.

$$\begin{aligned}
\alpha &= \log(V/N) - \log \lambda^3 \\
&= -\log p + \log\left[(kT)^{5/2}(2\pi m)^{3/2}/h^3\right]
\end{aligned} \tag{7.28}$$

This gives

$$g(T) = (kT)^{5/2}(2\pi m)^{3/2}/h^3 \equiv \gamma T^{5/2} \tag{7.29}$$

and

$$f_a = \frac{p}{p + \gamma T^{5/2}e^{-T_o/T}} \tag{7.30}$$

For fixed p, f_a goes from 1 to 0 as T goes from 0 to infinity.

7.3   QUANTUM IDEAL GASES

The distribution functions for Bose–Einstein and Fermi–Dirac ideal gases were derived in Chapter 2 by the method of determining the most probable macrostate. It will be instructive to apply the grand canonical ensemble to a study of the same systems. For both bosons and fermions, the many-particle quantum states are defined by a sequence of occupation numbers $N_1, N_2, \ldots$, where N_k is the number of particles in the kth single-particle quantum state. The total number of particles and the total energy of the state are given by

$$N = \sum_k N_k \tag{7.31}$$

and

$$E = \sum_k \varepsilon_k N_k \tag{7.32}$$

In using the grand canonical ensemble, the first thing one should do is calculate the grand partition function Λ. From Λ, the grand potential, $\psi = \log \Lambda$, can be easily calculated. Then the thermodynamic relations $N = -\partial \psi / \partial \alpha$, $E = -\partial \psi / \partial \beta$, and $\beta p = \partial \psi / \partial V$ will give the equations of state of the system. According to Eq. (7.9), Λ is equal to the sum of $\exp(-\alpha N - \beta E)$ over all values of N and, for each value of N, over all N-particle quantum states of the system. Expressing N and E by means of Eqs. (7.31) and (7.32), we can write Λ as

$$\begin{aligned} \Lambda &= \sum_{N=0}^{\infty} \sum_{\{N_k\}_N} \exp\left(-\alpha \sum_k N_k - \beta \sum_k \varepsilon_k N_k\right) \\ &= \sum_{N=0}^{\infty} \sum_{\{N_k\}_N} \exp\left(-\sum_k (\alpha + \beta \varepsilon_k) N_k\right) \end{aligned} \tag{7.33}$$

where $\{N_k\}_N$ indicates a sum over all all combinations of occupation numbers that add up to N. But summing over all values of N and, for each value of N, summing over all N-particle quantum states is equivalent to summing independently over all single-particle occupation numbers. Therefore, writing $\{N_k\}$ for the infinite sequence of occupation numbers $\{N_1, N_2, \ldots\}$, we get

$$\begin{aligned} \Lambda &= \sum_{\{N_k\}} \exp\left(-\sum_k (\alpha + \beta \varepsilon_k) N_k\right) \\ &= \sum_{N_1} e^{-(\alpha + \beta \varepsilon_1) N_1} \sum_{N_2} e^{-(\alpha + \beta \varepsilon_2) N_2} \sum_{N_3} e^{-(\alpha + \beta \varepsilon_3) N_3} \cdots \end{aligned} \tag{7.34}$$

and, because the logarithm of a product is equal to the sum of the logarithms of its factors

$$\psi(\alpha, \beta, V) = \log \Lambda = \sum_k \log\left(\sum_{N_k} e^{-(\alpha + \beta \varepsilon_k) N_k}\right) \tag{7.35}$$

For bosons, $N_k = 0, 1, 2, \ldots$, which means that the sum over N_k in Eq. (7.35) is of the form

$$\sum_{N_k} x^{N_k} = \frac{1}{1-x}$$

with $x = \exp[-(\alpha + \beta \varepsilon_k)]$. That is

$$\sum_{N_k=0}^{\infty} e^{-(\alpha + \beta \varepsilon_k) N_k} = \frac{1}{1 - e^{-(\alpha + \beta \varepsilon_k)}} \tag{7.36}$$

and therefore

$$\psi(\alpha, \beta, V) = -\sum_k \log\left[1 - e^{-(\alpha + \beta \varepsilon_k)}\right] \tag{7.37}$$

For fermions, the sum has only two terms, giving

$$\sum_{N_k=0}^{1} e^{-(\alpha + \beta \varepsilon_k) N_k} = 1 + e^{-(\alpha + \beta \varepsilon_k)} \tag{7.38}$$

and

$$\psi(\alpha, \beta, V) = \sum_k \log\left[1 + e^{-(\alpha + \beta \varepsilon_k)}\right] \tag{7.39}$$

Both cases can be treated simultaneously by the formula

$$\psi = \zeta \sum_k \log\left[1 + \zeta e^{-(\alpha + \beta \varepsilon_k)}\right] \tag{7.40}$$

where

$$\zeta = \begin{cases} +1, & \text{for FD} \\ -1, & \text{for BE} \end{cases} \tag{7.41}$$

N and E are then given, in terms of α and β, by the thermodynamic relations

$$N = -\frac{\partial \psi}{\partial \alpha} = \sum_k \frac{e^{-(\alpha + \beta \varepsilon_k)}}{1 + \zeta e^{-(\alpha + \beta \varepsilon_k)}} = \sum_k \frac{1}{e^{\alpha + \beta \varepsilon_k} + \zeta} \tag{7.42}$$

and

$$E = -\frac{\partial \psi}{\partial \beta} = \sum_k \frac{\varepsilon_k e^{-(\alpha + \beta \varepsilon_k)}}{1 + \zeta e^{-(\alpha + \beta \varepsilon_k)}} = \sum_k \frac{\varepsilon_k}{e^{\alpha + \beta \varepsilon_k} + \zeta} \tag{7.43}$$

These equations are in agreement with the Fermi–Dirac and Bose–Einstein distribution functions obtained in Chapter 2. The average number of particles that occupy any single-particle quantum state of energy ε is

$$f_{\text{BE}}(\varepsilon) = \frac{1}{e^{\alpha + \beta \varepsilon} - 1} \quad \text{or} \quad f_{\text{FD}}(\varepsilon) = \frac{1}{e^{\alpha + \beta \varepsilon} + 1} \tag{7.44}$$

for bosons and fermions, respectively.

7.4   BOLTZMANN STATISTICS

If the average occupation number of each single-particle quantum state is much less than one, then $e^{\alpha + \beta \varepsilon} \gg 1$, and both the Bose–Einstein and the Fermi–Dirac density formulas can be approximated by the Maxwell–Boltzmann formula

$$f_{\text{MB}}(\varepsilon) = e^{-\alpha - \beta \varepsilon} \tag{7.45}$$

When this is done, one says that one is using *Boltzmann statistics*. Because $f_{\text{FD}} < f_{\text{MB}} < f_{\text{BE}}$, quantities calculated using this approximation usually are intermediate between those for bosons and fermions at the same values of α and β.

7.5   THE IDEAL FERMI GAS AT LOW TEMPERATURE

In Chapter 2 it was shown that the quantum mechanical distribution functions reduced to their classical equivalents at sufficiently high temperatures. Therefore, the specifically quantum mechanical aspects of quantum gases are revealed by studying their low-temperature behavior. This must be done separately for Fermi–Dirac and Bose–Einstein systems, because their low-temperature properties are very different. In the next two sections the low-temperature properties of a system of nonrelativistic spin-$\frac{1}{2}$ fermions will be analyzed. Following this, we will consider an equivalent system of bosons.

It is assumed that the particles are in a cubic periodic box of side L. It was shown in Section 2.15 that the single-particle energy eigenstates are then plane waves with allowed momentum values that form a cubic lattice of spacing h/L in momentum space, where $h = 2\pi\hbar$. The energy of a particle with momentum $\mathbf{p}$ is $\varepsilon_p = p^2/2m$. Each allowed momentum eigenstate can be occupied by two particles of opposite spin. Therefore, the sum in Eq. (7.39) for the grand potential ψ can be converted to an integral of the form

$$\psi = 2(L/h)^3 \int \log[1 + e^{-(\alpha + \beta p^2/2m)}] d^3\mathbf{p} \tag{7.46}$$

By converting to an energy variable, $\varepsilon = p^2/2m$, and using the facts that $p = \sqrt{2m\varepsilon}$, $dp = d\varepsilon\sqrt{2m/\varepsilon}/2$, and $L^3 = V$, this integral can be transformed as follows.

$$
\begin{aligned}
\psi &= \frac{8\pi V}{h^3} \int_o^\infty \log[1 + e^{-(\alpha + \beta p^2/2m)}] p^2 \, dp \\
&= 2\gamma V \int_o^\infty \log[1 + e^{-(\alpha + \beta\varepsilon)}] \varepsilon^{1/2} d\varepsilon
\end{aligned}
\tag{7.47}
$$

where the constant $\gamma = 2\pi(2m/h^2)^{3/2}$. The same transformation of Eqs. (7.42) and (7.43) gives

$$N = 2\gamma V \int_o^\infty \frac{\varepsilon^{1/2} \, d\varepsilon}{e^{\alpha + \beta\varepsilon} + 1} \tag{7.48}$$

and

$$E = 2\gamma V \int_o^\infty \frac{\varepsilon^{3/2} \, d\varepsilon}{e^{\alpha + \beta\varepsilon} + 1} \tag{7.49}$$

The logarithmic function in Eq. (7.47) can be eliminated by a partial integration. If that is done, and the thermodynamic relation $\psi = \beta p V$ is used, one finds that the classical relationship between pressure and energy density, $p = \frac{2}{3}(E/V)$, is also valid for a quantum mechanical ideal gas (see Problem 7.8).

In analyzing the low-temperature properties of the Fermi–Dirac gas, it is best to change variables from the affinity α to the chemical potential $\mu = -\alpha\tau$, where $\tau = kT$. The integrals for N and E, in terms of μ and τ, are

$$\frac{N}{V} = 2\gamma \int_o^\infty \frac{\varepsilon^{1/2} \, d\varepsilon}{e^{(\varepsilon - \mu)/\tau} + 1} \tag{7.50}$$

and

$$\frac{E}{V} = 2\gamma \int_o^\infty \frac{\varepsilon^{3/2} \, d\varepsilon}{e^{(\varepsilon - \mu)/\tau} + 1} \tag{7.51}$$

These integrals are both of the general form

$$I = \int_o^\infty \frac{g(\varepsilon) \, d\varepsilon}{e^{(\varepsilon - \mu)/\tau} + 1} \tag{7.52}$$

where $g(\varepsilon)$ is a smooth function that is independent of τ. Therefore, we will present a general method of evaluating integrals of this type as a power series expansion in the temperature. It is easy to verify the following limit formula for the Fermi–Dirac function $f_{\mathrm{FD}}(\varepsilon)$.

$$\lim_{\tau \to 0} \left(\frac{1}{e^{(\varepsilon - \mu)/\tau} + 1} \right) = \begin{cases} 1, & \text{if } \varepsilon < \mu \\ 0, & \text{if } \varepsilon > \mu \end{cases} \tag{7.53}$$

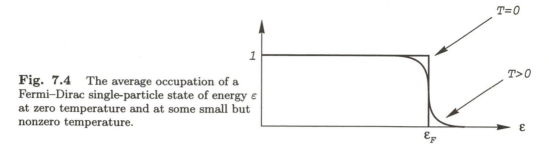

Fig. 7.4   The average occupation of a
Fermi–Dirac single-particle state of energy ε
at zero temperature and at some small but
nonzero temperature.

This has a simple physical interpretation. The Fermi–Dirac function gives the
probability that a single-particle state of energy ε is occupied. At zero temperature,
the N-particle system goes to a state in which the N single-particle states of
lowest energy are occupied while all higher states are unoccupied. Thus at zero
temperature the chemical potential has a value equal to the cutoff energy separating
the occupied states from the unoccupied ones. This is called the *Fermi energy* and
is written ε_F. The Fermi energy is defined by the relation [see Eq. (7.50)]

$$n = \frac{N}{V} = 2\gamma \int_{o}^{\varepsilon_F} \varepsilon^{1/2}\, d\varepsilon = \tfrac{4}{3}\gamma\varepsilon_F^{3/2} \tag{7.54}$$

which gives $\varepsilon_F = (3n/4\gamma)^{2/3}$. At finite temperature $f_{\text{FD}}(\varepsilon)$ deviates from its zero-
temperature limit, as shown in Fig. 7.4.

In the Mathematical Appendix, the following low-temperature expansion is
derived for the general Fermi–Dirac integral shown in Eq. (7.52).

$$\int_{o}^{\infty} \frac{g(\varepsilon)\, d\varepsilon}{e^{(\varepsilon-\mu)/\tau}+1} = \int_{o}^{\mu} g(\varepsilon)d\varepsilon + \frac{\pi^2}{6}g'(\mu)\tau^2 + \frac{7\pi^4}{360}g'''(\mu)\tau^4 + \cdots \tag{7.55}$$

where the primes indicate derivatives with respect to ε, evaluated at $\varepsilon = \mu$. Using
this expansion to order τ^2 in Eq. (7.50) gives the following relation for the particle
density, $n = N/V$.

$$\frac{3n}{4\gamma} \approx \mu^{3/2} + \frac{\pi^2\tau^2}{8\mu^{1/2}} \tag{7.56}$$

We can determine the dependence of the chemical potential μ on the temperature
τ for small τ as follows. In the second term, which is of order τ^2, we will make a
negligible error in replacing μ by ε_F. The left-hand side is equal to $\varepsilon_F^{3/2}$. Therefore

$$\varepsilon_F^{3/2} \approx \mu^{3/2} + \frac{\pi^2\tau^2}{8\varepsilon_F^{1/2}} \tag{7.57}$$

Dividing through by $\varepsilon_F^{3/2}$, solving for μ/ε_F, and expanding the solution as a power
series in τ^2 gives

$$\frac{\mu}{\varepsilon_F} \approx \left(1 - \frac{\pi^2\tau^2}{8\varepsilon_F^2}\right)^{2/3} \approx 1 - \frac{\pi^2\tau^2}{12\varepsilon_F^2} \tag{7.58}$$

or

$$\mu \approx \varepsilon_F - \frac{\pi^2\tau^2}{12\varepsilon_F} \tag{7.59}$$

which shows that, to first order in τ, the chemical potential is independent of
temperature for fixed density. The reason is that, in going from zero temperature

to some low temperature, the *change* in the Fermi–Dirac distribution function is an odd function of $\varepsilon - \mu$. Thus for fixed μ the same number of particles are added to the region $\varepsilon > \mu$ as are removed from the region $\varepsilon < \mu$ (see Fig. 7.4).

The expansion of the energy equation to second order in τ is

$$\frac{E}{2\gamma V} = \tfrac{2}{5}\mu^{5/2} + \frac{\pi^2}{4}\mu^{1/2}\tau^2 \tag{7.60}$$

The energy to order τ^2 at fixed n is obtained by using $\mu = \varepsilon_F - \pi^2\tau^2/12\varepsilon_F$ in this equation and retaining terms of order τ^2. Using the definition of ε_F, the result can be expressed in the form

$$\frac{E}{N} = \tfrac{3}{5}\varepsilon_F + \frac{\pi^2\tau^2}{4\varepsilon_F} \tag{7.61}$$

The pressure as a function of temperature and density is then given by the relation $p = \tfrac{2}{3}(E/V)$.

The dimensionless expansion parameter in these expressions is clearly τ/ε_F. If $\tau/\varepsilon_F \gg 1$, then the ideal Fermi gas behaves as a classical ideal gas. In that case it is said to be *nondegenerate*. For $\tau/\varepsilon_F \ll 1$ the gas is called a *degenerate* Fermi gas, and the exclusion principle is the most important factor determining the properties of the system.

7.6   THE ELECTRON GAS

A metallic crystal is distinguished from other types of crystals by the existence of a large system of *conduction electrons* within it. The number of conduction electrons is equal to the number of atoms in the sample times the valence of the atomic species (1 for sodium, potassium, and copper, 2 for zinc, etc.). The conduction electrons occupy single-particle quantum states in what is called the *conduction band*. Semiconductors also contain conduction electrons, but these are either electrons that have been thermally excited from the valence band to the conduction band (in which case the material is called an *intrinsic semiconductor*) or they are electrons that are associated with donor impurities (in a *doped semiconductor*). In either case, the number of conduction electrons is much smaller than it is for a typical metal. At zero temperature the states in the conduction band are filled (two electrons per state, with opposite spins) up to the Fermi energy, ε_F. Each state in the conduction band is also an eigenstate of crystal momentum with eigenvalue $\mathbf{p} = \hbar\mathbf{k}$, where $\mathbf{k}$ is the wave vector of the state. For some metals the relationship between the energy and momentum of a state in the conduction band can be approximated by that for a free particle.

$$\varepsilon_{\mathbf{k}} = \frac{\hbar^2 k^2}{2m^*} \tag{7.62}$$

where the *effective mass* m^* is a parameter that is of the order of an electron mass but is not equal to m_e.

We can get an estimate of the contribution made by the conduction electrons to the specific heat of a typical metal by taking $m^* = m_e$ and $V/N = a^3$, where $a = 4\,\text{Å}$ is the diameter of a sodium atom. Using these values to calculate γ and ε_F, we find that $\varepsilon_F \approx 1\,\text{eV}$. This energy corresponds to a temperature of about 10,000 K. Therefore, at room temperature, the expansion parameter $\tau/\varepsilon_F \approx 0.03$, which justifies neglecting all but the first nonvanishing term in the expansion for E/N. The specific heat per particle at room temperature is then given by

$$C_{\text{el}} = \frac{\partial(E/N)}{\partial T} = \frac{\pi^2 k^2 T}{2\varepsilon_F} \approx 0.15\,k \tag{7.63}$$

where k is Boltzmann's constant. The Debye temperature of sodium is about 158 K and, therefore, at room temperature we can use the Dulong–Petit law, which states that the vibrational specific heat per particle is equal to $3\,k$.

$$C_{\text{vib}} = 3\,k \tag{7.64}$$

Comparing these equations shows that at room temperature the electronic motion makes only a slight contribution to the specific heat of a metal. However, at temperatures lower than the Debye temperature of the metal, the vibrational specific heat goes to zero as T^3 while the electronic specific heat is proportional to T. Thus at sufficiently low temperature the electronic specific heat dominates.

7.7   THE IDEAL BOSE GAS AT LOW TEMPERATURE

As $T \to 0$ the Pauli exclusion principle becomes the most important factor determining the properties of a system of fermions. Since bosons do not satisfy an exclusion principle, their physical characteristics at low temperature will obviously be very different from those of fermions. In this section we will analyze the low-temperature properties of a gas of noninteracting spinless bosons. The basic equations for N and E in terms of α and β can be taken from the fermion case by dropping the factor of 2 that is due to the spin degeneracy, and making a change of sign in the Fermi–Dirac distribution function. When this is done, the equation for N [Eq. (7.50)] becomes

$$n = \frac{N}{V} = \gamma \int_o^\infty \frac{\varepsilon^{1/2}\,d\varepsilon}{e^{\alpha+\beta\varepsilon} - 1} \tag{7.65}$$

The temperature dependence of this equation can be made explicit by transforming the variable of integration to $x = \beta\varepsilon$.

$$n = \gamma\tau^{3/2} \int_o^\infty \frac{x^{1/2}\,dx}{e^{\alpha+x} - 1} \tag{7.66}$$

The affinity is limited to the range $0 \le \alpha < \infty$. If α is chosen to be negative, then $f_{\text{BE}}(\varepsilon)$ would be negative at small ε, which is physically meaningless, because it is the average occupation of a single-particle state. It is easy to see that $n(\alpha, \tau)$ is a monotonically decreasing function of α (at each value of x, the integrand decreases if α is increased). The maximum value, which occurs at $\alpha = 0$, is

$$n_c(\tau) = \gamma\tau^{3/2} \int_o^\infty \frac{x^{1/2}\,dx}{e^x - 1} \tag{7.67}$$

n_c is called the critical density. This integral is of a type that can be evaluated in terms of the gamma function $\Gamma(x)$ and the Riemann zeta function $\zeta(x)$. The general formula is

$$\int_o^\infty \frac{x^\nu\,dx}{e^x - 1} = \Gamma(\nu + 1)\zeta(\nu + 1) \tag{7.68}$$

The maximum particle density is therefore given by

$$n_c(\tau) = \Gamma\left(\tfrac{3}{2}\right)\zeta\left(\tfrac{3}{2}\right)\gamma\tau^{3/2} \tag{7.69}$$

But this result is clearly absurd! Since the particles do not interact, there is no physical mechanism preventing one from constructing a system with a density greater than n_c. A mistake has been made somewhere. The mistake is not hard to find. The lowest-energy single-particle state is the zero-momentum state (which has zero energy). According to the Bose–Einstein distribution, the number of particles in that state at affinity α is

$$N_{\mathbf{p}=0} = \frac{1}{e^\alpha - 1} \to \infty \quad \text{as } \alpha \to 0 \tag{7.70}$$

Thus, in making the transformation from a sum over discrete momentum states to an integral, the zero-momentum state must be singled out for special treatment. In Problem 7.22, the reader is asked to show that this is sufficient to fix the difficulty; it is not necessary to take into account, individually, the other very low-energy states neighboring the zero-momentum state. Therefore, Eq. (7.66) must be modified to read

$$n = \frac{1}{V(e^\alpha - 1)} + \gamma \tau^{3/2} \int_0^\infty \frac{x^{1/2}\, dx}{e^{\alpha+x} - 1} \tag{7.71}$$

With this modification, for any value of the density n, there is a corresponding value of the affinity α that satisfies this equation. When n is larger than the critical density, then α is of order $1/N$ and, in the integral part of Eq. (7.71), α may be set equal to zero. In this case, a finite fraction of all of the particles in the system occupy the discrete zero-momentum quantum state.

The collection of particles in the zero momentum state is called the Bose–Einstein *condensate*. It exists whenever $n > n_c(\tau)$. The number of particles in the condensate is just the difference between the total number of particles and $n_c V$. Since the condensate, when it exists, accounts for a finite fraction of the particles in the system, the gas with the condensate has macroscopically different properties than the gas without it. The change from one state to the other is a thermodynamic phase transition, called *Bose condensation*. The phase diagram is shown in Fig. 7.5. The phase without the condensate (that is, $n < n_c$) is called phase I, while the phase that contains a condensate ($n > n_c$) is called phase II.

In phase I the function $\alpha(n, \tau)$ must be determined by numerically solving the Eq. (7.71) in the form

$$\int_0^\infty \frac{x^{1/2}\, dx}{e^{\alpha+x} - 1} = \frac{n}{\gamma \tau^{3/2}} \tag{7.72}$$

In phase II the function $\alpha(n, \tau)$ has the simple form (see Problem 7.24)

$$\alpha(n, \tau) = \frac{1}{V[n - n_c(\tau)]} \qquad \text{Phase II} \tag{7.73}$$

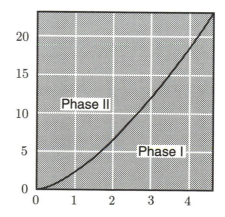

Fig. 7.5 The phase diagram for the ideal Bose gas. The vertical axis gives n/γ and the horizontal axis is $\tau = kT$. Recall that $\gamma = 2\pi(2m/h^2)^{3/2}$.

For phase II, in any integral, the value of α may be replaced by zero, since it is of order $1/N$. The condensate particles all have zero energy; therefore, the condensate makes no contribution to the energy and so the integral expression for E can be used in both phases.

$$E(n, \tau, V) = \gamma V \int_o^\infty \frac{\varepsilon^{3/2}\, d\varepsilon}{e^{\alpha + \beta \varepsilon} - 1} = \gamma V \tau^{5/2} \int_o^\infty \frac{x^{3/2}\, dx}{e^{\alpha + x} - 1} \qquad (7.74)$$

where α is understood to mean $\alpha(n, \tau)$. In the condensed phase, $\alpha \approx 0$, and thus, the energy becomes independent of n.

$$E = \gamma V \tau^{5/2} \int_o^\infty \frac{x^{3/2}\, dx}{e^x - 1} = \Gamma(\tfrac{5}{2})\zeta(\tfrac{5}{2})\gamma V \tau^{5/2} \qquad \text{Phase II} \qquad (7.75)$$

The constant volume specific heat (in rational units) in Phase II is

$$C_V = \frac{dE}{d\tau} = \tfrac{5}{2}\Gamma(\tfrac{5}{2})\zeta(\tfrac{5}{2})\gamma V \tau^{3/2} \qquad \text{Phase II} \qquad (7.76)$$

The entropy is the integral of dQ/τ.

$$S^o = \int_o^\tau \frac{C_V\, d\tau}{\tau} = \tfrac{5}{3}\Gamma(\tfrac{5}{2})\zeta(\tfrac{5}{2})\gamma V \tau^{3/2} \qquad \text{Phase II} \qquad (7.77)$$

For a given density, the transition from one phase to the other occurs at a *critical temperature* τ_c that is given by solving Eq. 7.69 for τ.

$$\tau_c = \left[n/\Gamma(\tfrac{3}{2})\zeta(\tfrac{3}{2})\gamma\right]^{2/3} \qquad (7.78)$$

In terms of τ_c, the specific heat per particle can be written in the simple form

$$\frac{C_V}{N} = \frac{15}{4}\frac{\zeta(\tfrac{5}{2})}{\zeta(\tfrac{3}{2})}\left(\frac{\tau}{\tau_c}\right)^{3/2} \qquad \text{Phase II} \qquad (7.79)$$

In Phase I, $E(n, \tau)$ and $C_V(n, \tau)$ must be calculated from Eq. (7.74) using the function $\alpha(n, \tau)$. A graph of C_V/N as a function of τ/τ_c is given in Fig. 7.6. Notice that Bose condensation is a very subtle phase transition. Across the phase transition line the density and the energy per particle are continuous. Even the specific heat is continuous. One must go to $dC_V/d\tau$ in order to find a quantity that jumps discontinuously at the phase transition.

7.8   FLUCTUATIONS IN THE IDEAL BOSE GAS*

A study of fluctuations in the ideal Bose gas below the critical temperature yields some surprising results. In phase II, $\alpha = 0$, and therefore, the fundamental relation for S^o becomes

$$S^o = \beta E + \beta p V = \frac{E + pV}{\tau} \qquad (7.80)$$

Noninteracting bosons also satisfy the ideal gas relation $pV = \tfrac{2}{3}E$ (see Problem 7.8), which gives

$$S^o = \tfrac{5}{3}E/\tau \qquad (7.81)$$

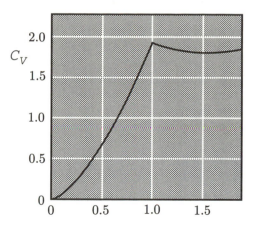

Fig. 7.6 The constant volume specific heat per particle as a function of τ/τ_c. In rational units the specific heat is dimensionless.

Equation (7.75) gives τ as a function of the energy density E/V, which can be used in Eq. (7.81) to write S^o in terms of E and V.

$$S^o = \frac{5}{3}\left(\Gamma(\tfrac{5}{2})\zeta(\tfrac{5}{2})\gamma\right)^{2/5}V^{2/5}E^{3/5} \tag{7.82}$$

We notice that this entropy function does not contain the variable N. This is reasonable, because $\alpha = (\partial S^o/\partial N)_E = 0$ in phase II. If more particles are added to the system, they all go into the condensate, leaving E and S^o unchanged. We will now see that this has disastrous consequences for the thermodynamic properties of the system. Letting s^o, n, and ε be the entropy, number of particles, and energy per unit volume, Eq. (7.82) can be written

$$s^o(n,\varepsilon) = \tfrac{5}{3}A\varepsilon^{3/5} \tag{7.83}$$

where $A = \left[\Gamma(\tfrac{5}{2})\zeta(\tfrac{5}{2})\gamma\right]^{2/5}$. The fact that s^o is not a function of n means that, in the situation depicted in Fig. 7.7, the condition of maximum entropy would not determine the equilibrium density in the two halves. This peculiar phenomenon

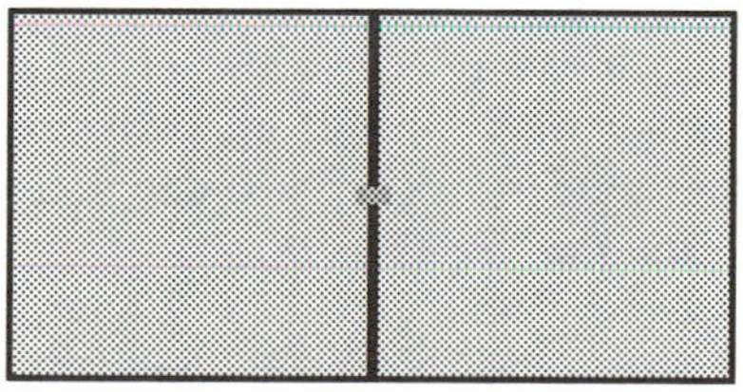

Fig. 7.7 Two volumes containing an ideal Bose gas are connected by a pinhole.

is easy to understand. Because the condensate particles all have zero momentum, they make no contribution to the pressure in a container. Thus, if the condensate densities in the left and right sides were unequal, this would not create a pressure difference that would bring the system back to equal densities. Because of this characteristic, the ideal Bose gas cannot be considered as an acceptable model of any physical system. An example of the bad features of the model is that, in the presence of a weak gravitational field in the z direction, if the temperature is below the critical temperature, all of the condensate, which contains a finite fraction of the particles in the system, becomes smeared against the bottom of the vessel in a single quantum state that has a scale height of $(\hbar^2/m^2 g)^{1/3}$. For the earth's gravitational field, which is weak for thermodynamic purposes, this scale height is about one micrometer (see Problem 7.28). Also, if the periodic boundary conditions are replaced by hard-wall boundary conditions, then a macroscopic change would take place in the particle density throughout the container, not only near the walls (see Problem 7.29). This extreme sensitivity of the particle density to the boundary conditions is never exhibited by real systems.

7.9   THE MEAN-FIELD INTERACTION*

In order to obtain a reasonable model of a system of Bose–Einstein particles, some interaction between the particles must be introduced. Even a very weak interaction eliminates the bizarre properties of the model. In order to construct a model that has normal physical properties without seriously complicating the analysis, we will introduce an interaction potential, $v(r)$, with the following characteristics:

1. $v(r)$ is a steadily decreasing positive function. Therefore, it creates purely repulsive interparticle forces.
2. The range of $v(r)$ is much larger than the average distance between particles in the system but the volume integral of $v(r)$, $4\pi \int_o^\infty v(r) r^2 dr = \sigma$, is finite.

An interaction potential with these characteristics, called a *mean-field interaction*, can be constructed by starting with any positive, decreasing potential function $V(r)$ that has a finite volume integral σ and reducing its amplitude while increasing its range by scaling it according to the formula

$$v(r) = \epsilon^3 V(r/\epsilon) \tag{7.84}$$

where ϵ is a dimensionless adjustable parameter. The integral of $v(r)$ is then independent of ϵ, but its range gets larger and larger as ϵ is made smaller and smaller.

For very small values of ϵ the interaction between any two particles is negligible but a particle finds itself in a very smooth potential field due to the combined interactions of thousands of particles in its large "interaction volume." If the density of particles is uniform, then that potential field will be constant. But adding a constant to the Hamiltonian will not change the energy eigenfunctions of the system. Thus the energy eigenstates for the uniform ideal Bose–Einstein gas are still energy eigenstates of the system, but their energies have been shifted by an amount that depends upon the particle density. Being careful not to double count the interaction energies, one can easily evaluate the total energy of a state with occupation numbers $N_\mathbf{p}$.

$$E = \sum_\mathbf{P} \frac{p^2}{2m} N_\mathbf{p} + \tfrac{1}{2}\sigma n N \equiv E_K + \tfrac{1}{2}\sigma n N \tag{7.85}$$

where the sum is over all the momentum eigenstates of the system, σ is the volume integral of $v(\mathbf{r})$, and $n = N/V$. With fixed N, a shift in all energy levels has no effect on their occupations. Therefore, E_K and S^o, as functions of T, are unaffected by the shift. Thus we can use Eq. (7.83) for S^o with E replaced by E_K.

$$S^o(N, E, V) = \frac{5}{3} A V^{2/5} (E - \tfrac{1}{2}\sigma n^2 V)^{3/5} \tag{7.86}$$

The entropy per unit volume is then

$$s^o(n, \varepsilon) = \frac{5}{3} A(\varepsilon - \tfrac{1}{2}\sigma n^2)^{3/5} \tag{7.87}$$

The reader will be asked to show (see Problem 7.30) that this entropy function is strictly convex and therefore leads to reasonable predictions for macroscopic properties. (Note that, because E_K is always positive, ε has the range $\tfrac{1}{2}\sigma n^2 < \varepsilon < \infty$.)

7.10 THE STATISTICS OF COMPOSITE PARTICLES

The elementary particles of nature are either bosons or fermions. Those that contribute to the structure of atoms, namely protons, neutrons, and electrons, are all spin-$\frac{1}{2}$ Fermi–Dirac particles. The atom, when treated as a single particle, inherits its statistics from the elementary particles that compose it. If it contains an even number of elementary fermions, then it is itself a Bose–Einstein particle; if it contains an odd number, it is a Fermi–Dirac particle. The reasoning behind this rule is easily understood by considering a simple example. A hydrogen molecule can be treated either as a system composed of two hydrogen atoms or as a system of two protons and two electrons. In the first case, it has a wave function $\phi(Q_A, Q_B)$, where Q_A and Q_B are each a set of coordinates, sufficient to define the position and state of a hydrogen atom. In the second case it has a wave function $\psi(x_A, y_A, x_B, y_B)$, where x_A and x_B are coordinates appropriate to the two protons and y_A and y_B are each space and spin coordinates for an electron. The transformation

$$\psi(x_A, y_A, x_B, y_B) \longrightarrow \psi(x_B, y_B, x_A, y_A) \tag{7.88}$$

which describes an exchange in position of the protons and the electrons, is equivalent to the combination of the two separate transformations

$$\psi(x_A, y_A, x_B, y_B) \longrightarrow \psi(x_B, y_A, x_A, y_B) \longrightarrow \psi(x_B, y_B, x_A, y_A) \tag{7.89}$$

But, because protons and electrons are Fermi–Dirac particles, each of the separate transformations multiplies the wave function ψ by -1. Thus, the combined transformation leaves ψ unchanged. When this fact is expressed in atomic coordinates it takes the form

$$\phi(Q_a, Q_B) = \phi(Q_B, Q_A) \tag{7.90}$$

This line of reasoning clearly leads to the conclusion that the exchange of the full sets of coordinates for two identical atoms multiplies the wave function by $+1$ or -1 depending upon whether the atoms contain an even or odd number of elementary fermions.

7.11 SUPERFLUIDITY AND LIQUID HELIUM

There is no real system that is well approximated by the ideal Bose gas. Photons are Bose–Einstein particles, and they are noninteracting to a very high degree, but the fact that the number of photons is not a conserved quantity has a strong effect on the properties of the photon gas. In particular, it eliminates the phase transition associated with Bose condensation, which is the most striking characteristic of the ideal Bose gas. The photon gas will be discussed in a later section.

Liquid helium is the system that is closest (although not very close) to the ideal Bose gas. Helium exists in two isotopic forms, ^{3}He and ^{4}He. The two isotopes, although they are chemically almost identical, have very different physical properties at low temperatures. The differences in their physical characteristics are not due to the difference in their masses, which are approximately in a ratio of 3 to 4. Their most fundamental difference is that the ^{3}He atom, composed of 5 fermions (2 protons, 1 neutron, and 2 electrons), is itself a Fermi–Dirac particle, while the ^{4}He atom is a Bose–Einstein particle. We will restrict our attention to the heavier isotope, which incidentally constitutes the bulk of naturally occurring helium, the lighter isotope contributing less than one part per million. The phase diagram of bulk ^{4}He is shown in Fig. 7.8. It differs substantially from that of other inert elements, such as neon and argon. It has no triple point, in the ordinary sense of having liquid, solid, and gas simultaneously in equilibrium, because the liquid phase extends down to zero temperature. The λ line separates two different liquid phases. (The isotopes of helium are the only pure substances that have more than one liquid phase.) The phase transition separating the two liquid phases, called the *lambda transition*, derives its name from the appearance of the specific heat singularity that occurs at the transition. As can be seen in Fig. 7.8, the specific heat curve resembles the Greek letter λ. The phase transition curve in the p-T plane is called the λ line and is written $T_\lambda(p)$. If no pressure is specified, then T_λ will always mean its lower endpoint, where the λ line meets the liquid–vapor phase transition curve. Extremely careful measurements have shown that, very close to $T_\lambda(p)$, along the whole length of the λ transition line, C_p can be accurately represented by a function of the form

$$C_p = B - At^{-\alpha}(1 + D\sqrt{t}) \quad \text{for } T > T_\lambda(p) \tag{7.91}$$

and

$$C_p = B' - A't^{-\alpha}(1 + D'\sqrt{t}) \quad \text{for } T < T_\lambda(p) \tag{7.92}$$

where $t = |T - T_\lambda|/T_\lambda$ is a dimensionless temperature variable. The parameters A, A', B, B', D, and D' have values that vary substantially along the λ transition line, but their ratios, A/A', B/B', and D/D', are independent of position on the λ line. The values of these *universal amplitude ratios* and the value of the *critical exponent* α (all given in Table 7.1) can be calculated using the renormalization group theory of critical phase transitions, which will be discussed in a later chapter. There is, at present, no way of calculating the individual amplitudes, $A(p)$, $B(p)$, and $D(p)$. They must be taken from experiment.

Table 7.1
Specific heat parameters for liquid He*

α	A/A'	B/B'	D/D'
$-.016$	1.07	1.00	1.03

* These values are "best estimates." There is still substantial experimental and theoretical uncertainty in them.

Although liquid helium is a system of strongly interacting particles, and therefore has properties very different from the ideal Bose gas, it has one thing in common with the ideal Bose gas; it is clear that the λ transition is associated with a Bose–Einstein condensation of the helium atoms into a macroscopically occupied single-particle quantum state. Liquid helium has another striking property, namely superfluidity, that the ideal Bose gas, if it existed, would not have. The phenomenon

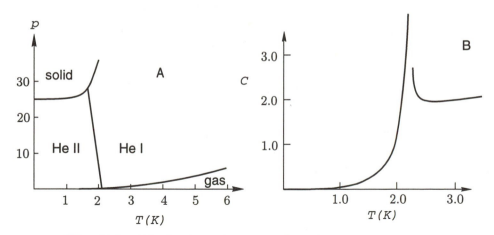

Fig. 7.8 (A) The phase diagram for ^{4}He. At zero temperature, it requires about 25 atmospheres of pressure to solidify helium. (B) The specific heat of liquid helium in equilibrium with its own vapor (in J/g-K).

of superfluidity, in which a fluid can flow through extremely narrow, crooked channels with absolutely no resistance or viscosity, seems to defy one's physical intuition. In this section we will explore the basic mechanism that leads to superfluidity.

In trying to elucidate the mechanism of superfluidity, we will consider the properties of the substance near zero temperature, where the theoretical analysis is simpler than it is close to the λ line. The most essential aspect of the analysis is the form of the quantum mechanical excitations near the ground state. According to Eq. (7.85), the total energy and momentum of a Bose–Einstein gas with mean-field interactions can be written in terms of a set of occupation numbers as

$$E = E_g + \sum_{\mathbf{P}} N_{\mathbf{p}} \varepsilon_p \tag{7.93}$$

and

$$\mathbf{P} = \sum_{\mathbf{P}} N_{\mathbf{p}} \mathbf{p} \tag{7.94}$$

where the ground-state energy, E_g, depends on the total density and $\varepsilon_p = p^2/2m$. Each occupation number can take the values $0, 1, 2, \ldots$. Any many-particle system whose total energy and momentum eigenvalues satisfy formulas of this general form is said to have a Bose–Einstein quasiparticle spectrum. For the ideal Bose gas the quasiparticles are simply the actual particles in the system, and the prefix "quasi" is therefore illogical. However, for liquid ^{4}He close to $T = 0$, it can be shown, both experimentally and theoretically, that the energy and momentum eigenstates are given by formulas like Eqs. (7.93) and (7.94) in which the quasiparticles are quantized sound waves or phonons. These excitations are similar to the phonon excitations that account for the vibrational specific heat of crystals at low temperatures, with the one difference that, for a liquid, there are only longitudinally polarized waves. The energy and momentum carried by a phonon are given in terms of its angular frequency and wave vector by $\varepsilon = \hbar\omega$ and $\mathbf{p} = \hbar\mathbf{k}$. For very long wavelengths (very small k) the usual dispersion relation for sound waves, $\omega = ck$, where c is the speed of sound, gives rise to a linear energy function.

$$\varepsilon_p = cp \tag{7.95}$$

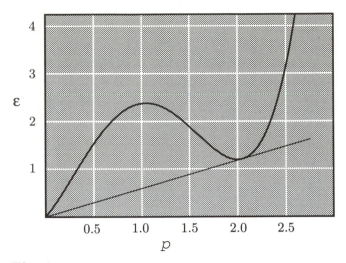

Fig. 7.9   The relation between energy and momentum for the elementary excitations in liquid helium near zero temperature. ε is given in units of 10^{-22}J and p in units of 10^{-24}kg-m/s. The slope of the straight line is defined as V_c.

As the wavelength decreases, becoming comparable to the average interparticle spacing in the liquid, the function ε_p begins to differ substantially from its long-wavelength linear form. The actual energy-momentum relation for quasiparticle excitations in liquid ^{4}He is shown in Fig. 7.9. The quasiparticle excitations in the linear part of the curve are called *phonons*; those near the local minimum are called *rotons*. There is no sharp borderline between phonons and rotons. In Fig. 7.9, a line is drawn whose slope defines what we will call the *critical velocity*, V_c. The critical velocity is the largest velocity that has the property that $V_c \leq \varepsilon_p/p$ for all quasiparticle excitations.

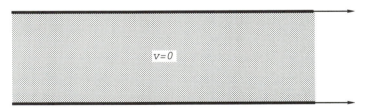

Fig. 7.10   A stationary sample of liquid helium at $T = 0$ fills an infinite pipe that is initially moving to the right.

To see how the details of the excitation spectrum are associated with the existence of superfluidity, we consider the system shown in Fig. 7.10. An infinite circular pipe, whose linear mass density is M, is filled with liquid ^{4}He in its ground state. The pipe, but not the fluid, is initially moving to the right at a speed v, less than V_c. We do not assume that the surface of the pipe is perfectly smooth. Rather, we assume that it has microscopic irregularities that would, because of the relative motion of the pipe and the fluid, be expected to transfer energy and momentum from the pipe to the fluid until both are moving with the same velocity. This would mean that, at some later time, the fluid would no longer be in its ground state but would be in some excited state of energy E and momentum P. Naturally, the energy and momentum picked up by the fluid must have come from the pipe, since the total system is assumed to be isolated. The pipe is assumed to be a heavy, classical system. We will consider what happens within a unit length of the pipe. Initially, the pipe has an energy $Mv^2/2$ and the fluid has energy E_g. At the later time, the pipe velocity has decreased to $v - \Delta v$ and the fluid is in an excited state described by occupation numbers $\{N_\mathbf{p}\}$. The energy and momentum conservation laws demand that

$$\tfrac{1}{2}Mv^2 + E_g = \tfrac{1}{2}M(v - \Delta v)^2 + E_g + \sum_\mathbf{p} N_\mathbf{p}\varepsilon_p \tag{7.96}$$

and

$$Mv = M(v - \Delta v) + \sum_\mathbf{p} N_\mathbf{p} p_x \tag{7.97}$$

If we assume that $\Delta v \neq 0$, then

$$\sum_\mathbf{p} N_\mathbf{p}\varepsilon_p = Mv\,\Delta v - \tfrac{1}{2}M(\Delta v)^2 < Mv\,\Delta v \tag{7.98}$$

and

$$M\,\Delta v = \sum_\mathbf{p} N_\mathbf{p} p_x \leq \sum_\mathbf{p} N_\mathbf{p} p \tag{7.99}$$

Using the second inequality to eliminate $M\,\Delta v$ in the first, we get

$$\sum_\mathbf{p} N_\mathbf{p}\varepsilon_p < v \sum_\mathbf{p} N_\mathbf{p} p \tag{7.100}$$

or

$$\sum_\mathbf{p} N_\mathbf{p}(\varepsilon_p - vp) < 0 \tag{7.101}$$

But this is clearly impossible, since $\varepsilon_p - vp > 0$ for all $\mathbf{p}$ and $N_\mathbf{p} \geq 0$. (Remember that we have assumed that v is less than V_c.) Thus, the only solution to the energy and momentum conservation laws is the trivial one, $\Delta v = 0$ and $N_\mathbf{p} = 0$ for all $\mathbf{p}$. In spite of its rough surface, there is no way that the pipe can transfer energy and momentum to the fluid without violating the fundamental conservation laws. Therefore, according to our argument, no energy and momentum will be transferred and the fluid will simply remain stationary, sitting there inside a rough, moving pipe! It would be very reasonable for the reader to react with extreme skepticism toward this bizarre conclusion. To overcome some of that skepticism, we describe the following real experiment.

A toroidal (doughnut-shaped) hollow tube is tightly packed with a fine powder, such as ground glass, saturated with liquid helium, and sealed (see Fig. 7.11). It is mounted on a fine wire so that it can oscillate about its axis of symmetry as a torsional pendulum within an evacuated, temperature-controlled chamber. The resonant frequency of the torsional pendulum is monitored as the temperature is gradually reduced. It remains constant until the temperature falls below T_λ. Then the oscillation frequency begins to increase and, near $T = 0$, reaches the value that it would have if there were no liquid helium present, that is, if only the container and the ground glass were contributing to the moment of inertia. The process is completely reversible. If the temperature is raised, the full moment of inertia returns. The only obvious explanation is that, below T_λ, only a part

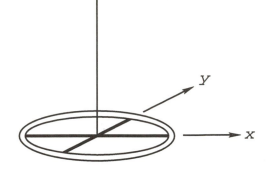

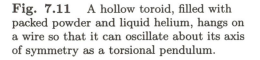

Fig. 7.11 A hollow toroid, filled with packed powder and liquid helium, hangs on a wire so that it can oscillate about its axis of symmetry as a torsional pendulum.

of the fluid (called the *normal fluid* component) is participating in the oscillatory motion, while the rest of it (the *superfluid* component) avoids moving along with the tube by winding its way, without resistance, through the narrow convoluted channels between the pieces of ground glass! One should not attempt to picture the superfluid motion as the ordinary motion of fluid particles. It is something closer to quantum tunneling on a massive scale. A way to appreciate just how "quantum mechanical" the superfluid motion is, is to remember that, if a helium atom is in a single-particle quantum state that extends throughout the free volume within the torus, then that atom is not at any particular place in the torus at all—it is simultaneously everywhere in that macroscopic volume! But, with Bose–Einstein condensation, a substantial fraction of the particles are in just such a macroscopically extended state.

Another real experiment, using the same ground glass filled torus, illustrates the wonders of superfluid motion even more clearly. By rotating the wire, the torus is spun like a top about its symmetry axis. This is done at a temperature above T_λ. Keeping the rotation constant, the temperature is gradually reduced to near 0 K. Then a small amount of friction is applied to the wire so that the torus slows down and stops. Everything looks completely quiescent. Let us call the symmetry axis (the wire axis) the z axis and choose perpendicular x and y axes in the plane of the torus. Now a small amount of torque is applied about the x axis. Due to that torque the torus rotates *about the y axis*, like a spinning top! The torus clearly carries angular momentum. The fluid inside has never stopped moving and is now flowing freely through the narrow channels. Remember, the fluid was put into rotation while it was normal ($T > T_\lambda$), kept rotating as it was cooled to $T = 0$, and only then was the torus slowed down.

7.12   QUANTIZED VORTICES

There is an important flaw in the theoretical analysis given here that we have to discuss now. The experiments that we have described all involved tubes filled with ground glass. This was not intended only to make the phenomenon seem still more spectacular. If the same experiments are done using a hollow tube, filled only with liquid helium, they will not work as described. In general, when liquid helium near zero temperature flows through a channel, it will behave as a superfluid as long as the flow velocity is less than some critical velocity V_c. But the critical velocity will not be equal to the slope of the line drawn in Fig. 7.9. Instead it will depend on the size of the channel, and for macroscopic channels it will be so small that the superfluid phenomena will be very difficult to observe. The source of the difficulty is not that the basic logic of the theoretical analysis is wrong, but rather that we

Fig. 7.12 A cylindrical bucket of liquid helium can be rotated about its symmetry axis. Actually, the bucket must be closed on top; otherwise it will empty by the liquid creeping up and down over the surface of the bucket—a characteristic of liquid helium that has not been discussed in the text.

have not taken into account all possible excited states of the system. The value of the critical velocity is determined by the ratio of the energy of an excitation to its momentum. There exist other excitations, whose nature we will describe shortly, that have much higher energies than the phonon or roton modes shown in Fig. 7.9 but that nonetheless have much lower energy-to-momentum ratios. The creation of those excitations allows energy and momentum to be transferred to the fluid at a much lower critical velocity than that determined from the phonon–roton spectrum.

To describe the nature of the excited states that we have missed, it is best to consider the case of helium within a small rotating bucket rather than the case of helium flowing through a tube (Fig. 7.12). If a bucket containing liquid helium at $T = 0$ is slowly brought into rotation, the liquid will behave in the following way. (We are now describing real experiments, not theoretical predictions, although the two things agree.) If the angular velocity of the bucket is maintained at some small value, the liquid will remain stationary as the bucket rotates around it. If the angular velocity of the bucket is increased beyond some critical angular velocity, then the stationary fluid will make an abrupt transition to a state in which there is a *vortex line* near the axis of rotation. Actually, there is a certain energy barrier (intermediate states of higher energy) between the stationary state and the state with the vortex line, so that the experimenter may be required to tap the experimental apparatus gently in order to help along the transition. But once the line has been created, further taps will not create new lines nor get rid of the one that is there. If the angular velocity of the bucket is below the critical angular velocity no amount of tapping will create a stable vortex line. The vortex line, running from top to bottom, is a very miniaturized version of the whirlpool created in an emptying sink or bathtub. The center of it, which is almost entirely empty, has a diameter of only a few Ångstroms. The fluid flows around the line with a velocity that decreases with distance from the line. The *vorticity*, which is defined as the line integral of the local fluid velocity along any path encircling the line, is exactly equal to h/m, where h is Planck's constant and m is the mass of a helium atom.

$$\oint \mathbf{v} \cdot d\ell = \frac{h}{m} \tag{7.102}$$

The existence of this quantum of vorticity is well understood. In fact, it was predicted theoretically long before it was experimentally observed. It can be derived by assuming that the superfluid velocity is related to the condensate wave function by the equation $m\mathbf{v}(\mathbf{r}) = \hbar\nabla\phi$, where the macroscopically occupied state is of the form $\psi = Re^{i\phi}$ (see Problem 7.32).

Returning to the experiment, if the angular velocity is increased further, at a second critical value a second vortex line appears. The two lines do not coalesce, but remain separated. Vortex lines of vorticity $2h/m$ are not fundamentally forbidden, but they are energetically unfavorable in comparison with two single lines. As the angular velocity is increased, more and more quantized vortex lines are created. Of course, the flow around each vortex line is in the direction of the bucket rotation. Once there are many lines, they will distribute themselves so that the average fluid velocity (averaged over a volume containing many vortex lines) will be the same as would exist in rigid body motion. Since an ordinary fluid would rotate (in a rotating bucket) with rigid body motion, the liquid helium, if not observed very carefully, would seem to be rotating like an ordinary fluid. Actually, once the first vortex line is created, the macroscopically occupied single-particle state is no longer a zero-momentum eigenstate. Rather, it is a quantum state with a macroscopic flow pattern that includes the vortex line. That is, the excited states that we have neglected in our analysis are not states in which a small vibration has been set up in an otherwise uniform stationary fluid, but ones in which all the fluid participates in a quantized flow with a velocity that is everywhere proportional to Planck's constant. For liquid helium flowing through a pipe, the excited states that are important in determining the critical flow velocity are states containing quantized *vortex rings*, as illustrated in Fig. 7.13. The critical velocity associated with the creation of quantized rings goes to zero as the channel size increases. However, for very narrow channels, V_c increases to about $50\,\mathrm{m/s}$, a value comparable to the slope of the line in Fig. 7.9.

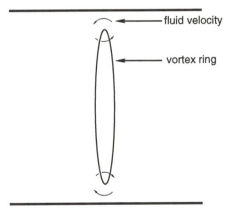

Fig. 7.13   If liquid helium is flowing through a pipe at a velocity larger than its critical velocity, it will transfer energy and momentum to the pipe by creating vortex rings, which are quantized flow patterns in which the singular vortex line is in the form of a closed ring, similar to a smoke ring (which the reader has probably never seen, since nobody smokes these days).

7.13   THE TWO-FLUID MODEL

Liquid ^{4}He, below the λ transition, behaves as if it were composed of two interpenetrating fluids that can pass through one another without friction. The first, a normal fluid, has properties similar to helium above the λ transition. That is, it has a small but finite viscosity and interacts normally with the surface of the container. The second, a superfluid, flows without friction in a flow pattern determined by the phase of the condensate wave function. The mass density of the helium, which

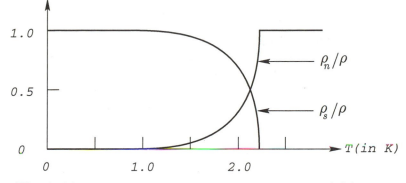

Fig. 7.14 The superfluid and normal fluid fractions, $\rho_s(T)/\rho$ and $\rho_n(T)/\rho$, between $T = 0$ and T_λ.

depends only slightly on temperature, is the sum of the mass densities of the normal and the superfluid components.

$$\rho = \rho_n(T) + \rho_s(T) \tag{7.103}$$

The superfluid and normal fractions are shown in Fig. 7.14 as functions of T at saturated vapor pressure. Notice that, below $1\,\mathrm{K}$, liquid helium is almost a pure superfluid.

Now we need to clarify a subtle and easily confused point. We called the collection of particles in the macroscopically occupied single-particle state the condensate. (For liquid helium at rest the condensate state is simply the zero-momentum state, but in the presence of superfluid flow it is more complicated.) No one has yet succeeded in measuring the fraction of particles in the condensate, but it is possible to make fairly reliable theoretical estimates of the condensate fraction. At zero temperature it is estimated that about 10% of the particles are in the zero-momentum state. But, at the same temperature, all of the particles are in the superfluid. Clearly, the superfluid is not to be identified with the condensate. At zero temperature the whole system of N particles becomes locked into a complicated N-particle ground-state wave function that extends over the full macroscopic volume. If that wave function is expanded in terms of single-particle plane waves, then the zero-momentum plane wave has an average occupation of about $0.1\,N$. But it is the complete system of particles in the N-particle ground state that moves coherently and constitutes the superfluid, not only those particles in the heavily occupied single-particle state.

For liquid helium, at close to zero temperature, it is not hard to answer the question: "What is the normal fluid?" The superfluid is then, by definition, everything else. We consider a system composed of liquid helium, in a very narrow stationary pipe, at equilibrium at some small but nonzero temperature. At time zero the pipe is instantaneously given some small velocity v that is less than the critical velocity. If the temperature were zero, then, according to our previous analysis, no change could take place in the fluid. However, at a finite temperature, thermal excitations already exist within the fluid.

If an excitation of momentum $\mathbf{p}$ is scattered, at the wall, into a state of momentum $\mathbf{p}'$, then the change in energy of the liquid will be $\Delta E = \varepsilon_{\mathbf{p}'} - \varepsilon_{\mathbf{p}}$ and the ratio of energy change to momentum change will be $(\varepsilon_{\mathbf{p}'} - \varepsilon_{\mathbf{p}})/|\mathbf{p}' - \mathbf{p}|$. This quantity is not necessarily larger than V_c. In fact, it can have any value at all. For example, if $\mathbf{p}' \approx -\mathbf{p}$, then the ratio of ΔE to $|\Delta \mathbf{p}|$ will be very small. Therefore the energy and momentum conservation laws do not prevent the scattering of existing thermal excitations, although they do still prevent the transfer of momentum from the pipe to the fluid by the creation of new excitations. The "quasiparticle gas" will absorb energy and momentum from the pipe, and its momentum distribution will shift until it comes into equilibrium with the pipe. Equilibrium will be established when the average group velocity of the quasiparticles is equal to the velocity of the pipe. But if the temperature is very low then the density of thermal excitations will also be very low, and equilibrium can be established with a very small transfer of energy and momentum from the pipe. Thus the system behaves as if the pipe were filled with a very low-density normal fluid. That is, the gas of thermal excitations *is* the normal fluid. The normal fluid mass density can be defined in terms of the density of momentum carried by the quasiparticle gas $\vec{\pi}$ and the velocity of the pipe **v**, which is the same as the average group velocity of the quasiparticles.

$$\vec{\pi} = \rho_n \mathbf{v} \tag{7.104}$$

A detailed calculation of ρ_n at low temperature is left to the exercises in the supplement to this chapter.

7.14   THERMODYNAMICS OF HELIUM NEAR $T = 0$

At low temperatures, the thermodynamic properties of liquid helium, and in particular its specific heat, can be calculated by treating the system as an ideal gas of Bose–Einstein quasiparticles with the energy–momentum relation shown in Fig. 7.9. The energies of quantized vortex states are so high that those states can be completely neglected at low temperatures. Assuming that the system is contained in a cubic box of side L, the density of allowed quasiparticle momentum values is $(L/h)^3 = V/h^3$. The grand potential of the system of quasiparticles is given by Eq. (7.40), with $\zeta = -1$ and the sum converted to an integral by using the momentum state density.

$$\psi = -\frac{V}{h^3} \int \log\left(1 - e^{-\alpha - \beta \varepsilon_p}\right) d^3\mathbf{p} \tag{7.105}$$

7.15   THE AFFINITY OF THE QUASIPARTICLE GAS

The greatest difference between the quasiparticle gas and an ordinary ideal Bose–Einstein gas is not the altered relationship between energy and momentum, but the fact that the number of quasiparticles is not a conserved quantity. The creation of a quantized sound wave does not change the number of helium atoms in the system. Thus these quantized vibrations can be freely created and absorbed at the boundary of the container. If a macroscopic variable is not constrained by a conservation law, then its value at equilibrium is determined by the maximum entropy principle of thermodynamics. That principle states that, at equilibrium,

$$\frac{\partial S}{\partial N} = 0 \tag{7.106}$$

But, by definition, $\partial S/\partial N = \alpha$, and thus, for a gas of nonconserved particles, such as the quasiparticles of liquid helium, the affinity is zero at equilibrium. (The same condition pertains to the set of quantized vibrations in a solid.) With $\alpha = 0$, the grand potential for the system is

$$\psi = -\frac{4\pi V}{h^3} \int \log\left(1 - e^{-\beta \varepsilon_p}\right) p^2 \, dp \tag{7.107}$$

The assumption that the quasiparticles do not interact is only valid when their density is small. As the temperature approaches T_λ, the density of quasiparticles increases to the point where their interactions strongly affect the excitation spectrum of the system. Near T_λ the system can no longer be treated as an ideal quasiparticle gas. Therefore, our present analysis is restricted to a range of temperature of about $0 < T < 2\,\text{K}$. The energy at the roton minimum is ε_o, where $\varepsilon_o = 8.6\,\text{K}$. Thus, below $2\,\text{K}$, only the states on the linear portion of the phonon curve and the roton states close to the minimum are significant. We can therefore make the approximation of replacing the single set of quasiparticles, which has a somewhat complicated spectrum, by two independent sets of quasiparticles. One set, which we will call phonons, has a purely linear energy–momentum curve with a slope given by the zero-temperature sound velocity. That is,

$$\varepsilon_p = cp \tag{7.108}$$

with $c = 244\,\text{m/s}$. The other set, the rotons, have a parabolic relationship between energy and momentum, with the parameters chosen to match the experimental dispersion curve at the roton minimum. This requires that, for the rotons,

$$\varepsilon_p = \varepsilon_o + \frac{(p - p_o)^2}{2\mu} \tag{7.109}$$

where

$$\varepsilon_o/k = 8.6\,\text{K}, \qquad p_o = 2.0 \times 10^{-24}\,\text{kg m/s}, \qquad \text{and} \qquad \mu = 1.0 \times 10^{-27}\,\text{kg}$$

The grand potential is the sum of the grand potentials due to the phonons and the rotons.

$$\psi = \psi_{\text{phon}} + \psi_{\text{rot}} \tag{7.110}$$

The integral for ψ_{phon} is easily calculated

$$
\begin{aligned}
\psi_{\text{phon}} &= -\frac{4\pi V}{h^3} \int_o^\infty \log(1 - e^{-\beta cp}) p^2 \, dp \\
&= -\frac{4\pi V}{(h\beta c)^3} \int_o^\infty \log(1 - e^{-x}) x^2 \, dx \quad (\text{with } x = \beta cp) \\
&= \frac{4\pi V}{3(h\beta c)^3} \int_o^\infty \frac{x^3 \, dx}{e^x - 1} \quad (\text{by partial integration}) \\
&= \frac{4\pi V}{3(h\beta c)^3} \Gamma(4)\zeta(4) \quad [\text{by Eq. (7.68)}] \\
&= \frac{4\pi^5}{45} \frac{V\tau^3}{(hc)^3} \qquad \left(\zeta(4) = \frac{\pi^4}{90} \text{ and } \tau = \beta^{-1}\right)
\end{aligned}
\tag{7.111}
$$

The roton grand potential is

$$\psi_{\text{rot}} = -\frac{4\pi V}{h^3} \int_o^\infty \log\left(1 - e^{-\beta\varepsilon_o} e^{-\beta(p-p_o)^2/2\mu}\right) p^2 \, dp \tag{7.112}$$

For $0 < T < 2\,\text{K}$, we have $e^{-\beta\varepsilon_o} < e^{-4} = 0.018$, and therefore one can use the approximation $\log(1 - x) \approx -x$, which is equivalent to using a Maxwell–Boltzmann distribution for the rotons, rather than a Bose–Einstein distribution.

$$\psi_{\text{rot}} = \frac{4\pi V}{h^3} e^{-\beta\varepsilon_o} \int_o^\infty e^{-\beta(p-p_o)^2/2\mu} p^2 \, dp \tag{7.113}$$

The maximum of the exponential factor in the integral, which occurs at $p = p_o$, is unity. If $p < 0$ and $T < 2\,\mathrm{K}$, then $\beta(p - p_o)^2/2\mu > 144$, which means that $\exp[-\beta(p - p_o)^2/2\mu] < e^{-144}$. Thus we can extend the integration to $-\infty$.

$$
\begin{aligned}
\psi_{\mathrm{rot}} &= \frac{4\pi V}{h^3} e^{-\varepsilon_o/\tau} \int_{-\infty}^{\infty} e^{-(p-p_o)^2/2\mu\tau} p^2 \, dp \qquad (\tau = kT = 1/\beta) \\
&= \frac{4\pi V}{h^3} e^{-\varepsilon_o/\tau} \int_{-\infty}^{\infty} e^{-q^2/2\mu\tau} (q^2 + 2p_o q + p_o^2) \, dq \qquad (q = p - p_o) \qquad (7.114) \\
&= \frac{4\pi V}{h^3} \sqrt{2\pi\mu\tau} (p_o^2 + \mu\tau) e^{-\varepsilon_o/\tau}
\end{aligned}
$$

These formulas for ψ_{phon} and ψ_{rot} will be used in Problem 7.31 to calculate the specific heat of liquid helium at low temperatures.

7.16   BLACKBODY RADIATION

If a cavity, whose walls are kept at a finite temperature, is completely evacuated of ordinary matter, its energy density does not become zero, because it is impossible to eliminate from the cavity the electromagnetic radiation that is constantly being emitted and absorbed by the walls. This section is devoted to a study of the physical properties of such cavity radiation. A perfectly black body (one that absorbed all radiation that was incident upon it) would, at equilibrium within the cavity, have to emit radiation with the same intensity and frequency distribution as the cavity radiation. Therefore, the type of electromagnetic radiation found within a cavity at equilibrium is also referred to as *blackbody radiation*. The electromagnetic energy within the cavity may be treated as an ideal gas of massless Bose–Einstein particles, called *photons*. Since the relation between the momentum and the wave vector, $\mathbf{p} = \hbar\mathbf{k}$, which is valid for nonrelativistic particles, is also true for zero mass photons, the allowed momentum states for photons in a periodic box of volume $V = L^3$ are the same as those for nonrelativistic particles. However, for each allowed momentum state, there are two possible values of the photon spin, corresponding to the two possible transverse polarization states of an electromagnetic wave. The energy of a photon of momentum $\mathbf{p}$ is

$$
\varepsilon_p = cp \qquad (7.115)
$$

Since photons are not a conserved particle species, as are electrons or neutrinos, the argument given in Section 7.15 applies, and one may conclude that the affinity α is exactly zero. Using the facts that the density of single-particle quantum states in momentum space is $2(L/h)^3$ and that $\alpha = 0$ for the photon gas, Eq. (7.40) for the grand potential can be written as

$$
\psi(\beta, V) = -\frac{2V}{h^3} \int \log[1 - e^{-\beta cp}] \, d^3\mathbf{p} \qquad (7.116)
$$

Except for the factor of 2, this is the integral that is worked out in Eq. (7.111).

$$
\psi(\beta, V) = \frac{8\pi^5}{45} \frac{V\tau^3}{(hc)^3} \qquad (7.117)
$$

The pressure and energy density of the radiation gas can now be obtained from the grand potential by the thermodynamic identities

$$
p = \tau \frac{\partial \psi}{\partial V} = \frac{8\pi^5}{45} \frac{\tau^4}{(hc)^3} \qquad (7.118)
$$

and

$$\frac{E}{V} = -\frac{1}{V}\frac{\partial \psi}{\partial \beta} = \frac{8\pi^5}{15}\frac{\tau^4}{(hc)^3} \tag{7.119}$$

Notice that, because of the modified relationship between the energy and the momentum of the particles, the photon gas obeys a relation $p = \frac{1}{3}E/V$, rather than $p = \frac{2}{3}E/V$.

Equation (7.119) shows that the electromagnetic energy density in a cavity is proportional to the fourth power of the temperature. This relationship is called *Stefan's law*.

7.17 THE FREQUENCY DISTRIBUTION

In order to determine the frequency distribution function of the radiation, we write $E/V = -(\partial \psi/\partial \beta)/V$ as an integral, using Eq. (7.116).

$$\begin{aligned}
\frac{E}{V} &= \frac{2}{h^3}\int \frac{cpe^{-\beta cp}}{1-e^{-\beta cp}}d^3\mathbf{p} \\
&= \frac{2}{h^3}\int \frac{cp}{e^{cp/\tau}-1}d^3\mathbf{p}
\end{aligned} \tag{7.120}$$

Changing to an angular frequency variable, $\hbar\omega = cp$, and using the fact that $d^3\mathbf{p} \to 4\pi p^2 dp = 4\pi(\hbar/c)^3\omega^2 d\omega$, gives

$$\frac{E}{V} = \frac{\hbar}{\pi^2 c^3}\int_o^\infty \frac{\omega^3}{e^{\hbar\omega/\tau}-1}d\omega \tag{7.121}$$

The frequency distribution function, $D(\omega, T)$ is defined by saying that the energy density of radiation with angular frequency within the interval $d\omega$ at temperature T is $D(\omega, T)\,d\omega$. It is clear from Eq. (7.121) that

$$D(\omega, T) = \frac{\hbar}{\pi^2 c^3}\frac{\omega^3}{e^{\hbar\omega/kT}-1} \tag{7.122}$$

At any given temperature, the dominant frequency is determined by finding the value of ω that maximizes $D(\omega, T)$. The equation $\partial D/\partial \omega = 0$ gives

$$\frac{e\omega^2}{e^{\hbar\omega/kT}-1} - \frac{(\hbar/kT)\omega^3}{(e^{\hbar\omega/kT}-1)^2} = 0 \tag{7.123}$$

which can be written in the form

$$\frac{\hbar\omega}{kT} = 3(1 - e^{-\hbar\omega/kT}) \tag{7.124}$$

The solution of the equation $x = 3(1 - e^{-x})$ is $x = 2.82144$, and therefore at temperature T the maximum in the frequency distribution function occurs at

$$\omega_{\max} = 2.82144\frac{kT}{\hbar} \tag{7.125}$$

That $\omega_{\max}$ is directly proportional to T is *Wien's displacement law*. These predictions are all in excellent agreement with experimental results.

PROBLEMS

7.1 Do the 'equivalent calculation' mentioned just prior to Eq. (7.8).

7.2 For a system of noninteracting one-dimensional classical particles in a harmonic oscillator potential: (a) Calculate the grand potential, $\psi(\alpha, \beta)$. Note that there is no fixed "volume" for this system. (b) Obtain N and E as functions of α and β and show that they satisfy the equipartition theorem.

7.3 (a) Using the grand canonical ensemble, verify the fluctuation formulas used in Chapter 5, namely $(\Delta N)^2 = \partial^2 \psi / \partial \alpha^2$ and $(\Delta E)^2 = \partial^2 \psi / \partial \beta^2$. (b) Show that, for a macroscopic system, $\Delta N / N$ and $\Delta E / E$ are both of order $1/\sqrt{N}$ and are therefore completely negligible.

7.4 If the adsorbed particles on a surface are free to move across the surface, then they may sometimes be approximated as a two-dimensional ideal gas in which the energy of a particle of momentum (p_x, p_y) is $\varepsilon(\mathbf{p}) = (p_x^2 + p_y^2)/2m - \varepsilon_o$, where ε_o is the binding energy of the particle to the surface. Using this approximation, calculate the surface density of adsorbed particles if the pressure in the three-dimensional gas, in equilibrium with the adsorbed particles on its surface, is p.

7.5 Diatomic molecules are sometimes decomposed, upon adsorption, into their constituent atoms, attached to the surface. Consider a gas of diatomic nitrogen (N_2), and assume that the vibrational excitations can be ignored, so that Eq. (4.65) is applicable. Assuming that the gas is in equilibrium with K *atomic* adsorption sites, derive the equivalent of Eq. (7.24) for the fraction of occupied sites.

7.6 For a surface with K adsorption sites and a single bound state at each site, show that the entropy (in rational units) of the adsorbed particles can be written in the form
$$S^o = -K\big[f \log f + (1 - f) \log(1 - f)\big]$$
where $f = N/K$ is the fraction of sites that are filled.

7.7 It is clear from Eq. (7.7) that the grand partition function is given by $\Lambda(\alpha, \beta, V) = \sum e^{-\alpha N} Z_N(\beta, V)$, where Z_N is the canonical partition function for a system of N particles. Using the fact that $Z_N(\beta, V) = \exp \phi(N, \beta, V)$, where ϕ is the canonical potential, approximate the sum by an integral and use the method described in Section A.9 to show that

$$\psi \equiv \log \Lambda = \phi - \alpha N = S^o - \alpha N - \beta E$$

where $N(\alpha, \beta, V)$ is the solution of $\alpha = \partial \phi(N, \beta, V)/\partial N$.

7.8 (a) Equation (7.47) gives the grand potential $\psi = \beta p V$ for a gas of spin-$\frac{1}{2}$ fermions. By a partial integration, show that $pV = \frac{2}{3}E$ for the Fermi–Dirac ideal gas. (b) Do the same for a gas of spinless bosons.

7.9 If a hydrogen atom is placed in a magnetic field, its nucleus (a proton) can have its spin, and therefore its magnetic moment, oriented either parallel or antiparallel to the field. The energy levels of the two orientations are $\pm \mu B$, where $\mu = 2.79255 \, \mu_N$. ($\mu_N = e\hbar/2M_p = 5.05 \times 10^{-27}$ J/T is the *Bohr nuclear magneton*.) (a) In a field of 2 Tesla, at a temperature of 300 K, what is the ratio $R(T) = (N_- - N_+)/(N_- + N_+)$.

(N_- and N_+ are the numbers of parallel and antiparallel spins, respectively.) (b) If a substance containing hydrogen is placed in a magnetic field and subjected to electromagnetic radiation at a frequency $\nu = 2\mu B/h$, energy will be absorbed from the radiation field, causing transitions between the two nuclear magnetic energy states. The absorption rate is proportional to $R(T)$. This is the phenomenon of *nuclear magnetic resonance.* (c) For a field of 2 T, in what range is the nuclear magnetic resonance frequency (infrared, microwave, etc.)?

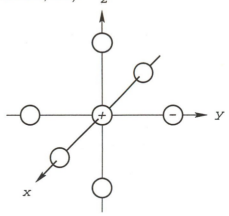

Fig. 7.15

7.10 Shown in Fig. 7.15 is an impurity atom in a cubic crystal of lattice constant a. The atom has become ionized by transferring an electron to one of its six nearest neighbors. Assuming that, in the absence of an electric field, the six positions are equivalent, determine the average polarization of the ion-electron system when an electric field **E** is imposed in the z direction.

7.11 In the grand canonical ensemble, the probability of finding the system with exactly N particles in some particular state is given by Eq. (7.6). Assume that the system is an ideal monatomic gas in a volume V, at temperature T. (a) Calculate $\bar{N}$, the average number of particles in the system, as a function of the affinity α. (b) Show that ΔN, the uncertainty in N, satisfies Eq. (5.73). That is, that $(\Delta N)^2 = -\partial\bar{N}/\partial\alpha$. (c) For one mole of gas at STP, calculate $\Delta N/\bar{N}$.

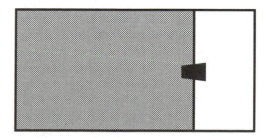

Fig. 7.16

7.12 Consider the system shown in Fig. 7.16. The volume on the left is $\frac{3}{4}$ of the total volume. In the left-hand side is a Fermi–Dirac gas at zero temperature. In the initial state $\varepsilon_F = 1\,\text{eV}$. The right-hand side is completely empty, and the system is totally isolated. At $t = 0$ the plug is pulled. What is the temperature when the system comes to equilibrium?

7.13 If the Fermi energy is much larger than the rest energy of an electron $(\varepsilon_F \gg m_e c^2)$, then the relationship between energy and momentum can be approximated by that for zero-mass particles; that is, $E = cp$. For such an ultrarelativistic electron gas, obtain an expansion for the energy and pressure up to order τ^2.

7.14 Massless particles have the energy–momentum relationship $\varepsilon = cp$. (a) For a system of spinless one-dimensional Fermi–Dirac massless particles, determine the exact relationship involving N, μ, and τ. (b) Show that, at $\tau = 0$, $\mu = \varepsilon_F$.

7.15 Neutrinos are zero-mass spin-$\frac{1}{2}$ Fermi–Dirac particles. Although neutrinos have spin-$\frac{1}{2}$, there is only one neutrino state for each momentum, because of the fact that the spin angular momentum of a neutrino is always antiparallel to its momentum vector. Treating the neutrino number as a strictly conserved quantity, obtain expansions for p and E/N, up to order τ^2, for a neutrino gas.

7.16 The antiparticle to the neutrino is the antineutrino, a particle with the same properties as the neutrino, but with its spin angular momentum always parallel to its linear momentum. In fact, the neutrino number N_ν and the antineutrino number $N_{\bar{\nu}}$ are not separately conserved, but only the *lepton number*, $L = N_\nu - N_{\bar{\nu}}$, is conserved. Due to creation of neutrino–antineutrino pairs, the total number of particles, $N = N_\nu + N_{\bar{\nu}}$, can change spontaneously. A neutrino–antineutrino gas may be treated as two interpenetrating Fermi–Dirac ideal gases with a chemical reaction, $\nu + \bar{\nu} \leftrightarrow 0$. Using the chemical equilibrium equation, calculate the pressure of such a system as an expansion in τ, separately, for the two cases $L = 0$ and $L > 0$. (Note: This problem requires the solution of Problem 7.15.)

7.17 Calculate the specific heat at low temperature of a two-dimensional Fermi–Dirac ideal gas of spin-$\frac{1}{2}$ particles.

7.18 The baloneyon is an imaginary Fermi–Dirac particle with spin-$\frac{1}{2}$ and the following relationship between energy and momentum.

$$E = B|\mathbf{p}|^4$$

where B is the baloney constant. What is the Fermi energy of a system of baloneyons as a function of the particle density?

7.19 Consider a mixture of N α particles and $2N$ electrons within a volume V. When the average kinetic energy of the electrons, due to the Pauli Exclusion Principle, is much larger than the binding energy of an electron bound to an α particle, then the electrons will be stripped off the α particles and the system can be approximated as an interpenetrating Fermi–Dirac electron gas and a classical gas of α particles, rather than as a gas of helium atoms, which it would become at lower densities. This situation exists in high-density stars. (a) Determine the mass density of the system when the average kinetic energy of the electrons (at $0\,K$) is ten times the binding energy of a single electron to an α particle in free space. (b) How does it compare with the estimated mass densities at the center of the sun $(10^5\,\mathrm{kg/m^3})$ and in a white dwarf $(2 \times 10^9\,\mathrm{kg/m^3})$? (c) The internal temperature of white dwarf stars is about $10^7\,K$. What is the ratio of τ to ε_F in a white dwarf? (This ratio determines whether the zero-temperature approximation is accurate.)

7.20 Copper has a density of about $8.9\,\mathrm{g/cm^3}$. Copper has one conduction electron per atom. (a) Calculate the pressure of the conduction electron gas at $0\,K$. (b) Calculate the increase in pressure when the temperature is raised to 300 K. (c) What is holding copper together against the pressures obtained in (a) and (b)?

7.21   The pressure in a finite-density Fermi–Dirac ideal gas does not go to zero as the temperature approaches zero. Show that the entropy satisfies the limit equations mentioned in Section 5.18; namely

$$\lim_{E \to E_o} \left(\frac{\partial S}{\partial V} \right) = \infty \quad \text{and} \quad \frac{\partial}{\partial V} \left(\lim_{E \to E_o} S \right) = 0$$

7.22   In this problem we want to show that, below the Bose–Einstein condensation temperature, it is not necessary to treat any state except the zero momentum state individually. Consider the thermodynamic limit, $N = nL^3 \to \infty$, with fixed n and β. Show that the number of particles in any fixed finite number of momentum states that does not include the zero-momentum state, divided by N, approaches zero, even for $\alpha = 0$. (Hint: The most heavily occupied nonzero momentum states are the ones next to the zero momentum state.)

7.23   Show that, in an ideal Bose gas, the Bose–Einstein transition occurs when the number of particles in a cubic thermal de Broglie wavelength is equal to 2.612. That is, when $n\lambda^3 = \zeta(\frac{3}{2})$.

7.24   Using Eqs. (7.68) and (7.71), derive the function $\alpha(n, \tau)$ in phase II of an ideal Bose–Einstein gas.

7.25   For an ideal Bose–Einstein gas, calculate the ratio of the condensate density to the total density as a function of τ/τ_c. [τ_c is defined in Eq. (7.78).]

7.26   (a) For a two-dimensional gas of spin-zero particles in a periodic box of area L^2, calculate the number of momentum eigenstates with energies within the interval ε to $\varepsilon + d\varepsilon$. (b) Use the result of (a) to obtain an integral formula for N/A as a function of α and τ for a 2D Bose–Einstein gas. (c) Show that the result obtained in (b) implies that no Bose–Einstein condensation occurs for the two-dimensional ideal Bose–Einstein gas.

7.27   (a) Consider a system of conserved bosons for which the relationship between energy and momentum is not $\varepsilon = p^2/2m$, but $\varepsilon = cp$, where $p = |\mathbf{p}|$. For this system, calculate the Bose–Einstein condensation temperature τ_c as a function of the particle density. (b) For a system of particles in which $\varepsilon = cp^3$, show that there is no Bose–Einstein condensation.

7.28   For a particle in a gravitational potential $U = mgz$, the Schrödinger equation for the ground state is

$$-\frac{\hbar^2}{2m} \frac{d^2\psi(z)}{dz^2} + U\psi(z) = E_o\psi(z)$$

Assume that a "hard floor" exists at $z = 0$, so that ψ has the boundary condition $\psi(0) = 0$. (a) Show that the ground-state wave function has the form $\psi(z) = u(z/s)$, where $s = (\hbar^2/m^2 g)^{1/3}$ and $u(x)$ satisfies an eigenvalue equation that does not contain the parameters $\hbar$, m, or g. Thus the parameter s gives the scale height for a quantum particle in a gravitational potential. For an ideal Bose–Einstein gas at zero temperature, all the particles will go into this single-particle ground state. (b) Evaluate s for the earth's gravitational field and m equal to the mass of a ^{4}He atom.

7.29   (a) Write the ground-state wave function for a single particle of mass m in a hard-walled box of volume L^3. (b) For an N-particle Bose–Einstein ideal gas in such a hard-walled box, determine the particle density as a function of position at temperature zero.

7.30*   (a) It is a mathematical theorem that a function $f(x,y)$ is strictly convex if and only if $\partial^2 f/\partial x^2 < 0$, $\partial^2 f/\partial y^2 < 0$, and $(\partial^2 f/\partial x^2)(\partial^2 f/\partial y^2) > (\partial^2 f/\partial x\,\partial y)^2$. Prove that $s^o(n,\varepsilon)$, given in Eq. (7.87), is a strictly convex function. (b) Show that the peculiar characteristic of the ideal Bose gas illustrated in Fig. 7.7 disappears for the Bose gas with mean-field interactions.

7.31   Using Eq.s 7.111 and 7.114, separately calculate and plot the contributions to the specific heat, $C = \partial E/\partial \tau$, of liquid helium due to phonons and rotons.

7.32   Assume that the condensate wave function has the form $\psi(\mathbf{r}) = R\exp[i\phi(\mathbf{r})]$ and that the superfluid velocity at point $\mathbf{r}$ is related to ψ by $m\mathbf{v} = \hbar\,\nabla\phi(\mathbf{r})$. From the single valuedness of ψ, derive the quantization of vorticity [Eq. (7.102)].

7.33   Calculate the number of photons in a 1 cm cube at 300 K.

7.34   Consider a small area dA on the inside surface of a cavity at temperature T. Show that the electromagnetic energy that falls upon that area, per second, can be written as $I(\omega,T)\,dA$, where the *spectral intensity function* $I(\omega,T) = (c/4)D(\omega,T)$. Thus, the frequency distribution function $D(\omega,T)$ also gives the frequency distribution of the radiation that would pour out of a hole in a cavity at temperature T. This is not true for the speed distribution for nonrelativistic particles. The speed distribution for particles within a box (see Problem 2.8) is not the same as the speed distribution of particles exiting from a small hole in the box. Explain what is the essential characteristic of photons that makes I proportional to D.

Chapter 8
Systems of Interacting Particles and Magnetism

8.1 INTRODUCTION

Up to this point, we have only treated various forms of ideal systems, that is, systems without interparticle interactions. The analysis of lattice vibrations may seem to be an exception to this statement, in that the particles in the solid interact with harmonic forces, but the system of lattice vibrations was completely equivalent to a collection of noninteracting harmonic oscillators. In any real system, the interactions are essential in determining the physical properties of the substance. It is the interactions between the molecules that cause a gas to condense to a liquid state and that create the varieties of crystal structure in the solid state. Unfortunately, with the exception of certain simple models in one and two dimensions, a complete statistical mechanical calculation cannot be carried through exactly for any interacting system. This is not surprising. In classical and quantum mechanics, even the three-body problem cannot be solved exactly, and in statistical mechanics we are faced with the N_A-body problem. What is remarkable is how much progress has been made in making reliable and accurate approximate calculations of the properties of interacting systems.

In this chapter we will discuss three important methods of dealing with interactions. The first is a method that generates a power series expansion, called the *cluster expansion*, that is particularly useful in deriving the properties of dense gases. It allows a systematic calculation of the equations of state of a gas in terms of the intermolecular interaction potential of the gas molecules. Often it is used in the reverse mode; that is, the equations of state are experimentally determined and that information is then used, via the cluster expansion, to calculate the interaction potential. The second method is called the *mean-field approximation*. It is a method of calculating, approximately, the properties of phase transitions in strongly interacting systems by neglecting the statistical correlations between neighboring particles. The third method is a numerical computational scheme, called the *Monte*

Carlo method. It has been vital in obtaining accurate results for many strongly interacting systems, such as liquids and solids. It is basically an efficient method of generating a sampling sequence, in the sense of Chapter 1, for the canonical ensemble probability density.

8.2   THE IDEAL GAS IN THE GRAND CANONICAL ENSEMBLE

In preparation for deriving the cluster expansion, it is useful to calculate the grand partition function and the grand potential of a classical ideal gas. The grand partition function is given in terms of H_N, the Hamiltonian function of the system with N particles, by

$$\Lambda = 1 + \sum_N \frac{e^{-\alpha N}}{h^{3N} N!} \int e^{-\beta H_N} \, d^{3N}x \, d^{3N}p \tag{8.1}$$

We define a function $\zeta(\alpha, \beta)$, called the *activity*, by

$$\begin{aligned}
\zeta(\alpha, \beta) &\equiv \frac{e^{-\alpha}}{h^3} \int e^{-\beta p^2/2m} \, d^3p \\
&= e^{-\alpha}(2\pi m/\beta h^2)^{3/2} \\
&= e^{-\alpha}/\lambda^3
\end{aligned} \tag{8.2}$$

where λ is the thermal de Broglie wavelength. For an ideal gas, $H_N = \sum p_n^2/2m$, and therefore the integral in Eq. (8.1) can be expressed easily in terms of the activity.

$$\frac{e^{-\alpha N}}{h^{3N}} \int e^{-\beta H_N} \, d^{3N}x \, d^{3N}p = V^N \zeta^N \tag{8.3}$$

Λ is then given as a well-known power series in the activity.

$$\Lambda = \sum_{N=0}^{\infty} \frac{V^N \zeta^N}{N!} = e^{V\zeta} \tag{8.4}$$

The grand potential is therefore

$$\psi = \log \Lambda = V\zeta \tag{8.5}$$

The thermodynamic relation $N = -\partial\psi/\partial\alpha$ and the fact that $\partial\zeta/\partial\alpha = -\zeta$ yields $N = V\zeta$, which shows that for the ideal gas ζ is equal to the particle density.

$$\zeta = \frac{N}{V} = n \tag{8.6}$$

This suggests that, for gases with interactions (usually called *real gases*), an expansion of ψ as a power series in the activity would be essentially an expansion in terms of the density. But at low density the ideal gas approximation is accurate. Therefore, such an expansion would yield systematic corrections to the ideal gas equations of state due to the interparticle interactions.

8.3   THE CLUSTER EXPANSION FOR A REAL GAS

A gas of N interacting particles has a Hamiltonian function that is a sum of a kinetic energy term and a potential energy term.

$$H_N = \sum_{i=1}^{N} \frac{p_i^2}{2m} + U_N(\mathbf{r}_1, \dots, \mathbf{r}_N) \tag{8.7}$$

Using Eqs. (8.1) and (8.2), the grand partition function Λ can be written in terms of the activity and the $3N$-dimensional integrals

$$I_N = \int e^{-\beta U_N} d^3\mathbf{r}_1 \cdots d^3\mathbf{r}_N \tag{8.8}$$

in the form

$$\Lambda = 1 + I_1\zeta + \tfrac{1}{2}I_2\zeta^2 + \frac{1}{3!}I_3\zeta^3 + \cdots \tag{8.9}$$

Using the following expansion for the logarithmic function

$$\log(1+x) = x - \tfrac{1}{2}x^2 + \tfrac{1}{3}x^3 - \tfrac{1}{4}x^4 + \cdots \tag{8.10}$$

the grand potential, $\psi = \log \Lambda$, can be expressed as a power series in the activity (only terms necessary to carry out the expansion to order ζ^3 are shown explicitly).

$$\psi = (I_1\zeta + \tfrac{1}{2}I_2\zeta^2 + \frac{1}{3!}I_3\zeta^3 + \cdots) - \tfrac{1}{2}(I_1\zeta + \tfrac{1}{2}I_2\zeta^2 + \cdots)^2 + \tfrac{1}{3}(I_1\zeta + \cdots)^3 - \cdots \tag{8.11}$$

Expanding out the brackets, we get, to third order in ζ,

$$\psi = I_1\zeta + \tfrac{1}{2}(I_2 - I_1^2)\zeta^2 + \tfrac{1}{6}(I_3 - 3I_1I_2 + 2I_1^3)\zeta^3 + \cdots \tag{8.12}$$

The coefficient of ζ^N divided by the volume is called the Nth *cluster integral* C_N. It is easy to see that the first cluster integral is equal to one. ($C_1 = I_1/V = 1$.) In terms of the cluster integrals,

$$\psi = V(\zeta + C_2\zeta^2 + C_3\zeta^3 + \cdots) \tag{8.13}$$

The thermodynamic identity $\psi = \beta pV$ shows that the grand potential is proportional to the system volume. Since ζ is a function of α and β, and therefore independent of V, Eq. (8.13) implies that for large systems the cluster integrals are functions only of the inverse temperature β. This will be confirmed immediately for C_2 and in the exercises for C_3. Comparing Eqs. (8.12) and (8.13), we see that

$$C_2 = \frac{1}{2V}(I_2 - I_1^2) = \frac{1}{2V}\int (e^{-\beta U_2} - 1)\, d^3\mathbf{r}_1\, d^3\mathbf{r}_2 \tag{8.14}$$

The interaction potential for two particles is a function only of the distance between them. That is, $U_2 = v(|\mathbf{r}_2 - \mathbf{r}_1|)$. In the integral, we can transform to center-of-mass and relative variables, $\mathbf{R} = (\mathbf{r}_1 + \mathbf{r}_2)/2$ and $\mathbf{r} = \mathbf{r}_2 - \mathbf{r}_1$. Because the interaction potential is negligible beyond a few Ångstroms, the integrand is zero unless $\mathbf{r}_1$ is very close to $\mathbf{r}_2$. Therefore, the relative variable can be extended to infinity and the center-of-mass variable can be integrated over the volume of the system. This gives

$$C_2 = \frac{1}{2} \int (e^{-\beta v(r)} - 1)\, d^3\mathbf{r} = 2\pi \int_o^\infty (e^{-\beta v(r)} - 1) r^2 dr \qquad (8.15)$$

If the interparticle interaction potential is known, then this equation allows one to compute $C_2(\beta)$.

Equation (8.12), when combined with the relations $\beta p = \psi/V$, $n = -\partial(\psi/V)/\partial\alpha$, and $\partial\zeta/\partial\alpha = -\zeta$, gives both the pressure and density as power series expansions in the activity.

$$\frac{p}{kT} = \zeta + C_2\zeta^2 + C_3\zeta^3 + \cdots \qquad (8.16)$$

$$n = \zeta + 2C_2\zeta^2 + 3C_3\zeta^3 + \cdots \qquad (8.17)$$

8.4   THE POTENTIAL ENERGY OF A REAL GAS

The justification for studying the ideal gas has been the assumption that, under ordinary conditions, the average value of the potential energy of most common gases is only a small fraction of the value of the kinetic energy. When that assumption is true, it is reasonable to ignore the potential energy term in the Hamiltonian function. Now that we have a method for dealing with interparticle interactions (namely, the cluster expansion), it is possible to check on the accuracy of the ideal gas approximation by actually calculating the expectation value of the potential energy for typical real gases.

Equation (8.12) gives an expansion for the grand potential as a power series in the activity.

$$\psi = V[\zeta + C_2(\beta)\zeta^2 + C_3(\beta)\zeta^3 + \cdots] \qquad (8.18)$$

The energy of the gas is given by the thermodynamic relation $E = -\partial\psi(\alpha, \beta, V)/\partial\beta$. Remembering that the activity ζ is a function of α and β, we can express E in the form

$$E = -V(1 + 2C_2\zeta + 3C_3\zeta^2 + \cdots)\frac{\partial\zeta}{\partial\beta} - V(C_2'\zeta^2 + C_3'\zeta^3 + \cdots) \qquad (8.19)$$

where $C_n' = \partial C_n(\beta)/\partial\beta$. From Eq. (8.2) we see that

$$\frac{\partial\zeta}{\partial\beta} = -\tfrac{3}{2}\beta^{-1}\zeta \qquad (8.20)$$

When this is used in Eq. (8.19), and one recognizes the series for the density $n = N/V$ from Eq. (8.17), the equation becomes

$$E = \tfrac{3}{2}NkT - V[C_2'(\beta)\zeta^2 + C_3'(\beta)\zeta^3 + \cdots] \qquad (8.21)$$

The first term is clearly the kinetic energy which, for a classical monatomic gas is always equal to $\frac{3}{2}NkT$. The expansion that follows that term must therefore give the potential energy in terms of the inverse temperature and the activity.

$$\frac{E_{\text{pot}}}{V} = -C_2'(\beta)\zeta^2 + C_3'(\beta)\zeta^3 - \cdots \qquad (8.22)$$

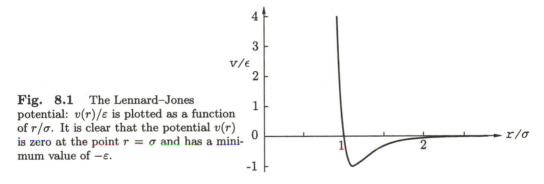

Fig. 8.1 The Lennard–Jones potential: $v(r)/\varepsilon$ is plotted as a function of r/σ. It is clear that the potential $v(r)$ is zero at the point $r = \sigma$ and has a minimum value of $-\varepsilon$.

For a low-density gas we can keep only the first term and also use the first term in Eq. (8.17) to replace ζ by n. When that is done, we obtain the following approximation for the average potential energy per particle.

$$\frac{E_{\text{pot}}}{N} \approx -C_2'(\beta)n \qquad (8.23)$$

Using Eq. (8.15), which gives $C_2(\beta)$ as a one-dimensional integral involving the interaction potential, this can be written as

$$\frac{E_{\text{pot}}}{N} \approx 2\pi n \int_o^\infty v(r)e^{-\beta v(r)}r^2 \, dr \qquad (8.24)$$

In order to proceed further, we must know the interaction potential $v(r)$. The interaction potentials of real atoms cannot be expressed exactly in any simple algebraic form. However, a simple function containing two parameters that gives a reasonably good fit to the interaction potentials of many common substances is the *Lennard–Jones potential.*

$$v(r) = 4\varepsilon[(\sigma/r)^{12} - (\sigma/r)^6] \qquad (8.25)$$

The parameters ε and σ, which must be chosen separately for each gas, have simple physical interpretations. ε is the depth of the potential at its lowest point. σ is the distance at which the potential is zero. As can be seen from Fig. 8.1, the distance σ is a reasonable definition of the diameter of the atom. Table 8.1 gives Lennard–Jones parameters and the ratio of the potential energy to the kinetic energy for four noble gases at standard temperature and pressure. One can see that, although the average value of the potential energy is much smaller than the average of the kinetic energy, it is by no means undetectably small.

	$\sigma(\text{Å})$	$\varepsilon(\text{K})$	$\dfrac{E(\text{pot})}{E(\text{kin})}$
Ne	2.8	35	$-.005$
A	3.4	120	$-.023$
Kr	3.6	180	$-.036$
Xe	4.1	225	$-.063$

TABLE 8.1 Lennard–Jones parameters for selected atoms

8.5 THE VIRIAL EXPANSION

In order to compare theory with experiment, one would like to eliminate the affinity (or the activity) and write the pressure as a function of the temperature and the density of the gas. The cluster expansions for p and n allow us to write the pressure as an expansion in powers of the density. That expansion is called the *virial expansion*, and the coefficients in it are known as the *virial coefficients*. It is of the form

$$\frac{p}{kT} = n + B_2(T)n^2 + B_3(T)n^3 + \cdots \tag{8.26}$$

It is fairly easy to use Eqs. (8.16), (8.17), and (8.26) to write the first few virial coefficients in terms of the cluster integrals. Working out higher ones becomes progressively more complicated. In Problem 8.1 the reader is asked to show that

$$B_2 = -C_2 \quad \text{and} \quad B_3 = -2C_3 + 4C_2^2 \tag{8.27}$$

Instead of going any further in the formal analysis, we will look at the effects, on the equation of state of a very simple two-particle interaction for which B_2 can be calculated exactly.

8.6 THE HARD-CORE PLUS STEP POTENTIAL

Realistic interactions between noble gas atoms are strongly repulsive at short distances but mildly attractive at larger distances. The interaction potential shown in Fig. 8.1 is typical. The simplest potential function with those characteristics is the combination of a hard-core potential plus an attractive step potential. It is defined here and illustrated in Fig. 8.2.

$$v(r) = \begin{cases} +\infty, & 0 < r < a \\ -V_o, & a < r < b \\ 0, & b < r < \infty \end{cases} \tag{8.28}$$

For this potential the second virial coefficient can be computed exactly.

$$\begin{aligned} B_2(T) &= -2\pi \int_0^\infty (e^{-v(r)/kT} - 1)r^2 \, dr \\ &= -2\pi\left(-\int_0^a r^2 dr + \int_a^b (e^{V_o/kT} - 1)r^2 \, dr\right) \\ &= \tfrac{2}{3}\pi a^3 - \tfrac{2}{3}\pi(e^{V_o/kT} - 1)(b^3 - a^3) \end{aligned} \tag{8.29}$$

The behavior of B_2 as a function of T in this very simple model is typical of what is found for more realistic interaction potentials. At low temperatures the second term, which is negative, becomes much larger than the first term, which is positive. Therefore, at low T the second virial coefficient is negative. This indicates that the pressure at low temperature is less than it would be for an ideal gas.

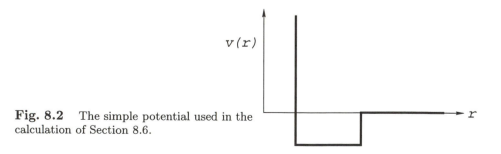

Fig. 8.2   The simple potential used in the calculation of Section 8.6.

Fig. 8.3 A graph of p/RT (in units of moles/cm^3) as a function of the density for a gas satisfying a two-term virial equation of state with $B_2(T)$ given by Eq. (8.29). The constants a, b, and V_o, have been chosen to approximately match the second virial coefficient of neon. For neon, $B_2(T)$ is negative for $T < 125$ K and positive for $T > 125$ K. Since $B_2(T) = 0$ at $T = 125$ K, the curve for that temperature is identical with the ideal gas equation, $p = nRT$. For positive B, the pressure increases, with increasing density, faster than that of an ideal gas. For negative B, it increases more slowly. For negative B the pressure would eventually become negative, but the two-term virial equation of state is not valid in that range of density.

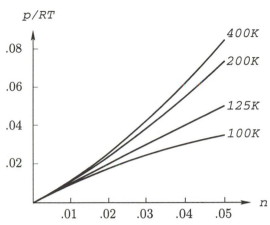

When the particles are moving more slowly, the weak but long-range attractive force pulls the particles together, reducing the pressure on the container walls. At high temperature, when kT is much larger than V_o, the particles have such large kinetic energies that the weak attractive interaction is negligible, and the only feature of the interparticle interaction that is significant is the hard core. The second term becomes much smaller than the first, and the value of the second virial coefficient approaches a positive constant equal to one-half of the volume of a hard-core particle of radius a. Thus the gas behaves like a gas with a simple hard-core interaction, which has a pressure that is larger than that of an ideal gas at the same temperature and density. Figure 8.3 shows the equation of state for a gas of these model particles. The reader should keep in mind that this two-term virial equation of state is only valid when the gas density is much less than the density of a liquid of the same substance.

8.7   THE MEAN-FIELD APPROXIMATION

The number and complexity of the terms appearing in the cluster expansion increase very rapidly with the order of the expansion. For that reason, the method is restricted to densities at which the first few terms give an adequate approximation to the entire series, and therefore it cannot be used to analyze phase transitions. In fact, it can be proved that neither the cluster expansion nor the virial expansion is a convergent expansion at any phase transition point. In contrast, the mean-field approximation, to be presented in this section, can be used in the presence of phase transitions. However, it is an inherently approximate method. There are no higher-order "corrections" that can be invoked if the basic approximation is not accurate enough. The cluster expansion and the virial expansion are both precisely defined approximation techniques. Once the interaction potential is known, there can be no ambiguity about what are the coefficients in an expansion of the equations of state, either as a power series in the activity or as a power series in the particle density. In contrast, the "mean-field approximation" is a phrase that describes any method in which, at some crucial step in computing the partition function, statistical correlations among the particles are ignored. For a single system, there may be more than one detailed implementation of the mean-field approximation that make somewhat different predictions.

As mentioned previously, the great advantage of the mean-field approximation is that it provides a way of analyzing phase transitions in systems with strong interactions. For reasons that we will not discuss, the mean-field approximation tends to become more accurate as the dimensionality of the system increases. For one-dimensional systems it is often qualitatively wrong, predicting phase transitions when an exact calculation shows that they do not occur. In two dimensions the method is usually qualitatively correct, predicting the phase transition at roughly the right place, but it is seldom quantitatively accurate. In three dimensions the quantitative accuracy of mean-field calculations usually improves, and the qualitative predictions are more reliable.

To illustrate the mean-field approximation, we will consider models that exhibit two different types of phase transitions. Each case will be done independently, so that the reader may skip either one without losing continuity.

8.8 NEMATIC LIQUID CRYSTALS

Nematic liquid crystals are solutions of long, rigid or partially rigid molecules that, in certain ranges of density and temperature, develop an *orientational ordering*, in which all the molecules in a large region in space tend to point in the same direction. That is not to say that they are all precisely parallel; they are not. In the *nematic phase*, the statistical distribution of their orientations is peaked in some direction, called the *director*. At lower densities or higher temperatures the same system exists in an *isotropic phase*, in which equal numbers of molecules are oriented in all directions. There are also other, more complicated types of liquid crystal phases, called *smectic* and *cholesteric* liquid crystals, that we will not consider. Figure 8.4 illustrates the two phases we want to study. The method that we are going to use to analyze the phase transition can be applied to realistic models of specific liquid crystals. However, when considering realistic liquid crystals, which are always composed of large complex organic molecules, the mathematical details become very complicated and tedious. Therefore, we will consider only an extremely simplified model, which is intended to capture the essential mechanism of the nematic phase transition, but which is obviously not intended to be a picture of any real liquid crystal.

The model is a two-dimensional gas of straight, infinitely thin, rodlike molecules, each of length ℓ. Instead of allowing the molecules to point in every possible direction, the molecules are constrained to point either horizontally or vertically. The horizontally and vertically pointing molecules will be treated as two different kinds of particles, called, respectively, x particles and y particles. The particles will not interact, except for the absolute constraint that no two particles can overlap. That is, the particles cannot touch one another. Since the particles are infinitely thin, the probability of two x particles or two y particles touching is zero, so that the constraint would have no effect on a gas composed purely of x particles or purely of y particles. As shown in Fig. 8.5, if a y particle has a given location, then no x particle can be located within an $\ell \times \ell$ square around its center of mass. (The "location" of a particle means the location of its center of mass.)

Using a canonical ensemble, we will consider a system composed of N_x x

isotropic

nematic

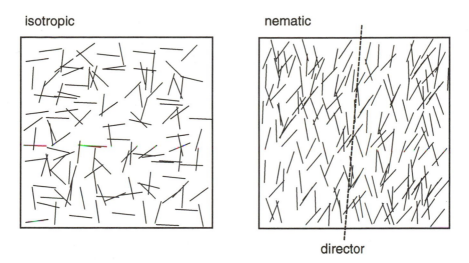

director

Fig. 8.4   The nematic phase transition. Above the phase transition temperature, the rodlike molecules are oriented in all directions with equal probabilities. Below the transition temperature, an orientation parallel to the director is favored over one perpendicular to it.

particles and N_y y particles, within a two-dimensional area A, where, of course, $A \gg \ell^2$. To avoid cumbersome notation, the collection of all the coordinates of the x particles will be called C_x and the set of all the coordinates of the y particles will be called C_y.

$$C_x \equiv (\mathbf{r}_1, \mathbf{r}_2, \ldots, \mathbf{r}_{N_x}) \qquad \text{and} \qquad C_y \equiv (\mathbf{r}_1, \mathbf{r}_2, \ldots, \mathbf{r}_{N_y}) \qquad (8.30)$$

The canonical partition function should involve an integral over all allowed configurations of the particles. We can extend the integral to an integral over all possible configurations of the particles, ignoring the no-overlap constraints, if we include a factor in the integrand that is zero whenever the configuration violates any constraint. We define such a constraint function by

$$F(C_x, C_y) = \begin{cases} 0, & \text{if any } x \text{ particle overlaps any } y \text{ particle} \\ 1, & \text{otherwise} \end{cases} \qquad (8.31)$$

If λ is the thermal de Broglie wavelength, then

$$Z(N_x, N_y, \beta, A) = \frac{1}{N_x! N_y! \lambda^{2(N_x + N_y)}} \int_A F(C_x, C_y)\, dC_x\, dC_y \qquad (8.32)$$

where $dC_x \equiv d^2\mathbf{r}_1 d^2\mathbf{r}_2 \cdots d^2\mathbf{r}_{N_x}$, with a similar definition for dC_y.

Fig. 8.5   The center of mass of any x particle is constrained to lie outside the square region shown in order to prevent an intersection of the two particles.

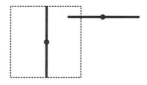

The canonical potential is given by

$$\phi(N_x, N_y, \beta, A) = \log Z \qquad (8.33)$$

Because there are two kinds of particles, there will be two affinities, which we will call α_x and α_y. The affinity for x particles can be written as (we do not indicate β and A explicitly)

$$\alpha_x = \phi(N_x+1, N_y) - \phi(N_x, N_y) = \log\left(\frac{Z(N_x+1, N_y)}{Z(N_x, N_y)}\right) \qquad (8.34)$$

We will see that, if we can calculate $\alpha_x(N_x, N_y)$, then we will be able to predict the exact location of the phase transition and the composition of the two phases. Therefore, we need to analyze carefully the ratio of $Z(N_x+1, N_y)$ to $Z(N_x, N_y)$. According to Eq. (8.32), this is given by [we will call the position of the (N_x+1)st particle, simply $\mathbf{r}$]

$$\frac{Z(N_x+1, N_y)}{Z(N_x, N_y)} = \frac{\int F(\mathbf{r}, C_y) F(C_x, C_y)\, d^2\mathbf{r}\, dC_x\, dC_y}{(N_x+1)\lambda^2 \int F(C_x, C_y)\, dC_x\, dC_y} \qquad (8.35)$$

The factor $F(\mathbf{r}, C_y)$ is zero whenever the (N_x+1)st x particle overlaps any y particle. For a given configuration of all the y particles (that is, for a given C_y), we define an observable $a(C_y)$ that is equal to the area that would be available to an x particle within the system (Fig. 8.6). This can be expressed as

$$a(C_y) = \int_A F(\mathbf{r}, C_y)\, d^2\mathbf{r} \qquad (8.36)$$

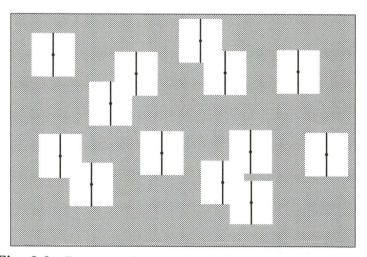

Fig. 8.6 For a given distribution of the y particles, the variable $a(C_y)$ is defined as the area available to an x particle—that is, the area outside all the squares.

Then the ratio of integrals in Eq. (8.35) can be written as

$$R = \frac{\int F(C_x, C_y) a(C_y) \, dC_x \, dC_y}{\int F(C_x, C_y) \, dC_x \, dC_y} \tag{8.37}$$

This can be interpreted as the expectation value of the observable $a(C_y)$ in an ensemble composed of all the allowed configurations of the system having $N_x + N_y$ particles, that is, in an ensemble with a probability density $P(C_x, C_y)$ that is proportional to $F(C_x, C_y)$. We will write the expectation value of any quantity in such an ensemble as $\langle \cdots \rangle_F$. Thus

$$R = \langle a(C_y) \rangle_F \tag{8.38}$$

Up to this point, everything has been exact. Now we will make the crucial approximation. We replace the value of $\langle a(C_y) \rangle_F$ by the expectation value of $a(C_y)$ in an ensemble where the N_y y particles are distributed at random within the area A. We will call this $\langle a(C_y) \rangle_{\text{random}}$. For any configuration of the y particles, $a(C_y)/A$ is equal to the probability that, if an x particle is placed at random within A, it will miss all the y particles. We will now calculate that probability for a random distribution of y particles. The probability that an x particle will miss a single y particle is

$$p = 1 - \frac{\ell^2}{A} \tag{8.39}$$

In the random distribution, the y particles are statistically uncorrelated, and therefore the probability that an x particle will miss all of them is just p^{N_y}. Therefore,

$$\frac{\langle a(C_y) \rangle_{\text{random}}}{A} = \left(1 - \frac{\ell^2}{A}\right)^{N_y} = \left(1 - \frac{\ell^2 n_y}{N_y}\right)^{N_y} = e^{-\ell^2 n_y} \tag{8.40}$$

where $n_y = N_y/A$ is the density of y particles and we have used the definition of the exponential function $e^x \equiv \lim_{N \to \infty} (1 + x/N)^N$. This gives for the ratio of the partition functions

$$\frac{Z(N_x + 1, N_y)}{Z(N_x, N_y)} = \frac{A e^{-\ell^2 n_y}}{(N_x + 1)\lambda^2} \tag{8.41}$$

and, for the x particle affinity,

$$\alpha_x = \phi(N_x + 1, N_y) - \phi(N_x, N_y) = -\log n_x - \log \lambda^2 - \ell^2 n_y \tag{8.42}$$

By the obvious symmetry between x particles and y particles, we can write an equivalent formula for the y particle affinity.

$$\alpha_y = -\log n_y - \log \lambda^2 - \ell^2 n_x \tag{8.43}$$

The x particles and y particles actually represent two possible orientations of a single type of molecule. Therefore, N_x and N_y should not be considered as separately conserved variables; only the sum, $N_x + N_y$, is fixed. The change of an x particle to a y particle and vice versa is best treated as a simple chemical reaction. $x \leftrightarrow y$. The condition of equilibrium in the presence of such a chemical reaction

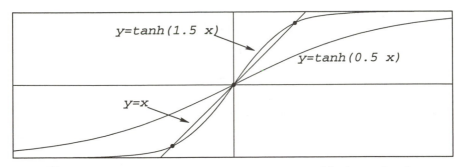

Fig. 8.7   In order to solve the equation $x = \tanh(ax)$, we plot the curves $y = x$ and $y = \tanh(ax)$. The value of x at their intersection is a solution to the equation. For $a < 1$ the only intersection occurs at $x = 0$. For $a > 1$ there are also two nonzero solutions.

is that $\alpha_x = \alpha_y$. Let n be the conserved total density. Then if we assume that at equilibrium,

$$n_x = \tfrac{1}{2}n(1+\gamma) \tag{8.44}$$

where γ is some parameter to be determined, then

$$n_y = n - n_x = \tfrac{1}{2}n(1-\gamma) \tag{8.45}$$

The condition that $\alpha_x = \alpha_y$ gives

$$
\begin{aligned}
-&\log\left(\frac{n}{2}(1+\gamma)\right) - \log \lambda^2 - \tfrac{1}{2}\ell^2 n(1-\gamma) \\
&= -\log\left(\frac{n}{2}(1-\gamma)\right) - \log \lambda^2 - \tfrac{1}{2}\ell^2 n(1+\gamma)
\end{aligned}
\tag{8.46}
$$

Cancelling identical terms from both sides, we get

$$-\log(1+\gamma) + \tfrac{1}{2}\ell^2 n\gamma = -\log(1-\gamma) - \tfrac{1}{2}\ell^2 n\gamma \tag{8.47}$$

which can be rearranged as

$$\log(1+\gamma) - \log(1-\gamma) = \ell^2 n\gamma \tag{8.48}$$

Using the identity, $\log(1+\gamma) - \log(1-\gamma) = 2\tanh^{-1}\gamma$, this can be written as

$$\gamma = \tanh(\ell^2 n\gamma/2) \tag{8.49}$$

This equation always has the trivial solution $\gamma = 0$, which describes a state with equal densities of x particles and y particles. When $\ell^2 n/2 > 1$ it also has two nontrivial solutions, describing states in which the symmetry between x and y particles is spontaneously broken. In one of these two states, which we will call the x phase, there is an excess of x particles; in the other there is an excess of y particles. In Problem 8.8 the reader is asked to show that, whenever these nontrivial solutions exist, the trivial solution is unstable and therefore does not really represent an equilibrium state.

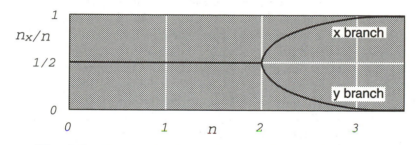

Fig. 8.8   The density of x particles as a function of the total density for the case $\ell = 1$. If the density is increased beyond the critical density, the macroscopic state of the system shifts, in an unpredictable way, to one of the two "broken symmetry" states.

If the density is gradually increased, the system will remain in the isotropic state, with $n_x = n_y = n/2$, until the density reaches the critical value $n = 2/\ell^2$. If it is increased beyond that value, the isotropic state becomes unstable, and the system will spontaneously transform into one of the two nematic liquid crystal phases, either the x phase, with an excess of x-oriented particles, or the y phase, with an excess of y-oriented particles. Which of these broken symmetry states will appear cannot be calculated within the model. In a real system, it would usually depend on small, neglected terms, such as unsymmetric interactions with the walls of the container. If a real system is kept very carefully symmetric, and the system is repeatedly cycled back and forth through the critical density, it will sometimes go into the x phase and sometimes go into the y phase, in a wholly unpredictable way that depends on the detailed microscopic state of the system at the time of the transition. As can be seen in Fig. 8.8, the value of n_x varies continuously through the nematic phase transition, but the slope of the function $n_x(n)$ is discontinuous. This indicates that the nematic phase transition is a second-order phase transition.

8.9   ADSORPTION WITH INTERACTIONS

The particles in an adsorbed surface layer are generally much closer to one another than those in the gas with which the adsorbed particles are in equilibrium. Because of their proximity, interactions between neighboring adsorbed particles are often significant. We will see that such interactions between neighboring particles can drastically change the adsorption characteristics by causing a phase transition in the adsorbed layer.

The model that will be used is similar to the one considered in Section 7.2. There is an $L \times L$ square lattice of adsorption sites, each of which can accommodate a single particle. For simplicity, it will be assumed that there is only a single adsorption state, of energy ε, at each site. The new element that will be introduced here is the assumption that two particles at neighboring sites interact with a potential energy v. If v is negative, then the interaction describes a short-range attraction between the adsorbed particles, in that it would require an amount of energy $|v|$ to separate a pair of neighboring particles. A positive value of v corresponds to a repulsive force between particles. This system is often referred to as a two-dimensional *lattice gas*. The lattice sites can be described by a two-dimensional vector with components $(x, y) = (Xa, Ya)$, where a is the lattice spacing and X and Y are integers in the range, $1, 2, \ldots, L$. Assuming that L is very large, we will always ignore the special case of particles at the very edge sites, that have only 3, rather than 4, neighboring sites. It will be convenient to work in a peculiar system of units in which the lattice spacing a is our unit of length. Then the total area of the surface A is equal to L^2, the number of adsorption sites.

In analyzing a macroscopic system, one is free to use whatever ensemble is most convenient. A change of ensemble cannot lead to a change in the thermodynamic relations for the system, but only to the same thermodynamic relations, expressed in a different representation. Although the adsorption calculation was done in Chapter 7 using the grand canonical ensemble, it is more convenient to do this calculation in the canonical ensemble.

The system consists of N indistinguishable particles, to be distributed among A sites. A configuration is defined by choosing N occupied sites from the A available sites. Thus the total number of configurations is

$$\binom{A}{N} = \frac{A!}{N!(A-N)!} \tag{8.50}$$

The energy of interaction between the particles and the adsorption sites is simply $N\varepsilon$. The energy of interaction between the particles themselves depends on the detailed configuration of the particles. In the mean-field approximation we calculate the interparticle interaction energy by assuming that the particles are randomly distributed over the sites. That is, instead of taking the proper canonical distribution of configurations, we take a uniform distribution, in which all possible configurations are equally likely, in spite of the fact that they do not all have the same energy.

At this point we could make a false argument that seems to prove that, in the thermodynamic limit, this mean-field approximation is perfectly accurate. It is useful to do so, because the refutation of the argument will point out certain properties of the canonical ensemble that are important in understanding the Monte Carlo method, which is the next topic in this chapter. The argument goes as follows: The interaction potential is a macroscopic observable. Therefore, in the limit of large N and A, almost all the configurations give the same value of the observable. Since the observable has the same value for almost all states, we will make a negligible error in using an equal probability ensemble of configurations rather than a canonical ensemble in calculating the expectation value of the interaction energy. What is wrong with this argument? Recall that it was shown that the equal probability ensemble would give the same value of $\langle A \rangle$ as any other probability function, *provided that the other probability function did not, in the thermodynamic limit, concentrate all the probability on the tiny fraction of exceptional states.* Now let us see that this is exactly what the canonical ensemble does. Suppose we consider the case of particles with attractive interactions. Then the low-energy configurations are those in which the particles are huddled together so that they can interact. The high-energy configurations are those in which the particles are spread out. As N becomes large, the difference in energy between an average energy state and a typical low-energy state, call it ΔE, is proportional to N. But the ratio of probabilities in the canonical ensemble is then

$$\frac{P(\text{low})}{P(\text{ave.})} = e^{\beta \Delta E} \sim e^N \tag{8.51}$$

Thus, in the thermodynamic limit, the canonical ensemble probability becomes extremely concentrated on a very small fraction of the possible states of the system.

It is this characteristic that invalidates the argument given, based on the law of large numbers.

The expectation value of the interaction energy for a collection of N particles, randomly distributed over A lattice sites, can be calculated as follows. We first concentrate our attention on any one of the particles. That particle has 4 nearest-neighbor sites. The probability that any one of those sites is filled is equal to N/A. Therefore, the average interaction energy of that particle is $4vN/A$. Adding up this number for each of the N particles actually double counts the interaction potential by counting the interaction between any pair of particles twice, once from each end. Therefore, the mean-field approximation for the total energy is

$$E = \varepsilon N + 2vN^2/A \qquad (8.52)$$

Using this for all configurations, the canonical partition function is equal to $\exp(-\beta E)$ times the number of configurations.

$$Z = \frac{A!}{N!(A-N)!}e^{-\beta(\varepsilon N + 2vN^2/A)} \qquad (8.53)$$

The canonical potential can be calculated using Stirling's approximation.

$$\phi = A\log A - N\log N - (A-N)\log(A-N) - \beta(\varepsilon N + 2vN^2/A) \qquad (8.54)$$

In order to study the equilibrium between the adsorbed particles and the gas, we must determine the relationship between the affinity and the surface density of adsorbed particles. We will do so for the special case of zero binding energy. The case of nonzero ε is left as a problem. The affinity of the adsorbed particles is

$$\alpha = \frac{\partial\phi}{\partial N} = -\log N + \log(A-N) - 4\beta vN/A \qquad (8.55)$$

This can be written in terms of the surface density, $n = N/A$.

$$\alpha = \log(1-n) - \log n - 4\beta vn \qquad (8.56)$$

If v is positive, indicating a repulsive interaction between nearest-neighbor particles, then the graph of $\alpha(n)$ has the general form shown in Fig. 8.9(A). In that figure, the graph of $\alpha(n)$ for the case of no interaction is also drawn for comparison. One can see that a repulsive interaction does not change the properties of the system of adsorbed particles in any qualitative way. As the affinity of the gas is decreased, by increasing the gas density, the surface density of adsorbed particles smoothly approaches the density of adsorption sites. The case of negative v is much more interesting. Graphs of $\alpha(n)$ are shown in Fig. 8.9(B) for values of the parameter βv equal to $-\frac{3}{4}$, -1, and $-\frac{4}{3}$, that is, for kT equal to $4|v|/3$, $|v|$, and $3|v|/4$. At the lowest temperature, the graph of $\alpha(n)$ has some obviously unphysical properties. For a range of α there are three different surface densities corresponding to the same value of α. Thus there seems to be no unique answer to the question: What would be the surface density in this system if it were in contact with a gas at affinity α? Between points B and C on the curve, the affinity increases with increasing particle number. But $\partial\alpha/\partial n$ is a second derivative of the entropy, which must be negative for a convex entropy function. If the entropy function is not convex,

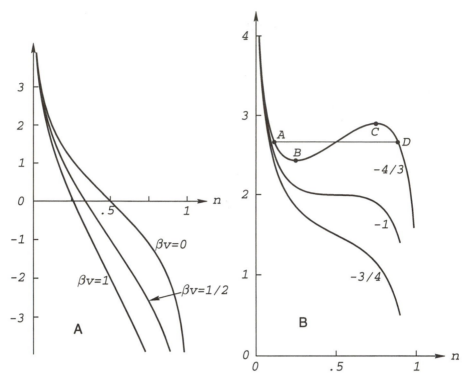

Fig. 8.9 (A) Graphs of the affinity, as a function of the particle density, for no interparticle interaction and for two positive values of βv. (B) Graphs of the affinity as a function of the particle density for three negative values of βv. When $\beta v < -1$ (that is, when $kT < |v|$), the mean-field method incorrectly predicts a van der Waals loop in the curve $\alpha(n)$. In a fairly ad hoc way, this is replaced by a horizontal section from point A to point D, where the points are chosen by the argument given in Section 8.10.

then the system could increase its entropy by separating into two separate phases at different densities. In fact, that is just what happens. But it does not begin to happen at point B, as one might expect. If the system is initially in a very low-density state and the affinity is gradually decreased, then the point describing the state of the system gradually moves down the curve on the left until it reaches some point A that we will calculate shortly. At that point, if new particles are added, the system breaks up into two phases. A fraction of the particles (initially, almost all of them) remains in the low-density phase described by point A, but the remainder condense into a high-density phase. Since the high-density phase must be a stable phase with the same affinity as the low-density phase (the two phases are in contact), it must be the phase represented by the point D on the graph. The affinity will remain constant until enough particles are added to bring the system completely into phase D. If still more particles are added, the system point begins to move down the curve on the right. Thus the true $\alpha(n)$ curve for $kT = 3|v|/4$ is the curve shown to the left of point A, a horizontal straight line from A to D, and the curve shown to the right of point D. Between A and D, the theoretical curve does not represent the actual state of the system. It gives the affinity of a system that is artificially constrained, by our use of the mean-field approximation, to an unstable uniform density state.

One thing should be made clear: the whole S-shaped part of the curve between points A and D is purely an artifact of the mean-field approximation. That portion of the curve is called a *van der Waals loop*, because it first appeared in the van der Waals equation of state for a real gas. If an exact calculation of $\alpha(n)$ could be made, without using the mean-field or any other approximation, it would show no van der Waals loop. This is confirmed in Fig. 8.10, where there is shown the results of an accurate numerical calculation of $\alpha(n)$, made by using the Monte Carlo method.

"It is clear that, if we are to find the pressure of the saturated vapor, then we must ... draw a horizontal line so that it cuts the isotherm at three points. But I have not succeeded in finding, in any of the properties of the saturated vapour, a characteristic one that I could use to determine where this line must be drawn."
——J. D. van der Waals *Doctoral Thesis in Physics, Leiden* (1875)

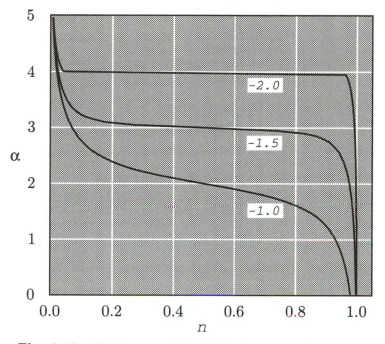

Fig. 8.10 The function $\alpha(n)$ calculated on a 5×5 lattice, by the Monte Carlo method for three negative values of βv. (See Section 8.11 for a description of the method.) For $\beta v = -2$, the curve shows a phase transition. That is, for nearly constant α, the density varies from a low "gas" value to a high "liquid" value. There is clearly no sign of a van der Waals loop. The two-phase portion of the curve, which should be exactly horizontal, is slightly tilted. This effect is due to the small size of the system used in the calculation. It is called a *finite-size effect*. One cannot obtain a true phase transition in any finite system. As the system size is made larger, that portion of the curve becomes closer and closer to being horizontal. Notice that the curve for $\beta v = -1.5$, which would show a phase transition in the mean-field approximation, does not actually show any phase transition. This is an example of the fact that the mean-field approximation is not quantitatively accurate.

8.10 PRESSURE EQUILIBRIUM

We are still faced with the problem of calculating the point A at which the phase separation starts. Since the low-density phase at point A has the same affinity as the high-density phase at point D, and since all points on the curve represent states of equal temperature, the two phases satisfy two of the three conditions necessary for two-phase equilibrium. The third condition is that the pressures of the two phases be equal. When that condition is added, it is possible to calculate the point at which the phase transition occurs. For a two-dimensional gas, the pressure has the same relation to the area of the system that, for a three-dimensional gas, it would have to the volume. That is, $\beta p = \partial \phi / \partial A$. Using Eq. (8.54), this gives

$$p = kT \log(1-n) + 2vn^2 \tag{8.57}$$

In a two-phase equilibrium state, if we call the densities of the "liquid" and the "gas" phases n_l and n_g, then the values of these quantities are determined by the simultaneous equations, $\alpha(n_l) = \alpha(n_g)$ and $p(n_l) = p(n_g)$. Solving these simultaneous nonlinear equations seems like a formidable job. However, using a clever geometrical construction due to Maxwell, it is actually not difficult to obtain an exact solution.

In this analysis, we must assume that the theoretical curve, calculated by the mean-field approximation, is an adequate description of the system to the left of point A and to the right of point D, although we know that it does not describe the system between points A and D. We recall that the Gibbs–Duhem equation of Section 5.19 was derived using only the extensivity property of the entropy function. It did not require the convexity property. Therefore, because the theoretical canonical potential is clearly an extensive function, although it does not have the correct convexity properties, it will satisfy the Gibbs–Duhem relation. That relation says that, along an isotherm [see Eq. (5.46)],

$$\beta \, dp = -n \, d\alpha \tag{8.58}$$

But $-n \, d\alpha$ is just the area between a piece of curve of height $d\alpha$ and the left coordinate axis. Therefore, if we integrate $-n \, d\alpha$ from A to D, remembering that the contribution to the integral is counted as positive if $d\alpha$ is negative and vice versa, we see that

$$\beta[p(D) - p(A)] = A_2 - A_1 \tag{8.59}$$

where A_1 and A_2 are the areas shown in Fig. 8.11.

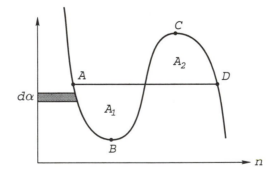

Fig. 8.11 The area of the shaded region is equal to $-n \, d\alpha$. The condition that the pressure in the gas phase be equal to the pressure in the liquid phase implies that the areas A_1 and A_2 are equal.

Thus the third condition for the equilibrium of the two phases implies that the areas A_1 and A_2 must be equal, which uniquely determines the point A. What makes the Maxwell construction so convenient in this problem is the fact that the curve $\alpha(n)$ is perfectly antisymmetric about the point drawn through $\alpha\left(\frac{1}{2}\right)$ (see Problem 8.10). Setting $n = \frac{1}{2}$ in Eq. (8.56) shows that the phase transition occurs at the value α_o, given by

$$\alpha_o = \alpha\left(\tfrac{1}{2}\right) = 2\beta|v| \tag{8.60}$$

The energy may be calculated from the thermodynamic relation

$$E = -\frac{\partial \phi}{\partial \beta} = \frac{2vN^2}{A} \tag{8.61}$$

which simply reproduces the mean-field approximation to the energy, assumed in Eq. (8.52). (Recall that ε has been set equal to zero.)

If a gas in equilibrium with an adsorption surface is slowly compressed at constant temperature, then the gas pressure will rise smoothly until the adsorbed particles begin to undergo their phase transition. At that point, until enough particles are removed from the gas to complete the phase transition on the surface, the gas pressure will remain constant (in order to keep the affinity constant). Such a constant pressure plateau will give the appearance of a phase transition within the gas itself.

8.11   THE MONTE CARLO METHOD

The Monte Carlo method is most easily described for a system with a finite number of possible states, although it is in no way limited to such systems. Once the method has been defined for finite-state systems, the generalization to a system with a continuum of possible states is quite easy.

We consider a system with a finite number of possible states. Each state will be called a "configuration" and denoted C. In the first chapter, the idea of probability was defined by means of a sampling sequence, which was an infinite sequence of configurations, $C_1, C_2, \ldots$, such that, for any particular configuration C the limit

$$\lim_{N \to \infty} \left(\frac{\text{the number of times } C_n = C \text{ for } n = 1, \ldots, N}{N} \right) \tag{8.62}$$

existed and was defined as $P(C)$, the probability of configuration C.

The Monte Carlo method is an efficient algorithm for constructing, using a computer, a sampling sequence whose associated probability function is the canonical probability density,

$$P_\beta(C) = e^{-\beta E(C)}/Z(\beta) \tag{8.63}$$

where β is any chosen positive number and $E(C)$ is the energy of the system in state C. The objective in constructing a canonical sampling sequence is that the canonical average of any observable, $A(C)$, defined by

$$\langle A \rangle_\beta = \sum_C P_\beta(C) A(C) \tag{8.64}$$

can be calculated by averaging $A(C)$ over a long but finite section of the sequence.

$$\langle A \rangle_N = \frac{1}{N} \sum_{n=1}^{N} A(C_n) \tag{8.65}$$

Of course, the Monte Carlo average, $\langle A \rangle_N$, only becomes equal to the true canonical average as N goes to infinity, but, by making N sufficiently large, one can achieve any desired finite accuracy.

One obvious objection to the idea of using a sampling sequence for calculating thermal averages is that, since the system has a finite number of possible states, one could use Eqs. (8.63) and (8.64) to calculate $\langle A \rangle_\beta$ directly and obtain the result with zero error. To see that this direct method is completely impractical, let us consider using it for a $10 \times 10 \times 10$ cubic crystal of atoms, each of which have only two possible states. Certainly this is an unrealistically small and simple system. Such a system has 2^{1000} possible states. A computer that was capable of summing up a million states per second would take about 10^{288} years of steady running to carry out the sum over all of them. No future improvements in computer speed are likely to make the direct method practical. In contrast, to achieve 1% accuracy, using the Monte Carlo method, about ten thousand terms must be summed, because, according to the law of large numbers, the uncertainty will be proportional to $1/\sqrt{N}$. (The reader is warned that there are important exceptions to this rough estimate of the uncertainty.)

In preparation for defining the Monte Carlo algorithm, we picture a system that makes abrupt transitions from one configuration to another by a probabilistic, rather than a deterministic, rule. Once each second, the system makes a jump either to the same or to a different configuration. If it is in configuration C, it has a *transition probability* $T(C{\to}C')$ of making a jump to configuration $C' \neq C$. It is assumed that $T(C{\to}C')$ does not depend upon the past history of the states that the system has been in prior to configuration C. Let $P_n(C)$ be the probability that the system is in configuration C immediately after the nth step. We want to derive an equation for $P_n(C)$. The probability that the system was in state C before the nth jump is $P_{n-1}(C)$. The probability that, if it were in state C, it would remain in that state is $1 - \sum' T(C{\to}C')$, where the sum is over all the configurations C' and the prime on the sum indicates that the term $C' = C$ is omitted. The probability that, on the nth jump, the system would jump into state C from some other state is $\sum' P_{n-1}(C')T(C'{\to}C)$. Therefore, the desired equation for $P_n(C)$ is

$$P_n(C) = P_{n-1}(C)\Big[1 - \sum_{C'}' T(C{\to}C')\Big] + \sum_{C'}' P_{n-1}(C')T(C'{\to}C) \qquad (8.66)$$

or

$$P_n(C) - P_{n-1}(C) = \sum_{C'}' \Big[P_{n-1}(C')T(C'{\to}C) - P_{n-1}(C)T(C{\to}C')\Big] \qquad (8.67)$$

A probabilistic transition process of this sort, in which the probability of making a particular transition depends only on the present state of the system, is called a *Markov process*. The above equation for $P_n(C)$ is known as the *master equation*.

Since the transition rate $T(C{\to}C')$ is time-independent, it is reasonable to assume (and it can be mathematically proved) that the probability function will eventually approach some time-independent function $P(C)$. The steady-state function will satisfy the equation obtained by setting $P_n - P_{n-1}$ equal to zero in Eq. (8.67).

$$\sum_{C'}' P(C)T(C{\to}C') = \sum_{C'}' P(C')T(C'{\to}C) \qquad (8.68)$$

This equation simply says that at equilibrium the total rate of jumping out of configuration C is equal to the total rate of jumping into that configuration.

The normal problem in studying a Markov process is to calculate $P(C)$, given the transition probability $T(C{\to}C')$. In the Monte Carlo method, we look at the inverse problem, that is, what transition rate will give the canonical probability function $P_\beta(C)$ as the time-independent probability distribution? The solution is not unique, but that does not matter, what we need is any convenient solution. Certainly one solution is any transition probability that satisfies, for every pair of configurations C and C', the relation

$$P_\beta(C)T(C{\to}C') = P_\beta(C')T(C'{\to}C) \tag{8.69}$$

This is called the condition of *detailed balance*. If it is satisfied, then the terms in Eq. (8.68) will cancel in pairs. Thus, to get the canonical probability density, we want T to satisfy the relation,

$$e^{-\beta E(C)}T(C{\to}C') = e^{-\beta E(C')}T(C'{\to}C) \tag{8.70}$$

8.12   A SIMPLE EXAMPLE

At this point, in order to define the detailed computational method, we have to specialize our discussion to some concrete problem. We will first illustrate the method by considering a system that is so simple that one could, in fact, use the direct method for calculating any expectation value $\langle A \rangle_\beta$ by summing over all the possible configurations of the system. We consider a single particle in one dimension whose coordinate x is confined to the points of a one-dimensional lattice of spacing a (Fig. 8.12). That is, x must be equal to one of the values, $x_\ell = \ell a$, where ℓ is any (positive, negative, or zero) integer. The particle is subjected to a harmonic oscillator potential, which at point x_ℓ has the value

$$E(\ell) = \tfrac{1}{2}kx_\ell^2 = \varepsilon\ell^2 \tag{8.71}$$

where $\varepsilon = ka^2/2$. We want to construct a sampling sequence for the canonical probability function,

$$P(\ell) = Ce^{-\beta\varepsilon\ell^2} \tag{8.72}$$

Fig. 8.12   A single particle is confined to the points of a one-dimensional lattice. At the nth lattice point it has the potential energy εn^2.

Since the procedure is a numerical one, the parameters β and ε must have some definite values, but we will not specify now what they are. With this simple system, the "configuration" of the system is just the value of ℓ. Therefore, a sampling sequence means a sequence of integers, $\ell_1, \ell_2, \ldots$, that have the probability distribution $C\exp(-\beta\varepsilon\ell^2)$. We picture the sequence of configurations as a particle that hops from lattice point to lattice point. We use the following rule, which will be justified in the following.

At each step, the particle can either stay in the same place, jump one lattice space to the right, or jump one lattice space to the left. If it is at lattice point ℓ, it has a probability of $R(\ell)$ to jump to the right, a probability $L(\ell)$ to jump to the left, and a probability $1 - R(\ell) - L(\ell)$ (which must be non-negative for the scheme to make sense) to stay put. This means that, in our former notation, the transition probability $T(\ell \rightarrow k)$ has the values

$$T(\ell \rightarrow k) = \begin{cases} R(\ell), & \text{if } k = \ell + 1 \\ L(\ell), & \text{if } k = \ell - 1 \\ 1 - R(\ell) - L(\ell), & \text{if } k = \ell \\ 0, & \text{otherwise} \end{cases} \tag{8.73}$$

Consider the possible transitions between states ℓ and $\ell + 1$ for some particular ℓ. The transition probability in the forward direction $(\ell \rightarrow \ell + 1)$ is $R(\ell)$. In the backward direction $(\ell + 1 \rightarrow \ell)$ it is $L(\ell + 1)$. In order to satisfy the condition of detailed balance it is necessary that

$$e^{-\beta E(\ell)} R(\ell) = e^{-\beta E(\ell+1)} L(\ell + 1) \tag{8.74}$$

We can satisfy this equation as follows: If a jump to the right would increase the energy [that is, if $E(\ell + 1) > E(\ell)$] then that jump is made with probability $\frac{1}{2} \exp(-\beta \Delta E)$, where $\Delta E = E(\ell + 1) - E(\ell)$. Otherwise the move is made with probability $1/2$. The same rule is used for jumps to the left; if the move would increase the energy then it has probability $(1/2) \exp(-\beta \Delta E)$, otherwise it has probability $\frac{1}{2}$. It is easy to check that with this rule Eq. (8.74) is satisfied for all values of ℓ. This rule also guarantees that the probability of staying put, namely $1 - L(\ell) - R(\ell)$, is never negative.

There is another condition that is satisfied by the above rule that has not been stated but is obviously necessary for any Monte Carlo scheme to work. That is, that it is *possible*, by some series of Monte Carlo moves, to go from any configuration of the system to any other configuration. Such would not have been the case if we had made all jumps have size ± 2. In that case we would have stayed on the odd-numbered or the even-numbered sites. Essentially, we would have inadvertently added a new conservation law to the system. Although no one would be so stupid as to choose jumps of ± 2, in treating more complex systems by the Monte Carlo method, people have sometimes accidentally imposed subtle new conservation laws (not valid in the actual system under study) that have caused erroneous results.

The reader should note that the rule introduced above makes no detailed reference to the fact that the energy function is $E(\ell) = \varepsilon \ell^2$. It would therefore work with other functions also.

At this point we must discuss the question of how to carry out a probabilistic process on a computer, which is, one hopes, a completely predictable device. All programming languages have random number generators. They are procedures that, if called repeatedly, will generate a random sequence of numbers*, uniformly distributed within the interval $0 < r < 1$. From now on, when we say: "Do

---

* The question of how to define a "random" sequence is actually a deep and difficult one. One possible definition of randomness is that the sequence of two-dimensional points

A with probability p_A, B with probability p_B, and C with probability p_C, where $p_A + p_B + p_C = 1$." what we mean is: "Generate a random number x. If $0 < x < p_A$, then do A. If $p_A < x < p_A + p_B$, then do B. Otherwise do C." This is something that can be done with a computer.

Given any two functions $R(\ell)$ and $L(\ell)$ that satisfy Eq. (8.74), the following simple algorithm will generate a sample sequence with a probability distribution $P(\ell) = C \exp[-\beta E(\ell)]$.

1. Let $n = 0$ and let $\ell_o = 0$ (or any other starting value) and go to Step 2.
2. Let $\ell_{n+1} = \ell_n + 1$ with probability $R(\ell_n)$, let $\ell_{n+1} = \ell_n - 1$ with probability $L(\ell_n)$, and let $\ell_{n+1} = \ell_n$ with probability $1 - R(\ell_n) - L(\ell_n)$. Go to Step 3.
3. Increase n by one and go to Step 2.

With this algorithm, for any function $F(x)$,

$$\lim_{N\to\infty}\left(\frac{F(x_1) + \cdots + F(x_N)}{N}\right) = Z^{-1}(\beta) \sum_{\ell=-\infty}^{\infty} F(a\ell)e^{-\beta E(\ell)} \qquad (8.75)$$

where $x_n = a\ell_n$.

8.13   THE ISING MODEL

In order to illustrate the application of the Monte Carlo method to a problem that cannot be done by the direct method, we will choose the two-dimensional version of the Ising model, introduced briefly in Chapter 3. It is a system of atoms, fixed at the points of an $L \times L$ square lattice. Each atom has only a spin degree of freedom. The spin value at $\mathbf{r} = (x, y)$ can be "up," denoted by $\sigma(\mathbf{r}) = +1$, or it can be "down," denoted by $\sigma(\mathbf{r}) = -1$. (The lattice coordinates x and y are integers in the range 1 to L.) A configuration C is a set of spin values for all lattice points. The energy of a configuration is assumed to be of the form

$$E(C) = -H \sum_{\mathbf{r}} \sigma(\mathbf{r}) - V \sum_{NN} \sigma(\mathbf{r})\sigma(\mathbf{r}') \qquad (8.76)$$

The first term is what one would expect if each atom had a magnetic moment $\mathbf{m}$ parallel to its spin and there existed an external magnetic field giving an interaction energy $-\mathbf{m} \cdot \mathbf{B}$ with $H = mB$. The second term is summed over all pairs of lattice points that are nearest neighbors. If the nearest-neighbor interaction energy is negative (if V is positive), then two neighboring spins will have an interaction energy $-V$ if they agree (either both up or both down) and an interaction energy $+V$ if they disagree. A model with positive V is called a *ferromagnetic Ising model*. A model with negative V is an antiferromagnetic model. In an antiferromagnetic model, at low temperature, one tends to get a checkerboard pattern of up and down spins so that nearest neighbors mostly disagree.

Using this expression for $E(C)$, we can construct a solution to Eq. (8.70) as follows. Two configurations, C and C', are called *neighboring* configurations (this

(r_1, r_2), (r_2, r_3), $(r_3, r_4),\ldots$ be uniformly distributed in the unit square, the sequence of three-dimensional points (r_1, r_2, r_3), (r_2, r_3, r_4), $(r_3, r_4, r_5),\ldots$ be uniformly distributed in the unit cube, and so forth. We will not discuss this problem any further, although the fact that the sequences generated by most random number generators are not really random has, at times in the past, caused errors in Monte Carlo calculations.

has nothing to do with neighboring lattice sites) if they differ by the value of a single spin. It is obvious that any configuration is connected to any other configuration by a sequence of neighboring configurations. That is, we can go from any configuration to any other configuration by changing a finite number of spins, one spin at a time. We now choose $T(C{\to}C') = T(C'{\to}C) = 0$ for any pair of configurations that are not neighboring configurations. If C and C' are neighboring configurations, and $E(C) \geq E(C')$, then we choose

$$T(C{\to}C') = L^{-2} \tag{8.77}$$

and

$$T(C'{\to}C) = L^{-2}e^{-\beta[E(C)-E(C')]} = L^{-2}e^{-\beta\,\Delta E} \tag{8.78}$$

If $E(C') > E(C)$ we use these equations with C and C' exchanged. In other words, for any pair of neighboring configurations, the transition probability for a downhill jump is L^{-2}, while the transition probability for the reverse uphill jump is $L^{-2}e^{-\beta\,\Delta E}$.

A random process having these transition rates can be carried out on a computer, using the following algorithm.

1. Let $n = 0$ and choose any state C_0 as the starting state.
2. Pick two random integers x and y between 1 and L. This can be done with a random number generator. Let C be the configuration obtained from C_n by changing the value of $\sigma(x,y)$. Evaluate $\Delta E = E(C) - E(C_n)$.
3. Increase n by 1. If $\Delta E \leq 0$, set $C_n = C$ and go to Step 2. If $\Delta E > 0$, then set $C_n = C$ with probability $e^{-\beta\,\Delta E}$; otherwise set $C_n = C_{n-1}$. Go to Step 2.

This algorithm produces a sequence of configurations $C_0, C_1, C_2, \ldots$, which eventually approaches a sampling sequence for the canonical probability function $P_\beta(C) = \text{const.} \times e^{-\beta E(C)}$. One of the problems associated with this method is that it is difficult to predict just how long it will take for the sequence to reach its time-independent asymptotic state. This is usually determined empirically by seeing whether averages of quantities over large time intervals ("time" means the subscript n on C_n) are still gradually shifting, which would indicate that the steady state has not yet been reached.

Notice that at each iteration only ΔE has to be calculated. For a large lattice this is a major convenience in that ΔE involves only a few lattice points, namely the nearest neighbors of the point (x,y), chosen in Step 2, but an evaluation of $E(C)$ would involve a sum over the complete lattice.

8.14   THE CALCULATION OF C AND χ

Both the specific heat per site, C, and the magnetic susceptibility, χ, are defined as derivatives. In particular, $L^2 C = \partial E/\partial\tau$ and $\chi = \partial m/\partial H$, where

$$m = L^{-2}\left\langle \sum_{i=1}^{L^2} \sigma_i \right\rangle \tag{8.79}$$

is the average polarization per site. (The average $\langle \cdots \rangle$ means the canonical ensemble average.) The calculation of these physical quantities is very important in the study of phase transitions and critical phenomena. The obvious way to calculate the specific heat is to calculate E at two nearby temperatures and then to take the

Fig. 8.13 A Monte Carlo calculation would produce quantities such as $E(T)$ at particular points and with finite error bars. But there is then a wide range of possible tangents to the curve that are consistent with the calculated data within the error bars. That is, the derivative of a curve that is only approximately known is usually very poorly defined.

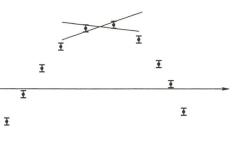

ratio $[E(\tau_2) - E(\tau_1)]/(\tau_2 - \tau_1)$. However, even small errors in $E(\tau_1)$ and $E(\tau_2)$ can lead to very large errors in the derivative, as can be seen by looking at Fig. 8.13. Errors are tremendously amplified when numerical data points, containing random errors, are differentiated.

Fortunately, the fluctuation formulas, discussed in the last three sections of Chapter 5, offer a quite different method of calculating C and χ that is both convenient and free of the difficulties inherent in numerical differentiation. For example, the specific heat per site may be expressed in terms of the canonical potential by the formula

$$L^2 C = \frac{\partial E}{\partial \tau} = -\beta^2 \frac{\partial E}{\partial \beta} = \beta^2 \frac{\partial^2 \phi}{\partial \beta^2} \tag{8.80}$$

But, according to Eq. (5.68), the mean-square fluctuation in the energy is given by

$$(\Delta E)^2 = \langle E^2 \rangle - \langle E \rangle^2 = \frac{\partial^2 \phi}{\partial \beta^2} \tag{8.81}$$

Since $\langle E^2 \rangle$ and $\langle E \rangle^2$ may be easily calculated by the Monte Carlo method, C can be obtained from the relation

$$L^2 C = \beta^2 (\langle E^2 \rangle - \langle E \rangle^2) \tag{8.82}$$

The magnetic susceptibility may be computed using a similar formula, whose derivation is left to Problem 8.15.

$$L^2 \chi = \beta (\langle M^2 \rangle - \langle M \rangle^2) \tag{8.83}$$

with $M = \sum \sigma_i$.

8.15 MAGNETISM

Many atoms in their ground states have permanent magnetic dipole moments. When such atoms combine to form a solid, the magnetic degrees of freedom interact with one another and with any external magnetic field, giving rise to interesting and easily observed macroscopic magnetic phenomena. The study of the magnetic behavior of solids has for a long time been one of the most active subfields of statistical mechanics, both because of its fundamental physical significance and because of its technological importance. Most substances fall into one of four broad classes with regard to their magnetic properties. They are (1) diamagnetic, (2) paramagnetic, (3) ferromagnetic, or (4) antiferromagnetic. There is also a fifth class, called *ferrimagnets*, which are generally more complex solids, such as two-component alloys. Some cases of antiferromagnetism and ferrimagnetism will be covered in the supplement, but we will not consider these phenomena any further in this chapter. These five classes are not, by any means, an exhaustive list of possible magnetic behaviors. The full range of magnetic phenomena is much too complex to be covered in this short introductory presentation.

Diamagnetism is exhibited by materials composed of atoms or molecules that do not have permanent magnetic moments. It is a very weak effect. The introduction of a diamagnetic substance into a region in which a magnetic field exists will typically cause a change in the magnetic field strength of one part in 10^5 or 10^6. The permeability μ of a diamagnetic material is slightly smaller than μ_o, the permeability of vacuum. This indicates that the small magnetic dipole moments that are induced in the atoms of the substance by the external field point antiparallel to the inducing field. The calculation of those induced dipole moments is purely a problem in quantum mechanics, and, therefore, diamagnetism will not be considered in any detail in this book.

8.16 PARAMAGNETISM

Paramagnetism describes the interaction of atoms that have permanent magnetic dipole moments with an imposed magnetic field in the case where interaction effects between the atoms are relatively weak. This can occur in a dilute gas of magnetic atoms, but the far more important case is that of complex paramagnetic salts, which have crystal structures in which the magnetic atoms (or ions) are well separated from one another by intervening lattice sites containing nonmagnetic elements. A paramagnetic atom is like a small magnetic dipole. The source of the atom's dipole moment may be either the intrinsic magnetic dipole moment of the electrons, due to their spin, or the electrical currents implied by the orbital motion of the electrons around the nucleus. The magnetic moment of an isolated atom is a vector $\mathbf{m}$ that is antiparallel to the atom's angular momentum vector. The z component of the angular momentum vector can be written in terms of the magnetic quantum number M as $M\hbar$, where $-J \leq M \leq J$. Since the vector $\mathbf{m}$ is antiparallel to the angular momentum, its z component is also proportional to the magnetic quantum number. That is, $m_z = -\alpha M$. The proportionality constant α is different for different atoms and is a number of the order of the *Bohr magneton*, $\mu_B = e\hbar/2m_e = 0.927 \times 10^{-23}$ J/T. For a particular atom, the ratio of α to μ_B is called the *gyromagnetic ratio*. If we choose the z axis in the direction of the external magnetic field, then the energy associated with the interaction of the atom's magnetic dipole moment and the magnetic field is given by

$$E_M = -m_z B = \alpha M B \tag{8.84}$$

Using a canonical ensemble, the magnetic energy per atom at temperature T is

$$E = \langle E_M \rangle = \sum_M E_M e^{-\beta E_M} / Z \tag{8.85}$$

where

$$Z(\beta) = \sum_{M=-J}^{J} e^{-\beta E_M} \tag{8.86}$$

The sum defining the magnetic partition function $Z(\beta)$ can be calculated exactly (see Problem 8.21), giving

$$Z(\beta) = \frac{\sinh\left[\beta\alpha B(J + \frac{1}{2})\right]}{\sinh(\beta\alpha B/2)} \tag{8.87}$$

From Eq. (8.85) we can see that $E = -\partial \log Z / \partial\beta$, as expected. Therefore, the magnetic energy per particle is

$$E = -\frac{\partial \log Z}{\partial\beta} = -\alpha B(J + \tfrac{1}{2}) \coth\left[\beta\alpha B(J + \tfrac{1}{2})\right] + \tfrac{1}{2}\alpha B \coth(\beta\alpha B/2) \tag{8.88}$$

For small values of β (that is, at high T) or for small values of the magnetic field, one can expand the hyperbolic cotangent function in a power series ($\coth x \approx 1/x + x/3$), obtaining

$$E \approx -\tfrac{1}{3}\beta\alpha^2 B^2 J(J + 1) \tag{8.89}$$

The magnetic susceptibility is usually defined as the ratio of the magnetization of the substance to the imposed H field in the limit of small field strength. However, for the purposes of statistical mechanics and thermodynamics, it is much more convenient to define the magnetic susceptibility χ_m as the ratio of the magnetization to the magnetic field strength B. With that definition, we can express χ_m in terms of the density of magnetic dipoles n as

$$\chi_m = \lim_{B\to 0} \frac{n\langle m_z\rangle}{B} \tag{8.90}$$

Using the fact that $E = -B\langle m_z\rangle$ and the small-B expansion for E [Eq. (8.89)], one gets the following expression for the magnetic susceptibility.

$$\chi_m = \frac{n\alpha^2 J(J + 1)}{3kT} \tag{8.91}$$

Thus the magnetic susceptibility of a paramagnetic substance is predicted to be proportional to the inverse of the absolute temperature. That characteristic is called *Curie's law*. For any paramagnetic material, the ratio of χ_m to $1/T$ is called the *Curie constant* of the substance. A comparison of the predictions of Eq. (8.91) with experimental results is given in Fig. 8.14.

8.17 THE SPIN-$\frac{1}{2}$ PARAMAGNET

The simplest possibility is a paramagnetic atom or ion with a spin-$\frac{1}{2}$ ground state. Then $J = \frac{1}{2}$ and Eq. (8.87) can be simplified by using the identity $\sinh(A + B) = \sinh(A)\cosh(B) - \cosh(A)\sinh(B)$ in the numerator, and cancelling one of the factors with the denominator, giving

$$Z(\beta) = 2\cosh(\beta\alpha B/2) \tag{8.92}$$

Of course, this formula could have easily been obtained directly from Eq. (8.86). Being a two-level system, the spin-$\frac{1}{2}$ paramagnet has a magnetic contribution to its specific heat that exhibits the Schottky specific heat anomaly shown in Fig. 4.9. However, because the energy spacing between the two levels, $\varepsilon = \alpha B$, is proportional

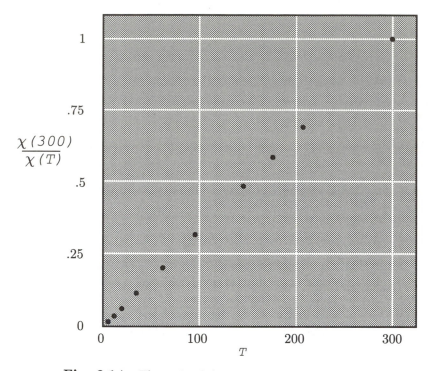

Fig. 8.14   The ratio of the magnetic susceptibility χ at 300 K to that at temperature T for the paramagnetic crystal $CuSO_4 \cdot 5H_2O$. According to Curie's law, the points should lie on a straight line through the origin. (Data from J. Crangle, *The Magnetic Properties of Solids*, Edward Arnold, London, 1977.)

to the magnetic field, the temperature of the specific heat maximum can be tuned by changing the magnetic field strength. For larger values of J, the sharpness of the specific heat maximum is reduced (see Problem 8.22).

8.18   COOLING BY ADIABATIC DEMAGNETIZATION

A particular characteristic of how the entropy of a paramagnet depends on the temperature and the magnetic field strength is the source of an important method of cooling substances to very low temperatures. The technique relies only on the fact that, according to Eq. (8.87), $\phi = \log Z$ is a function only of the combination βB. That is, $\phi(\beta, B) = f(\beta B)$. The entropy function can be shown to have the same property.

$$S^o = \phi + \beta E = \phi - \beta \frac{\partial \phi}{\partial \beta} = f(\beta B) - \beta B \frac{df(\beta B)}{d(\beta B)} \tag{8.93}$$

The method works as follows: (1) In a constant magnetic field B_o, one cools the sample to a temperature T_o. (2) The sample is then insulated so that no heat can flow in or out, and the magnetic field is gradually reduced to some small final value $B \ll B_o$. Because no heat can flow in or out of the sample during stage (2), the entropy of the sample remains constant. But if S_o is to remain constant while B decreases, then T must also decrease in order to maintain the value of $\beta B = B/kT$.

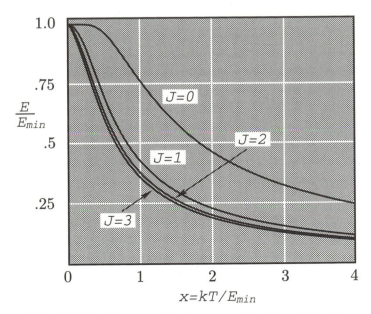

Fig. 8.15 The magnetic energy as a function of the temperature for paramagnets of various total angular momentum values. $E_{min} = E(T=0) = \alpha\beta J$. For large J the curves approach the result given by a classical analysis. (Compare with Fig. 4.6.)

Thus, according to Eq. (8.93), the final temperature should be given by

$$T = \frac{B}{B_o}T_o \tag{8.94}$$

This equation seems to show that, by turning off the magnetic field entirely, one could bring the sample to zero temperature. That, of course, is not true. First, this analysis is all based on the assumption that the magnetic dipoles are not interacting. But when kT becomes smaller than the interaction energy, which must exist in reality, then the model leading to Eq. (8.93) is no longer adequate to describe the real substance. Second, the analysis is also based on the assumption that the system is perfectly insulated. But it is clearly impossible to insulate the magnetic degrees of freedom from the other sources of energy in the substance, such as lattice vibrations. Therefore, as the magnetic field is reduced, energy will be transferred from other degrees of freedom of the sample, keeping the temperature finite. Nevertheless, the effect is an important and effective means of reaching low temperatures.

8.19 FERROMAGNETISM

When there is a significant interaction between nearby atoms in a substance, and the interaction favors parallel magnetic moments, then the substance exhibits ferromagnetic behavior. Ferromagnetism is such a strong and striking effect that it is, in fact, the only class of magnetic phenomena observable to the nonscientist. The detection of diamagnetism or paramagnetism requires delicate and careful experimental procedures. In common parlance, the word *magnetism* means ferromagnetism. Because ferromagnetism is an effect of strong interatomic interactions, its theoretical

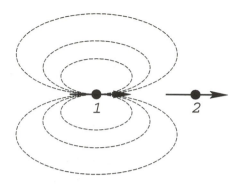

Fig. 8.16 Dipole 1 produces a magnetic field that tends to cause a parallel alignment of dipole 2, and vice versa.

analysis is very challenging. It is a collective effect, in which vast numbers of atoms act in a synchronized, coherent way.

The first question that must be considered is what is the source of the interaction that leads to parallel spin alignment. It is easy to picture a direct magnetic dipole–dipole interaction that would favor alignment. Let us first convince ourselves that this direct magnetic dipole interaction is actually completely insignificant at the temperatures at which ferromagnetic phenomena really occur. In Fig. 8.16 we see two magnetic dipoles separated by a distance a. The magnetic field due to dipole 1 at the location of dipole 2 is $B = \mu_o m / 2\pi a^3$. The energy difference between antiparallel and parallel orientations of the dipoles is $\Delta E = 2mB = \mu_o m^2 / \pi a^3$. For an electron spin, m is equal to one Bohr magneton, μ_B. A reasonable value of a is 2 Å. This gives $\Delta E = 4.3 \times 10^{-24}\,\mathrm{J} = 0.3\,\mathrm{K}$. Such energy differences would be important at a temperature of about 1 K. But the typical transition temperatures of real ferromagnets are about 1000 K. The magnetic dipole interaction is too weak by a factor of about 1000. Doing a more exact analysis by considering the effects on a given dipole of a whole cubic lattice of magnetic dipoles only makes the discrepancy worse (see Problem 8.23). The true source of the electron spin alignment of neighboring atoms is what is called the *exchange force*. This is also the source of the spin alignment within a single atom that has a partially filled shell. In a single atom, the tendency for spins to align themselves in incompletely filled shells is called *Hund's rule*.

To understand the exchange force, we consider two electrons on neighboring atoms in orbitals $\phi_1(\mathbf{r})$ and $\phi_2(\mathbf{r})$ that overlap slightly. If the electrons have identical spin orientations, then their spatial wave function must be antisymmetric. Thus the wave function of the pair will be (except for normalization) $\psi(\mathbf{r}_1, \mathbf{r}_2) = \phi_1(\mathbf{r}_1)\phi_2(\mathbf{r}_2) - \phi_2(\mathbf{r}_1)\phi_1(\mathbf{r}_2)$. For any value of $\mathbf{r}$, $\psi(\mathbf{r}, \mathbf{r}) = 0$. Thus this antisymmetric combination gives a vanishing probability for the electrons to be in exactly the same place and a very small probability for them to be very near one another. But because of the repulsive Coulomb force between electrons, such a repulsive short-range correlation in their positions is just what is needed to lower the average potential energy. If the two electrons have different spins, then they are distinguishable, and there is no reason for their spatial wave function to be antisymmetric. Therefore, it is the electrostatic interaction between electrons that, in combination with the antisymmetry requirement on many-electron wave functions, favors parallel spin alignment for electrons in neighboring atoms.

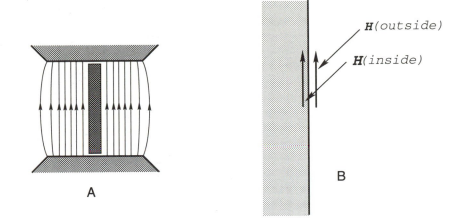

Fig. 8.17 (A) A long rod is placed parallel to the external field. (B) At the boundary between two different materials, the tangential component of the H field is continuous. This guarantees that $\mathbf{H}$(inside) = $\mathbf{H}$(outside). But, the outside material is vacuum, so $\mathbf{B} = \mu_o\mathbf{H}$ there. Thus $\mathbf{B}$(outside) = $\mu_o\mathbf{H}$(inside).

8.20 DOMAINS AND HYSTERESIS

Before making any attempt to relate the theoretical predictions of any model to the experimental properties of any real ferromagnet, there is another effect that must be discussed. It is one that greatly complicates the interpretation of experiments on ferromagnetic substances.

We consider a rod of unmagnetized iron placed between the poles of an electromagnet that is initially turned off [see Fig. 8.17(A)]. The field between the pole pieces is gradually increased to some vey large maximum value B_o, then gradually brought back to zero and passed through zero and brought to $-B_o$ (that is, B_o in the opposite direction). After that, it is slowly cycled between $-B_o$ and B_o. While this is being done, the magnetization M of the iron is measured, simultaneously with the external magnetic field. As shown in Fig. 8.17(B), the value of the internal H field is directly proportional to that of the applied B field. If the values of M and H are plotted on an M–H plane, one obtains a hysteresis curve such as that shown in Fig. 8.18. The fact that when the external field is returned to its initial zero value the magnetization does not return to its initial value indicates that the states being observed are not true equilibrium states. The values of the pressure (which, at moderate values, is more or less irrelevent), the temperature, and the external field should constitute a complete set of thermodynamic variables. Their values would therefore determine the values of all other macroscopic variables at equilibrium. In fact, the value obtained for the magnetization at given external field depends on the past history of the field values. One can reach any point within the two bounding curves on the hysteresis diagram simply by stopping the increase of H some time before saturation (the limiting value of M) is reached and then reducing the value of H.

What prevents the easy establishment of full equilibrium in most ferromagnetic samples is the formation of *magnetic domains*, which are regions that are very large on an atomic scale (one hundredth of a millimeter is typical) but quite small in comparison with the size of the sample. Each domain is internally in equilibrium, with a magnetization that is larger than the net magnetization of the

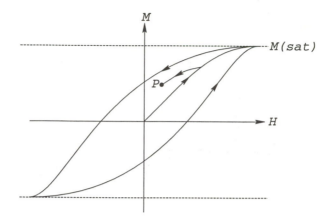

Fig. 8.18   Beginning at the origin, as H is increased, the magnetization approaches some maximum value, called the *saturated* magnetization. If H is then reduced to zero, the magnetization is left at some positive value (the point at which the curve cuts the positive M axis). If H is repeatedly cycled, the magnetization follows the outside *hysteresis* curve. By stopping at various points on the curve and "backing up" the external field, one can reach any point on the M–H plane within the hysteresis curve, such as point P.

sample. However, the direction of magnetization varies from domain to domain. To understand the reason for domain formation, let us picture a cube of some ferromagnetic material whose internal crystal structure makes it advantageous for the magnetization vector to lie either in the $\pm x$ or the $\pm y$ direction. At equilibrium, the Helmholtz free energy of the system is a minimum. The free energy is composed of the internal free energy of the substance, which is minimized when the complete sample is magnetized in a single direction, either $\pm x$ or $\pm y$, plus the field energy of the magnetic field outside the sample. When the internal free energy is at its minimum, the external free energy is fairly large. By modifying the internal state, as shown in Fig. 8.19(B), the external field energy can be greatly reduced at the small cost of introducing some narrow regions, called *domain walls*, of high free energy density. The result is a lowering of the total free energy. Thus, although the internal free energy alone favors a simple homogeneous state, the actual equilibrium state is a more complicated state of nonuniform magnetization. This analysis is appropriate for single perfect crystals. In such samples the domains are quite large, and their detailed shape could be calculated by theoretical analysis. However, real samples are almost always polycrystalline, with the axes of easy magnetization varying from one crystallite to the next, and with preexisting *zone boundaries* between the crystallites. They also contain crystal imperfections and impurities. All these act as anchors for the magnetic domain walls, causing the magnetization pattern to get stuck in configurations that are local, but not true minima of the total free energy, and are therefore nonequilibrium states. We do not intend to discuss the complex subject of domain theory any further, and therefore we will always consider only perfect crystals and will ignore the external magnetic field energy. Our analysis is adequate to describe the properties of small perfect crystal samples (because the importance of the external field energy diminishes as the sample size decreases) or to describe the situation within a single domain.

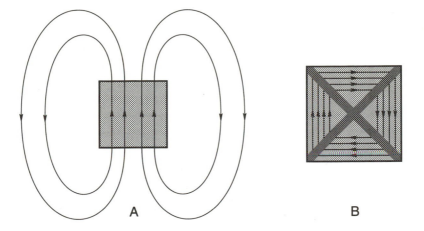

Fig. 8.19   In state A, the Helmholtz free energy of the material itself is a minimum. But due to the contribution of the external field energy, the total free energy of the system is not a minimum. In state B, the external field energy has been greatly reduced, but, due to the transition regions (zone boundaries), where the internal magnetization changes from one value to another, the internal free energy is no longer at its minimum. For large samples, the increase in free energy that is caused by introducing zone boundaries is more than compensated by the decrease in free energy due to eliminating, or greatly reducing, the external field. Thus the total free energy of state B is smaller than that of state A.

8.21   THE 3D ISING MODEL

A realistic model of a magnetic substance must treat the crystal as a system of interacting electrons and atomic nuclei. The magnetic properties are then determined, along with the other physical properties of the crystal, by a very complex quantum mechanical calculation. Such a calculation would involve concepts well beyond the level of this book. Instead of attempting such a thing, we will look for a model that can capture the essential physics of ferromagnetic phase transitions and yet can be analyzed, at least approximately, without a great deal of complication. In our attempt to obtain a simple model for ferromagnetism, our greatest sacrifice will be to restrict our analysis to *uniaxial* ferromagnets. These are a subset of ferromagnetic crystals in which, due to their anisotropic crystal structure, there is a fixed axis along which, in either direction, the atomic magnetic moments are strongly favored to be aligned. The magnetic behavior of such a crystal can be reasonably approximated by that of a three-dimensional Ising model (Fig. 8.20).

The 3D Ising model is the obvious generalization of the 2D Ising model, discussed in the section on the Monte Carlo method. It is a cubic lattice, with lattice points defined by three integers X, Y, and Z that each take values in the range $1, 2, \ldots, L$, giving a total of $N = L^3$ lattice sites. If the lattice constant is a, then the lattice sites are located at positions

$$\mathbf{r} = (aX, aY, aZ) \tag{8.95}$$

The z axis is taken to be the axis of easy magnetization. At each site there is a spin variable $\sigma(\mathbf{r})$ with the possible values of ± 1, indicating that the spin at that

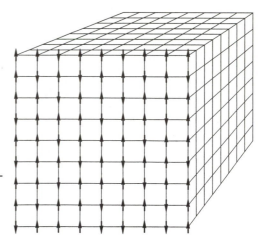

Fig. 8.20 The 3D Ising model is a three-dimensional cubic array of spins, each of which can have two possible values, up or down.

site points in the positive or negative z direction. The z component of the magnetic moment of the atom at site $\mathbf{r}$ is $m_z(\mathbf{r}) = m_o\sigma(\mathbf{r})$.

A given spin interacts with its six neighbors in the x, y, and z directions and with an external magnetic field $\mathbf{B}$ pointing in the z direction. A configuration C is defined by choosing spin values at every site. The energy of the system in a given configuration is assumed to be of the form

$$E(C) = -m_oB \sum_{\mathbf{r}} \sigma(\mathbf{r}) - V \sum_{\mathrm{NN}} \sigma(\mathbf{r})\sigma(\mathbf{r}') \qquad (8.96)$$

The terms in the first sum describe the interaction between the spin magnetic moments and the external field. The product of the magnetic moment and the magnetic field m_oB will, in the future, be written as H. The terms in the second sum describe the interaction between all nearest-neighbor pairs of spins—that is, all pairs of spins that are separated by only one lattice distance. The interaction constant V is positive for a ferromagnetic model, in which the favored, low-energy configurations have most nearest-neighbor spins pointing in the same direction.

The 3D Ising model being considered here is the *isotropic* Ising model. This means that the interactions between neighboring particles in the x, y, and z directions are equal. A more general model is the *anisotropic* Ising model, in which there are three different interaction parameters, describing interactions between nearest-neighbor spins in the three coordinate directions. For the anisotropic Ising model,

$$E(C) = -H \sum_{\mathbf{r}} \sigma(\mathbf{r}) - \sum_{\mathbf{r}} \sigma(\mathbf{r})\big[V_x\sigma(\mathbf{r} + \mathbf{a}_x) + V_y\sigma(\mathbf{r} + \mathbf{a}_y) + V_z\sigma(\mathbf{r} + \mathbf{a}_z)\big] \quad (8.97)$$

where $\mathbf{a}_x$, $\mathbf{a}_y$, and $\mathbf{a}_z$ are lattice vectors in the x, y, and z directions. In the next section, we will obtain an approximate solution of the isotropic Ising model. The reader will be asked to extend the calculation to the anisotropic model. (See Problem 8.25.)

8.22   THE MEAN-FIELD SOLUTION

The 2D Ising model, with $H = 0$, was exactly solved in 1944 by Lars Onsager. Both the 2D model with finite H and the 3D model with any value of H have resisted exact solution for about 50 years. In order to investigate the properties of the 3D Ising model, we will first obtain an approximate solution of the model, using the mean-field approximation and then compare the predictions of that method with the results of accurate numerical solutions, obtained with the Monte Carlo method and by other methods. It will require very little extra work to obtain simultaneously the mean-field solution on any type of lattice (triangular, hexagonal, etc.) in any number of dimensions. In order to facilitate the simultaneous treatment of all Ising systems with nearest-neighbor interactions, we introduce a parameter c, called the *coordination number* of the lattice, that is equal to the number of nearest neighbors of any spin in the lattice under consideration. For the three-dimensional cubic lattice, $c = 6$.

The mean-field calculation is made by first expressing each spin variable $\sigma(\mathbf{r})$ as an average value plus a fluctuation. That is

$$\sigma(\mathbf{r}) = \bar{\sigma} + \delta\sigma(\mathbf{r}) \tag{8.98}$$

where $\bar{\sigma} = \langle\sigma(\mathbf{r})\rangle$, which is independent of $\mathbf{r}$, because of the translational symmetry of the model. The fluctuating variables each have zero average value, $\langle\delta\sigma(\mathbf{r})\rangle = 0$. The basic assumption made in the mean-field method is that the statistical correlation of neighboring spins is negligible. Therefore, a term such as $\langle\delta\sigma(\mathbf{r})\delta\sigma(\mathbf{r}')\rangle$ will give a very small value, because

$$\langle\delta\sigma(\mathbf{r})\delta\sigma(\mathbf{r}')\rangle \approx \langle\delta\sigma(\mathbf{r})\rangle\langle\delta\sigma(\mathbf{r}')\rangle = 0 \tag{8.99}$$

If, in the interaction term in Eq. (8.96), we replace all σ's by $\bar{\sigma} + \delta\sigma$ and ignore products of $\delta\sigma$s [because, according to Eq. (8.99), they make negligible contributions to average values], we obtain

$$\begin{aligned} E(C) &= - H \sum_{\mathbf{r}} \sigma(\mathbf{r}) - V \sum_{NN} \left[\bar{\sigma} + \delta\sigma(\mathbf{r})\right]\left[\bar{\sigma} + \delta\sigma(\mathbf{r}')\right] \\ &\approx -H \sum_{\mathbf{r}} \sigma(\mathbf{r}) - \tfrac{1}{2}cNV\bar{\sigma}^2 - cV\bar{\sigma}\sum_{\mathbf{r}} \delta\sigma(\mathbf{r}) \end{aligned} \tag{8.100}$$

Having eliminated the second-order terms in $\delta\sigma$, it is now better to rewrite our approximate expression for $E(C)$ in terms of the original σ variables, by replacing $\delta\sigma(\mathbf{r})$ by $\sigma(\mathbf{r}) - \bar{\sigma}$.

$$\begin{aligned} E(C) &= \tfrac{1}{2}cNV\bar{\sigma}^2 - (H + cV\bar{\sigma})\sum_{\mathbf{r}}\sigma(\mathbf{r}) \\ &= \tfrac{1}{2}cNV\bar{\sigma}^2 - H'\sum_{\mathbf{r}}\sigma(\mathbf{r}) \end{aligned} \tag{8.101}$$

where $H' = H + cV\bar{\sigma}$. The last line is, except for the constant term $\tfrac{1}{2}cNV\bar{\sigma}^2$, the energy expression for a collection of *noninteracting* magnetic moments in an external field H'. For such a system, the probability that the spin $\sigma(\mathbf{r})$ has the value ± 1 is proportional to $\exp(\pm\beta H')$. When that probability is normalized,

$$P(\pm 1) = \frac{e^{\pm\beta H'}}{e^{\beta H'} + e^{-\beta H'}} \tag{8.102}$$

Therefore, the condition that $\langle \sigma \rangle = \bar{\sigma}$ gives

$$(+1)P(+1) + (-1)P(-1) = \frac{e^{\beta H'} - e^{-\beta H'}}{e^{\beta H'} + e^{-\beta H'}} = \tanh \beta(H + cV\bar{\sigma}) = \bar{\sigma} \qquad (8.103)$$

This is a nonlinear equation for the value of $\bar{\sigma}$, given the external field H, the interaction parameter V, and the inverse temperature β. We will call $\bar{\sigma}$ the magnetization, although the actual macroscopic magnetization is m_o times the particle density times $\bar{\sigma}$. There are two ways of treating this equation. The first is to picture solving it by a graphical technique. We let $x = \beta(H + cV\bar{\sigma})$ and note that $\bar{\sigma} = (kTx - H)/cV$, so that the equation to be solved is

$$\tanh x = \frac{kTx - H}{cV} \equiv s(x - x_o) \qquad (8.104)$$

where the slope of the linear function on the right-hand side is $s = kT/cV$ and the displacement $x_o = H/kT$.

If $s \geq 1$, then the equation has a unique solution x for each value of x_o. The case $s < 1$ is more interesting. As can be seen in Fig. 8.21, for large x_o (large H) there is a unique solution for x, but for smaller values of x_o there are generally three possible solutions of the equation. Let us assume that x_o, and therefore H, is positive. It is easy to see that only one of the solutions will have $x > x_o$. For that solution, $\bar{\sigma}$ and H have the same sign. But we can prove by the following argument that, in the equilibrium state, $\bar{\sigma}$ and H must have the same sign. Therefore, for $x_o > 0$ it is the rightmost solution that is the physical one. Similarly, for $x_o < 0$ it is the leftmost solution that represents the equilibrium state.

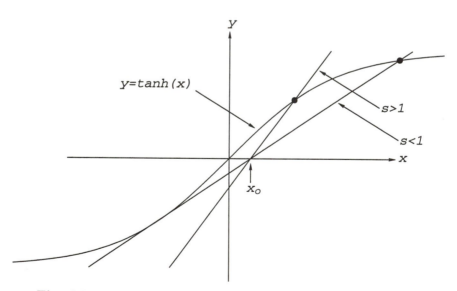

Fig. 8.21 The curve is the graph of the function $y = \tanh(x)$. The two straight lines represent the function $y = s(x - x_o)$ for two different values of s. As x_o (which is proportional to the external field H) goes to zero, the positive solution of the equation $\tanh(x) = s(x - x_o)$ (and it is only the positive solution that is physically acceptable) approaches zero when $s > 1$ but remains at a finite value of x when $s < 1$. Actually, for $x_o = 0$, the equation to be solved is equivalent to Eq. (8.49), and the analysis given there could have been used also.

We must now show that $\bar{\sigma}$ and H must have the same sign at equilibrium. We can prove this without using the mean-field approximation. For any configuration $C = (\sigma_1, \ldots, \sigma_N)$, let C^* be the configuration that is obtained by flipping all of the spins. That is, $C^* = (-\sigma_1, \ldots, -\sigma_N)$. Let Ω_1 be the set of all the configurations that have $\sigma_1 = +1$. Then, if C ranges over all the configurations in Ω_1, the combination of C and C^* will give all possible configurations. According to Eq. (8.96), the energy of a configuration is

$$E(C) = E_H(C) + E_V(C) \tag{8.105}$$

where

$$E_H = -H \sum_{\mathbf{r}} \sigma(\mathbf{r}) \tag{8.106}$$

and

$$E_V = -\sum_{NN} \sigma(\mathbf{r})\sigma(\mathbf{r}') \tag{8.107}$$

It is easy to see that $E_H(C^*) = -E_H(C)$ and $E_V(C^*) = E_V(C)$. The probability of any configuration is $P(C) = Z^{-1} \exp[-\beta E_H(C)] \exp[-\beta E_V(C)]$. Therefore, the average value of E_H is

$$\langle E_H \rangle = Z^{-1} \sum_{C \in \Omega_1} \left[E_H(C) e^{-\beta E(C)} + E_H(C^*) e^{-\beta E(C^*)} \right]$$

$$= Z^{-1} \sum_{C \in \Omega_1} \left[E_H(C) e^{-\beta E_H(C)} - E_H(C) e^{\beta E_H(C)} \right] e^{-\beta E_V(C)} \tag{8.108}$$

$$= -2Z^{-1} \sum_{C \in \Omega_1} E_H(C) \sinh[\beta E_H(C)] e^{-\beta E_V(C)}$$

Using the fact that $x \sinh(\beta x) \geq 0$ for all x, one can see that $\langle E_H \rangle < 0$. But it is also easy to see that $\langle E_H \rangle = -NH\bar{\sigma}$. Therefore, H and $\bar{\sigma}$ must have the same sign.

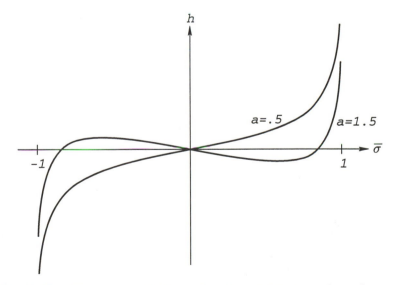

Fig. 8.22   Plots of $h = H/cV$ as a function of $\bar{\sigma}$ for two values of the parameter $a = kT/cV$.

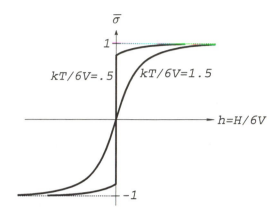

Fig. 8.23   Plots of $\bar{\sigma}$ as a function of the parameter $h = H/cV$ for temperatures above and below the ferromagnetic phase transition.

The second method of dealing with the nonlinear relation between H and $\bar{\sigma}$ is to stand it on its head by solving for H as a function of $\bar{\sigma}$. Taking the arctanh of both sides of Eq. (8.103), one easily obtains the relation

$$H = kT \tanh^{-1}\bar{\sigma} - cV\bar{\sigma} \tag{8.109}$$

This function is plotted in Fig. 8.22. It can be seen that, for $kT < cV$, the function cannot be simply inverted to give $\bar{\sigma}(H)$, but again the "same sign" rule relating $\bar{\sigma}$ and H picks out the correct solution. (Correct, that is, within the mean-field approximation, which is itself not very accurate.) Using this rule to choose the unique physical solution when multiple solutions exist, $\bar{\sigma}(H, T)$ has been plotted in Fig. 8.23 for the cases $kT/cV = 0.5$ and $kT/cV = 1.5$.

8.23   SPONTANEOUS MAGNETIZATION
Figure 8.23 exhibits the essential feature of ferromagnetism—when the strength of the external field H is reduced to zero, the magnetization of the sample is not also brought to zero. When T is less than the *Curie temperature*, $T_c = cV/k$, there exist two possible equilibrium states at $H = 0$. They have values of $\bar{\sigma}$ that are solutions of Eq. (8.109) with $H = 0$. Using the definition of T_c, this can be written as

$$\tanh\left(\frac{T_c}{T}\bar{\sigma}\right) = \bar{\sigma} \tag{8.110}$$

The phenomenon of nonzero magnetization in the presence of zero external field is called *spontaneous magnetization*. It is a special case of spontaneous symmetry breakdown, which is said to exist whenever the possible equilibrium states of a system do not exhibit the symmetries of the Hamiltonian function of the system. In the case of the Ising model, when $H = 0$, the Hamiltonian $E(C)$ has an obvious symmetry with respect to a simultaneous flipping of all the spins in the system. But each of the equilibrium states is either magnetized up or down and is therefore not unchanged under spin flip. Notice, however, that the collection of all the possible equilibrium states *is* invariant under spin flip. This is a general phenomenon. Whenever the equilibrium states have spontaneous symmetry breaking, then they are also not unique and the set of equivalent equilibrium states transform into one another under the symmetry operation.

8.24   THE PHASE DIAGRAM

In discussing the thermodynamics of our model of a ferromagnetic substance, the natural independent variables are the external field H and the temperature T. On the T–H plane, there is a line of first-order phase transitions at $H = 0$ and $0 < T < T_c$ (Fig. 8.24). The phase transition is first order because there is a finite jump in the magnetization across the phase transition line. However, the endpoint of the line (called the *critical point*), at $H = 0$ and $T = T_c$, is a point of second-order phase transition. This is best seen by considering what happens to the magnetization as we approach the critical point from the left. (However, because we are using the mean-field approximation in this analysis, all our results must be viewed with some scepticism. Later we will see that, although the order of the transition is predicted correctly, most quantitative aspects of the solution must be significantly modified.)

Just below T_c, one can get an analytic solution of Eq. (8.110) for $\bar{\sigma}(T)$ by using the fact that $\bar{\sigma}$ is very small and expanding $\tanh^{-1}\bar{\sigma}$ as a power series. Keeping the first two terms, the equation becomes

$$\tanh^{-1}\bar{\sigma} \approx \bar{\sigma} + \tfrac{1}{3}\bar{\sigma}^3 = \frac{T_c}{T}\bar{\sigma} \tag{8.111}$$

Cancelling a factor of $\bar{\sigma}$, this equation can be written as

$$\frac{1}{3}\bar{\sigma}^2 = \frac{T_c}{T} - 1 = \frac{T_c - T}{T} \approx \frac{T_c - T}{T_c} \tag{8.112}$$

where the last step is valid for $T \approx T_c$. This has the solution

$$\bar{\sigma} = \pm\sqrt{3}\left(\frac{T_c - T}{T_c}\right)^{1/2} \tag{8.113}$$

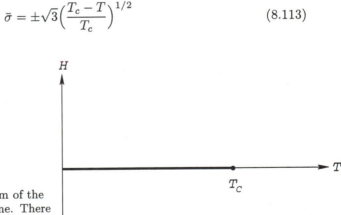

Fig. 8.24   The phase diagram of the 3D Ising model in the H–T plane. There is a line of first-order phase transitions, ending in a critical point.

This analysis is all for $T < T_c$. For $T > T_c$, $\bar{\sigma}$ is zero. Therefore, we can see that $\bar{\sigma}$ is a continuous (but not a differentiable) function of T across the phase transition point at $T = T_c$. Keep in mind that this is all for $H = 0$. If H is kept at a fixed nonzero value while T is varied, there is no phase transition at all because the phase transition line is never intercepted.

$\bar{\sigma}$ is the quantity that exhibits the spontaneous symmetry breaking. In general, whatever variable is nonzero in the broken symmetry states, when it would be expected to be zero just by looking at the Hamiltonian, is called the *order parameter* associated with the spontaneous symmetry breaking. Equation (8.113) says that, as we approach the critical point, the order parameter goes to zero as $(-t)^{1/2}$, where $t = (T - T_c)/T_c$ is a dimensionless variable that measures how far T deviates from T_c. As we will see in the next chapter, this behavior of the order parameter near a second-order phase transition is a completely general prediction of the mean-field approximation when it is applied to any system that has spontaneous symmetry breakdown. In reality, this prediction does not agree with experimental results for uniaxial ferromagnets (nor with numerical calculations, or more sophisticated theoretical analysis), which show that, as T approaches T_c, $\bar{\sigma}$ does go to zero as $(-t)^\beta$, but that the parameter β called a *critical exponent*, has the value

$$\beta = 0.324 \pm 0.006 \tag{8.114}$$

rather than the value $\frac{1}{2}$.

8.25   THERMODYNAMICS OF THE 3D ISING MODEL

In this section, we will specialize to the particular case of a 3D cubic lattice, that is, to the case $c = 6$. The equations of state of the system should give the magnetization and the energy as functions of the two independent variables, T and H. The magnetization has already been determined. The energy is calculated by using the fact that, in the mean-field approximation, any two spin variables are statistically uncorrelated. Therefore, if $\mathbf{r} \neq \mathbf{r}'$, then

$$\langle \sigma(\mathbf{r})\sigma(\mathbf{r}') \rangle = \langle \sigma(\mathbf{r}) \rangle \langle \sigma(\mathbf{r}') \rangle = \bar{\sigma}^2 \tag{8.115}$$

From this, one can easily calculate the thermal energy in terms of the magnetization.

$$\begin{aligned} E &= \langle E(C) \rangle \\ &= -NH\bar{\sigma} - \tfrac{1}{2}cVN\bar{\sigma}^2 \\ &= -N(H\bar{\sigma} + 3V\bar{\sigma}^2) \end{aligned} \tag{8.116}$$

This equation, when combined with Eq. (8.109), which defines $\bar{\sigma}(T, H)$, gives the energy as a function of the temperature and the external magnetic field.

The specific heat per spin is the partial derivative of E/N with respect to T at constant H.

$$C = \frac{\partial(E/N)}{\partial T} = -(H + 6V\bar{\sigma})\frac{\partial \bar{\sigma}}{\partial T} \tag{8.117}$$

It is interesting to evaluate C along the line $H = 0$ and $0 < T < \infty$, because this passes through the critical point at T_c. Along this line, to the right of the critical point, $\bar{\sigma} = 0$ and therefore $C = 0$. Just to the left of the critical point we can use the approximate equation, Eq. (8.113), for $\bar{\sigma}$, which says that

$$\bar{\sigma}^2 \approx 3\left(1 - \frac{T}{T_c}\right) \tag{8.118}$$

But with $H = 0$, Eq. (8.117) can be written as $C = 3V\partial\bar{\sigma}^2/\partial T$. Therefore, for $T \approx T_c$,

$$C \approx 9V/T_c = \tfrac{3}{2}k \tag{8.119}$$

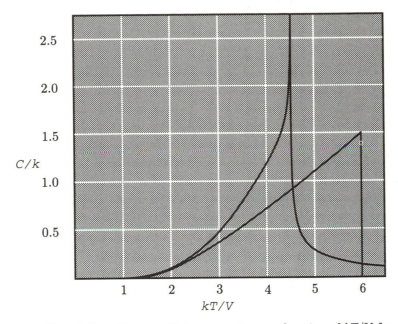

Fig. 8.25   The specific heat per site as a function of kT/V for a 3D Ising model. The true specific heat curve was obtained from an analysis of high-order perturbation calculations, similar to the cluster expansion presented in Section 8.3. It goes to infinity at the critical temperature $(kT_c/V \approx 4.51)$ with an integrable singularity of the form $|T - T_c|^{-\alpha}$, where α is very close to 1/8. The mean-field result has no singularity, but has a finite discontinuity at $kT_c/V = 6$.

A plot of C on the line $H = 0$ is given in Fig. 8.25, along with the results of an accurate numerical calculation for comparison. It is clear from a study of that figure that the mean-field approximation does not make an accurate prediction of the critical temperature, which is actually equal to about $4.51\,V/k$, rather than $6\,V/k$. Nor does it give the correct behavior of the specific heat close to T_c, even if both the mean-field result and the accurate numerical result are plotted as functions of $x = T/T_c$, so that their critical points are forced to coincide at $x = 1$.

PROBLEMS

8.1   Derive Eq. (8.27) for B_2 and B_3 in terms of C_2 and C_3.

8.2   In Exercise 4.17 it was shown that the equation of state of a one-dimensional gas of hard-core particles is $\beta p = n/(1 - an)$, where a is the hard-core diameter. (a) Calculate the cluster integrals C_2 and C_3, for one-dimensional hard-core particles, using Eqs. (8.8), (8.12), and (8.13). (b) By using Eq. (8.27) and your result for C_2 and C_3, calculate the virial coefficients B_2 and B_3. (c) By expanding the exact equation of state as a power series in the density n, verify the result that you obtained in (b).

8.3   A Joule–Thompson process is defined and discussed in Problem 6.13. The necessary condition for Joule–Thompson cooling is that $\beta_p T > 1$. The Joule-

Thompson inversion curve is the curve, in the p-T plane, separating those states in which the substance is cooled by a Joule–Thompson process from those states in which it is heated. For a gas that satisfies the approximate equation of state, $\beta p = n + B_2 n^2$, show that the Joule–Thompson inversion curve is the straight line $T = T_i$, where T_i is the solution of the equation $B_2(T_i) = T_i B_2'(T_i)$. On which side of the curve does one obtain cooling?

8.4 (a) Determine the cluster integral, $C_2(T)$, for particles with a repulsive interaction potential, $v(r) = a/r^\lambda$, where $\lambda > 3$. [Hint: $\int_o^\infty \exp(-x) x^s \, dx = \Gamma(s - 1)$.] (b) Write expressions for βp and E/N as expansions in the particle density n up to second order.

8.5 Determine the virial coefficients, $B_2(T)$ and $B_3(T)$, for a gas that satisfies Dieterici's equation of state, $p(v-b) = RT \exp(-a/vT)$, where a and b are constants and v is the molar volume.

8.6 (a) Using a two-term virial equation of state, $\beta p = n + B_2(T) n^2$, and Eq. (8.23) for the average potential energy, calculate the constant volume molar specific heat for a gas whose virial coefficient is given by Eq. (8.29). (b) For the same system, calculate the constant pressure molar specific heat.

8.7 A one-dimensional gas of "soft-core" particles have interaction potentials of the form $v(x_{ij}) = V\theta(a - x_{ij})$, where $x_{ij} = |x_i - x_j|$ and θ is the unit step function. (a) Calculate the cluster integrals C_2 and C_3. (b) Write the virial equation of state up to order n^3.

8.8 In the model of a nematic liquid crystal, the parameter γ is related to the densities of x particles and y particles by $\gamma = (n_x - n_y)/n$. Equation (8.49) for γ always has the trivial solution $\gamma = 0$. Show that, when $\gamma = 0$, the equilibrium condition $\partial\phi/\partial\gamma = 0$ is satisfied but the extra condition $\partial^2\phi/\partial\gamma^2 < 0$, which guarantees that the stationary point is actually a maximum of ϕ, is only satisfied when $\ell^2 n < 2$.

8.9 Derive the equivalent of Eq. (8.56) in the case in which there is a nonzero interaction energy ε between a particle and any surface binding site. Does this modification cause a significant change in the physical characteristics of the system?

Fig. 8.26

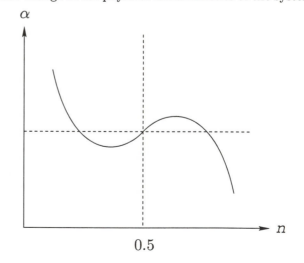

0.5

8.10   (a) Show that the function $\alpha(n)$, given in Eq. (8.56), is antisymmetrical with respect to the dashed axes shown in Fig. 8.26. (b) Equation (8.56) was derived using the mean-field approximation. Show that the antisymmetry property would also hold for the exact solution.

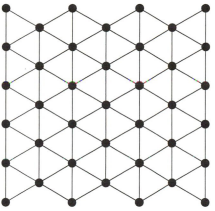

Fig. 8.27

8.11   A triangular Ising lattice is shown in Fig. 8.27. At each lattice point there is a spin that can take the two possible values ±1. With no external field, the energy of a configuration is

$$E = -V \sum_{\text{NN}} \sigma_i \sigma_j \tag{8.120}$$

where the sum is over all pairs of nearest neighbor spins (indicated by "bond lines" in the figure). Consider a very large lattice, so that edge effects can be neglected. For a given configuration, let $p(+,+)$ be the fraction of bonds with $+$ spins on both ends, $p(-,-)$ the fraction of bonds with $-$ spins on both ends, and $p(+,-) = 1 - p(+,+) - p(-,-)$ the fraction of bonds with different spins on the two ends. (a) Show that the energy per spin is given by $E = -3V[p(+,+)+p(-,-)-p(+,-)]$ (b) For $V > 0$ (ferromagnetic coupling), show that the ground-state energy is $-3NV$, where N is the number of spins. (c) For $V < 0$ (antiferromagnetic coupling), show that the ground-state energy is $E_g = NV$. (d) Let $S(N, E)$ be the entropy of the system. For $V < 0$, show that, as $N\rightarrow\infty$, $S(N, NV)/N$ does not appoach zero, that is, that this system does not satisfy Nernst's law.

8.12   Consider a gas of classical particles with the purely repulsive interaction potential,

$$v(r) = \frac{\varepsilon}{1 + (r/a)^4} \tag{8.121}$$

where r is the interparticle distance and ε and a are parameters with the dimensions of energy and length, respectively. (a) Select some particle and, by ignoring any statistical correlation between the selected particle and the other particles, calculate the average interaction energy involving that particle and all the other particles in the system. (Hint: A necessary integral can be found in any table of definite integrals.) Being careful to avoid double counting, use your result to obtain a "mean-field approximation" to the average potential energy of the system. (b) Using the mean-field approximation in place of the actual potential function, calculate the partition function $Z(N, \beta, V)$ and the canonical potential ϕ. (c) Write the energy and pressure equations of state for the gas in this mean field approximation.

8.13 (a) Write a computer program to carry out a Monte Carlo calculation of $\bar{E}(\tau)$ for a system with nondegenerate quantum states of energy $E_n = \varepsilon n + \gamma n^2$, where $n = 0, 1, \ldots$ and $\tau = kT$. (b) Use the program to draw rough graphs of $\bar{E}(\tau)$ for the case $\varepsilon = 1$ and $\gamma = 0.1$ in the range $0 \leq \tau \leq 2$. Hint: Simply modify the program HotBall.for.

8.14 The condition of detailed balance is that, for every pair of configurations C and C',

$$e^{-\beta E(C)} T(C \to C') = e^{-\beta E(C')} T(C' \to C) \tag{8.122}$$

This condition does not uniquely determine the Monte Carlo transition rates, $T(C \to C')$. (a) For the simple example defined in Section 8.12, show that the transition rates

$$T(k \to k+1) = \frac{e^{-\beta E(k+1)}}{2(e^{-\beta E(k)} + e^{-\beta E(k+1)})} \tag{8.123}$$

and

$$T(k \to k-1) = \frac{e^{-\beta E(k-1)}}{2(e^{-\beta E(k)} + e^{-\beta E(k-1)})} \tag{8.124}$$

satisfy detailed balance but differ from the values used in that section. (b) Calculate the transition rate, $T(k \to k)$, and show that it is not negative.

8.15 Using the canonical ensemble, derive Eq. (8.83) for the magnetic susceptibility of an Ising system. (Hints: (1) If $E_o(C)$ is the energy of configuration C without an external field, then $E(C) = E_o(C) - HM(C)$ is the energy with external field H. (2) $L^2 \chi \equiv \partial \langle M \rangle / \partial H$, evaluated at $H = 0$.)

8.16 In the detailed Monte Carlo algorithm described in Section 8.13, a spin is picked at random from an $L \times L$ lattice and flipped with probability P, where $P = 1$ if the flip would lower the energy and $P = \exp(-\beta \Delta E)$ if the flip would increase the energy. This can easily be seen to satisfy the principle of detailed balance. In Ising 2, however, each lattice point is considered in turn in a raster pattern, and the spin is flipped with the same probability as given above. Consider two configurations, C_1 and C_2, that differ only by the value of $\sigma(i, j)$. Suppose the system is in state C_1. In a particular move, what is the probability of a transition to state C_2? If the system *got into* C_1 on the last move by flipping spin $\sigma(i, j-1)$, then the probability of jumping to C_2 is fairly large. But if the system got into state C_1 by flipping spin $\sigma(i, j+1)$, then the probability of jumping into C_2 on the very next move is zero. Thus the algorithm does not seem to produce a Markov process (as defined in Section 8.11). But the Markovian nature of the process is essential for justifying the Monte Carlo method. By taking a complete sweep of the lattice as your definition of a "move," show that the process is a Markov process and that it satisfies detailed balance.

8.17 Run Hot Spring 2 (Display)* three times with the value of *Range* $= 15$ and the maximum jump set to 1, 15, and 150. Judging by eye, determine which case gives the best approximation to the exact Gaussian curve and try to explain your results. Hint: If there is a strong tendency to remain at the same point for a long time, then the effective number of sample points is reduced.

---

*Descriptions of the computer programs included in the Program Disk are given at the end of the Supplement to Chapter 8.

8.18   Run the program HotBall.for* three times, with the scale height set at 20, the number of moves set at 10,000, and the maximum jump size set at 1, 20, and 200. For each run, calculate the error

$$\text{error} = \sum_{\ell=0}^{\infty} |F(\ell) - P(\ell)| \tag{8.125}$$

where $F(\ell)$ is the fraction of time spent at point ℓ and $P(\ell) = s^{-1} \exp(-\ell/s)$ is the theoretical probability function.

8.19   Consider a 2D Ising system at inverse temperature β. According to Eq. (5.69), the probability distribution in its energy is given by $P(E) = C \exp \phi^*(\beta, E)$ where $\phi^*(\beta, E) = S(E) - \beta E$. (a) Choose some value of τ between 3 and 4. Using Ising_2.for, with $L = 10$, $V = 1$, $H = 0$, and $N_{\text{sweeps}} = 1000$, determine an approximate value of $\bar{E}$ and ΔE. (b) Break up the interval from $\langle E \rangle - 2\Delta E$ to $\langle E \rangle + 2\Delta E$ into 15 equal subintervals, which we will call *energy bins*. Run the program with the same parameters as above, but with $N_{\text{sweeps}} = 10000$. Let K_1, $K_2, \ldots, K_{15}$ be the number of times that the system is found in each energy bin. After each sweep of the lattice, augment the appropriate K_i by one (unless E falls outside of the complete interval). From the values of K_1, K_2, etc., plot $S(E) - S(\bar{E})$ over the interval $\langle E \rangle - 2\Delta E$ to $\langle E \rangle + 2\Delta E$. Is it a convex function? (c) Explain how you could determine the conditional entropy function $S^*(E, M)$, where M is the magnetization, in an analogous way.

8.20   (a) Using Ising_1.for, with $V = 1$, $H = 0$, $L = 20$, and $N_{\text{moves}} = 100,000$, draw rough plots of the magnetic susceptibility per spin and the specific heat per spin in the temperature range $1 < \tau < 10$. (b) Repeat (a) with $H = 2$. (c) Plot the predictions of the mean-field approximation on the same graphs as (a) and (b). When does mean-field theory work best?

8.21   Using the formula for a finite geometrical sum, evaluate the sum given in Eq. (8.86) and confirm Eq. (8.87).

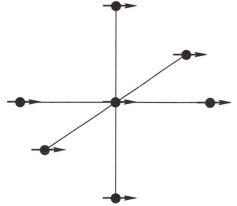

Fig. 8.28

8.22   (a) Using Eq. (8.88), obtain a formula for the specific heat per particle, in rational units, for a paramagnetic substance. (b) Let $x = \tau/\alpha B$ and plot the specific heat, as a function of x, from $x = 0$ to 10, for the three cases $J = \frac{1}{2}$, 1, and $\frac{3}{2}$. (c) What happens to the width and the location of the specific heat peak as J increases?

8.23 In Section 8.19 it was shown that two collinear magnetic dipoles, each of dipole moment $m = \mu_B$ (one Bohr magneton), have a lower energy if their dipole moments are parallel than if they are antiparallel, but that the difference in energy is only about 0.3 K, and therefore is much too small to explain the phenonemon of ferromagnetism, which occurs typically at hundreds or thousands of Kelvin. To show that taking more than two atoms into account will not help matters, calculate the energy difference for the central atom in Fig. 8.28 between parallel and antiparallel arrangements of its magnetic moment, assuming that all seven atoms have magnetic moments equal to μ_B and that the cubic lattice spacing is 2 Å.

8.24 (a) Using the mean-field approximation given in Eq. (8.101) for the energy of a configuration, calculate the canonical potential of a 3D Ising model with N spins for the case $H = 0$. Hint: See Eq. (8.53). (b) Show that the equation $\partial\phi/\partial\bar{\sigma} = 0$ always has the solution $\bar{\sigma} = 0$. (c) Show that ϕ has a maximum at $\bar{\sigma} = 0$ if $T \geq T_c = 6V/k$ and that ϕ has a minimum at $\bar{\sigma} = 0$ if $T < T_c$.

8.25 Using the mean-field approximation, derive an equation for the average polarization $\bar{\sigma}$ for the anisotropic Ising model, defined at the end of Section 8.21. Hint: One should obtain Eq. (8.103) with V replaced by $\bar{V} = (v_x + V_y + V_z)/3$. That the solution depends only on $\bar{V}$ is an artifact of the mean-field approximation. It would not be true if a more accurate method were used to solve the problem.

8.26 In the program Ising_3.for, set Ileft = +1, Iright = 0, $K = 10$, and $L = 20$. This will produce a 2D Ising model in the form of a tube of length 21 and circumference 10, with all the spins on the left edge fixed at +1 and a free boundary on the right end (see Fig. 8.29). Let $\beta V = 0.3$, which gives a temperature that is well above the transition temperature. [For a 2D square lattice, $(\beta V)_c = 0.4407$.] The output of Ising 3 is $P(i)$, the average polarization of the spins in the ith ring, where i is measured from the left end. By plotting $\log[P(i)]$, confirm that, near the left end, $P(i) \approx \exp(-i/\ell)$, and determine the correlation length ℓ.

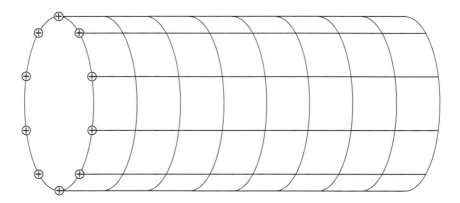

FIGURE 8.29

Chapter 9
Critical Phenomena and Renormalization Theory

9.1 WHAT IS A PHASE TRANSITION?

This chapter will be devoted to the study of second-order phase transitions, but, before considering the particular case of second-order phase transitions, it is best to look in more detail at the question of how one defines a phase transition in general. The most familiar phase transitions, such as those between water, ice, and steam, are almost misleadingly clear, in that the properties of the phases involved are so strikingly different. However, even in this seemingly obvious case, the meaning of the word *phase* begins to become more subtle when we realize that we can convert a sample of pure liquid to pure gas with no phase transition, simply by moving the system along a trajectory around the critical point in the p-T plane (see Fig. 9.1). The phase transition lines do not necessarily separate the thermodynamic space into disjoint regions of different phases. Although it is possible to define precisely a phase transition point in the p-T plane, and we will do so presently, it is not possible to separate clearly the liquid phase points from the gas phase points.

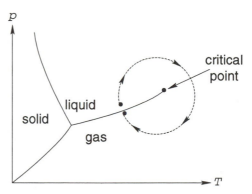

Fig. 9.1 By varying the values of p and T to move the system along the curve shown, it is possible to transform a liquid to a gas without encountering a phase transition line.

"The ordinary gaseous and liquid states are, in short, only widely separated forms of the same condition of matter and can be made to pass into one another by a series of gradations so gentle that the passage shall nowhere present any interuption or break of continuity."
——   Thomas   Andrews   *Bakerian   Lecture   to   the   Royal   Society.*(1869)
Andrews was the first person to demonstrate clearly the existence of a critical point in the liquid–gas transition.

The phase transition encountered in the ideal Bose gas is typical of many, more subtle, phase transitions. Across the phase transition line, the density is continuous, the energy density is continuous, and even the specific heat is continuous. One must go to the derivative of C_V with respect to T in order to find a property that changes discontinuously. If phase transitions can be so delicate, what is the defining characteristic of a phase transition point?

The unambiguous definition of a phase transition involves the mathematical concept of analyticity. A mathematical function of many variables (we will consider the case of two variables) is called *analytic* at a point (x, y) if its Taylor expansion converges to the value of the function at every point within some finite circle about (x, y). That is, if

$$f(x + \Delta x, y + \Delta y) = f(x, y) + \sum_{m,n=0}^{\infty} f_{mn} \frac{(\Delta x)^m (\Delta y)^n}{m! n!} \tag{9.1}$$

whenever $\Delta x^2 + \Delta y^2 < a^2$ for some number $a > 0$, where $f_{mn} = \partial^{m+n} f / \partial x^m \partial y^n$. Single-phase regions in the p-T plane (or, for magnetic systems, in the H-T plane) are regions in which the appropriate thermodynamic potential [$G(T, p)$ for the p-T plane and $\phi(H, T)$ for the H-T plane] is an analytic function. They are bounded by curves along which the function is not analytic.

Equations (7.72) and (7.73) for $\alpha(n, \tau)$ the affinity of an ideal Bose gas, illustrate the point. This function must be calculated separately for the two different phases. Its value in one phase cannot be determined by smoothly extrapolating its values across the phase transition line. (See Fig. 9.2.)

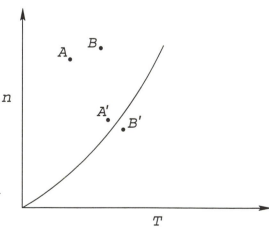

Fig. 9.2   One can calculate the value of $\alpha(n, T)$ at point B by expanding the function in a Taylor expansion centered at point A, but one cannot evaluate $\alpha(B')$ by using a Taylor expansion centered at A'.

This definition of a phase transition, in terms of the analyticity of the thermodynamic potential, presents a deep problem for the theory of statistical mechanics. To see what the problem is, let us consider a three-dimensional Ising model on a cubic lattice with $N = L^3$ points. The partition function is

$$Z(N, \beta, H) = \sum_C \exp\left(-\beta H \sum \sigma(\mathbf{r}) - \beta V \sum_{NN} \sigma(\mathbf{r})\sigma(\mathbf{r}') \right) \qquad (9.2)$$

Each term in the sum over configurations is an analytic function of β and H. There are 2^N configurations. The sum of a finite number of analytic functions is analytic. Where are the phase transition points? The function defined in Eq. (9.2) is analytic for all values of β and H. This difficulty for many years caused knowledgeable people to have some doubts as to whether the formalism of statistical mechanics could properly treat phase transitions without some modification. Of course, models such as the Ising model, when treated with mean-field theory, predict phase transitions. However, it was not clear whether or not those predictions were simply artifacts of the mean-field approximation that would vanish if an exact calculation could be made. The doubt was strengthened by the fact that the one-dimensional Ising model, which can be solved exactly, gives no phase transition in the exact solution but does give a phase transition in the mean-field approximation. It was only when Onsager was able to solve the two-dimensional Ising model exactly and show that it did predict a phase transition that these questions about the adequacy of statistical mechanics to treat phase transitions were finally put to rest. The resolution of the analyticity paradox can be given along the following lines. Recall that the laws of thermodynamics, and those of statistical mechanics, can only be used with complete confidence in the limit of large system size. Therefore, rather than looking at the total canonical potential for a finite system, we should be looking at something like the canonical potential per particle in the thermodynamic limit. That is,

$$\phi(\beta, H) = \lim_{N \to \infty} \left[\phi(N, \beta, H)/N \right] \qquad (9.3)$$

By some fairly complicated mathematical analysis, it can be proved that the limit defined here exists for all reasonable Hamiltonian functions. But it is a fact of mathematics that the statement that a sequence of functions $f_N(x)$ converges, as $N \to \infty$, to a limit function $f(x)$ does *not* imply that df_N/dx converges to df/dx. Thus, even though $\phi(N, \beta, H)/N$ is analytic for every finite N, it is still possible for $\phi(\beta, H)$ to be nonanalytic. This may seem like mathematical quibbling, but Fig. 9.3 shows that it is this very effect (the fact that taking a derivative does not commute with taking a limit) that produces the nonanalytic behavior in the thermodynamic functions. This is just another example of the general rule that the proper subjects of thermodynamics and statistical mechanics are those characteristics of matter that appear in the thermodynamic limit.

9.2   SECOND-ORDER PHASE TRANSITIONS

A first-order phase transition involves a restructuring of the substance on a microscopic level. Looking at a small bit of the substance through a powerful microscope, one could easily detect the change that takes place in the material when T goes from just above a first order transition to just below it. In a gas-liquid transition, the density undergoes a finite jump. In a liquid-solid transition, a crystal lattice forms when none existed above the transition temperature and there is also a finite change in the density.

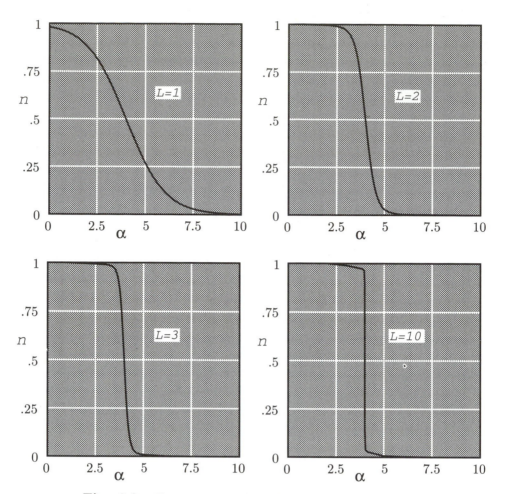

Fig. 9.3 The average number of particles per lattice site for
the two-dimensional lattice gas with attractive nearest-neighbor
interactions. This is the system that was analyzed in Section 8.9. All
the graphs are for a value of $\beta v = -2$. Calculations have been made
using the Monte Carlo method. Shown is the relationship between
the affinity and the average occupation number for $L \times L$ lattices of
various sizes, with periodic boundary conditions. It is clear that $n(\alpha)$
becomes a truly discontinuous function only in the limit of large lattice
size. The curve for $L = 10$, although it looks discontinuous, is actually
slightly rounded and not quite vertical.

In contrast, a second-order phase transition could not be detected by observing
a microscopic sample of the substance. Most second-order phase transitions separate
a more symmetrical phase from a less symmetrical one. For example, the equilibrium
state of the Ising model with no external field has up–down symmetry above T_c
but does not have that symmetry below T_c. Of course, many (but not all) first-
order transitions also separate more symmetrical from less symmetrical phases. The
distinction between the two cases is that in first-order transitions there is a finite
change in structure over an infinitesimal temperature interval—the substance is
a translationally invariant liquid at any temperature above the transition and a
noninvariant crystal at any temperature below it. In second-order transitions, the

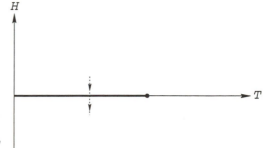

Fig. 9.4   If, at fixed T, the value of H is varied along the curve shown, there is a discontinuous finite change in the value of $\langle\sigma\rangle$ as the point $H = 0$ is passed. In a real system, this could be detected by observing a finite number of spins for a finite amount of time. Thus the transition is of first order, although it has no latent heat.

unsymmetrical state connects smoothly to the symmetrical state as T approaches T_c from below. (Usually, but not always, the unsymmetrical phase is the low-temperature phase.) Again, the Ising model is one of the simplest examples—the magnetization is nonzero below T_c and zero above T_c, but it is continuous, although not differentiable, through the transition. Because of this property, one could not detect a second-order transition by observing a microscopic sample. In a microscopic sample, the uncertainty in $\bar{\sigma}$ is always fairly large, because one is observing only a few fluctuating spins, and so one cannot distinguish between a zero value and a very small nonzero value.

On the other hand, if we move through the phase transition line on the trajectory shown in Fig. 9.4, by imposing a weak positive external field H and then slowly varying the value of H through zero to negative values, the average polarization of a single spin behaves in the following way. As $H \to 0$ from positive values, $\langle\sigma\rangle \to M(T)$, where $M(T)$ is the *finite* positive value of the spontaneous magnetization. A finite value of $\langle\sigma\rangle$ is not completely obscured by the random fluctuations in a finite microscopic sample, and can thus be observed. As we move from tiny positive values of H to tiny negative values, the average polarization $\langle\sigma\rangle$ changes *discontinuously* to $-M(T)$. Since this finite jump can be detected on a microscopic level, we say that all the points on the phase transition line, *except the endpoint of the line*, are points of first-order phase transitions. The endpoint of the line (the critical point in the Ising model) is a point of second-order phase transition.

Most first-order phase transitions have a finite latent heat associated with them. However, it is easy to see that the first-order transition that has just been described has no latent heat. Therefore, the existence or nonexistence of a latent heat is not a reliable criterion for distinguising first-order from second-order transitions. The general criterion is that any phase transition that could be detected by observing a microscopically finite sample for a finite time is first order. All others are called second-order.

In a second-order transition from a more symmetrical to a less symmetrical phase, the quantity whose nonzero value distinguishes the unsymmetrical from the symmetrical state is called the *order parameter*. The order parameter for a uniaxial ferromagnet is the magnetization M, which is a real scalar, because the magnetization axis is fixed. For a ferromagnetic material in which the magnetization can point in any direction (called an isotropic ferromagnet), the order parameter is a vector magnetization $\mathbf{M}$. For the λ transition in liquid helium, the order parameter is usually taken to be the superfluid wave function $\psi(\mathbf{r})$ normalized so that $\psi^*\psi$ is equal to the superfluid density. The order parameter in that case is a complex scalar. In our analysis of second-order phase transitions, we will, for the most part, restrict ourselves to the three-dimensional Ising model, which has a real scalar order parameter, and only mention equivalent results for the the more complicated cases.

9.3   UNIVERSALITY AND CRITICAL EXPONENTS

The points in thermodynamic space at which second-order phase transitions occur are called *critical points*, and the behavior of systems at second-order phase transitions is generally referred to as *critical phenomena*. One of the striking characteristics of critical phenomena is the fact that certain detailed quantitative measures of a system's behavior near a critical point are quite independent of the details of the interactions between the particles making up the system. This characteristic of critical phenomena is called *universality*. The universal features are not only independent of the numerical details of the interparticle interactions, but are also independent of the most fundamental aspects of the structure of the system. For example, we mentioned that the order parameter $\bar{\sigma}$ for the 3D Ising model on a cubic lattice approached zero as the temperature approached the critical temperature from below as $(-t)^{\beta}$, where $\beta \approx 0.324$. What we mean by universality is illustrated by the fact that the value of β is independent of:

1. The numerical values of V_x, V_y, and V_z, in the case of an anisotropic Ising model (see the end of Section 8.21), so long as they are greater than zero. (If one of them is zero, then the system is not a three-dimensional system, but is a collection of completely noninteracting two-dimensional systems, and one thing that β *does* depend upon is the dimensionality of the system.)

2. The structure of the lattice. If a ferromagnetic Ising model is defined on a face-centered cubic lattice, a hexagonal lattice, or, in fact, on any of the fourteen possible Bravais lattices in three dimensions, and the value of $\bar{\sigma}$ is determined very close to the ferromagnetic transition temperature for that lattice, then it is found that $\bar{\sigma}$ is proportional to $(-t)^{\beta}$ as $t \to 0$ with the same value of β as is shown by a cubic lattice.

3. The details of the interactions. The Ising model is defined as having only nearest-neighbor interactions. If longer-range interactions (next-nearest-neighbor, third-nearest-neighbor, etc.) are added, then, as long as they are chosen so that a ferromagnetic transition still occurs, $\bar{\sigma}$ will go to zero as $(-t)^{\beta}$ at the Curie point for the system, with the same value of β.

4. Whether the system is a lattice system or a three-dimensional gas with continuous coordinates and momenta. Consider any ordinary gas, even one with complex molecules, which have rotational and vibrational degrees of freedom, such as nitrogen, hexane, octane, etc. Measure the parameter, $\Delta n = n(\text{liquid}) - n(\text{gas})$, where n is the particle density, as a function of $t = (T - T_c)/T_c$ along the coexistence curve in the p-T plane. [The coexistence curve ends at the critical point. At that point $n(\text{liquid})$ becomes equal to $n(\text{gas})$.] As $T \to T_c$ from below, Δn approaches zero as $(-t)^{\beta}$ with the same value of β as the 3D Ising model. For the liquid–gas phase transition, the value of Δn is the order parameter. It is zero above the critical temperature and nonzero below it. That this system exhibits the same exponent as the 3D Ising model is really an astonishing fact since there seems to be nothing in common between a magnetic transition in a three-dimensional lattice model and the liquid–gas phase transition in a continuous system. (Also see Fig. 9.5.)

The order parameter exponent β is only one of a number of different exponents that can be defined at a critical phase transition. We will define some others shortly. They all share the characteristic of universality. Their values are independent of

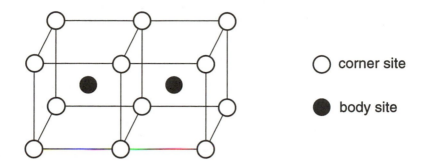

Fig. 9.5   Beta-brass is a binary alloy made up of equal numbers of zinc and copper atoms. It has a body-centered cubic (bcc) crystal structure. A bcc lattice can be defined by beginning with a cubic lattice, all of whose sites we will call *corner sites*, and adding an additional lattice site, called a *body site*, at the center of each cube. There are equal numbers of corner sites and body sites. In fact, there is no fundamental distinction between the set of corner sites and the set of body sites. The set of body sites forms a cubic lattice that could just as well have been called the corner sites. We define the order parameter $\Delta n = |Z_B - Z_C|/(Z_B + Z_C)$, where Z_B is the number of zinc atoms at the body sites and Z_C is the number of zinc atoms at the corner sites. In beta-brass, there is a phase transition at a temperature $T_c \approx 741\,\text{K}$. Above T_c, $\Delta n = 0$, indicating that a zinc atom has an equal probability of being at a body or a corner site. Below T_c, $\Delta n > 0$ and, as $T \to T_c$, $\Delta n \sim (-t)^\beta$, with $\beta \approx 0.324$. The quantity Δn can be measured by X-ray diffraction studies.

almost all detailed properties of the system. In fact, they generally depend on only two characteristics of the phase transition, namely, the number of components in the order parameter (one for a real number, two for a complex number or a two-dimensional vector, three for a vector, etc.) and the dimensionality of the system.

At zero external field, the specific heat of the three-dimensional Ising model, as a function of temperature, has an integrable singularity* of the form

$$C \sim \begin{cases} A_-(-t)^{-\alpha}, & \text{as } T \to T_c \text{ from below} \\ A_+ t^{-\alpha}, & \text{as } T \to T_c \text{ from above} \end{cases} \tag{9.4}$$

where, as usual, $t = (T - T_c)/T_c$ and the *specific heat exponent* α has the value

$$\alpha \approx 0.11 \pm 0.008 \tag{9.5}$$

A_- and A_+ are constants with the appropriate units. Since the right-hand side of Eq. (9.4) goes to infinity at $t = 0$, there is some question about what is actually meant, in a mathematical sense, by the equation. What the equation signifies is that, as $T \to T_c$ from below,

$$\lim_{T \to T_c} \left[C(T)(-t)^\alpha \right] = A_- \tag{9.6}$$

with a corresponding statement for the limit as $T \to T_c$ from above.

---

* A function $f(x)$ has an integrable singularity at x_o if $f(x) \to \infty$ as $x \to x_o$ but the integral $\int_a^b f(x)\, dx$ for $a < x_o < b$ is finite. A good example is the function $\log(|x|)$, which has an integrable singularity at $x = 0$.

The magnetic susceptibility χ_m of the 3D Ising model at zero external field also shows a singularity at T_c. The form of the singularity defines a *susceptibility exponent* γ.

$$\chi_m(T, H = 0) \sim \begin{cases} B_-(-t)^{-\gamma}, & \text{as } T \to T_c \text{ from below} \\ B_+ t^{-\gamma}, & \text{as } T \to T_c \text{ from above} \end{cases} \tag{9.7}$$

where

$$\gamma = 1.24 \pm 0.004 \tag{9.8}$$

The critical exponent β describes how $\bar{\sigma}$ goes to zero as one approaches the critical point along the direction 1, shown in Fig. 9.6. It is defined by saying that, as $T \to T_c$ from below, $\bar{\sigma} \sim (-t)^\beta$

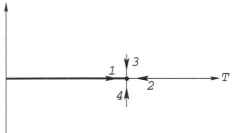

Fig. 9.6   The behavior of $\bar{\sigma}$ as one approaches the critical point in direction 1 defines the critical exponent β. The same thing along path 3 or 4 defines the exponent δ.

There is another critical exponent δ that indicates how $\bar{\sigma} \to 0$ along the directions 3 and 4. That is, at $T = T_c$ exactly, and for $H \to 0$,

$$|\bar{\sigma}| \sim C|H|^{1/\delta} \tag{9.9}$$

(That the exponent is written as $1/\delta$, rather than δ, is just an accident of history.) The same exponent is exhibited by a fluid system as one approaches the critical point by varying the pressure along the critical isotherm.

$$|n - n_c| \sim C|p - p_c|^{1/\delta} \tag{9.10}$$

where n_c and p_c are the particle density and pressure at the critical point. For that reason, δ is called the *critical isotherm exponent*. Its value is

$$\delta = 4.82 \pm 0.006 \tag{9.11}$$

When the same numbers appear in a variety of very different physical situations there must be some simple mechanism as work. But what can the magnetism of a uniaxial ferromagnet have in common with the condensation of carbon dioxide or the ordering transition in beta-brass? The Hamiltonian functions for the three systems are utterly different in structure. But ultimately it is the Hamiltonian function of each system that must determine the characteristics of all the phase transitions involving the system, including any critical exponents.

9.4   LANDAU THEORY

The physics of critical phenomena is subtle. Therefore, we will approach it in stages. We will begin with a simple thermodynamic analysis, due to L. D. Landau, that gives the same predictions for all the critical exponents as is given by the mean-field approximation. It has the advantage over mean-field theory of being based on a few general thermodynamic assumptions that do not involve the detailed microscopic model of the system. It can therefore be used, without significant modifications, to analyze the critical behavior of microscopically different systems. Since it predicts exactly the same behavior for each system, it explains the puzzling property of universality. Unfortunately, the universal behavior that the Landau theory predicts is wrong. For example, we have seen that mean-field theory predicts that, near T_c, $\bar{\sigma} \sim \sqrt{3}(-t)^{1/2}$, and therefore that the exponent $\beta = \frac{1}{2}$. But this disagrees with experiments and with more adequate theoretical analyses that do not use the mean-field approximation. They all show that $\beta \approx 0.324$. However, since a modification of the Landau theory plays a central role in the modern theory of critical phenomena, the time spent in presenting the theory will not be wasted. We will use the Landau theory to analyze a uniaxial ferromagnetic substance and then indicate how it can be applied to other systems.

In discussing the thermodynamics of a magnetic substance, one is free to use, as the "magnetic" variable, the magnetic field B, the magnetic displacement H, or the magnetic polarization of the substance M. The use of M, rather than B or H, has two advantages. First, it is completely determined by the microscopic state of the substance, without reference to external agents such as coils or pole pieces. The more important advantage is that the thermodynamic states of the substance are more adequately represented by the points in the T–M plane, as opposed to the points in the T–H or T–B planes. The shaded region of the T–M plane, shown in Fig. 9.7(A), describes two-phase states, in which a portion of the

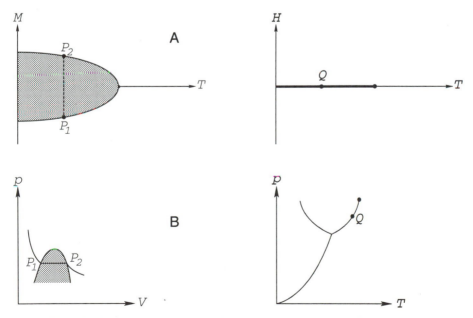

Fig. 9.7   (A) For a 3D Ising lattice, the set of thermodynamic states lying on the straight line between points P_1 and P_2 in the M–T plane are all mapped into the single point Q in the H–T plane. (B) A similar thing happens in liquid–gas transitions—the set of points between P_1 and P_2 in the p-V plane are mapped into the point Q in the p-T plane.

substance is polarized in the up direction while the remainder is polarized in the down direction. This whole region is mapped into the single line, $H = 0, 0 < T < T_c$, in the T–H plane. Thus single points in the T–H plane represent a multitude of possible macroscopic equilibrium states of the substance. A similar thing happens for fluids, where the region of two-phase (liquid and gas) states shown in the p-V plane collapses to the single-phase transition line in the p-T plane.

The Landau theory begins with some simple assumptions about the form of the canonical potential $\phi(T, M)$. The potential $\phi(T, M)$ is meant to be interpreted as the conditional canonical potential for a system of fixed magnetization *in the absence of an external magnetic field*. Recall that the conditional canonical potential is defined so that the probability of finding the system with magnetization M is proportional to $\exp \phi(M)$. From a microscopic point of view,

$$\phi(T, M) = \log\Big(\sum_C \delta\Big(\sum \mu_i - MV\Big)e^{-\beta E(C)}\Big) \tag{9.12}$$

where the sum on C is over all configurations of the system and μ_i is the magnetic moment of the ith particle in configuration C. The delta function limits the sum over configurations to those configurations that have a total magnetic moment equal to MV—that is, a magnetization equal to M. With no external field, this function is clearly symmetric in M. The Landau theory assumes that, for small values of M, it can be expanded as a power series in M.

$$\phi(T, M) = \phi_o(T) - VA(T)M^2 - VB(T)M^4 \tag{9.13}$$

The explicit factors of V in the second and third terms are included in order to make A and B independent of the system size. The minus signs, which suggest that the corresponding terms are negative, are included so that the stationary point at $M = 0$ is a maximum when A and B are positive.

Since we are interested in analyzing the thermodynamic properties of the system at the phase transition, where the thermodynamic functions are known to be nonanalytic, it may seem to be a fundamental error to assume that $\phi(T, M)$ is analytic in M. However, it is the equilibrium potential $\phi(T)$ that is nonanalytic. It will be shown that an analytic conditional potential $\phi(T, M)$ can still produce a nonanalytic equilibrium potential $\phi(T)$.

In the Landau theory, the critical temperature T_c is interpreted as the point at which $A(T)$ changes sign. That is, $A(T) > 0$ for $T > T_c$ and $A(T) < 0$ for $T < T_c$. $B(T)$ must remain positive or else the maximum of $\phi(T, M)$ would occur at $M = \pm\infty$. Since we are interested in the properties of the system in the vicinity of T_c, it is adequate to treat $A(T)$ as a linear function with a zero at T_c and to treat $B(T)$ as a constant. Thus we assume that

$$A(T) = at \quad \text{and} \quad B(T) = b \tag{9.14}$$

where a and b are constants and $t = (T - T_c)/T_c$, as usual.

The quantity $\phi(T, M)$ is plotted as a function of M for temperatures less than, equal to, and greater than T_c in Fig. 9.8. For $T < T_c$, the maximum of $\phi(T < M)$, which should describe the equilibrium state, occurs at either of the two nonzero values shown in the figure. At those temperatures the system will exhibit a finite magnetization, even in the absence of any external magnetic field. Using Eqs. 9.13

Fig. 9.8 If the function $\phi(T, M)$ is assumed to have the form given in Eq. (9.13), then, for T above T_c, $\phi(T, M)$ has single quadratic maximum at $M = 0$. One would therefore expect normal (that is, Gaussian) fluctuations about that most probable value of M. For $T = T_c$, the most probable value of M is still zero, but the probability distribution is much flatter, indicating a much larger probability of observing nonzero values. For $T < T_c$, we have spontaneous symmetry breakdown— the most probable values are the two nonzero values indicated. Notice that the fluctuations about those nonzero values would again be Gaussian, because the function is locally quadratic.

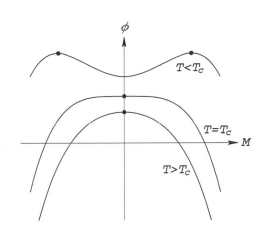

and 9.14, the equilibrium value of M as a function of t is easy to calculate. Setting $\partial\phi(T, M)/\partial M = 0$ in the expression

$$\phi = \phi_o - VatM^2 - VbM^4 \tag{9.15}$$

gives the cubic equation

$$atM + 2bM^3 = 0 \tag{9.16}$$

This has the three solutions

$$M = 0 \quad \text{and} \quad \pm\sqrt{\frac{a}{2b}}(-t)^{1/2} \tag{9.17}$$

When t is positive, the $M = 0$ solution describes a maximum of $\phi(T, M)$ and the nonzero solutions are imaginary and therefore physically irrelevant. But for negative t the zero solution describes a local minimum of $\phi(T, M)$ and it is the now real nonzero solutions that describe the possible equilibrium values of the magnetization. It is obvious from Eq. (9.17) that the Landau prediction for the magnetization exponent is that $\beta = \frac{1}{2}$.

In order to calculate the specific heat below T_c, we must transform to the regular canonical potential $\phi(T)$. This is done by replacing the independent variable M in $\phi(T, M)$ by its equilibrium value as a function of T. For T less than but close to T_c, this gives

$$\phi(T) = \phi_o(T) - VatM^2(T) - VbM^4(T)$$
$$= \phi_o(T) + V\frac{a^2}{4b}\frac{(T - T_c)^2}{T_c^2} \tag{9.18}$$

The energy as a function of T is given by

$$E = -\frac{\partial\phi}{\partial\beta}$$
$$= kT^2\frac{\partial\phi}{\partial T} \tag{9.19}$$
$$= kT^2\left(\phi_o'(T) + V\frac{a^2(T - T_c)}{2bT_c^2}\right)$$

The specific heat for $T < T_c$ is

$$C = \frac{\partial E}{\partial T} = k\frac{\partial(T^2\phi_o')}{\partial T} + V\left(\frac{ka^2T^2}{2bT_c^2} + \frac{ka^2Tt}{bT_c}\right) \tag{9.20}$$

The limit of C as $T \to T_c$ from below is

$$C = k\frac{\partial(T^2\phi_o')}{\partial T}\bigg|_{T_c} + V\frac{ka^2}{2b} \tag{9.21}$$

Above T_c, M is zero, and therefore the specific heat is given by the first term of Eq. (9.21). Thus in the Landau theory there is no singularity, but only a finite discontinuity in $C(T)$ at T_c of magnitude

$$\Delta C = V\frac{ka^2}{2b} \tag{9.22}$$

In order to make this agree with Eq. (9.4), which defines the specific heat exponent α, we must choose $\alpha = 0$. This prediction disagrees with the experimental result given in Eq. (9.11).

In order to calculate the magnetic susceptibility, it is necessary to have $\phi(T, M)$ in the presence of an external magnetic field. If a magnetic field B is present, then the energy of a configuration C is changed from $E(C)$ to $E(C) - B\sum \mu_i$. In that case

$$\begin{aligned}
\phi(T, M) &= \log\left[\sum_C \delta\left(\sum\mu_i - MV\right)e^{-\beta E(C)+\beta B\sum\mu_i}\right] \\
&= \log\left[e^{\beta BMV}\sum_C \delta\left(\sum\mu_i - MV\right)e^{-\beta E(C)}\right] \\
&= \phi_{B=0}(T, M) + \beta BMV \\
&= \phi_o(T) + \beta BMV - VatM^2 - VbM^4
\end{aligned} \tag{9.23}$$

Assuming that B is small and positive and that $T < T_c$, the plot of $\phi(T, M)$ as a function of M now has an asymmetrical double maximum. The unique equilibrium state is given by the higher maximum at positive M. The maximum occurs at the solution of $\partial\phi(T, M)/\partial M = 0$. This gives the equation

$$\beta B - 2atM - 4bM^3 = 0 \tag{9.24}$$

The point $B = 0$ and $t = 0$ is a singular point for the thermodynamic functions and therefore one must be very careful about the order in which limits and derivatives are taken. To calculate the susceptibility exponent γ, one must separately look at the limit of $(\partial M/\partial B)|_{B=0}$ as $t \to 0$ from below and above. We will first do the calculation for negative t. Equation (9.24) is an equation for $M(t, B)$. Taking the derivative of the equation with respect to B, setting B equal to zero, and defining χ_m as $\partial M/\partial B$ gives

$$\beta - 2at\chi_m - 12bM^2(t, 0)\chi_m = 0 \tag{9.25}$$

Replacing $M^2(t, 0)$ by its value $-at/2b$ obtained from Eq. (9.24) gives

$$\chi_m = \frac{\beta_c}{4a}(-t)^{-1} \tag{9.26}$$

where $\beta_c = 1/kT_c$.

For positive values of t, M is proportional to B as $B \to 0$, and therefore the cubic term in Eq. (9.24) can be dropped. The magnetic susceptibility is then given by

$$\chi_m = \frac{M}{B} = \frac{\beta}{2a} t^{-1} \tag{9.27}$$

Comparing these results with Eq. (9.7) shows that the Landau theory prediction is that

$$\gamma = 1 \quad \text{and} \quad \frac{B_+}{B_-} = 2 \tag{9.28}$$

Notice that γ and the ratio B_+/B_- are both independent of the numerical values of the parameters a and b that appear in the thermodynamic function. (This is not true for the separate amplitudes, B_+ and B_-.) Thus, according to the Landau theory, these quantities should have the same value at the Curie point of all uniaxial ferromagnets. This prediction of universality for γ and the amplitude ratio B_+/B_- is supported by experimental observation, but the specific numbers predicted in Eq. (9.28) are not.

The isotherm exponent δ is calculated by setting $t = 0$ in Eq. (9.24). This gives

$$M = \left(\frac{\beta}{4b}\right)^{1/3} B^{1/3} \tag{9.29}$$

A comparison of this with the definition of δ given in Eq. (9.9) yields the estimate $\delta = 3$ for the isothermal exponent.

In Section 8.24 the mean-field approximation was used to calculate the order-parameter exponent β. The mean-field result was that $\beta = \frac{1}{2}$, in agreement with the Landau theory. Similar mean-field calculations of all the other exponents would show that all of the critical-point predictions of the mean field approximation are in agreement with those of Landau theory. Since mean field theory, which neglects correlation effects, is obviously a relatively crude approximation to reality, it is not surprising that it predicts critical exponents that do not agree with experiment or with more careful calculations. However, the Landau theory is a thermodynamic theory, and its basic assumption [that is, Eq. (9.13)] seems so reasonable that it is much more difficult to see why it fails so badly. In the next section we will see that the essential weakness of the Landau theory is that it neglects thermodynamic fluctuations.

9.5   CRITICAL-POINT FLUCTUATIONS

There are two ways of seeing that the careful treatment of large-scale fluctuations is important in any analysis of critical phenomena. The first is to consider an experimental effect, called *critical opalescence*. When a transparent fluid, such as water, is brought to its critical point, it becomes milky white, although it is quite clear at all neighboring pressures and temperatures. It is known that the effect is caused by the light being scattered, as it passes through the fluid, by small, disordered, spatially varying fluctuations in the fluid density. A spatially varying fluctuation in the density will create a similar fluctuation in the index of refraction, which will distort the wave fronts of a plane wave and cause a scattering of its energy flux.

The second way of seeing the importance of fluctuations at the critical point is to recall the relationship between energy fluctuations and specific heat that is expressed by the equation

$$\beta \, \Delta E = C^{1/2} \tag{9.30}$$

where $(\Delta E)^2 = \langle E^2 \rangle - \langle E \rangle^2$ and C is the specific heat of the system. But, as $T \to T_c$, the specific heat of any system with a positive exponent, α, approaches infinity as $1/|t|^\alpha$. Equation (9.30), which is based on a Gaussian approximation to the function $\exp(\phi(T, M))$, is not really applicable right at the critical point, but it clearly indicates that fluctuations will be very large there. The fact that the magnetic susceptibility χ_m diverges at the critical point, combined with the relationship between χ_m and ΔM, shows that the magnetization will also exhibit large fluctuations at the critical point.

9.6   THE FLUCTUATION PROBABILITY

The equilibrium states of a macroscopic 3D Ising model show uniform magnetization. The magnetization M is zero above the transition temperature and finite below that temperature, but it is always uniform throughout the sample. However, the fluctuations about the uniform average magnetization are, by their nature, nonuniform. In order to incorporate fluctuations into the thermodynamic analysis, it is necessary to construct some expression that describes the probability that the sample will be found with a particular spatially varying magnetization $M(\mathbf{r})$. It is easiest to begin at a temperature above T_c, so that the equilibrium value of M is zero. Above T_c the parameter at is positive, and the probable values of M are very small (of order $N^{-1/2}$). Therefore, the term proportional to M^4 can be neglected. That term becomes important only when t is very small or negative.

In the Landau theory, the magnetization is always assumed to be uniform and, with no external field and neglecting the M^4 term, the probability of obtaining a value M for the uniform magnetization is taken as

$$P(M) = C \exp(-atVM^2) \tag{9.31}$$

A natural way of extending the Landau theory to treat nonuniform patterns of magnetization would be to replace the term VM^2 by a term of the form $\int M^2(\mathbf{r}) \, d^3\mathbf{r}$. That is, the probability of finding the sample with a magnetization pattern $M(\mathbf{r})$ would be taken as

$$P[M(\mathbf{r})] = C \exp\left(-at \int M^2(\mathbf{r}) \, d^3\mathbf{r}\right) \tag{9.32}$$

This reduces to Eq. (9.31) when $M(\mathbf{r})$ is constant. Equation (9.32) would imply that the fluctuations in different macroscopic regions were statistically uncorrelated. That is, if we decompose the total volume V into two disjoint regions, V_1, and V_2, then, because

$$\int_V M^2 \, d^3\mathbf{r} = \int_{V_1} M^2 \, d^3\mathbf{r} + \int_{V_2} M^2 \, d^3\mathbf{r} \tag{9.33}$$

the probability of simultaneously getting a magnetization $M_1(\mathbf{r})$ within V_1 and a magnetization $M_2(\mathbf{r})$ within V_2 would be equal to the exponential of the sum of the integrals and therefore equal to the product of the separate probabilities.

9.7   THE GRADIENT EFFECT

The modification of Landau theory proposed neglects an important effect, namely the *gradient effect*. It would state that the probability of the two magnetization patterns shown in Fig. 9.9 were equal, which seems clearly unreasonable. The rapid back and forth changes in $M(\mathbf{r})$ shown in the pattern of Fig. 9.9(B) would require high-energy domain walls that would very much suppress the probability of occurrence of such a magnetization pattern. In the expression for the probability of

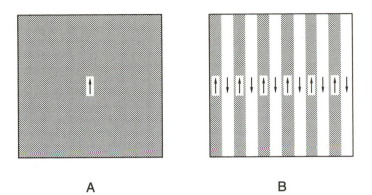

Fig. 9.9 (A) A magnetization pattern in which $M(\mathbf{r})$ is constant throughout the sample. In (B) the magnitude of $M(\mathbf{r})$ is constant, but its sign varies discontinuously from region to region. If the probability of any given pattern of magnetization is assumed to have the form given in Eq. (9.32), then the magnetization patterns shown in (A) and (B) would be equally probable, which seems obviously unreasonable.

a particular macrostate, some account must be taken of the rate of spatial variation (the gradient) of the magnetization. To determine the specific term that is necessary, one can refer to the results of the computer experiment carried out in Problem 8.26. It is done on a two-dimensional Ising lattice, but the same result would be obtained in three dimensions. An Ising lattice in the form of a long tube, with a circumference of ten lattice units, is considered. The spins at one end of the tube are maintained at a fixed magnetization, $M = \langle\sigma\rangle = +1$. The average of the magnetization of a circular column of spins is then measured as a function of distance from the fixed end. $M(x) = \langle\sigma\rangle$, where the average is taken over all ten spins in the column at x and over a long Monte Carlo run at fixed temperature. It is found that the equilibrium value of $M(x)$ has an exponential form. That is, near the fixed column at $x = 0$,

$$M(x) = me^{-x/\ell} \tag{9.34}$$

where the *correlation length* ℓ increases as T approaches T_c. The exponential function is a solution of the equation

$$-\ell^2 \nabla^2 M + M = 0 \tag{9.35}$$

This is the equation that would be obtained by maximizing the following expression for the conditional canonical potential associated with a magnetization distribution $M(\mathbf{r})$. (See Problem 9.2.)

$$\phi(M) = \phi_o - \int_V \left[atM^2(\mathbf{r}) + c|\nabla M(\mathbf{r})|^2 \right] d^3\mathbf{r} \tag{9.36}$$

where $\ell^2 = c/at$. This expression for the value of the canonical potential to be assigned to any given pattern of magnetization seems to be adequate for $T > T_c$, where t is positive. In order to treat the case of zero or negative t, it will be necessary to again add a term of the form bM^4. The expression one then obtains is known as the *Landau–Ginsburg potential*.

$$\phi(T, M) = \phi_o - \int_V \left(atM^2 + bM^4 + c|\nabla M|^2 \right) d^3\mathbf{r} \tag{9.37}$$

The term proportional to $|\nabla M|^2$ will make the value of $\phi(M)$ large and negative for any function $M(\mathbf{r})$ that has very rapid variations in space. The probability of such a magnetization distribution, which is equal to $\exp\phi(M)$, will therefore be very small.

9.8 FUNCTIONAL INTEGRALS

The interpretation of the expression

$$P = C \exp\left[-\int (atM^2 + bM^4 + c|\nabla M|^2) d^3\mathbf{r}\right] \qquad (9.38)$$

as the probability of finding the system with the magnetization pattern $M(\mathbf{r})$ requires some careful mathematical analysis. In order for the expression to have any meaning as a probability, it must be normalized so that the sum of P over all possible distributions of magnetization $M(\mathbf{r})$ is one. But $M(\mathbf{r})$ is a function. How can one possibly take a sum over the vast set of conceivable functions, $M(\mathbf{r})$? Actually, there is a mathematical structure, called a *functional integral*, that was introduced by the mathematician Norbert Wiener to handle just such problems in probability theory. It was later rediscovered by Richard Feynman and used in a reformulation of fundamental quantum mechanics. Functional integration is a rather complex and difficult subject. Fortunately, the very fact that the Landau–Ginzburg theory must be interpreted as a macroscopic, thermodynamic theory of critical phenomena will allow us to avoid entirely the complexities of functional integration.

In attempting to give precise meaning to Eq. (9.38), we will consider a system that is in the form of a cube of side L, write the Cartesian components of $\mathbf{r}$ as x_1, x_2, and x_3, and assume that $M(x_1, x_2, x_3)$ satisfies periodic boundary conditions. It will simplify many of the later formulas if, for every integer K, we define the function

$$u_K(x) = \begin{cases} \sqrt{2}\cos kx, & \text{if } K > 0 \\ 1, & \text{if } K = 0 \\ \sqrt{2}\sin kx, & \text{if } K < 0 \end{cases} \qquad (9.39)$$

where $k = 2\pi|K|/L$. Then the functions u_K satisfy the simple orthonormality relations

$$\int_o^L u_M(x)u_N(x)\,dx = L\delta_{MN} \qquad (9.40)$$

Any real periodic function $f(x)$ with a period L can be expanded as the Fourier series

$$f(x) = \sum_K f_K u_K(x) \qquad (9.41)$$

where

$$f_K = L^{-1}\int_o^L f(x)u_K(x)\,dx \qquad (9.42)$$

Without this artifice, the special case $K = 0$ would require separate treatment in most of the following analysis. Introducing a single symbol $\mathbf{K}$ for the triplet of integers (K_1, K_2, K_3), we will expand M as a triple Fourier series, involving the *fluctuation amplitudes* $M_{\mathbf{K}}$.

$$M(x_1, x_2, x_3) = \sum_{\mathbf{K}} M_{\mathbf{K}}\, u_{K_1}(x_1)\, u_{K_2}(x_2)\, u_{K_3}(x_3) \qquad (9.43)$$

The integers (K_1, K_2, K_3) are called the *mode numbers* corresponding to the wave vector $\mathbf{k} = 2\pi\mathbf{K}/L$. Clearly, the term with a wave vector $\mathbf{k}$ describes a fluctuation in the magnetization with a wavelength of size $2\pi/k$. But $M(\mathbf{r})$ is defined as the average magnetization within a volume, centered at $\mathbf{r}$, that contains very many individual spins. Variations in this function on any scale comparable to the lattice spacing have no meaning. They are microscopic details that have already been incorporated into the term $\phi_o(T)$. We must, in a somewhat arbitrary way, choose a minimum length scale Λ (and therefore a maximum wave vector $k_m = 2\pi/\Lambda$) for macroscopic fluctuations and declare that all details on a scale smaller than Λ are included in the ϕ_o term. That our arbitrary choice of Λ will not affect the outcome of any calculations very close to the critical point is shown by the results of Problem 9.7, where the reader is asked to plot the magnetic susceptibility per spin for systems of increasing size as a function of temperature. The sharp, high peak in χ_m, that is symptomatic of the critical phase transition, only becomes evident as L, the size of the system, becomes large. The fluctuations in M that are contributing to the divergence of χ_m are large-scale fluctuations. Such large-scale fluctuations are possible only in a large system. In a very large system, the small-scale fluctuations, whose calculation might be affected by our choice of Λ, make a contribution to χ_m that is negligible in comparison to the divergent contribution made by the large wavelength fluctuations, once we are sufficiently close to the critical point. Therefore, for the purpose of calculating the critical-point properties of the system, we can restrict the sum on $\mathbf{K}$ in Eq. (9.43) to the finite number of terms with $|\mathbf{K}| < L/\Lambda$.

The partition function Z is the sum of $\exp(-\beta E)$ over all possible configurations of the system. By assuming that the conditional canonical potential is given by Eq. (9.37), we have assumed that the sum of $\exp(-\beta E)$ over all those microscopic configurations that have a given set of fluctuation amplitudes $\{M_\mathbf{K}\}$ is equal to

$$Z(\{M_\mathbf{K}\}) = \exp\left(\phi_o - \int (atM^2 + bM^4 + c|\nabla M|^2) d^3\mathbf{r}\right) \qquad (9.44)$$

where the function $M(\mathbf{r})$ is given in terms of the $M_\mathbf{K}$ by Eq. (9.43). The full partition function is the integral of $Z(\{M_\mathbf{K}\})$ over all values of the fluctuation amplitudes.

$$Z = e^{\phi_o} \int \exp\left[-\int (atM^2 + bM^4 + c|\nabla M|^2) d^3\mathbf{r}\right] \prod_\mathbf{K} dM_\mathbf{K} \qquad (9.45)$$

Let us first evaluate this integral for $T > T_c$ by neglecting the term proportional to M^4. This is called the *Gaussian approximation*. Our first task is to evaluate the integral $\int (atM^2 + c|\nabla M|^2) d^3\mathbf{r}$ as an explicit function of the fluctuation amplitudes. This can easily be done by using the orthonormality relations [Eq. (9.40)] for the functions u_K and the identity

$$\int \nabla M \cdot \nabla M \, d^3\mathbf{r} = -\int M \, \nabla^2 M \, d^3\mathbf{r} \qquad (9.46)$$

The result is that

$$\int (atM^2 + c|\nabla M|^2) \, d^3\mathbf{r} = L^3 \sum_\mathbf{K} (at + ck^2) M_\mathbf{K}^2 \qquad (9.47)$$

where $k^2 = k_1^2 + k_2^2 + k_3^2 = (2\pi/L)^2(K_1^2 + K_2^2 + K_3^2)$. The integral for Z then factors into independent integrals over each of the variables $M_\mathbf{K}$.

$$Z = e^{\phi_o} \prod_\mathbf{K} \int_{-\infty}^{\infty} \exp\left[-L^3(at + ck^2)M_\mathbf{K}^2\right] dM_\mathbf{K}$$
$$= e^{\phi_o} \prod_\mathbf{K} \left(\frac{\pi/L^3}{at + ck^2}\right)^{1/2} \tag{9.48}$$

The canonical potential, including the contribution of the fluctuations, is

$$\phi = \phi_o + \frac{1}{2}\sum_\mathbf{K} \log\left(\frac{\pi/L^3}{at + ck^2}\right) \tag{9.49}$$

The energy is given by $-\partial\phi/\partial\beta$.

$$E = E_o - \frac{1}{2}\frac{\partial}{\partial\beta}\sum_\mathbf{K} \log\left(\frac{\pi/L^3}{at + ck^2}\right)$$
$$= E_o + \frac{1}{2}\frac{T^2}{T_c}\frac{\partial}{\partial t}\sum_\mathbf{K} \log\left(\frac{\pi/L^3}{at + ck^2}\right) \tag{9.50}$$
$$= E_o - \frac{1}{2}\frac{T^2}{T_c}\sum_\mathbf{K} \frac{a}{at + ck^2}$$

Since we are only interested in the singularity in the specific heat that occurs close to the critical point, we can set the smooth function T^2/T_c equal to T_c, its value at $T = T_c$. The specific heat, close to T_c, is equal to $\partial E/\partial T = (1/T_c)\partial E/\partial t$.

$$C = C_o + \frac{1}{2}\sum_\mathbf{K} \frac{a^2}{(at + ck^2)^2} \tag{9.51}$$

To evaluate the specific heat at the critical point, we must set $t = 0$, obtaining

$$C = C_o + \frac{1}{2}(a/c)^2 \sum_\mathbf{K} k^{-4} \tag{9.52}$$

Clearly, the term with $\mathbf{k} = 0$ diverges. What is important to note, however, is that all of the long-wavelength modes give effectively divergent contributions to the specific heat. Consider the contribution to the sum of, let us say, the 1000 modes of smallest, but nonzero, k. They all have wave vectors of order L^{-1}. Thus each of them would contribute a term of order L^4. Since the specific heat is an extensive variable, and thus of order L^3, these contributions all diverge in the thermodynamic limit. Because of this *infrared divergence*, in the limit $t \to 0$, the term proportional to M^4 in the Landau–Ginsburg potential cannot be neglected.

9.9   THE CRITICAL DIMENSION

When the term BM^4 is included in Eq. (9.45), the integral no longer factors into independent integrals over each of the variables $M_\mathbf{K}$. It then becomes a very high-dimensional integral of a non-Gaussian exponential function, which cannot be evaluated. One can get some idea of the effect of the fourth-order term by calculating the partition function for an artificial situation in which all of the fluctuation modes except a single mode with mode numbers $\mathbf{K}$ are suppressed. In order to see an important effect of the dimensionality of the space, we will do this analysis for a space of unspecified dimension D. With only a single mode operating, the magnetization pattern is of the form

$$M(x_1, \ldots, x_D) = M_{\mathbf{K}} u_{K_1}(x_1) \cdots u_{K_D}(x_D) \tag{9.53}$$

This calculation will *not* allow us to determine the contributions to the canonical potential of the fluctuations in the macroscopic magnetization, because the integral for Z does not factor, and therefore the separate fluctuations do not give independent contributions to ϕ. Near the critical point, the fluctuations of different wavelengths strongly interact with one another. The contribution to ϕ of the fluctuations at various wavelengths is not equal to the sum of the contributions of the wavelengths taken separately. However, this calculation will exhibit a certain characteristic of the problem that has been very important in finding a practical method of calculating critical exponents and other characteristics of critical phenomena.

With $M(\mathbf{r})$ given by Eq. (9.53), the spatial integral that appears in the exponent in Eq. (9.45) can be easily evaluated. At $T = T_c$, t is zero, and therefore, the terms we need to integrate are the $|\nabla M|^2$ term and the M^4 term. We are most interested in determining how these integrals depend upon the size of the system L and the spatial dimension D. In the following analysis, $k^2 = k_1^2 + \cdots + k_D^2 = 4\pi^2 K^2 L^{-2}$. First we note that

$$\begin{aligned}
\int \nabla M \cdot \nabla M \, dx_1 \cdots dx_D &= -\int M \, \nabla^2 M \, dx_1 \cdots dx_D \\
&= k^2 L^D M_{\mathbf{K}}^2 \\
&= 4\pi^2 K^2 L^{D-2} M_{\mathbf{K}}^2
\end{aligned} \tag{9.54}$$

The M^4 integral is

$$\begin{aligned}
\int M^4 dx_1 \cdots dx_D &= M_{\mathbf{K}}^4 \int_o^L u_{K_1}^4 dx_1 \int_o^L u_{K_2}^4 dx_2 \cdots \int_o^L u_{K_D}^4 dx_D \\
&\equiv I_{\mathbf{K}} L^D M_{\mathbf{K}}^4
\end{aligned} \tag{9.55}$$

The calculation of $I_{\mathbf{K}}$, which depends on the mode numbers, but is never far from one, is left as an exercise for the reader (see Problem 9.8). Using these formulas, one can evaluate the necessary integral

$$\int (c|\nabla M|^2 + bM^4) \, d^D x = 4\pi^2 K^2 c L^{D-2} M_{\mathbf{K}}^2 + b I_{\mathbf{K}} L^D M_{\mathbf{K}}^4 \tag{9.56}$$

With all other modes frozen at zero, the partition function is given by integrating $\exp[\phi_o - \int (c|\nabla M|^2 + bM^4) d^D x]$ over all possible values of the mode amplitude $M_{\mathbf{K}}$ [see Eq. (9.45)].

$$Z = e^{\phi_o} \int_{-\infty}^{\infty} \exp\left(-4\pi^2 K^2 c L^{D-2} M_{\mathbf{K}}^2 - b I_{\mathbf{K}} L^D M_{\mathbf{K}}^4\right) dM_{\mathbf{K}} \tag{9.57}$$

In order to isolate the L dependence, we make a transformation of variables, $m = L^{(D-2)/2} M_{\mathbf{K}}$. Then

$$Z = e^{\phi_o} L^{-(D-2)/2} \int \exp\left(-4\pi^2 K^2 c m^2 - b I_{\mathbf{K}} L^{4-D} m^4\right) dm \tag{9.58}$$

If $D > 4$ and $\mathbf{K} \neq 0$, then, because of the factor L^{4-D}, the second term in the exponent becomes negligible in comparison with the first in the thermodynamic limit, $L \to \infty$. But it is the M^4 term that produces the troublesome interactions between different modes. The same analysis is valid when any number of nonuniform modes exist simultaneously. In a space of more than four dimensions, the interaction terms are negligible in the thermodynamic limit, and therefore the partition function integral becomes a multiple Gaussian integral that can be evaluated.

Recall that, according to Eq. (9.52), the contribution to the specific heat, at $T = T_c$, of any nonuniform mode, calculated with the Gaussian approximation, was proportional to L^4, and therefore divergent in comparison with the ordinary specific heat, which is proportional to L^3. But, in dimension D, the ordinary specific heat is proportional to L^D. Thus, for $D > 4$, the separate terms in the Gaussian approximation each give a negligible contribution to the specific heat and, once the $\mathbf{K} = 0$ mode has been separated out, the discrete sum may be converted to an integral, using the usual rule that

$$\sum_{\mathbf{K}} (\cdots) \to (L/2\pi)^D \int (\cdots) \, d^D k \tag{9.59}$$

The $\mathbf{K} = 0$ mode must, however, be treated separately. But the $\mathbf{K} = 0$ mode describes a uniform magnetization—just the sort of thing that is included in the simple Landau theory. Therefore, using the simple Landau theory to calculate the contribution of the uniform mode and the Gaussian approximation for all the nonuniform fluctuations would allow one to calculate all of the critical exponents for systems in dimension D larger than four. This must seem to the reader like a completely empty result, since no real system is more than three dimensional. The point in looking at higher dimensional systems is that they are the closest solvable systems to the three-dimensional systems whose critical behavior we would like to calculate.

9.10 THE EPSILON EXPANSION

In contrast to textbook problems, most real problems that come up in physics cannot be solved exactly. One of the standard techniques for obtaining an approximate solution to an intractable mathematical problem is to find a solvable problem that is, in some sense, close to the given, unsolvable problem. The most common sense in which the solvable problem is close to the unsolvable one is that the two problems differ in the value of some numerical parameter λ. For example, it is impossible to calculate the orbits of the planets in the solar system, because of gravitational interactions between the planets. However, if the interplanetary interactions are turned off, then the problem of a planet moving in the gravitational field of the sun can be completely solved. We can relate the solvable problem to the unsolvable one as follows. We consider a sun of mass M and nine planets of masses λm_1, λm_2, $\ldots, \lambda m_9$, where λ is an adjustable parameter and m_i is the actual mass of the ith planet. For fixed λ and some given initial positions and velocities, let $\mathbf{R}_i(t, \lambda)$ be the trajectory of the ith planet. What we can calculate is $\mathbf{R}_i(t, 0)$. (When λ is very small, the planets move in the same way as "dust particles" in the field of the sun and their trajectories are simple ellipses, independent of their masses.) What we would like to know is $\mathbf{R}_i(t, 1)$. The *perturbation method* assumes that the function $\mathbf{R}_i(t, \lambda)$ can be expanded as a power series in λ.

$$\mathbf{R}_i(t, \lambda) = \mathbf{R}_i(t, 0) + \mathbf{R}'_i(t)\lambda + \mathbf{R}''_i(t)\lambda^2 + \cdots \tag{9.60}$$

It is possible to write the functions $\mathbf{R}_i'$, $\mathbf{R}_i''$, etc. as integrals involving the known functions $\mathbf{R}_i(t, 0)$. If t is not too large, the sizes of the terms in the series decrease rapidly, and it is a good approximation to cut off the infinite series after a few terms. In this way, one can construct useful, but not exact, analytic solutions to the problem of planetary motion.

In a similar way, it is possible to construct equations for critical-point parameters, such as exponents and amplitude ratios. The equations involve the parameter D, the space dimension, in such a way that they are meaningful for noninteger D. The parameter D is written in terms of an expansion parameter ϵ as

$$D = 4 - \epsilon \tag{9.61}$$

The equations can be solved exactly for $\epsilon = 0$, and the desired quantities (critical exponents, etc.) can be written as power-series expansions in ϵ. The next few sections will present the fundamental ideas needed to construct such equations.

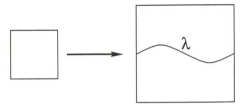

Fig. 9.10   When the size of a system at its critical temperature is increased, the important new degrees of freedom are fluctuations of a wavelength that could not have fit into the smaller system.

9.11   RENORMALIZATION THEORY

Although the Landau–Ginsburg theory is basically an adequate theory of critical phenomena, the complete intractability of the integral over fluctuation amplitudes for many years prevented any use of the theory to make calculations of critical parameters. The essential elements of a tractable calculational method were supplied, in the early 1970s, by Kenneth Wilson of Cornell University. Wilson's analysis, which is called *renormalization theory*, begins with a study of what happens to the distribution of configurations of a system when one increases the size of the system. One can see why such a peculiar study might be relevant to critical phenomena by considering what new elements are introduced if we double the dimensions of a sample of ferromagnetic material at its critical temperature. (See Fig. 9.10.)

By doubling the size of the system, we have increased the number of microscopic degrees of freedom, but that is a trivial change, because the microscopic degrees of freedom in the larger system are basically identical to those in the smaller one. The new elements that have been added are fluctuations of a size that could not be accommodated in the smaller system. But it is exactly such large-scale fluctuations that are responsible for critical behavior. Thus, in looking only at the *changes* in going from a smaller to a larger system, all the details that are irrelevant to the critical point cancel out. The theoretical device used in the study is a two-step process, called a *renormalization transformation*.

1. In the first step, we scale up the size of the system by some factor larger than one. (This is all done theoretically—renormalization is not an experimental procedure.) What we aim to do is to compare the probability distribution for the configurations of the larger system with that of the smaller one. But this cannot be done immediately, because the larger system has more coordinates, and therefore many more possible configurations. The probability distributions

would therefore be functions of very different numbers of variables and thus not directly comparable. To circumvent that difficulty, one carries out the second step.

2. We separate the coordinates in the larger system into a set of short-range coordinates and a set of long-range coordinates, in such a way that the number of long-range coordinates is equal to the total number of coordinates in the smaller system. We now compute the probability distribution for the long-range coordinates by integrating over the short-range coordinates. This leaves us with a probability function for the larger system that can be compared with that of the smaller system.

The procedure can be made clear by carrying it out in detail for the system shown in Fig. 9.11; a one-dimensional chain of coupled harmonic oscillators held fixed at its left end. The system is described by the $N+1$ coordinates, $x_o, x_1, \ldots,$ x_N, where x_o is fixed at zero. The value of βE for a given configuration is assumed to be of the form

$$\beta E = K_1(x_1^2 + \cdots + x_{N-1}^2) + \tfrac{1}{2}K_1 x_N^2 + K_2 \sum_{n=1}^{N}(x_n - x_{n-1})^2 \qquad (9.62)$$

The spring constant K_1 has been reduced on the last particle to $K_1/2$ for a technical reason that will be made clear shortly. Such a modification would have no effect on the thermodynamic behavior of a very large system. The chain can be considered to be made up of N elementary *links*, such as shown in Fig. 9.11(B). To make up the chain, the mass points of the links are pasted together, thus increasing the interior spring constants from $K_1/2$ to K_1. In the canonical ensemble, the probability that the two ends of a single isolated link have coordinates x and y is proportional to the function

$$g_1(x,y) = \exp[-\tfrac{1}{2}K_1(x^2 + y^2) - K_2(x - y)^2] \qquad (9.63)$$

The partition function for the system is the integral of $\exp(-\beta E)$ over the N free coordinates, $x_1, \ldots, x_N$. It can be written as

$$Z_N = \int g_1(0,x_1)g_1(x_1,x_2) \cdots g_1(x_{N-1},x_N)\, dx_1 \cdots dx_N \qquad (9.64)$$

We now increase the number of links to $2N$, writing the coordinates in the longer chain as $x_o, y_1, x_1, y_2, x_2, \ldots, y_N, x_N$. (Remember, x_o is defined as zero.) The partition function of the longer chain is

$$\begin{aligned} Z_{2N} &= \int g_1(0,y_1)g_1(y_1,x_1) \cdots g_1(y_N,x_N)\, dy_1 \cdots dy_N\, dx_1 \cdots dx_N \\ &= \int g_2(0,x_1)g_2(x_1,x_2) \cdots g_2(x_{N-1},x_N)\, dx_1 \cdots dx_N \end{aligned} \qquad (9.65)$$

where

$$g_2(x,y) = \int_{-\infty}^{\infty} g_1(x,z)g_1(z,y)\, dz \qquad (9.66)$$

$g_2(x,y)$ is the unnormalized probability function for the end coordinates of a double link. With this transformation, the unnormalized configurational probability for the larger system, namely $g_2(x_o,x_1)g_2(x_1,x_2) \cdots g_2(x_{N-1},x_N)$, is now in a form in which it can be compared with that of the smaller system. However, the coordinates

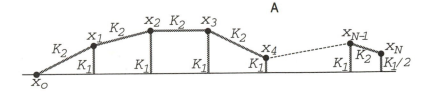

A

Fig. 9.11   (A) A system of $N+1$ particles that move only in the vertical direction. The potential energy has terms proportional to the square of the distance of a particle from its equilibrium position and terms proportional to the squares of the differences in the vertical coordinates of neighboring particles. (B) The system can be considered as being composed of N "links" of the form shown. What this really means is that the potential energy function is the sum of N terms of the form, $(K_1/2)(x_n^2 + x_{n-1}^2) + K_2(x_n - x_{n-1})^2$.

B

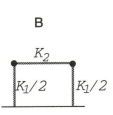

in the larger system represent the endpoints of larger pieces (that is, double links) than those in the smaller system. The detailed integral for g_2 is not difficult to carry out.

$$g_2(x,y) = \int \exp[-\tfrac{1}{2}K_1(x^2 + y^2) - K_1 z^2 - K_2(x - z)^2 - K_2(z - y)^2]\, dz \quad (9.67)$$

The integral is done by expanding the quadratic polynomial, completing the square in z, and writing the rest in terms of x, y, and $x - y$.

$$\begin{aligned}
&\tfrac{1}{2}K_1(x^2 + y^2) + K_1 z^2 + K_2(x - z)^2 + K_2(z - y)^2 \\
&= \tfrac{1}{2}(K_1 + 2K_2)(x^2 + y^2) + (K_1 + 2K_2)z^2 - 2K_2 z(x + y) \quad (9.68) \\
&= \tfrac{1}{2}(K_1 + 2K_2)(x^2 + y^2) + (K_1 + 2K_2)\Big(z - \frac{K_2(x+y)}{K_1 + 2K_2}\Big)^2 - \frac{K_2^2}{K_1 + 2K_2}(x+y)^2 \\
&= \tfrac{1}{2}K_1\Big(\frac{K_1 + 4K_2}{K_1 + 2K_2}\Big)(x^2 + y^2) + \frac{K_2^2}{K_1 + 2K_2}(x-y)^2 + (K_1 + 2K_2)\Big(z - \frac{K_2(x+y)}{K_1 + 2K_2}\Big)^2
\end{aligned}$$

With this we see that

$$g_2(x,y) = \Big(\frac{\pi}{K_1 + 2K_2}\Big)^{1/2} \exp[-\tfrac{1}{2}\tilde{K}_1(x^2 + y^2) - \tilde{K}_2(x - y)^2] \quad (9.69)$$

where $\tilde{K}_1 = K_1(K_1 + 4K_2)/(K_1 + 2K_2)$ and $\tilde{K}_2 = K_2^2/(K_1 + 2K_2)$. The factor $\sqrt{\pi/(K_1 + 2K_2)}$ is important in computing the partition function, but not in discussing the probability distribution, which would have to be normalized anyway. The configurational probability of the larger system has the same form as that of the smaller system, except for the fact that the parameters K_1 and K_2, have undergone the transformation

$$K_1 \to K_1\Big(\frac{K_1 + 4K_2}{K_1 + 2K_2}\Big) \quad \text{and} \quad K_2 \to \frac{K_2^2}{K_1 + 2K_2} \quad (9.70)$$

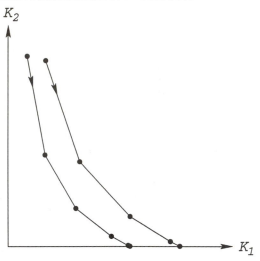

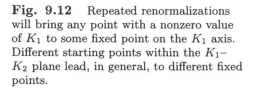

Fig. 9.12 Repeated renormalizations will bring any point with a nonzero value of K_1 to some fixed point on the K_1 axis. Different starting points within the K_1–K_2 plane lead, in general, to different fixed points.

If we repeat this transformation, over and over again, we obtain the probability distribution for the endpoints of longer and longer links in a larger and larger system. The parameters K_1 and K_2 move as shown in the *renormalization flow diagram* in Fig. 9.12. If we begin with any nonzero value of K_1, then after repeated renormalizations the value of K_2 approaches zero and the value of K_1 approaches some point on the K_1 axis that depends on our starting values. If we begin with any point $K_1 > 0$ and $K_2 = 0$, then the point does not change under renormalization. Thus all the points on the K_1 axis are called *fixed points* of the renormalization flow. These are called trivial fixed points, because they describe systems in which $K_2 = 0$ and therefore the two ends of every link are statistically independent. That is, they describe systems of noninteracting harmonic oscillators. Thus any system with a nonzero value of K_1, after it is scaled a sufficient number of times, resembles a system with noninteracting coordinates. That after many renormalization transformations neighboring particles become noninteracting is not surprising—the coordinates of neighboring particles in the final system represent the positions of quite distant particles in the original system.

9.12   THE NONTRIVIAL FIXED POINTS

If, in the initial system, $K_1 = 0$ exactly, then the renormalization flow equations say that K_1 remains zero and that $K_2 \to K_2/2$. This does not look like a fixed point, but that impression is wrong. We have not yet used up all of our freedom in mapping the larger system onto the smaller one. In comparing the configurations of the larger system with those of the smaller one, we are free to scale the coordinates, defining coordinates $\tilde{x}_1, \ldots, \tilde{x}_N$ in the larger system by the relations $x_n = \sqrt{2}\,\tilde{x}_n$. Then in terms of $\tilde{x}$ and $\tilde{y}$, with $K_1 = 0$,

$$g_2(\tilde{x}, \tilde{y}) = \left(\frac{\pi}{2K_2}\right)^{1/2} \exp[-K_2(\tilde{x} - \tilde{y})^2] \tag{9.71}$$

which means that the new probability distribution is the same as the original one. Since $g_2(\tilde{x}, \tilde{y})$ is not equal to a function of $\tilde{x}$ times a function of $\tilde{y}$, the probability distribution for the configurations of the system will not approach one for a system of noninteracting elements as the system size grows. The set of fixed points along the K_2 axis are not trivial fixed points. Nontrivial fixed points describe states of the system that have *scale-invariant* probability distributions. Let us confirm that this is so for the harmonically coupled chain that we are considering.

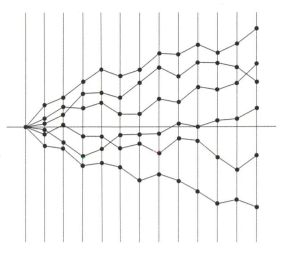

Fig. 9.13 A small ensemble of reasonably probable configurations of the system with $K_1 = 0$. The probability distribution for the system is the same as that for a random walk of N steps. As N increases, the fluctuations of the right end increase as $\sqrt{N}$. In contrast, when K_1 is not zero, the configurations remain close to the horizontal axis. Then the fluctuations about that axis provide a fixed scale of length and, as $N \to \infty$, a picture of the ensemble would show the configurations more and more concentrated on the axis.

9.13   SCALE INVARIANCE

If $K_1 = 0$, then the probability of a configuration $(x_1, x_2, \ldots, x_N)$ is equal to

$$P(x_1, x_2, \ldots, x_N) = C g_1(0, x_1) g_1(x_1, x_2) \cdots g_1(x_{N-1}, x_N) \qquad (9.72)$$

where $g_1(x, y) = \exp[-K_2(x - y)^2]$. A small ensemble of configurations for such a system is shown in Fig. 9.13. It is clear from the picture that the probability distribution for the system configurations resembles that of a random walk of length N. (See Problems 1.24 to 1.26 for a definition of the random walk problem.) It is easy to confirm that it is precisely the probability distribution for a random walk of individual step size $a = 1/\sqrt{2K_2}$. Imagine that in the picture N is made so large that the individual steps are well beyond the limit of discernibility. Then the individual curves in the ensemble become examples of fractals. In that limit, an ensemble of random walks is known to have the characteristic that, if we take the piece of the picture between 0 and ℓ, blow it up by a factor L/ℓ in the horizontal direction, and by a factor of $\sqrt{L/\ell}$ in the vertical direction, then it will match exactly the original picture of the ensemble between 0 and L. Whenever the probability distribution for the configurations of a complex system has the property that, by scaling a few parameters, one can make the ensemble for a smaller system exactly match that of a larger system, then the probability distribution is said to be scale invariant. It can be shown that all nontrivial fixed points of renormalization transformations lead to scale-invariant probability distributions.

The two fundamental assumptions of Wilson's analysis of critical phenomena were that

1. At the critical point, the probability distribution for the fluctuations in the order parameter (the magnetization for ferromagnetic transitions) is scale invariant.

2. The scale-invariant distribution could be obtained as a nontrivial fixed point of a properly formulated renormalization transformation.

9.14   STABILITY OF FIXED POINTS

Looking at the renormalization flow diagram, one can see that a system point located close to, but not at, a nontrivial fixed point behaves, under repeated transformations, very differently from one located close to a trivial fixed point. The flow lines are all directed away from the nontrivial fixed points and toward the trivial ones. An initial point (K_1, K_2) that has K_2 very small but nonzero would be close to the trivial fixed point $(K_1, 0)$. Under repeated renormalizations, it would move toward that fixed point. However, an initial point with K_1 very small but nonzero, which would be close to the nontrivial fixed point $(0, K_2)$, would move steadily further from the K_2 axis, eventually coming to a trivial fixed point on the K_1 axis. For this reason, the trivial fixed points are called stable, in analogy with a stable equilibrium position of a mechanical system, while the nontrivial fixed points are called unstable fixed points. This is a general property of trivial and nontrivial fixed points. The nontrivial fixed points describe critical points of the associated thermodynamic system. At a critical point, the correlation functions for local macroscopic observables, such as the order parameter correlation function,

$$c(\mathbf{r}, \mathbf{r}') = \langle M(\mathbf{r})M(\mathbf{r}') \rangle \tag{9.73}$$

fall off as some power of $|\mathbf{r} - \mathbf{r}'|$. That is, *

$$\langle M(\mathbf{r})M(\mathbf{r}') \rangle \sim \frac{1}{|\mathbf{r} - \mathbf{r}'|^\lambda} \tag{9.74}$$

This can be seen to be a direct consequence of the assumption that the critical probability distribution is scale invariant. If a system of volume V is scaled to a volume $s^3 V$, where s is some scaling parameter larger than one, then it should be possible, by comparing some scaled order parameter γM at the "corresponding points" in the new system (namely $s\mathbf{r}$ and $s\mathbf{r}'$) with the original order parameter at $\mathbf{r}$ and $\mathbf{r}'$, to obtain exactly the same correlation function. This implies that

$$\langle \gamma M(s\mathbf{r})\gamma M(s\mathbf{r}') \rangle = \langle M(\mathbf{r})M(\mathbf{r}') \rangle \tag{9.75}$$

With the correlation function given in Eq. (9.74), it is clear that this is satisfied for $\gamma = s^{\lambda/2}$.

$$\frac{\gamma^2}{|s\mathbf{r} - s\mathbf{r}'|^\lambda} = \frac{1}{|\mathbf{r} - \mathbf{r}'|^\lambda} \Rightarrow \gamma = s^{\lambda/2} \tag{9.76}$$

In any thermodynamic state that is not at the critical point, the correlation function at large distances has an exponential behavior.

$$c(\mathbf{r}, \mathbf{r}') \sim \frac{e^{-|\mathbf{r}-\mathbf{r}'|/\ell}}{r^\lambda} \tag{9.77}$$

where the parameter ℓ is the correlation length. There is no way of rescaling the order parameter by a factor γ that will make

$$\frac{\gamma^2}{s^\lambda} e^{-s|\mathbf{r}-\mathbf{r}'|/\ell} = e^{-|\mathbf{r}-\mathbf{r}'|/\ell} \tag{9.78}$$

for all values of $|\mathbf{r}-\mathbf{r}'|$. This shows that the probability distribution is not scale invariant. The correlation length ℓ acts as a standard of length that does not scale with the system size.

    If the basic correlations in the system decay exponentially with a fixed correlation length, then after repeated renormalization transformations we will arrive

---

*Usually the power λ in the following equation is written as $\lambda = D - 2 + \eta$, where D is the space dimension and η is yet another critical exponent.

at a set of long-range variables that are statistically independent. In the harmonically coupled chain those variables describe the endpoints of very long links. In a Landau–Ginsburg theory they will describe local magnetizations $M(\mathbf{r})$ that are averages over larger and larger domains. As the domains become much larger than the correlation length ℓ, the averages over different domains become statistically independent. When the renormalization process has produced statistically independent variables, then the set of parameters in the probability distribution (for the harmonically coupled chain, simply K_1 and K_2) will have reached a trivial fixed point. Thus the stability properties of the fixed points are closely associated with the decay of the correlation function in the states described by those fixed points. Unstable fixed points give power-law decay, while stable fixed points describe states with exponentially decaying correlation functions.

The renormalization flow diagram of the harmonically coupled chain has one feature that is not a general characteristic of such diagrams. That is that both the trivial and the nontrivial fixed points in the diagram form continuous lines (in this case, the K_1 and K_2 axes). It is very common for them to be, instead, discrete isolated points. However, there do exist other systems with continuous curves of fixed points.

9.15   THE CALCULATION OF EXPONENTS

In contrast to the Landau–Ginsburg theory, in which all calculations of critical properties were prevented by the impossibility of evaluating the partition function integral, the renormalization theory has allowed the accurate computation of the universal properties of critical phenomena, such as exponents and amplitude ratios. The reason for this practical success is that the desired quantities are entirely determined by the details of the renormalization flow diagram close to the nontrivial fixed point that represents the critical state of the system. As an example of how one relates critical exponents to flow diagrams, we will calculate two critical exponents that are associated with the correlation function of the harmonic chain. Before we do so, we will make a simple modification in our definition of the system. Instead of fixing the left end of the chain by demanding that $x_o = 0$, we will use periodic boundary conditions, defined by taking $x_o = x_N$, where N is the number of particles in the system. With periodic boundary conditions, the two-particle correlation function, to be defined, will be translationally invariant. This modification of the boundary conditions has no effect on the renormalization flow equations, $(K_1, K_2) \to (\tilde{K}_1, \tilde{K}_2)$, as can be easily confirmed by the reader.

In a uniaxial ferromagnet, the two-point correlation function is defined as the expectation value of the product of the magnetization at two different points, $\mathbf{r}_1$ and $\mathbf{r}_2$. Because of translational invariance, it is a function only of the relative variable $\mathbf{r} = \mathbf{r}_2 - \mathbf{r}_1$. Another way of writing this is that

$$c(\mathbf{r}) = \langle M(\mathbf{R}) M(\mathbf{R} + \mathbf{r}) \rangle \tag{9.79}$$

where $\mathbf{R}$ is any point within the system. The equivalent function for the harmonic chain is

$$c(r | K_1, K_2) = \langle x_n x_{n+r} \rangle_K \tag{9.80}$$

where r is an integer, x_n is the coordinate of the nth mass point, and $\langle \ldots \rangle_K$ means the expectation value in a system with parameters K_1 and K_2. The dependence of the correlation function on the parameters of the system, K_1 and K_2, is shown explicitly because it will play an important part in the analysis.

We have seen that the probability distribution for the coordinates $\tilde{x}_n$ in the larger system with parameters K_1 and K_2 is identical to the probability distribution for the coordinates x_n in the smaller system with parameters $\tilde{K}_1$ and $\tilde{K}_2$, given in Eq. (9.70). This means that

$$\langle \tilde{x}_m \tilde{x}_n \rangle_K = \langle x_m x_n \rangle_{\tilde{K}} \tag{9.81}$$

But if in the larger system one uses an ordinary numbering of the coordinates, so that x_k is the coordinate of the kth particle (recall that in the previous analysis we called the particle coordinates $x_o, y_1, x_1, \ldots$), then

$$\tilde{x}_n = \frac{1}{\sqrt{2}} x_{2n} \tag{9.82}$$

In the renormalization transformation for the harmonic chain the factor s by which the size of the system is scaled up is 2. The factor γ by which the variables x_{2n} are scaled down to make the variables $\tilde{x}_n$ is $1/\sqrt{2}$. In order to facilitate the generalization of the following analysis to other possible scaling factors, we will rewrite Eq. (9.82) as

$$\tilde{x}_n = \gamma x_{sn} \tag{9.83}$$

Putting this into Eq. (9.81) immediately gives the important renormalization relation

$$\gamma^2 c(sr|K_1, K_2) = c(r|\tilde{K}_2, \tilde{K}_2) \tag{9.84}$$

If $K^* \equiv (K_1, K_2)$ is a fixed point, then $\tilde{K} = K^*$ and this relation can be written

$$\gamma^2 c(sr|K^*) = c(r|K^*) \tag{9.85}$$

This implies that, at the nontrivial fixed point, K^*, $c(r) = A/r^\lambda$ with λ and γ related by

$$\gamma = s^{\lambda/2} \tag{9.86}$$

Thus we see that γ, the scaling factor for the field, in combination with the scaling factor s, determines the power in the correlation function at the critical point. For the harmonic chain, $s = 2$ and $\gamma = 1/\sqrt{2}$, which shows that $\lambda = -1$.

The renormalization relation, Eq. (9.84), can be used to calculate another critical exponent that is associated with the correlation function. If K is not a fixed point, then the correlation function at large r takes an exponential form

$$c(r|K) \sim \frac{e^{-r/\ell}}{r^\lambda} \tag{9.87}$$

As $K_1 \to 0$, which means that K approaches a nontrivial fixed point, the correlation length ℓ, which depends on the values of K_1 and K_2, goes to infinity as some negative power of K_1.

$$\ell(K_1, K_2) \sim \ell_o K_1^{-\nu} \tag{9.88}$$

Using Eq. (9.87), the relation $\gamma^2 c(sr|K) = c(r|\tilde{K})$ can be written as

$$\frac{\gamma^2}{s^\lambda} e^{-sr/\ell(K)} = e^{-r/\ell(\tilde{K})} \tag{9.89}$$

Clearly λ must have the same value as it has at the critical point in order for γ^2/s^λ to be equal to one. The other relation we obtain is that

$$\frac{s}{\ell(K)} = \frac{1}{\ell(\tilde{K})} \tag{9.90}$$

or

$$\frac{\ell(K)}{\ell(\tilde{K})} = s \tag{9.91}$$

Using Eq. (9.88) and the fact that $s = 2$ gives

$$\lim_{K_1 \to 0} \frac{\tilde{K}_1^\nu}{K_1^\nu} = 2 \tag{9.92}$$

When the rescaling, $\tilde{x}_n = x_{2n}/\sqrt{2}$, is taken into account, the flow equations near a nontrivial fixed point take the form

$$K_1 \to 2K_1\left(\frac{K_1 + 4K_2}{K_1 + 2K_2}\right) \quad \text{and} \quad K_2 \to \frac{2K_2^2}{K_1 + 2K_2} \tag{9.93}$$

Near $K_1 = 0$, the K_1 flow equation gives

$$\tilde{K}_1 \approx 4K_1 \tag{9.94}$$

Thus Eq. (9.92) says that $4^\nu = 2$ or

$$\nu = \tfrac{1}{2} \tag{9.95}$$

The harmonic chain is merely a toy model. It is of no intrinsic importance, but has been presented in order to illustrate the concepts and methods of renormalization theory in a situation where everything can be worked out exactly. For realistic systems, such as those described by the Landau–Ginsburg theory, the integrations necessary to determine the renormalization flow equations are quite complicated, and only the results of those integrations will be given here.

9.16   MOMENTUM-SPACE RENORMALIZATION

The renormalization transformation that has been used in analyzing the harmonic chain is an example of a *real-space renormalization* calculation, which means that the variables used in the calculation (that is, the x_ns) describe properties of the system at locations in real space. It is often more convenient to carry out the calculation in terms of variables that are Fourier components of the real-space variables. The fluctuation amplitudes $M_\mathbf{K}$ are a good example of such variables. One then says that the calculation is being done in momentum space, a terminology taken from quantum mechanics, where the momentum-space and real-space wave functions are just Fourier transforms of one another. To show how the renormalization transformation is done in momentum space, we will consider the renormalization theory analysis of the Landau–Ginsburg model in three dimensions.

The magnetization at point $\mathbf{r}$ is given in terms of the fluctuation amplitudes by the equation

$$M(\mathbf{r}) = \sum_\mathbf{K} M_\mathbf{K}\, u_{K_1}(x_1)\, u_{K_2}(x_2)\, u_{K_3}(x_3) \tag{9.96}$$

where the functions $u_K(x)$ are defined in Eq. (9.39) and the mode numbers, $\mathbf{K} = (K_1, K_2, K_3)$ are restricted to the range

$$|\mathbf{K}| < L/\Lambda \tag{9.97}$$

This restriction guarantees that the wavelengths of the fluctuations remain larger than Λ. We will only consider the case of zero external field. Then the probability distribution for the fluctuations in $M(\mathbf{r})$ is of the form

$$P[\{M_\mathbf{K}\}] = C \exp\left(-\int (AM^2 + bM^4 + c|\nabla M|^2)\, d^3\mathbf{r}\right) \tag{9.98}$$

In the simple Landau theory, the critical point occurs when $A = 0$. We will see that, when more accurate calculations are made using renormalization theory, the nontrivial fixed point, in fact, occurs at a slightly negative value of A, not at $A = 0$. That is why we have not written A as at.

The probability distribution is determined by the parameters A, b, and c. These are the equivalents, in the Landau–Ginsburg theory, of the parameters K_1 and K_2 for the harmonic chain. The momentum-space renormalization cycle has the following steps.

1. The size of the system in each direction is increased from L to $L' = sL$, where the scale factor s is any number larger than one. The number of fluctuation amplitudes will therefore increase. There will be one variable for each set of mode numbers (K_1, K_2, K_3) in the range

$$|\mathbf{K}| < L'/\Lambda \tag{9.99}$$

2. In order to reduce the number of variables to the previous value, the minimum allowable wavelength is increased from Λ to $\Lambda' = s\Lambda$. This restricts the allowable mode numbers to the range

$$|\mathbf{K}| < L'/\Lambda' = L/\Lambda \tag{9.100}$$

To obtain the new probability distribution, one must integrate over the eliminated fluctuation amplitudes. That is, one integrates over all those variables $M_\mathbf{K}$ with $\mathbf{K}$ in the range

$$L'/\Lambda' < |\mathbf{K}| < L'/\Lambda \tag{9.101}$$

3. The magnetization $M(\mathbf{r})$ is rescaled by multiplying it by some as yet undetermined scale factor γ.

4. The new probability distribution is for a magnetization field $M(\mathbf{r})$, defined within the volume $V' = (L')^3$. The old field was defined within the volume $V = L^3$. The new distribution can be mapped into the old volume by a scale transformation. Combining steps 3 and 4, the new magnetization variable is

$$\tilde{M}(\mathbf{r}) = \gamma M(\mathbf{r}/s) \tag{9.102}$$

The difficult step is Step 2, the integration. It cannot be done exactly, but it can be carried out using the ϵ expansion (or other approximation methods that we have not discussed). The details of the calculation are quite complicated and will not be given. We will only quote the results.

9.17   FIXED POINTS OF THE LANDAU–GINSBURG THEORY

What we are trying to determine are the fixed points, if any exist, in a fluctuation probability distribution of the Landau–Ginsburg form. A fixed point would mean a value of A, b, and c with the property that, if one started with a Landau–Ginsburg distribution having those values of the parameters and carried out the renormalization transformation defined, then the resulting probability distribution would still be of the Landau–Ginsburg form with the same parameter values.

One case in which the integration is not difficult is the case $b = 0$. In that case the integral in the exponent of Eq. (9.98) is

$$\int (AM^2 + c|\nabla M|^2)\, d^3\mathbf{r} = \sum_{\mathbf{K}} (A + ck^2) M_{\mathbf{K}}^2 \tag{9.103}$$

Integrating over the variables $M_{\mathbf{K}}$ with $\mathbf{K}$ in the shell $L'/\Lambda' < |\mathbf{K}| < L'/\Lambda$ simply produces a factor in front of the exponential that can be absorbed into the normalization constant. The probability distribution, after the integration of step 2, is of the same form as the starting one with the same value of the parameters. However, one must still complete steps 3 and 4. When that is done the completed transformation has the form

$$\int_{V'} (AM^2 + c|\nabla M|^2)\, d^3\mathbf{r} \to \frac{s^3}{\gamma^2} \int_V [A\tilde{M}^2(\mathbf{x}) + s^{-2}c|\nabla \tilde{M}(\mathbf{x})|^2]\, d^3\mathbf{x} \tag{9.104}$$

where $\mathbf{x} = \mathbf{r}/s$. The parameter A will remain unchanged if we choose $\gamma = s^{3/2}$, but then the parameter c will remain constant only if its value is zero. Thus there is a line of fixed points defined by the equations

$$b = 0, \quad c = 0, \quad \text{and} \quad 0 < A < \infty \tag{9.105}$$

The fact that $c = 0$ means that the values of the magnetization at different points $\mathbf{r}_1$ and $\mathbf{r}_2$ are statistically independent in the thermodynamic limit. The states described by these fixed points are noncritical states in which the correlation function decays exponentially with some correlation length ℓ. After many renormalizations [recall that, in each renormalization, we keep only the long-wavelength components of $M(\mathbf{r})$] the function $M(\mathbf{r})$ varies very slowly with $\mathbf{r}$. It is an average of the magnetization over a subvolume of size Λ. When $\Lambda \gg \ell$, these subvolumes become statistically independent. Another way of saying the same thing is that two points with a separation $r = |\mathbf{r}_1 - \mathbf{r}_2|$ in the starting volume are mapped into points with a separation sr after the renormalization. If r is not zero, then sr eventually becomes much larger than the correlation length ℓ and the magnetizations at the two points become statistically independent. As expected, the renormalization flow equations show that these fixed points are stable. Any initial state close to this line of fixed points flows into it.

According to Eq. (9.104), one can obtain another fixed point by letting $A = 0$ and $\gamma = s^{1/2}$. However, with $b = 0$, that set of parameter values does not describe a normalizable probability distribution. It is actually outside the physically allowable parameter space.

9.18   THE WILSON FIXED POINT

If b is not zero, then the integration over the fluctuation variables $M_{\mathbf{K}}$ that is required in the renormalization transformation becomes very difficult. In fact, it is impossible to carry out exactly. We have seen that if the space dimension D is larger than 4, then the bM^4 term becomes negligibly small. It is therefore possible to expand the required integral as a power series in $\epsilon = 4 - D$. In order to determine the fixed points, it is enough to calculate the renormalization flow equations for an expansion factor s only marginally larger than one. Assuming that $s = 1 + \delta$, we will simply quote the results of the renormalization transformation, given to first order in ϵ and δ. The renormalization equation for the parameter c is unchanged from Eq. (9.104). That is

$$c \to \frac{s}{\gamma^2} c \qquad (9.106)$$

In order to find a fixed point with a nonzero value of c, we must choose the order parameter scaling factor, $\gamma = s^{1/2}$. The renormalization flow equations for the other two parameters, A and b, take a simpler form if they are written in terms of the equivalent variables

$$x = \frac{\Lambda^2}{2\pi^2} A \quad \text{and} \quad y = \frac{3}{2\pi^2} b \qquad (9.107)$$

They are

$$x \to \tilde{x} = x + (2x + y - xy)\delta \qquad (9.108)$$

and

$$y \to \tilde{y} = y + (\epsilon y - 3y^2)\delta \qquad (9.109)$$

Of course, γ has been chosen so that $\tilde{c} = c$, and, therefore, that parameter will no longer be explicitly considered.

To determine the fixed point, we set $\tilde{x} = x$, $\tilde{y} = y$, and recall that, for a three-dimensional system, $\epsilon = 1$. The resulting equations have two solutions. The first, $(x^*, y^*) = (0, 0)$, is the unphysical solution that was rejected previously. The other, called the Wilson fixed point, is at

$$(x^*, y^*) = (-\tfrac{1}{5}, \tfrac{1}{3}) \qquad (9.110)$$

To first order in ϵ, this gives the correct distribution of fluctuations at the critical point for any system with a scalar order parameter in three dimensions, such as the 3D Ising model.

9.19   DIFFERENTIAL FLOW EQUATIONS
Using the fact that $\log s = \log(1 + \delta) \approx \delta$, one can write the renormalization equations given in differential form.

$$\frac{dx}{d(\log s)} = \lim_{\delta \to 0} \frac{\tilde{x} - x}{\delta} = 2x + y - xy \qquad (9.111)$$

and

$$\frac{dy}{d(\log s)} = \lim_{\delta \to 0} \frac{\tilde{y} - y}{\delta} = \epsilon y - 3y^2 \qquad (9.112)$$

The flow lines produced by these differential equations are shown in Fig. 9.14.

In the vicinity of the Wilson fixed point (x^*, y^*), the differential equations can be linearized by writing x as $x^* + \Delta x$ and y as $y^* + \Delta y$ and then ignoring second-order terms in Δx and Δy. One then obtains the following linearized flow equations

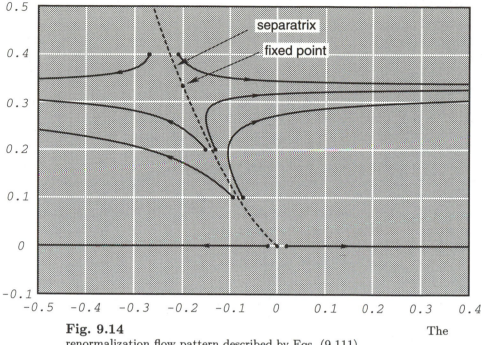

Fig. 9.14   The
renormalization flow pattern described by Eqs. (9.111)
and (9.112). All points to the right of the dotted line, called the
separatrix, flow, as $s \to \infty$, to $x = +\infty$. The points to the left of
the separatrix flow to $x = -\infty$. The points on the separatrix flow to
the Wilson fixed point. Those points describe fluctuation probability
distributions for a system at a critical point. One thing that must be
kept in mind is that this renormalization flow diagram depends on the
fact that we have chosen the scaling parameter γ equal to $s^{1/2}$. For
those points that flow to $x = \pm\infty$ this is clearly wrong. Had we chosen
$\gamma = s^{3/2}$, those points would flow to noncritical fixed points with
$c = 0$ and A either negative or positive, indicating that the system
was either below or above the critical temperature.

for the variables Δx and Δy. They are valid only close to the fixed point, but that
is exactly where we are most interested in the details of the renormalization flow.

$$\frac{d(\Delta x)}{d(\log s)} = \tfrac{5}{3}\Delta x + \tfrac{6}{5}\Delta y \tag{9.113}$$

and

$$\frac{d(\Delta y)}{d(\log s)} = -\Delta y \tag{9.114}$$

Let us first look at the equation for Δy. It can be written in the form
$d\log(\Delta y) = d\log(s^{-1})$, which has the solution $\Delta y(s) = \Delta y(1)/s$. Thus, if the
parameter y is initially displaced from its fixed point value y^*, it will monotonically
move to that value under renormalization. A parameter with that characteristic is
called an *irrelevant parameter*. The logic behind calling a parameter irrelevant if it
flows to its fixed point value under renormalization is that, insofar as the study of
critical phenomena is concerned, such a parameter can simply be set equal to its
fixed point value. One is only interested in very large systems, and, therefore, it is
reasonable to assume that a number of renormalizations have already been done,
which would be enough to bring all irrelevant parameters to their fixed point values.

The Landau–Ginsburg distribution is actually only an approximation to the true critical point probability distribution. Once the renormalization integrals are calculated to more than first order in ϵ, it is found that the strict Landau–Ginsburg form has no nontrivial fixed point. After one renormalization, a distribution function of the Landau–Ginsburg form is converted to one of a more general form, with terms such as M^6, M^8, $M^2|\nabla M|^2$, $|\nabla M|^4$, etc.. Therefore, in order to find a fixed point, one must start with a completely general probability distribution for the fluctuations in the order parameter. The coefficients of all the terms (actually infinite in number, although their size at the fixed point drops rapidly as the power of M increases) are completely determined by the fixed point condition that the distribution function must reproduce itself under a renormalization transformation. Also, these extra parameters are all irrelevant parameters, so that they automatically flow to their fixed point values. For that reason, one can discuss all the essential physics of the critical point in terms of the simple Landau–Ginsburg distribution. However, these terms do affect the values of critical properties when things are calculated to higher than first order in ϵ.

If $y = y^*$ but x is displaced slightly from its fixed point value, then the flow equation for Δx is

$$\frac{d(\Delta x)}{d \log s} = \tfrac{5}{3} \Delta x \tag{9.115}$$

It is clear that the point will move farther and farther from the fixed point. Eventually, it will come to the line of trivial fixed points. For this reason, x is called a *relevant parameter*. We will see that the only other relevant parameter in the theory is the magnetic field B, which we have set equal to its fixed point value of zero in this discussion.

Renormalization theory is such an abstract construction that it is necessary, even at the hazard of appearing repetitious, to review periodically the physical meaning of the mathematical operations. The flow equations clearly do not describe the physical movement of anything. They describe a purely intellectual procedure in which one looks at more and more macroscopic variables in larger and larger systems. What is involved is obviously a form of thermodynamic limit, as outlined in Chapter 3. The rescaling of the order parameter M at each renormalization step is analogous to something that is done in the proof of the Central Limit Theorem. In that proof, one must look at the variable $T = (x_1 + \cdots + x_N)/\sqrt{N}$. If one looks at the average, $x = (x_1 + \cdots + x_N)/N$, then the fluctuations disappear as $N \to \infty$. If one looks at the sum, $S = x_1 + \cdots + x_N$, then they blow up in that limit. The scaling factor $1/\sqrt{N}$ is just enough to give a nontrivial probability distribution in the thermodynamic limit. In the same way, one must rescale M with just the right factor in order to keep the fluctuations in view as one takes the thermodynamic limit. In fact, renormalization theory can be considered as a major extension of the Central Limit Theorem from systems of independent variables to systems with significant interactions. The trivial fixed points, where the observables become statistically independent in the thermodynamic limit, give Gaussian probability distributions that correspond exactly to the usual Central Limit Theorem. However, because of the interactions and the statistical correlations that they produce, completely different limiting probability distributions are also possible. They are described by the nontrivial fixed points, which have much more complex fluctuation probability distributions. Renormalization theory shares with the Central Limit Theorem the marvelous characteristic that the detailed mathematical form of the probability distributions in the thermodynamic limit depend very little on the microscopic details of the system. It is certainly remarkable that the complicated distribution at the fixed point, which cannot even be written down in any finite formula, is not determined by the physical details of the real system, but by a purely mathematical fixed point condition. In the macroscopic limit, where thermodynamics becomes valid, statistical effects dominate everything else and limit the fluctuation distribution function to those few forms (including the Gaussian distribution) that satisfy the fixed point condition.

The essential characteristic of the flow equations, for determining critical exponents, is the rate at which all relevant parameters diverge from the fixed point as s is increased. Equation (9.115) for Δx can be written as

$$d(\log \Delta x) = \tfrac{5}{3} d(\log s) \qquad (9.116)$$

This has the solution

$$\Delta x(s) = \Delta x(1) s^{5/3} \qquad (9.117)$$

The parameter x is proportional to the parameter A. In the Landau theory, the critical point occurred at $A = 0$, and, therefore, near the critical point, A could be assumed to be proportional to $t = (T - T_c)/T_c$. Now that we have determined the correct value for A at the critical point (at least to order ϵ), we can again express A in terms of t. We do this by writing x in the form $\Delta x = x - x^* = Ct$, where the constant C cannot be determined by renormalization theory. Putting this into Eq. (9.117) gives

$$t = s^{5/3} t_o \qquad (9.118)$$

where $t_o = \Delta x(1)/C$. Again we must ask ourselves what this equation means. It does not mean that, as we look at larger and larger systems, we change the temperature as a function of the system size. All the systems in the thermodynamic limit sequence are at the same temperature. What is meant is that the probability distribution for the scaled-up variables in the scaled-up system, at temperature t_o, is the same as the probability distribution for the starting variables in the starting system at temperature $t_o s^{5/3}$. This is reasonable. As one looks at larger and larger scale variables at fixed temperature, they become less correlated. As one looks at a fixed set of variables at larger and larger values of t, they also become less correlated. Equation (9.118) gives an equivalence between the two effects.

9.20   THE MAGNETIC FIELD

If we ignore the irrelevant parameter y, then, close to the critical point, for sufficiently large systems, the fluctuation distribution is determined by the single parameter t, the deviation of the temperature from its critical value. However, this is true only because we have been assuming that there is no external magnetic field. It is known that the critical point occurs only at $B = 0$. Therefore B must be a relevant parameter. The calculation of the renormalization flow equations in the presence of an external magnetic field is even more complicated than in the zero-field case, and thus we will again only quote the results. To first order in ϵ, near the critical point, a nonzero magnetic field scales as

$$B = s^{5/2} B_o \qquad (9.119)$$

The specific numbers $\tfrac{5}{3}$ and $\tfrac{5}{2}$, in Eqs. (9.118) and (9.119) are obtained when the ϵ expansion is cut off after the linear term. If the expansion is carried further, then the form of the equations remains the same but the numerical values of the exponents are modified. Therefore, we will write the equations in the more general form

$$t = s^{y_t} t_o \quad \text{and} \quad B = s^{y_B} B_o \qquad (9.120)$$

and use the first-order values, $y_t = \frac{5}{3}$ and $y_B = \frac{5}{2}$, only when we want specific numbers for a calculation.

9.21   THE SCALING FORMULAS

Very close to the critical point, those properties that show nonanalytic behavior, such as the susceptibility and the specific heat, are dominated by the large-scale fluctuations in the order parameter. For that reason, the part of the canonical potential that is contributed by the microscopic degrees of freedom [for example, the ϕ_o term in Eq. (9.37)] may be ignored in the calculation of specifically critical properties. That fact, when combined with the scaling properties of t and B, will be seen to yield simple algebraic formulas for the various critical exponents in terms of the two parameters y_t and y_B.

Consider a large ferromagnetic system of, let us say, 10^{25} microscopic coordinates. Choose some number K, such as 10^{20}, that is very large but still much smaller than the total number of coordinates. Let $\hat{\phi}(t, B, V)$ be that part of the canonical potential that is due to the K fluctuation modes of largest wavelength. Very close to the critical point we can treat $\hat{\phi}$ as the total canonical potential. The distribution of the K largest modes for a system of volume $s^3 V$ at temperature t and field B is the same as it is for a system of volume V at temperature $s^{y_t} t$ with a field $s^{y_B} B$. Integrating over that distribution to obtain $\hat{\phi}$ gives the relationship

$$\hat{\phi}(t, B, s^3 V) = \hat{\phi}(s^{y_t} t, s^{y_B} B, V) \qquad (9.121)$$

Taking a derivative of Eq. (9.121) with respect to V and defining the canonical potential density $f(t, B) = \phi(t, B, V)/V = \partial\phi(t, B, V)/\partial V$ gives the basic scaling relation that we will use to derive the exponent formulas.

$$s^3 f(t, B) = f(s^{y_t} t, s^{y_B} B) \qquad (9.122)$$

We are actually making an assumption in this analysis, namely that the singular part of the canonical potential is itself an extensive quantity (i.e. proportional to V). This is called the *hyperscaling* property. There are some statistical mechanical models in which this hyperscaling assumption is not valid.

In order to obtain the magnetization exponent β in terms of the two basic fixed point parameters y_t and y_B, we differentiate this equation with respect to B and use the thermodynamic relation $M(t, B) = kT\partial f(t, B)/\partial B$ with T set equal to T_c (since t is assumed to be very small).

$$s^3 M(t, B) = s^{y_B} M(s^{y_t} t, s^{y_B} B) \qquad (9.123)$$

The exponent β is defined by saying that, for very small negative t, $M(t, 0) \approx A(-t)^\beta$. Setting B to zero in Eq. (9.123) and assuming that t is a small negative number, gives the relation

$$s^3(-t)^\beta = s^{y_B + \beta y_t}(-t)^\beta \qquad (9.124)$$

which implies that

$$y_B + \beta y_t = 3 \qquad (9.125)$$

or

$$\beta = \frac{3 - y_B}{y_t} \tag{9.126}$$

Using the values to first order in ϵ, $y_t = \frac{5}{3}$ and $y_B = \frac{5}{2}$, gives the exponent value

$$\beta = 0.3 \tag{9.127}$$

which is much closer to the experimentally observed value of 0.324 than is the mean-field prediction, $\beta = 0.5$. Taking higher powers in the ϵ expansion of y_B and y_t eliminates the remaining discrepancy.

9.22   THE SUSCEPTIBILITY EXPONENT

To evaluate the susceptibility exponent in terms of y_t and y_B, we take a derivative of Eq. (9.123) with respect to B and then set B equal to zero.

$$s^3 \chi_m(t, 0) = s^{2y_B} \chi_m(s^{y_t} t, 0) \tag{9.128}$$

The exponent γ was defined by Eq. (9.7), which says that the magnetic susceptibility in zero field, close to the critical temperature, has the form $\chi_m \sim B_\pm |t|^{-\gamma}$. Using this in Eq. (9.128) gives

$$s^3 |t|^{-\gamma} = s^{2y_B - \gamma y_t} |t|^{-\gamma} \tag{9.129}$$

which shows that

$$2y_B - \gamma y_t = 3 \tag{9.130}$$

or

$$\gamma = \frac{2y_B - 3}{y_t} \tag{9.131}$$

Using the values to first order in ϵ for y_B and y_t gives $\gamma = 1.2$—quite close to the experimental value of 1.24. Again, higher-order terms bring the predictions completely in line with the experimental results. Other cases will be left to the problems and exercises.

9.23   REAL-SPACE RENORMALIZATION

The Landau–Ginsburg potential is a thermodynamic object, and therefore the renormalization theory that is based on it has all the advantages and disadvantages of any thermodynamic calculation. It is not based on any detailed microscopic model of the system, and it is therefore applicable to a wide class of systems that have only some broad features in common. The disadvantage in this is that the theory provides no method of relating the detailed features of the microscopic Hamiltonian, such as the spin–spin interaction strength, to the thermodynamic properties of the system, like the phase transition temperature. It is obvious that any theory that does not distinguish between a ferromagnetic solid and a nonmagnetic fluid must leave out many physically important details.

There is another version of renormalization theory, illustrated by the calculation that was made involving the harmonic chain, that does begin from a detailed microscopic model of the system. This real-space renormalization method, being

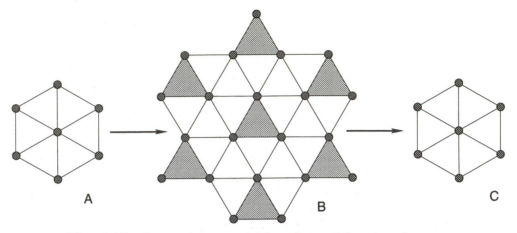

Fig. 9.15   System A is expanded by a factor of three to make System B. The site spins in System B are grouped into sets of three, called blocks. Associated with each block, is a block spin variable, whose value is ±1, depending upon whether the sum of the vertex spins is positive or negative. The set of block spins is equivalent to another seven-spin system (System C), which differs from System A only in having an altered Hamiltonian function.

based on a microscopic picture of the system, is properly a part of statistical mechanics, rather than thermodynamics. It has become a particularly powerful technique for analyzing systems with discrete variables, such as the Ising model.

As an example of a real-space renormalization calculation for a discrete system, we consider the arrays of two-valued spins shown in Fig. 9.15. The leftmost array, which we will call System A, has seven spins in a hexagonal pattern on a triangular lattice. The renormalization procedure consists of the following two steps.

1. The dimensions of System A are increased by a factor of $\sqrt{3}$, and each spin is replaced by three spins to produce a larger, related system of 21 spins (System B) on a similar triangular lattice. The spin–spin interactions in the second system are assumed to be the same as those in the first, so that System B is simply a larger version of System A.

2. The 21 spins in System B, shown by filled dots, will be called it site spins. They can each take the values of ±1. The probability of any configuration is given by

$$P(\sigma_1, \ldots, \sigma_{21}) = Z^{-1} \exp H(\sigma_1, \ldots, \sigma_{21}) \qquad (9.132)$$

where $H = -\beta E$ and E is the energy of the spin system. Although it is dimensionless and is proportional to minus energy, the function H is usually called the Hamiltonian of the system. (It is really a conditional canonical potential, but we will follow the more standard nomenclature.) We will not assume that E necessarily has the simple Ising model form that involves only nearest neighbor interactions. Each of the shaded triangles is called a block, and associated with each block is a two-valued *block spin* variable $\tilde{\sigma}$, defined as follows.

$$\tilde{\sigma} = S(\sigma + \sigma' + \sigma'') \qquad (9.133)$$

where σ, σ', and σ'' are the site spins at the three vertices of the block and

$S(n)$ is the *sign function*,

$$S(n) = \begin{cases} -1, & \text{for } n < 0 \\ 0, & \text{for } n = 0 \\ +1, & \text{for } n > 0 \end{cases} \tag{9.134}$$

The procedure of assigning the values ± 1 to the block spin variables, according to whether most of the site spins are up or down, is called *majority rule*. As can be seen, there are seven block spins, arranged in a hexagonal array, similar to System A. We define a new block spin Hamiltonian, $\tilde{H}(\tilde{\sigma}_1, \ldots, \tilde{\sigma}_7)$, by the relation

$$\exp \tilde{H}(\tilde{\sigma}_1, \ldots, \tilde{\sigma}_7) = {\sum}' \exp H(\sigma_1, \ldots, \sigma_{21}) \tag{9.135}$$

where the prime on the summation sign indicates that the sum is taken only over those configurations that give the set of block spin values that are indicated on the left hand side. Another way of writing this is

$$\exp \tilde{H}(\tilde{\sigma}_1, \ldots, \tilde{\sigma}_7) = \sum \delta(\tilde{\sigma}_1 - S_1) \cdots \delta(\tilde{\sigma}_7 - S_7) \exp H(\sigma_1 \ldots, \sigma_{21}) \tag{9.136}$$

where $S_i \equiv S(\sigma_i + \sigma_i' + \sigma_i'')$ and σ_i, σ_i', and σ_i'' are the site spins at the vertices of the ith block. With this definition of the block spin Hamiltonian, it is easy to see that the probability for the block spin variables is

$$\tilde{P}(\tilde{\sigma}_1, \ldots, \tilde{\sigma}_7) = \tilde{Z}^{-1} \exp \tilde{H}(\tilde{\sigma}_1, \ldots, \tilde{\sigma}_7) \tag{9.137}$$

This renormalization transformation maps the 21-spin system with the Hamiltonian $H(\sigma_1, \ldots, \sigma_{21})$ into a 7-spin system, System C, similar to the starting system, but with a Hamiltonian $\tilde{H}(\tilde{\sigma}_1, \ldots, \tilde{\sigma}_7)$. If the Hamiltonian in the starting system had some particularly simple form, such as

$$H = K \sum_{\text{NN}} \sigma_i \sigma_j \tag{9.138}$$

there is no reason to expect that form to be retained in the mapping. In fact, the renormalized Hamiltonian will usually be of a more complicated form. It will generally have next-nearest-neighbor terms, three-spin terms, etc.

$$\tilde{H} = \tilde{K}_1 \sum_{\text{NN}} \tilde{\sigma}_i \tilde{\sigma}_j + \tilde{K}_2 \sum_{\text{NNN.}} \tilde{\sigma}_i \tilde{\sigma}_j + \tilde{K}_3 \sum_{\text{triangles}} \tilde{\sigma}_i \tilde{\sigma}_j \tilde{\sigma}_k + \cdots \tag{9.139}$$

This effect is known as a *proliferation of interactions*. The same effect occurs in the momentum-space renormalization of the Landau–Ginsburg potential when the calculation is done beyond first order in epsilon.

The aim of the calculation is to find a starting Hamiltonian

$$H = K_1 \sum_{\text{NN}} \sigma_i \sigma_j + K_2 \sum_{\text{NNN.}} \sigma_i \sigma_j + K_3 \sum_{\text{tri.}} \sigma_i \sigma_j \sigma_k + \cdots \tag{9.140}$$

with the property that its form and the specific values of its parameters are preserved under the transformation. In practical calculations one starts with a Hamiltonian with a finite (and fairly small) number of parameters, such as K_1, K_2, and K_3. One then makes some approximation to the exact renormalization transformation that prevents the proliferation of interactions and thus gives a renormalized Hamiltonian of the same form but with new parameters. That is,

$$(K_1, K_2, K_3) \rightarrow (\tilde{K}_1, \tilde{K}_2, \tilde{K}_3) \tag{9.141}$$

The solutions of the equation

$$(K_1, K_2, K_3) = (\tilde{K}_1, \tilde{K}_2, \tilde{K}_3) \tag{9.142}$$

define the fixed points of the transformation. For Ising systems, there are usually two trivial fixed points and a nontrivial one that describes the critical properties of the system. One trivial fixed point has the values $K_1 = K_2 = K_3 = 0$. Since the Hamiltonian is related to the energy by $H = -\beta E$, having $H = 0$ is equivalent to having infinite temperature. At this trivial fixed point the site spins are totally uncorrelated. But if the site spins are uncorrelated, then the block spins will also be. That is why $H = 0$ is always a fixed point. The other trivial fixed point occurs at zero temperature, or infinite interaction strength. In that case, the site spins are perfectly correlated—if one of them is up, then they are all up; if one of them is down, then they are all down. But again, if that is true of the site spins, then it will also be true of the block spins, which explains why that case is a fixed point. Both trivial fixed points are stable. The interesting fixed point, which one hopes will at least approximately describe the critical properties of the system, is the third, unstable one.

There is a wide variety of different approximation schemes, designed to prevent the proliferation of interactions. We will illustrate some of them in the exercises that are associated with this chapter. The properties of the renormalization flow equations near the nontrivial fixed point give the critical exponents of the system.

Although the calculational details of the method will be left to the exercises, there are a number of important general points that should be discussed here.

1. What is really needed is a renormalization transformation for an infinite (or at least very large) system, rather than one for a system like a seven-spin hexagon. However, no one has been able to compute the renormalization transformation for even a mildly complex Hamiltonian on an infinite lattice. Fortunately, although most properties of small systems are very different than the corresponding properties of very large systems, the renormalization transformation itself, which can be used to calculate the critical behavior of an infinite system, is affected very little by the size of the system. Thus the scheme has the marvelous characteristic of allowing one to make reasonably accurate calculations on very large systems by doing a calculation on a very small one.

2. It was no accident that our introduction to the real-space renormalization method was based on a triangular Ising lattice. In a square lattice a new problem appears that would have significantly complicated the analysis. The problem, called *block spin ambiguity*, is illustrated in Fig. 9.16. In a square lattice, one would usually form blocks from sets of four site spins. If the block spin variables are to take the values ± 1, so that they have the same form as the site spin variables, then it becomes very unclear what block spin value should be assigned to the six block configurations that have zero total spin. This problem can be resolved in a number of different ways. Some people have simply assigned a block spin value of $+1$ to three of the configurations, such as (1), (2), and (3) and a value of -1 to the others. This can be done in four different ways if one adds the natural restriction that reversing all of the spins

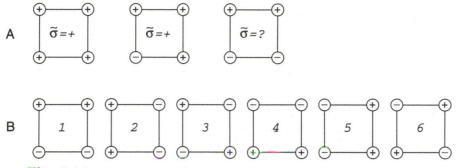

Fig. 9.16 It is obvious that a block spin value of +1 should be assigned to the first two configurations in (A), but it is not clear what block spin value should be assigned to the third. The six-site spin configurations that give an ambiguous block spin value are shown in (B).

in a configuration should reverse the value of the block spin. Another way of treating the problem is to allow the block spin variables to have the three possible values, ±1 and 0. This procedure has the drawback that it maps the original system of two-valued spins onto a system of three-valued spins, but all subsequent renormalizations will take the system of three-valued spins into itself. In the exercises, we will illustrate still other methods of resolving the problem of block spin ambiguity.

3. In contrast to the momentum-space renormalization method, the real-space renormalization method allows one to make a prediction of the critical temperature T_c. To illustrate how this is done, let us imagine making a calculation in which only two Hamiltonian parameters, K_1 and K_2, are used, where K_1 is the nearest-neighbor interaction strength and K_2 is the next-nearest-neighbor strength, as in Eq. (9.140). Using some particular finite system, one can compute a renormalization flow diagram, as shown in Fig. 9.17. We will assume that the physical system that we are interested in is an Ising lattice with only nearest-neighbor interactions. For such a system,

$$E = -V \sum_{\text{NN}} \sigma_i \sigma_j \tag{9.143}$$

with some particular fixed value of V. Therefore, in the physical system, $K_1 = V/kT$ and $K_2 = 0$. From the renormalization flow diagram, one can find the point on the K_1 axis, call it K_1^o, that flows to the nontrivial fixed point under repeated renormalizations. The equation

$$K_1^o = V/kT_c \tag{9.144}$$

then allows one to calculate T_c.

4. The real-space renormalization method is particularly valuable in analyzing two-dimensional systems because the alternative method of momentum space renormalization relies upon the epsilon expansion. Power series expansions work well only when the expansion parameter is small, but in two dimensions epsilon is equal to two.

The following argument, although it is highly simplified, should indicate how the real-space renormalization procedure is related to the previous renormalization

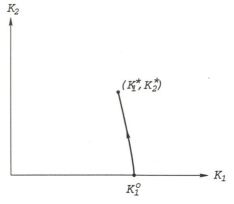

Fig. 9.17 The point $(K_1^o, 0)$ flows, under renormalization, to the fixed point (K_1^*, K_2^*). This means that the physical system, with interaction strength K_1^o will be at its critical point. Since $K_1 = V/\tau$, the critical temperature is given by $\tau_c = V/K_1^o$.

procedure that used continuous magnetization variables. We consider a rectangular 2D lattice in which we have introduced only two magnetization variables, M_1 and M_2, which represent the sums of the spins in the respective squares. It is assumed that the number of spins is so large that M_1 and M_2 can be treated as continuous variables, ranging from $-\infty$ to ∞. Under a single renormalization, the system is converted to a larger system with corresponding magnetization variables, M_1' and M_2'. (See Fig. 9.18.) When the possibility of scaling is taken into account, the existence of a fixed point would imply that

$$\mathbf{P}[M_1 = a, M_2 = b] = \mathbf{P}[M_1' = \gamma a, M_2' = \gamma b] \tag{9.145}$$

Discrete block spin variables, σ_1, σ_2, σ_1', and σ_2', defined by majority rule, are related to the magnetization variables by $\sigma(M) = \text{sgn}(M)$. (With continuous variables, we can ignore the possibility of M being exactly zero.) The joint probability distribution of σ_1 and σ_2 is related to that for M_1 and M_2 by

$$\mathbf{P}[\sigma_1 = +1, \sigma_2 = +1] = \mathbf{P}[M_1 > 0, M_2 > 0] \tag{9.146}$$

with similar equations for the other three possible values of σ_1 and σ_2. It is now easy to verify that, if Eq. (9.145) is satisfied with any positive value of the scaling parameter γ, then

$$\mathbf{P}[\sigma_1 = +1, \sigma_2 = +1] = \mathbf{P}[\sigma_1' = +1, \sigma_2' = +1] \tag{9.147}$$

because the scaling parameter will have no effect on the integrated probability that M_1 and M_2 are positive. Thus the equality of the continuous probability distribution, *with scaling*, implies the simple equality of the discrete probability distributions. The search for fixed points in the block spin formalism is basically equivalent to the search for fixed points in the scaled magnetization formalism.

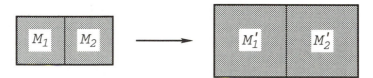

Fig. 9.18 Under renormalization, the magnetization variables of the smaller system are mapped into those of the larger one.

PROBLEMS

9.1 Equation (8.100) is the basic equation of the mean-field approximation applied to the 3D Ising model. With no external field, it states that the energy of a configuration is

$$E(C) = 3NV\bar{\sigma}^2 - 6V\bar{\sigma}\sum \sigma(\mathbf{r}) \qquad (9.148)$$

It is not at all obvious that this approximation is equivalent to the Landau theory, which states that the conditional canonical potential for a fixed value of the uniform magnetization M is

$$\phi(T, M) = \phi_o - VatM^2 - VbM^4 \qquad (9.149)$$

The objects of this problem are to show that, close to the critical point of the mean-field approximation ($kT_c = 6V$), for small values of M, the value of $\phi(T, M)$ calculated from Eq. (9.148) agrees with the Landau theory and to identify the Landau parameters ϕ_o, a, and b. In comparing the two theories, the Landau order parameter M is identified with the average spin polarization $\bar{\sigma}$. (a) Replace $\bar{\sigma}$ by M in Eq. (9.148) and evaluate $\phi(M)$ by summing $\exp(-\beta E(C))$ over all 2^N configurations. (b) Using the definition, $t = (T - T_c)/T_c$, and the fact that $kT_c = 6V$, eliminate β in the expression you obtained in (a), in favor of t. (c) Expand the expression obtained in (b) as a power series in the small quantities, M^2 and t, retaining all terms through second-order. [That is, one keeps the second-order terms, $M^2 t$ and $(M^2)^2$, but drops the third order terms, $M^2 t^2$ and $(M^2)^2 t$. At the critical point, $M \sim (-t)^{1/2}$, so that M^2 and t are of the same order.] (d) Comparing the result obtained in (c) with Eq. (9.149), determine the Landau parameters ϕ_o, a, and b for the 3D Ising model.

9.2 Consider a two-dimensional Ising system in the form of an $L \times L$ square. We assume that the lattice spacing is so small in comparison with L that the state of the system can be described by a continuous magnetization variable $M(x, y)$, where x and y have the range 0 to L. The conditional canonical potential associated with any magnetization pattern $M(x, y)$ is assumed to have the two-dimensional equivalent of the form given in Eq. (9.37).

$$\phi(M) = \phi_o - \int\int \left[atM^2(x, y) + bM^4 + c|\nabla M(x, y)|^2\right] dx\, dy \qquad (9.150)$$

We also assume that the values of $M(x, y)$ at the boundaries are some known, fixed, functions, let us say

$$M(0, y) = M_o \quad \text{and} \quad M(x, 0) = M(x, L) = M(L, y) = 0 \qquad (9.151)$$

Since the probability of obtaining any particular magnetization pattern is proportional to $\exp \phi(M)$ and, away from the critical point, the fluctuations from the most probable pattern are negligible, the pattern that would actually be observed is the one that maximizes $\phi(M)$. We call the function that maximizes $\phi(M)$ and also satisfies the boundary conditions [Eq. (9.151)] $\bar{M}(x, y)$. Let $\Delta M(x, y)$ be any function that is zero on all of the boundaries. Then $M(x, y, \lambda) = \bar{M}(x, y) + \lambda \Delta M(x, y)$ is a function that satisfies the boundary conditions for any value of the parameter λ and

$$\phi(\lambda) \equiv \phi_o - \int\int \left[atM^2(x, y, \lambda) + bM^4(x, y, \lambda) + c|\nabla M(x, y, \lambda)|^2\right] dx\, dy \qquad (9.152)$$

has a maximum at $\lambda = 0$. (a) Using the condition that $\Delta M(x, y)$ vanishes on all the boundaries, prove the identity

$$\int \int \nabla M(x, y) \cdot \nabla[\Delta M(x, y)] \, dx \, dy = - \int \int \nabla^2 M(x, y) \Delta M(x, y) \, dx \, dy \quad (9.153)$$

(b) Using the identity proved in (a), show that the condition that $d\phi(\lambda)/d\lambda = 0$ at $\lambda = 0$ for any $\Delta M(x, y)$ that vanishes on all the boundaries implies that $\bar{M}(x, y)$ satisfies the equation

$$-c\nabla^2 \bar{M}(x, y) + at\bar{M}(x, y) + 2b\bar{M}^3(x, y) = 0 \qquad (9.154)$$

9.3 (a) Repeat the calculation described in Problem 8.26 for the values, $\beta V = 0.3$, 0.35, and 0.4. (b) For a 2D Ising model, the critical value of βV is known from the Onsager solution to be $(\beta V)_c = 0.4407 \ldots$. Assume that, as $\beta V \to (\beta V)_c$, the correlation length goes to infinity as some negative power of $(\beta V)_c - \beta V$, that is, that

$$\ell = \ell_o (0.4407 - \beta V)^{-\nu}$$

Plotting $\log \ell$ versus $\log(0.4407 - \beta V)$, try to determine the value of ν for a 2D Ising model.

9.4 (a) Using Eq. (9.122), derive a scaling formula that gives the specific heat exponent α [see Eq. (9.4)], in terms of the fixed point parameters y_t and y_B. (b) Use the epsilon expansion formulas given in Exercise 9.4, to estimate α for a 3D Ising model. Because of the fact that α is quite small for the 3D Ising model, the result you obtain will have a large fractional error. The exponent α, because of its small size, is difficult to determine accurately, both experimentally and theoretically.

9.5 In deriving the scaling laws in Section 9.21, explicit use was made of the fact that, as the dimensions of a system are scaled by a factor s, the volume increases as $s^3 V$. This is true only in three dimensions, and therefore the conclusions drawn there are restricted to 3D systems. (a) Rederive the scaling formulas for β and γ for the general case of a D-dimensional system. (b) In Exercise 9.4, an expansion is given, through second-order in $\varepsilon = 4 - D$, of the fundamental exponents y_t and y_B. Use that expansion, and the results of (a), to show that in four dimensions one obtains mean-field (that is, simple Landau theory) values for β and γ.

9.6 (a) Use Eq. (9.121) and the thermodynamic relation $\partial\phi/\partial B = \beta\langle M \rangle$ to derive a scaling formula for the critical isotherm exponent δ for a 3D system. (b) Use the epsilon expansion given in Exercise 9.4 to estimate the value of δ for a 3D uniaxial ferromagnet.

9.7 The aim of this problem is to show that the sharp spike in the magnetic susceptibility per spin at the critical temperature, $\tau_c \approx 4.511$, appears only in the thermodynamic limit. Using the Fortran program Ising_3D.for, with $H = 0$ and $V = 1$, calculate and plot the magnetic susceptibility per spin for an $L \times L \times L$ lattice in the range $4 < \tau < 5$ for lattices of size $L = 2$, 4, and 8.

9.8 Calculate the constant $I_{\mathbf{K}}$ defined in Eq 9.55.

9.9 Extend the renormalization flow diagram shown in Exercise 9.7 to negative values of K and "explain" your result.

9.10   Using the program Ising_2.for, with $V = 1$, on an 8×8 lattice, plot the magnetic susceptibility per spin between $\tau = 1$ and 4 for the external field values, $H = 0$, 0.05, and 0.1. Use at least 10,000 sweeps for each point.

9.11   In this problem, we will investigate the number of iterations needed for a Monte Carlo process to come to equilibrium. In the program Ising_2.for, after each sweep through the lattice, the current values of $|M|$, M^2, E, and E^2 are added to Sum_M, Sum_M2, Sum_E, and Sum_E2, to be used later in calculating $\langle |M| \rangle$, $\langle M^2 \rangle$, $\langle E \rangle$, and $\langle E^2 \rangle$. (a) Modify the program in the following ways: (1) Drop all instructions used in calulating $\langle |M| \rangle$, $\langle M^2 \rangle$, $\langle E \rangle$, and $\langle E^2 \rangle$. (2) Add an integer array, Mag(1),..., Mag(10000). (3) Let $N_{\text{sweeps}} = 10000$ and, after the kth sweep, store the total magnetization in Mag(k). (3) For each value of K from 0 to 1000, calculate the *magnetization autocorrelation function*

$$C(K) = \left(\sum \text{Mag}(i)\text{Mag}(i+K) \right) \Big/ (10000 - K) \qquad (9.155)$$

where the sum on i is from 1 to $10000 - K$. (b) For an 8×8 lattice, with $H = 0$ and $V = 1$, calculate and plot $C(k)$ for the temperatures, $\tau = 3$, 2.5, and 2.25. (c) Explain why and how the function $C(k)$ is related to the number of sweeps necessary to bring the lattice to equilibrium.

9.12   This problem will illustrate some of the difficulties one encounters in trying to determine τ_c, the transition temperature for an Ising system, by making Monte Carlo calculations. It will also illustrate two of the methods used to circumvent those difficulties. For an $L \times L$ Ising lattice, the specific heat per spin C is given by $L^2 C = (\langle E^2 \rangle - \langle E \rangle^2)/\tau^2$. We also define a quantity Q by $L^2 Q = 2L^2 C - (\langle E^3 \rangle - 3\langle E^2 \rangle \langle E \rangle + 2\langle E \rangle^3)/\tau^3$. In the limit $L \to \infty$, $C(\tau)$ becomes infinite at the critical temperature τ_c but for finite L, C only develops a finite maximum at some temperature $\tau_{\text{max}}(L)$, which depends on the size of the lattice. Of course, $\tau_{\text{max}}(L) \to \tau_c$ as $L \to \infty$. (a) Modify the program Ising_2.for so that it calculates and saves the values of C and Q for each value of τ. (It may be helpful to take out the instructions that calculate and save the values of $\langle M \rangle$ and χ.) (b) With $L = 8$, $H = 0$, $V = 1$, $\tau_{\text{min}} = 2$, $\tau_{\text{max}} = 3$, $N_{\text{runs}} = 11$, and $N_{\text{sweeps}} = 10000$, run the program and show that statistical fluctuation in $C(\tau)$ prevent one from determining $\tau_{\text{max}}(8)$ with reasonable accuracy. (c) By taking first, second, and third derivatives of the canonical potential $\phi(\beta)$, prove that another criterion for τ_{max} is that $Q(\tau_{\text{max}}) = 0$ and show that this criterion does allow a reasonably accurate determination of $\tau_{\text{max}}(8)$ by interpolation. In a similar way, determine $\tau_{\text{max}}(4)$ and $\tau_{\text{max}}(6)$. (d) It can be proved that, for large values of L, $\tau_{\text{max}}(L) \approx \tau_c + A/L$ for some constant A. Plotting the three values of τ_{max} that you have determined as a function of $x = 1/L$, try to determine the value of τ_c for a two-dimensional Ising model. (The exact value is about 2.2692.)

9.13   Use the program LatGas.for to obtain data for the curves shown in Fig. 9.3.

9.14   The spin–spin correlation function in the x direction on an Ising lattice is defined as

$$C(n) = \left\langle \sum_x \sum_y \sigma(x,y)\sigma(x+n,y) \right\rangle \Big/ L_x L_y - \left(\langle \sum \sigma \rangle / L_x L_y \right)^2 \qquad (9.156)$$

where the expectation value is taken over a canonical ensemble or, in Monte Carlo calculations, over a long Monte Carlo run. (a) Modify Ising_2.for so that it can take different values for L_x and L_y. Take $L_x = 20$ and $L_y = 15$. With $H = 0$, $V = 1$, and $N_{\text{sweeps}} = 10,000$, calculate $C(3)$, $C(5)$, and $C(7)$ for the temperatures, $\tau = 2.0$, 2.25, and 2.5. (b) At the critical temperature, $C(n)$ has the form $C(n) = A/n^\lambda$ for some constants A and λ. In that case $\log C(n) = a - \lambda \log n$. That is, $\log C(n)$ is a linear function of $\log n$. For each temperature, determine values of a and λ that minimize the error E, defined as

$$E = \sum_n [\log C(n) - a - \lambda \log n]^2 \qquad (9.157)$$

where n is summed over the values 3, 5, and 7. The value of E is a measure of how far $C(n)$ deviates from having a power-law form. For what temperature is E smallest?

9.15 Using the display program AB Model, describe qualitatively the equilibrium phases for the nine combinations of α and βV that occur by taking $\alpha = -1$, 0, and $+1$ and $\beta V = -0.5$, 0, and $+0.5$.

9.16 For the AB Model, we define an observable Γ by

$$\Gamma = \sum_m \sum_n mn \, \text{Lat}(m, n) \qquad (9.158)$$

Modify the program AB.for so that it calculates $\langle \Gamma^2 \rangle$ and, with $\beta V = 0.5$, $L = 20$, and $N_{\text{sweeps}} = 10,000$, plot $\langle \Gamma^2 \rangle$ as a function of α in the range $-2 < \alpha < 2$ and give a physical interpretation of your results.

9.17 Using Ising_4.for with $L = 16$, plot the magnetic susceptibility, with $V = 1$ and τ going from 2 to 3, for the vacancy rates 0, 0.1, and 0.2. Qualitatively describe the effects of random vacancies on the magnetic properties of this finite system.

9.18 The program Ising_4.for can be used to study the effects of randomly distributed vacancies on the thermodynamic properties of a two-dimensional Ising model and, one hopes, on real magnetic solids. Regarding the effects of random defects, such as vacancies, there are two seemingly reasonable arguments that lead to contradictory conclusions. They are:

Argument 1. The randomness of the vacancy distribution will cause different parts of the system to have different critical temperatures. Thus the magnetic phase transition in an infinite system will be rounded over, as it is for a finite-size lattice. There will be no perfectly sharp critical temperature.

Argument 2. As the critical temperature is approached, the fluctuations that are important in the phase transition are of such long wavelength that they sample very large portions of the system. On those large scales, the vacancy distribution is effectively uniform, due to the law of large numbers. Thus, although the vacancies may cause quantitative shifts in the properties of the system, such as the value of the critical temperature or the critical exponents, at least for low enough density, they will have no qualitative effect on the phase transition—it will still be perfectly sharp.

Try to use the program Ising_4.for to obtain some useful information about this fundamental qualitative question on the effects of random defects on phase transitions. Warning: This is a very difficult assignment.

Chapter 10
Nonequilibrium Statistical Mechanics

10.1 INTRODUCTION

With some few exceptions, everything in the preceding chapters has been devoted to determining the properties of systems at equilibrium. Very little attention has been given to the obviously important processes that bring a system from its initial state to an equilibrium state. We have also avoided any discussion of systems in time-independent but nonequilibrium states, such as an electrical conductor that is connected to a constant source of potential difference or a piece of material that is conducting heat between two large reservoirs. There was a good reason for avoiding any discussion of nonequilibrium phenomena. It is that the fundamental principle that allowed us to make so much progress in describing the equilibrium properties of complex systems is that *equilibrium properties are entirely dominated by statistics rather than by detailed dynamical mechanisms.* For example, we could determine the equilibrium values of the ionization of a hydrogen gas, as a function of temperature and pressure, without ever looking at any details of the ionization process for a single hydrogen atom. We needed only to determine which value of the ionization corresponded to the largest number of microstates. We never had to ask how the system would get itself into that most probable macrostate. The equilibrium state is so overwhelmingly most probable that, unless some conservation law prevents the movement to that macrostate, we could be confident that the system would eventually reach it. As soon as one makes any attempt to calculate the detailed intermediate states between some given initial nonequilibrium macrostate and the final equilibrium state of a system, then the dominance of statistics disappears and one is forced to deal with all of the complicated details of time-dependent microscopic processes. To realize how sensitive the rate of approach to equilibrium is to the details of a system one might note that a container filled with a mixture of hydrogen and oxygen at room temperature will remain unchanged for centuries until there is a tiny spark, at which point it will explode. Except for some special circumstances, some of which will be explored in this chapter, calculating rates of approach to equilibrium is much more difficult than calculating the properties of the equilibrium state itself.

Nonequilibrium phenomena are currently an area of very active research. Unfortunately, it would require a major expansion of this already large book to introduce the reader to any significant fraction of the varied methods used in nonequilibrium studies. Instead, we will restrict ourselves to only a few topics in the field. For systems that are close to equilibrium, there are two important general theorems that will be presented and explored. The Fluctuation-Dissipation Theorem gives a close relationship between the spontaneous thermal fluctuations of a system at equilibrium and the equations of motion of the same system away from equilibrium. The Onsager Reciprocity Theorem gives certain discrete symmetries of the near-equilibrium equations of motion of any system. They are a consequence of the fact that the fundamental microscopic dynamical laws of the system are symmetrical under time reversal.

The third topic of the chapter will be an important equation, derived by Ludwig Boltzmann, that describes the approach to equilibrium of the momentum density function of an ideal gas. This will lead to a discussion of the increase of entropy in the universe and the directionality of time.

10.2   EQUILIBRIUM FLUCTUATIONS IN SIMPLE CIRCUITS

In preparation for studying time-dependent and nonequilibrium phenomena in general thermodynamic systems, we will look at the charge and current fluctuations in two simple electrical circuits. This will allow the reader to relate the somewhat abstract formalism of nonequilibrium statistical mechanics to the concrete physical characteristics of certain simple and familiar devices. The first question that we will consider is the calculation of charge fluctuations in a simple RC circuit. We assume that the device shown in Fig. 10.1 is completely isolated and that C and R are sufficiently large that inertial effects, due to inductance in the wires, can be neglected. We will idealize the capacitor to a device with no thermal energy, and therefore no entropy. All of the entropy of the system resides in the resistor. If the conserved total energy of the system is E, then, at a time when the charge on the capacitor is Q, the thermal energy of the resistor is $E - Q^2/2C$. Thus, the entropy of the system can be expressed as

$$S(E, Q) = S_R(E - Q^2/2C) \tag{10.1}$$

where S_R is the entropy function of the resistor. The equilibrium value of Q, the charge on the capacitor, is zero. However, at any finite temperature, the charge on the capacitor will not remain at zero precisely, but will show spontaneous thermal fluctuations. Because of the fact that the energy associated with the capacitor charge is the purely quadratic function $Q^2/2C$ one would expect, using the equipartition theorem, that

$$\langle Q^2/2C \rangle = \tfrac{1}{2}kT \tag{10.2}$$

However, a rigorous application of the equipartition theorem would require that one actually write the Hamiltonian function of the complete system as a function of Q and some set of other variables. A more convincing way of arriving at the same result is to use the fact that the probability of finding the isolated system with the charge Q on the capacitor is proportional to the exponential of the conditional entropy,[*]

$$P(Q) = \text{const.} \times e^{S(E, Q)} \tag{10.3}$$

[*]Throughout this chapter, we will always write thermodynamic potentials, such as the entropy, in rational units.

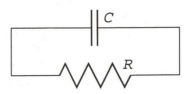

Fig. 10.1 An RC circuit.

Since the energy associated with the spontaneous charge fluctuations is much smaller than the thermal energy of the resistor, we can use the first-order expansion

$$S(E, Q) \approx S^o(E) - \beta Q^2/2C \tag{10.4}$$

where $\beta = \partial S^o/\partial E = 1/kT$. Then $\langle Q^2 \rangle$ can be computed from the general formula

$$\begin{aligned}
\langle Q^2 \rangle &= \int_{-\infty}^{\infty} Q^2 P(Q) \, dQ \\
&= \int Q^2 \exp(-\beta Q^2/2C) \, dQ \Big/ \int \exp(-\beta Q^2/2C) \, dQ \\
&= C/2kT
\end{aligned} \tag{10.5}$$

which simply confirms the prediction of the equipartition theorem. In the LCR circuit shown in Fig. 10.2, the thermal energy, expressed in terms of the conserved total energy of the system and the instantaneous values of the charge and current, is

$$E_{\text{th}} = E - Q^2/2C - LI^2/2 \tag{10.6}$$

In an LCR circuit, specifying the charge does not determine the current, nor vice versa. Thus Q and I are independent variables. At equilibrium, with total energy E, the joint probability for finding the circuit to have charge and current, Q and I, is

$$\begin{aligned}
P(Q, I) &= \text{const.} \times \exp S_R(E - Q^2/2C - LI^2/2) \\
&\approx \text{const.} \times \exp(-\beta Q^2/2C - \beta LI^2/2)
\end{aligned} \tag{10.7}$$

Using this probability distribution, one can easily confirm the results, expected from the equipartition theorem, that

$$\langle Q^2/2C \rangle = \langle LI^2/2 \rangle = \tfrac{1}{2}kT \tag{10.8}$$

and that Q and I are statistically independent variables.

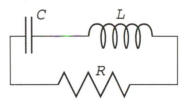

Fig. 10.2 An LCR circuit.

10.3 TIME-DEPENDENT CORRELATION FUNCTIONS

The formulas given for the mean-square fluctuations of the charge and current in an LCR circuit are very simple consequences of the principles of equilibrium statistical mechanics. In this chapter it will be shown that, by using certain general aspects of Hamiltonian dynamics, it is possible to obtain much more detailed information about the thermal fluctuations in the observables of any thermodynamic system. In particular, if $Q(t)$ and $I(t)$ are the instantaneous values of the charge and current in an LCR circuit at temperature T, it will be possible to obtain formulas for the time-dependent correlation functions, $\langle Q(0)Q(t)\rangle$, $\langle Q(0)I(t)\rangle$, $\langle I(0)Q(t)\rangle$, and $\langle I(0)I(t)\rangle$. The formulas will involve only the easily measurable circuit parameters L, C, and R.

In defining the time-dependent correlation functions, the average, indicated by the brackets, $\langle\ldots\rangle$, is either a microcanonical or a canonical ensemble average. However, in actual experimental determinations of quantities such as $\langle Q(0)Q(t)\rangle$, the experimentalist cannot construct an "ensemble of RC circuits" but must instead carry out the determination of $\langle Q(0)Q(t)\rangle$ on one circuit by using the alternative definition,

$$\langle Q(0)Q(t)\rangle \equiv \lim_{T\to\infty}\left(\frac{1}{T}\int_o^T Q(\tau)Q(\tau+t)\,d\tau\right) \tag{10.9}$$

The experimentally observed quantity is therefore a time-averaged correlation function, rather than an ensemble-averaged one. In the absence of extra conservation laws, the ergodic theorem, mentioned in Section 3.16, assures us that the two quantities will agree. If the system does not satisfy the requirements of the ergodic theorem, then, in general, it is only the ensemble-averaged correlation fuction that is well defined. The value obtained for the time-averaged correlation function might depend upon the particular initial values of the extra conserved variables. These would generally differ from one experimental measurement to another. In this chapter, only ensemble-averaged correlation fuctions will be considered. In the following sections, we will generalize the definition of a time-dependent correlation function to include arbitrary physical observables within any system obeying Hamilton's equations.

10.4   MACROSCOPIC DYNAMICAL LAWS

We consider a system with a $2K$-dimensional phase space and a Hamiltonian function $H(x_1,\ldots,x_K,p_1,\ldots,p_K)$. It will be convenient to introduce a compact notation for a point in the phase space.

$$z \equiv (x_1,\ldots,x_K,p_1,\ldots,p_K) \tag{10.10}$$

The phase-space volume element will be written as

$$d^{2K}z \equiv dx_1\cdots dx_K\,dp_1\cdots dp_K \tag{10.11}$$

The entropy, in rational units, is given by the integral

$$e^{S^o(E)} = \int \delta(E - H(z))\,d^{2K}z/N!h^K \tag{10.12}$$

Let us choose any n observables, $A_1(z)$, $A_2(z),\ldots$, $A_n(z)$, and indicate by the boldface symbol $\mathbf{A}(z)$ the set of the n chosen observables.

$$\mathbf{A}(z) \equiv (A_1(z), A_2(z),\ldots,A_n(z)) \tag{10.13}$$

For any n-dimensional real vector $\mathbf{a} = (a_1,a_2,\ldots,a_n)$, the n-dimensional delta function $\delta(\mathbf{a}-\mathbf{A}(z))$ is defined as the product of n ordinary delta functions.

$$\delta(\mathbf{a}-\mathbf{A}(z)) \equiv \delta(a_1 - A_1(z))\cdots\delta(a_n - A_n(z)) \tag{10.14}$$

The conditional entropy $S(\mathbf{a}, E)$ (often we will not explicitly indicate its dependence upon E) is given by the integral

$$e^{S(\mathbf{a},E)} \equiv \int \delta(\mathbf{a} - \mathbf{A}(z))\delta(E - H(z))d^{2K}z/N!h^K \qquad (10.15)$$

It is related to the thermodynamic entropy $S^o(E)$ by the usual identity

$$\int e^{S(\mathbf{a},E)} d^n\mathbf{a} = e^{S^o(E)} \qquad (10.16)$$

The vector $\mathbf{a}$ will be considered to be a point in an n-dimensional space that will be referred to as the *macrostate space* of the system. That is, the values of the n observables, $A_1, A_2, \ldots, A_n$, will be said to define the macrostate of the system. The microstate of the system is, as usual, defined by the values of the $2K$ phase-space coordinates.

Given any function of the n observables $F(\mathbf{A})$, the microcanonical expectation value of F may be converted to an integral over the macrostate space of the system.

$$\langle F \rangle = e^{-S^o} \int F(\mathbf{A}(z))\, \delta(E - H(z))\, d^{2K}z/N!h^K$$
$$= e^{-S^o} \int d^n\mathbf{a}\, F(\mathbf{a}) \int \delta(\mathbf{a} - \mathbf{A}(z))\, \delta(E - H(z))\, d^{2K}z/N!h^K \qquad (10.17)$$
$$= \int e^{S(\mathbf{a})-S^o} F(\mathbf{a})\, d^n\mathbf{a}$$

To confirm that the second line of Eq. (10.17) is equal to the first line, one need only carry out the integration over $\mathbf{a}$ using the n-dimensional delta function. The third line follows from taking the integral over the $2K$-dimensional phase space and using the definition of $S(\mathbf{a})$ given in Eq. (10.15). This equation shows that $\exp[S(\mathbf{a}) - S^o]$ is the probability density to be associated with the point $\mathbf{a}$ in the macrostate space of the system.

For any point z^o in phase space, there is a unique function $z(t, z^o)$ that is the solution of Hamilton's equations satisfying the initial condition $z(0, z^o) = z^o$. It will be convenient to define a *time evolution operator* T_t by the condition that T_t operating on any point z^o gives the phase-space point $z(t, z^o)$ that would result from allowing z^o to move according to Hamilton's equations for a time interval t.

$$T_t z^o \equiv z(t, z^o) \qquad (10.18)$$

For any initial point $\mathbf{a}^o$ in the macrostate space, we define a function $\mathbf{a}(t, \mathbf{a}^o)$ as follows:
1. At time zero, we construct an ensemble containing all the points on the energy surface that have $\mathbf{A}(z) = \mathbf{a}^o$.
2. We let the points in the ensemble evolve, according to Hamilton's equations, for a time interval t to give a new ensemble.
3. With this new ensemble, we calculate the expectation value of $\mathbf{A}(z)$. That value we call $\mathbf{a}(t, \mathbf{a}^o)$.

One can see that this prescription is satisfied by the formula

$$\mathbf{a}(t, \mathbf{a}^o) = e^{-S(\mathbf{a}^o)} \int \mathbf{A}(T_t z)\delta(\mathbf{a}^o - \mathbf{A}(z))\, \delta(E - H(z))\, d^{2K}z/N!h^K \qquad (10.19)$$

The value $\mathbf{a}(t, \mathbf{a}^o)$ is the average value that would be obtained if one measured $\mathbf{A}(z)$ at time 0 and time t for all the members of a microcanonical ensemble, threw out all the cases in which the original measurement did not give $\mathbf{a}^o$, and took an average of the second measurement over the remaining cases. It is the ensemble average of the set of observables $\mathbf{A}$, at time t, given that their values at time zero are known to be $\mathbf{a}^o$. The function $\mathbf{a}(t, \mathbf{a}^o)$ defines what will prove to be a very useful equation of motion within the macrostate space.

10.5   MACROSCOPIC DETERMINISM

Suppose that the system under consideration is a macroscopic RC circuit and that the set of observables $\mathbf{A}(z)$ is simply the charge on the capacitor. Setting the value of Q certainly does not fix the immensely complicated microscopic state of the system. Thus the initial value of the charge $Q(0)$ does not exactly determine its subsequent values $Q(t)$. However, it is well known that, if a capacitor in an RC circuit is repeatedly charged to the same initial value and then allowed to discharge through the resistor, the variations in $Q(t)$ from one run to the next will be exceedingly small. For all practical purposes, the initial value of the observable Q determines its subsequent behavior. We say that the variable Q exhibits *macroscopic determinism*.

In general, for a particular energy E and a particular set of observables $\mathbf{A}(z)$, a large system is said to exhibit macroscopic determinism if almost all microstates on the energy surface that have the same initial values of the observables will give the ensemble-average time dependence with only very small errors. That is, for almost all z^o satisfying $H(z^o) = E$ and $\mathbf{A}(z^o) = \mathbf{a}^o$,

$$\mathbf{A}(T_t z^o) \approx \mathbf{a}(t, \mathbf{a}^o) \tag{10.20}$$

Clearly, a precise definition of macroscopic determinism would involve some sort of thermodynamic limit. However, the definition of the thermodynamic limit for time-dependent phenomena is much more complex and subtle than it is for equilibrium properties. Because we do not want to discuss that difficult question in this book, we will make no attempt to give a more logically precise definition of macroscopic determinism than has been given previously.

10.6   THE INCREASE OF ENTROPY

The concept of macroscopic determinism is useful in obtaining a qualitative understanding of why the entropy of a system, not initially in its equilibrium macrostate, monotonically increases as it approaches that state. We will treat the case in which there is only one relevant macroscopic observable, $A(z)$. The extension to any finite number of observables is straightforward. As usual, we define the entropy associated with a particular value of the observable by the integral

$$e^{S(a,E)} = \int \delta(a - A(z))\, \delta(E - H(z))\, d^{2K} z / N! h^K \tag{10.21}$$

The integral defines the $(2K - 2)$-dimensional "volume" of the set of points with the given values of the energy and the observable A. Let us assume that the system exhibits macroscopic determinism. Then, if we start the system in a microstate z^o that has an initial value $A(z^o) = a^o$ of the observable A, the subsequent behavior of $A(T_t z^o)$ is almost certain to be the same as the average behavior of all states with that initial value.

$$A(T_t z^o) = a(t, a^o) \tag{10.22}$$

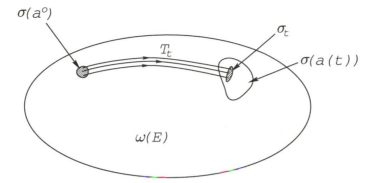

Fig. 10.3   The initial set $\sigma(a^o)$ goes to σ_t, which is a subset of $\sigma(a(t, a^o))$ if the system has macroscopic determinism.

What we intend to show is that

$$S(a^o, E) \leq S(a(t, a^o), E) \tag{10.23}$$

which means that the entropy function increases as the system proceeds toward its equilibrium state. Consider the set of points on the energy surface, $\omega(E)$, that have $A(z) = a^o$. Call that set $\sigma(a^o)$. [In general, for any a, we call $\sigma(a)$ the set of points on the energy surface that have $A(z) = a$.] Let the points of $\sigma(a^o)$ evolve according to Hamilton's equations for a time interval, t. The resulting set of points we will call σ_t. (See Fig. 10.3.)

In Exercise 3.10, a theorem was proved, called Liouville's theorem, that says that the volume of any $2K$-dimensional set of points in phase space remains constant if those points move according to Hamilton's equations. By an extension of Liouville's theorem, it is possible to prove that the $(2K - 2)$-dimensional volume of the set of points σ_t also remains constant as those points move according to Hamilton's equations. But the value of $S(a^o, E)$ is just the logarithm of the volume of $\sigma(a^o)$ divided by $N!h^K$.

$$S(a^o, E) = \log\left[V(\sigma(a^o))/N!h^K\right] \tag{10.24}$$

By the property of macroscopic determinism, almost every point in σ_t has $A(z) = a(t, a^o)$. Thus σ_t is a subset of $\sigma(a(t, a^o))$ and must therefore have a volume less than or equal to $V(\sigma(a(t, a^o)))$. By Liouville's theorem, $V(\sigma_t) = V(\sigma(a^o))$, and therefore

$$V(\sigma(a^o)) \leq V(\sigma(a(t, a^o))) \tag{10.25}$$

Taking the logarithm of this inequality gives the desired inequality for the entropy [Eq. (10.23)].

This pictorial argument is not by any means a rigorous mathematical proof of the increase of entropy theorem. It was only intended to give the reader a qualitative understanding of the physical principles behind the phenomenon of entropy increase. For systems that do not exhibit macroscopic determinism, there is no doubt that the entropy still increases, although our "proof" of that fact is not applicable.

10.7 THERMODYNAMIC FORCES

For each of the macroscopic variables a_i, we define a thermodynamically conjugate variable, that we call the corresponding thermodynamic force, as the derivative of the entropy with respect to that variable.

$$\beta_i = \frac{\partial S(\mathbf{a})}{\partial a_i} \tag{10.26}$$

Since the goal of our analysis is to obtain formulas for the thermodynamic fluctuations in the observables $A_1(z), \ldots, A_n(z)$, it is reasonable to assume that none of the chosen observables is a conserved quantity (which, by definition, would have zero fluctuation in an isolated system). For any nonconserved quantity, its equilibrium value is that value that maximizes the entropy function. Thus the equilibrium values of the variables $\mathbf{A}$ are found by solving the n equations

$$\beta_i = \frac{\partial S(\mathbf{a})}{\partial a_i} = 0 \tag{10.27}$$

That is, the equilibrium condition is that all components of the thermodynamic force be zero. The analogy with the equilibrium condition for a mechanical system explains why the β_i are called forces. The n-component vector field, $\vec{\beta}(\mathbf{a}) = (\beta_1, \ldots, \beta_n)$, may be pictured as an n-dimensional force field in the macrostate space. Because of the convexity property of the entropy, it has a unique equilibrium point. It is convenient to make a simple shift in the definition of each observable $A_i(z)$ by subtracting its constant equilibrium value. This will bring the equilibrium values of all the chosen observables to the origin in the macrostate space. That is, $\vec{\beta}(\mathbf{a}) = 0$ will occur at $\mathbf{a} = 0$.

10.8   THE FLUCTUATION-DISSIPATION THEOREM
The $n \times n$ time-dependent correlation matrix $C_{ij}(t)$ is defined as the microcanonical average of the product of observables, $A_i(T_t z) A_j(z)$. This can be transformed into an average within the macrostate space in the following way:

$$
\begin{aligned}
C_{ij}(t) &= \langle A_i(T_t z) A_j(z) \rangle \\
&= e^{-S^\circ} \int A_i(T_t z) A_j(z) \, \delta(E - H(z)) \, d^{2K}z / N! h^K \\
&= e^{-S^\circ} \int d^n \mathbf{a}^\circ \int \delta(\mathbf{a}^\circ - \mathbf{A}(z)) A_i(T_t z) A_j(z) \, \delta(E - H(z)) \, d^{2K}z / N! h^K \\
&= e^{-S^\circ} \int d^n \mathbf{a}^\circ a_j^\circ \int \delta(\mathbf{a}^\circ - \mathbf{A}(z)) A_i(T_t z) \, \delta(E - H(z)) \, d^{2K}z / N! h^K \\
&= \int e^{S(\mathbf{a}^\circ) - S^\circ} a_i(t, \mathbf{a}^\circ) a_j^\circ \, d^n \mathbf{a}^\circ \tag{10.28}
\end{aligned}
$$

The last line uses the definition of the macrostate equation of motion given in Eq. (10.19).

For a single system drawn from the microcanonical ensemble, if the initial values of all the chosen observables are their equilibrium values [that is, if $\mathbf{A}(z^\circ) = 0$], then, as a function of time, they will not remain exactly zero, but will execute small erratic fluctuations about their zero equilibrium values. However, if we take an ensemble average of the time dependence of all the points that have $\mathbf{A}(z^\circ) = 0$, then the random fluctuations will completely cancel and the value of the ensemble average will remain zero with no thermal fluctuations. This is expressed by the fact that $\mathbf{a}(t, 0) = 0$ for all t. In general, although $\mathbf{A}(t) = \mathbf{A}(z(t, z^\circ))$ is a very irregular function when viewed with high resolution, the macrostate equation of motion $\mathbf{a}(t, \mathbf{a}^o)$ is quite smooth and regular. Therefore, close to $\mathbf{a}^o = 0$ it would be reasonable to expand the function $\mathbf{a}(t, \mathbf{a}^o)$ to first order in $\mathbf{a}^o$. Such an expansion would have the form

$$a_i(t, \mathbf{a}^o) = \sum_k G_{ik}(t)a_k^o \tag{10.29}$$

However, at $\mathbf{a}^o = 0$, the thermodynamic forces β_k are also zero. We will see that it is more convenient near equilibrium to expand $a_i(t, \mathbf{a}^o)$ to first order in the thermodynamic forces $\beta_k(\mathbf{a}^o)$ rather than in the coordinates a_k^o directly. That expansion would be of the form

$$a_i(t, \mathbf{a}^o) = \sum_k F_{ik}(t)\beta_k(\mathbf{a}^o) \tag{10.30}$$

Since thermal fluctuations are generally quite small, it should be an excellent approximation to use this linearized macrostate equation of motion in the last line of Eq. (10.28) for $C_{ij}(t)$. When that is done, one can, by a simple partial integration, obtain an important identity, known as the Fluctuation-Dissipation Theorem.

$$
\begin{aligned}
C_{ij}(t) &= \int e^{S(\mathbf{a}^o)-S^o} \sum_k F_{jk}(t)\beta_k(\mathbf{a}^o)a_j^o \, d^n\mathbf{a}^o \\
&= \sum_k F_{ik}(t) \int e^{S(\mathbf{a}^o)-S^o} \frac{\partial S(\mathbf{a}^o)}{\partial a_k^o} a_j^o \, d^n\mathbf{a}^o \\
&= \sum_k F_{ik}(t) \int \frac{\partial}{\partial a_k^o}\left(e^{S(\mathbf{a}^o)-S^o}\right) a_j^o \, d^n\mathbf{a}^o \\
&= -\sum_k F_{ik}(t)\delta_{jk} \int e^{S(\mathbf{a}^o)-S^o} \, d^n\mathbf{a}^o \\
&= -F_{ij}(t)
\end{aligned} \tag{10.31}
$$

In going from the third to the fourth line, we have used the fact that $\exp[S(\mathbf{a}^o) - S^o]$ must go to zero for large $|\mathbf{a}^o|$ since the integral of that function over all $\mathbf{a}^o$ is equal to one. This theorem gives a close relationship between the spontaneous thermal fluctuations, described by the correlation matrix $C_{ij}(t)$ and the macroscopic dynamics of the system, described by the coefficients $F_{ij}(t)$ in the linearized macroscopic equation of motion. The macroscopic equation of motion describes the rate at which any initial deviation from equilibrium decays (that is, dissipates) as the system returns to equilibrium—hence, the name "fluctuation-dissipation theorem." The content of the theorem will become clearer when, in the next section, it is applied to the dynamics of an *LCR* circuit. When the definitions of the correlation function and the macroscopic equation of motion are properly extended to a quantum mechanical system, the theorem can be shown to be valid, without change, in quantum mechanics also (see Exercise 10.8).

10.9   CORRELATION FUNCTIONS OF THE *LCR* CIRCUIT

In applying the fluctuation-dissipation theorem to an *LCR* circuit, the macroscopic observables to be considered will naturally be the charge Q and the current I. One must define precise observables, $A_Q(z)$ and $A_I(z)$, corresponding to Q and I. There are many ways of doing this, and the final results are not sensitive to the details of how A_Q and A_I are defined, but the fact that they should have *some* precise

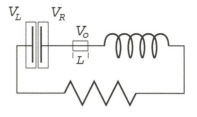

Fig. 10.4 The volumes V_L, V_R, and V_o used to define A_Q and A_I.

definition is important for understanding the theorem. In Fig. 10.4 are shown three volumes: V_L and V_R surround the two capacitor plates, and V_o includes a section of the wire. We define A_Q as $(Q_R - Q_L)/2$, where Q_R and Q_L are the total electric charges in V_R and V_L, respectively. A_I is defined as

$$A_I = \frac{1}{L} \sum_{V_o} q_i v_{ix} \tag{10.32}$$

where L is the length of V_o , as shown in the figure, and q_i and v_{ix} are the charge and x component of velocity of the ith particle. The sum is taken over all the particles in V_o. With these definitions, it is clear that $Q(t)$ and $I(t)$ will be, if measured with arbitrary precision and time resolution, extremely erratic functions. Each time that an electron enters one of the volumes, Q or I will change discontinuously. Certainly $I(t)$ will not be equal to dQ/dt. In fact, dQ/dt will not be well defined.

The next question that must be explored is how to interpret the circuit equations for the system. It is obvious that the circuit equations do not describe the detailed behavior of the very complex functions, $Q(t)$ and $I(t)$. The natural interpretation is that the circuit equations describe the curves that would result if $Q(t)$ and $I(t)$ were somehow "smoothed out." But that smoothing out is exactly what happens when one transforms from the detailed behavior of a single system to the ensemble-averaged dynamics described by the macroscopic equation of motion, $\mathbf{a}(t, \mathbf{a}^o)$. It is the smooth function $\mathbf{a}(t, \mathbf{a}^o)$ that we will assume is described by the circuit equations. In fact, if this function were not described by the circuit equations, then there would exist other equations that would, on an average, better describe the time dependence of the system than the circuit equations. The ensemble-averaged quantities a_1 and a_2, corresponding to $A_Q(z)$ and $A_I(z)$, will be written as $q(t)$ and $i(t)$.

The circuit equations are

$$L\frac{di}{dt} + Ri + \frac{1}{C}q = 0 \tag{10.33}$$

and

$$\frac{dq}{dt} = i \tag{10.34}$$

Assuming a solution of the form

$$q(t) = e^{-\gamma t}(A \cos \omega t + B \sin \omega t) \tag{10.35}$$

one obtains, after some algebra, the relations

$$\gamma = R/2L \text{ and } \omega = (\omega_o^2 - \gamma^2)^{1/2} \tag{10.36}$$

for the decay constant and the angular frequency, where $\omega_o^2 = 1/LC$. In order to be sure that ω is real, we must assume that $\gamma < \omega_o$. This is called the *underdamped oscillator condition.* The opposite case, $\gamma > \omega_o$, will be done in the exercises. The initial conditions give A and B in terms of q^o and i^o, the initial values of $q(t)$ and $i(t)$.

$$A = q^o \quad \text{and} \quad B = \frac{\gamma}{\omega}q^o + \frac{1}{\omega}i^o \tag{10.37}$$

Thus the functions $q(t, q^o, i^o)$ and $i(t, q^o, i^o)$ are

$$q(t) = e^{-\gamma t}\left(\cos\omega t + \frac{\gamma}{\omega}\sin\omega t\right)q^o + \left(\frac{1}{\omega}e^{-\gamma t}\sin\omega t\right)i^o \tag{10.38}$$

and

$$i(t) = -\left(\frac{\omega_o^2}{\omega}e^{-\gamma t}\sin\omega t\right)q^o + e^{-\gamma t}\left(\cos\omega t - \frac{\gamma}{\omega}\sin\omega t\right)i^o \tag{10.39}$$

One can evaluate the thermodynamic forces β_q^o and β_i^o, corresponding to q^o and i^o, from the entropy function given in Eq. (10.7).

$$S(E, q, i) = S_R\left(E - \frac{1}{2C}q^2 - \frac{L}{2}i^2\right) \tag{10.40}$$

For the case of spontaneous thermal fluctuations, $q^2/2C$ and $Li^2/2$ are each of order kT, while the complete thermal energy E is very much larger. Thus we can evaluate the derivatives to first order in q^o and i^o.

$$\beta_q^o = \frac{\partial S}{\partial q^o} = -\frac{q^o}{CkT} \tag{10.41}$$

and

$$\beta_i^o = \frac{\partial S}{\partial i^o} = -\frac{Li^o}{kT} \tag{10.42}$$

In order to put the circuit equations in the standard form given in Eq. (10.30), we must write $q(t)$ and $i(t)$ in term of β_q^o and β_i^o. Using the substitutions $q^o = -kTC\beta_q^o$ and $i^o = -(kT/L)\beta_i^o$ in Eqs. 10.38 and 10.39, one finds that

$$q(t) = F_{qq}(t)\beta_q^o + F_{qi}(t)\beta_i^o \tag{10.43}$$

and

$$i(t) = F_{iq}(t)\beta_q^o + F_{ii}(t)\beta_i^o \tag{10.44}$$

where

$$F_{qq} = -kTCe^{-\gamma t}\left(\cos\omega t + \frac{\gamma}{\omega}\sin\omega t\right) \tag{10.45}$$

$$F_{ii} = -\frac{kT}{L}e^{-\gamma t}\left(\cos\omega t - \frac{\gamma}{\omega}\sin\omega t\right) \tag{10.46}$$

and

$$F_{qi} = -F_{iq} = -\frac{kTC}{\omega}e^{-\gamma t}\sin\omega t \tag{10.47}$$

The fact that $F_{qi} = -F_{iq}$ is no accident. It is an example of the *Onsager Reciprocity Theorem*, which will be proven shortly. These formulas, together with the Fluctuation-Dissipation Theorem, give us the time-dependent correlation functions, $\langle Q(0)Q(t)\rangle$, $\langle Q(0)I(t)\rangle$, $\langle I(0)Q(t)\rangle$, and $\langle I(0)I(t)\rangle$ in terms of the easily measured circuit parameters, L, C, and R. They show us that the seemingly random and structureless thermal fluctuation functions actually contain hidden dynamical relations. Notice that, because of the exponential decay factor, the values of $Q(t)$ and

$I(t)$ become statistically independent of $Q(0)$ and $I(0)$ for large t. Notice also that $Q(0)$ and $I(0)$ are statistically independent. That is, the thermal fluctuations in the charge at any given time are independent of the fluctuations in the current at the same time (but not independent of the fluctuations in the current at a different time).

10.10   THE RECIPROCITY THEOREM

The Reciprocity Theorem, first proven by Lars Onsager, shows that the matrix of coefficients $F_{ij}(t)$ that appears in the linearized macroscopic equations of motion must have certain symmetry properties. The source of the symmetries is the well-known symmetry of mechanics under the operation of time reversal. If the trajectory $(\mathbf{r}_1(t), \ldots, \mathbf{r}_N(t))$ for an N-particle system satisfies Newton's equations of motion during the time interval $-T < t < T$, then the "time-reversed" trajactory, $(\tilde{\mathbf{r}}_1(t), \ldots, \tilde{\mathbf{r}}_N(t)) = (\mathbf{r}_1(-t), \ldots, \mathbf{r}_N(-t))$ (which just describes the particles running backwards in time over the same paths with the negative of their original velocities), will also satisfy Newton's equations. When expressed in terms of Hamilton's equations, the time-reversal symmetry transformation must take into account the fact that the momenta are reversed on the time-reversed path. Thus it says that, if the phase-space trajectory $(x(t), p(t))$ is a solution of Hamilton's equations, then the time-reversed trajectory, $(\tilde{x}(t), \tilde{p}(t)) = (x(-t), -p(-t))$, is also a solution of the same equations.

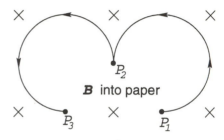

Fig. 10.5   If, after a positively charged particle moves from P_1 to P_2, its velocity is reversed, it does not return to P_1, but rather, moves to P_3. In the presence of a magnetic field, the orbits are not time reversible.

The time-reversal symmetry property does not hold for charged particles moving in an external magnetic field. (See Fig. 10.5.) This lack of time-reversal symmetry in magnetic fields has a simple explanation. Suppose that the external field is being produced by an electromagnet. If we extend the definition of our system to include the charged particles within the coils of the electromagnet, then the extended system will exhibit time-reversal symmetry. However, when we look at the time-reversed motion, because of a reversal of the currents in the coils of the electromagnet, the magnetic field will also be reversed. For the case of charged particles moving in an external magnetic field, the correct formulation of the time-reversal symmetry property of mechanics is to say that, if $(x(t), p(t))$ is a solution of Hamilton's equations with the magnetic field $\mathbf{B}(\mathbf{r})$, then $(x(-t), -p(-t))$ is a solution of Hamilton's equations with the magnetic field $\tilde{\mathbf{B}}(\mathbf{r}) = -\mathbf{B}(\mathbf{r})$. For now, we will exclude magnetic fields from our considerations while we prove the Onsager theorem. In an exercise in the supplement to this chapter, we will extend the theorem to include systems with external fields.

We will use the same notation as was used in the proof of the Fluctuation-Dissipation Theorem. The time-dependent correlation matrix for n observables, $A_1(z), \ldots, A_n(z)$, is defined as the microcanonical expectation value

$$C_{ij}(t) = \langle A_i(T_t z) A_j(z) \rangle$$
$$= e^{-S^\circ} \int A_i(T_t z) A_j(z) \, \delta(E - H(z)) \, d^{2K} z / N! h^K \qquad (10.48)$$

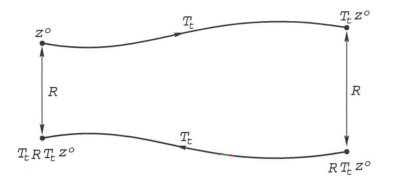

Fig. 10.6 Allowing the point z^o to evolve for a time interval t, then reversing all momenta and allowing it to evolve for another time interval t, will get the system back to its initial state with all momenta reversed.

We define a momentum-reversal operator R, which, like the time evolution operator, takes any point in phase space into another point in phase space. In particular, if $z = (x_1, \ldots, x_K, p_1, \ldots, p_K)$, then

$$Rz = (x_1, \ldots, x_K, -p_1, \ldots, -p_K) \tag{10.49}$$

The time-reversal symmetry property of mechanics implies that, if we start a system in a state z^o, let it run for a time interval t, then reverse all of the momenta and again let it run for a time interval t, it will be left in the state Rz^o. (See Fig. 10.6.)
This says that, for any point z^o,

$$T_t R T_t z^o = R z^o \tag{10.50}$$

The inverse of the time evolution operator T_t is an operator that we write as T_{-t}. If $z(t, z^o)$ is a solution of Hamilton's equations for all t, then $T_{-t} z^o \equiv z(-t, z^o)$. For any z^o, $T_{-t} z^o$ gives the state that the system must have been in at time $-t$ in order to be in the state z^o at time zero. It is easy to confirm that T_{-t} is the inverse of T_t by noting that, for any z^o,

$$T_{-t} T_t z^o = T_{-t} z(t, z^o) = z^o \tag{10.51}$$

Multiplying Eq. (10.50) by T_{-t} on the left, we see that, for any z^o,

$$R T_t z^o = T_{-t} R z^o \tag{10.52}$$

With this basic identity, we are ready to prove the theorem. In Eq. (10.48), which defines the correlation matrix element $C_{ij}(t)$ we make a transformation of variables, $z \to \tilde{z}$, where the new variables are defined in terms of the old by

$$\tilde{z}(z) = R T_t z \tag{10.53}$$

The inverse transformation that gives z in terms of $\tilde{z}$ is $z = T_{-t} R \tilde{z}$. Also, because $R^2 = I$, where I is the identity operator, $R\tilde{z} = T_t z$. Thus

$$C_{ij}(t) = e^{-S^o} \int A_i(R\tilde{z}) A_j(T_{-t} R \tilde{z}) \, \delta(E - H(T_{-t} R \tilde{z})) \, d^{2K}\tilde{z}/N!h^K \tag{10.54}$$

In writing Eq. (10.54), we have used a property of the transformation that we have yet to show, namely that the Jacobian of the transformation is equal to one. Whenever a transformation of variables is made in an integral, one must replace the original volume element $d^{2K}z$ by the expression $J\,d^{2K}\tilde{z}$, where the Jacobian J is the ratio of an infinitesimal volume in the old variables to the value of the transformed volume. That is, if we take some very small volume V made up of the set of points S in phase space, and we transform each point in S by the transfomation $z \to \tilde{z} = R\,T_t z$ to get a new set of points $\tilde{S}$ with a new volume $\tilde{V}$, then $J = V/\tilde{V}$. In Exercise 3.10 it was shown that the motion of a set of points in phase space satisfying Hamilton's equations does not change the volume of the set of points (Liouville's Theorem). Thus the transformation from z to $T_t z$ would cause no change in volume. But the momentum-reversal operation R can be pictured as a sequence of K ordinary reflections, the first across the plane $p_1 = 0$, the second across the plane $p_2 = 0$, etc. Each reflection has no effect on the volume of a set of points. Therefore, $\tilde{V} = V$, and the Jacobian, being equal to one, can be omitted.

Using the identity given in Eq. (10.52), we can replace the term $A_i(T_{-t}R\tilde{z})$ in Eq. (10.54) by $A_i(RT_t\tilde{z})$. The Hamiltonian function is quadratic in the momentum variables, and thus $H(Rz) = H(z)$ for any z. Also the Hamiltonian is a conserved quantity, and thus $H(T_t z) = H(z)$ for any values of t and z. These two facts imply that

$$H(T_{-t}R\tilde{z}) = H(R\tilde{z}) = H(\tilde{z}) \tag{10.55}$$

Therefore, we can write $C_{ij}(t)$ in the form

$$C_{ij}(t) = e^{-S^o} \int A_i(RT_t\tilde{z})A_j(R\tilde{z})\,\delta(E - H(\tilde{z}))\,d^{2K}\tilde{z}/N!h^K \tag{10.56}$$

Now we will assume that each of the n observables chosen has either odd or even parity under a reversal of momentum. That is, that

$$A_i(Rz) = \epsilon_i A(z) \tag{10.57}$$

where ϵ_i is either $+1$ or -1. In reference to the case of the LCR circuit, the variable $A_Q(z)$ has even parity while the variable $A_I(z)$ has odd parity. A little thought will convince the reader that almost any "reasonable" observable will have either odd or even parity under momentum reversal. With that assumption, one obtains the Onsager symmetry relation

$$\begin{aligned} C_{ij}(t) &= \epsilon_i\epsilon_j e^{-S^o} \int A_j(T_t\tilde{z})A_i(\tilde{z})\,\delta(E - H(\tilde{z}))\,d^{2K}\tilde{z}/N!h^K \\ &= \epsilon_i\epsilon_j C_{ji}(t) \end{aligned} \tag{10.58}$$

Using the Fluctuation-Dissipation Theorem, this can be written as an equivalent symmetry relation for the coefficients $F_{ij}(t)$ in the linearized macroscopic equations of motion.

$$F_{ij}(t) = \epsilon_i\epsilon_j F_{ji}(t) \tag{10.59}$$

10.11   THERMOELECTRIC EFFECTS

The Reciprocity Theorem has important consequences for any system in which there are simultaneous flows of several conserved quantities. An excellent example is the simultaneous flow of electric charge and energy through a body within which a temperature gradient and an electric field both exist. It is an experimental fact that the electric current is affected, not only by the electric field, but also by the temperature gradient. Conversely, the heat flux is affected by both the temperature gradient and the electric field. Such interaction effects between charge and energy flow are called *thermoelectric effects*. If it is assumed that the flow equations are linear (which in most cases is an excellent approximation), then, with two flows and two force fields, there are four linear coefficients in the equations. The Reciprocity Theorem allows one of the coefficients to be related to another, reducing the number of independent coefficients to three.

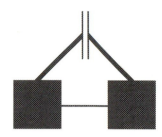

Fig. 10.7   A capacitor, two reservoirs, and a thin wire are all made of one substance.

The system that we will consider is shown in Fig. 10.7. It consists of a capacitor, the plates of which are strongly connected to two thermal reservoirs. Between the reservoirs is a fine wire. Everything is constructed of one substance. Because of the thinness of the wire, the time required for the system to transfer heat or charge from one reservoir to the other is much larger than the time required for either of the reservoirs to come into internal equilibrium. The system is isolated. At complete equilibrium, the charge on the capacitor would be zero and the energies of the left and right reservoirs, which we assume to be identical, would both be E_o. The two macroscopic observables, defining nonequilibrium states of the system, are the capacitor charge Q and the energy difference

$$\varepsilon = (E_L - E_R)/2 \tag{10.60}$$

where E_L and E_R are the actual energies of the left and right reservoirs. Since we are interested only in states of total energy $2E_o$, the energy conservation equation states that

$$E_L + E_R + \frac{Q^2}{2C} = 2E_o \tag{10.61}$$

These two equations can easily be solved for E_L and E_R.

$$E_L = E_o + \varepsilon - \frac{Q^2}{4C} \quad \text{and} \quad E_R = E_o - \varepsilon - \frac{Q^2}{4C} \tag{10.62}$$

Assuming that the entropy of the thin wire is negligible, the entropy of the system is

$$S = S_L(E_o + \varepsilon - Q^2/4C) + S_R(E_o - \varepsilon - Q^2/4C) \tag{10.63}$$

where S_L and S_R are actually the same function. The thermodynamic forces conjugate to Q and ε are

$$\beta_Q = \frac{\partial S}{\partial Q} = \tfrac{1}{2}(\beta_L + \beta_R)\frac{Q}{C} \approx \frac{V}{kT} \tag{10.64}$$

and

$$\beta_\varepsilon = \frac{\partial S}{\partial \varepsilon} = \beta_L - \beta_R \approx -\frac{\Delta T}{kT^2} \tag{10.65}$$

where, in the last step of each line, we have neglected terms of higher than first order in $\Delta T = T_L - T_R$. This "near-equilibrium" approximation is completely adequate for analyzing the linearized flow equations. In the last step of Eq. (10.64), V is the potential difference between the two ends of the wire.

The charge and energy currents in the wire, I_Q and I_E, are assumed to be linear functions of the electric potential and temperature differences.

$$I_Q = AV + a\,\Delta T \tag{10.66}$$

and

$$I_E = B\,\Delta T + bV \tag{10.67}$$

Using Eqs. (10.64) and (10.65) to eliminate ΔT and V in favor of the thermodynamic forces, and using the facts that $\dot{Q} = -I_Q$ and $\dot{\varepsilon} = -I_E$, the evolution equations for Q and ε can be written in the form,

$$\dot{Q} = -kTA\beta_Q + kT^2 a\beta_\varepsilon \tag{10.68}$$

and

$$\dot{\varepsilon} = -kTb\beta_Q + kT^2 B\beta_\varepsilon \tag{10.69}$$

Although these equations are not in exactly the standard form, given in Eq. (10.30), it is not difficult to prove that the Onsager theorem implies that the matrix of coefficients on the right hand side of these equations must be symmetric. (See Problem 10.3.) Thus, the off-diagonal terms, $kT^2 a$ and $-kTb$, must be equal, showing that

$$b = -Ta \tag{10.70}$$

We will now consider the question of how the three coefficients A, B, and a are related to standard experimental quantities. Using the Onsager relation to eliminate the parameter b, we can rewrite Eqs. (10.66) and (10.67) as

$$I_Q = AV + a\,\Delta T \quad \text{and} \quad I_E = B\,\Delta T - TaV \tag{10.71}$$

The electrical resistance of the wire is defined as the ratio of V to I_Q, measured at uniform temperature ($\Delta T = 0$). Thus $A = 1/R$. The thermal conductivity of the wire, κ, is defined as the ratio of I_E to ΔT, measured with no electric current flowing, that is, with the wire electrically, but not thermally, insulated. But $I_Q = 0$ implies that a potential difference exists between the ends of the wire, given by $AV + a\,\Delta T = 0$, or

$$V = -aR\,\Delta T \tag{10.72}$$

In that case, Eq. (10.71) becomes

$$I_E = (B + Ra^2 T)\,\Delta T \tag{10.73}$$

which shows that

$$\kappa = B + Ra^2 T \tag{10.74}$$

Clearly, we need one more parameter in order to determine B and a separately. The third standard parameter, called the *thermoelectric power* of the material composing

Fig. 10.8   If two ends of a wire are held at different temperature, a potential difference develops that is proportional to the temperature difference.

the wire, is related to the just-mentioned potential difference that exists between the ends of an insulated wire if those ends are kept at different temperatures.

In Fig. 10.8 is shown an electrically insulated wire whose ends are in temperature reservoirs at the slightly different temperatures T_1 and T_2. It is found that a potential difference is produced between the ends that is proportional to ΔT and independent of the dimensions of the wire. The ratio of $V = \phi_2 - \phi_1$ to $T_1 - T_2 = -\Delta T$ is called the thermoelectric power T_p of the substance. It is clear from Eq. (10.72) that

$$T_p = -\frac{V}{\Delta T} = Ra \qquad (10.75)$$

which allows a determination of all the linear coefficients in terms of R, κ, and T_p.

Fig. 10.9   A potential difference also develops in the voltmeter wires, obscuring the reading of the potential difference that we are trying to measure.

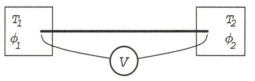

A certain serious problem occurs when one attempts to measure the potential difference indicated in Fig. 10.8, in order to determine the thermoelectric power of a substance. The natural thing to do would be to connect a voltmeter across the leads, as shown in Fig. 10.9. However, the voltmeter leads are necessarily made of *some* substance that, when placed in two reservoirs at different temperatures, would exhibit its own thermoelectric effect, obviously interfering with the measurement. If one knows the thermoelectric power of any one substance, then one can correct for this interfering effect (see Problem 10.4) and measure the same parameter for any other substance. A great help is the fact that the thermoelectric power of any superconductor is exactly zero. Thus at all temperatures below the highest attainable superconducting temperature, accurate measurements of T_p are relatively easy. Further investigation of this problem may be pursued in the references.

10.12   THE RESPONSE FUNCTION

The Fluctuation-Dissipation Theorem presented in Section 10.7 relates the correlation function for thermal fluctuations to the linearized macroscopic equation of motion in the same system. Another form of the same theorem relates the correlation function to a slightly different quantity, namely, the *linear response function* of the system.

The response function of a system describes the effect on an observable B when the system is "perturbed" by another observable A. A perfect example of a response function is the electric susceptibility. If an initially unpolarized substance is placed in a weak electric field, then it developes a polarization proportional to the field strength. In this case, the perturbing observable A is the electric field and the perturbed observable B is the polarization. The response function (that is, the susceptibility) gives the ratio of B to A. Because in this chapter we are interested in studying time-dependent phenomena, we will have to generalize the definition of a response function to include the time-dependent response of a system to a time-dependent perturbation. Now let us define a response function more precisely.

We consider a canonical ensemble for a quantum mechanical system with a Hamiltonian operator H_o. We choose two time-independent Hermitian operators A and B. By "perturbing the system with observable A," we mean that we change the Hamiltonian from the time-independent operator H_o, to a time-dependent Hamiltonian

$$H(t) = H_o + \epsilon(t)A \tag{10.76}$$

where $\epsilon(t)$ is some small function that we will assume is zero for $t < 0$. For example, this is what would happen if we took a system that was initially at equilibrium without a magnetic field and began to turn on a magnetic field. In general, we make some small change in the Hamiltonian, starting at time zero. It is convenient to choose the operator B so that in the initial equilibrium state its expectation value is zero. At some time, $t > 0$, we measure the average value of B for the system. This we call $B(t)$. For very small values of $\epsilon(t)$, the value of $B(t)$ will depend linearly on the values of $\epsilon(t')$ for all $t' < t$. That is, $B(t)$ will be given by an equation of the form

$$B(t) = \int_{-\infty}^{t} R_{BA}(t - t')\epsilon(t')\, dt' \tag{10.77}$$

This formula defines the time-dependent response function $R_{BA}(t)$. Clearly, $R_{BA}(t - t')$ is zero for $t < t'$. The system will not respond before it is disturbed. A familiar example of a time-dependent response function is the admittance function discussed in Exercise 10.2.

With the same two observables A and B we can define a time-dependent correlation function for the fluctuations in the system when it is in equilibrium at temperature T.

$$C_{BA}(t) \equiv \langle B(t)A \rangle \tag{10.78}$$

where the average is taken over a quantum mechanical canonical ensemble. The quantum mechanical Fluctuation-Dissipation Theorem is a simple relationship between the Fourier transforms of $R_{BA}(t)$ and $C_{BA}(t)$. For any function $F(t)$ the Fourier transform $\tilde{F}(\omega)$ is defined by the equation

$$\tilde{F}(\omega) = \int_{-\infty}^{\infty} F(t)e^{i\omega t}\, dt \tag{10.79}$$

The theorem, first proved by H. B. Callen and T. A. Welton, is that, for any two observables A and B,

$$\tilde{R}_{BA}(\omega) = \frac{1 - e^{-\beta\hbar\omega}}{i\hbar}\tilde{C}_{BA}(\omega) \tag{10.80}$$

The proof of the theorem requires some fairly fancy quantum mechanical manipulations. It will therefore be omitted. However, certain basic quantum mechanical facts are necessary in order for the reader to appreciate the meaning of the theorem. Certainly, if the theorem is to have any significance, one must precisely define the expectation value $\langle B(t)A \rangle$. For a classical system, if observables A and B are represented by the phase-space functions $A(z)$ and $B(z)$ and $P(z)$ is the ensemble probability density function, then

$$\langle B(t)A \rangle \equiv \int B(T_t z)A(z)P(z)\, dz \tag{10.81}$$

For a quantum mechanical system, $A(z)$ is replaced by some Hermitian operator A. The more difficult question is: What is the operator equivalent of $B(T_t z)$? Suppose we have a machine with a button that measures $B(z)$ whenever we push the button. How do we make a machine to measure the observable $B(T_t z)$? We construct a new machine with a button. When we push the button on our new machine, it waits for a time interval t and then pushes the button on the old machine. In other words, for any state z, the observable $B(T_t z)$ is defined to have, *now*, the value that $B(z)$ will have in t seconds from now. In Problem 10.9, the reader is asked to show that, for any quantum mechanical operator B, the operator that is measured by the composite machine, which first waits for t seconds and then measures B, is

$$B(t) = e^{iHt/\hbar} B e^{-iHt/\hbar} \tag{10.82}$$

Therefore, for a quantum mechanical system in a canonical ensemble given any two Hermitian operators A and B we define $C_{BA}(t)$ by

$$
\begin{aligned}
C_{BA}(t) &= \langle B(t)A \rangle \\
&= \sum_n e^{-\beta E_n} \langle \phi_n | B(t)A | \phi_n \rangle / Z
\end{aligned}
\tag{10.83}
$$

where $B(t)$ has been defined and ϕ_n is the nth energy eigenfunction. $C_{BA}(t)$ is the expectation value of the product of A, now, and B, in t seconds.

The response function $R_{BA}(t)$ can also be written as an expectation value of operators. In the exercises, it is shown that

$$
\begin{aligned}
i\hbar R_{BA}(t) &= \langle [B(t), A] \rangle \\
&= \sum_n e^{-\beta E_n} \langle \phi_n | [B(t), A] | \phi_n \rangle / Z
\end{aligned}
\tag{10.84}
$$

where $[B(t), A] = B(t)A - AB(t)$ is the usual commutator of the two operators. Comparing the operator representations of C_{BA} and R_{BA}, it is not surprising that the two functions are simply related. In the exercises it is shown that, converting from the commutator $[B(t), A]$ to the simple product $B(t)A$, gives the factor of $1 - e^{-\beta \hbar \omega}$ in the Fourier transform.

10.13 BOLTZMANN'S EQUATION

One of the most important developments in the history of statistical mechanics was the derivation by Ludwig Boltzmann of an integrodifferential equation that describes the approach to equilibrium of the one-particle phase-space density function $F(\mathbf{r}, \mathbf{p})$ of a dilute gas. A number of fundamental results flowed directly from Boltzmann's equation.

1. It could be shown that the only time-independent solution of the equation is the Maxwell–Boltzmann distribution $F(\mathbf{r}, \mathbf{p}) = C \exp\{-[p^2/2m + U(\mathbf{r})]/kT\}$. (Maxwell himself had never given a completely satisfactory derivation of his velocity distribution function.)
2. It was possible to define an entropy function with the characteristics that:
   a. The entropy function is equal to the logarithm of the phase-space volume available to a system with a given phase-space density.
   b. The entropy increases steadily until the phase-space density function becomes equal to the Maxwell–Boltzmann distribution.
   c. At equilibrium, the Boltzmann entropy function is equal to the known thermodynamic entropy of an ideal gas.

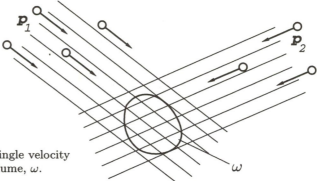

Fig. 10.10   Two dilute single velocity beams intersect within a volume, ω.

For simplicity, only the case of a spatially uniform gas will be considered here. The phase-space density function then becomes independent of $\mathbf{r}$ and is simply the momentum distribution function $F(\mathbf{p})$. In preparation for deriving Boltzmann's equation, we will look at certain general characteristics of two-particle scattering in dilute, intersecting beams of point particles.* We consider a region of volume ω, through which two very dilute single velocity beams of identical particles are passing. (See Fig. 10.10.) A single velocity beam is a collection of particles, all moving with the same velocity, but randomly distributed with a density that is uniform over the volume ω. The particles interact with a two-body force that is derivable from a potential $\phi(r_{ij})$, where $r_{ij} = |\mathbf{r}_i - \mathbf{r}_j|$. We assume that the potential, and therefore also the force, vanishes whenever r_{ij} is larger than some interaction distance d. We also assume that, in both beams, the average distance between nearest-neighbor particles is much larger than the interaction distance. We say that a collision has occurred whenever two particles (one from each beam) come closer than d. We want to calculate the rate at which certain special types of collisions occur within the volume ω. Because of the extreme diluteness of the beams, collisions are infrequent events and we can ignore, over the volume ω, the reduction in density of the beams due to particles being lost by collisions. The collection of particles that have momentum $\mathbf{p}_1 = m\mathbf{v}_1$ we call beam 1; the other set, of momentum, $\mathbf{p}_2 = m\mathbf{v}_2$, we call beam 2. The two beams have respective densities, n_1 and n_2. In any particular collision, a particle from beam 1 has its momentum altered from $\mathbf{p}_1$ to some value $\mathbf{p}_1'$ while its scattering partner has its momentum changed from $\mathbf{p}_2$ to $\mathbf{p}_2'$. Although $\mathbf{p}_1$ and $\mathbf{p}_2$ are given, $\mathbf{p}_1'$ and $\mathbf{p}_2'$ can take some continuous range of possible values in different collisions. It is a restricted range—for example, it is not possible for $p_1'^2/2m$ or $p_2'^2/2m$ to be greater than the total kinetic energy of the initial pair, $(p_1^2 + p_2^2)/2m$.

We choose some fixed momentum vector $\mathbf{p}_1'$ and some infinitesimal range $d^3\mathbf{p}_1'$ centered at $\mathbf{p}_1'$. We now ask: What is the rate at which collisions occur within ω, in which the final momentum of the particle from beam 1 falls within the range $d^3\mathbf{p}_1'$? Following the terminology of Boltzmann, we call such collisions "special collisions." The rate will be proportional to $d^3\mathbf{p}_1'$. Note that there is no need to specify the final momentum of the particle from beam 2. If $\mathbf{p}_1$, $\mathbf{p}_2$, and $\mathbf{p}_1'$ are given, then the momentum conservation law guarantees that $\mathbf{p}_2' = \mathbf{p}_1 + \mathbf{p}_2 - \mathbf{p}_1'$. Since the particles in beam 1 are statistically independent, the rate of special collisions within ω is proportional to the number of beam 1 particles within that volume, namely ωn_1.

---

*At this point it would be helpful for the reader to read or review Exercises 2.9 through 2.17.

The probability that a beam 1 particle within ω will have a special collision during a time interval Δt will obviously be doubled if we double the number of particles in beam 2. That is, the rate is also proportional to n_2. The rate of special collisions can therefore be written in the form

$$\text{rate of special collisions} = \omega n_1 n_2 R(\mathbf{p}_1, \mathbf{p}_2, \mathbf{p}_1')\, d^3\mathbf{p}_1' \qquad (10.85)$$

where $R(\mathbf{p}_1, \mathbf{p}_2, \mathbf{p}_1')\, d^3\mathbf{p}_1'$ is the rate of special collisions within a unit volume for beams of unit density. (If the reader is bothered by the sheer absurdity of considering an atomic beam with a density of one atom per cubic meter, he or she may imagine working in some system of units in which a unit density beam is a more feasible object.) Equation (10.85) has the disadvantageous property that it is not symmetric with respect to particles in the two beams, although the basic scattering process certainly is. In order to write it in a symmetrical fashion, we use the momentum conservation law, and define a function of four momentum variables, $P(\mathbf{p}_1, \mathbf{p}_2|\mathbf{p}_1', \mathbf{p}_2')$, by

$$P(\mathbf{p}_1, \mathbf{p}_2|\mathbf{p}_1', \mathbf{p}_2') = R(\mathbf{p}_1, \mathbf{p}_2, \mathbf{p}_1')\delta(\mathbf{p}_1 + \mathbf{p}_2 - \mathbf{p}_1' - \mathbf{p}_2') \qquad (10.86)$$

Then $P(\mathbf{p}_1, \mathbf{p}_2|\mathbf{p}_1', \mathbf{p}_2')d^3\mathbf{p}_1' d^3\mathbf{p}_2'$ is the rate at which collisions take place within a unit volume that send the momentum of a particle from beam 1 into $d^3\mathbf{p}_1'$ and that of a particle from beam 2 into $d^3\mathbf{p}_2'$.

It is possible to write $P(\mathbf{p}_1, \mathbf{p}_2|\mathbf{p}_1', \mathbf{p}_2')$ in terms of the differential scattering cross section for the two-particle scattering process. If we intended to use the Boltzmann equation for making practical calculations, such as calculating the viscosity or heat conductivity of a gas of particles with known interparticle interactions, then it would be advisable to write $P(\mathbf{p}_1, \mathbf{p}_2|\mathbf{p}_1', \mathbf{p}_2')$ in that way. No such calculations will be attempted in this book, although they are very important in their practical results and in verifying the basic correctness of the Boltzmann theory. Instead, we will concentrate our attention on determining the basic symmetries of the special collision rate, $P(\mathbf{p}_1, \mathbf{p}_2|\mathbf{p}_1', \mathbf{p}_2')$.

Because the particles in the two beams are identical, the function P has the symmetry property

$$P(\mathbf{p}_1, \mathbf{p}_2|\mathbf{p}_1', \mathbf{p}_2') = P(\mathbf{p}_2, \mathbf{p}_1|\mathbf{p}_2', \mathbf{p}_1') \qquad (10.87)$$

which corresponds to simply renaming beams 1 and 2. If we keep the two beams fixed, but reflect the x, y, and z axes, then the components of all the momentum vectors go into their negatives. Because the interparticle potential is spherically symmetric, the force law is unchanged under reflection. This shows that P is unchanged under the simultaneous reflection of all momentum variables.

$$P(\mathbf{p}_1, \mathbf{p}_2|\mathbf{p}_1', \mathbf{p}_2') = P(-\mathbf{p}_1, -\mathbf{p}_2|-\mathbf{p}_1', -\mathbf{p}_2') \qquad (10.88)$$

The function $P(\mathbf{p}_1, \mathbf{p}_2|\mathbf{p}_1', \mathbf{p}_2')$ is determined by the collection of all the two-particle space–time orbits that satisfy Newton's equation of motion with the given force law. But that collection of orbits is unchanged under time reversal. Time reversal will take any orbit with input momenta $\mathbf{p}_1$ and $\mathbf{p}_2$ and output momenta $\mathbf{p}_1'$ and $\mathbf{p}_2'$ and transform it into an orbit with input momenta $-\mathbf{p}_1'$ and $-\mathbf{p}_2'$ and output momenta $-\mathbf{p}_1$ and $-\mathbf{p}_2$. Thus, by time-reversal invariance,

$$P(\mathbf{p}_1, \mathbf{p}_2|\mathbf{p}_1', \mathbf{p}_2') = P(-\mathbf{p}_1', -\mathbf{p}_2'|-\mathbf{p}_1, -\mathbf{p}_2) \qquad (10.89)$$

$$S = k \log W$$

—— Epitaph ingraved on the tombstone of Ludwig Boltzmann in Vienna

Boltzmann, after a reasonably successful career as a Professor in Graz and Vienna, committed suicide in 1906. It is an open question, how much that act was influenced by the lack of acceptance he encountered to his ideas in continental Europe. (They were widely accepted in England and the United States.) Particularly in Germany and Austria, his theoretical concepts met with strong opposition for a number of related reasons. One was the popularity of a rather extreme form of philosophical scepticism, called Positivism, which attempted to eliminate from scientific discussions, all concepts that were not experimentally verifiable. At the time, one such concept was the atomic nature of matter. There was also a growing school of scientists, called energists, who opposed the view that energy was not a fundamental substance in itself, but was simply the sum of the kinetic and potential energies of the particles composing a substance. The third development was the fact that, due to the growing interest in electromagnetic theory and other fields, kinetic theory and statistical mechanics became somewhat unfashionable, so that Boltzmann's work, rather than being actively opposed, was simply ignored. Unfortunately, it was only shortly after his death that almost all elements of this intellectual climate were changed dramatically.

Combining Eqs. (10.88) and (10.89), we see that

$$P(\mathbf{p}_1, \mathbf{p}_2 | \mathbf{p}_1', \mathbf{p}_2') = P(\mathbf{p}_1', \mathbf{p}_2' | \mathbf{p}_1, \mathbf{p}_2) \tag{10.90}$$

10.14   BOLTZMANN'S EQUATION FOR A UNIFORM SYSTEM

Boltzmann's equation is a master equation for the momentum distribution function of a dilute gas. (The master equation for a system with discrete states is derived in Section 8.11.) A master equation is an equation that describes how the probability density for a system evolves in time due to transitions from one state to another. In a dilute gas, the transitions are caused by collisions between pairs of gas particles.

We consider a unit volume of a gas whose spatial density is constant. At time t, the momentum distribution function of the gas is $F(\mathbf{p})$. We choose momentum volume elements, $d^3\mathbf{p}_1$, $d^3\mathbf{p}_2$, $d^3\mathbf{p}_1'$, and $d^3\mathbf{p}_2'$, centered around the momenta $\mathbf{p}_1$, $\mathbf{p}_2$, $\mathbf{p}_1'$, and $\mathbf{p}_2'$. We then ask the question: At what rate do collisions occur that knock one particle from the momentum range $d^3\mathbf{p}_1$ into the momentum range $d^3\mathbf{p}_1'$ while its collision partner has its momentum changed from within the range $d^3\mathbf{p}_2$ to within the range $d^3\mathbf{p}_2'$? In answering this, we can imagine that all the particles whose momenta are not in momentum ranges $d^3\mathbf{p}_1$ or $d^3\mathbf{p}_2$ have instantaneously vanished, because they could not contribute to this process anyway. We are then left with two single velocity beams of densities $F(\mathbf{p}_1) d^3\mathbf{p}_1$ and $F(\mathbf{p}_2) d^3\mathbf{p}_2$. From our previous analysis, we can see that the rate of collisions of the special type just described is

$$\begin{aligned} \text{rate}(d^3\mathbf{p}_1 &\to d^3\mathbf{p}_1', d^3\mathbf{p}_2 \to d^3\mathbf{p}_2') \\ &= (F(\mathbf{p}_1)d^3\mathbf{p}_1)(F(\mathbf{p}_2)d^3\mathbf{p}_2)P(\mathbf{p}_1, \mathbf{p}_2 | \mathbf{p}_1', \mathbf{p}_2') \, d^3\mathbf{p}_1' \, d^3\mathbf{p}_2' \end{aligned} \tag{10.91}$$

We now ask: What is the total rate at which particles are knocked out of the momentum range $d^3\mathbf{p}_1$? Clearly, we must take the answer to the last question and integrate over $\mathbf{p}_1'$, $\mathbf{p}_2$, and $\mathbf{p}_2'$.

$$
\begin{aligned}
&\text{rate}(d^3\mathbf{p}_1 \to \text{anything}) \\
&= (F(\mathbf{p}_1)\, d^3\mathbf{p}_1) \int F(\mathbf{p}_2) P(\mathbf{p}_1, \mathbf{p}_2 | \mathbf{p}_1', \mathbf{p}_2')\, d^3\mathbf{p}_1'\, d^3\mathbf{p}_2\, d^3\mathbf{p}_2'
\end{aligned}
\tag{10.92}
$$

In order to calculate the rate of change of the momentum distribution function $F(\mathbf{p})$, we will also need to know the rate at which particles are knocked *into* the momentum element $d^3\mathbf{p}_1$ by two-particle collisions. To calculate that rate, we interchange the symbols used to denote the input and output momenta in a collision. We consider collisions of the type $(d^3\mathbf{p}_1' \to d^3\mathbf{p}_1, d^3\mathbf{p}_2' \to d^3\mathbf{p}_2)$. Summing these over all values of $\mathbf{p}_1'$, $\mathbf{p}_2'$, and $\mathbf{p}_2$ gives the total rate at which particles are knocked into the momentum element $d^3\mathbf{p}_1$.

$$
\begin{aligned}
&\text{rate}(\text{anything} \to d^3\mathbf{p}_1) \\
&= d^3\mathbf{p}_1 \int (F(\mathbf{p}_1')\, d^3\mathbf{p}_1')(F(\mathbf{p}_2')\, d^3\mathbf{p}_2') P(\mathbf{p}_1', \mathbf{p}_2' | \mathbf{p}_1, \mathbf{p}_2)\, d^3\mathbf{p}_2
\end{aligned}
\tag{10.93}
$$

The Boltzmann equation is derived by noting that $F(\mathbf{p}_1)\, d^3\mathbf{p}_1$ is the number of particles with momenta in the range $d^3\mathbf{p}_1$ and, therefore,

$$
\frac{\partial}{\partial t}\big[F(\mathbf{p}_1)\, d^3\mathbf{p}_1\big] = \text{rate}(\text{anything} \to d^3\mathbf{p}_1) - \text{rate}(d^3\mathbf{p}_1 \to \text{anything})
\tag{10.94}
$$

This gives the equation

$$
\begin{aligned}
\frac{\partial F(\mathbf{p}_1, t)}{\partial t} = \int &\big[F(\mathbf{p}_1') F(\mathbf{p}_2') P(\mathbf{p}_1', \mathbf{p}_2' | \mathbf{p}_1, \mathbf{p}_2) \\
&- F(\mathbf{p}_1) F(\mathbf{p}_2) P(\mathbf{p}_1, \mathbf{p}_2 | \mathbf{p}_1', \mathbf{p}_2') \big]\, d^3\mathbf{p}_2\, d^3\mathbf{p}_1'\, d^3\mathbf{p}_2'
\end{aligned}
\tag{10.95}
$$

Using the time-reversal symmetry described by Eq. (10.90), this can be written as

$$
\begin{aligned}
\frac{\partial F(\mathbf{p}_1, t)}{\partial t} = \int &\big[F(\mathbf{p}_1') F(\mathbf{p}_2') - F(\mathbf{p}_1) F(\mathbf{p}_2) \big] \\
&\times P(\mathbf{p}_1, \mathbf{p}_2 | \mathbf{p}_1', \mathbf{p}_2')\, d^3\mathbf{p}_2\, d^3\mathbf{p}_1'\, d^3\mathbf{p}_2'
\end{aligned}
\tag{10.96}
$$

This is Boltzmann's equation for the momentum distribution of a system with a spatially uniform density.

10.15 THE BOLTZMANN ENTROPY

Boltzmann identified a certain expression (which we will present shortly) as the entropy to be associated with any given momentum distribution function $F(\mathbf{p})$ and then showed that, if F satisfies the Boltzmann equation, then the associated entropy increases monotonically until the momentum distribution becomes the Maxwell distribution, $F(\mathbf{p}) = C \exp(-\beta p^2/2m)$, after which the entropy remains constant and is equal to the equilibrium entropy for an ideal gas.

Our first task in presenting the Boltzmann theory is to decide on the appropriate form for the entropy of a system of particles with a known momentum distribution function. In fact, we will answer a somewhat more difficult question, namely: In a system of weakly interacting particles with an external potential $U(\mathbf{r})$, what is the entropy to be associated with a given one-particle phase-space distribution function $F(\mathbf{r}, \mathbf{p})$? The entropy of a spatially uniform system with a given momentum distribution will then be a special case. Since an arbitrary phase-space distribution certainly does not describe an equilibrium state, the entropy that we are looking for must be some conditional entropy function—it cannot be the simple thermodynamic entropy. As always, what we mean when we say that the particles are weakly interacting is that the interaction potential energy is almost certain to be negligible in comparison with the total energy. In that case, giving the phase-space distribution function $F(\mathbf{r}, \mathbf{p})$ already determines the energy of the system. That is, all microstates with the given phase-space distribution have the same energy, namely

$$E = \int \left(\frac{p^2}{2m} + U(\mathbf{r}) \right) F(\mathbf{r}, \mathbf{p}) \, d^3r \, d^3\mathbf{p} \tag{10.97}$$

Therefore, in calculating the conditional entropy, we can leave out the energy fixing delta function $\delta(E - H)$ because all states will automatically be on the energy surface. The principle that we use is that the entropy is proportional to the logarithm of the volume of all the states in the $6N$-dimensional phase space that are consistent with the given conditions on the system. We break up the one-particle phase space into numbered six-dimensional boxes, each with a phase-space volume, ω. Let N_i be the number of particles in the ith box. First we consider states in which particles numbered 1 to N_1 are in box 1, those numbered $N_1 + 1$ to $N_1 + N_2$ are in box 2, etc. The total phase space of those configurations is ω^N, which is obtained by integrating the phase-space coordinates of each particle over the box it is in. The total phase-space volume available to the system for given values of the occupation numbers, $N_1, N_2, \ldots$, is

$$\Omega(\{N_i\}) = \frac{\omega^N N!}{\prod_i N_i!} \tag{10.98}$$

The logarithm of this quantity, in the limit of small ω, should give us the conditional entropy.

$$\log \Omega = N \log \omega + N(\log N - 1) - \sum_i N_i(\log N_i - 1)$$

$$= N \log N - \sum_i N_i \log\left(\frac{N_i}{\omega}\right)$$

$$= N \log N - \omega \sum_i \frac{N_i}{\omega} \log\left(\frac{N_i}{\omega}\right) \tag{10.99}$$

$$\sim N \log N - \int d^3\mathbf{r} \, d^3\mathbf{p} \, F(\mathbf{r}, \mathbf{p}) \log F(\mathbf{r}, \mathbf{p})$$

Since $N \log N$ is a constant, it can be dropped,* giving the following expression for the Boltzmann entropy (in rational units).

$$S_B = - \int F(\mathbf{r}, \mathbf{p}) \log F(\mathbf{r}, \mathbf{p}) \, d^3\mathbf{r} \, d^3\mathbf{p} \tag{10.100}$$

For a spatially uniform distribution $F(\mathbf{p})$, the integral over $\mathbf{r}$ in Eq. (10.100) gives a factor of the volume. Thus the entropy per unit volume takes the form

$$s_B = (S_B/V) = - \int F(\mathbf{p}) \log F(\mathbf{p}) \, d^3\mathbf{p} \tag{10.101}$$

*Actually, the correct expression for the entropy is $S = \log(\Omega/N!h^K)$. Thus the $N \log N$ term would be cancelled exactly by the $\log N!$. There is no way that Boltzmann could have known anything about the h^K term.

The rate of change of s_B is obtained by differentiating the integrand.

$$\frac{ds_B}{dt} = -\int (1 + \log F)\frac{\partial F}{\partial t}d^3\mathbf{p} \tag{10.102}$$

Using the fact that $\int(\partial F/\partial t)\,d^3\mathbf{p} = d(N/V)/dt = 0$, changing the name of the variable of integration from $\mathbf{p}$ to $\mathbf{p}_1$, and using Eq. (10.96) for $\partial F(\mathbf{p}_1)/\partial t$, we can write ds_B/dt as

$$\begin{aligned}\frac{ds_B}{dt} = -\int &\log F(\mathbf{1})[F(\mathbf{1}')F(\mathbf{2}') - F(\mathbf{1})F(\mathbf{2})] \\ &\times P(\mathbf{1},\mathbf{2}|\mathbf{1}',\mathbf{2}')\,d^3\mathbf{p}_1\,d^3\mathbf{p}_2\,d^3\mathbf{p}_1'\,d^3\mathbf{p}_2'\end{aligned} \tag{10.103}$$

where, for compactness, $\mathbf{p}_1$, $\mathbf{p}_2$, $\mathbf{p}_1'$, and $\mathbf{p}_2'$ have been written as $\mathbf{1}$, $\mathbf{2}$, $\mathbf{1}'$, and $\mathbf{2}'$. By renaming the variables in the following way: $\mathbf{p}_1 \to \mathbf{p}_2$, $\mathbf{p}_2 \to \mathbf{p}_1$, $\mathbf{p}_1' \to \mathbf{p}_2'$, $\mathbf{p}_2' \to \mathbf{p}_1'$, and by using the symmetry relation $P(\mathbf{1},\mathbf{2}|\mathbf{1}',\mathbf{2}') = P(\mathbf{2},\mathbf{1}|\mathbf{2}',\mathbf{1}')$, one can also write ds_B/dt as

$$\begin{aligned}\frac{ds_B}{dt} = -\int &\log F(\mathbf{2})[F(\mathbf{1}')F(\mathbf{2}') - F(\mathbf{1})F(\mathbf{2})] \\ &\times P(\mathbf{1},\mathbf{2}|\mathbf{1}',\mathbf{2}')\,d^3\mathbf{p}_1\,d^3\mathbf{p}_2\,d^3\mathbf{p}_1'\,d^3\mathbf{p}_2'\end{aligned} \tag{10.104}$$

Taking the average of these two expressions, and using the identity $\log F(\mathbf{1}) + \log F(\mathbf{2}) = \log[F(\mathbf{1})F(\mathbf{2})]$, gives

$$\begin{aligned}\frac{ds_B}{dt} = -\frac{1}{2}\int &\log[F(\mathbf{1})F(\mathbf{2})][F(\mathbf{1}')F(\mathbf{2}') - F(\mathbf{1})F(\mathbf{2})] \\ &\times P(\mathbf{1},\mathbf{2}|\mathbf{1}',\mathbf{2}')d^3\mathbf{p}_1 d^3\mathbf{p}_2 d^3\mathbf{p}_1' d^3\mathbf{p}_2'\end{aligned} \tag{10.105}$$

Now we again rename the variables, making $\mathbf{p}_1 \to \mathbf{p}_1'$, $\mathbf{p}_2 \to \mathbf{p}_2'$, $\mathbf{p}_1' \to \mathbf{p}_1$, and $\mathbf{p}_2' \to \mathbf{p}_2$. Recalling that $P(\mathbf{1},\mathbf{2}|\mathbf{1}',\mathbf{2}') = P(\mathbf{1}',\mathbf{2}'|\mathbf{1},\mathbf{2})$, we see that

$$\begin{aligned}\frac{ds_B}{dt} = \frac{1}{2}\int &\log[F(\mathbf{1}')F(\mathbf{2}')][F(\mathbf{1}')F(\mathbf{2}') - F(\mathbf{1})F(\mathbf{2})] \\ &\times P(\mathbf{1},\mathbf{2}|\mathbf{1}',\mathbf{2}')d^3\mathbf{p}_1 d^3\mathbf{p}_2 d^3\mathbf{p}_1' d^3\mathbf{p}_2'\end{aligned} \tag{10.106}$$

Again, averaging Eqs. (10.105) and (10.106), we obtain

$$\begin{aligned}\frac{ds_B}{dt} = \frac{1}{4}\int &\{\log[F(\mathbf{1}')F(\mathbf{2}')] - \log[F(\mathbf{1}')F(\mathbf{2}')]\} \\ &\times [F(\mathbf{1}')F(\mathbf{2}') - F(\mathbf{1})F(\mathbf{2})]P(\mathbf{1},\mathbf{2}|\mathbf{1}',\mathbf{2}')\,d^3\mathbf{p}_1\,d^3\mathbf{p}_2\,d^3\mathbf{p}_1'\,d^3\mathbf{p}_2'\end{aligned} \tag{10.107}$$

But it is easy to verify that, for any values of x and y,

$$(\log x - \log y)(x - y) \geq 0 \tag{10.108}$$

because both factors have the same sign unless they are zero. Using this inequality, and taking $x = F(\mathbf{1}')F(\mathbf{2}')$ and $y = F(\mathbf{1})F(\mathbf{2})$, one sees that the product of the first two factors in the integrand in Eq. (10.107) is nonnegative. The function $P(\mathbf{1},\mathbf{2}|\mathbf{1}',\mathbf{2}')$ is defined as the rate at which special collisions take place per unit volume for unit density beams and thus it cannot be negative. The integrand is therefore nonnegative for all values of $\mathbf{p}_1$, $\mathbf{p}_2$, $\mathbf{p}_1'$, and $\mathbf{p}_2'$, showing that

$$\frac{ds_B}{dt} \geq 0 \qquad (10.109)$$

The Boltzmann entropy can never decrease!

Now we will look at the question of when ds_B/dt is zero. Suppose that for some momentum density function $F_o(\mathbf{p})$ the Boltzmann entropy remains constant. Then for the function $F_o(\mathbf{p})$ the integrand in Eq. (10.107) must be zero for all values of $\mathbf{p}_1$, $\mathbf{p}_2$, $\mathbf{p}_1'$, and $\mathbf{p}_2'$, because, since it is nonnegative, some range in which it has positive values cannot be cancelled by some other range in which it is negative.

Because two-particle scattering by an interaction potential conserves the energy and momentum of the pair of particles, the scattering function $P(\mathbf{p}_1, \mathbf{p}_2 | \mathbf{p}_1', \mathbf{p}_2')$ is zero unless

$$\mathbf{p}_1 + \mathbf{p}_2 = \mathbf{p}_1' + \mathbf{p}_2' \qquad (10.110)$$

and

$$p_1^2 + p_2^2 = p_1'^2 + p_2'^2 \qquad (10.111)$$

For realistic interatomic potentials, $P > 0$ for any set of four momenta that satisfy the preceding two conservation laws. (See Problem 10.10.) Therefore, one must have $F_o(1')F_o(2') = F_o(1)F_o(2)$ for any values of their variables that satisfy Eqs. (10.110) and (10.111). This is equivalent to the condition that, if we let $G_o(\mathbf{p}) = \log F_o(\mathbf{p})$, then

$$G_o(\mathbf{p}_1) + G_o(\mathbf{p}_2) = G_o(\mathbf{p}') + G_o(\mathbf{p}_2') \qquad (10.112)$$

whenever the energy and momentum conservation laws are satisfied. This condition has three obvious solutions. They are

$$G_o(\mathbf{p}) = a, \quad G_o(\mathbf{p}) = \mathbf{b} \cdot \mathbf{p}, \quad \text{and} \quad G_o(\mathbf{p}) = cp^2 \qquad (10.113)$$

where a and c are arbitrary constants and $\mathbf{b}$ is an arbitrary vector. In fact, it is possible to show that the only function that satisfies Eq. (10.112) whenever the four variables satisfy Eqs. (10.110) and (10.111) is a linear combination of these three solutions. Therefore, any function that gives $ds_B/dt = 0$ is of the form

$$\log F_o(\mathbf{p}) = a + \mathbf{b} \cdot \mathbf{p} + cp^2 \qquad (10.114)$$

By completing the square and exponentiating, this can be written as

$$F_o(\mathbf{p}) = \exp\left[a + c(\mathbf{p} - \mathbf{p}_o)^2\right] \qquad (10.115)$$

It is clear that the constant c must be negative for a normalizable momentum density function. We will write the constant as $-\beta/2m$. Also, because the momentum density function is symmetric about the vector $\mathbf{p}_o$, that vector is the average momentum of the particles in the system. But for a system of particles in a closed stationary container, the average momentum must be zero for a spatially uniform solution. (This can be shown mathematically by taking into account the collisions of the particles with the walls of the container, but it entails a great deal of work for a fairly obvious result.) Therefore, the only functions that yield constant entropy are of the form of the Maxwell distribution.

$$F_o(\mathbf{p}) = C \exp(-\beta p^2/2m) \qquad (10.116)$$

10.16   A DEEPER LOOK AT THE BOLTZMANN THEORY

From Newton's laws of mechanics, we seem to have derived, at least for gases, the law of increase of entropy. But this presents a fundamental logical paradox. The law of entropy increase is not symmetrical with respect to time reversal. It would give one a means of orienting a loose strip of movie film that showed the time evolution of a gas. The "beginning" is the end with the lowest entropy; the "finale" is the end with the highest. But the laws of mechanics are completely invariant under time reversal—they offer no way of orienting the filmstrip. If, run in one direction, the particles are seen to move according to Newton's laws, then, run in the other direction, they do also. This is often expressed by saying that Newton's laws offer no way of orienting the *arrow of time*. It seems, and in fact it is, utterly impossible to derive correctly a time-asymmetric result from a fundamental theory that is time-symmetric. Boltzmann's theory was repeatedly attacked for having what his opponents felt was this basic flaw. But before we go into this deeper question, we should go back and point out a few more mundane weaknesses in our presentation of Boltzmann's analysis. Without any explicit mention of them, a number of significant approximations and assumptions have been made in the derivation of Boltzmann's equation. What we intend to show is that, as the ratio of the interaction length d to the average nearest-neighbor distance [this is approximately given by $D = (V/N)^{1/3}$] becomes smaller and smaller, the approximations that have been made become more and more accurate. Thus, although these approximations introduce a certain finite error into any specific calculation made with the Boltzmann equation (for example, a calculation of the heat conductivity of argon at STP), they could hardly be the root cause of the time-asymmetry paradox, since, by making a gas sufficiently dilute, the errors can be made arbitrarily small.

In deriving the Boltzmann equation, two unmentioned approximations were made.

1. The time evolution of the momentum distribution was assumed to be caused entirely by complete two-body collisions. The possibility that, while two particles were in the process of colliding, a third particle would disturb the collision was completely ignored. This could be important, because contact with a third particle during the collision eliminates the necessity of conserving the momentum and energy of the original two particles. The third particle could carry off some of the energy and momentum of the original pair.

2. It was assumed, without discussion, that the scattering rate calculated using dilute single-velocity beams could be applied without change to the subsets of particles with momenta $\mathbf{p}_1$ and $\mathbf{p}_2$ within a gas. But, before their collision, the particles in the two single-velocity beams are naturally completely uncorrelated in position because they come from some unspecified sources that are distant from one another. They are like the photons reaching the earth from two different stars. This assumed lack of correlation was essential in showing that the scattering rate was proportional to the product of the densities, $n_1 n_2$. But in a gas the same two particles might interact repeatedly, which could produce strong statistical correlations in their positions. Such statistical correlations would make the result derived from uncorrelated single-velocity beams irrelevant to the scattering processes in a gas.

We can estimate the importance of the "three-body effect" mentioned in point 1 by asking the question: During the time that any two colliding particles are interacting, what is the probability that one of them will come into interaction with a third particle? First, we should note that, for a billiard ball model of gas particles, the interaction time is zero and thus the probability that three particles will simultaneously be interacting is also zero. For other particles, one can roughly estimate the time for a collision to take place (that is, the time interval during which the two particles are actually interacting) as the interaction distance divided by the average speed of particles in the gas, $t_c \approx d/\bar{v}$. The average time between collisions is approximately the mean free path (calculated in Exercise 2.17) divided by the average velocity. We will call it the mean free time t_{mf}. If we drop factors of order one, then $t_{\mathrm{mf}} \approx \ell/\bar{v} \approx 1/n\bar{v}d^2$. For a single particle, if the intervals during which it is colliding are distributed at random along the time axis, then the fraction of them that will overlap, meaning that the particle is colliding with two particles at once, is approximately equal to the ratio of t_c to t_{mf}. This would give the fraction of collisions that are three-body, rather than two-body, collisions.

$$F = \frac{t_c}{t_{\mathrm{mf}}} \approx d^3 n \qquad (10.117)$$

Looking at Table 8.1, we see that, for typical noble gas atoms, $d \approx 5 \times 10^{-10}$ m. At STP, $n \approx 3 \times 10^{25}$ m^{-3} and therefore $F \approx 4 \times 10^{-3}$. This is fairly small, but the important point is that the relative importance of three-body versus two-body collisions is proportional to n and can therefore be made arbitrarily small by reducing the density of the gas.

Fig. 10.11   It is clear that in a dilute gas the probability of finding a stationary particle at point 1, directly behind a fast moving particle, is much less that the probability of finding one at point 2

The second potential source of error listed is the assumed lack of statistical correlation between particles prior to their collision. Now the set of particles with momenta within the element $d^3\mathbf{p}_1$ form an extremely dilute beam; in fact, the density of that set of particles goes to zero with $d^3\mathbf{p}_1$. Therefore, it is perfectly reasonable to assume that those particles are statistically uncorrelated in position. The questionable assumption that was made is the following. We assumed that, *given that a particle is at position* $\mathbf{r}_1$ *with momentum* $\mathbf{p}_1$, the probability of finding another particle at a position $\mathbf{r}_2$ with momentum $\mathbf{p}_2$ is the same as it would be without that condition. That is, for any two positions and momenta that will lead to a collision, we assumed that

$$P_2(\mathbf{r}_1, \mathbf{p}_1, \mathbf{r}_2, \mathbf{p}_2) = P_1(\mathbf{r}_1, \mathbf{p}_1)P_1(\mathbf{r}_2, \mathbf{p}_2) \qquad (10.118)$$

where P_2 is the joint probability of finding particles at those phase-space points and P_1 is the one-particle phase-space probability density. It is an important point that we needed to make this assumption only for pairs of phase-space points that would lead to collisions, that is, for pairs of particles that are approaching one another. It is easy to show that Eq. (10.118) is not true for arbitrary pairs of phase-space points (see Fig. 10.11). If we consider two particles that are about to collide, then, on an average, each of them has been traveling for a distance ℓ since its last interaction with any particle. Thus the "sources" of the two particles are separated by a distance of order ℓ. But for small n ℓ is proportional to $1/n$ and thus goes to infinity as the density goes to zero. In a gas, at least one at equilibrium, there is a distance, called

"In my opinion it would be a great tragedy for science if the theory of gases were temporarily thrown into oblivion because of a momentary hostile attitude toward it, as was for example the wave theory because of Newton's authority.

I am conscious of being only an individual struggling weakly against the stream of time. But it still remains in my power to contribute in such a way that, when the theory of gases is again revived, not too much will have to be rediscovered."

—— L. Boltzmann, in *Theory of Gases* (translated by S. Brush)

the correlation length ℓ_c, over which pairs of particles are correlated in position. At low densities, the correlation length is not much larger then the interaction length d. As the density is made smaller and smaller, the sources of the colliding particles become further and further separated, and thus it is reasonable to assume that the particle positions become less and less correlated. Notice that this argument only holds for pairs of particles that are in an approaching configuration.

10.17   THE ARROW OF TIME

There is no confusing yesterday and tomorrow. We consider as mentally unbalanced any person who promises to meet us yesterday afternoon or remembers what will happen tomorrow morning. Yet the origin of this obvious temporal directionality has been a longstanding problem in physics. The laws of mechanics, quantum mechanics, and electromagnetic theory are perfectly symmetrical with respect to time reversal. The laws of particle physics are not. They have only the combined symmetry, called TCP: (time reversal) × (charge conjugation) × (parity reflection). That is, if we take any situation that satisfies these laws, imagine it being run backwards, interpret all particles as their antiparticles, and reverse the directions of our x, y, and z axes, then we get another situation that satisfies the same laws. The laws of particle physics connect the puzzle of time asymmetry with the puzzle of charge asymmetry. That is: Why are there not as many antiprotons as protons, positrons as electrons, etc.? However, the properties of matter at ordinary temperatures and densities are so totally dominated by mechanics, quantum mechanics, and electromagnetic interactions, that we will assume that it is possible to discuss the question of time asymmetry within the confines of these three time-symmetrical laws. The fact that we will ultimately not really solve the problem within this restricted context might be taken as evidence that a broader range of physics is required.

The first question that will be considered is whether the Boltzmann theory really predicts time asymmetry. In discussing this question, it is best first to look at Boltzmann's equation for entropy change [Eq. (10.107)], rather than at the Boltzmann equation itself. When making a time reversal, one must remember to replace $F(\mathbf{p})$ by $F(-\mathbf{p})$. Therefore, it will simplify matters if we restrict ourselves to momentum density functions that are only functions of p^2. If $F(\mathbf{p})$ has that property initially, it will certainly be preserved by the Boltzmann equation.

If, in Eq. (10.107), we replace the variable t by a new variable $t^* = -t$, then, on the left, $\partial s_B / \partial t \rightarrow -\partial s_B / \partial t^*$ but the integral on the right-hand side remains a nonnegative integral. Thus the time-reversed equation predicts that the entropy increases as a function of t^*, which would mean that it decreased as a function of t. This seems to show that the Boltzmann theory is clearly asymmetrical under time reversal. Boltzmann himself argued that the preceding argument was too naïve to answer the question of whether the Boltzmann theory is symmetrical

"In reading Clausius we seem to be reading mechanics; in reading Maxwell and in much of Boltzmann's most valuable work, we seem rather to be reading in the theory of probabilities."

—— J. W. Gibbs, in an obituary to Clausius (1889)

under time reversal. In order to determine the symmetries of a theory, it is not enough just to look at the major equations—one must take into account the whole theory, including the proper physical interpretation of the functions that appear in the equations. For example, the Schrödinger equation is a first-order differential equation in time, just as the Boltzmann equation is. It therefore changes under the time reversal, $t \rightarrow -t$. Also, the physical interpretation involves setting up a system in a given state and then taking a measurement at some later time, a process that seems intrinsically asymmetrical in time. However, one can show that nonrelativistic quantum mechanics cannot be used to orient the arrow of time. (See Problem 10.11 for a simplified version of the proof.)

In order to determine the time-reversal symmetry of the Boltzmann theory, we first have to ask: What question does the Boltzmann equation answer? The question is the following. Given that the momentum distribution of the system at time 0 is $F(\mathbf{p}, 0)$, and given that the system is undisturbed during the time interval $(0, t)$, what is the most probable momentum distribution at time t, *where probability is calculated using the ensemble of microstates that have the specified momentum distribution at time* 0. Boltzmann's answer is that the most probable momentum distribution at time t is $F(\mathbf{p}, |t|)$, where $F(\mathbf{p}, t)$ is the solution of the Boltzmann equation for positive time that satisfies the initial conditions at $t = 0$. This result is manifestly symmetrical in time. In other words, the time variable in the Boltzmann equation is the absolute value of the time interval between 0 and t. Although certain approximations are made in deriving the Boltzmann equation, there is no serious doubt that, for dilute gases, the equation provides a highly accurate answer to the question posed.

This interpretation of the Boltzmann theory seems still to have some problems. Suppose that a system has a momentum distribution $F(\mathbf{p}, 0)$ at $t = 0$ and one uses the Boltzmann equation to calculate the momentum distribution at times $t = 1$ and 2, obtaining $F(\mathbf{p}, 1)$ and $F(\mathbf{p}, 2)$. If the momentum distribution $F(\mathbf{p}, 1)$ is now used as initial data and projected forward in time for one second, using the Boltzmann equation, then the result will agree with $F(\mathbf{p}, 2)$. However, if the prescription described is used to make a *retrodiction* by calculating $F(\mathbf{p}, 0)$ from $F(\mathbf{p}, 1)$, then the result will be the patently false statement $F(\mathbf{p}, 0) = F(\mathbf{p}, 2)$. Why one obtains this false result is easy to see by considering Fig. 10.12. In that figure, $S(0)$ represents the large set of microstates in the $6N$-dimensional phase space that have the specified momentum distribution $F(\mathbf{p}, 0)$. Carried forward in time for one second (by Hamilton's equations), these states go into some set $S(+1)$. Almost all of the states in $S(+1)$ have the same momentum distribution, approximately equal to the solution of the Boltzmann equation, $F(\mathbf{p}, 1)$. If the states are carried backwards in time for one second, then, because of the time reversibility of Hamilton's equations, they will go into another set, $S(-1)$, that has the same momentum distribution as $S(+1)$. The set, S^*, is the set of all microstates that have the momentum distribution, $F(\mathbf{p}, 1)$. It is a much larger set of states than either $S(+1)$ or $S(-1)$, which are both tiny subsets of it. If we are told that at time $t = 1$ the system has the momentum distribution $F(\mathbf{p}, 1)$, and that is all that we know about the state

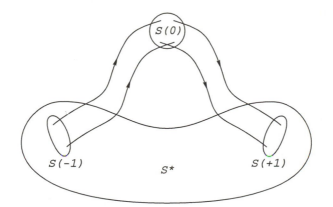

Fig. 10.12   The set $S(0)$ is the set of all the phase-space points that have the momentum density function $F(\mathbf{p}, 0)$. If we assume that $F(\mathbf{p}, 0) = F(-\mathbf{p}, 0)$, then, because of the time reversibility of mechanics, the sets $S(+1)$ and $S(-1)$ will have the same momentum density functions.

of the system, then we know only that at $t = 1$ the microstate is part of S^*. If we want to use that information to guess what the state of the system was at $t = 0$, it is clear that we should take the set S^* and carry it backwards in time for one second. It would be obviously wrong to expect that a typical member of the set S^*, when carried backwards in time, will behave, with respect to its momentum distribution, like a member of the tiny subset $S(+1)$, which is not a random subset of S^* but rather a subset that was specifically defined by the property that its members come from $S(0)$. If, instead of knowing only that the system was in S^* at $t = 1$, we also know that the system was in $S(0)$ at $t = 0$, then, in any attempt to calculate the momentum distribution at $t = 0$, the second piece of information is highly relevant and cannot be ignored without producing incorrect results.

Another question naturally arises from contemplating Fig. 10.12. When we use the Boltzmann equation to go from $F(\mathbf{p}, 0)$ to $F(\mathbf{p}, 2)$, we are, at least approximately, projecting the set $S(0)$ forward by two seconds, which is the same as projecting the set $S(+1)$ forward by one second. But when we use the Boltzmann equation to go from $F(\mathbf{p}, 1)$ to $F(\mathbf{p}, 2)$, we are projecting the set S^* forward by one second. Why does one obtain the same results using either S^* or $S(+1)$ in a forward projection, in spite of the fact that the two sets give very different results in a backward projection? It is a consequence of the Markov nature of two-particle collisions in a dilute gas. (A Markov process is described in Section 8.11.) The probability that two particles of momenta $\mathbf{p}_1$ and $\mathbf{p}_2$ will collide is independent of what collisions they have had in the past. This lack of effective memory is something that is only valid for dilute gases (and even there it is only a good approximation). For more dense systems, what a particle will do in the future is more closely correlated with what it has done in the past. For such systems, the future projections of $S(+1)$ and S^* would have measureable differences.

What the Boltzmann theory says is that Newton's equations will, with overwhelming probability, take a system with a non-Maxwellian momentum distribution into one whose momentum distribution is closer to the Maxwell distribution, where closeness may be measured by the size of the Boltzmann entropy function. Now that we see that the Boltzmann theory, properly interpreted, really says nothing about the origin of time asymmetry, because its predictions are symmetrical in time, we are still left with the question of what *is* the source of the very obvious time asymmetry of the real world. A natural way to approach this problem is to look for aspects of physics that are asymmetrical under time reversal. We must look for things other than the increase of entropy, since that is essentially what we are trying to explain.

10.18   THE ELECTROMAGNETIC ARROW

One such asymmetrical aspect of classical physics is the outgoing radiation condition used in electrodynamics. Consider the "bremstrahlung" process illustrated in Fig. 10.13. Two charged particles are scattered by their repulsive Coulomb forces. Because of their acceleration during the scattering process, approximately spherical outgoing electromagnetic waves are produced. If this process is viewed in a time-reversed mode, one sees spherical waves coming in from infinity that are carefully correlated so that they arrive at the location of the scattering particles just in time to deliver the appropriate amounts of energy and momentum. The time-reversed process satisfies all the laws of electrodynamics, but still appears to be highly artificial and unnatural. In the early 1900s it was argued, particularly by the German physicist W. Ritz (in a famous contraversy with A. Einstein), that the nonexistence of these converging wave solutions was a fundamental law of nature, just as basic to electromagnetic theory as Maxwell's equations. The requirement of only outgoing wave solutions can be expressed in a precise mathematical form. When it is added to Maxwell's equations, the resulting theory is not symmetrical under time reversal. Ritz argued that this lack of time symmetry was the fundamental cause of time irreversibility in the world. Unfortunately, this proposal cannot be adequately analyzed without a fairly extensive study of the mathematical details of electrodynamics—something that would be quite out of place in this book. The reader will have to take on faith the statement that Ritz's proposal has not held up well. The modern formulations of classical and quantum electrodynamics are quite symmetrical under time reversal. The outgoing wave condition seems to be based only on the special situations of most laboratory experiments, where the energies of the charged particles are much larger than the typical photon energies of the ambient radiation field. If one looks at a collection of approximately stationary particles within a cavity at equilibrium, then one finds that incoming spherical waves are as common as outgoing ones. It is purely a matter of choice in that time-symmetrical situation whether one solves the electromagnetic equations with one type of solution or the other.

10.19   THE COSMIC ARROW

Currently, the most widely accepted explanation of the observed time asymmetry is one proposed in 1962 by T. Gold. This cosmological explanation connects the direction of entropy increase with the expansion of the universe. Many independent pieces of evidence point to the "fact" that the universe, along with the space–time manifold in which it is imbedded, had a definite creation event, the big bang, some 15 billion years ago. In the beginning, the universe was extremely compressed, hot, and more or less homogeneous, but in a state of rapid expansion.

This expansion should not be viewed as an initially dense congregation of matter exploding out into a preexisting empty space. It is the space that is expanding, not the matter. The matter is, on an average, at rest. In the early stages, the temperature was so high that no matter could exist in the structured, aggregated forms that we find it today. There were no atoms or molecules, but only an unimaginably dense, high-temperature plasma in which particle creation and annihilation

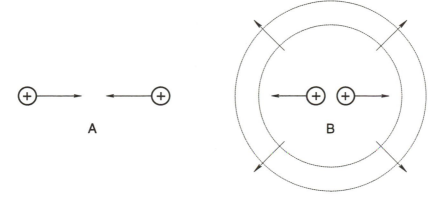

Fig. 10.13 (A) Before their collision, two charged particles approach one another in empty space. (B) After their collision, the particles move apart with smaller velocities. Because of the acceleration of the particles during their collision, some of their energy has been converted to electromagnetic radiation in the form of outgoing waves. The time-reversed version of this process looks totally unnatural.

processes maintained an equilibrium among all the varieties of elementary particles. There are reasons to believe that in the early stages of the universe there was a symmetry between each type of particle and its antiparticle; there were equal numbers of electrons and positrons, quarks and antiquarks, etc. The strong asymmetry that now exists between, for example, protons, which are the common constituents of matter, and antiprotons, which are extremely rare in the universe, is the result of a symmetry-breaking phase transition that took place once the expanding, cooling universe reached a temperature at which the symmetrical state was no longer a stable equilibrium state. It is analogous to the symmetry-breaking phase transition that occurs spontaneously when a piece of unmagnetized ferromagnetic material is cooled through its Curie temperature. Because of the attractive gravitational interaction, as the universe cooled, the uniform density state became unstable and matter condensed into the stars, galaxies, and clusters of galaxies seen today. At each important stage in the cosmic evolution, the disequilibrium between the various constituents became worse. While atoms were still largely ionized, the temperatures of the electrons, protons, and the surrounding blackbody radiation remained equal because the interaction between electromagnetic radiation and charged particles was effective in producing rapid transfer of energy between the three modes. Once neutral atoms formed, the electromagnetic interaction became much less effective and the background radiation decoupled from the matter and cooled, due to the expansion of space, on its own schedule. Nuclear processes in stars cause a stellar evolution with a natural time scale of billions of years—comparable to the total age of the universe. Because of such bottlenecks in energy transfer between its parts, the universe is now a hodgepodge of pieces at wildly different temperatures—a system that is vastly out of equilibrium. Since the disequilibrium of all the parts has a common cause, namely the homogeneous expansion of space, their time directions are thereby synchronized. This is in contrast to what would be the situation if the local disequilibria were due to random fluctuations. In that case, some parts would be fluctuating toward equilibrium while other parts were fluctuating away from equilibrium. If one defined an arrow of time locally as being in the direction toward more equilibrium, then each localized fluctuation would have its own arrow.

Of course, this explanation of the origin of time asymmetry simply transfers the problem to someone else. It is now the duty of cosmologists to explain the origin of the big bang (the origin of the origin?).

"Now it seems more likely that the universe is an eternally existing, self-producing entity, that is divided into many mini-universes much larger than our observable portion, and that the laws of low-energy physics and even the dimensionality of spacetime may be different in each of those mini-universes."
— A. D. Linde, in *Inflation and Quantum Cosmology* (1990)

With this extravagant quotation we will end the book, before we go off the deep end entirely.

PROBLEMS

10.1 From the probability function, given in Eq. (10.7), show that, for an *LCR* circuit at temperature T,

$$\langle Q^2/2C \rangle = \langle LI^2/2 \rangle = kT/2 \quad \text{and} \quad \langle QI \rangle = 0 \quad (10.119)$$

10.2 A pendulum hangs in a gas of density n and temperature T. When the pendulum swings, it has, for small amplitudes, a linear restoring force and a frictional force, due to the gas, that is proportional to its velocity. When just hanging at equilibrium, due to being struck by the gas molecules, the angle θ and the angular velocity Ω of the pendulum fluctuate. Calculate the correlation function, $\langle \Omega(t)\Omega(0) \rangle$.

10.3 Show that the Reciprocity Theorem [Eq. (10.59)] implies that the matrix of coefficients in Eqs. (10.68) and (10.69) is symmetric.

Fig. 10.14

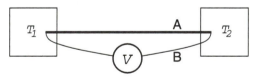

10.4 The system shown in Fig. 10.14, in which the wires A and B are made of different materials, is called a *thermocouple*. Assuming that the thermoelectric power of wire A is T_A and that of wire B is T_B, determine the reading on the voltmeter as a function of $\Delta T = T_2 - T_1$.

10.5 For any classical observable $A(x,p)$, the conditional canonical potential is defined by

$$e^{\phi(a,\beta)} = \int \delta\big(a - A(x,p)\big)e^{-\beta H(x,p)}\, d^K x\, d^K p / N! h^K \quad (10.120)$$

If $\langle F \rangle$ is the canonical expectation value of some function $F(A(x,p))$, derive the canonical ensemble equivalent of Eq. (10.17).

10.6 Beginning with the *LCR* circuit equation [Eq. (10.33)], derive Eqs. (10.36), (10.38), and (10.39).

10.7 For a canonical ensemble, the macroscopic equation of motion for a single observable $A(z)$ [see Eq. (10.19)] is defined by

$$a(t, a^o) = e^{-\phi(a^o)} \int A(T_t z)\, \delta(a^o - A(z)) e^{-\beta H(z)}\, d^{2K} z / N! h^K \tag{10.121}$$

where $\phi(a^o)$ is given in Problem 10.5. For a system composed of a single harmonic oscillator of mass m and spring constant k, derive the detailed macroscopic equation of motion $x(t, x^o)$ for the case where A is just the coordinate of the oscillator.

10.8 A particle of mass m and charge e sits at the center of a harmonic oscillator potential, $U(\mathbf{r}) = kr^2/2$. When it is displaced from equilibrium and released, it satisfies the equation of motion

$$m\dot{\mathbf{v}} = -k\mathbf{r} + m\tau\ddot{\mathbf{v}} \tag{10.122}$$

where the second term on the right is a *radiation reaction force*, due to the fact that the particle, when accelerating, radiates energy in the form of electromagnetic waves. The parameter $\tau = e^2/6\pi\epsilon_o mc^3$ has the units of time and, for an electron, a value of about 6×10^{-24} s. (a) Try a solution of the equation of motion of the form,

$$\mathbf{r}(t) = e^{-\gamma t}\left(\mathbf{A}\cos\omega t + \mathbf{B}\sin\omega t\right) \tag{10.123}$$

but, in evaluating derivatives, consistently drop terms of order γ^2, which is justified when $\gamma^2 \ll k/m$, an inequality that is usually well satisfied. Determine γ and ω in terms of the parameters in the equation of motion and determine $\mathbf{A}$ and $\mathbf{B}$ in terms of the initial values of the position and momentum of the particle. (b) At equilibrium, at temperature T, the particle fluctuates in position and momentum due to the thermally fluctuating electric field. Using the Fluctuation-Dissipation Theorem, evaluate the correlation function $\langle x(t)p_x(0)\rangle$. Hint: The x, y, and z components of the equation of motion are independent.

10.9 The wave function of an isolated quantum mechanical system with a Hamiltonian operator H satisfies the time-dependent Schrödinger equation $i\hbar\, \partial\psi(x,t)/\partial t = H\psi(x,t)$. For any operator A with eigenvalues a_n and eigenfunctions $\phi_n(x)$, the probability of getting the eigenvalue a_n if a measurement of the operator is made at time t is

$$P_n(t) = \left|\int \phi_n^*(x)\psi(x,t)\, dx\right|^2 \tag{10.124}$$

Consider a new operator, $A(t) = \exp(iHt/\hbar) A \exp(-iHt/\hbar)$. (a) Show that $A(t)$ has the same eigenvalues as A and determine the eigenfunctions. (b) Show that the probability of getting the eigenvalue a_n in a measurement of $A(T)$, taken at time 0, is the same as the probability of getting that eigenvalue in a measurement of A taken at time T.

Fig. 10.15

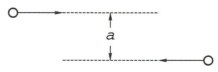

10.10 The object of this problem is to show that, for a "reasonable" interparticle potential, the scattering cross-section is nonzero for any set of incoming and outgoing momenta that satisfy the conservation laws expressed in Eqs. (10.110) and (10.111). Assume that the interaction potential $\phi(r)$ satisfies the conditions that $\phi(r) \to \infty$ as $r \to 0$ and $\phi(r)$ goes smoothly to zero as $r \to d$, the interaction distance. (a) Show that the theorem is satisfied in an arbitrary inertial frame if and only if it is satisfied in the center of mass frame. (b) In the center of mass frame, the scattering event is of the form, $(\mathbf{p}, -\mathbf{p}) \to (\mathbf{p}', -\mathbf{p}')$. Show that the conservation laws are satisfied if and only if $|\mathbf{p}'| = |\mathbf{p}|$.
(C) By considering what happens to the scattering angle as the impact parameter varies from 0 to d, show that the theorem is satisfied in the center of mass frame. The impact parameter a is defined in Fig. 10.15.

10.11 The object of this problem is to show that, in a particular situation, the experimental predictions of quantum mechanics are invariant under time reversal, even though the usual physical interpretation of the mathematical formalism involves a definite directionality of time. (That is, the state of a system must be set up before anything is measured.) Consider two Hermitian operators, A and B. Each of the operators has a complete set of nondegenerate eigenfunctions, which means that there exist two sets of functions, $|u_i\rangle$, and $|v_i\rangle$, such that

$$A|u_i\rangle = a_i|u_i\rangle \quad \text{and} \quad B|v_i\rangle = b_i|v_i\rangle \tag{10.125}$$

where no two a_is are equal and no two b_is are equal. The following measurements are made:

At time $t = 0$ we measure A.   At time $t = 1$ we measure B.
At time $t = 2$ we measure A.   At time $t = 3$ we measure B.
          . . .   . . .
At time $t = N-1$ we measure A.   At time $t = N$ we measure B.

where N is a very large odd number. The results of the measurements are: $A(0)$, $B(1)$, $A(2)$, ..., $B(N)$, where each $A(n)$ is one of the eigenvalues of A and each $B(n)$ is one of the eigenvalues of B. We know the Hamiltonian of the system H. By comparing the pairs of measurements $[A(0), B(1)]$, $[B(1), A(2)]$, $[A(2), B(3)]$, etc. with the predictions of the Schrödinger theory, we can determine whether they agree with that theory. Now consider the set of measurements as viewed by someone moving backwards in time. They would be $B(N)$, $A(N-1)$, $B(N-2)$, ..., $A(0)$. Show that this sequence of measurements will agree with the Schrödinger theory if and only if the original one did. One cannot use a sequence of measurements plus the Schrödinger equation to orient time's arrow.

Supplement to Chapter 1

REVIEW QUESTIONS

1.1 Describe the meaning of the terms: (a) microstate, (b) microstate space, (c) sampling sequence.

1.2 How is the probability function $P(x)$ defined, in the frequency interpretation of probability?

1.3 In the normalization condition $\mathbf{P}[\Omega] = 1$, what is the set Ω?

1.4 Given a statement $q(x)$, what is the corresponding subset $Q \subset \Omega$?

1.5 How is the probability $\mathbf{P}[Q]$ of an arbitrary subset $Q \subset \Omega$ defined in the frequency interpretation of probability?

1.6 In the measure theory interpretation of probability, what are the requirements for an acceptable probability distribution $\mathbf{P}[A]$?

1.7 How many different sets of n objects can be constructed from a set of N distinct objects?

1.8 How many ways can N distinct objects be grouped into three sets containing n_1, n_2, and n_3 objects? (Naturally, $n_1 + n_2 + n_3 = N$.)

1.9 Calculate the probability of getting n heads in a throw of N coins.

1.10 Calculate the Poisson distribution for the probability of finding exactly n particles within a small subvolume of an ideal gas.

1.11 What is a statistical ensemble, and how is it related to a sampling sequence?

1.12 Given a statistical ensemble, how do we define the probability $\mathbf{P}[A]$ of a subset $A \subset \Omega$?

1.13 If the microstates are points in a three-dimensional space and A is a three-dimensional region, how is the probability $\mathbf{P}[A]$ related to the probability density function $P(x, y, z)$?

1.14 Given a probability distribution $\mathbf{P}[Q]$ for every subset $Q \subset \Omega$ write formulas for: (a) $\mathbf{P}[q_1 \text{ or } q_2]$, (b) $\mathbf{P}[q_1 \text{ and } q_2]$, and $\mathbf{P}[\text{not } q]$, where q_1 and q_2 are statements.

1.15 If q and q' are statements in terms of the sampling sequence, what is the meaning of the conditional probability $\mathbf{P}[q'|q]$?

1.16 How is the conditional probability $\mathbf{P}[q'|q]$ related to the joint probability $\mathbf{P}[q' \text{ and } q]$?

1.17 What does it mean to say that two statements are statistically independent?

1.18 For statistically independent statements, what is the form of $\mathbf{P}[q \text{ and } q']$?

1.19 What is a random variable?

1.20 Suppose N is an integer random variable with a probability distribution $P_N(n)$, and $X(N)$ is some function of N. What is the probabilty distribution for X?

1.21 Suppose $\mathbf{R}$ is a three-dimensional random variable with a probability density $P_{\mathbf{R}}(\mathbf{r})$, and $Y(\mathbf{R})$ is some scalar function of $\mathbf{R}$. What is the probability density for Y?

1.22 If the continuous variable x has the probability density $P(x)$, how are $\bar{x}$ and Δx defined?

1.23 Prove Chebyshev's inequality, $\mathbf{P}[|x - \bar{x}| > a] \leq (\Delta x/a)^2$

1.24 Suppose $x_1, x_2, \ldots, x_N$ are N independent random variables with uncertainties $\Delta x_1, \Delta x_2, \ldots, \Delta x_N$. What is the uncertainty in their average, $x = (x_1 + x_2 + \cdots + x_N)/N$?

1.25 Prove the answer you gave in the last question.

1.26 With the same conditions as Question 1.24 and the added conditions that N is a very large number and that $\bar{x}_1 = \cdots = \bar{x}_N = 0$, what is the form of the probability density associated with the random variable x?

EXERCISES

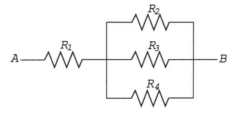

Fig. S1.1   A resistor network.

Exercise 1.1   The probability that resistor R_n $(n = 1, 2, 3, 4)$ is an open circuit $(R_n = \infty)$ is p_n. What is the probability that the resistance between points A and B is infinite?

Solution
$$
\begin{aligned}
R_{AB} &= R_1 + (R_2^{-1} + R_3^{-1} + R_4^{-1})^{-1} \\
&= R_1 + R'
\end{aligned}
\tag{S1.1}
$$

R' will be infinite if and only if R_2, R_3, and R_4 are all infinite. The probability of that is $p_2 p_3 p_4$. R_{AB} will be finite only if R_1 and R' are both finite. The probability of that is $(1 - p_1)(1 - p_2 p_3 p_4)$. Thus

$$
\mathbf{P}[R_{AB} = \infty] = 1 - (1 - p_1)(1 - p_2 p_3 p_4)
\tag{S1.2}
$$

Exercise 1.2   Let A and B be any two sets. Show that the three sets $S_1 = A$, $S_2 = \overline{A} \cap B$, and $S_3 = \overline{A \cup B}$ are mutually disjoint and that their sum is Ω.

Solution   An element x is in S_1 if it is in A. x is in S_2 if it is in B but not in A. (Remember, A and B may intersect.) x is in S_3 if it is in neither A nor B. It is clear that no x can satisfy two of these conditions and that every x must satisfy one of them.

Exercise 1.3   A cube that is painted blue is cut into 64 equal cubes. What is the probability P_n that a little cube, picked at random, has n painted faces, where $n = 0, 1, 2, 3$?

Solution   Of the $4^3 = 64$ cubes, there are $2^3 = 8$ inside cubes that have no paint on them. Thus $P_0 = 8/64 = 1/8$. In each of the six faces of the original cube there are 4 small cubes that will end up with one face painted. Thus $P_1 = 6 \times 4/64 = 3/8$. On each of the 12 edges of the original cube there are 2 small cubes that have two painted faces. Thus $P_2 = 12 \times 2/64 = 3/8$. Finally, the 8 corner cubes will have 3 painted faces, so $P_3 = 8/64 = 1/8$. Notice that $P_0 + P_1 + P_2 + P_3 = 1$.

Exercise 1.4 Two faces of a cube, chosen at random, are painted blue. The cube is then cut into 27 equal cubes. What is the probability that one of the little cubes, picked at random, has two blue faces?

Solution The probability that the two painted faces have a common edge is 4/5. This can be seen by imagining that one of the faces has been painted, leaving 5 unpainted faces, 4 of which share an edge with the painted one. If the two pointed faces do not have a common edge, then none of the small cubes will have two blue faces. If they do have a common edge, then 3 of the 27 cubes will have two blue faces. Thus the answer to the question is

$$P = \frac{4}{5}\frac{3}{27} = \frac{4}{45} \tag{S1.3}$$

Exercise 1.5 Assuming that the birth rate is constant throughout the year (which is definitely not true), what is the probability that a person born in this century was born on the 29th of February?

Solution The 29th of February has occurred every 4 years since 1900. The probability that a day, picked at random is the 29th of February is $1/(4 \times 365 + 1) = 0.000684$.

Exercise 1.6 Ten people throw dice, once per minute, at ten tables. When any person throws 12 he leaves. What is the probability that anyone will be left after one hour?

Solution There are 36 possible states for two dice and only one of them gives 12. The probability of *not* hitting 12 on one throw is 35/36. The probability that a given person will be left after one hour is $(35/36)^{60}$. The probability that a given person will be gone after one hour is $1 - (35/36)^{60}$. The probability that everyone will be gone after one l hour is $[1 - (35/36)^{60}]^{10}$. The answer to the question is

$$\mathbf{P}[\text{someone left}] = 1 - [1 - (35/36)^{60}]^{10} = 0.87 \tag{S1.4}$$

Exercise 1.7 A person is given 4 coins, each having equal and independent probabilities of being a penny, a nickel, a dime, or a quarter. What is the probability that the person has been given 37¢?

Solution Imagine that the coins are given one at a time. There are $4^4 = 256$ different possible sequences of four coins. Each has a probability of 1/256. In order to get 37¢, one must have a quarter, a dime, and two pennies. There are 12 ways of arranging those coins. (Imagine that the coins are numbered 1 to 4 as they are given. The quarter can have 4 possible numbers. The dime can then have any of the 3 remaining numbers. The numbers of the pennies are then determined.) Thus the probability of getting 37¢ is

$$P = \frac{12}{256} = \frac{3}{64} \tag{S1.5}$$

Exercise 1.8   If the integer N is chosen at random from a very large interval, what is the probability that the last digit of N^3 is 1?

Solution   It is always possible to write N as $N = 10K + d$, where $0 \leq d \leq 9$ and K is an integer. ($10K$ is what you get by setting the last digit of N to zero.) Then

$$N^3 = K^3 \times 10^3 + 3dK^2 \times 10^2 + 3d^2 K \times 10 + d^3 \qquad (S1.6)$$

Thus the last digit of N^3 is the same as the last digit of d^3. The last digit of d^3 is 1 only if $d = 1$. Thus the answer to the question is $P = 1/10$.

Exercise 1.9   Given a long table of squares of integers, what is the probability distribution P_n ($n = 0, 1, \ldots, 9$) of the last digits in the table?

Solution   By the argument of the previous problem, we can look at the distribution of the last digits in d^2, where $d = 0, 1, \ldots, 9$. But $0^2 = 0$, $1^2 = 1$, $2^2 = 4$, $3^2 = 9$, $4^2 = 16$, $5^2 = 25$, $6^2 = 36$, $7^2 = 49$, $8^2 = 64$, and $9^2 = 81$. Thus $P_0 = 1/10$, $P_1 = 2/10$, $P_2 = 0$, $P_3 = 0$, $P_4 = 2/10$, $P_5 = 1/10$, $P_6 = 2/10$, $P_7 = 0$, $P_8 = 0$, and $P_9 = 2/10$.

Exercise 1.10   What is the probability that a random number from 1 to 1000, inclusive, will be divisible by 13?

Solution $1000/13 = 76 + 12/13$. Thus, there are 76 multiples of 13 between 1 and 1000. Therefore the probability of hitting one of those multiples is $P = 76/1000 = 0.076$. (Note that $1/13 = 0.0769230769230 \ldots$.)

Exercise 1.11 N black circles of radius a are printed with their centers placed randomly on a sheet of paper of area A. (Note: The circles are not prevented from overlapping.) Assuming N and A are large enough so that edge effects can be neglected, calculate the average value of the total blackened area.

Solution   Because the circles may overlap, the blackened area is less than $N\pi a^2$. Consider an infinitesimal square of area $dA = dx\, dy$ on the paper. dA will be white if the centers of all N circles are further than a from dA. The probability that a particular black circle has its center further than a is $p = (A - \pi a^2)/A = 1 - \pi a^2/A$. The probability that all N centers fall outside the circle of radius a around dA is $P = p^N = (1 - \pi a^2/A)^N$. To write this in a more useful form, let $x = N\pi a^2/A$. Then $P = (1 - x/N)^N$. But $\lim(1 - x/N)^N = e^{-x}$ as $N \to \infty$. Thus the probability that an area dA, picked at random on the sheet, is white is $\exp(-N\pi a^2/A)$. The total blackened area is

$$A' = A\left(1 - e^{-N\pi a^2/A}\right) \qquad (S1.7)$$

There are two things to notice about this result.

1. If $N\pi a^2 \ll A$, then we can ignore overlaps and $A' \approx N\pi a^2$.
2. No matter how large N is, there is always a finite probability of being left with a small white area on the sheet. That is $A' < A$.

Exercise 1.12   The integers from 1 to 30 are written on 30 cards. The cards are shuffled and three cards are drawn. What is the probability that the numbers on them are equal to the lengths of the sides of a right triangle?

Solution   There is a total of $\binom{30}{3} = 4060$, equally probable, sets of 3 cards. There are basically two integer right triangles, namely (3,4,5) and (5,12,13). Counting multiples of those, the only sets that form right triangles are the 8 sets, (3,4,5), (6,8,10), (9,12,15), (12,16,20), (15,20,25), (18,24,30), (5,12,13), and (10,24,26). The probability of drawing one of those is $8/4060 = 1/508$.

Exercise 1.13   A gas contains a fraction x of isotope A and a fraction $1 - x$ of isotope B. What is the probability that n particles, chosen at random, contain at least one A particle?

Solution   The probability that a set of n particles are all B particles is $(1 - x)^n$. The negative of the foregoing statement is the statement that the set contains at least one A particle. Thus the probability of that is $P = 1 - (1 - x)^n$.

Exercise 1.14   In the same situation as was described in the last question, what is the probability that a set of n particles contains at least two A particles?

Solution   Let q_0 be the statement that the set of n particles contains no A particles and q_1 be the statement that the set contains exactly one A particle. From the previous answer, we know that $\mathbf{P}[q_0] = (1 - x)^n$. The probability that the first of the n particles is an A and all the rest are B's is $x(1 - x)^{n-1}$. Thus the probability that any one, but only one, is an A particle is $nx(1 - x)^{n-1}$. The set of n particles will contain at least two As if and only if both q_0 and q_1 are false. Noting that q_0 and q_1 are mutually exclusive, we see that

$$\mathbf{P}[\text{at least 2 } A\text{s}] = 1 - (1 - x)^n - nx(1 - x)^{n-1} \tag{S1.8}$$

Exercise 1.15   A random number generator is a computer program that generates a sequence of random real numbers uniformly distributed over the interval from 0 to 1. What is the probability that the product of two numbers from a random number generator is larger than $\frac{1}{2}$?

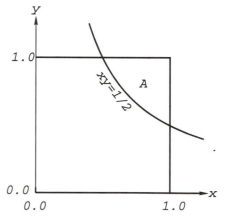

Fig. S1.2   A plot of the function $y = 1/2x$.

Solution   Let the two numbers generated be x and y. The point (x, y) is uniformly distributed on the unit square. The points that have a product larger that $1/2$ are those that lie in the region A, above the curve $xy = 1/2$, or equivalently $y = 1/2x$. (See Fig. S1.2.) The area of that region is the integral

$$A = \int_{1/2}^{1} (1 - 1/2x) \, dx$$
$$= \tfrac{1}{2} - \tfrac{1}{2} \Big[\log x \Big]_{1/2}^{1} \qquad\qquad (S1.9)$$
$$= \tfrac{1}{2}(1 - \log 2) = 0.153$$

which is equal to the probability of getting a pair of numbers whose product is greater than $1/2$.

Exercise 1.16   An integer n is drawn at random from the set 1 through 15. Are the statements, $q_1 =$ "n is odd' 'and $q_2 =$ "n is greater than 10" statistically independent?

Solution   It is easy to see that $\mathbf{P}[q_1] = 8/15$, $\mathbf{P}[q_2] = 5/15$, and $\mathbf{P}[q_1 \text{ and } q_2] = P(11) + P(13) + P(15) = 3/15$. Thus

$$\mathbf{P}[q_1|q_2] = \frac{\mathbf{P}[q_1 \text{ and } q_2]}{\mathbf{P}[q_2]} = \frac{3/15}{5/15} = \frac{3}{5} \neq \mathbf{P}[q_1] \qquad\qquad (S1.10)$$

The statements are not statistically independent.

Exercise 1.17   An ensemble consists of a large collection of points in the x–y plane, uniformly distributed within a circle of radius 2, whose center is the origin. Are the statements $q_1 =$ "$x > 1$" and $q_2 =$ "$y > 1$" statistically independent?

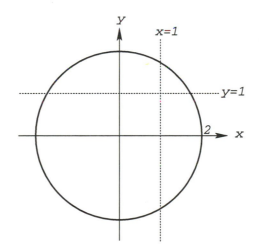

Fig. S1.3

Solution   The circle $x^2 + y^2 = 4$ is shown in Fig. S1.3. The area within the circle, lying to the right of the line $x=1$ we call A_x. That above the line $y=1$ we call A_y. The equation of the upper boundary of the circle is $y = \sqrt{4 - x^2}$. Therefore, the area A_x shown is given by

$$A_x = 2 \int_1^2 \sqrt{4 - x^2} \, dx$$
$$= \left[x\sqrt{4 - x^2} + 4\sin^{-1}\left(\frac{x}{2}\right) \right]_1^2 \qquad \text{(S1.11)}$$
$$= 4\pi/3 - \sqrt{3}$$

The area of the intersection, $A' = A_x \cap A_y$ is

$$A' = \int_1^{\sqrt{3}} (\sqrt{4 - x^2} - 1) \, dx = \pi/3 - \sqrt{3} + 1 \qquad \text{(S1.12)}$$

The area of the circle is 4π. Thus

$$\mathbf{P}[q_1] = \mathbf{P}[q_2] = \frac{A_x}{4\pi} \quad \text{and} \quad \mathbf{P}[q_1 \text{ and } q_2] = \frac{A'}{4\pi} \qquad \text{(S1.13)}$$

It is easy to see that $\mathbf{P}[q_1 \text{ and } q_2] < \mathbf{P}[q_1]\mathbf{P}[q_2]$ and , therefore, q_1 and q_2 are not independent.

Exercise 1.18   A particle moves on the x–y plane. The probability density for its velocity is $P(v_x, v_y)$. From the two assumptions (1) $P(v_x, v_y)$ is a function of $v^2 = v_x^2 + v_y^2$ only, and (2) v_x and v_y are statistically independent, show that $P(v_x, v_y) = (\beta/\pi) \exp(-\beta v^2)$, where β is a constant. This theorem, which can easily be extended to three dimensions, gives a "quick and dirty" derivation of the Maxwell velocity distribution for particles in a gas. It is the one that was originally given by Maxwell. Of course, there is no physical foundation for the second assumption. It does not follow from the first assumption. The first derivation of the Maxwell distribution that was based on physical principles was given by Boltzmann.

Solution   From assumption (1), we can write $P(v_x, v_y) = F(v_x^2 + v_y^2)$, where F is some unknown function. From assumption (2), we can write that $P(v_x, v_y) = p(v_x)p(v_y)$, where

$$p(v_x) = \int_{-\infty}^{\infty} P(v_x, v_y) \, dv_y \qquad \text{(S1.14)}$$

is the probability density for one velocity component. That $p(v_x)$ is equal to $p(v_y)$ follows from assumption (1). Thus

$$p(v_x)p(v_y) = F(v_x^2 + v_y^2) \qquad \text{(S1.15)}$$

Setting v_x to zero, we get $p(0)p(v_y) = F(v_y^2)$. Using this in Eq. (S1.15) gives

$$F(s + t) = C^2 F(s)F(t) \qquad \text{(S1.16)}$$

with $s = v_x^2$, $t = v_y^2$, and $C = 1/p(0)$. Differentiating this equation with respect to t and setting t equal to 0, we get

$$F'(s) = C^2 F'(0)F(s) \equiv -\beta F(s) \qquad \text{(S1.17)}$$

or

$$\frac{d}{ds} \log F(s) = -\beta \tag{S1.18}$$

which has the solution $F(s) = \text{const.} \times e^{-\beta s}$. Therefore

$$P(v_x, v_y) = \text{const.} \times e^{-\beta(v_x^2 + v_y^2)} \tag{S1.19}$$

β cannot be negative because $P(v_x, v_y)$ must be normalizable. The normalization condition for the probability density is

$$\int \int P(v_x, v_y) \, dv_x \, dv_y = \text{const.} \times \left(\int_{-\infty}^{\infty} e^{-\beta v^2} \, dv \right)^2$$
$$= \text{const.} \times \left(\frac{\pi}{\beta} \right) = 1 \tag{S1.20}$$

The integral is given in the Table of Integrals. This gives the value of the constant factor and completes the solution.

Exercise 1.19   A one-dimensional gas of noninteracting, statistically uncorrelated particles has an average density of n particles per unit length. What is the probability density function for the distance between any particle and its nearest-neighbor to the right?

Solution   Take the location of one of the particles as the origin of our coordinate system. Because the particles are uncorrelated, the fact that there is a particle at the origin does not affect the probability distribution for the other particles. Let $p(x)$ be the probability that there is no particle in the interval from 0 to x. The probability of finding a particle in the interval $(x, x + dx)$ is $n \, dx$. Thus the probability that the *nearest particle* will be in the interval $(x, x + dx)$ is $p(x) \, n \, dx$, which is the answer to our question. We must now calculate $p(x)$. There will be no particle in the interval from 0 to $x + dx$ iff there is no particle in the interval 0 to x and there is no particle in the interval from x to $x + dx$. This means that $p(x + dx) = p(x)(1 - n \, dx)$. Using the expansion, $p(x + dx) = p(x) + p'(x) \, dx$, we get the differential equation

$$p'(x) = -n \, p(x) \tag{S1.21}$$

The probability that there is no particle in the interval from 0 to x approaches 1 as x approaches 0. Therefore $p(0) = 1$. With this initial condition, the solution of the differential equation is $p(x) = e^{-nx}$. Therefore, the probability density for the nearest neighbor distance is

$$P(x) = ne^{-nx} \tag{S1.22}$$

Notice that $\int_o^{\infty} P(x) \, dx = 1$.

Exercise 1.20   In the previous exercise, if we choose a particle at random, what is the probability distribution for the distance to its nearest neighbor, which may be left or right?

Solution   Because the particles are independent, the distance to the right neighbor is statistically independent of the distance to the left neighbor. Thus the probability density that the particle's left neighbor is at distance x_L and its right neighbor is at distance x_R is

$$P(x_L, x_R) = n^2 e^{-n(x_L + x_R)} \tag{S1.23}$$

The quantity whose probability distribution we want to calculate is

$$x \equiv \min(x_L, x_R) \tag{S1.24}$$

Using Eq. (1.20), we obtain

$$
\begin{aligned}
P(x) &= n^2 \int_0^\infty \int_0^\infty \delta(x - \min(x_L, x_R)) e^{-n(x_L + x_R)} \, dx_L \, dx_R \\
&= 2n^2 \int_0^\infty dx_1 \, \delta(x - x_1) e^{-n x_1} \int_{x_1}^\infty dx_2 \, e^{-n x_2} \\
&= 2n \int_0^\infty dx_1 \, \delta(x - x_1) e^{-2n x_1} \\
&= 2n e^{-2nx}
\end{aligned}
\tag{S1.25}
$$

Exercise 1.21   A particle traveling through a dilute gas has a probability of $\gamma \, dt$ of being struck by a gas molecule during any very short time interval dt. What is the probability $P(T)$ that, beginning at time 0, the particle survives for a time interval T without being struck?

Solution   The particle will avoid being struck during the time interval $0 < t < T + \Delta T$ iff it is not struck during the time interval $0 < t < T$ and it is not struck during the time interval $T < t < T + \Delta T$. The probability of satisfying the first condition is $P(T)$, and the probability of satisfying the second condition is $1 - \gamma \Delta T$. Therefore

$$P(T + \Delta T) = P(T)(1 - \gamma \, \Delta T) \tag{S1.26}$$

This can be written as

$$\frac{P(T + \Delta T) - P(T)}{\Delta T} = -\gamma P(T) \tag{S1.27}$$

which leads in the limit $\Delta T \to 0$, to the differential equation

$$\frac{dP(T)/dT}{P(T)} = -\gamma \tag{S1.28}$$

The probability that it will survive for a zero time interval is one. That is, $P(0) = 1$. The solution of Eq. (S1.28) that satisfies this initial condition is

$$P(T) = e^{-\gamma T} \tag{S1.29}$$

Exercise 1.22   An ensemble of points in the x–y plane has a probability density $P(x, y) = C \exp(-r^2)$, where $r^2 = x^2 + y^2$. What is the probability that three points, chosen at random from the ensemble, all lie on the same side of the line $x = 1$?

Solution   Because $P(x, y) = (\sqrt{C}e^{-x^2})(\sqrt{C}e^{-y^2})$, it is clear that x and y are statistically independent and that the probability distribution for x is $P(x) = \sqrt{C}e^{-x^2}$. In order for $P(x)$ to be normalized, $\sqrt{C}$ must be equal to $1/\sqrt{\pi}$. The probability that a given point will lie to the right of $x = 1$ is given in terms of the complementary error function.

$$\mathbf{P}[x > 1] = \frac{1}{\sqrt{\pi}} \int_1^\infty e^{-x^2} dx = \tfrac{1}{2}\text{erfc}(1) = 0.07865 \tag{S1.30}$$

The probability that a chosen point lies to the left of the line is then

$$\mathbf{P}[x < 1] = 1 - 0.07865 = 0.92135 \tag{S1.31}$$

The probability that three points all lie on the same side is the sum of the probabilities that they all lie to the right of the line and that they all lie to the left of the line.

$$\mathbf{P}[\text{same side}] = (0.07865)^3 + (0.92135)^3 = 0.7826 \tag{S1.32}$$

Exercise 1.23   Two points A and B are chosen at random on the bottom side of a square. Two points A' and B' are chosen at random on the top side. What is the probability that the lines AA' and BB' intersect inside the square? (See Fig. S1.4.)

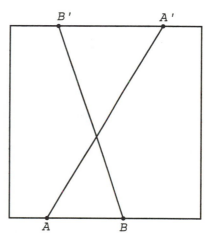

Fig. S1.4   What is the probability that the two lines intersect within the square?

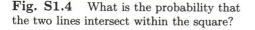

Solution   The lines will intersect inside the square iff $A < B$ and $A' > B'$ or $A > B$ and $A' < B'$. The probability of each of the four separate conditions is clearly $1/2$. Thus the probability of intersection is

$$P = (\tfrac{1}{2})^2 + (\tfrac{1}{2})^2 = \tfrac{1}{2} \tag{S1.33}$$

What is the probability that three lines, so drawn, will all intersect?

Exercise 1.24   A particle is distributed along the positive x axis with a probability density $P(x) = a^{-1}e^{-x/a}$. What is the probability density associated with the variable $u = x^3$?

Solution   For a given positive number α, the probability that $u < \alpha$ is

$$
\begin{aligned}
\mathbf{P}[u < \alpha] &= \frac{1}{a} \int_o^{\alpha^{1/3}} e^{-x/a}\,dx \\
&= \int_o^{\alpha^{1/3}/a} e^{-y}\,dy \\
&= \left(1 - e^{-\alpha^{1/3}/a}\right)
\end{aligned}
\qquad (S1.34)
$$

But the probability density $P_u(\alpha)$ is the derivative of $\mathbf{P}[u < \alpha]$ with respect to α. Thus

$$
P_u(\alpha) = \frac{e^{-\alpha^{1/3}/a}}{3\alpha^{2/3}a}
\qquad (S1.35)
$$

Another way of doing this problem is to use the general formula given in Eq. (1.50). Then

$$
P_u(\alpha) = \frac{1}{a} \int_o^\infty e^{-x/a}\,\delta(\alpha - x^3)\,dx
\qquad (S1.36)
$$

The general rule for doing an integral involving a delta function of a function is that

$$
\int_a^b F(x)\,\delta(\alpha - f(x))\,dx = \sum_r \frac{F(r)}{|f'(r)|}
\qquad (S1.37)
$$

where the sum is over all solutions of the equation $f(x) = \alpha$ that lie in the interval (a, b). In the case at hand, the only solution to the equation $x^3 = \alpha$ in the interval $0 < x < \infty$ is $x = \alpha^{1/3}$, and $f'(x) = d(x^3)/dx = 3x^2$.

Exercise 1.25   The particles in a gas in a uniform gravitational field g at temperature T have a probability density $P(z) = h^{-1}e^{-z/h}$, where the scale height $h = kT/mg$ and k is Boltzmann's constant. If two particles are picked at random, what is the probability density associated with the absolute difference in their heights $Z = |z_1 - z_2|$?

Solution   By Eq. 1.50,

$$
P_Z(x) = \frac{1}{h^2} \int \int \delta(x - |z_1 - z_2|)e^{-(z_1+z_2)/h}\,dz_1\,dz_2
\qquad (S1.38)
$$

Using the fact that the integrand is symmetric under exchange of z_1 and z_2, we can restrict the region of integration to $z_1 > z_2$ and multiply the result by 2.

$$
P_Z(x) = \frac{2}{h^2} \iint_{z_1 > z_2} \delta(x - (z_1 - z_2))e^{-(z_1+z_2)/h}\,dz_1\,dz_2
\qquad (S1.39)
$$

Changing variables from z_1 to $u = z_1 - z_2$ and using the fact that $z_1 + z_2 = u + 2z_2$, we get

$$
\begin{aligned}
P_Z(x) &= \frac{2}{h^2} \int_o^\infty du \int_o^\infty dz_2\,\delta(x - u)e^{-(u+2z_2)/h} \\
&= \frac{e^{-x/h}}{h}
\end{aligned}
\qquad (S1.40)
$$

Exercise 1.26   A two-digit number is uniformly distributed from 00 to 99. What is the probability function for the sum of the two digits?

Solution   There is one combination that gives a sum of 0. There are 2 that give a sum of 1, 3 that give a sum of 2, 4 that give a sum of 3, ..., 10 that give a sum of 9, 9 that give a sum of 10, 8 that give a sum of 11, ..., 1 that gives a sum of 18. If s is the sum of the digits, then $0 \le s \le 18$ and

$$P(s) = \frac{10 - |s - 9|}{100} \tag{S1.41}$$

Exercise 1.27   10,000 tickets are numbered from 0000 to 9999. What is the probability of drawing a ticket in which the sum of the first two digits equals the sum of the last two?

Solution   The first two digits are statistically independent of the last two digits. Therefore $P(s, s') = P(s)P(s')$. The probability distribution for the sum of either pair of digits was given in the last exercise. The probability that $s = s'$ is

$$
\begin{aligned}
P = \sum_{s=0}^{18} P(s, s) &= \sum_{s=0}^{18} \frac{(10 - |s - 9|)^2}{10,000} \\
&= 2 \sum_{s=0}^{8} \frac{(1 + s)^2}{10,000} + \frac{1}{100} = 0.067
\end{aligned}
\tag{S1.42}
$$

Exercise 1.28   Let $P_1(x_1)$, $P_2(x_2)$, and $P_3(x_3)$ be the probability densities associated with three independent random variables. If $S = x_1 + x_2 + x_3$, show that

$$P_S(s) = \int P_1(s - v)P_2(v - w)P_3(w)\, dv\, dw \tag{S1.43}$$

Solution   We know that

$$P_S(s) = \int P_1(x_1)P_2(x_2)P_3(x_3)\, \delta(s - x_1 - x_2 - x_3)\, dx_1\, dx_2\, dx_3 \tag{S1.44}$$

Introducing variables $u = x_1 + x_2 + x_3$, $v = x_2 + x_3$, and $w = x_3$. The Jacobian of the transformation is

$$\frac{\partial(u, v, w)}{\partial(x_1, x_2, x_3)} = \begin{vmatrix} 1 & 1 & 1 \\ 0 & 1 & 1 \\ 0 & 0 & 1 \end{vmatrix} = 1 \tag{S1.45}$$

The inverse transformation is $x_1 = u - v$, $x_2 = v - w$, and $x_3 = w$. Thus

$$
\begin{aligned}
P_S(s) &= \int P_1(u - v)P_2(v - w)P_3(w)\delta(s - u)\, du\, dv\, dw \\
&= \int P_1(s - v)P_2(v - w)P_3(w)\, dv\, dw
\end{aligned}
\tag{S1.46}
$$

The integral on the right-hand side is called the *convolution* of the functions P_1, P_2, and P_3. The result can be extended to any number of random variables.

Exercise 1.29   A particle is distributed along the positive x axis with a probability density $P(x) = Cx^2 \exp[-(x/a)^2]$. Determine its average coordinate $\bar{x}$ and the fluctuation Δx.

Solution   First we must determine the normalization constant C.

$$C \int_o^\infty x^2 e^{-(x/a)^2} \, dx = Ca^3 \int_o^\infty u^2 e^{-u^2} \, du = C \frac{a^3 \sqrt{\pi}}{4} = 1 \qquad \text{(S1.47)}$$

Therefore $C = 4/a^3 \sqrt{\pi}$. Then

$$
\begin{aligned}
\bar{x} &= \frac{4}{a^3 \sqrt{\pi}} \int_o^\infty x^3 e^{-(x/a)^2} \, dx \\
&= \frac{4a}{\sqrt{\pi}} \int_o^\infty u^3 e^{-u^2} \, du = \frac{2a}{\sqrt{\pi}}
\end{aligned}
\qquad \text{(S1.48)}
$$

and $(\Delta x)^2 = \overline{x^2} - \bar{x}^2$ where

$$
\begin{aligned}
\overline{x^2} &= \frac{4}{a^3 \sqrt{\pi}} \int_o^\infty x^4 e^{-(x/a)^2} \, dx \\
&= \frac{4a^2}{\sqrt{\pi}} \int_o^\infty u^4 e^{-u^2} \, du = \tfrac{3}{2} a^2
\end{aligned}
\qquad \text{(S1.49)}
$$

Therefore

$$(\Delta x)^2 = (\frac{3}{2} - \frac{4}{\pi}) a^2 = 0.227 a^2 \qquad \text{(S1.50)}$$

Exercise 1.30   The coordinate of a particle is distributed over the whole x axis with a probability density $P(x) = C/(a^2 + x^2)^{3/2}$. Calculate C, $\bar{x}$, and Δx.

Solution

$$C^{-1} = \int_{-\infty}^\infty \frac{dx}{(a^2 + x^2)^{3/2}} = \left[\frac{x}{a^2 \sqrt{a^2 + x^2}} \right]_{-\infty}^\infty = \frac{2}{a^2} \qquad \text{(S1.51)}$$

$\bar{x} = 0$ because $P(x) = P(-x)$. Therefore

$$(\Delta x)^2 = \overline{x^2} = \frac{a^2}{2} \int_{-\infty}^\infty \frac{x^2 \, dx}{(a^2 + x^2)^{3/2}} = \infty \qquad \text{(S1.52)}$$

For this probability distribution, the mean-square fluctuation diverges!

Exercise 1.31   The Fourier transform of the probability density,

$$\tilde{P}(k) = \int_{-\infty}^\infty e^{ikx} P(x) \, dx \qquad \text{(S1.53)}$$

is called the *characteristic function* of the random variable x. Let $F(k) = \log \tilde{P}(k)$ and prove that: (1) $F(0) = 0$, (2) $F'(0) = i\bar{x}$, and (3) $F''(0) = -(\Delta x)^2$.

Solution $\tilde{P}(0) = \int P(x)\, dx = 1$. Therefore $F(0) = \log 1 = 0$.

$$\tilde{P}'(0) = i \int x P(x)\, dx = i\bar{x} \tag{S1.54}$$

$$F'(0) = \frac{\tilde{P}'(0)}{\tilde{P}(0)} = i\bar{x} \tag{S1.55}$$

$$\tilde{P}''(0) = -\int x^2 P(x)\, dx = -\overline{x^2} \tag{S1.56}$$

But $F'(k) = \tilde{P}'(k)/\tilde{P}(k)$ and therefore

$$F''(k) = \frac{\tilde{P}''(k)}{\tilde{P}(k)} - \frac{\left[\tilde{P}'(k)\right]^2}{\left[\tilde{P}(k)\right]^2} \tag{S1.57}$$

which implies that

$$F''(0) = -\overline{x^2} + \bar{x}^2 \tag{S1.58}$$

Exercise 1.32 N points are randomly and independently distributed over an interval L. Calculate the probability that some subinterval ℓ will contain exactly n points.

Solution   Assume that the N points are laid down one after the other. The probability that a particular point will fall in ℓ is ℓ/L and the probability that it will fall outside ℓ is $(L-\ell)/L$. Therefore, the probability that the first n points will fall inside ℓ and the last $N - n$ points will fall outside ℓ is $\ell^n (L - \ell)^{N-n}/L^N$. The probability that any set of n points will fall in ℓ and all other points outside ℓ is

$$P = \binom{N}{n} \frac{\ell^n (L - \ell)^{N-n}}{L^N} \tag{S1.59}$$

Exercise 1.33   A pair of two-dimensional vector variables, $\mathbf{x} = (x_1, x_2)$ and $\mathbf{y} = (y_1, y_2)$, are statistically independent and have probability densities $P_1(\mathbf{x})$ and $P_2(\mathbf{y})$. Let $\mathbf{u} = \mathbf{x} + \mathbf{y} = (u_1, u_2)$ and show that

$$\langle \mathbf{u} \rangle = \langle \mathbf{x} \rangle + \langle \mathbf{y} \rangle \tag{S1.60}$$

and

$$\langle (u_i - \bar{u}_i)(u_j - \bar{u}_j) \rangle = \langle (x_i - \bar{x}_i)(x_j - \bar{x}_j) \rangle + \langle (y_i - \bar{y}_i)(y_j - \bar{y}_j) \rangle \tag{S1.61}$$

where $i, j = 1, 2$. The quantity $\langle (u_i - \bar{u}_i)(u_j - \bar{u}_j) \rangle$ is called the *covariance matrix*. For a one-dimensional variable it has only one component, equal to $(\Delta u)^2$.

Solution   Since the variables are statistically independent, their joint probability is $P(\mathbf{x}, \mathbf{y}) = P_1(\mathbf{x}) P_2(\mathbf{y})$.

$$\begin{aligned}
\langle \mathbf{u} \rangle &= \int (\mathbf{x} + \mathbf{y}) P_1(\mathbf{x}) P_2(\mathbf{y})\, d^2x\, d^2y \\
&= \int \mathbf{x} P_1(\mathbf{x})\, d^2x \int P_2(\mathbf{y})\, d^2y + \int P_1(\mathbf{x})\, d^2x \int \mathbf{y} P_2(\mathbf{y})\, d^2y \\
&= \langle \mathbf{x} \rangle + \langle \mathbf{y} \rangle
\end{aligned} \tag{S1.62}$$

Also,

$$\langle(u_i - \bar{u}_i)(u_j - \bar{u}_j)\rangle = \langle(x_i - \bar{x}_i + y_i - \bar{y}_i)(x_j - \bar{x}_j + y_j - \bar{y}_j)\rangle \qquad \text{(S1.63)}$$

But,

$$\langle(x_i - \bar{x}_i)(y_j - \bar{y}_j)\rangle = \int P_1(\mathbf{x})(x_i - \bar{x}_i)\,d^2x \int (y_j - \bar{y}_j)P_2(\mathbf{y})\,d^2y = 0 \qquad \text{(S1.64)}$$

Similarly $\langle(y_i - \bar{y}_i)(x_j - \bar{x}_j)\rangle = 0$, which easily gives the desired result for the covariance matrix.

Exercise 1.34   Given a random variable x with average $\bar{x}$ and uncertainty Δx and a smooth function $y(x)$, show that, in the case that Δx is very small, if we keep only first-order terms in Δx, then

$$\bar{y} = y(\bar{x}) \quad \text{and} \quad \Delta y = |y'(\bar{x})|\,\Delta x \qquad \text{(S1.65)}$$

Solution   For any particular value of x, let $\delta x \equiv x - \bar{x}$. Then $x = \bar{x} + \delta x$, $\langle \delta x \rangle = 0$, and $(\Delta x)^2 = \langle(\delta x)^2\rangle$. Using these relations, we see that

$$\bar{y} = \langle y(x)\rangle = \langle y(\bar{x} + \delta x)\rangle \approx y(\bar{x}) + y'(\bar{x})\langle \delta x\rangle + \tfrac{1}{2}y''(\bar{x})\langle(\Delta x)^2\rangle \approx y(\bar{x}) \qquad \text{(S1.66)}$$

and

$$\begin{aligned}
(\Delta y)^2 &= \langle[y(\bar{x} + \delta x) - y(\bar{x})]^2\rangle \\
&\approx [y'(\bar{x})]^2\langle(\delta x)^2\rangle \qquad\qquad\qquad \text{(S1.67)} \\
&= [y'(\bar{x})]^2(\Delta x)^2
\end{aligned}$$

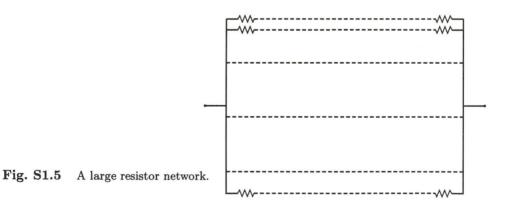

Fig. S1.5   A large resistor network.

Exercise 1.35   The rectangular array of resistors shown in Fig. S1.5 has K parallel rows of L resistors in series. The resistors are drawn from a lot with an average resistance of $1\,\Omega$ and an uncertainty of 10%. Assuming that K and L are large numbers, calculate the average of and the uncertainty in the equivalent resistance of the array.

Solution   The single series branches have, according to the law of large numbers, a mean resistance of $L\,\Omega$ and an uncertainty of $\Delta R = 0.1\sqrt{L}\,\Omega$. The conductance of a branch is given by $C = 1/R$. Using the result of the previous exercise, we see that

$$\langle C \rangle = \frac{1}{\langle R \rangle} = L^{-1} \tag{S1.68}$$

and

$$\Delta C = \frac{\Delta R}{\langle R \rangle^2} = 0.1\,L^{-3/2} \tag{S1.69}$$

The conductance of the whole array, C_A, is the sum of the conductances of the branches. Therefore

$$\langle C_A \rangle = K \langle C \rangle = K/L \tag{S1.70}$$

and

$$\Delta C_A = \sqrt{K}\,\Delta C = 0.1\,\sqrt{K}/L^{3/2} \tag{S1.71}$$

The resistance of the array is $R_A = 1/C_A$. Using the result of the previous exercise again, we get

$$\langle R_A \rangle = \frac{L}{K} \quad \text{and} \quad \Delta R_A = \frac{\Delta C_A}{\langle C_A \rangle^2} = 0.1\,L^{1/2}K^{-3/2} \tag{S1.72}$$

Notice that

$$\frac{\Delta R_A}{\langle R_A \rangle} = \frac{0.1}{\sqrt{LK}} \tag{S1.73}$$

which is the uncertainty in one element divided by the square root of the total number of elements, even though R_A is not simply a sum of the resistances of all the elements.

Exercise 1.36   Consider an $N \times N$ array of resistors, similar to that shown in the previous exercise. Assume that each resistor has a probability of $1\,\%$ of being an open circuit ($R = \infty$). What is the probability that the whole array is an open circuit? Evaluate the result for $N = 3$ and 1000.

Solution   The probability that any particular series branch has a finite resistance is $(0.99)^N$. Therefore, the probability that the branch has an infinite resistance is $1 - (0.99)^N$. The whole circuit will have an infinite resistance only if every branch has an infinite resistance. The probability of that is

$$P = \left[1 - (0.99)^N\right]^N \tag{S1.74}$$

For $N = 3$, $P \approx 0.000026$ and, for $N = 1000$, $P \approx 0.96$.

Exercise 1.37   For a random variable X, with a range of $-\infty < X < \infty$, it is known that

$$\mathbf{P}[X < x] = A + B \tan^{-1} x \tag{S1.75}$$

Determine the values of A and B and the probability density $P_X(x)$.

Solution The limit as $x \to -\infty$ of $\mathbf{P}[X < x]$ must be zero. But $\tan^{-1}(-\infty) = -\pi/2$. Therefore $A = \pi B/2$. The limit as $x \to \infty$ of $\mathbf{P}[X < x]$ must be one. Thus

$$\frac{\pi B}{2} + \frac{\pi B}{2} = 1 \tag{S1.76}$$

or $B = 1/\pi$. Therefore

$$\mathbf{P}[X < x] = \tfrac{1}{2} + \pi^{-1}\tan^{-1}x \tag{S1.77}$$

But

$$P_X(x) = \frac{d}{dx}\mathbf{P}[X < x] = \frac{1}{\pi(1+x^2)} \tag{S1.78}$$

Exercise 1.38 An atom has a magnetic moment that is a vector $\mathbf{m}$ of fixed length m. All possible directions of $\mathbf{m}$ are equally likely. What is the probability density associated with the z component of $\mathbf{m}$?

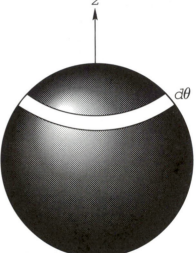

Fig. S1.6 The magnetic moment vector is equally distributed over the sphere.

Solution Picture a very large number, N, of such atoms. If we plot the magnetic moment vectors of all N atoms in a three-dimensional vector space, they will be uniformly distributed on a sphere of radius m (Fig. S1.6). Let $m_z = m\cos\theta$. Clearly, the probability that θ lies in some small interval $d\theta$ is equal to the area on the sphere associated with the angular range $d\theta$ (that is, $2\pi m^2 \sin\theta\, d\theta$) divided by the total area of the sphere, $4\pi m^2$. That is, $dP = \tfrac{1}{2}\sin\theta\, d\theta$. But $\cos\theta = m_z/m$, and therefore $\sin\theta\, d\theta = dm_z/m$ and $dP = dm_z/2m$. This implies that $P(m_z) = 1/2m$. That is, the component m_z is uniformly distributed in its range, $-m \le m_z \le m$.

Exercise 1.39 A molecule has a finite number K of nondegenerate rotational states of energies E_1, E_2, $\ldots, E_K$. The molecule is bathed in electromagnetic radiation with a continuous spectrum. If the molecule is in state k, then, during the short time interval dt, it has a probability $A(k \to \ell)\, dt$ of absorbing a photon and making a transition to state ℓ and it also has a probability $E(k \to m)\, dt$ of emitting a photon and making a transition from state k to state m. Suppose that, at $t = 0$, the molecule was in state k_o. Construct an explicit equation for calculating $P_n(t)$, the probability that, at time t, the molecule is in state n.

Solution   First, let us define a $K \times K$ *transition rate matrix* $T_{k\ell}$ by saying that

$$T_{k\ell} = \begin{cases} E(k \to \ell), & \text{if } E_\ell < E_k \\ 0, & \text{if } k = \ell \\ A(k \to \ell), & \text{if } E_\ell > E_k \end{cases} \tag{S1.79}$$

Then, during the time interval dt, a molecule in state k has a probability $T_{k\ell}\, dt$ of making a transition to state ℓ, where ℓ is any of the other $K - 1$ states. At time $t + dt$ the molecule will be in state k if it was in state k at time t and made no transition during the time interval dt *or* it was in some other state ℓ and made a transition into state k during the time interval dt (we can, for very small dt, neglect the possibility of the molecule's making two or more transitions). The probability that it was in state k at time t and made no transition during the time interval dt is $P_k(t)(1 - \sum_{\ell \neq k} T_{k\ell}\, dt)$. The probability that it was in some other state ℓ at time t and made a transition to state k during the time interval dt is $\sum_{\ell \neq k} P_\ell(t) T_{\ell k}\, dt$. Therefore, the probability that, at time $t + dt$, it will end up in state k is

$$P_k(t + dt) = P_k(t)\left(1 - \sum_{\ell \neq k} T_{k\ell}\, dt\right) + \sum_{\ell \neq k} P_\ell(t) T_{\ell k}\, dt \tag{S1.80}$$

This equation can be rearranged to read

$$\frac{P_k(t + dt) - P_k(t)}{dt} = \sum_{\ell \neq k}\left[P_\ell(t) T_{\ell k} - P_k(t) T_{k\ell}\right] \tag{S1.81}$$

The limit $dt \to 0$ obviously gives the set of differential equations

$$\frac{dP_k(t)}{dt} = \sum_{\ell \neq k}\left[P_\ell(t) T_{\ell k} - P_k(t) T_{k\ell}\right] \tag{S1.82}$$

which must be solved with the initial condition

$$P_k(0) = \begin{cases} 1, & \text{for } k = k_o \\ 0, & \text{otherwise} \end{cases} \tag{S1.83}$$

This equation is called the *master equation* for a statistical process of this sort.

Exercise 1.40   The following question, called the *Monty Hall Problem* after the television game show host, Monty Hall, was correctly answered by the newspaper columnist Marilyn vos Savant in her weekly column. Her solution stimulated thousands of letters, many from professors of mathematics and statistics, that claimed that her answer was incorrect. Here is the problem:

A game show contestant is presented with three closed doors. Behind two of the doors are goats and behind the third is a new car. The contestant chooses a door without opening it. Monty Hall, who knows which door has the car, always goes to one of the other doors and opens it to reveal a goat. The contestant is then given the opportunity of switching to the other unopened door. Is it to the contestant's advantage to switch, to remain with the original choice, or does it make no difference?

Solution Let us call the doors A, B, and C and assume that door A has been chosen and door C has been opened. Knowing that door C has been opened, we are then faced with a problem of conditional probability. After A has been chosen but before any door was opened there were four possible sequences with the following probabilities (we are assuming that, when the contestant chooses the car, Monty Hall chooses one of the remaining two doors randomly)

  1. The car is behind A and Monty Hall opens B. $P[A$ and $B] = \frac{1}{6}$.

  2. The car is behind A and Monty Hall opens C. $P[A$ and $C] = \frac{1}{6}$.

  3. The car is behind B and Monty Hall opens C. $P[B$ and $C] = \frac{1}{3}$.

  4. The car is behind C and Monty Hall opens B. $P[C$ and $B] = \frac{1}{3}$.

Given that Monty Hall has opened C, what we need to calculate are the two conditional probabilities

$$P[A|C] = \frac{P[A \text{ and } C]}{P[C]} = \frac{1/6}{1/6 + 1/3} = \frac{1}{3}$$

$$P[B|C] = \frac{P[B \text{ and } C]}{P[C]} = \frac{1/3}{1/6 + 1/3} = \frac{2}{3}$$

The contestant who consistently switches wins two thirds of the time. An easy way of seeing this is to imagine a new game in which there is a second contestant who always gets whatever is behind the two doors that the first contestant did not pick and in which there are no door openings by Monty Hall and no switches allowed. Clearly, the second contestant would get twice as many cars as the first. But the second contestant in this new game would get exactly as many cars as the first contestant in the old game who consistently switched.

Supplement to Chapter 2

REVIEW QUESTIONS

2.1 For a spatially uniform ideal gas, how is the velocity distribution $f(\mathbf{v})$ defined?

2.2 Derive the relation $p = \frac{2}{3}E/V$ for an ideal gas.

2.3 In deriving the Maxwell–Boltzmann distribution, how are microstates and macrostates defined?

2.4 For a given set of occupation numbers N_{kl}, how many microstates correspond to that macrostate?

2.5 What is the assumption of "equal a priori probability"?

2.6 What is the function that must be maximized in the derivation of the Maxwell–Boltzmann distribution?

2.7 What are the constraints on the set of variables in the function to be maximized in the previous question?

2.8 Describe Lagrange's procedure for finding the maximum of a function $F(x, y, z)$ on the constraint surface $g(x, y, z) = 1$.

2.9 Using the results of Questions 4–8, derive the Maxwell–Boltzmann distribution.

2.10 What is the form of the Maxwell–Boltzmann distribution?

2.11 For ideal gas particles in an external potential $U(\mathbf{r})$ that is equal to zero at some location $\mathbf{r}_o$, what is the equilibrium particle density?

2.12 For a uniform ideal gas of density n, calculate the normalization constant C in the Maxwell velocity distribution $f(\mathbf{v}) = C \exp(-\frac{1}{2}mv^2/kT)$.

2.13 How is the speed distribution $F(v)$ related to the velocity distribution $f(\mathbf{v})$?

2.14 How are the average speed $\bar{v}$ and v_{rms} defined?

2.15 For an ideal gas at temperature T, what is v_{rms}?

2.16 How is the mean free path defined?

2.17 For a quantum mechanical system of N particles in a three-dimensional potential $U(\mathbf{r})$, what is the Hamiltonian operator?

2.18 How are the single-particle energy eigenfunctions defined?

2.19 For a system of N Bose–Einstein particles without interactions, describe the form of the N-particle energy eigenfunctions.

2.20 What are the allowed values for the quantum state occupation numbers N_k

for Bose–Einstein particles?

2.21 For a system of N Fermi–Dirac particles without interactions, describe the form of the N-particle energy eigenfunctions.

2.22 What are the allowed values for the quantum state occupation numbers N_k for Fermi–Dirac particles?

2.23 In deriving the Bose–Einstein and Fermi–Dirac distribution functions, how were microstates and macrostates defined?

2.24 How many ways can ν fermions be arranged in K quantum states?

2.25 How many Fermi–Dirac microstates are there for a macrostate defined by the bin occupation numbers $\nu_1, \nu_2, \ldots$ if the kth bin has K_k states?

2.26 Using the result of the previous question, derive the Fermi–Dirac distribution function, $f_{\text{FD}}(\varepsilon) = (e^{\alpha + \beta \varepsilon} + 1)^{-1}$.

2.27 What is the form of the Bose–Einstein distribution function $f_{\text{BE}}(\varepsilon)$?

2.28 For a particle with no external potential in three dimensions, write the Schrödinger energy eigenvalue equation.

2.29 For a particle in a cubic box of side L, what are "periodic boundary conditions"?

2.30 What are the eigenfunctions and eigenvalues of the Schrödinger equation for a free particle with periodic boundary conditions?

2.31 For a particle in a cubic periodic box of side L, calculate the number of energy eigenstates with energy less than E.

2.32 Under what condition do the quantum mechanical distribution functions reduce to the Maxwell distribution?

EXERCISES

Exercise 2.1 Draw a realistic picture of the situation in a typical noble gas, say argon, at STP by drawing two spheres with diameters proportional to the "interaction diameters" of argon atoms separated by a distance that is proportional, with the same proportionality constant, to the typical distance between particles, namely $d = 1/n^{1/3}$. On the same scale, how large is the mean free path?

Fig. S2.1

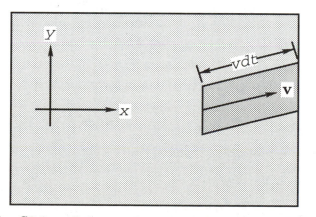

Solution At STP, the particle density is $2.7 \times 10^{25}\,\mathrm{m}^{-3}$. Thus $d = 3.3 \times 10^{-9}\,\mathrm{m}$. The *Handbook of Chemistry and Physics* gives an approximate diameter for Ar, based on crystal structure, of $3.1\,\text{Å} = 3.1 \times 10^{-10}\,\mathrm{m}$. Thus the ratio of the atomic diameter to d is about one to ten.

The mean free path of hard spherical particles is calculated approximately in Problem 2.8 and accurately in Exercises 2.15 through 2.17. It is given by $\ell = 1/(\sqrt{2}\,\pi D^2 n)$, where D is the diameter of the particles and n is the particle density. For argon at STP, $\ell = 8.7 \times 10^{-8}$. In Fig. S2.1, in which d is represented by two inches, ℓ would be about 4.4 feet (1.34 m).

Exercise 2.2 A two-dimensional ideal gas is a collection of noninteracting particles that move on a plane. Suppose a two-dimensional ideal gas of N particles, with a total energy E, is confined by one-dimensional "walls" to an area A. The pressure on a wall is defined as the force per unit length. Show that $p = E/A$.

Fig. S2.2 All the particles with velocity **v** that are in the shaded region will strike the wall within the time interval dt.

Solution   Let $f(v_x, v_y)\, dv_x\, dv_y$ be the number of particles per unit area that have velocities in the range $dv_x\, dv_y$. Consider a length ℓ of the container wall, which we assume is in the y direction. A particle whose velocity is in the range $dv_x\, dv_y$ will strike the wall in that length, during the time interval dt, if it is within the shaded region shown in Fig. S2.2. The area of that region is $\ell v_x\, dt$, and therefore the number of such particles is

$$dN = \ell v_x\, dt\, f(v_x, v_y)\, dv_x\, dv_y \qquad (S2.1)$$

The momentum delivered to the section of wall by the rebounding particle is $2mv_x$. Thus the total rate at which momentum is being delivered to that section of the wall is

$$\begin{aligned}
\frac{dP}{dt} \equiv F &= 2\ell m \int_{v_x > 0} f(v_x, v_y) v_x^2\, dv_x\, dv_y \\
&= \ell m \int f(v_x, v_y) v_x^2\, dv_x\, dv_y \\
&= \ell \int f(v_x, v_y) \left[\frac{m}{2}(v_x^2 + v_y^2) \right] dv_x\, dv_y \\
&= \ell E / A
\end{aligned} \qquad (S2.2)$$

Exercise 2.3   Consider two very dilute gases of slightly different densities and temperatures, separated by a thin wall with a small hole in it of area A. Calculate the rate at which particles and energy are transferred through the hole as a function of the temperature and density differences, to first order in ΔT and Δn. (See Fig. S2.3.)

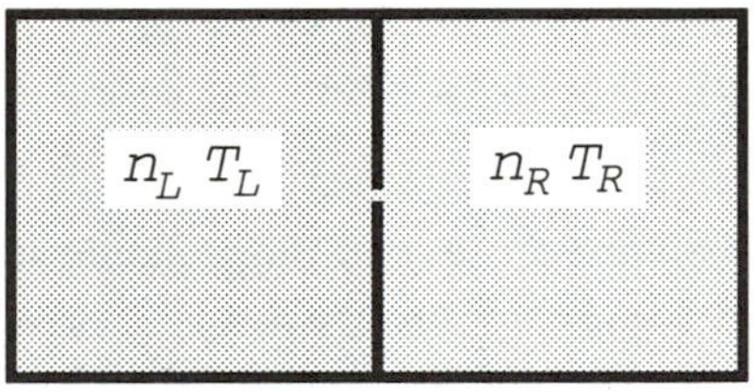

Fig. S2.3 n_L, T_L, n_R, and T_R are the densities and temperatures of the dilute gases on the two sides.

Solution   The rate at which particles pass through the hole, from left to right, is

$$R(L \to R) = A n_L (m/2\pi k T_L)^{3/2} \int_{v_x > 0} e^{-mv^2/2kT_L} v_x\, d^3v \qquad (S2.3)$$

Using spherical coordinates with the polar axis in the x direction

$$\begin{aligned}
R(L \to R) &= A n_L (m/2\pi k T_L)^{3/2} 2\pi \int_o^{\pi/2} \cos\theta \sin\theta\, d\theta \int_o^\infty e^{-mv^2/2kT_L} v^3\, dv \\
&= \pi A n_L (m/2\pi k T_L)^{3/2} \tfrac{1}{2} (2kT_L/m)^2 \\
&= \frac{1}{\sqrt{2\pi}} A n_L (kT_L/m)^{1/2} \\
&= \frac{1}{4} n_L \bar{v}_L A
\end{aligned}$$

$$(S2.4)$$

where $\bar{v}_L$ is the average speed of the particles on the left, given in Eq. (2.30). The rate of transfer of particles from right to left is

$$R(R \to L) = \frac{An_R T_R^{1/2}}{\sqrt{2\pi m/k}} \tag{S2.5}$$

Assuming that $n_L = n + \Delta n/2$, $T_L = T + \Delta T/2$, $n_R = n - \Delta n/2$, and $T_R = T - \Delta T/2$, the net rate of particle transfer from left to right is

$$
\begin{aligned}
\frac{dN_R}{dt} &= R(L \to R) - R(R \to L) \\
&= \frac{A}{\sqrt{2\pi m/k}} \Big[(n + \Delta n/2)(T + \Delta T/2)^{1/2} \\
&\qquad - (n - \Delta n/2)(T - \Delta T/2)^{1/2} \Big] \\
&\approx \frac{A}{\sqrt{2\pi m/k}} \Big(T^{1/2} \Delta n + \frac{1}{2}\frac{n}{T^{1/2}} \Delta T \Big)
\end{aligned}
\tag{S2.6}
$$

A particle of speed v that goes through the hole carries an amount of energy $mv^2/2$. Therefore the rate at which energy is carried through the hole from left to right can be obtained from Eq. (S2.3) by putting a factor of $mv^2/2$ in the integrand.

$$
\begin{aligned}
R_E(L \to R) &= An_L(m/2\pi kT_L)^{3/2} \int_{v_x > 0} e^{-mv^2/2kT_L}(mv^2/2)v_x\, d^3v \\
&= \pi m A n_L (m/2\pi kT_L)^{3/2} \int_o^{\pi/2} \cos\theta \sin\theta\, d\theta \cdot \int_o^\infty e^{-mv^2/2kT_L} v^5\, dv \\
&= \frac{1}{2\sqrt{\pi}} m A n_L (2kT_L/m)^{3/2} \\
&= \Big(\frac{2k^3}{\pi m} \Big)^{1/2} An_L T_L^{3/2}
\end{aligned}
\tag{S2.7}
$$

with a similar formula for $R_E(R \to L)$. The net rate of transfer of energy is

$$
\begin{aligned}
\frac{dE_R}{dt} &= \Big(\frac{2k^3}{\pi m} \Big)^{1/2} A \Big[(n + \Delta n/2)(T + \Delta T/2)^{3/2} - (n - \Delta n/2)(T - \Delta T/2)^{3/2} \Big] \\
&= \Big(\frac{2k^3}{\pi m} \Big)^{1/2} A \big(T^{3/2} \Delta n + \tfrac{3}{2} n T^{1/2} \Delta T \big)
\end{aligned}
\tag{S2.8}
$$

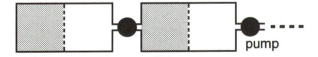

Fig. S2.4   A series of diffusion chambers, each with many small holes through which the gas diffuses.

Exercise 2.4   This exercise concerns a simplified model of the process of isotope separation by diffusion. Two chambers are separated by a thin wall containing many small holes, as shown in Fig. S2.4. In the left chamber is a very dilute mixed ideal gas containing a density n_A of particles of mass m_A and a density n_B of particles of mass m_B. The right chamber is pumped so that its density can be approximated as zero. (a) What is the ratio of the number of B particles to A particles in the gas that is being pumped out of the right chamber? (b) Assume that the gas being pumped out of the right chamber is being pumped into the left chamber of a second diffusion stage similar to this one. Assume that the process is repeated for 20 stages, that $n_A = n_B$ in the initial chamber, and that $m_A/m_B = 0.8$. What is n_A/n_B in the final chamber?

Solution   (a) According to Eq. (S2.3), the rate at which A particles pass through the wall is

$$\frac{dN_A}{dt} = \frac{A}{4}\bar{v}_A n_A \tag{S2.9}$$

where $\bar{v}_A = \sqrt{8kT/\pi m_A}$. Similarly, the rate at which B particles pass through the wall is

$$\frac{dN_B}{dt} = \frac{A}{4}\bar{v}_B n_B \tag{S2.10}$$

The ratio of the number of B particles to that of A particles in the gas extracted from the right chamber is

$$\frac{dN_B/dt}{dN_A/dt} = \frac{\bar{v}_B n_B}{\bar{v}_A n_A} = \left(\frac{m_A}{m_B}\right)^{1/2}\frac{n_B}{n_A} \tag{S2.11}$$

(b) In each stage, the ratio n_B/n_A is multiplied by the factor $(m_A/m_B)^{1/2}$. If n_B/n_A is initially one, then after 20 stages

$$\frac{n_B}{n_A} = \left(\frac{m_A}{m_B}\right)^{10} = (.8)^{10} = 0.107 \tag{S2.12}$$

Exercise 2.5   In Section 2.2 we calculated the force per unit area exerted on the walls of a container filled with an ideal gas by assuming that the walls were perfectly smooth, so that each gas particle underwent simple reflection when it hit the wall. When we remember that the walls of a real container are made of atoms that are roughly the same size as the gas particles, it becomes clear that the assumption that the walls are smooth on an atomic scale is completely unrealistic. Assuming that the piston shown in the figure is in equilibrium and taking, as our system, everything within the volume shown by the dashed line in Fig. S2.5, use the momentum conservation law to derive the relation $p = (2/3)E/V$ without the unrealistic "smooth wall" assumption.

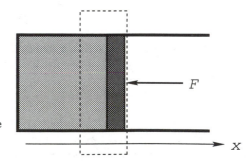

Fig. S2.5   A cylinder with a movable piston. The system is everything within the dotted lines.

Solution Our "system" is everything that is contained in a cylindrical volume that includes the piston plus a certain amount of the gas, to the left of the piston. At the left surface of the system, particles carry momentum into and out of the system. Let us choose our x axis as shown in Fig. S2.5. Then, because momentum is a conserved quantity, the x component of the momentum carried into the system by particles entering from the left minus the x component of the momentum carried out of the system by particles leaving at the left during a time interval Δt must equal the momentum delivered to the system by the force, namely $F\Delta t$, if the total momentum of the system is to remain constant. If we take a small area ΔA on the imaginary surface that bounds the system on the left, the number of particles entering the system through ΔA during time interval Δt that have velocities in the range d^3v, centered at $\mathbf{v}$, is $[f(\mathbf{v})\,d^3v]v_x\,\Delta A\,\Delta t$. The x component of momentum carried into the system by those particles is $m[f(\mathbf{v})\,d^3v]v_x^2\,\Delta A\,\Delta t$. The x momentum carried into the system by all the particles entering through ΔA is

$$m\,\Delta A\,\Delta t \int_{v_x>0} f(\mathbf{v})v_x^2\,d^3v \tag{S2.13}$$

The total rate at which particles carry x momentum into the system at the left is

$$\frac{dP_x(\text{in})}{dt} = mA \int_{v_x>0} f(\mathbf{v})v_x^2\,d^3v \tag{S2.14}$$

The particles that leave the system at the left carry with them *negative* x momentum, since their values of v_x must be negative. The loss of negative P_x counts as a gain of P_x by the system. It is easy to see that the rate at which the system gains momentum by the loss of those particles is

$$\frac{dP_x(\text{out})}{dt} = mA \int_{v_x<0} f(\mathbf{v})v_x^2\,d^3v \tag{S2.15}$$

Combining the two effects, we get

$$\begin{aligned}
\frac{dP_x}{dt} &= mA \int f(\mathbf{v})v_x^2\,d^3v \\
&= \tfrac{2}{3}A \int f(\mathbf{v})\Big(\frac{m}{2}(v_x^2 + v_y^2 + v_z^2)\Big)d^3v \\
&= \tfrac{2}{3}AE/V
\end{aligned} \tag{S2.16}$$

The force delivers negative P_x to the system at a rate F. Setting this equal to dP_x/dt gives the desired result. Notice that, in this analysis, what goes on at the left surface of the piston is a process internal to the system and therefore of no relevance to the conservation of total momentum.

Exercise 2.6 In computing the pressure by calculating the momentum carried by particles across an imaginary surface, we are ignoring another possible mode of momentum transfer across the surface. That is, that the particles on one side of the surface, without moving across the surface, can exert forces on the particles on the other side, due to the direct interparticle force. In a dense system, such as a liquid, most of the pressure force across any surface is transmitted by interparticle forces rather than by particle transfer. In fact, in a solid, where the particles are not free to move across the surface at all, all the pressure force is transmitted in that way. In order to estimate how much force is transmitted in a dilute gas by interparticle forces, assume that the particles in a gas at STP are distributed randomly with a density n and have an interparticle potential that depends on the separation of the particles r as

$$\phi(r) = \phi_o e^{-r^2/a^2} \tag{S2.17}$$

where the range of the potential, $a = 2\,\text{Å}$ and $\phi_o = 1\,\text{eV}$. Calculate the pressure transmitted by the interparticle force across a surface of area A and compare it with the kinetic pressure force (the momentum transferred by moving particles).

Solution   We choose as our surface, dividing the gas into two parts, the surface $x = 0$. We calculate the potential energy at a point $(-x_o, 0, 0)$ to the left of the surface, due to all the particles that lie to the right of the surface. That is

$$\phi(-x_o) = \int_o^\infty dx \int_{-\infty}^\infty dy \int_{-\infty}^\infty dz\, n\phi_o \exp\left(-\frac{(x+x_o)^2}{a^2} - \frac{y^2}{a^2} - \frac{z^2}{a^2}\right)$$
$$= n\phi_o \pi a^2 \int_o^\infty dx\, e^{(x+x_o)^2/a^2} \tag{S2.18}$$

Letting $u = (x + x_o)/a$, this can be written as

$$\phi(-x_o) = n\phi_o \pi a^3 \int_{x_o/a}^\infty e^{-u^2}\, du \tag{S2.19}$$

The force on a particle toward the left, due to all the particles on the right, is

$$F = -\frac{\partial \phi(-x_o)}{\partial x_o} = \pi n\phi_o a^2 e^{-x_o^2/a^2} \tag{S2.20}$$

The total force on all the particles on the left is the integral of F times the density of particles over all the region on the left. The y and z integrals give a factor of A, the area of the surface separating the two halves. Thus

$$F_{\text{total}} = \pi n^2 \phi_o a^2 A \int_o^\infty e^{-x_o^2/a^2}\, dx_o$$
$$= \pi^{3/2} n^2 \phi_o a^3 A \tag{S2.21}$$

The pressure exerted by the direct interparticle forces is

$$p_{\text{int}} = \frac{F_{\text{total}}}{A} = \pi^{3/2} n^2 \phi_o a^3 \tag{S2.22}$$

Putting in $n = n(\text{STP}) = 2.7 \times 10^{25}/\text{m}^3$, $\phi_o = 1\,\text{eV} = 1.6 \times 10^{-19}\,\text{J}$, and $a = 2 \times 10^{-10}\,\text{m}$, we get $p_{\text{int}} = 5.2 \times 10^3\,\text{N/m}^2$ But, at STP, $p_{\text{kin}} \approx 10^5\,\text{N/m}^2$. Thus $p_{\text{int}}/p_{\text{kin}} = 0.05$.

Exercise 2.7   The rate of a certain chemical reaction of the form $A + A \to A_2$ in an ideal gas is given by

$$R = \alpha \times \left(\begin{array}{c}\text{the density of pairs of particles with}\\ \text{c.m. kinetic energy greater than } \varepsilon_o\end{array}\right) \tag{S2.23}$$

where R is the number of reactions per unit volume per second and α is some constant. Write a one-dimensional integral expression for R as a function of particle density and temperature.

Solution   Consider a unit volume of the gas. The number of particles with velocity in the range $d^3\mathbf{v}$ is $f(\mathbf{v})\, d^3\mathbf{v}$, where $f(\mathbf{v})$ is the Maxwell velocity distribution function. The number of pairs of particles, where the first particle has velocity in $d^3\mathbf{v}$ and the second in $d^3\mathbf{v}'$ is $f(\mathbf{v})f(\mathbf{v}')\, d^3\mathbf{v}\, d^3\mathbf{v}'$. We must now transform from the absolute velocities of the two particles to center-of-mass and relative velocities, defined by

$$\mathbf{V} = (\mathbf{v} + \mathbf{v}')/2 \qquad \text{and} \qquad \mathbf{u} = \mathbf{v} - \mathbf{v}' \tag{S2.24}$$

In terms of $\mathbf{u}$ and $\mathbf{V}$, the kinetic energy of the pair is

$$K = \frac{m}{2}(v^2 + v'^2) = \frac{m}{4}u^2 + mV^2 \tag{S2.25}$$

The term involving $\mathbf{u}$ is called the center-of-mass kinetic energy. Thus

$$K_{\text{cm}} = \frac{m}{4}u^2 \tag{S2.26}$$

The significance of K_{cm} is that $K_{\text{cm}} + \phi(r)$ remains constant while the two particles interact, where ϕ is the interaction potential energy associated with the interaction force between the particles. K_{cm} is the energy available in the center-of-mass frame (where the two particles always have equal and opposite velocities) to do work against a repulsive two-particle interaction force.

The number of pairs of particles within the unit volume with $K_{\text{cm}} > \varepsilon_o$ is therefore

$$N = \frac{1}{2}\int f(\mathbf{v})f(\mathbf{v}')\theta\left(\frac{m}{4}u^2 - \varepsilon_o\right) d^3\mathbf{v}\, d^3\mathbf{v}' \tag{S2.27}$$

where θ is the unit step function. The factor of $1/2$ is necessary to compensate for the fact that the integral "double counts" the pairs of particles. To see what is meant, let us try to calculate the number of pairs of students in a class whose total ages add up to 39 years. If a_x is the age of student x (always an integer), then we might sum the quantity $\delta(a_x + a_y - 39)$, where $\delta(0) = 1$ and $\delta(n) = 0$ for $n \neq 0$, and x and y independently take on the values of all the students in the class. But this would count "Mary and John" and "John and Mary" as two pairs although they are really only one.

We must now transform the variables of integration from $d^3\mathbf{v}\, d^3\mathbf{v}'$ to $d^3\mathbf{u}\, d^3\mathbf{V}$. Whenever integration variables are changed, one must include the Jacobian determinant of the transformation. According to Eq. (S2.24), u_x and V_x are functions of v_x and v_x' and similarly for the y and z components. The transformation formula is

$$dv_x\, dv_x' \rightarrow |J|\, du_x\, dV_x \tag{S2.28}$$

This is the same transformation that is shown in the Mathematical Appendix to have $|J| = 1$. Also

$$f(\mathbf{v})f(\mathbf{v}') = n^2(m/2\pi kT)^3 \exp\left(-\frac{m(v^2 + v'^2)}{2kT}\right) \tag{S2.29}$$

where

$$\frac{m}{2}(v^2 + v'^2) = \frac{m}{4}u^2 + mV^2 \tag{S2.30}$$

Therefore

$$N = \tfrac{1}{2}n^2(m/2\pi kT)^3 \int e^{-mV^2/kT}d^3V \int \theta\left(\frac{m}{4}u^2 - \varepsilon_o\right)e^{-mu^2/4kT}d^3u \qquad (S2.31)$$

The first integral can be done easily

$$I_1 = \int e^{-mV^2/kT}d^3V = (\pi kT/m)^{3/2} \qquad (S2.32)$$

The second integral can be reduced to a one-dimensional integral by transforming to spherical coordinates in **u** space.

$$I_2 = \int \theta\left(\frac{m}{4}u^2 - \varepsilon_o\right)e^{-mu^2/4kT}d^3u$$
$$= 4\pi \int_{u_o}^{\infty} e^{-mu^2/4kT}u^2\,du \qquad (S2.33)$$

where $u = 2\sqrt{\varepsilon_o/m}$. Introducing a variable $x = (m/4kT)^{1/2}u$ with a lower limit, $x_o = (m/4kT)^{1/2}u_o = \sqrt{\varepsilon_o/kT}$, we can write I_2 as

$$I_2 = 32\pi \left(\frac{kT}{m}\right)^{3/2} \int_{x_o}^{\infty} e^{-x^2}x^2\,dx \qquad (S2.34)$$

Putting these expressions for I_1 and I_2 into Eq. (S2.31), we get

$$R = \frac{8\alpha n^2}{\sqrt{2\pi}} \int_{x_o}^{\infty} e^{-x^2}x^2\,dx \qquad (S2.35)$$

If kT is much smaller than the activation energy for the reaction ε_o, then x_o is large and, because of the factor of $\exp(-x^2)$ in the integrand, R is very small.

Exercise 2.8   Consider an ideal gas composed mostly of A particles of mass m_A but with a low density of B particles of mass m_B. If a steady force is applied to the B particles, but not to the A particles (one could imagine that only the B particles carry an electric charge and there is a weak electric field present), then the B particles will drift through the gas at a speed that is proportional to the force.

$$v = \mu F \qquad (S2.36)$$

The proportionality constant μ is called the *mobility constant*. Because of the drift, there will be a flux of the B particles equal to

$$\phi = n_B v = \mu n_B F \qquad (S2.37)$$

In the absence of any external force, if the density of B particles is not uniform (let us say that it depends on the coordinate z), then there will be a net flux of the B particles from the regions of high B particle density to those of low density. This diffusion of B particles will eventually bring about a uniform equilibrium state. It is observed that the flux is proportional to the density gradient.

$$\phi = -D\frac{dn_B}{dz} \qquad (S2.38)$$

The constant of proportionality D is called the *diffusion constant*. It is reasonable (and correct) to assume that, if there are both a force and a density gradient, then the net flux is given by the sum of the two terms. Make that assumption and use the equilibrium distribution [Eq. (2.20)] in a gravitational field, to derive a relationship between the mobility constant μ and the diffusion constant D.

Solution   For a gravitational field pointing in the negative z direction, the force on a B particle is $F_z = -m_B g$. The flux of B particles is given by

$$\phi = -\mu n_B m_B g - D\frac{dn_B}{dz} \qquad (S2.39)$$

At equilibrium the flux must vanish, giving the following differential equation for the equilibrium distribution.

$$\mu g m_B n_B = -D\frac{dn_B}{dz} \qquad (S2.40)$$

This can be written as

$$\frac{d[\log n_B(z)]}{dz} = -\frac{\mu g m_b}{D} \equiv -\frac{1}{h} \qquad (S2.41)$$

which has a solution

$$n_B(z) = n_B(0)e^{-z/h} \qquad (S2.42)$$

with $h = D/\mu g m_B$. But the Maxwell–Boltzmann formula for the scale height is $h = kT/m_B g$. Thus we see that, in order for these two formulas to agree, the diffusion constant and the mobility constant must be related by

$$D = \mu kT \qquad (S2.43)$$

This formula was first derived by A. Einstein.

   The next few exercises involve calculations using single-velocity beams. A single-velocity beam is a shower of particles, all with velocity $\mathbf{v}_o$, but distributed randomly throughout space with density n_o. First, for simplicity, we will consider two-dimensional single-velocity beams.

Exercise 2.9   In two dimensions, a smooth hard wall has a normal $\mathbf{n}$ and moves with velocity $\mathbf{u}$. A single-velocity beam is being reflected from the wall (Fig. S2.6). What is the force per unit length on the wall?

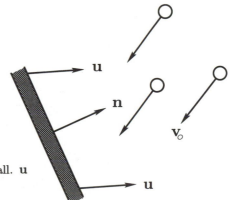

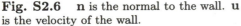

Fig. S2.6 $\mathbf{n}$ is the normal to the wall. $\mathbf{u}$ is the velocity of the wall.

Solution   We transform to the inertial frame in which the wall is at rest. In that frame the beam particles have velocity $\mathbf{v}_o - \mathbf{u}$. The rate at which they strike a unit length of the wall is equal to their density times the inward normal component of their velocity. That is

$$\text{Rate per length} = -n_o \mathbf{n} \cdot (\mathbf{v}_o - \mathbf{u}) \tag{S2.44}$$

Each particle has a change in momentum equal to $-2m[\mathbf{n} \cdot (\mathbf{v}_o - \mathbf{u})]\mathbf{n} \equiv \Delta\mathbf{p}$. The momentum delivered to a unit length of the wall is equal to $\Delta\mathbf{p}$ times the collision rate. This is defined as the force per unit length on the wall.

$$\frac{d\mathbf{F}}{d\ell} = -2mn_o[\mathbf{n} \cdot (\mathbf{v}_o - \mathbf{u})]^2\mathbf{n} \tag{S2.45}$$

As one would expect for a smooth wall, it is a normal force. One should note that this equation is valid only if, in the wall frame, the particles are actually directed at the wall. That is, if

$$\mathbf{n} \cdot (\mathbf{v}_o - \mathbf{u}) < 0 \tag{S2.46}$$

Exercise 2.10   A heavy smooth hard circular particle of radius R moves with velocity $\mathbf{u}$ through a single-velocity beam (Fig. S2.4). What is the net force on the particle?

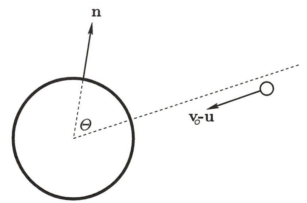

Fig. S2.7   The circle moves with velocity $\mathbf{u}$ through a single-velocity beam in two dimensions.

Solution   We let θ be the angle that a point on the surface of the circle makes with an axis drawn antiparallel to the vector $\mathbf{v}_o - \mathbf{u}$. Then $\mathbf{n}(\theta) \cdot (\mathbf{v}_o - \mathbf{u}) = -|\mathbf{v}_o - \mathbf{u}| \cos\theta$. The restriction that $\mathbf{n} \cdot (\mathbf{v}_o - \mathbf{u}) < 0$ says that we should integrate θ from $-\pi/2$ to $\pi/2$.

$$\mathbf{F} = \int_{-\pi/2}^{\pi/2} \frac{d\mathbf{F}}{d\ell} \frac{d\ell}{d\theta} d\theta$$

$$= -2mn_o|\mathbf{v}_o - \mathbf{u}|^2 R \int_{-\pi/2}^{\pi/2} \cos^2\theta\, \mathbf{n}(\theta)\, d\theta \tag{S2.47}$$

Certainly the net force $\mathbf{F}$ is parallel to $\mathbf{v}_o - \mathbf{u}$. Therefore it is sufficient to calculate

$$\mathbf{F} \cdot (\mathbf{v}_o - \mathbf{u}) = 2mn_o |\mathbf{v}_o - \mathbf{u}|^3 R \int_{-\pi/2}^{\pi/2} \cos^3 \theta \, d\theta \qquad (\text{S2.48})$$

$$= \tfrac{8}{3} m R n_o |\mathbf{v}_o - \mathbf{u}|^3$$

or

$$\mathbf{F} = \tfrac{8}{3} m R n_o |\mathbf{v}_o - \mathbf{u}|(\mathbf{v}_o - \mathbf{u}) \qquad (\text{S2.49})$$

Exercise 2.11   A disk moves at a very low velocity $\mathbf{u} = u\hat{\mathbf{x}}$ through a very dilute two-dimensional ideal gas. Calculate the net force on the disk.

Solution   The collection of particles with velocities within the range d^2v, centered at $\mathbf{v}$, can be considered as a single-velocity beam with a density $f(\mathbf{v}) \, d^2v$, where $f(\mathbf{v})$ is the two-dimensional Maxwell distribution function

$$f(v_x, v_y) = n(m/2\pi kT)e^{-mv^2/2kT} \qquad (\text{S2.50})$$

The net force on the disk due to those particles is, by Eq. (S2.49),

$$d\mathbf{F} = \tfrac{8}{3} m R n(m/2\pi kT)e^{-mv^2/2kT}|\mathbf{v} - \mathbf{u}|(\mathbf{v} - \mathbf{u}) \, d^2v \qquad (\text{S2.51})$$

The net force due to all the gas particles is obtained by integrating $d\mathbf{F}$ over all velocities.

$$\mathbf{F} = \frac{4m^2 Rn}{3\pi kT} \int e^{-mv^2/2kT}|\mathbf{v} - \mathbf{u}|(\mathbf{v} - \mathbf{u}) \, d^2v \qquad (\text{S2.52})$$

We make a transformation to the variable $\mathbf{V} = \mathbf{v} - \mathbf{u}$ and use the following approximations, which are valid for very small $\mathbf{u}$.

$$v^2 = (\mathbf{V} + \mathbf{u})^2 \approx V^2 + 2\mathbf{V} \cdot \mathbf{u} \qquad (\text{S2.53})$$

$$e^{-\alpha v^2} \approx e^{-\alpha V^2} e^{-2\alpha \mathbf{V} \cdot \mathbf{u}} \approx e^{-\alpha V^2}(1 - 2\alpha \mathbf{V} \cdot \mathbf{u}) \qquad (\text{S2.54})$$

Then

$$\mathbf{F} = \frac{4m^2 Rn}{3\pi kT} \int e^{-mV^2/2kT}(1 - \frac{m}{kT}\mathbf{V} \cdot \mathbf{u})|\mathbf{V}|\mathbf{V} \, d^2V \qquad (\text{S2.55})$$

The first term in the parentheses vanishes by symmetry. Since $\mathbf{u}$ is in the x direction, it is clear that the net force must have only an x component. The value of that component is

$$F_x = -\frac{4m^3 Rnu}{3\pi k^2 T^2} \int_o^\infty \int_{-\pi}^\pi e^{-mV^2/2kT} V^4 \cos^2 \theta \, dV \, d\theta \qquad (\text{S2.56})$$

$$= -2Rn\sqrt{2\pi mkT}\, u$$

Exercise 2.12   By reviewing the solution to Exercise 2.9, it is easy to see that Eq. (S2.45) is also valid in three dimensions with the trivial modification that $d\ell$ must be replaced by an element of area dA on the wall surface and n_o is the three-dimensional density of particles in the beam. Use this fact to redo Exercise 2.11

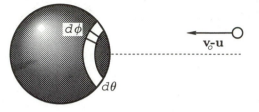

Fig. S2.8   In three dimensions, a hard sphere moves with velocity **u** through a single-velocity beam.

for the case of a heavy hard spherical particle in three dimensions, as shown in Fig. S2.8.

Solution   We construct a system of spherical coordinates with the polar axis in a direction opposite to the vector $\mathbf{v}_o - \mathbf{u}$. The area on the sphere associated with the range of spherical angles $d\theta$ and $d\phi$ is equal to $dA = R^2 \sin\theta\, d\theta\, d\phi$. The force $d\mathbf{F}$ imparted to that area by the particles in the single-velocity beam is given by

$$\frac{d\mathbf{F}}{R^2 \sin\theta\, d\theta\, d\phi} = -2mn_o[\mathbf{n} \cdot (\mathbf{v}_o - \mathbf{u})]^2 \mathbf{n} \tag{S2.57}$$

Again, as in two dimensions, $\mathbf{n} \cdot (\mathbf{v}_o - \mathbf{u}) = -|\mathbf{v}_o - \mathbf{u}|\cos\theta$. The restriction that $\mathbf{n} \cdot (\mathbf{v}_o - \mathbf{u}) < 0$ restricts θ to the range $0 < \theta < \pi/2$. Therefore, the total force is given by the integral

$$\mathbf{F} = -2mn_o|\mathbf{v}_o - \mathbf{u}|^2 R^2 \int_o^{2\pi} d\phi \int_o^{\pi/2} d\theta \cos^2\theta \sin\theta\, \mathbf{n}(\theta, \phi) \tag{S2.58}$$

By viewing the process in the rest frame of the sphere, it is obvious that the net force will be in the direction of $\mathbf{v}_o - \mathbf{u}$. Thus we need only calculate

$$\begin{aligned}
\mathbf{F} \cdot (\mathbf{v}_o - \mathbf{u}) &= -2mn_o|\mathbf{v}_o - \mathbf{u}|^3 R^2 \int_o^{2\pi} d\phi \int_o^{\pi/2} \cos^3\theta \sin\theta\, d\theta \\
&= -4\pi mn_o|\mathbf{v}_o - \mathbf{u}|^3 R^2 \left[\frac{\cos^4\theta}{4}\right]_o^{\pi/2} \\
&= \pi mn_o R^2 |\mathbf{v}_o - \mathbf{u}|^3
\end{aligned} \tag{S2.59}$$

which implies that

$$\mathbf{F} = \pi mn_o R^2 |\mathbf{v}_o - \mathbf{u}|(\mathbf{v}_o - \mathbf{u}) \tag{S2.60}$$

a result that differs only slightly from the two-dimensional case.

Exercise 2.13   What is the mobility constant of a hard sphere in a very dilute ideal gas?

Solution   The mobility constant is the constant μ in the equation $\mathbf{u} = \mu\mathbf{F}$, relating the steady-state velocity of an object moving through a fluid to the force being exerted on the object. But, for steady-state motion, the external force must balance the average force that the fluid particles are exerting on the object. Therefore, what we have to calculate is the average force exerted by the particles of a very dilute ideal gas on a sphere moving through the gas at velocity **u**.

We view the gas as being composed of a very large number of single-velocity beams. The collection of all the gas particles with velocities in the element d^3v

forms a single-velocity beam of density $f(\mathbf{v})\,d^3v$. That beam exerts an average force on the sphere of

$$dF = \pi m R^2 f(\mathbf{v})\,d^3v\,|\mathbf{v} - \mathbf{u}|(\mathbf{v} - \mathbf{u}) \tag{S2.61}$$

The total force on the sphere is

$$\mathbf{F} = \pi m R^2 \int f(\mathbf{v})|\mathbf{v} - \mathbf{u}|(\mathbf{v} - \mathbf{u})\,d^3v \tag{S2.62}$$

It is clear that the net force will be opposed to the velocity of the sphere $\mathbf{u}$, so that it is sufficient to calculate

$$\mathbf{F} \cdot \mathbf{u} = \pi m R^2 n (m/2\pi kT)^{3/2} \int e^{-mv^2/2kT}|\mathbf{v} - \mathbf{u}|(\mathbf{v} - \mathbf{u}) \cdot \mathbf{u}\,d^3v \tag{S2.63}$$

Transforming the variable of integration to the relative velocity $\mathbf{V} = \mathbf{v} - \mathbf{u}$, using the approximations

$$e^{-m(\mathbf{V}+\mathbf{u})^2/2kT} \approx e^{-mV^2/2kT}(1 - m\mathbf{V} \cdot \mathbf{u}/kT) \tag{S2.64}$$

and introducing spherical coordinates with the polar axis antiparallel to $\mathbf{u}$, we can write the integral for $\mathbf{F} \cdot \mathbf{u}$ as

$$\begin{aligned}
\mathbf{F} \cdot \mathbf{u} =&\, 2\pi^2 m R^2 n (m/2\pi kT)^{3/2} \int_0^\pi \sin\theta\,d\theta \\
&\times \int_0^\infty V^2 dV e^{-mV^2/2kT}\left(1 - \frac{m}{kT}uV\cos\theta\right)V^2 u\cos\theta
\end{aligned} \tag{S2.65}$$

Because the integral $\int_0^\pi \cos\theta\sin\theta\,d\theta = 0$, the first term in the parentheses vanishes. Thus $\mathbf{F} \cdot \mathbf{u}$ can be written in terms of the integrals

$$\int_0^\pi \cos^2\theta\sin\theta\,d\theta = \tfrac{2}{3} \tag{S2.66}$$

and

$$\int_0^\infty e^{-mV^2/2kT}V^5 dV = (2kT/m)^3 \tag{S2.67}$$

when this is done, one gets

$$\mathbf{F} = -\tfrac{8}{3}(2\pi mkT)^{1/2}R^2 n\mathbf{u} \tag{S2.68}$$

Note that, from this and Einstein's relation [Eq. (S2.43)], we can get the diffusion constant for hard spheres in a very dilute ideal gas.

$$D = \tfrac{8}{3}(2\pi m)^{1/2}(kT)^{3/2}R^2 n \tag{S2.69}$$

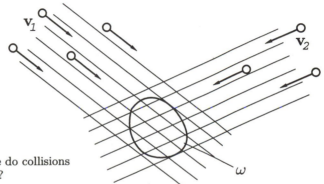

Fig. S2.9 At what rate do collisions occur within the volume ω?

Exercise 2.14   Two single-velocity beams are passing through the same volume ω. They have densities n_1 and n_2 and velocities $\mathbf{v}_1$ and $\mathbf{v}_2$. The particles in both beams are hard spheres of diameter d. At what rate do collisions occur within the volume ω? (See Fig. S2.9.)

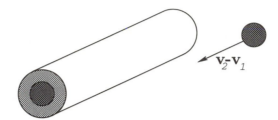

Fig. S2.10   In the frame of the first beam, the particles in the other beam have velocity $\mathbf{v}_2 - \mathbf{v}_1$. The length of the cylinder shown is $|\mathbf{v}_2 - \mathbf{v}_1|\, dt$. Its cross-sectional area is πd^2.

Solution   We observe the situation within the inertial frame of the first beam. Then there is a density n_1 of stationary particles and a density n_2 of particles with velocity $\mathbf{v}_2 - \mathbf{v}_1$. The number of stationary particles in ω is $n_1\omega$. The probability that one of those stationary particles will be hit within a time interval dt is equal to the probability that the center of one of the moving particles occupies the cylinder, of volume $\pi d^2 |\mathbf{v}_2 - \mathbf{v}_1|\, dt$, shown in Fig. S2.10. That probability is $n_2\pi d^2 |\mathbf{v}_2 - \mathbf{v}_1|\, dt$. Thus the number of collisions within ω during time interval dt is

$$dN = \omega n_1 n_2 \pi d^2 |\mathbf{v}_2 - \mathbf{v}_1|\, dt \qquad (\text{S2.70})$$

and the rate of collisions within ω is

$$R_{\text{coll}} = \omega n_1 n_2 \pi d^2 |\mathbf{v}_2 - \mathbf{v}_1| \qquad (\text{S2.71})$$

Notice that the rate is symmetric with respect to exchange of $\mathbf{v}_1$ and $\mathbf{v}_2$, which shows that we would have obtained the same result by making the calculation in the rest frame of the second beam.

Exercise 2.15   Within a volume ω in an ideal gas of spherical particles of diameter d, at what rate do collisions occur?

Solution   Within ω, the rate of collisions between pairs of particles, of which one particle has velocity in the range d^3v_1 and the other has velocity in the range d^3v_2 is

$$R_{\mathbf{v}_1\mathbf{v}_2} = \pi \omega d^2 f(\mathbf{v}_1)\, d^3v_1\, f(\mathbf{v}_2)\, d^3v_2\, |\mathbf{v}_1 - \mathbf{v}_2| \qquad (\text{S2.72})$$

By integrating $\mathbf{v}_1$ and $\mathbf{v}_2$ over their full ranges, we obtain twice the total collision rate within ω because we count collisions between particles of velocities $\mathbf{v}$ and $\mathbf{v}'$ when $\mathbf{v}_1 = \mathbf{v}$ and $\mathbf{v}_2 = \mathbf{v}'$ and also when $\mathbf{v}_1 = \mathbf{v}'$ and $\mathbf{v}_2 = \mathbf{v}$. Thus the total collision rate is

$$R = \tfrac{1}{2}\pi \omega d^2 \int f(\mathbf{v}_1) f(\mathbf{v}_2) |\mathbf{v}_1 - \mathbf{v}_2|\, d^3v_1\, d^3v_2 \qquad (\text{S2.73})$$

We transform to center-of-mass and relative velocities, $\mathbf{u} = \mathbf{v}_1 - \mathbf{v}_2$ and $\mathbf{V} = (\mathbf{v}_1 + \mathbf{v}_2)/2$.

$$f(\mathbf{v}_1)f(\mathbf{v}_2) = n^2(m/2\pi kT)^3 \exp[-m(v_1^2 + v_2^2)/2kT]$$
$$= n^2(m/2\pi kT)^3 \exp\left[-\left(\frac{m}{4}u^2 + mV^2\right)/kT\right] \tag{S2.74}$$

With this transformation,

$$R = \tfrac{1}{2}\pi\omega d^2 n^2 (m/2\pi kT)^3 \int e^{-mV^2/kT} d^3V \int e^{-mu^2/4kT}|\mathbf{u}| \, d^3u$$
$$= 2\omega d^2 n^2 \sqrt{\pi kT/m} \tag{S2.75}$$

Exercise 2.16   In an ideal gas of density n at temperature T, what is the total distance traveled by all the particles within a volume ω during one second?

Solution   The number of particles is $n\omega$. Their average speed is $\bar{v} = \sqrt{8kT/\pi m}$. Therefore, the total distance traveled by all particles together is

$$s = n\omega\bar{v} = n\omega\sqrt{8kT/\pi m} \tag{S2.76}$$

Exercise 2.17   Given the results of the previous two exercises, what is the average distance between collisions of a particle in an ideal gas?

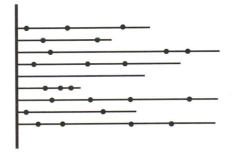

Fig. S2.11   Each line represents the absolute distances traveled by one particle. The dots are points where the particle collided with another particle.

Solution   Let us imagine that Fig. S2.11 shows a graph of the absolute distances traveled by each of the particles in a volume ω in an ideal gas during one second. For each particle, a dot has been drawn at the location of each collision involving that particle. Note that each collision creates two dots on the diagram because it involves two particles. The total length of the lines is $s = n\omega\sqrt{8kT/\pi m}$. The total number of dots is $2R = 4\omega d^2 n^2 \sqrt{\pi kT/m}$. Therefore, the average distance between dots on the lines (that is, the mean free path of the particles) is

$$\ell = \frac{s}{2R} = \frac{1}{\sqrt{2}\pi d^2 n} \tag{S2.77}$$

Exercise 2.18   An ideal gas of A particles contains a low density of B particles. The diffusion constant for the B particles, defined in Eq. (S2.38), is D. Assume that the initial density of the B particles, $n(x, 0)$, depends only on the variable x. Derive a differential equation, the *diffusion equation*, for $n(x, t)$.

Solution   Consider a cylinder of unit cross-sectional area whose axis is parallel to the x axis. Let $N(x_o, t)$ be the total number of B particles within the cylinder to the left of the plane $x = x_o$. By the meaning of the B particle flux, it is clear that

$$\frac{\partial N(x, t)}{\partial t} = -J(x, t) = D\frac{\partial n(x, t)}{\partial x} \tag{S2.78}$$

where J is the flux at position x at time t. But, by the definition of the particle density, it is also true that $n(x, t) = \partial N(x, t)/\partial x$. Differentiating Eq. (S2.78) with respect to x and using the fact that $\partial^2 N/\partial x \partial t = \partial n/\partial t$ gives the desired equation.

$$\frac{\partial n}{\partial t} = D\frac{\partial^2 n}{\partial x^2} \tag{S2.79}$$

Exercise 2.19   A *Brownian particle* is a particle in a fluid that is large enough to be observed in an ordinary microscope (which means that it must be much larger and heavier than the fluid molecules) but is still small enough that its gradual drift due to the impulses delivered to it by rebounding fluid particles is observable in the same microscope. The thermal motion of a truly macroscopic object is much too small to be observable. The resultant random drifting motion of such a particle is called *Brownian motion*. Consider a dilute mixture, sometimes called a suspension, of identical Brownian particles in water. Assume that the particles, in water, are known to have a diffusion constant D. A particle begins at time 0 with x coordinate 0. What is the probability distribution for its x coordinate at time t?

Solution   Imagine a system of pure water with a large collection of the Brownian particles randomly distributed over the plane $x = 0$. The initial density of Brownian particles would be

$$n(x, 0) = C\,\delta(x) \tag{S2.80}$$

where C is some positive constant equal to the two-dimensional density of the particles in the y–z plane. After time zero, the density of Brownian particles will satisfy the diffusion equation

$$\frac{\partial n(x, t)}{\partial t} = D\frac{\partial^2 n(x, t)}{\partial x^2} \tag{S2.81}$$

We are now faced with a mathematical problem of finding a solution of the diffusion equation that satisfies the initial condition given by Eq. (S2.80). In order to guess the form of the solution, it is best to think about the physics of the Brownian motion process. The particle drifts a noticeable amount during a given time interval because it is being knocked randomly left and right by individually unnoticeable amounts very many times during the time interval. Its final x displacement is the sum of a very large number of separate x displacements. It is natural to guess that the probability distribution for an individual particle's x coordinate will be a Gaussian function of the form predicted by the central limit theorem. The density of particles at time t will certainly be proportional to the probability density for an individual particle. Therefore we try a solution of the form

$$n(x, t) = C\sqrt{\gamma(t)/\pi}\, e^{-\gamma(t)x^2} \tag{S2.82}$$

where $\gamma(t)$ is, as yet, unknown. The square root factor is necessary in order to maintain normalization. Since the particles are conserved, it is necessary that

$$\int_{-\infty}^{\infty} n(x,t)\, dx = C \int_{-\infty}^{\infty} \delta(x)\, dx = C \qquad \text{(S2.83)}$$

Putting this form into the differential equation, we get

$$(\tfrac{1}{2}\gamma^{-1/2} - \gamma^{1/2}x^2)\frac{d\gamma}{dt} = -4D(\tfrac{1}{2}\gamma^{3/2} - \gamma^{5/2}x^2)$$
$$= -4D(\tfrac{1}{2}\gamma^{-1/2} - \gamma^{1/2}x^2)\gamma^2$$

which would be satisfied if

$$\frac{d\gamma/dt}{\gamma^2} = -\frac{d(1/\gamma)}{dt} = -4D \qquad \text{(S2.85)}$$

or $1/\gamma = 4Dt$. Thus the solution of Eq. (S2.81) is

$$n(x,t) = \frac{C}{\sqrt{4\pi Dt}}\, e^{-x^2/4Dt} \qquad \text{(S2.86)}$$

The probability density for a single particle differs from the particle density for the ensemble of Brownian particles only by its normalization. Therefore

$$P(x,t) = \frac{1}{\sqrt{4\pi Dt}}\, e^{-x^2/4Dt} \qquad \text{(S2.87)}$$

Historical note: In the year 1828, the botanist Robert Brown detected, in a suspension of fine pollen grains in water, an irregular "swarming motion." He attributed it to a somewhat mystical primitive life force. A long period of controversy followed regarding the question of whether the Brownian motion was a genuine effect or was only an artifact of his experiment, due to vibrations of the microscope, currents in the water caused by evaporation, etc. By 1905, when Einstein's theoretical explanation and mathematical analysis of the effect appeared, it had been confirmed unambiguously that an irregular diffusive motion with no apparent energy source continued indefinitely within a closed system of small suspended particles. Stimulated by Einstein's analysis, a number of experimental physicists, but most notably the French physicist Perrin, made detailed quantitative experimental studies of Brownian motion. These studies had two important effects.

1. They completely confirmed the fundamental concept, which was controversial at that time, that heat was simply microscopic kinetic energy.
2. They allowed, using Einstein's theoretical analysis, a determination, for the first time, of the mass of individual molecules and, from that, of Avogadro's number.

A more recent study (see *Scientific American*, August 1991) has concluded that, with Brown's experimental techniques, the size and density of pollen grains, and the microscopes available at the time of Brown's experiments, Brown could not possibly have seen what is now called Brownian motion. What Brown saw really was an artifact of his experimental technique. He has immortalized his name by a lucky fluke!

Exercise 2.20   A harmonic oscillator potential is a potential of the form $U(x) = \frac{1}{2}kx^2$. A single particle of mass m in a harmonic oscillator potential has quantum mechanical energy levels

$$E_n = \hbar\omega(n + \tfrac{1}{2}) \tag{S2.88}$$

where $n = 0, 1, 2, \dots$ and $\omega = \sqrt{k/m}$, which is the angular frequency of vibration of a classical particle in the same potential. What is the energy and degeneracy of the ground state of a system of five noninteracting particles in a harmonic oscillator potential in the cases that (a) the particles are spin-0 bosons, (b) the particles are spin-$\frac{1}{2}$ fermions, (c) the particles are spin-$\frac{1}{2}$ bosons, and (d) the particles are spin-0 fermions? Note: cases (c) and (d) are known to be impossible. (See Fig. S2.12.)

Solution   (a) For bosons, any number of particles may occupy a given single-particle quantum state. The five-particle ground state is obtained by putting all the particles in the single-particle state with quantum number $n = 0$. Thus $E_g = \frac{5}{2}\hbar\omega$. The ground-state is clearly unique; that is, it has degeneracy one. (b) For spin-$\frac{1}{2}$ fermions, each single-particle state can hold two particles of opposite spin. The spin of the particle in the $n = 2$ state can be either up or down, giving the five-particle ground state a degeneracy of two. The ground-state energy is $E_g = 2(\hbar\omega/2) + 2(3\hbar\omega/2) + (5\hbar\omega/2) = 13\hbar\omega/2$. (c) This has the same energy as case (a), but the number of spin-up particles can vary from 0 to 5, giving the ground state a degeneracy of 6. (d) Only one particle can go into each single-particle state. Thus $E_g = 25\hbar\omega/2$ and the state is nondegenerate.

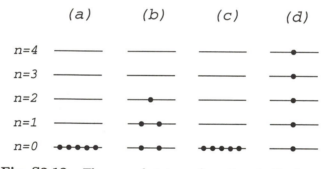

Fig. S2.12   The ground-state configurations in the four cases.

Exercise 2.21   For particles in a one-dimensional hard-walled box of length π, the single-particle energy eigenstates are $u_n(x) = \sqrt{2/\pi}\sin(nx)$, where $n = 1, 2, \dots$. If we assume that the particles are bosons, and ignore the problem of normalization, what is the explicit form of the three-particle wave function associated with the occupation numbers $N_1 = 2$, $N_2 = 1$, $N_3 = N_4 = \dots = 0$?

Solution

$$\psi(x_1, x_2, x_3) = A \sum_{\text{perm}} u_1(x_i)u_1(x_j)u_2(x_k) \tag{S2.89}$$

where A is a normalization constant and in the sum (i, j, k) take the values (1,2,3), (1,3,2), (3,1,2), (3,2,1) (2,3,1), and (2,1,3). The result is

$$\begin{aligned}
\psi = {}& A(\sin(x_1)\sin(x_2)\sin(2x_3) + \sin(x_1)\sin(x_3)\sin(2x_2) \\
& + \sin(x_2)\sin(x_3)\sin(2x_1))
\end{aligned} \tag{S2.90}$$

Exercise 2.22 For a particle of mass m in a two-dimensional periodic box of area L^2: (a) how many single-particle quantum states are there with energies less than E? (b) What is the density of single-particle quantum states as a function of energy? (c) If $m = m_e$, and $L = 1\,\mathrm{cm}$, what is the numerical value of the density of single-particle quantum states?

Solution (a) The Schrödinger energy equation for the system is

$$-\frac{\hbar^2}{2m}\left(\frac{\partial^2 u}{\partial x^2} + \frac{\partial^2 u}{\partial y^2}\right) = Eu \qquad (\text{S2.91})$$

with the periodic boundary conditions

$$u(0, y) = u(L, y) \qquad \text{and} \qquad u(x, 0) = u(x, L) \qquad (\text{S2.92})$$

The solutions of this equation are plane waves of the form

$$u(x, y) = A\exp[i(k_1 x + k_2 y)] \qquad (\text{S2.93})$$

where

$$(k_1, k_2) = \left(\frac{2\pi}{L}K_1, \frac{2\pi}{L}K_2\right) \qquad (\text{S2.94})$$

and K_1 and K_2 are integers. The corresponding energy eigenvalue is

$$E(k_1, k_2) = \frac{\hbar^2(k_1^2 + k_2^2)}{2m} \qquad (\text{S2.95})$$

If we plot these allowed wave vectors on a two-dimensional $\mathbf{k}$ plane, they form a square lattice with a spacing of $2\pi/L$ (see Fig. 2.5). For a given value of energy E, the states with energy less than E all fall within a circle whose radius is $k = \sqrt{2mE}/\hbar$. The number of such states is

$$N(E) = \frac{\pi k^2}{(2\pi/L)^2} = \frac{mEL^2}{2\pi\hbar^2} \qquad (\text{S2.96})$$

(b) The density of eigenstates is

$$\frac{dN(E)}{dE} = \frac{mL^2}{2\pi\hbar^2} \qquad (\text{S2.97})$$

(c) For $m = m_e$ and $L = 10^{-2}\,\mathrm{m}$, $dN(E)/dE = 1.3 \times 10^{33}$ states/joule.

Exercise 2.23 1200 particles are to be distributed among three energy states with energies $\varepsilon_1 = 1$ eV, $\varepsilon_2 = 2$ eV, and $\varepsilon_3 = 3$ eV. Assume that the total energy is 2400 eV and that each possible microstate is equally probable. (a) What is the probability distribution for the number of particles in state 1 if the particles are distinguishable? (b) What is the probability distribution for the number of particles in state 1 if the particles are Bose–Einstein particles?

Solution (a) For distinguishable particles, a microstate is defined by giving the energy value of each particle. That is E_1, E_2, $\ldots$, E_{1200}, where E_k can be ε_1, ε_2, or ε_3. Let N_1, N_2, and N_3 represent the number of particles in each energy state. The number of microstates with given values of N_1, N_2, and N_3 is

$$K = \frac{1200!}{N_1!N_2!N_3!} \qquad (S2.98)$$

The restriction that $E = 2400$ eV will be satisfied if and only if $N_3 = N_1$. But also, $N_1 + N_2 + N_3 = 1200$. Thus only N_1 is independent. For any N_1 between 0 and 600,

$$N_3 = N_1 \qquad \text{and} \qquad N_2 = 1200 - 2N_1 \qquad (S2.99)$$

When we take the restrictions into account, the number of microstates with a given value of N_1 is

$$K(N_1) = \frac{1200!}{(N_1!)^2(1200 - 2N_1)!} \qquad (S2.100)$$

To find the most likely value of N_1 we let

$$F(N_1) = \log K(N_1) \approx 1200\log(1200) - 2N_1\log N_1 \\ - (1200 - 2N_1)\log(1200 - 2N_1) \qquad (S2.101)$$

and set $dF(N_1)/dN_1 = 0$, getting

$$-2\log N_1 + 2\log(1200 - 2N_1) = 0 \qquad (S2.102)$$

or $(1200 - 2N_1)/N_1 = 1$, whose solution is $N_1 = 1200/3 = 400$.

We will calculate $P(N_1)/P(400) = \exp[F(N_1) - F(400)]$. A bit of algebra and use of the expansion, $\log(1 + \varepsilon) \approx \varepsilon - \varepsilon^2/2$ will show that

$$F(400 + n) - F(400) \approx -3n^2/400 \qquad (S2.103)$$

or

$$\frac{P(400 + n)}{P(400)} = e^{-3n^2/400} \qquad (S2.104)$$

(b) For Bose–Einstein particles, specifying N_1, N_2, and N_3 completely specifies the 1200-particle quantum state. But Eq. (S2.99) gives N_2 and N_3 in terms of N_1. For any N_1 between 0 and 600, there is exactly one microstate, and therefore all values of N_1 in that range are equally probable.

Exercise 2.24 What is the maximum value of the function

$$F(x_1, x_2, \ldots, x_{10}) = \sum_{n=1}^{10} n\,x_n^2 \qquad (S2.105)$$

subject to the constraint

$$\sum_{n=1}^{10} x_n = 1 \qquad (S2.106)$$

Solution Using Lagrange's method, we look at the function

$$G = \sum_{n=1}^{10} n\,x_n^2 - \lambda \sum_{n=1}^{10} x_n \qquad (S2.107)$$

The unconstrained maximum of G is given by

$$\frac{\partial G}{\partial x_n} = 2nx_n - \lambda = 0 \tag{S2.108}$$

or $x_n = \lambda/2n$. The value of λ is determined by the constraint equation

$$\sum_{n=1}^{10} x_n = \frac{\lambda}{2} \sum_{n=1}^{10} \frac{1}{n} = 1 \tag{S2.109}$$

which gives

$$\lambda = 2\left(\sum_{n=1}^{10} \frac{1}{n}\right)^{-1} = 0.68283 \tag{S2.110}$$

Then

$$F = \sum_{n=1}^{10} n\left(\frac{\lambda}{2n}\right)^2 = \frac{\lambda^2}{4} \sum_{n=1}^{10} \frac{1}{n} = \frac{\lambda}{2} = 0.341417 \tag{S2.111}$$

Exercise 2.25   In Exercise 2.22 it was shown that the number of single-particle quantum states of a two-dimensional quantum gas with energies within the interval dE is equal to $(mL^2/2\pi\hbar^2)\,dE$. (a) What is the spatial density of particles in a two-dimensional Fermi–Dirac gas as an explicit function of temperature and affinity? (b) For particles of electronic mass and $T = 300\,\mathrm{K}$, what value of α will give a density of 1 particle per square Ångstrom?

Solution   (a) For given values of α and $\beta = 1/kT$, the average number of particles in a single quantum state of energy ε is $(e^{\alpha+\beta\varepsilon}+1)^{-1}$. The number of quantum states in energy interval $d\varepsilon$ is $(mL^2/2\pi\hbar^2)\,d\varepsilon$. Thus the average number of particles in all quantum states is

$$\begin{aligned}
N &= \int_o^\infty \frac{(mL^2/2\pi\hbar^2)\,d\varepsilon}{e^{\alpha+\beta\varepsilon}+1} \\
&= \frac{mL^2}{2\pi\hbar^2} \int_o^\infty \frac{e^{-\beta\varepsilon}\,d\varepsilon}{e^\alpha + e^{-\beta\varepsilon}}
\end{aligned} \tag{S2.112}$$

Changing the integration variable to $x = e^{-\beta\varepsilon}$ and $dx = -\beta e^{-\beta\varepsilon}\,d\varepsilon$, we can write N as

$$\begin{aligned}
N &= \frac{kTmL^2}{2\pi\hbar^2} \int_o^1 \frac{dx}{e^\alpha + x} \\
&= \frac{kTmL^2}{2\pi\hbar^2} \Big[\log(e^\alpha + x)\Big]_o^1 \\
&= \frac{kTmL^2}{2\pi\hbar^2} \Big[\log(1 + e^\alpha) - \alpha\Big]
\end{aligned} \tag{S2.113}$$

The spatial density of particles is N/L^2, or

$$n = \frac{mkT}{2\pi\hbar^2} \Big[\log(1 + e^\alpha) - \alpha\Big] \tag{S2.114}$$

(b) Setting $n = 10^{20}$ particles/m^2 and $T = 300\,\mathrm{K}$, we get

$$\log(1 + e^\alpha) - \alpha = 1821 \tag{S2.115}$$

The function on the left becomes large for large negative values of α. If α is large and negative, then e^α is very small and

$$\log(1 + e^\alpha) - \alpha \approx e^\alpha - \alpha \approx -\alpha \tag{S2.116}$$

Therefore, $\alpha \approx -1821$.

Exercise 2.26   What is the density n as a function of T and α if the particles of the previous question are bosons?

Solution   For Bose–Einstein particles, Eq. (S2.112) must be replaced by

$$
\begin{aligned}
N &= \frac{mL^2}{2\pi\hbar^2} \int_o^\infty \frac{d\varepsilon}{e^{\alpha+\beta\varepsilon} - 1} \\
&= \frac{mkTL^2}{2\pi\hbar^2} \int_o^1 \frac{dx}{e^\alpha - x} \\
&= \frac{mkTL^2}{2\pi\hbar^2} \left[-\log(e^\alpha - x) \right]_o^1 \\
&= \frac{mkTL^2}{2\pi\hbar^2} \left[\alpha - \log(e^\alpha - 1) \right]
\end{aligned}
\tag{S2.117}
$$

or

$$n = \frac{mkT}{2\pi\hbar^2} \left[\alpha - \log(e^\alpha - 1) \right] \tag{S2.118}$$

Supplement to Chapter 3

REVIEW QUESTIONS

3.1 For a system of N particles in three dimensions, what are the variables $x_1, x_2, \ldots, x_K$ and $p_1, p_2, \ldots, p_K$?

3.2 What is the meaning of the Hamiltonian function $H(x, p)$?

3.3 What is the form of the Hamiltonian function for a system of N three-dimensional particles in an external potential $U(\mathbf{r})$ and with two-particle interactions $\phi(r_{ij})$?

3.4 What are Hamilton's equations?

3.5 How is the energy surface $\omega(E)$ defined?

3.6 What is the dimensionality of $\omega(E)$?

3.7 How are one-body observables defined? Give one example.

3.8 How are two-body observables defined? Give one example.

3.9 What are the necessary mathematical properties of a microstate probability density?

3.10 What is the probability density function for the uniform ensemble?

3.11 Given that the volume of an N-dimensional sphere of radius R is $\pi^{N/2} R^N / (N/2)!$, calculate the normalization constant for a uniform ensemble for a system of noninteracting particles in a box of volume V.

3.12 What is the microcanonical ensemble probability density?

3.13 Given the microstate probability density $P(x, p)$, how can we calculate $n(\mathbf{r})$, the average particle density at position $\mathbf{r}$?

3.14 Given the microstate probability density $P(x, p)$, how can we calculate $F_1(\mathbf{r}, \mathbf{p})$, the phase-space density function?

3.15 Given the microstate probability density $P(x, p)$, how can we calculate $F_2(\mathbf{r}, \mathbf{p}, \mathbf{r}', \mathbf{p}')$, the two-particle phase-space density function?

3.16 Express the average value of a one-body observable in terms of $F_1(\mathbf{r}, \mathbf{p})$.

3.17 Explain what is meant by the "thermodynamic limit."

3.18 What is a uniform intensive observable?

3.19 What is the property of asymptotic factorization?

3.20 Show that $\Delta A^2 \to 0$ for any uniform intensive observable if a system has the property of asymptotic factorization.

3.21 With respect to a uniform intensive observable, what is an exceptional point on the energy surface?

3.22 How does Chebbyshev's inequality show that the fraction of points on the energy surface that are exceptional goes to zero in the thermodynamic limit?

3.23 Why does the property that was discussed in the last question justify the use of the microcanonical ensemble?

3.24 Describe a three-dimensional Ising model, giving an expression for the energy in terms of the spin variables.

3.25 How is the mean magnetization defined for an Ising model?

3.26 What happens to the probability distribution for the mean magnetization m at low temperatures?

3.27 How is the phenomenon discussed in the previous question related to the no-fluctuation theorem?

3.28 Describe the Standard Calculational Procedure.

3.29 How was an "extra" conservation law defined?

3.30 Why would an extra conservation law make calculations based on a microcanonical ensemble unreliable?

3.31 What does the ergodic theorem say?

3.32 What is a "complex system" and why does such a system take an extremely long time to come to equilibrium?

3.33 If two thermodynamic systems are in contact, what is the condition that determines how the total energy will be shared between them?

3.34 A system is composed of two weakly interacting parts, so that its Hamiltonian can be approximated by $H = H_1(x_1, p_1) + H_2(x_2, p_2)$. Using a microcanonical ensemble for the total system, calculate $P_1(E_1)$, the probability that subsystem 1 has energy E_1.

3.35 What is the condition that $P_1(E_1)$ be a maximum (consistent with energy conservation)?

3.36 How do the results of the previous three questions relate the thermodynamic entropy and the microcanonical partition function $Q(E)$?

3.37 How are the absolute temperature and the mechanical pressure related to the entropy function?

EXERCISES

Exercise 3.1   Two particles, each of mass m, are attached by a massless spring of force constant k and equilibrium length l. The particles are constrained to the x–y plane. Using center-of-mass and relative coordinates, write the Hamiltonian function and Hamilton's equations.

Solution   Let the coordinates of the two particles be (x_1, y_1) and (x_2, y_2). The kinetic energy of the system is

$$K = \frac{m}{2}(\dot{x}_1^2 + \dot{y}_1^2 + \dot{x}_2^2 + \dot{y}_2^2) \tag{S3.1}$$

where, as usual, $\dot{x} \equiv dx/dt$. Center-of-mass and relative coordinates are defined by the equations

$$X = \frac{x_1 + x_2}{2}, \ Y = \frac{y_1 + y_2}{2}, \ x = x_1 - x_2, \ \text{and} \ y = y_1 - y_2. \tag{S3.2}$$

In terms of the center-of-mass and relative velocities,

$$\dot{x}_1 = \dot{X} + \frac{\dot{x}}{2}, \ \dot{y}_1 = \dot{Y} + \frac{\dot{y}}{2}, \ \dot{x}_2 = \dot{X} - \frac{\dot{x}}{2}, \ \text{and} \ \dot{y}_2 = \dot{Y} - \frac{\dot{y}}{2} \tag{S3.3}$$

Using these relations in order to write the kinetic energy in terms of center-of-mass and relative velocities, we get

$$K = \tfrac{1}{2}M(\dot{X}^2 + \dot{Y}^2) + \tfrac{1}{2}\mu(\dot{x}^2 + \dot{y}^2) \tag{S3.4}$$

where $M = 2m$ and $\mu = m/2$. The momenta that are canonical to X and Y are

$$P_X = \frac{\partial K}{\partial \dot{X}} = M\dot{X} \quad \text{and} \quad P_Y = \frac{\partial K}{\partial \dot{Y}} = M\dot{Y} \tag{S3.5}$$

The momenta canonical to x and y are

$$p_x = \frac{\partial K}{\partial \dot{x}} = \mu\dot{x} \quad \text{and} \quad p_y = \frac{\partial K}{\partial \dot{y}} = \mu\dot{y} \tag{S3.6}$$

The potential energy is proportional to the square of the displacement from equilibrium of the spring.

$$V = \tfrac{1}{2}k(\sqrt{x^2 + y^2} - l)^2 \tag{S3.7}$$

The Hamiltonian function is simply $K + V$, written in terms of the coordinates and canonical momenta.

$$H = \frac{1}{2M}(P_X^2 + P_Y^2) + \frac{1}{2\mu}(p_x^2 + p_y^2) + \frac{k}{2}(\sqrt{x^2 + y^2} - l)^2 \tag{S3.8}$$

Hamilton's equations for this system are

$$\dot{X} = \frac{\partial H}{\partial P_X} = \frac{P_X}{M}$$

$$\dot{x} = \frac{\partial H}{\partial p_x} = \frac{p_x}{\mu}$$

$$\dot{P}_X = -\frac{\partial H}{\partial X} = 0 \qquad\qquad\qquad\qquad (\text{S3.9})$$

$$\dot{p}_x = -\frac{\partial H}{\partial x} = -k\frac{x}{\sqrt{x^2 + y^2}}(\sqrt{x^2 + y^2} - l)$$

and similar equations for Y, y, P_Y, and p_y.

Exercise 3.2   A hanging pendulum of mass m and length L is free to move in two dimensions. (See Fig. S3.1.) (a) Using a system of spherical angles, θ and ϕ, write the Hamiltonian function, $H(\theta, \phi, p_\theta, p_\phi)$. (b) Write Hamilton's equations for the system.

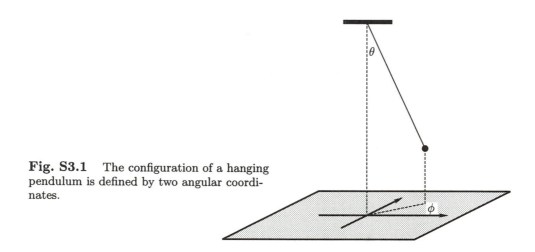

Fig. S3.1   The configuration of a hanging pendulum is defined by two angular coordinates.

Solution   (a) A small change in the polar angle $d\theta$ gives a displacement of the mass equal to $ds = L\,d\theta$. A small change in the azimuthal angle, $d\phi$, gives a displacement of the mass equal to $ds = L\sin\theta\,d\phi$. Since the two displacements are in perpendicular directions, a simultaneous change of both angular coordinates gives a displacement $ds = (L^2\,d\theta^2 + L^2\sin^2\theta\,d\phi^2)^{1/2}$. But the kinetic energy is given by $K = (m/2)(ds/dt)^2$. Therefore

$$K = \tfrac{1}{2}mL^2\left[\left(\frac{d\theta}{dt}\right)^2 + \sin^2\theta\left(\frac{d\phi}{dt}\right)^2\right]$$

$$= \tfrac{1}{2}I(\dot{\theta}^2 + \sin^2\theta\,\dot{\phi}^2) \qquad\qquad (\text{S3.10})$$

The momenta that are canonical to the coordinates θ and ϕ are

$$p_\theta = \frac{\partial K}{\partial \dot{\theta}} = I\dot{\theta} \qquad \text{and} \qquad p_\phi = \frac{\partial K}{\partial \dot{\phi}} = I\sin^2\theta\,\dot{\phi} \qquad (\text{S3.11})$$

The Hamiltonian function is just the energy expressed in terms of the coordinates and canonical momenta. The potential energy is equal to mgz. As we have defined our coordinates, $z = -L\cos\theta$. Making the substitutions $\dot\theta = p_\theta/I$ and $\dot\phi = p_\phi/I\sin^2\theta$ in the kinetic energy, we obtain

$$H = \frac{1}{2I}\left(p_\theta^2 + \frac{p_\phi^2}{\sin^2\theta}\right) - mgL\cos\theta \qquad (S3.12)$$

(b) Hamilton's equations are

$$\frac{d\theta}{dt} = \frac{\partial H}{\partial p_\theta} = \frac{p_\theta}{I}$$

$$\frac{dp_\theta}{dt} = -\frac{\partial H}{\partial\theta} = \frac{\cos\theta\, p_\phi^2}{I\sin^3\theta} - mgL\sin\theta \qquad (S3.13)$$

$$\frac{d\phi}{dt} = \frac{\partial H}{\partial p_\phi} = \frac{p_\phi}{I\sin^2\theta}$$

and $\qquad \dfrac{dp_\phi}{dt} = -\dfrac{\partial H}{\partial\phi} = 0$

Exercise 3.3 For a simple pendulum: (a) Write the Hamiltonian function $H(\theta, p_\theta)$. (b) In the two-dimensional phase space, indicate some of the one-dimensional equal energy "surfaces," noting the change in their structure from lower to higher energies. (c) Describe the "topology" or connectedness of the phase space. (See Fig. S3.2.)

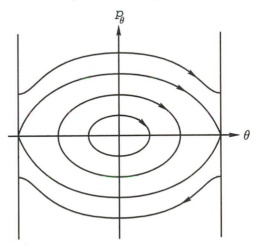

Fig. S3.2   The phase-space trajectories of a simple pendulum.

Solution   (a) The kinetic energy is given by $K = \frac{1}{2}ml^2\dot\theta^2$, where l is the length of the pendulum. The canonical momentum is $p_\theta = \partial K/\partial\dot\theta = ml^2\dot\theta$. The potential energy is $V = -mgl\,\cos\theta$. The Hamiltonian is therefore

$$H = \frac{p_\theta^2}{2ml^2} - mgl\,\cos\theta \qquad (S3.14)$$

(b) The range of θ is $-\pi \leq \theta \leq \pi$. The range of p_θ is $-\infty < p_\theta < \infty$. The phase space can be shown as a vertical strip. (c) Since $\theta = -\pi$ and π really describe identical configurations of the system the representation shown maps, in some cases, a single physical state of the system into two points on the plane. In order to avoid such one-to-two mappings, one must bend the strip around and connect it so as to form an infinite cylinder. That is, the phase space of this system has the topology of a cylinder.

Exercise 3.4   For a system of N one-dimensional particles, write an observable whose value is equal to "the number of particles to the right of the origin."

Solution   Let us first try the simplest case, namely, $N = 1$. Since whether the particle is to the right of the origin depends only on its position and not its momentum, the observable will be only a function of x. The number of particles to the right of the origin can only be 0 or 1. Thus, for this single-particle system we want the observable to have the values

$$A(x) = \begin{cases} 1, & \text{if } x > 0 \\ 0, & \text{if } x < 0 \end{cases} \tag{S3.15}$$

But this equation defines the unit step function. That is, for $N = 1$ the derived observable is $A(x) = \theta(x)$. Now that we have the trick, it is easy to do the same thing for an N-particle system. The one-body observable

$$A(x_1, \ldots, x_N) = \sum_{n=1}^{N} \theta(x_n) \tag{S3.16}$$

just counts the number of particles to the right of the origin.

Exercise 3.5   For a system of N one-dimensional massless particles in a one-dimensional "box" of length L, calculate the uniform ensemble normalization integral $C(E)$ and the entropy function of the system $S(N, E, L)$.

Solution   For massless particles, $E = c|p|$, where c is the speed of light. Therefore, Eq. (3.18) for $C(E)$ must be replaced by

$$C(E) = L^N \int \theta\left(E - c\sum_n |p_n|\right) dp_1 \cdots dp_N \tag{S3.17}$$

Making a transformation of variables, $q_1 = cp_1$, $q_2 = cp_2$, $\ldots, q_N = cp_N$ and using the fact that the integrand is an even function of each variable we can write $C(E)$ as

$$C(E) = (2L/c)^N \int_0^\infty \theta\left(E - \sum_n q_n\right) dq_1 \cdots dq_N \tag{S3.18}$$

The problem boils down to evaluating the N-dimensional integral

$$I_N(a) = \int_0^\infty dx_N \cdots \int_0^\infty dx_2 \int_0^\infty dx_1 \, \theta\left(a - \sum_n x_n\right) \tag{S3.19}$$

since the integral is zero if any $x_n > a$, we can write $I_n(a)$ as

$$I_N(a) = \int_o^a dx_N \cdots \int_o^a dx_2 \int_o^a dx_1 \, \theta\left(a - \sum_n x_n\right) \qquad \text{(S3.20)}$$

Using the simple identity

$$\sum_n^N x_n = x_N + \sum_n^{N-1} x_n \qquad \text{(S3.21)}$$

we can derive the recursion relation

$$I_N(a) = \int_o^a dx_N \, I_{N-1}(a - x_N) \qquad \text{(S3.22)}$$

Making the change of variables, $y = a - x_N$, with $dy = -dx_N$, we can rewrite the recursion relation as

$$\begin{aligned} I_N(a) &= -\int_a^o dy \, I_{N-1}(y) \\ &= \int_o^a I_{N-1}(y) \, dy \end{aligned} \qquad \text{(S3.23)}$$

It is obvious that $I_1(a) = a$. But, then

$$\begin{aligned} I_2(a) &= \int_o^a y \, dy = \frac{a^2}{2} \\ I_3(a) &= \frac{1}{2} \int_o^a y^2 \, dy = \frac{a^3}{3!} \end{aligned} \qquad \text{(S3.24)}$$

and it is easy to see that $I_N(a) = a^N/N!$. Using this result in Eq. (S3.18) we get

$$C(E) = \frac{(2LE/c)^N}{N!} \qquad \text{(S3.25)}$$

The entropy function of the system can be calculated using Eq. (3.86)

$$\begin{aligned} S &= k \log\left[C/h^N N!\right] \\ &= k \log\left[(2LE/hc)^N/(N!)^2\right] \\ &= kN\left[\log(E/N) + \log(L/N) + \text{const.}\right] \end{aligned} \qquad \text{(S3.26)}$$

Exercise 3.6   The position of a two-dimensional diatomic molecule, with fixed distance between the atoms, can be described by the three coordinates, (x, y, θ), where x and y are the coordinates of the molecule's center of mass and θ gives the orientation of the molecular axis with respect to the x axis. The momenta canonically conjugate to the coordinates are p_x, p_y, and p_θ where the first two are the Cartesian components of the center-of-mass momentum and the third is the angular momentum of the molecule about its center of mass. The energy of the molecule is

$$E = (p_x^2 + p_y^2)/2m + p_\theta^2/2I \tag{S3.27}$$

Where I is the moment of inertia about the center of mass. (a) For a system of N noninteracting two-dimensional diatomic molecules confined to a two-dimensional area A, use the uniform ensemble to calculate the entropy function $S(N, E, A)$. (b) Using the entropy function, derive the equations of state of the system—that is, the equations that give the pressure and the energy per particle as functions of the temperature and density. Note: In two dimensions, the "pressure" is the force per unit length that is required to confine the particles to an area A. It is given by a formula analogous to the three-dimensional formula for p/T, namely $p/T = \partial S/\partial A$. (c) Calculate the constant volume (actually constant area) specific heat per particle, defined by $C_V = N^{-1} \partial E(N, T, A)/\partial T$.

Solution   (a) S is related to the normalization integral for the uniform ensemble. For a system of N molecules, that integral is given by

$$C = \int \theta\Big(E - \sum^N (p_{nx}^2 + p_{ny}^2)/2m - \sum^N p_{n\theta}^2/2I\Big) \tag{S3.28}$$
$$\times \, dx_1 \, dy_1 \, d\theta_1 \, dp_{1x} \, dp_{1y} \, dp_{1\theta} \cdots dx_N \, dy_N \, d\theta_N \, dp_{Nx} \, dp_{Ny} \, dp_{N\theta}$$

For each molecule, (x_k, y_k) are integrated over the area A. These integrals give a factor of A^N. For each molecule, θ_k is integrated over 2π. Those integrals give a factor of $(2\pi)^N$. To do the momentum integrals, we define $3N$ variables, $u_1, \ldots, u_N,$ $v_1, \ldots, v_N, w_1, \ldots, w_N$, by

$$u_k = p_{kx}/\sqrt{2m}, \quad v_k = p_{ky}/\sqrt{2m}, \quad w_k = p_{k\theta}/\sqrt{2I} \tag{S3.29}$$

for $k = 1, \ldots, N$. The result is that

$$C = (2\pi A)^N (8m^2 I)^{N/2} \int \theta\Big(E - \sum^N (u_n^2 + v_n^2 + w_n^2)\Big) d^N u \, d^N v \, d^N w \tag{S3.30}$$

The remaining integral is just the volume of a $3N$-dimensional sphere of radius $\sqrt{E}$, which is equal to $(\pi E)^{3N/2}/(3N/2)!$. Thus

$$C = (2\pi A)^N (8m^2 I)^{N/2} (\pi E)^{3N/2}/(3N/2)!$$
$$= (2\pi m I^{1/2} A)^N (2\pi E)^{3N/2}/(3N/2)! \tag{S3.31}$$

In using the formula that relates C to S, we must remember that each molecule has three coordinates, and therefore $K = 3N$. Thus

$$S = k \log(C/h^{3N} N!)$$
$$= kN\big(\log A + \tfrac{3}{2} \log E - \tfrac{5}{2} \log N + \text{const.}\big) \tag{S3.32}$$

where the constant term has not been written explicitly because it will not contribute to the equations of state. (b) The relation

$$\frac{1}{T} = \frac{\partial S}{\partial E} = \frac{3kN}{2E} \tag{S3.33}$$

gives the energy equation

$$E = \tfrac{3}{2}NkT \tag{S3.34}$$

Looking back over the derivation, one can see that each of the momentum variables p_x, p_y, and p_θ contribute $\tfrac{1}{2}kT$ to the energy per particle. This is an example of a general theorem that will be derived in Chapter 4, called the equipartition theorem. It states that any term in the Hamiltonian function that is a simple square contributes $\tfrac{1}{2}kT$ to the thermal energy at temperature T. The pressure is given by

$$p = T\frac{\partial S}{\partial A} = kT\frac{N}{A} = nkT \tag{S3.35}$$

which is the same as that for a gas without rotational degrees of freedom. (c) The specific heat per particle is easily calculated from Eq. (S3.34).

$$C_V = N^{-1}\frac{\partial E}{\partial T} = \tfrac{3}{2}k \tag{S3.36}$$

Exercise 3.7   A one-dimensional particle in a harmonic oscillator potential has an energy $E = p^2/2m + k_o x^2/2$. A system of N particles in the same potential, using a microcanonical ensemble, has a probability function

$$P(x_1, p_1 \ldots, x_N, p_N) = Q^{-1}\delta\Big(E - \sum_n p_n^2/2m - \sum_n k_o x_n^2/2\Big) \tag{S3.37}$$

(a) For such a system, calculate the one-particle phase-space density function $F_1(x,p)$. (b) Assume that $E = N\varepsilon$, where ε is fixed, and show that, for large N, $F_1(x,p)$ is equal to the Maxwell–Boltzmann distribution function that was derived in the last chapter.

Solution   By Eq. (3.34),

$$F_1(x,p) = \sum_{i=1}^{N}\int P(x_1, p_1, \ldots, x_N, p_N)\delta(x - x_i)\delta(p - p_i)\, d^N x\, d^N p \tag{S3.38}$$

Because the probability function P is symmetric under the exchange of the space and momentum coordinates of any two particles, each term in the sum over i makes the same contribution. Therefore, we can keep only the $i = 1$ term and multiply the result by N. One can then use the delta functions to do the integrals over x_1 and p_1, obtaining

$$F_1(x,p) = N\int P(x,p,x_2,p_2,\ldots,x_N,p_N)\, dx_2\, dp_2 \cdots dx_N\, dp_N \tag{S3.39}$$

This is really an obvious result. The integral, without the factor of N, is the probability of finding the first particle at phase-space point (x,p), regardless of the states of the other particles. Multiplying that by N gives the probability of finding *any* particle at (x,p). If we define E' by

$$E' = E - p^2/2m - k_o x^2/2 \tag{S3.40}$$

then

$$F_1(x,p) = N\frac{\int \delta(E' - \sum_2^N p_n^2/2m - \sum_2^N k_o x_n^2/2)\, dx_2 \cdots dp_N}{\int \delta(E - \sum_1^N p_n^2/2m - \sum_1^N x_n^2/2)\, dx_1 \cdots dp_N} \tag{S3.41}$$

$$= N\frac{Q_{N-1}(E')}{Q_N(E)}$$

By making the transformation of coordinates $u_n = p_n/\sqrt{2m}$ and $v_n = x_n\sqrt{k_o/2}$, the integral $Q_N(E)$ can be written in terms of the surface area of a $2N$-dimensional sphere, given in the Mathematical Appendix.

$$Q_N(E) = \frac{(2\pi/\omega)^N E^N}{N!} \tag{S3.42}$$

where $\omega = \sqrt{k_o/m}$ is the oscillation frequency of a particle of mass m in the potential. The integral in the numerator can be done in a similar way.

$$Q_{N-1}(E') = \frac{(2\pi/\omega)^{N-1}(E - p^2/2m - k_o x^2/2)^{N-1}}{(N-1)!} \tag{S3.43}$$

Using these in Eq. (S3.41) gives

$$F_1(x,p) = \frac{\omega N^2}{2\pi}\frac{(E - p^2/2m - k_o x^2/2)^{N-1}}{E^N} \tag{S3.44}$$

(b) Putting $E = N\varepsilon$, we can write this as

$$F_1(x,p) = \frac{\omega N}{2\pi\varepsilon}\left[1 - \frac{1}{N}\left(\frac{p^2}{2m\varepsilon} + \frac{k_o x^2}{2\varepsilon}\right)\right]^{N-1} \tag{S3.45}$$

But

$$\lim_{N\to\infty}\left(1 - \frac{x}{N}\right)^{N-1} = \lim_{N\to\infty}\left(1 - \frac{x}{N}\right)^N = e^{-x} \tag{S3.46}$$

Therefore, for large N,

$$F_1(x,p) \approx \frac{\omega N}{2\pi kT}\exp\left(-\frac{p^2/2m + k_o x^2/2}{kT}\right) \tag{S3.47}$$

where $kT \equiv \varepsilon$. This is exactly the Maxwell–Boltzmann distribution given in Eq. (2.16), with $\phi(x) = k_o x^2/2$.

Exercise 3.8   In Chapter 3 it is stated that the entropy may be calculated by Eq. (3.84), using either the function $C(E)$, associated with the uniform ensemble, or the function $Q(E)$, associated with the microcanonical ensemble. Let S_U be the entropy, calculated with the uniform ensemble, and S_M be that calculated with the microcanonical ensemble. Show that, for large N, $(S_M - S_U)/S_U$ is of order $1/N$.

Solution

$$S_M = k\log(Q/h^k N!) \tag{S3.48}$$

and

$$S_U = k \log(C/h^k N!) \tag{S3.49}$$

where

$$Q = \int \delta(E - H) \, d^k x \, d^k p \tag{S3.50}$$

and

$$C = \int \theta(E - H) \, d^k x \, d^k p \tag{S3.51}$$

We can see that $Q(E) = \partial C(E)/\partial E$, which means that

$$S_M - S_U = k \log \left(\frac{\partial C/\partial E}{C} \right) \tag{S3.52}$$

But

$$\frac{\partial C/\partial E}{C} = \frac{\partial}{\partial E} \log C = \frac{\partial(S_U/k)}{\partial E} = \frac{1}{kT} \tag{S3.53}$$

Thus

$$S_M - S_U = -k \log(kT) \tag{S3.54}$$

For a large system, the right-hand side is independent of N, but S_U is proportional to N. Therefore $(S_M - S_U)/S_U \sim 1/N$.

Exercise 3.9   Suppose the Hamiltonian function has one term that is proportional to some scalar parameter λ. That is,

$$H(x, p) = H_o(x, p) + \lambda h(x, p) \tag{S3.55}$$

An example is a set of particles in an external gravitational field. Then the parameter is g, the gravitational acceleration, and $h = m \sum z_i$. In that case the entropy will depend on the value of the parameter. Show that the microcanonical average of the observable $h(x, p)$ is given by the formula

$$\langle h \rangle = -T \frac{dS}{d\lambda} \tag{S3.56}$$

Solution

$$\begin{aligned} S &= k \log \left(\frac{1}{h^K N!} \int \theta(E - H_o - \lambda h) \, d^K x \, d^K p \right) \\ &= k \log \left(\int \theta(E - H_o - \lambda h) \, d^K x \, d^K p \right) - k \log(h^K N!) \end{aligned} \tag{S3.57}$$

Using the fact that $d\theta(E - H_o - \lambda h)/d\lambda = -h\delta(E - H_o - \lambda h)$, we see that

$$\begin{aligned} \frac{dS}{d\lambda} &= k \frac{\left[\frac{d}{d\lambda} \int \theta(E - H_o - \lambda h) \, d^K x \, d^K p \right]}{\int \theta(E - H_o - \lambda h) \, d^K x \, d^K p} \\ &= -k \frac{\int h(x, p)\delta(E - H_o - \lambda h) \, d^K x \, d^K p}{\int \theta(E - H_o - \lambda h) \, d^K x \, d^K p} \\ &= -k \frac{\int h\delta(E - H) \, d^K x \, d^K p}{\int \delta(E - H) \, d^K x \, d^K p} \frac{\int \delta(E - H) \, d^K x \, d^K p}{\int \theta(E - H) \, d^K x \, d^K p} \end{aligned} \tag{S3.58}$$

But the microcanonical average of h is

$$\langle h \rangle = \frac{\int h\delta(E-H)\,d^K x\,d^K p}{\int \delta(E-H)\,d^K x\,d^K p} \tag{S3.59}$$

and

$$\begin{aligned}\frac{1}{T} &= k\frac{\partial}{\partial E}\log\left[\int\theta(E-H)\,d^K x\,d^K p\right]\\ &= k\frac{\int\delta(E-H)\,d^K x\,d^K p}{\int\theta(E-H)\,d^K x\,d^K p}\end{aligned} \tag{S3.60}$$

which gives the desired result.

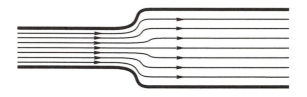

Fig. S3.3   The flow lines of a two-dimensional fluid.

Exercise 3.10   Fig. S3.3 shows the flow lines for the steady (that is, time-independent) flow of a two-dimensional fluid. The flow velocity at position (x_1, x_2) is $\mathbf{u}(x_1, x_2)$. If the volume occupied by a certain amount of fluid remains constant under the flow, then the flow is called *incompressible*. The mathematical condition for incompressible flow is that the divergence of the velocity field vanish everywhere.

$$\nabla\cdot\mathbf{u} = \frac{\partial u_1}{\partial x_1} + \frac{\partial u_2}{\partial x_2} = 0 \tag{S3.61}$$

(a) For a two-dimensional phase space, the variables (x, p) could be identified with a pair of Cartesian coordinates (z_1, z_2). Then Hamilton's equations would define a velocity field

and

$$\begin{aligned}u_1 &= \frac{dz_1}{dt} = \frac{dx}{dt} = \frac{\partial H}{\partial p} = \frac{\partial H(z_1, z_2)}{\partial z_2}\\ u_2 &= \frac{dz_2}{dt} = \frac{dp}{dt} = -\frac{\partial H}{\partial x} = -\frac{\partial H(z_1, z_2)}{\partial z_1}\end{aligned} \tag{S3.62}$$

Show that this Hamiltonian flow is incompressible. (b) using the fact that $\nabla\cdot\mathbf{u} = 0$ is the condition for incompressible flow in any number of dimensions, generalize (a) to a $2K$-dimensional phase space. The result is known as Liouville's theorem.

Solution   (a)

$$\nabla\cdot\mathbf{u}(z_1, z_2) = \frac{\partial u_1}{\partial z_1} + \frac{\partial u_2}{\partial z_2} = \frac{\partial^2 H}{\partial z_1\,\partial z_2} - \frac{\partial^2 H}{\partial z_2\,\partial z_1} = 0 \tag{S3.63}$$

(b) For a $2K$-dimensional phase space, we identify the $2K$ variables $(z_1, z_2, \ldots, z_{2K})$ with the phase-space variables $(x_1, \ldots, x_K, p_1, \ldots, p_K)$. Then Hamilton's equations give the $2K$-dimensional velocity field defined as follows. For $k = 1$ to K,

$$u_k = \frac{dz_k}{dt} = \frac{dx_k}{dt} = \frac{\partial H}{\partial p_k} = \frac{\partial H}{\partial z_{K+k}} \tag{S3.64}$$

and

$$u_{K+k} = \frac{dz_{K+k}}{dt} = \frac{dp_k}{dt} = -\frac{\partial H}{\partial x_k} = -\frac{\partial H}{\partial z_k} \tag{S3.65}$$

Then

$$\nabla \cdot \mathbf{u} = \sum_{k=1}^{K} \left(\frac{\partial u_k}{\partial z_k} + \frac{\partial u_{K+k}}{\partial z_{K+k}} \right)$$

$$= \sum_{k=1}^{K} \left(\frac{\partial^2 H}{\partial z_k \partial z_{K+k}} - \frac{\partial^2 H}{\partial z_{K+k} \partial z_k} \right) = 0 \tag{S3.66}$$

Exercise 3.11   According to Liouville's theorem, the motion of phase-space points defined by Hamilton's equations conserves phase-space volume. In this exercise we will illustrate this conservation of phase-space volume with a particular case. The Hamiltonian for a single particle in one dimension, subjected to a constant force F, is

$$H(x,p) = p^2/2m - Fx \tag{S3.67}$$

Consider the phase-space rectangle defined by $0 < x < A$ and $0 < p < B$. (a) Letting the points in the rectangle move for a time t according to Hamilton's equations, show the region into which the rectangle develops. (b) Show that the area of the region is AB, which is the area of the initial rectangle.

Solution   First we have to determine the phase-space position at time t of a point that starts out at (x_o, p_o). Hamilton's equations are

$$\frac{dx}{dt} = \frac{\partial H}{\partial p} = \frac{p(t)}{m} \tag{S3.68}$$

and

$$\frac{dp}{dt} = -\frac{\partial H}{\partial x} = F \tag{S3.69}$$

The second equation can be easily integrated to give

$$p(t) = p_o + Ft \tag{S3.70}$$

Using this result, we can integrate the first equation.

$$x(t) = x_o + \frac{1}{m} \int_o^t p(t)\, dt = x_o + \frac{p_o}{m}t + \frac{F}{2m}t^2 \tag{S3.71}$$

In Fig. S3.4 the four sides of the intial rectangle are numbered. Side 1 consists of the phase-space points $(x_o, 0)$, where $0 < x_o < A$. An initial point $(x_o, 0)$ goes to a point

$$(x_o, 0) \rightarrow \left(x_o + \frac{F}{2m}t^2, Ft \right) \tag{S3.72}$$

Side 2 consists of the points $(0, p_o)$, where $0 < p_o < B$. These points are carried, by Hamilton's equations, into

$$(0, p_o) \rightarrow \left(\frac{p_o}{m}t + \frac{F}{2m}t^2, p_o + Ft \right) \tag{S3.73}$$

Side 3 consists of the points (x_o, B), where $0 < x_o < A$. These go to

$$(x_o, B) \rightarrow \left(x_o + \frac{B}{m}t + \frac{F}{2m}t^2, B + Ft \right) \tag{S3.74}$$

Side 4 consists of the points (A, p_o), with $0 < p_o < B$. These go to

$$(A, p_o) \rightarrow \left(A + \frac{p_o}{m}t + \frac{F}{2m}t^2, p_o + Ft \right) \tag{S3.75}$$

In particular, the four corners of the rectangle are transformed as follows:

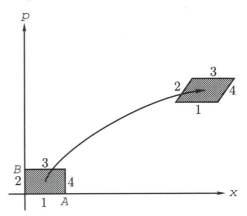

Fig. S3.4   The motion of a region in phase space under Hamiltonian flow.

$$
\begin{aligned}
(0,0) &\rightarrow \left(\frac{F}{2m}t^2, Ft \right) \\
(A,0) &\rightarrow \left(A + \frac{F}{2m}t^2, Ft \right) \\
(0,B) &\rightarrow \left(\frac{B}{m}t + \frac{F}{2m}t^2, B + Ft \right) \\
\text{and}\quad (A,B) &\rightarrow \left(A + \frac{B}{m}t + \frac{F}{2m}t^2, B + Ft \right)
\end{aligned}
\tag{S3.76}
$$

The rectangle is transformed into a parallelogram with a base of A and an altitude of B, which has an area AB.

Exercise 3.12   Given any probability density in phase space $P(x, p)$ a statistical ensemble associated with that probability density is a very large but finite collection of points in phase space that has the property that the density of ensemble points $N(x, p)$, is, at every location, proportional to the probability function $P(x, p)$. The ensemble may be pictured as a vast collection of exact replicas of the system, each in a different physical state (having different values of particle positions and/or velocities). For example, if the system is a simple pendulum, then the ensemble is a vast warehouse of identical pendula, each with its own combination of angle and rotational velocity. But this picture of an ensemble immediately suggests the question: What happens to the ensemble density as all the pendula move according to Hamilton's equations? The answer is given by the following remarkable theorem.

Theorem: As the system points move according to Hamilton's equations, the ensemble density at the location of any moving phase space-point remains constant. (a) Prove the theorem, which is also due to Liouville. (b) Show that, as the system points move, the ensemble density $N(x, p, t)$ satisfies *Liouville's equation*,

$$\frac{\partial N(x, p, t)}{\partial t} + \sum_{k=1}^{K} \left(\frac{\partial N}{\partial x_k} \frac{\partial H}{\partial p_k} - \frac{\partial N}{\partial p_k} \frac{\partial H}{\partial x_k} \right) = 0 \qquad (S3.77)$$

Solution   (a) Let us consider the case of a simple pendulum. As each pendulum moves, its system point in the two-dimensional phase space moves like a speck of dust caught in the Hamiltonian flow. Suppose at time 0 we consider a small region δV_o in the phase space. That small region will contain a number of ensemble points, δN. Now let all the points of δV_o (not only the ensemble points) move for a time t according to Hamilton's equations. The region δV_o will then be transformed into the region δV_t. But, because Hamiltonian flow is incompressible, the volume of δV_t will equal the volume of δV_o. The ensemble points, like dust particles, remain in δV as it moves. Thus at time t the number of ensemble points in δV_t will be the same as the original number of ensemble points in δV_o. Since the volume of the region has not changed, the density of ensemble points in δV_t at time t will equal the density of ensemble points in δV_o at time 0. That is, the density at the location of a moving ensemble point remains constant. (b) Let $N(x, p, t)$ be the ensemble density at time t and let $(x(t), p(t))$ be a solution of Hamilton's equations of motion. What we have just shown can be written as

$$N(x(t), p(t), t) = N(x(0), p(0), 0) \qquad (S3.78)$$

Taking a time derivative of this equation gives

$$\frac{\partial N}{\partial t} + \sum_{k=1}^{K} \left(\frac{\partial N}{\partial x_k} \frac{dx_k}{dt} + \frac{\partial N}{\partial p_k} \frac{dp_k}{dt} \right) = 0 \qquad (S3.79)$$

Using Hamilton's equations to eliminate dx_k/dt and dp_k/dt gives Liouville's equation.

Exercise 3.13   For a uniform ensemble, the ensemble density is $N(x, p) = N_o \theta(E - H(x, p))$, where N_o is a constant that is proportional to the number of system points in the ensemble. For a microcanonical ensemble, the density is of the form $N(x, p) = N_o \delta(E - H(x, p))$. If these density functions are to describe the time-independent equilibrium state of a complex system, then one characteristic that we would certainly want them to have is that the ensemble density function would not change in time as the ensemble points moved according to Hamilton's equations. Show that a time-independent ensemble density that is *any* function of the system Hamiltonian [that is, $N(x, p, t) = F(H(x, p))$] is a solution of Liouville's equation.

Solution   If $N(x, p, t) = F(H(x, p))$ then

$$\frac{\partial N}{\partial t} + \sum_{k=1}^{K} \left(\frac{\partial N}{\partial x_k} \frac{\partial H}{\partial p_k} - \frac{\partial N}{\partial p_k} \frac{\partial H}{\partial x_k} \right) = \frac{dF}{dH} \sum_{k=1}^{K} \left(\frac{\partial H}{\partial x_k} \frac{\partial H}{\partial p_k} - \frac{\partial H}{\partial p_K} \frac{\partial H}{\partial x_k} \right) \qquad (S3.80)$$

$$= 0$$

Exercise 3.14 For any observable, let $\langle A \rangle$ represent the time average of $A(t)$ over the time period $0 < t < \infty$. That is,

$$\langle A \rangle \equiv \lim_{T \to \infty} \frac{1}{T} \int_o^T A(t)\, dt \tag{S3.81}$$

For any isolated system of point particles in a fixed volume, prove the *Virial Theorem of Clausius*,

$$\langle E_K \rangle = \left\langle -\frac{1}{2} \sum \mathbf{r}_i \cdot \mathbf{F}_i \right\rangle \tag{S3.82}$$

where E_K is the kinetic energy of the system and $\mathbf{F}_i$ is the instantaneous force on the ith particle.

Solution Consider the observable

$$S(t) = \sum_i \mathbf{r}_i \cdot \mathbf{p}_i \tag{S3.83}$$

The time derivative of S is

$$\begin{aligned}
\frac{dS}{dt} &= \sum_i \mathbf{v}_i \cdot \mathbf{p}_i + \sum_i \mathbf{r}_i \cdot \mathbf{F}_i \\
&= 2E_K + \sum_i \mathbf{r}_i \cdot \mathbf{F}_i
\end{aligned} \tag{S3.84}$$

Averaging this equation over the time interval $(0, T)$ gives

$$\frac{1}{T} \int_o^T \frac{dS}{dt}\, dt = \frac{S(T) - S(0)}{T} = \frac{1}{T} \int_o^T \left(2E_K + \sum \mathbf{r}_i \cdot \mathbf{F}_i \right) dt \tag{S3.85}$$

For a system in a fixed volume, $S(T)$ is bounded and, therefore, $[S(T) - S(0)]/T \to 0$ as $T \to \infty$. In the same limit, the last term approaches $2\langle E_K \rangle + \langle \sum \mathbf{r}_i \cdot \mathbf{F}_i \rangle$, which proves the theorem.

Exercise 3.15 For a gas with no interparticle interactions, in a rectangular volume, show that the virial theorem leads to the well-known result, $pV = \frac{2}{3} E_K$.

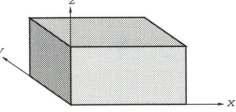

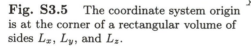

Fig. S3.5 The coordinate system origin is at the corner of a rectangular volume of sides L_x, L_y, and L_z.

Solution Assume that the dimensions of the volume are L_x, L_y, and L_z (Fig. S3.5). Take a coordinate system whose origin is located at one corner of the volume. Let X_i, Y_i, and Z_i be the rectangular components of the force on particle i. Then

$$\langle E_K \rangle = -\frac{1}{2} \left\langle \sum (x_i X_i + y_i Y_i + z_i Z_i) \right\rangle \tag{S3.86}$$

In the absence of interparticle interactions, forces are exerted on the particles only when they are microscopically close to one of the walls. Thus, at any instant of time, the sum on the right-hand side of Eq. (S3.86) has six contributions associated with the sets of particles that, at that instant, are very close to each of the six walls.

For every particle near the wall at $x=0$, the x coordinate is essentially zero. The total force that those particles exert on that wall is, by definition of the pressure, equal to the instantaneous pressure times the area of the wall, namely $L_y L_z$. Thus, for that wall,

$$\sum_i x_i X_i = 0 \cdot \sum_i X_i = 0 \cdot L_y L_z = 0 \qquad \text{(S3.87)}$$

For simplicity, we will assume that the forces exerted on the particles by the walls are always normal to the walls. Then, the y and z components of the forces exerted on the particles by the wall at $x=0$ are zero. (This assumption could be replaced by the weaker assumption that the tangential components of the forces were statistically independent of position on the wall. Their sum would then be zero by cancellation.) In a similar way, the contribution of the particles near the wall at $x=L_x$ is

$$\sum (x_i X_i + y_i Y_i + z_i Z_i) = L_x(-L_y L_z p) = -L_x L_y L_z p \qquad \text{(S3.88)}$$

Taking into account the other four walls, one obtains

$$-\sum (x_i X_i + y_i Y_i + z_i Z_i) = 3 L_x L_y L_z p = 3V p \qquad \text{(S3.89)}$$

giving

$$2\langle E_K \rangle = 3V \langle p \rangle \qquad \text{(S3.90)}$$

If we neglect the fluctuations in the macroscopic observables E_K and p, we may drop the time average symbols.

Exercise 3.16   In the last exercise, it was shown that, for a gas without interparticle interactions, $pV = \frac{2}{3} E_K$. Assume that a nonideal gas has interparticle forces derivable from a two-body interaction potential $v(r_{ij})$. That is, that the Hamiltonian function is

$$H = \sum_i \frac{p_i^2}{2m} + \frac{1}{2} \sum_{i \neq j} v(r_{ij}) \qquad \text{(S3.91)}$$

The quantity $pV - \frac{2}{3} E_K$ we will call the deviation from the ideal gas law. (a) Using the Virial Theorem, and assuming that the system has no extra conservation laws, so that time averages can be replaced by microcanonical ensemble averages, show that the deviation from the ideal gas law can be written as the average value of a two-particle observable involving the interaction potential $v(r)$. (b) The function

$$n_2(\mathbf{r}_1, \mathbf{r}_2) = \int F_2(\mathbf{r}_1, \mathbf{p}_1, \mathbf{r}_2, \mathbf{p}_2) \, d^3\mathbf{p}_1 \, d^3\mathbf{p}_2 \qquad \text{(S3.92)}$$

gives the probability density for finding a particle at $\mathbf{r}_1$ and another particle at $\mathbf{r}_2$, regardless of their momenta. (F_2 is the two-particle phase space density function, defined in Section 3.10.) In a rotationally and translationally invariant system, $n_2(\mathbf{r}_1, \mathbf{r}_2)$ is a function of $r_{12} = |\mathbf{r}_1 - \mathbf{r}_2|$ only. We write it as

$$n_2(\mathbf{r}_1, \mathbf{r}_2) = n^2 g(r_{12}) \tag{S3.93}$$

where n is the particle density and $g(r)$ is called the *two-particle correlation function*. Using the result of (a), write $pV - \frac{2}{3}E_K$ as a one-dimensional integral involving $v(r)$ and $g(r)$.

Solution   (a) According to the Virial Theorem, $\langle E_K \rangle = \langle -\frac{1}{2} \sum \mathbf{r}_i \cdot \mathbf{F}_i \rangle$. Assuming that the system has no extra conservation laws, we can interpret the averages as microcanonical averages. The force on the ith particle is a sum of the forces exerted on that particle by all the other particles in the system plus the force on that particle by the walls (in case it is sufficiently close to one of the walls). Let us write the force of the walls on the ith particle as $\mathbf{W}_i$. The force exerted on the ith particle by the jth particle is given by

$$\mathbf{F}_{ij} = -v'(r_{ij})\mathbf{n}_{ij} \tag{S3.94}$$

where $\mathbf{n}_{ij} = \mathbf{r}_{ij}/r_{ij}$ is a unit vector, pointing from particle j to particle i. Then, ignoring fluctuations in E_K,

$$E_K = \left\langle -\frac{1}{2} \sum_i \mathbf{r}_i \cdot \mathbf{W}_i \right\rangle + \frac{1}{2} \left\langle \sum_{i \neq j} \mathbf{r}_i \cdot \mathbf{n}_{ij} v'(r_{ij}) \right\rangle \tag{S3.95}$$

The first term on the right, by the argument of the previous exercise, gives $\frac{3}{2}pV$. The pair of particles labeled 1 and 2 contribute two terms to the double sum in Eq. (S3.95). Using the fact that $\mathbf{n}_{ij} = -\mathbf{n}_{ji}$, they can be written as

$$(\mathbf{r}_1 - \mathbf{r}_2) \cdot \mathbf{n}_{12} v'(r_{12}) = r_{12} v'(r_{12}) \tag{S3.96}$$

Doing the same for the other pairs of particles, we get

$$\begin{aligned} E_K &= \frac{3}{2}pV + \frac{1}{2}\left\langle \sum_{i<j} r_{ij} v'(r_{ij}) \right\rangle \\ &= \frac{3}{2}pV + \frac{1}{4}\left\langle \sum_{i \neq j} r_{ij} v'(r_{ij}) \right\rangle \end{aligned} \tag{S3.97}$$

From the definition of $n_2(\mathbf{r}_1, \mathbf{r}_2)$,

$$\left\langle \sum_{i \neq j} r_{ij} v'(r_{ij}) \right\rangle = \int d^3\mathbf{r}_1 \, d^3\mathbf{r}_2 \, r_{12} v'(r_{12}) n_2(\mathbf{r}_1, \mathbf{r}_2) \tag{S3.98}$$

We transform the integral to center-of-mass and relative coordinates, $\mathbf{R} = (\mathbf{r}_1 + \mathbf{r}_2)/2$ and $\mathbf{r} = \mathbf{r}_1 - \mathbf{r}_2$, and use the fact that $n_2(\mathbf{r}_1, \mathbf{r}_2) = n^2 g(r)$. The term $v'(r)$ is zero unless r is very small; thus we can extend the $\mathbf{r}$ integral to infinity. The integral over $\mathbf{R}$ gives a factor of V. Introducing polar coordinates in the $\mathbf{r}$ integral gives

$$\left\langle \sum_{i \neq j} r_{ij} v'(r_{ij}) \right\rangle = 4\pi V n^2 \int_0^\infty r^3 v'(r) g(r) \, dr \tag{S3.99}$$

and therefore,

$$E_K - \frac{3}{2}pV = \pi V n^2 \int_o^\infty r^3 v'(r) g(r)\, dr \tag{S3.100}$$

In the next chapter it will be shown that, for a classical system, even one with interparticle interactions, $E_K = \frac{3}{2}NkT$. When this is used in Eq. (S3.100), one obtains an exact equation for the pressure of a nonideal gas in terms of the two-particle correlation function.

$$p = nkT - \frac{2}{3}\pi n^2 \int_o^\infty r^3 v'(r) g(r)\, dr \tag{S3.101}$$

Exercise 3.17   The three-dimensional Ising model described in Section 3.13, in which the energy depends on the relative orientation of nearest-neighbor spins, cannot be solved exactly. That is, no one has been able to derive an analytic expression for the thermodynamic functions, such as the entropy, associated with the model. There is a simpler model that can easily be solved exactly. It is called the noninteracting Ising model. The configurations of the model are still described by N spin variables, $\sigma_1, \ldots, \sigma_N$, but the model attempts to describe a system of N noninteracting magnetic moments in an external magnetic field. For a given configuration, the energy is given by

$$E = H(\sigma_1, \sigma_2, \ldots, \sigma_N) = \sum_{i=1}^{N} mB\sigma_i \tag{S3.102}$$

where m is the magnetic moment of the particle, B is the magnetic field strength, and σ_i tells whether the ith magnetic moment is parallel or antiparallel to the field. For any energy value of the form $E = mBK$, where K is an integer, we can construct a microcanonical ensemble with a probability density

$$P(\sigma_1, \sigma_2, \ldots, \sigma_N) = Q^{-1}\delta(E - H(\sigma_1, \sigma_2, \ldots, \sigma_N)) \tag{S3.103}$$

where $\delta(0) = 1$ and $\delta(x) = 0$ for $x \neq 0$. (a) Calculate the microcanonical normalization sum

$$Q(N, E) = \sum_{\sigma_1} \cdots \sum_{\sigma_N} \delta(E - H(\sigma_1, \sigma_2, \ldots, \sigma_N)) \tag{S3.104}$$

(b) Using the relation, $S = k \log(Q/h^N)$, calculate the thermal energy per particle as a function of the temperature. Note: For particles that are fixed in place in a lattice, there is no possibility of particle exchange and therefore there is no need to include the $N!$ factor in the entropy expression. Such particles are said to be *distinguishable*.

Solution   (a)

$$Q = \sum_{\sigma_1} \cdots \sum_{\sigma_N} \delta\left(E - mB\sum \sigma_i\right) \tag{S3.105}$$

where $E = mBK$. For a given configuration, let N_+ be the number of up spins and N_- be the number of down spins. Then $mB\sum \sigma_i = E$ iff $N_+ - N_- = K$. But we also know that $N_+ + N_- = N$. Therefore, $N_+ = (N+K)/2$ and $N_- = (N-K)/2$. Q is equal to the number of configurations with given values of N_+ and N_-. That is

$$Q = \frac{N!}{N_+!N_-!} = \frac{N!}{[(N+K)/2]![(N-K)/2]!} \quad \text{(S3.106)}$$

(b) Using Stirling's approximation

$$S/k = N\log N - \frac{N+K}{2}\log\left(\frac{N+K}{2}\right) - \frac{N-K}{2}\log\left(\frac{N-K}{2}\right) - N\log h \quad \text{(S3.107)}$$

The relation $E = mBK$ allows us to express S in terms of N and E.

$$\begin{aligned}\frac{S}{k} =& N\log N - \frac{1}{2}\left(N + \frac{E}{mB}\right)\log\left(N + \frac{E}{mB}\right)\\ &- \frac{1}{2}\left(N - \frac{E}{mB}\right)\log\left(N - \frac{E}{mB}\right) - N\log\left(\frac{h}{2}\right)\end{aligned} \quad \text{(S3.108)}$$

The temperature is related to the energy by

$$\frac{1}{kT} = \frac{\partial(S/k)}{\partial E} = \frac{1}{2mB}\log\left(\frac{N - E/mB}{N + E/mB}\right) \quad \text{(S3.109)}$$

or

$$\frac{N + E/mB}{N - E/mB} = e^{-2mB/kT} \quad \text{(S3.110)}$$

Letting $x = E/mBN$ and $y = e^{-2mB/kT}$, this equation can be written as

$$1 + x = y(1 - x) \quad \text{(S3.111)}$$

which has the solution $x = (y-1)/(y+1)$, or

$$\begin{aligned}\frac{E}{N} &= mB\frac{e^{-2mB/kT} - 1}{e^{-2mB/kT} + 1}\\ &= -mB\frac{e^{mB/kT} - e^{-mB/kT}}{e^{mB/kT} + e^{mB/kT}}\\ &= -mB\tanh(mB/kT)\end{aligned} \quad \text{(S3.112)}$$

Exercise 3.18 Consider a set of N particles in two dimensions in a symmetrical harmonic oscillator potential $V(x,y) = \frac{1}{2}k(x^2 + y^2)$. Because of the perfect rotational symmetry of the potential the angular momentum about the origin is an extra conserved quantity. Assume that the angular momentum L and the total energy E are given. (a) Use the method of Chapter 2 to determine the phase-space particle density $f(x,y,p_x,p_y)$ at equilibrium. (b) Determine the spatial particle density $n(r)$ as a function of distance from the origin. (c) Determine the average velocity of particles at position (x,y) and show that it is consistent with the velocity field of rigid body rotation. (Warning: This is a very long and difficult exercise.)

Solution   (a) We break up the four-dimensional phase space (x,y,p_x,p_y) into little four-dimensional cubes, which we number with an integer $i = 1,2,\dots$. The center of the ith cube is at $(x_i, y_i, p_{xi}, p_{yi})$. A particle in the ith cube has an energy E_i and an angular momentum L_i, where

$$E_i = \frac{1}{2m}\left(p_{xi}^2 + p_{yi}^2\right) + \frac{k}{2}\left(x_i^2 + y_i^2\right) \tag{S3.113}$$

and

$$L_i = x_i p_{yi} - y_i p_{xi} \tag{S3.114}$$

We describe a macrostate by assigning occupation numbers, $N_1, N_2, \ldots$, to all cubes. We describe a microstate by saying which cube each particle is in. For a given macrostate, there correspond I microstates where

$$I = \frac{N!}{\prod_i N_i!} \tag{S3.115}$$

We maximize $\log I$, subject to the three constraints

$$\sum_i N_i = N \tag{S3.116}$$

$$\sum_i N_i E_i = E \tag{S3.117}$$

and

$$\sum_i N_i L_i = L \tag{S3.118}$$

This is done by maximizing the function

$$\begin{aligned} F &= \log I - \alpha \sum_i N_i - \beta \sum_i N_i E_i - \gamma \sum_i N_i L_i \\ &= \log N! - \sum_i N_i(\log N_i - 1) - \sum_i (\alpha + \beta E_i + \gamma L_i) N_i \end{aligned} \tag{S3.119}$$

with no constraints. Setting $\partial F / \partial N_i = 0$ gives the equation

$$\log N_i = -\alpha - \beta E_i - \gamma L_i \tag{S3.120}$$

If ω is the four-dimensional volume of one of the phase-space cubes, then the definition of the phase-space distribution function is

$$f(x_i, y_i, p_{xi}, p_{yi}) = \frac{N_i}{\omega} \tag{S3.121}$$

This shows that

$$f(x, y, p_x, p_y) = C\exp\left[-\beta\left(\frac{1}{2m}(p_x^2 + p_y^2) + \frac{k}{2}(x^2 + y^2)\right) - \gamma(xp_y - yp_x)\right] \tag{S3.122}$$

Next, we must evaluate the constants C, β, and γ in terms of N, E, and L.

$$\begin{aligned} C^{-1}N = \int \exp\Big[&-\beta\Big(\frac{1}{2m}(p_x^2 + p_y^2) + \frac{k}{2}(x^2 + y^2)\Big) \\ &- \gamma(xp_y - yp_x)\Big]\, dx\, dy dp_x\, dp_y \end{aligned} \tag{S3.123}$$

Notice that the integral factors into equivalent two-dimensional integrals involving (x, p_y) and (y, p_x).

$$\frac{N}{C} = \left[\int \exp(-\beta p_y^2/2m - \beta k x^2/2 - \gamma x p_y) \, dy \, dp_x \right]^2 \tag{S3.124}$$

The two-dimensional integral needed is of the form

$$I = \int e^{-(Ax^2 + 2Bxp + Cp^2)} \, dx \, dp \tag{S3.125}$$

But

$$Ax^2 + 2Bxp + Cp^2 = A\left(x + \frac{B}{A}p\right)^2 + \left(C - \frac{B^2}{A}\right)p^2 \tag{S3.126}$$

Changing variables from x to $u = x + (B/A)p$, the integral separates into two Gaussian integrals, giving

$$I = \frac{\pi}{\sqrt{A(C - B^2/A)}} = \frac{\pi}{\sqrt{AC - B^2}} \tag{S3.127}$$

In the case at hand, $A = \beta k/2$, $B = \gamma/2$, and $C = \beta/2m$. The final result is

$$C^{-1}N = \frac{4\pi^2}{\beta^2\omega^2 - \gamma^2} \tag{S3.128}$$

where $\omega = \sqrt{k/m}$. This gives the constant C in terms of N, β, and γ.

$$C = \frac{\beta^2\omega^2 - \gamma^2}{4\pi^2} N \tag{S3.129}$$

To do the integrals for the energy and angular momentum, it is best to use the following trick:

$$\left\langle \frac{p_x^2 + p_y^2}{2m} + \frac{k}{2}(x^2 + y^2) \right\rangle$$

$$= -C\frac{\partial}{\partial \beta} \int \exp\left[-\beta\left(\frac{1}{2m}(p_x^2 + p_y^2) + \frac{k}{2}(x^2 + y^2)\right) - \gamma(xp_y - yp_x)\right] dx \, dy \, dp_x \, dp_y \tag{S3.130}$$

$$= -N\frac{\beta^2\omega^2 - \gamma^2}{4\pi^2}\frac{\partial}{\partial \beta}\left(\frac{4\pi^2}{\beta^2\omega^2 - \gamma^2}\right)$$

$$= N\frac{2\beta\omega^2}{\beta^2\omega^2 - \gamma^2}$$

Also

$$\langle xp_y - yp_x \rangle = -N\frac{\beta^2\omega^2 - \gamma^2}{4\pi^2}\frac{\partial}{\partial \gamma}\left(\frac{4\pi^2}{\beta^2\omega^2 - \gamma^2}\right)$$

$$= -N\frac{2\gamma}{\beta^2\omega^2 - \gamma^2} \tag{S3.131}$$

Letting $\varepsilon = E/N$ and $l = L/N$ be the energy and angular momentum per particle, we get the following two equations for β and γ.

$$\frac{2\beta\omega^2}{\beta^2\omega^2 - \gamma^2} = \varepsilon \qquad (S3.132)$$

and

$$\frac{2\gamma}{\beta^2\omega^2 - \gamma^2} = -l \qquad (S3.133)$$

It is easy to see that these equations imply that

$$\gamma\varepsilon + \beta\omega^2 l = 0 \qquad (S3.134)$$

and

$$\gamma l + \beta\varepsilon = 2 \qquad (S3.135)$$

which are two linear equations for γ and β, with the solution

$$\gamma = -\frac{2\omega^2 l}{\varepsilon^2 - \omega^2 l^2} \quad \text{and} \quad \beta = \frac{2\varepsilon}{\varepsilon^2 - \omega^2 l^2} \qquad (S3.136)$$

(b) Because the system has rotational symmetry about the origin, it is sufficient to calculate the spatial density along the x axis.

$$n(x,0) = C \int \exp\left(-\beta\left[(p_x^2 + p_y^2)/2m + kx^2/2\right] - \gamma x p_y\right) dp_x \, dp_y \qquad (S3.137)$$

The integral over p_x gives a factor of $(2\pi m/\beta)^{1/2}$. The integral over p_y can be done by completing the square, as was done previously. It then gives a factor of $(2\pi m/\beta)^{1/2} \exp[(m\gamma^2/2\beta)x^2]$. Using Eq. (S3.108) for C, we get

$$\begin{aligned} n(x,0) &= N\frac{m}{2\pi\beta}(\beta^2\omega^2 - \gamma^2)\exp\left[-\left(\frac{\beta k}{2} - \frac{m\gamma^2}{2\beta}\right)x^2\right] \\ &= \frac{N}{\pi}\frac{k}{\varepsilon}e^{-(k/\varepsilon)x^2} \end{aligned} \qquad (S3.138)$$

where we have used Eq. (S3.111) and the fact that $\omega^2 = k/m$. Since both k and ε must be positive, it is clear that the density decreases at large distances, as it must for normalization. (c) The average velocity at position (x, y) is defined by saying that

$$mn(x,y)\mathbf{u}(x,y) = \int f(x,y,p_x,p_y)\mathbf{p}\, dp_x \, dp_y \qquad (S3.139)$$

Again we can utilize the rotational symmetry and calculate only u_y on the x axis, knowing that the velocity field must point in circles around the origin.

$$\begin{aligned} mn(x,0)u_y(x) &= \int f(x,0,p_x,p_y)p_y\, dp_x \, dp_y \\ &= -(C/x)\frac{\partial}{\partial\gamma}\int \exp\left[-\beta(p_x^2 + p_y^2)/2m - \beta kx^2/2 - \gamma x p_y\right] dp_x \, dp_y \end{aligned} \qquad (S3.140)$$

The integral factors into an integral over p_x times an integral over p_y.

$$\int e^{-\beta p_x^2/2m} \, dp_x = (2\pi m/\beta)^{1/2}$$

$$\int e^{-\beta p_y^2/2m - \gamma x p_y} \, dp_y = \exp\left(\frac{m\gamma^2}{2\beta}x^2\right)(2\pi m/\beta)^{1/2} \tag{S3.141}$$

Using this in Eq. (S3.119) and using Eq. (S3.117) for $n(x,0)$, we get

$$\begin{aligned}
\frac{m^2 N}{2\pi\beta}(\beta^2\omega^2 - \gamma^2)\exp\left[-\left(\frac{\beta k}{2} - \frac{m\gamma^2}{2\beta}\right)x^2\right]u(x) \\
- \frac{N}{4\pi^2}(\beta^2\omega^2 - \gamma^2)\frac{m\gamma}{\beta}x\frac{2\pi m}{\beta}\exp\left[-\left(\frac{\beta k}{2} - \frac{m\gamma^2}{2\beta}\right)x^2\right]
\end{aligned} \tag{S3.142}$$

which gives the velocity field

$$u(x) = -(\gamma/\beta)x = (l\omega^2/\varepsilon)x \equiv \Omega x \tag{S3.143}$$

This is the velocity field of a rigid body that rotates about the origin with an angular velocity $\Omega = l\omega^2/\varepsilon$.

Supplement to Chapter 4

REVIEW QUESTIONS

4.1 What is a statistical ensemble and how is it related to the probability density?

4.2 Describe the probability density for the uniform ensemble for a quantum system, defining any symbols you use.

4.3 For a quantum system, what is the meaning of $\Omega(E)$, the normalization constant for the uniform ensemble?

4.4 How is $\Omega(E)$ related to $C(E)$, the uniform ensemble normalization constant for the equivalent classical system?

4.5 How is $\Omega(E)$ related to the entropy of the system?

4.6 For a quantum system, what is the microcanonical probability distribution?

4.7 Describe the energy spectrum (the eigenvalues and their degeneracies) of a system of K quantized harmonic oscillators.

4.8 Calculate $\Omega'(E)$ for a system of K harmonic oscillators.

4.9 For a system of oscillators, how is the density of normal-modes function $D(\omega)$ defined?

4.10 Given $D(\omega)$, write a formula for the thermal energy of a lattice as a function of the temperature.

4.11 For $kT \gg \hbar\omega$, what is the average thermal energy of a normal mode of angular frequency ω?

4.12 For $kT \ll \hbar\omega$, what is the average thermal energy of a normal mode of angular frequency ω?

4.13 Why is it true that, at low T, only the long-wavelength vibrations of a lattice contribute to the specific heat?

4.14 Describe Einstein's approximation for the vibrational specific heat of a solid.

4.15 Describe the Debye approximation for the vibrational specific heat of a solid.

4.16 How are the Debye frequency and the Debye temperature defined?

4.17 Explain why the Debye theory gives accurate values for the vibrational specific heat of solids at temperatures much larger and much smaller than T_D.

4.18 Prove that the probability of finding a system that is in contact with a much larger thermal reservoir in energy eigenstate ψ_n is const. $\times \exp(-\beta E_n)$. You may assume a microcanonical ensemble for the combined system.

4.19 For a quantum mechanical system, how is the partition function Z defined? What parameters is it a function of?

4.20 How is the canonical potential related to the partition function?

4.21 How is the canonical potential related to the entropy of the system in practical units (Kelvins, etc.)?

4.22 What is the probability density in phase space for the classical canonical ensemble?

4.23 How is the canonical partition function defined for a classical system?

4.24 Using the dumbbell, or rigid rotor, model of a diatomic molecule, describe the full set of canonical coordinates and momenta needed to define the state of one molecule.

4.25 Write the Hamiltonian function for a rigid rotor molecule, explaining any nonobvious terms in it.

4.26 Calculate the internal (rotational) partition function of a rigid rotor molecule.

4.27 What extra term is added to the Hamiltonian of a diatomic molecule if it is placed in an electric field?

4.28 Explain the meaning of the terms in the formula $N\bar{\mu}_z = kT\,\partial\phi/\partial\mathcal{E}$.

4.29 Derive the formula that was given in the last question.

4.30 Describe the "orientational dipole moment" and the "induced dipole moment" contributions to the average dipole moment of a molecular gas.

4.31 How does the orientational contribution to the electric susceptibility of a diatomic gas depend on the temperature?

4.32 Explain, in words, what the Schottky anomaly is.

4.33 Derive the specific heat for a system of two-level atoms.

4.34 State the equipartition theorem.

4.35 Derive the equipartition theorem.

EXERCISES

Exercise 4.1   Consider a quantum system that has energy eigenstates $\psi_1, \psi_2, \ldots$ with energies $E_1, E_2, \ldots$. Let $P_1, P_2, \ldots$ be any probability distribution with the interpretation that P_n is the probability of finding the system in the state ψ_n. We define a quantity, called the *Boltzmann–Gibbs entropy*, by

$$\tilde{S} = -\sum_n P_n \log P_n \tag{S4.1}$$

(a) Show that the maximum value of $\tilde{S}$, subject to the constraints

$$\sum_n P_n = 1 \quad \text{and} \quad \sum_n E_n P_n = E \tag{S4.2}$$

is given by the canonical probability distribution $P_n = e^{-\beta E_n}/Z$. (b) Show that, for the canonical distribution, the value of $\tilde{S}$ is equal to the entropy in rational units, S^o.

Solution   (a) To maximize $\tilde{S}$, subject to the two constraints, we use Lagrange's method and set $\partial F/\partial P_n = 0$, where

$$F = -\sum_n P_n \log P_n - \alpha \sum_n P_n - \beta \sum_n P_n E_n \tag{S4.3}$$

This gives

$$\log P_n + 1 + \alpha + \beta E_n = 0 \tag{S4.4}$$

which has the solution

$$P_n = \text{const.} \times e^{-\beta E_n} \tag{S4.5}$$

In order to satisfy the first constraint, we must set the constant equal to Z^{-1}. (b) Letting $P_n = e^{-\beta E_n}/Z$, we see that $\log P_n = -\beta E_n - \log Z$ and

$$\tilde{S} = \sum_n P_n(\log Z + \beta E_n) = \log Z + \beta E \tag{S4.6}$$

But, by Eq. (4.45), $\log Z = S^o - \beta E$, which shows that $\tilde{S} = S^o$.

Exercise 4.2   Calculate the thermal energy of a system of K one-dimensional classical harmonic oscillators, using a canonical ensemble.

Solution   For a single harmonic oscillator of angular frequency ω the Hamiltonian is

$$H = p^2/2m + m\omega^2 x^2/2 \tag{S4.7}$$

and the partition function is

$$Z = \frac{1}{h} \int e^{-\beta p^2/2m} \, dp \int e^{-\beta m \omega^2 x^2/2} \, dx$$
$$= (2\pi m/\beta)^{1/2}(2\pi/\beta m \omega^2)^{1/2}/h \tag{S4.8}$$
$$= 1/\beta \hbar \omega$$

The canonical potential of a single oscillator is

$$\phi = \log z = -\log \beta - \log(\hbar \omega) \tag{S4.9}$$

and the thermal energy is

$$\varepsilon = -\frac{\partial \phi}{\partial \beta} = kT \tag{S4.10}$$

Since the K oscillators are independent, the energy of the system is

$$E = K\varepsilon = KkT \tag{S4.11}$$

The simplicity of this calculation illustrates the advantage of the canonical versus the microcanonical or uniform ensembles.

Exercise 4.3 A system is composed of K one-dimensional classical oscillators. Assume that the potential for the oscillators contains a small quartic "anharmonic term." That is

$$V(x) = \frac{k_o}{2}x^2 + \alpha x^4 \tag{S4.12}$$

where $\alpha \langle x^4 \rangle \ll kT$. To first order in the parameter α, derive the anharmonic correction to the Dulong–Petit law.

Solution Since the oscillators are noninteracting, the canonical potential of the system is K times the canonical potential of a single oscillator. The partition function of a single oscillator is

$$Z = \frac{1}{h} \int_{-\infty}^{\infty} e^{-p^2/2mkT} \, dp \int_{-\infty}^{\infty} e^{-V(x)/kT} \, dx$$
$$= \frac{\sqrt{2\pi mkT}}{h} \int_{-\infty}^{\infty} e^{-k_o x^2/2kT} e^{-\alpha x^4/kT} \, dx \tag{S4.13}$$
$$\approx \frac{\sqrt{2\pi mkT}}{h} \int_{-\infty}^{\infty} e^{-k_o x^2/2kT}(1 - \alpha x^4/kT) \, dx$$

But

$$\int_{-\infty}^{\infty} e^{-k_o x^2/2kT} \, dx = \sqrt{2\pi kT/k_o} \tag{S4.14}$$

and

$$\int_{-\infty}^{\infty} e^{-k_o x^2/2kT} x^4 \, dx = \frac{3}{4}\sqrt{\pi}(2kT/k_o)^{5/2} \tag{S4.15}$$

and therefore

$$Z \approx \frac{kT}{\hbar \omega}(1 - 3\alpha kT/k_o^2) \tag{S4.16}$$

where $\omega = \sqrt{k_o/m}$.

Using the expansion $\log(1 - x) \approx -x$, we get, for the system of K oscillators,

$$
\begin{aligned}
\phi &= K[\log kT - \log(\hbar\omega) - 3\alpha kT/k_o^2] \\
&= -K[\log\beta + \log(\hbar\omega) + 3\alpha/k_o^2\beta]
\end{aligned}
\tag{S4.17}
$$

The thermal energy per oscillator is given by

$$
\begin{aligned}
\frac{E}{K} &= -\frac{\partial(\phi/K)}{\partial\beta} = \frac{1}{\beta} - \frac{3\alpha}{k_o^2\beta^2} \\
&= kT - \frac{3\alpha}{k_o^2}(kT)^2
\end{aligned}
\tag{S4.18}
$$

The anharmonic correction tends to reduce the energy per oscillator in comparison to a perfectly harmonic oscillator, as T is increased.

Exercise 4.4 What is the thermal energy, at temperature T, of a particle of mass m in the one-dimensional potential $V(x) = \varepsilon_o|x/a|^n$? Note that the harmonic oscillator corresponds to $n = 2$. ε_o and a are parameters with the units of energy and length, respectively.

Solution   The partition function is

$$
\begin{aligned}
Z &= \frac{1}{h}\int_{-\infty}^{\infty} e^{-\beta p^2/2m}\, dp \int_{-\infty}^{\infty} e^{-\beta V(x)}\, dx \\
&= 2\frac{\sqrt{2\pi m/\beta}}{h}\int_o^{\infty} e^{-\beta\varepsilon_o(x/a)^n}\, dx
\end{aligned}
\tag{S4.19}
$$

The integral over x can be done by the transformation of variables $z = \beta\varepsilon_o(x/a)^n$. This gives $x = a(z/\beta\varepsilon_o)^{1/n}$ and $dx = (1/n)az^{(1-n)/n}/(\beta\varepsilon_o)^{1/n}dz$.

$$
\begin{aligned}
\int_o^{\infty} e^{-\beta\varepsilon_o(x/a)^n}\, dx &= \frac{a}{n(\beta\varepsilon_o)^{1/n}}\int_o^{\infty} e^{-z}z^{(1-n)/n}\, dz \\
&= \frac{a}{n(\beta\varepsilon_o)^{1/n}}\Gamma(1/n)
\end{aligned}
\tag{S4.20}
$$

Therefore

$$
Z = \text{const.} \times \beta^{-(1/2+1/n)}
\tag{S4.21}
$$

$$
\phi = -(\tfrac{1}{2} + 1/n)\log\beta + \text{const.}
\tag{S4.22}
$$

and

$$
E = -\frac{\partial\phi}{\partial\beta} = (1/2 + 1/n)kT
\tag{S4.23}
$$

What is surprising is that the thermal energy is exactly proportional to kT for any n. Only the proportionality constant is affected by the value of n. This means that, for any one-dimensional pure power potential, the particle has a constant specific heat.

Exercise 4.5   A quantum harmonic oscillator has energy levels $E_n = \hbar\omega(n + \frac{1}{2})$. What is the probability of finding the oscillator in its nth quantum state at temperature T?

Solution   The oscillator can be viewed as a small system coupled to a thermal reservoir at temperature T. The probability of finding the system in a quantum state of energy E_n is given by the canonical probability density

$$P_n = Ce^{-\beta E_n} \tag{S4.24}$$

Since $E_n = \hbar\omega(n + \frac{1}{2})$, the normalization constant is

$$C^{-1} = \sum_{n=0}^{\infty} e^{-\hbar\omega/2kT} e^{-n\hbar\omega/kT}$$

$$= e^{-\hbar\omega/2kT} \sum_{n=0}^{\infty} x^n \qquad (x \equiv e^{-\hbar\omega/kT}) \tag{S4.25}$$

$$= \frac{e^{-\hbar\omega/2kT}}{1 - e^{-\hbar\omega/kT}}$$

and therefore

$$P_n = (1 - e^{\hbar\omega/kT})e^{-n\hbar\omega/kT} \tag{S4.26}$$

Exercise 4.6   The Hamiltonian function of a two-dimensional dipolar molecule of mass m, moment of inertia I, and dipole moment μ, in an electric field $\mathcal{E}$, is (see Fig. S4.1)

$$H = \frac{1}{2m}(p_x^2 + p_y^2) + p_\theta^2/2I - \mu\mathcal{E}\cos\theta \tag{S4.27}$$

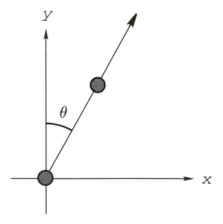

Fig. S4.1   The angle θ gives the orientation of a two-dimensional dipolar molecule.

(a) Calculate the canonical potential of a system of N classical dipolar molecules confined to an area A. Hint: The modified Bessel function $I_o(z)$ is defined by

$$I_o(z) = \frac{1}{\pi} \int_o^\pi e^{-z\cos\theta}\, d\theta \tag{S4.28}$$

(b) Use the power series expansion

$$I_o(z) = \sum_{k=0}^{\infty} (z/2)^{2k}/(k!)^2 \qquad (S4.29)$$

and the asymptotic approximation for large z

$$I_o(z) \sim e^z/\sqrt{2\pi z} \qquad (S4.30)$$

to obtain low- and high-temperature approximations for the specific heat.

Solution   (a) The partition function of a gas with internal degrees of freedom is a product of the translational partition function and the internal partition function.

$$Z = Z(\text{trans})Z(\text{int}) \qquad (S4.31)$$

where, for a two-dimensional gas,

$$\begin{aligned} Z(\text{trans}) &= \frac{A^N}{N!h^{2N}} \left(\int_{-\infty}^{\infty} e^{-\beta p^2/2m} \, dp \right)^{2N} \\ &= \frac{A^N(2\pi m/\beta)^N}{N!h^{2N}} \end{aligned} \qquad (S4.32)$$

and $Z(\text{int}) = z^N$, where

$$z = \frac{1}{h} \int_{-\infty}^{\infty} e^{-\beta p_\theta^2/2I} \, dp_\theta \int_{-\pi}^{\pi} e^{\beta\mu\mathcal{E}\cos\theta} \, d\theta \qquad (S4.33)$$

is the internal partition function for a single molecule. Using Eq. (S4.28), we obtain

$$z = \frac{\sqrt{2\pi I/\beta}}{h} 2\pi I_o(\beta\mu\mathcal{E}) \qquad (S4.34)$$

The canonical potential is

$$\phi = N\left[\log(A/N) - \tfrac{3}{2}\log\beta + \log I_o(\beta\mu\mathcal{E}) + C\right] \qquad (S4.35)$$

(b) As $T \to 0$, $\beta \to \infty$ and $I_o \approx e^{\beta\mu\mathcal{E}}/\sqrt{2\pi\beta\mu\mathcal{E}}$

$$\phi \approx N[\log(A/N) - 2\log\beta + \mu\mathcal{E}\beta + \text{const.}] \qquad (S4.36)$$

Therefore

$$E = -\frac{\partial\phi}{\partial\beta} \approx \frac{2N}{\beta} - \mu\mathcal{E}N \qquad (S4.37)$$

and

$$C = \frac{\partial E}{\partial T} = 2Nk \qquad (S4.38)$$

(At low T, the dipoles in the electric field act as one-dimensional harmonic oscillators.) As $T \to \infty$, $\beta \to 0$, $I_o \approx 1 + \tfrac{1}{4}(\beta\mu\mathcal{E})^2$, and, using the identity $\log(1+x) \approx x$, one obtains the high-temperature approximations,

$$\phi \approx N\left[\log(A/N) - \tfrac{3}{2}\log\beta + \tfrac{1}{4}(\beta\mu\mathcal{E})^2 + C\right] \qquad (S4.39)$$

$$E \approx \tfrac{3}{2}NkT - N\frac{(\mu\mathcal{E})^2}{2kT} , \tag{S4.40}$$

and

$$C \approx \tfrac{3}{2}Nk + \tfrac{1}{2}N(\mu\mathcal{E}/kT)^2 \tag{S4.41}$$

(Notice that the dipole-field interaction becomes negligible when $kT \gg \mu\mathcal{E}$.)

Exercise 4.7 The probability of finding a three-dimensional dipolar molecule in a state described by the phase-space variables $(x, y, z, \theta, \phi, p_x, p_y, p_z, p_\theta, p_\phi) \equiv (x, p)$ is proportional to $e^{-\beta H(x,p)}$. Show that, if the electric field is not uniform, then the molecular gas is more dense where the field strength is higher.

Solution For a canonical ensemble, the single-particle phase space probability density is

$$P(x, p) = Ce^{-\beta H(x,p)} \tag{S4.42}$$

The Hamiltonian function is

$$H = \frac{1}{2m}(p_x^2 + p_y^2 + p_z^2) + \frac{1}{2I}\left(p_\theta^2 + \frac{p_\phi^2}{\sin^2\theta}\right) - \mu\mathcal{E}\cos\theta \tag{S4.43}$$

To get the spatial probability density we must integrate Eq. (S4.42) over all variables except x, y, and z. The integrals over p_x, p_y, p_z, and p_θ only give factors that are independent of x, y, and z and can therefore be lumped together with the normalization constant C. The integral of over p_ϕ is

$$\int_{-\infty}^{\infty} e^{-\beta p_\phi^2/2I\sin^2\theta}\, dp_\phi = \sqrt{2\pi I/\beta}\, \sin\theta \tag{S4.44}$$

The square root factor can be combined with the normalization constant. This leaves only the integral over θ.

$$P(x, y, z) = \text{const.} \times \int_0^\pi e^{\beta\mu\mathcal{E}\cos\theta} \sin\theta\, d\theta \tag{S4.45}$$

But

$$
\begin{aligned}
\int_0^\pi e^{a\cos\theta}\sin\theta\, d\theta &= \int_{-1}^{1} e^{a\cos\theta}\, d(\cos\theta) \\
&= a^{-1}(e^a - e^{-a}) \\
&= 2\sinh(a)/a
\end{aligned}
\tag{S4.46}
$$

Therefore

$$P(x, y, z) = \text{const.} \times \frac{\sinh[\beta\mu\mathcal{E}(x, y, z)]}{\mathcal{E}(x, y, z)} \tag{S4.47}$$

Since $\sinh(x)/x$ is a monotonically increasing function of x, $P(x, y, z)$ is greatest where $\mathcal{E}(x, y, z)$ is greatest.

Exercise 4.8 In Problem 4.19 the reader is asked to evaluate the rotational partition function $Z(\text{rot})$ by using the approximation

$$\sum_{n=0}^{\infty} f(n) \approx \int_o^\infty f(x)\, dx \tag{S4.48}$$

where $f(n) = (2n + 1) \exp[-(\beta \hbar^2 / 2I) n(n + 1)]$. By considering the left-hand side of Eq. (S4.48) as an approximation to the integral on the right-hand side and by using an improved approximation to the integral, obtain an estimate of the error in the original approximation.

Solution We treat the integral from zero to infinity as a sum of integrals over unit intervals.

$$\int_0^\infty f(x)\, dx = \sum_{n=0}^\infty \int_n^{n+1} f(x)\, dx \qquad (S4.49)$$

We now want to construct an approximation for an integral over a unit interval. Let us look at the first interval. Within the interval $0 \le x \le 1$ we approximate $f(x)$ by a polynomial $P(x)$ that satisfies the conditions that $P(0) = f(0)$, $P'(0) = f'(0)$, $P(1) = f(1)$, and $P'(1) = f'(1)$. In order to satisfy four conditions, we need four coefficients in the polynomial. We therefore choose a cubic polynomial.

$$P(x) = f(0) + f'(0)x + ax^2 + bx^3 \qquad (S4.50)$$

It is a straightforward calculation to use the two conditions at $x = 1$ to determine the coefficients a and b. The results are that

$$a = 3f(0) - 2f'(0) - 3f(1) - f'(1) \qquad (S4.51)$$

and

$$b = 2f(0) + f'(0) - 2f(1) + f'(1) \qquad (S4.52)$$

Our approximation for the integral from 0 to 1 is

$$\int_0^1 f(x)\, dx \approx \int_0^1 P(x)\, dx = f(0) + \tfrac{1}{2}f'(0) + \tfrac{1}{3}a + \tfrac{1}{4}b$$
$$= \frac{f(0) + f(1)}{2} + \frac{f'(0) - f'(1)}{12} \qquad (S4.53)$$

Because the integral has the geometrical meaning of the area under the curve $f(x)$, it is clear that there is an equivalent formula that is valid for any unit interval.

$$\int_n^{n+1} f(x)\, dx \approx \frac{f(n) + f(n+1)}{2} + \frac{f'(n) - f'(n+1)}{12}$$

Using this in Eq. (S4.49) gives

$$\int_0^\infty f(x)dx \approx \tfrac{1}{2}(f(0) + f(1)) + \tfrac{1}{12}(f'(0) - f'(1))$$
$$+ \tfrac{1}{2}(f(1) + f(2)) + \tfrac{1}{12}(f'(1) - f'(2))$$
$$+ \cdots \qquad (S4.54)$$
$$= \tfrac{1}{2}f(0) + \sum_{n=1}^\infty f(n) + \tfrac{1}{12}f'(0)$$

Rearranging this formula, to give the sum of $f(n)$ from 0 to ∞, we get

$$\sum_{n=0}^\infty f(n) \approx \int_0^\infty f(x)\, dx + \tfrac{1}{2}f(0) - \tfrac{1}{12}f'(0) \qquad (S4.55)$$

These are the first three terms in what is known as the *Euler summation formula.* Applying this to the function

$$f(x) = (2x + 1) \exp[-\alpha x(x + 1)]$$

with $\alpha = \beta\hbar^2/2I$, gives

$$\sum_{n=0}^{\infty} f(n) \approx \frac{1}{\alpha} + \tfrac{1}{2}(1) - \tfrac{1}{12}(2 - \alpha) \approx \frac{1}{\alpha} + \tfrac{1}{3} \qquad (S4.56)$$

Therefore, the integral approximation is accurate when $2IkT/\hbar^2 \gg \tfrac{1}{3}$, or

$$T \gg \hbar^2/6kI \qquad (S4.57)$$

As can be seen in Table 6.1, $\hbar^2/6kI$ is typically less than 1 K. The full Euler summation formula is

$$\sum_{n=0}^{\infty} f(n) = \int_o^{\infty} f(x)\,dx + \tfrac{1}{2}f(0) - \frac{B_2}{2!}\frac{df}{dx}\Big|_0 - \frac{B_4}{4!}\frac{d^3f}{dx^3}\Big|_0 - \frac{B_6}{6!}\frac{d^5f}{dx^5}\Big|_0 - \cdots \quad (S4.58)$$

where the B_ns are the Bernoulli numbers, which can be found in any mathematical handbook.

Fig. S4.2   A polymer chain, attached at one end.

Exercise 4.9   A polymer is a molecule composed of a long chain of identical molecular units, called monomers. Imagine a long polymer, made of N rodlike units, each of length l, attached end-to-end. Assume that the connections between the monomers are completely flexible so that the rods can make any angle with respect to one another. One end of the polymer is held fixed while a constant force F, in the x direction, is applied to the other end (Fig. S4.2). Except for an arbitrary constant, which can be ignored, the potential energy of any configuration of the polymer can be written as

$$U(\theta_1, \phi_1, \ldots, \theta_N, \phi_N) = -\sum_{i=1}^{N} Fl \cos\theta_i \qquad (S4.59)$$

where θ_i and ϕ_i are polar angles defining the orientation, in three dimensions, of the ith monomer with respect to an axis parallel to the x axis. Assume that the configurational probability density,

$$P(\theta_1, \phi_1, \ldots, \theta_N, \phi_N) = Z^{-1} \exp(-\beta U)$$

where

$$\int P(\theta_1, \phi_1, \ldots, \theta_N, \phi_N) \prod_{i=1}^{N} \sin \theta_i \, d\theta_i \, d\phi_i = 1 \tag{S4.60}$$

and $\beta = 1/kT$. (a) Calculate the configurational partition function Z. (b) Calculate $\langle U \rangle$ as a function of F and β. (c) Make the assumption, which is true for a classical system, that the average kinetic energy is proportional to T and use the thermodynamic relation

$$S = \int \frac{\partial E}{\partial T} \frac{dT}{T} = \int \frac{\partial E}{\partial \beta} \beta \, d\beta \tag{S4.61}$$

where the integrals are indefinite integrals, to calculate the entropy.

Solution   (a)

$$Z = \int \exp\left(\beta \sum Fl \cos \theta_i\right) \sin \theta_1 \, d\theta_1 \, d\phi_1 \cdots \sin \theta_N \, d\theta_N \, d\phi_N = z^N \tag{S4.62}$$

where

$$z = 2\pi \int_0^{\pi} e^{\beta Fl \cos \theta} \sin \theta \, d\theta$$

$$= 2\pi \int_{-1}^{1} e^{\beta Fl \cos \theta} \, d(\cos \theta) \tag{S4.63}$$

$$= 4\pi \frac{\sinh(\beta Fl)}{\beta Fl}$$

(b)

$$\langle U \rangle = \left\langle -\sum Fl \cos \theta_i \right\rangle$$

$$= \frac{-\partial Z/\partial \beta}{Z}$$

$$= -\frac{\partial \log Z}{\partial \beta}$$

$$= -N \frac{\partial \log z}{\partial \beta} \tag{S4.64}$$

$$= -N \frac{\partial}{\partial \beta} (\log \sinh(\beta Fl) - \log \beta + \text{const.})$$

$$= N[\beta^{-1} - Fl \coth(\beta Fl)]$$

(c) Let $a \equiv Fl$. Assuming that the kinetic energy is given by $K = N\gamma kT$, we can write the total thermal energy per monomer as

$$\frac{E}{N} = (\gamma + 1)/\beta - a \coth(a\beta) \tag{S4.65}$$

Then

$$\frac{\partial(E/N)}{\partial \beta} = -(\gamma + 1)/\beta^2 + a^2 \text{csch}^2(a\beta) \tag{S4.66}$$

and, by Eq. (S4.61),

$$\frac{S}{N} = -(\gamma+1)\int \frac{d\beta}{\beta} + a^2 \int \mathrm{csch}^2(a\beta)\beta\, d\beta \tag{S4.67}$$

The first integral gives $-(\gamma+1)\log\beta$. The second can be calculated by the identity

$$\int \mathrm{csch}^2(a\beta)\beta\, d\beta = -\frac{\partial}{\partial a}\int \coth(a\beta)\, d\beta$$

$$= -\frac{\partial}{\partial a}[\log\sinh(a\beta)/a] \tag{S4.68}$$

$$= -\frac{\beta\coth(a\beta)}{a} + \frac{\log\sinh(a\beta)}{a^2}$$

giving

$$\frac{S}{N} = -(\gamma+1)\log\beta + \log\sinh(a\beta) - a\beta\coth(a\beta) + C \tag{S4.69}$$

Exercise 4.10   A system is composed of a large number N of one-dimensional quantum harmonic oscillators whose angular frequencies are distributed over the range $\omega_a \le \omega \le \omega_b$ with a frequency distribution function $D(\omega) = A/\omega$. (a) Calculate the specific heat per oscillator at temperature T. (b) Evaluate the result in the limits of high and low T.

Solution   (a) We will use a canonical ensemble. The partition function of a single oscillator of angular frequency ω, is

$$z_\omega = \sum_{n=0}^{\infty} \exp\left(-\beta\hbar\omega(n+\tfrac{1}{2})\right)$$

$$= e^{-\beta\hbar\omega/2}\sum_{n=0}^{\infty}\left(e^{-\beta\hbar\omega}\right)^n \tag{S4.70}$$

$$= e^{-\beta\hbar\omega/2}/(1-e^{-\beta\hbar\omega})$$

$$= \tfrac{1}{2}\sinh(\beta\hbar\omega/2)$$

The canonical potential of the oscillator is

$$\phi_\omega = -\log[2\sinh(\beta\hbar\omega/2)] \tag{S4.71}$$

The average thermal energy of the oscillator is

$$E_\omega = -\frac{\partial\phi}{\partial\beta} = \frac{\hbar\omega}{2}\coth(\beta\hbar\omega/2) \tag{S4.72}$$

The energy of the system of N oscillators is given in terms of the frequency distribution function by

$$E = \int_{\omega_a}^{\omega_b} D(\omega)E_\omega\, d\omega$$

$$= A\frac{\hbar}{2}\int_{\omega_a}^{\omega_b}\coth(\beta\hbar\omega/2)\, d\omega$$

$$= (A/\beta)\log\left(\frac{\sinh(\beta\hbar\omega_b/2)}{\sinh(\beta\hbar\omega_a/2)}\right) \tag{S4.73}$$

$$= AkT\log\left(\frac{\sinh(\hbar\omega_b/2kT)}{\sinh(\hbar\omega_a/2kT)}\right)$$

The constant A can be written in terms of N by using the normalization condition

$$N = \int_{\omega_a}^{\omega_b} D(\omega)\, d\omega = A \int_{\omega_a}^{\omega_b} \frac{d\omega}{\omega} = A \log(\omega_b/\omega_a) \tag{S4.74}$$

Therefore, the thermal energy per oscillator $\varepsilon = E/N$ is

$$\varepsilon = \frac{kT}{\log(\omega_b/\omega_a)} \log\!\left(\frac{\sinh(\hbar\omega_b/2kT)}{\sinh(\hbar\omega_a/2kT)}\right) \tag{S4.75}$$

The specific heat per oscillator is $\partial\varepsilon/\partial T$.

$$C = \frac{k}{\log(\omega_b/\omega_a)} \log\left[\frac{\sinh(\hbar\omega_b/2kT)}{\sinh(\hbar\omega_a/2kT)}\right] \\ + \frac{\hbar\omega_a \coth(\hbar\omega_a/2kT)}{2T\log(\omega_b/\omega_a)} - \frac{\hbar\omega_b \coth(\hbar\omega_b/2kT)}{2T\log(\omega_b/\omega_b)} \tag{S4.76}$$

(b) If $2kT \gg \hbar\omega$, then $\sinh(\hbar\omega/2kT) \approx \hbar\omega/2kT$ and $\coth(\hbar\omega/2kT) \approx 2kT/\hbar\omega$. Therefore, in the high temperature limit, the first term in Eq. (S4.76) approaches k while the second and third terms cancel one another.

$$C \approx k \tag{S4.77}$$

This is the expected Dulong–Petit result. In the low-temperature limit $\hbar\omega/2kT$ becomes very large. But for large values of x,

$$\log(e^x - e^{-x}) = \log\!\left[e^x(1 - e^{-2x})\right] \approx x - e^{-2x} \tag{S4.78}$$

and

$$\coth x = \frac{1 + e^{-2x}}{1 - e^{-2x}} \approx 1 + 2e^{-2x} \tag{S4.79}$$

These approximations give

$$C \approx \frac{k}{\log(\omega_b/\omega_a)}\left(\frac{\hbar(\omega_b - \omega_a)}{2kT} + e^{-\hbar\omega_a/kT} - e^{-\hbar\omega_b/kT}\right) \\ + \frac{\hbar}{2T\log(\omega_b/\omega_a)}\left(\omega_a + 2\omega_a e^{-\hbar\omega_a/kT} - \omega_b - 2\omega_b e^{-\hbar\omega_b/kT}\right) \tag{S4.80}$$

Using the facts that $\hbar\omega/kT \gg 1$ and $e^{-\hbar\omega_a/kT} \gg e^{-\hbar\omega_b/kT}$, we get

$$C \approx \frac{\hbar\omega_a e^{-\hbar\omega_a/kT}}{T\log(\omega_b/\omega_a)} \tag{S4.81}$$

This is an example of the general rule that if a system of oscillators has a lowest frequency that is larger than zero, then the specific heat always has a factor of $\exp(-\hbar\omega_{\min}/kT)$ at very low temperatures.

Exercise 4.11   In Eq. (4.32), the vibrational energy of a solid is expressed in terms of the Debye integral

$$I(\lambda) = \int_0^\lambda \frac{x^2\, dx}{e^x - 1} \tag{S4.82}$$

Work out approximations to this integral, giving at least two terms, for the two extremes $\lambda \gg 1$ and $\lambda \ll 1$.

Solution   For $\lambda \gg 1$, it is useful to decompose the integral as

$$I(\lambda) = \int_o^\infty \frac{x^3\, dx}{e^x - 1} - \int_\lambda^\infty \frac{x^3\, dx}{e^x - 1} \tag{S4.83}$$

The first integral is given in the Mathematical Appendix. It is equal to $\pi^4/15$. In the second integral, since $x \geq \lambda$, $e^x \gg 1$ and the -1 in the denominator can be neglected. This leads to an integral that can be done.

$$\begin{aligned}
I(\lambda) &\approx \frac{\pi^4}{15} - \int_\lambda^\infty e^{-x} x^3\, dx \\
&= \frac{\pi^4}{15} - (1 + \lambda + \tfrac{1}{2}\lambda^2 + \tfrac{1}{6}\lambda^3)e^{-\lambda}
\end{aligned} \tag{S4.84}$$

For $\lambda \ll 1$, one can use the expansion of the exponential function in the denominator of the integrand.

$$e^x - 1 \approx x + \tfrac{1}{2}x^2 \tag{S4.85}$$

Then $x^3/(e^x - 1) \approx x^2(1 + x/2)^{-1} \approx x^2 - x^3/2$ and

$$I(\lambda) \approx \int_0^\lambda (x^2 - x^3/2)\, dx = \tfrac{1}{3}\lambda^3 - \tfrac{1}{8}\lambda^4 \tag{S4.86}$$

Exercise 4.12   Consider a system of N classical particles in a potential field $U(\mathbf{r})$. Using the canonical ensemble, show that the equilibrium particle density is of the Maxwell–Boltzmann form

$$n(\mathbf{r}) = n_o e^{-U(\mathbf{r})/kT} \tag{S4.87}$$

Solution   The Hamiltonian function for the N-particle system is

$$H = \sum_{i=1}^N p_i^2/2m + \sum_{i=1}^N U(\mathbf{r}_i) \tag{S4.88}$$

The probability density for finding the system in state $(\mathbf{r}_1, \mathbf{p}_1, \ldots, \mathbf{r}_N, \mathbf{p}_N)$ is, according to Eq. (4.52),

$$P(\mathbf{r}_1, \mathbf{p}_1, \ldots, \mathbf{r}_N, \mathbf{p}_N) = I^{-1} e^{-\beta H} \tag{S4.89}$$

The normalization integral I factors into N identical terms

$$\begin{aligned}
I &= \int e^{-\beta H}\, d^3 r_1\, d^3 p_1 \cdots d^3 r_N\, d^3 p_N \\
&= \left(\int e^{-\beta U(\mathbf{r})}\, d^3 r \int e^{-\beta p^2/2m}\, d^3 p \right)^N \\
&= \left(2\pi m/\beta \right)^{3N/2} \left(\int e^{-\beta U}\, d^3 r \right)^N
\end{aligned} \tag{S4.90}$$

The probability of finding particle number 1 at location $\mathbf{r}$, regardless of its momentum and regardless of the positions and momenta of all the other particles, is calculated by setting $\mathbf{r}_1$ equal to $\mathbf{r}$ and integrating $P(\mathbf{r}, \mathbf{p}_1, \ldots, \mathbf{r}_N, \mathbf{p}_N)$ over all the irrelevant variables. The integrations simply cancel equivalent terms in the normalization constant, leaving

$$P_1(\mathbf{r}) = \frac{e^{-\beta U(\mathbf{r})}}{\int e^{-\beta U} d^3 r} \tag{S4.91}$$

The probability of finding particle number 1 inside $d^3 r$ is $P_1(\mathbf{r}) d^3 r$. Since all particles have the same distribution, the average number of particles inside the volume element $d^3 r$ is $N P_1(r) d^3 r$. But that defines the particle density function. Thus

$$n(\mathbf{r}) = \frac{N e^{-\beta U(\mathbf{r})}}{\int e^{-\beta U} d^3 r} \tag{S4.92}$$

which is of the Maxwell–Boltzmann form.

Exercise 4.13   Calculate the specific heat of a quantized square drumhead (a square elastic membrane with zero boundary conditions on the edges).

Solution   The wave function for a drumhead, $u(x, y, t)$, represents the displacement, in the z direction at time t, of that point on the drumhead with coordinates (x, y). If μ is the mass per unit area and σ is the surface tension (the pulling force per unit length within the membrane), then $u(x, y, t)$ satisfies the wave equation

$$\mu \frac{\partial^2 u}{\partial t^2} - \sigma \left(\frac{\partial^2 u}{\partial x^2} + \frac{\partial^2 u}{\partial y^2} \right) = 0 \tag{S4.93}$$

For an $L \times L$ drumhead, the boundary conditions are

$$u(0, y, t) = u(L, y, t) = u(x, 0, t) = u(x, L, t) = 0 \tag{S4.94}$$

The normal-mode solutions are of the form

$$u(x, y, t) = A_{mn}(t) \sin(m\pi x / L) \sin(n\pi y / L) \tag{S4.95}$$

Putting this into Eq. (S4.93), we see that A_{mn} satisfies a harmonic oscillator equation

$$\frac{d^2 A_{mn}}{dt^2} + \omega_{mn}^2 A_{mn} = 0 \tag{S4.96}$$

where

$$\omega_{mn}^2 = (\sigma \pi^2 / \mu L^2)(m^2 + n^2) \tag{S4.97}$$

The integers m and n must be positive. Let $N(\omega)$ be the number of normal modes with angular frequencies less then ω. Then $N(\omega)$ is the number of pairs of positive integers with $\sqrt{m^2 + n^2} < \sqrt{\mu/\sigma} \, L\omega/\pi \equiv R$. For large L we can approximate this by the area shown in Fig. S4.3.

$$N(\omega) = \frac{\pi}{4} R^2 = \frac{\mu L^2}{4\pi\sigma} \omega^2 \tag{S4.98}$$

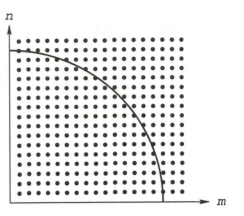

Fig. S4.3 Each pair of integers gives one normal mode of the square drumhead.

This gives the normal mode frequency distribution function

$$D(\omega) = \frac{dN(\omega)}{d\omega} = \frac{\mu L^2}{2\pi\sigma}\omega \qquad \text{(S4.99)}$$

According to Eq. (4.25), the thermal energy of the system at temperature T is

$$E = \frac{\mu L^2}{2\pi\sigma} \int_o^\infty \frac{\hbar\omega^2\, d\omega}{e^{\hbar\omega/kT} - 1} \qquad \text{(S4.100)}$$

Changing to the integration variable $x = \hbar\omega/kT$ and using the Table of Integrals, we get

$$\begin{aligned} E &= \frac{\mu L^2 (kT)^3}{2\pi\sigma\hbar^2} \int_o^\infty \frac{x^2\, dx}{e^x - 1} \\ &= \frac{\mu L^2 (kT)^3}{\pi\sigma\hbar^2}\zeta(3) \end{aligned} \qquad \text{(S4.101)}$$

The specific heat per unit area of membrane is

$$\frac{C}{L^2} = \frac{3\mu k^3 \zeta(3)}{\pi\sigma\hbar^2}T^2 \qquad \text{(S4.102)}$$

This specific heat goes to infinity as T increases, which is unrealistic. The source of the problem is that we have used the wave equation to generate an infinite number of normal modes. In fact, the wave equation is only valid for normal modes with wavelengths much larger than microscopic dimensions.

Exercise 4.14 A piston of mass M, in a frictionless vertical cylinder of area A, encloses an N-particle ideal gas. (a) Describing the state of the piston by a single coordinate z_o and momentum p_o (which is not realistic, since it actually has a microscopic structure), and using a canonical ensemble, calculate the probability distribution $P(z_o)$ for the piston height. Neglect the effect of the gravitational field on the ideal gas particles. (b) Show that, for large N, it is a Gaussian function. (c) Show that the fluctuation in the gas volume is given by

$$\frac{\Delta V}{V} = \frac{1}{\sqrt{N}} \qquad \text{(S4.103)}$$

Solution   (a) The Hamiltonian function for the system is

$$H = \frac{p_o^2}{2M} + Mgz_o + \sum_{i=1}^{N} p_i^2/2m \tag{S4.104}$$

The x and y coordinates of the gas particles are restricted to the area A. Their z coordinates must be less than z_o. The canonical probability density function is

$$P(z_o, p_o, \mathbf{r}_1, \mathbf{p}_1, \ldots, \mathbf{r}_N, \mathbf{p}_N) = I^{-1}e^{-\beta H} \tag{S4.105}$$

The normalization integral is

$$
\begin{aligned}
I = &\int_o^\infty dz_o\, e^{-\beta M g z_o} \int_o^{z_o} dz_1 \cdots dz_N \\
&\times \int_A dx_1\, dy_1 \ldots dx_N\, dy_N (2\pi M/\beta)^{1/2}(2\pi m/\beta)^{3N/2} \\
= &(2\pi M/\beta)^{1/2}(2\pi m/\beta)^{3N/2} A^N \int_o^\infty z_o^N e^{-\beta M g z_o}\, dz_o \\
= &(2\pi M/\beta)^{1/2}(2\pi m/\beta)^{3N/2} A^N N!/(\beta M g)^{N+1}
\end{aligned}
\tag{S4.106}
$$

where we have made use of the integral formula for $N!$ given in the Mathematical Appendix.

The probability distribution for z_o is obtained by integrating over all other coordinates and over all momenta. Most of the integrals cancel factors in I, leaving

$$P(z_o) = \frac{(\beta W)^{N+1}}{N!} z_o^N e^{-\beta W z_o} \tag{S4.107}$$

with $W = Mg$. (b) The most probable value of z_o, which we will call z^*, is given by $P'(z^*) = 0$. This gives

$$\frac{N}{z^*} - \beta W = 0 \tag{S4.108}$$

or

$$z^* = \frac{NkT}{W} \tag{S4.109}$$

$P(z_o)$ can be written as

$$P(z_o) = \frac{(\beta W)^{N+1}}{N!} e^{-g(z_o)} \tag{S4.110}$$

with $g(z) = \beta W z - N \log z$. Expanding $g(z)$ to second order about z^*, we get (with $\zeta = z - z^*$)

$$g \approx \beta W z^* - N \log z^* + \tfrac{1}{2} N(\zeta/z^*)^2 \tag{S4.111}$$

Therefore

$$P(z^* + \zeta) = \frac{(\beta W)^{N+1}(z^*)^N e^{-\beta W z^*}}{N!} e^{-N(\zeta/z^*)^2/2} \tag{S4.112}$$

Using the fact that $\beta W z^* = N$ and Stirling's approximation in the form $N! \approx N^N e^{-N}(2\pi N)^{1/2}$, we can write this as

$$P(z^* + \zeta) = \frac{\sqrt{N/2\pi}}{z^*} e^{-N(\zeta/z^*)^2/2} \tag{S4.113}$$

(c) The mean-square fluctuation in the height of the piston is

$$(\Delta z_o)^2 = \frac{\sqrt{N/2\pi}}{z^*} \int_{-\infty}^{\infty} e^{-N(\zeta/z^*)^2/2} \zeta^2 \, d\zeta = \frac{(z^*)^2}{N} \qquad \text{(S4.114)}$$

But, because V is proportional to z^*,

$$\frac{\Delta V}{V} = \frac{\Delta z_o}{z^*} = \frac{1}{\sqrt{N}} \qquad \text{(S4.115)}$$

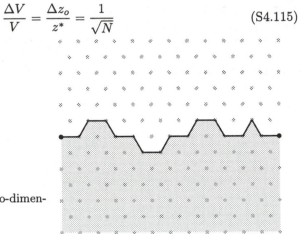

Fig. S4.4   The surface of a two-dimensional liquid.

Exercise 4.15   Shown in Fig. S4.4 is the surface of a two-dimensional "liquid." In order to make the possible surface configurations discrete, we assume that the surface is confined to a triangular lattice. Assume that the surface is pinned at the two points shown and that the energy of the system is proportional to the surface length. The two pins are L lattice distances apart. Assume also that overhangs (places where the outward-pointing normal to the surface points downward, rather than upward) are forbidden. Use the microcanonical ensemble to calculate the length of the surface as a function of temperature.

Solution   Let N_o be the number of horizontal surface elements, N_+ be the number of surface elements that rise from left to right, and N_- be the number of surface elements that fall from left to right. The length of the surface in lattice units is $N_o + N_+ + N_-$. If we take the zero-energy configuration as the flat, lowest-energy state, then the energy of any configuration can be written in terms of N_o, N_+, and N_- as

$$E = \varepsilon(N_o + N_+ + N_- - L) \qquad \text{(S4.116)}$$

In order for the surface to reach the fixed point on the right, it is necessary that $N_+ = N_-$. Because the projection of the surface onto the line between the pins always has length L, it is also necessary that $N_o + (N_+ + N_-)/2 = L$. From these two equations we can easily write N_+ and N_- in terms of N_o.

$$N_+ = N_- = L - N_o \qquad \text{(S4.117)}$$

For a system with discrete states, $\Omega(E)$ is the number of states that have energy E. But, combining Eqs. (S4.100) and (S4.101), we can write E in terms of N_o.

$$E = \varepsilon(L - N_o) \qquad \text{(S4.118)}$$

Therefore, the value of E determines the value of N_o and hence the values of N_+ and N_- also. For given N_o, N_+, and N_- we can get all possible surface configurations by permuting the three types of surface elements. The number of such configurations is $(N_o + N_+ + N_-)!/N_o!N_+!N_-!$. The spacing between neighboring energy values is ε. Therefore

$$\Omega'(E) = \frac{(N_o + N_+ + N_-)!}{N_o!N_+!N_-!}\frac{1}{\varepsilon} = \frac{(2L - N_o)!}{N_o!((L - N_o)!)^2}\frac{1}{\varepsilon} \tag{S4.119}$$

S^o is the logarithm of Ω'. Using Stirling's formula and the fact that $L - N_o = E/\varepsilon$, we see that

$$\begin{aligned} S^o =&(2L - N_o)\log(2L - N_o) - N_o \log N_o \\ &- 2(L - N_o)\log(L - N_o) - \log\varepsilon \\ =&\left(L + \frac{E}{\varepsilon}\right)\log\left(L + \frac{E}{\varepsilon}\right) - \left(L - \frac{E}{\varepsilon}\right)\log\left(L - \frac{E}{\varepsilon}\right) \\ &- 2\frac{E}{\varepsilon}\log\left(\frac{E}{\varepsilon}\right) - \log\varepsilon \end{aligned} \tag{S4.120}$$

The thermodynamic relation $\partial S^o/\partial E = 1/kT$ gives

$$\frac{1}{kT} = \frac{1}{\varepsilon}\left[\log\left(L + \frac{E}{\varepsilon}\right) + \log\left(L - \frac{E}{\varepsilon}\right) - 2\log\left(\frac{E}{\varepsilon}\right)\right] \tag{S4.121}$$

or

$$\frac{\varepsilon}{kT} = \log\left[\frac{L^2\varepsilon^2 - E^2}{E^2}\right] \tag{S4.122}$$

which gives

$$e^{\varepsilon/kT} = \frac{L^2\varepsilon^2 - E^2}{E^2} \tag{S4.123}$$

This equation can easily be solved for E.

$$E = \frac{L\varepsilon}{(e^{\varepsilon/kT} + 1)^{1/2}} \tag{S4.124}$$

Since E is equal to ε times the length of the surface minus its minimum length, we see that the length of the surface at temperature T is

$$l(T) = L[1 + (e^{\varepsilon/kT} + 1)^{-1/2}] \tag{S4.125}$$

For $kT \ll \varepsilon$ $l(T) \approx L$ and for $kT \gg \varepsilon$ $l(T) \approx L(1 + 1/\sqrt{2})$.

Exercise 4.16   For zero-mass particles, or for ordinary particles when $kT \gg mc^2$, one can approximate the relativistic relationship between energy and momentum by the simple extreme relativistic relation

$$E(p) = c|\mathbf{p}| \tag{S4.126}$$

Calculate the pressure and energy of a classical extremely relativistic gas as functions of density and temperature and show that $pV = E/3$.

Solution The partition function for a system of N particles in a volume V is

$$Z = \frac{V^N}{h^{3N} N!} \left(4\pi \int_o^\infty e^{-\beta c p} p^2 \, dp \right)^N \tag{S4.127}$$

But

$$\int_o^\infty e^{-\beta c p} p^2 \, dp = \frac{1}{(\beta c)^3} \int_o^\infty e^{-x} x^2 \, dx = \frac{2}{(\beta c)^3} \tag{S4.128}$$

Therefore

$$Z = \frac{(8\pi)^N V^N}{(\beta c h)^{3N} N!} \tag{S4.129}$$

and

$$\phi = N[\log(V/N) - 3\log\beta + \text{const.}] \tag{S4.130}$$

The pressure and energy are given by

$$\beta p = \frac{\partial \phi}{\partial V} = \frac{N}{V} \tag{S4.131}$$

and

$$E = -\frac{\partial \phi}{\partial \beta} = \frac{3N}{\beta} \tag{S4.132}$$

or

$$pV = NkT \quad \text{and} \quad E = 3NkT \tag{S4.133}$$

Fig. S4.5   Hard-core particles on a line.

Exercise 4.17   The system shown in Fig. S4.5 consists of N impenetrable beads of diameter a on a frictionless wire. The rightmost bead is subjected to a constant force F toward the origin. (a) Calculate the partition function of the system as a function of N, β, and F. Note: Since the beads cannot pass through one another, they should be taken as distinguishable and the $N!$ should be omitted in the definition of Z. (b) Calculate the specific heat of the system and the average position of the rightmost particle as functions of T. (c) For a one-dimensional system, the pressure is simply the force being applied to the last particle (there is no "area" to divide by). Show that the pressure equation of state is $p = nkT/(1 - na)$, where the density, $n \equiv N/\langle x_N \rangle$.

Solution   (a) We let x_n be the coordinate of the left side of the nth bead. The force on the Nth particle is derivable from the potential $U = F(x_N - (N-1)a)$ where the constant term in the potential has been chosen so that $U = 0$ when the system is in its lowest-energy state. That is, the state with $x_N = (N-1)a$. Therefore, the Hamiltonian function is

$$H = \sum_{n=1}^{N} p_n^2/2m + F(x_N - (N-1)a) \tag{S4.134}$$

The coordinates are restricted by the N inequalities,

$$0 < x_1, \ x_1 + a < x_2, \ x_2 + a < x_3, \ldots, x_{N-1} + a < x_N \qquad \text{(S4.135)}$$

Therefore, the partition function is

$$
Z = \left(\frac{1}{h} \int_{-\infty}^{\infty} e^{-\beta p^2 / 2m} \, dp \right)^N \int_{-\infty}^{\infty} dx_1
$$
$$
\times \int_{x_1+a}^{\infty} dx_2 \cdots \int_{x_{N-1}+a}^{\infty} dx_N \, e^{-\beta F(x_N - (N-1)a)} \qquad \text{(S4.136)}
$$

We define a coordinate transformation from the coordinates $(x_1, x_2, \ldots, x_N)$ to new coordinates $(u_1, u_2, \ldots, u_N)$ by the equations $u_1 = x_1$, $u_2 = x_2 - (x_1 + a)$, $u_3 = x_3 - (x_2 + a), \ldots, u_N = x_N - (x_{N-1} + a)$. We notice that

$$u_1 + u_2 + \ldots + u_N = x_N - (N-1)a \qquad \text{(S4.137)}$$

Written in terms of the new coordinates

$$
Z = (2\pi m / \beta h^2)^{N/2} \int_0^{\infty} e^{-\beta F u_1} \, du_1 \int_0^{\infty} e^{-\beta F u_2} \, du_2 \cdots \int_0^{\infty} e^{-\beta F u_N} \, du_N
$$
$$
= (2\pi m / \beta h^2)^{N/2} (\beta F)^{-N} \qquad \text{(S4.138)}
$$

(b) The canonical potential is

$$\phi = N\left[\tfrac{1}{2} \log(2\pi m / h^2) - \tfrac{3}{2} \log \beta - \log F \right] \qquad \text{(S4.139)}$$

The energy of the system at temperature T is

$$E = -\frac{\partial \phi}{\partial \beta} = \tfrac{3}{2} kT \qquad \text{(S4.140)}$$

which gives, for the specific heat,

$$C = \frac{\partial E}{\partial T} = \tfrac{3}{2} k \qquad \text{(S4.141)}$$

The average position of the Nth particle is $\langle x_N \rangle$. From Eq. (S4.136), one can see that

$$
\langle x_N - (N-1)a \rangle = -kT \frac{\partial Z / \partial F}{Z}
$$
$$
= -kT \frac{\partial \phi}{\partial F} \qquad \text{(S4.142)}
$$
$$
= NkT / F
$$

Thus

$$\langle x_N \rangle = (N-1)a + NkT / F \qquad \text{(S4.143)}$$

(c) For large N, we can replace the $N-1$ in the previous equation by N. If we do that and use the fact that $p = F$, we obtain the pressure equation of state.

$$p = \frac{Nkt}{x_N - Na} = \frac{nkT}{1 - na} \qquad \text{(S4.144)}$$

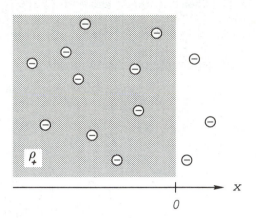

Fig. S4.6   Electrons diffuse out of a metal surface.

Since $\langle U \rangle = F\langle x_N - (N-1)a \rangle$, we see that, of the $\frac{3}{2}NkT$ of thermal energy, two-thirds of it is in the form of potential energy and only $\frac{1}{2}NkT$ is kinetic energy. This agrees with the equipartition theorem.

Exercise 4.18   A metal may be pictured as a rigid collection of positively charged ions plus a set of mobile conduction electrons. The ions are composed of the atomic nuclei plus their inner-shell electrons, which are tightly bound to the nuclei and play no role in electrical conductivity. A more simplified model of a metal, sometimes called the *jellium model*, is obtained by replacing the system of ions by a uniform positive background charge density of magnitude en_o within which moves a gas of electrons whose number is sufficient to make the whole system electrically neutral. Consider a classical jellium model of a metal surface. That is, take the background charge density to be

$$\rho_+(x) = \begin{cases} en_o, & \text{if } x < 0 \\ 0, & \text{if } x > 0 \end{cases} \tag{S4.145}$$

The charge density of the electrons is $\rho_- = -en(x)$, where $n(x)$ is the electron density—to be determined. Assume that the electrostatic potential $V(x)$ approaches zero deep inside the metal. Use Poisson's equation,

$$\frac{d^2V(x)}{dx^2} = -\frac{\rho(x)}{\epsilon_o} \tag{S4.146}$$

and the Maxwell–Boltzmann equilibrium density function (remember that $U = -eV$)

$$n(x) = Ce^{eV(x)/kT} \tag{S4.147}$$

along with the assumption that $kT \gg eV$, to determine $V(x)$ and $n(x)$ everywhere.

Solution   First let us determine the constant C in Eq. (S4.147). Deep inside the metal $V(x) \to 0$ and the background charge is en_o. But the total charge density must go to zero there. Thus $C = n_o$. Using the inequality $kT \gg eV$ we can expand the exponential and keep only the first two terms.

$$n(x) \approx n_o[1 + eV(x)/kT] \tag{S4.148}$$

Inside the metal (that is, for $x < 0$) Poisson's equation is

$$\frac{d^2V(x)}{dx^2} = -\frac{1}{\epsilon_o}(\rho_+ + \rho_-) = \frac{n_o e^2}{\epsilon_o kT}V(x) \qquad \text{(S4.149)}$$

The solution to this equation that goes to zero as $x \to -\infty$ is

$$V(x) = V_o e^{\lambda x} \qquad \text{(S4.150)}$$

where $\lambda = e\sqrt{n_o/\epsilon_o kT}$. Outside the metal $\rho_+ = 0$ and Poisson's equation takes the form

$$\frac{d^2V(x)}{dx^2} = \frac{e}{\epsilon_o}n(x) = \frac{en_o}{\epsilon_o}\left(1 + \frac{e}{kT}V(x)\right) \qquad \text{(S4.151)}$$

The way to solve this equation is to notice that one can eliminate the constant term on the right-hand side by defining $f(x) = V(x) + kT/e$. Then $f(x)$ satisfies the equation

$$\frac{d^2f(x)}{dx} = \frac{e^2 n_o}{\epsilon_o kT}f(x) = \lambda^2 f(x) \qquad \text{(S4.152)}$$

The only solution to this equation that does not diverge as $x \to \infty$ is

$$f(x) = f_o e^{-\lambda x} \qquad \text{(S4.153)}$$

The constants V_o and f_o are determined by demanding that $V(x)$ and $V'(x)$ be continuous at $x = 0$. These conditions give the equations

$$V_o = f_o - kT/e \quad \text{and} \quad V_o = -f_o \qquad \text{(S4.154)}$$

which have the solution

$$V_o = -\frac{kT}{2e} \quad \text{and} \quad f_o = \frac{kT}{2e} \qquad \text{(S4.155)}$$

Thus

$$V(x) = \begin{cases} -(kT/2e)e^{\lambda x}, & \text{if } x < 0 \\ -(kT/2e)(2 - e^{-\lambda x}), & \text{if } x > 0 \end{cases} \qquad \text{(S4.156)}$$

The electron density is then given by Eq. (S4.148).

$$n(x) = \begin{cases} n_o(1 - \frac{1}{2}e^{\lambda x}), & \text{if } x < 0 \\ \frac{1}{2}n_o e^{-\lambda x}, & \text{if } x > 0 \end{cases} \qquad \text{(S4.157)}$$

Notice that the total charge density is an antisymmetric function, $\rho(-x) = -\rho(x)$.

Exercise 4.19   A perfect atomic crystal has exactly one atom at each lattice site. Such a configuration is only stable at zero temperature. At any finite temperature, due to the thermal fluctuations, various types of crystal imperfections develop. One of the simplest, called a *vacancy*, is an empty lattice point. In this example we want to determine the temperature dependence of the density of vacancies. We take the energy of the perfect crystal as zero and assume that the energy of the crystal with n vacancies is $n\varepsilon$. We also assume that n is much less than N, the number of atoms in the crystal. In doing this calculation we must be careful because there are two small quantities available. One of them is n/N, the density of vacancies. The other is N_s/N, where N_s is the number of sites on the surface of the perfect crystal. Assuming that $N_s/N \ll n/N \ll 1$, calculate n/N as a function of T. (Since $N_s/N \sim 1/N^{1/3}$ is an extremely small number for a macroscopic crystal, this is the condition that usually holds.)

Solution Imagine that we begin with a perfect crystal, with N particles filling N sites, and begin to warm it. Occasionally a particle near the surface will jump to the surface, leaving behind a vacancy. The vacancies will then migrate throughout the volume of the crystal. Because we are assuming that, at equilibrium, the total number of vacancies n is much larger than the original number of surface sites, many new surface layers will have to be created in order to make room for the particles that come from the vacancies. The number of lattice sites occupied by the new, imperfect crystal will be $N + n$. We can view the system a being composed of N particles and n vacancies, distributed over $N + n$ sites. The number of possible configurations is just the binomial coefficient.

$$K = \frac{(N+n)!}{N!n!} \tag{S4.158}$$

The entropy is the logarithm of the number of configurations. Using Stirling's formula,

$$S^o = (N+n)\log(N+n) - N \log N - n \log n \tag{S4.159}$$

Since $E = \varepsilon n$, $\beta = \partial S^o / \partial E = (1/\varepsilon)\partial S^o / \partial n$. Thus

$$\frac{\varepsilon}{kT} = \frac{\partial S^o}{\partial n} = \log\left(\frac{N+n}{n}\right) \tag{S4.160}$$

or

$$\frac{N+n}{n} = e^{\varepsilon/kT} \tag{S4.161}$$

which can easily be solved for n/N.

$$\frac{n}{N} = \frac{1}{e^{\varepsilon/kT} - 1} \approx e^{-\varepsilon/kT} \tag{S4.162}$$

where the last step uses the fact that $e^{\varepsilon/kT}$ must be much larger than one if $n/N \ll 1$.

Exercise 4.20 Another type of crystal defect is an *interstitial atom*, that is, an atom squeezed into a space between filled lattice spaces. Such interstitial defects cause a local distortion of the crystal structure and therefore have fairly high energy. To study such defects let us consider a perfect crystal of N lattice sites that is rigidly constrained to its original volume and warmed to temperature T. At equilibrium, n particles will have jumped from their original positions to interstitial positions, creating n vacancies and n interstitial defects. Assuming that the energy of the configuration described is proportional to n and that there are N interstitial positions, calculate n/N as a function of T.

Solution The vacancies can be distributed among the N lattice sites in $N!/n!(N-n)!$ ways. The interstitial atoms can be distributed among the N interstitial sites in $N!/n!(N-n)!$ ways. Therefore, the total number of configurations of the system is

$$K = \left[N!/n!(N-n)!\right]^2 \tag{S4.163}$$

The entropy is, using Stirling's approximation,

$$S^o = \log K = 2[N \log N - n \log n - (N-n) \log(N-n)] \tag{S4.164}$$

The energy is given by $E = \varepsilon n$, and therefore $\beta = \partial S^o / \partial E = (1/\varepsilon) \partial S^o / \partial n$.

$$\frac{\varepsilon}{kT} = \frac{\partial S^o}{\partial n} = 2 \log \frac{N-n}{n} \tag{S4.165}$$

This equation can be solved for n/N, giving

$$\frac{n}{N} = \frac{1}{e^{\varepsilon/2kT} + 1} \approx e^{-\varepsilon/2kT} \tag{S4.166}$$

Exercise 4.21   The molecules of an imaginary ideal gas have internal energy levels that are equally spaced so that the nth energy eigenvalue is $E_n = n\varepsilon$, where $n = 0, 1, 2, \ldots$. The degeneracy of the nth energy level is $n+1$. Calculate the contribution to the thermal energy of the internal energy states.

Solution   For a single molecule, the internal partition function is

$$
\begin{aligned}
z_{\text{int}} &= \sum_{n=0}^{\infty} (n+1) e^{-\beta \varepsilon n} \\
&= \sum_{n=0}^{\infty} (n+1) x^n \qquad (x = e^{-\beta \varepsilon}) \\
&= \frac{d}{dx} \sum_{m=0}^{\infty} x^m \qquad (m = n+1) \\
&= \frac{d}{dx} (1-x)^{-1} \\
&= (1-x)^{-2} \\
&= (1 - e^{-\beta \varepsilon})^{-2}
\end{aligned}
\tag{S4.167}
$$

The contribution of the internal states to the canonical potential of the molecule is

$$\phi_{\text{int}} = -2 \log(1 - e^{-\beta \varepsilon}) \tag{S4.168}$$

The internal energy of the molecule is

$$E_{\text{int}} = -\frac{\partial \phi_{int}}{\partial \beta} = \frac{2\varepsilon}{e^{\beta \varepsilon} - 1} \tag{S4.169}$$

The total energy of the N-particle gas would be

$$E = \frac{3}{2} NkT + \frac{2N\varepsilon}{e^{\varepsilon/kT} - 1} \tag{S4.170}$$

Notice that Eq. (S4.169) has the same form as the thermal energy of a harmonic oscillator with angular frequency $\omega = \varepsilon/\hbar$ except for the factor of 2. This is easy to understand. If the internal energy states were composed of two independent harmonic oscillators with the same angular frequency, then the energy levels, measured from the ground state, would be given by

$$E(n_1, n_2) = \hbar\omega(n_1 + n_2) \tag{S4.171}$$

But it is easy to see that there are just $n + 1$ combinations of n_1 and n_2 that add up to n. Thus the spectrum and degeneracies of the molecule are those of a pair of harmonic oscillators and so the thermal energy is twice that of a single oscillator.

Exercise 4.22   The total dipole moment of a gas of dipolar molecules in an electric field directed along the z axis is defined as $P = N\bar{\mu}_z$. In Eq. (4.78) it is shown that $P = kT\, \partial\phi/\partial\mathcal{E}$, where ϕ is the canonical potential and $\mathcal{E}$ is the electric field strength. Derive a similar (but different) formula for $(\Delta P)^2$, the mean-square fluctuation in P.

Solution   We know that

$$(\Delta P)^2 = \langle P^2 \rangle - \langle P \rangle^2 \tag{S4.172}$$

But, $P = \sum \mu \cos\theta_i$, and therefore

$$\langle P^2 \rangle = \left\langle \mu^2 \sum_{i,j} \cos\theta_i \cos\theta_j \right\rangle \tag{S4.173}$$

Also, $H = H_o - \mathcal{E}\mu \sum \cos\theta_i$. If we define the normalization integral

$$I = \int \exp\left(-\beta H_o - \beta\mu\mathcal{E}\sum_n \cos\theta_n\right) d^K x\, d^K p \tag{S4.174}$$

then

$$\langle P^2 \rangle = \frac{\int (\mu\sum\cos\theta_i)^2 \exp[-\beta H_o - \beta\mu\mathcal{E}\sum\cos\theta_n]\, d^K x\, d^K p}{I}$$
$$= (kT)^2 \frac{\partial^2 I/\partial\mathcal{E}^2}{I} \tag{S4.175}$$

and

$$\langle P^2 \rangle - \langle P \rangle^2 = (kT)^2 \left[\frac{\partial^2 I/\partial\mathcal{E}^2}{I} - \left(\frac{\partial I/\partial\mathcal{E}}{I}\right)^2\right]$$
$$= (kT)^2 \frac{\partial}{\partial\mathcal{E}}\left(\frac{\partial I/\partial\mathcal{E}}{I}\right) \tag{S4.176}$$
$$= (kT)^2 \frac{\partial^2\phi}{\partial\mathcal{E}^2}$$

where, in the final step, we have used the fact that

$$\phi = \log(I/h^K N!) = \log I - \log(h^K N!) \tag{S4.177}$$

which shows that $\partial\phi/\partial\mathcal{E} = \partial(\log I)/\partial\mathcal{E}$. Notice that this analysis does not really depend on the detailed form of the observable P. In fact, if any observable A and some parameter λ satisfy the relation $\langle A \rangle = \partial \phi / \partial \lambda$, then the fluctuation in A is given by $(\Delta A)^2 = \partial^2 \phi / \partial \lambda^2$.

Exercise 4.23 This exercise concerns another way of deriving the canonical ensemble for a quantum mechanical system. Consider a system that is composed of a very large number N of weakly interacting identical subsystems. The total system is isolated and has energy E. Each subsystem has discrete energy eigenstates with eigenvalues $E_n (n = 1, 2, \ldots)$. Let N_n be the number of subsystems that are in the nth quantum state. Calculate the most probable distribution of the subsystems among their possible quantum states. That is, calculate the distribution N_n that corresponds to the maximum number of states for the complete system.

Solution The microstate of the complete system is defined by specifying the exact quantum state of each subsystem. This can be done by giving the quantum numbers $n_1, n_2, \ldots, n_N$ that tell the quantum states of each subsystem. The macrostates of the system are defined by specifying how many subsystems are in each possible quantum state, that is, by giving $N_1, N_2, \ldots$. The number of microstates that correspond to a particular macrostate is given by the usual formula

$$I = \frac{N!}{N_1! N_2! \cdots} \tag{S4.178}$$

There are two restrictions on the possible choice of macrostates. The sum over all the N_is must give the total number of subsystems N and the sum of N_i times E_i must give the fixed total energy of the complete system. Thus

$$\sum_i N_i = N \quad \text{and} \quad \sum_i N_i E_i = E \tag{S4.179}$$

Maximizing $\log I$, using Lagrange's method to handle the constraints and Stirling's approximation for $\log(N_i!)$, leads to the problem of maximizing the function

$$F = N(\log N - 1) - \sum_i N_i (\log N_i - 1) - \alpha \sum N_i - \beta \sum N_i E_i \tag{S4.180}$$

Setting $\partial F / \partial N_i = 0$ gives the relation

$$\log N_i = -\alpha - \beta E_i \tag{S4.181}$$

or

$$N_i = C e^{-\beta E_i} \tag{S4.182}$$

where $C = e^{-\alpha}$. The normalization condition $\sum N_i = N$ fixes the value of the constant C.

$$N_i = N \frac{e^{-\beta E_i}}{\sum e^{-\beta E_k}} \tag{S4.183}$$

But the probability of finding any particular subsystem in its nth quantum state is N_n/N. Thus

$$P_n = Z^{-1} e^{-\beta E_n} \tag{S4.184}$$

which is exactly the canonical probability density. What we have done is simply to choose, as our reservoir, $(N - 1)$ copies of the sample system.

Exercise 4.24 A solid is composed of N atoms whose nuclei have angular momentum $\hbar/2$. Associated with the nuclear angular momentum is a nuclear magnetic moment of magnitude μ. When the solid is placed in a magnetic field the component of the magnetic moment of any nucleus, in the direction of the field, can have the two values $\pm\mu$. The interactions between the nuclear magnetic moments are completely negligible. In many real solids, the interaction between the nuclear magnetic moments and all other degrees of freedom of the solid, such as lattice vibrations and electronic motions, is so weak that the collection of nuclei can be considered as an isolated system.

Using the canonical ensemble, calculate the energy and entropy of the system of nuclear magnetic moments in a magnetic field B at temperature T.

Solution For the ith nucleus we introduce a variable σ_i that is equal to -1 if the magnetic moment is parallel to B and $+1$ if it is antiparallel. The configuration of the system is then defined by the set of variables $(\sigma_1, \sigma_2, \ldots, \sigma_N)$. The energy associated with a given configuration is

$$E = \mu B \sum_n \sigma_n \tag{S4.185}$$

Since the energy is a sum of independent terms, the partition function is the Nth power of the partition function for one nucleus. Thus $Z = z^N$, where

$$
\begin{aligned}
z &= \sum_\sigma e^{-\beta E(\sigma)} \\
&= \sum_{\sigma=\pm 1} e^{-\beta \mu B \sigma} \\
&= 2\cosh(\beta\mu B)
\end{aligned}
\tag{S4.186}
$$

The canonical potential is

$$\phi = N\log[2\cosh(\beta\mu B)] \tag{S4.187}$$

and the energy at inverse temperature β is

$$E = -\frac{\partial\phi}{\partial\beta} = -N\mu B \tanh(\beta\mu B) \tag{S4.188}$$

The entropy is related to the canonical potential and the energy by $\phi = S^o - \beta E$. Therefore

$$S^o = N\big\{\log[2\cosh(\beta\mu B)] - \beta\mu B \tanh(\beta\mu B)\big\} \tag{S4.189}$$

This system is exactly the noninteracting Ising model that was analyzed in Exercise 3.14. A comparison of this short calculation with that much longer one shows again the advantage of the canonical versus the microcanonical ensemble.

Exercise 4.25 We can define an unrealistic but solvable model of a molecular gas by assuming that the gas consists of N diatomic molecules and that the two atoms in the ith molecule have masses m and m' and are described by position and momentum vectors $\mathbf{r}_i$, $\mathbf{r}'_i$, $\mathbf{p}_i$, and $\mathbf{p}'_i$. The atoms in any given molecule are bound together with a harmonic oscillator potential $v(|\mathbf{r}_i - \mathbf{r}'_i|) = k_o|\mathbf{r}_i - \mathbf{r}'_i|^2/2$, but they do not interact with the atoms in any other molecule. With such a model, the Hamiltonian is

$$H = \sum_{i=1}^{N}\left(\frac{p_i^2}{2m} + \frac{p_i'^2}{2m'} + \frac{k_o}{2}|\mathbf{r}_i - \mathbf{r}'_i|^2\right) \tag{S4.190}$$

Calculate the thermal energy of the system, using classical mechanics, and show that it agrees with the equipartition theorem.

Solution   The canonical partition function is

$$Z = \frac{1}{h^{6N}N!}\int e^{-\beta H}\,d^{3N}p\,d^{3N}p'\,d^{3N}r\,d^{3N}r' \tag{S4.191}$$

The momentum integrals are straightforward, giving a factor of $(2\pi\sqrt{mm'}/\beta)^{3N}$. The coordinate integrals factor into N six-dimensional integrals. The ith integral involves the coordinates of the ith molecule only. Each of the integrals is of the form

$$I = \int_V e^{-\beta k_o|\mathbf{r}-\mathbf{r}'|^2/2}\,d^3r\,d^3r' \tag{S4.192}$$

We transform to center-of-mass and relative variables, $\mathbf{R} = (\mathbf{r}+\mathbf{r}')/2$ and $\mathbf{q} = \mathbf{r}-\mathbf{r}'$. We have seen before that the Jacobian of this transformation is one. The integral is then

$$I = \int e^{-\beta k_o q^2/2}d^3R\,d^3q \tag{S4.193}$$

The exponential factor goes rapidly to zero as q becomes large. Therefore, we can extend the q integral to infinity, making negligible error. The molecular center of mass must be integrated over the system volume. Using the fact that $q^2 = q_x^2+q_y^2+q_z^2$, the integral over the relative variable can be factored into three Gaussian integrals,

$$I = V(2\pi/\beta k_o)^{3/2} \tag{S4.194}$$

Putting these all together, the partition function is

$$Z = \frac{(2\pi\sqrt{mm'}/\beta)^{3N}V^N(2\pi/\beta k_o)^{3N/2}}{h^{6N}N!} \tag{S4.195}$$

$$= \left(8\pi^3 mm'/h^4\sqrt{k_o}\right)^{3N/2}V^N\Big/\beta^{9N/2}N!$$

and

$$\phi = N\left[\log(V/N) - \tfrac{9}{2}\log\beta + \text{const.}\right] \tag{S4.196}$$

The thermal energy is

$$E = -\frac{\partial\phi}{\partial\beta} = \tfrac{9}{2}NkT \tag{S4.197}$$

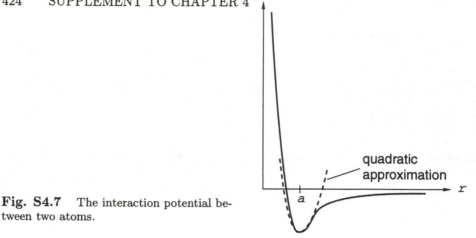

Fig. S4.7   The interaction potential between two atoms.

Since the Hamiltonian function is composed of $9N$ simple squares (remember that $p^2 = p_x^2 + p_y^2 + p_z^2$, etc.), this result could have been predicted by the equipartition theorem.

Exercise 4.26   The most unrealistic characteristic of the model of a diatomic ideal gas used in the last example is the fact that the equilibrium distance between the atoms in the molecule is zero. In reality the interaction potential between two atoms in a diatomic molecule is similar to the function shown in Fig. S4.7. For such an interaction potential, if we still ignore the interactions between atoms in different molecules, the Hamiltonian function is

$$H = \sum_{i=1}^{N} \left(\frac{p_i^2}{2m} + \frac{p_i'^2}{2m'} + v(|\mathbf{r}_i - \mathbf{r}_i'|) \right) \qquad (S4.198)$$

(a) Treating the system by classical mechanics, derive a formula for the thermal energy per molecule in terms of one-dimensional integrals involving $v(r)$. (b) Approximate $v(r)$ by a displaced harmonic oscillator potential with a minimum $-\varepsilon_o$ at $r = a$ and a force constant k_o

$$v(r) = -\varepsilon_o + \tfrac{1}{2}k_o(r - a)^2 \qquad (S4.199)$$

Assume that $kT \ll ka^2/2$ and show that the simple modification of displacing the minimum of the potential has an effect on the temperature dependence of the thermal energy. Actually, the Hamiltonian in Eq. (S4.198), combined with the displaced harmonic oscillator approximation, gives a reasonably accurate description of many diatomic molecules if the vibrational degrees of freedom are treated quantum mechanically.

Solution   (a) All of the analysis used in the preceding example is still valid, but the integral for the internal partition function of a single molecule must be changed to

$$I = V \int e^{-\beta v(q)} d^3q = 4\pi V \int_o^\infty e^{-\beta v(q)} q^2 \, dq \qquad (S4.200)$$

With this modification, Eq. (S4.195) for the partition function of the system becomes

$$Z = \frac{(2\pi\sqrt{mm'}/\beta)^{3N} V^N (4\pi \int_o^\infty e^{-\beta v} q^2 \, dq)^N}{h^{6N} N!} \qquad (S4.201)$$

and the equation for the canonical potential becomes

$$\phi = N\left[\log\left(\frac{V}{N}\right) - 3\log\beta + \log\left(\int_o^\infty e^{-\beta v}q^2\,dq\right) + C\right] \tag{S4.202}$$

The energy per molecule is

$$\frac{E}{N} = -\frac{\partial(\phi/N)}{\partial\beta} = 3kT + \frac{\int ve^{-\beta v}q^2\,dq}{\int e^{-\beta v}q^2\,dq} \tag{S4.203}$$

This is an obvious formula. The $3kT$ comes, via the equipartition theorem, from the six momentum degrees of freedom per molecule. The term involving the interaction potential simply shows that the probability distribution for the interatomic distance within a molecule is the Maxwell–Boltzmann function, $Ce^{-\beta v(q)}$. (b) Using the displaced harmonic oscillator approximation, the integral in the denominator in Eq. (S4.203) becomes

$$\int_o^\infty e^{-\beta v(q)}q^2\,dq = e^{\beta\varepsilon_o}\int_{-a}^\infty e^{-\beta k_o x^2/2}(x+a)^2\,dx$$
$$\approx e^{\beta\varepsilon_o}\int_{-\infty}^\infty e^{-\beta k_o x^2/2}(x^2+a^2)\,dx \tag{S4.204}$$
$$= \sqrt{2}e^{\beta\varepsilon_o}\left[(\beta k_o)^{-3/2} + a^2(\beta k_o)^{-1/2}\right]$$

In going from the first to the second line, we have used the fact that $\beta k_o a^2/2 \gg 1$ to extend the integral to $-\infty$ and the fact that the term $2ax$ gives a zero integral. The other necessary integral is

$$\int_o^\infty v(q)e^{-\beta v(q)}q^2\,dq = -\frac{d}{d\beta}\int_o^\infty e^{-\beta v(q)}q^2\,dq$$
$$= -\frac{d}{d\beta}\left\{\sqrt{2}e^{\beta\varepsilon_o}\left[(\beta k_o)^{-3/2} + a^2(\beta k_o)^{-1/2}\right]\right\} \tag{S4.205}$$
$$= -\sqrt{2}\varepsilon_o e^{\beta\varepsilon_o}\left[(\beta k_o)^{-3/2} + a^2(\beta k_o)^{-1/2}\right]$$
$$+ \sqrt{2}k_o e^{\beta\varepsilon_o}\left[\tfrac{3}{2}(\beta k_o)^{-5/2} + \tfrac{1}{2}a^2(\beta k_o)^{3/2}\right]$$

Using these values for the integrals in Eq. (S4.203), we get, after some simple algebra,

$$\frac{E}{N} = 3kT - \varepsilon_o + \tfrac{1}{2}kT\frac{1+3kT/k_o a^2}{1+kT/k_o a^2} \tag{S4.206}$$

If $kT \ll k_o a^2$, as we have already assumed, then

$$\frac{E}{N} \approx \tfrac{7}{2}kT - \varepsilon_o \tag{S4.207}$$

The $-\varepsilon_o$ comes from a trivial shift in the zero of our interatomic potential. The change of the factor in front of kT from 9/2 to 7/2 shows that displacing the harmonic oscillator potential has changed it from a three-dimensional to a one-dimensional harmonic oscillator, thus eliminating two contributions of $\tfrac{1}{2}kT$. Notice that letting $a \to 0$ in Eq. (S4.206) makes the two suppressed terms reappear.

Exercise 4.27   In this example we want to show that the temperature coefficient of expansion of a solid is related to the anharmonic terms in the interatomic potential. We take as our model a one-dimensional solid in which only nearest-neighbor particles interact and they interact with a potential

$$v(x) = \frac{k_o}{2}(x-a)^2 - \gamma(x-a)^3 \tag{S4.208}$$

where the second term is assumed to be much smaller than the first. The minimum of the potential occurs at $x = a$, and thus at zero temperature the configuration of the system will be a perfect lattice of spacing a. The Hamiltonian for the system is

$$H = \sum_{n=1}^{N} \frac{p_n^2}{2m} + \sum_{n=1}^{N} v(x_n - x_{n-1}) \tag{S4.209}$$

In order to fix one end of the crystal, we assume that the first particle interacts with a completely stationary particle at $x_o = 0$. That accounts for the $n = 1$ term in the potential sum in Eq. (S4.209). Using classical mechanics, determine the length of the N-particle crystal as a function of temperature and show that it is constant if $\gamma = 0$.

Solution   The probability of a given configuration is

$$P(x_1, x_2, \ldots, x_N) = I^{-1} \exp\left(-\beta \sum v(x_n - x_{n-1})\right) \tag{S4.210}$$

where $x_o = 0$. The normalization integral is

$$I = \int_{-\infty}^{\infty} \exp\left(-\beta \sum v(x_n - x_{n-1})\right) dx_1 \cdots dx_N \tag{S4.211}$$

We make a transformation of variables, $y_1 = x_1$, $y_2 = x_2 - x_1$, $y_3 = x_3 - x_2$, ..., $y_N = x_N - x_{N-1}$. Then

$$I = \left(\int \exp[-\beta v(y)]\, dy\right)^N \tag{S4.212}$$

The length of the crystal is the average value of the position of the last particle. That is $L = \langle x_N \rangle$. Using the fact that, with the transformation defined, $y_1 + y_2 + \cdots + y_N = x_N$, we get

$$L(T) = \left\langle \sum y_n \right\rangle \tag{S4.213}$$

Written in terms of the y variables

$$P(x_1, x_2, \ldots, x_N) = I^{-1} \prod_{n=1}^{N} \exp[-\beta v(y_n)] \tag{S4.214}$$

Using this fact, it is easy to see that

$$\left\langle \sum y_n \right\rangle = N \frac{\int y e^{-\beta v(y)}\, dy}{\int e^{-\beta v(y)}\, dy} \tag{S4.215}$$

The integral in the denominator is

$$
\begin{aligned}
I_{\text{den}} &= \int_{-\infty}^{\infty} \exp\left[-\beta k_o (y-a)^2/2 + \beta\gamma(y-a)^3\right] dy \\
&= \int_{-\infty}^{\infty} \exp(-\beta k_o q^2/2) \exp(\beta\gamma q^3) \, dq \\
&\approx \int_{-\infty}^{\infty} \exp(-\beta k_o q^2/2)(1 + \beta\gamma q^3) \, dq \\
&= \sqrt{2\pi/\beta k_o}
\end{aligned}
\tag{S4.216}
$$

where, in the second line, we have used the definition $q = y - a$, and in the third line we have expanded the exponential to first order in the small parameter γ. The integral in the numerator is done in a similar way.

$$
\begin{aligned}
I_{\text{num}} &= \int_{-\infty}^{\infty} \exp\left[-\beta k_o (y-a)^2/2 + \beta\gamma(y-a)^3\right] y \, dy \\
&= \int_{-\infty}^{\infty} \exp(-\beta k_o q^2/2) \exp(\beta\gamma q^3)(a+q) \, dq \\
&\approx \int_{-\infty}^{\infty} \exp(-\beta k_o q^2/2)(1 + \beta\gamma q^3)(a+q) \, dq \\
&= \int_{-\infty}^{\infty} \exp(-\beta k_o q^2/2)(a + \beta\gamma q^4) \, dq \\
&= a\sqrt{2\pi/k_o\beta} + 3\sqrt{2\pi}\gamma(kT)^{3/2}/k_o^{5/2}
\end{aligned}
\tag{S4.217}
$$

$L(T)$ is equal to $NI_{\text{num}}/I_{\text{den}}$.

$$
L(T) = N\left[a + 3\gamma(kT/k_o)^2\right]
\tag{S4.218}
$$

Notice that if the anharmonic term is eliminated by setting $\gamma = 0$, then the length of the crystal becomes independent of temperature.

Exercise 4.28 A nonideal gas has an equation of state $p = p(n, T)$, where n is the particle density. The gas is at equilibrium in a uniform gravitational field g in the negative z direction. (a) Using hydrostatic arguments, obtain a differential equation for the density at height z. (b) Show that, for an ideal gas, the equation predicts the usual exponential atmosphere, $n(z) = n_0 \exp(-mgz/kT)$.

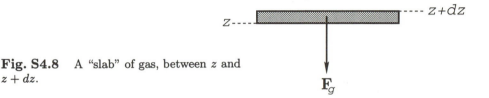

Fig. S4.8 A "slab" of gas, between z and $z + dz$.

Solution (a) We consider a horizontal slab of the gas (see Fig. S4.8) with its top surface at $z + dz$ and its bottom surface at z. The area of the slab is A. The z

component of the pressure forces exerted on the top and bottom surfaces of the slab add up to

$$F_p = A\Big[p(n(z), T) - p(n(z + dz), T)\Big] \qquad \text{(S4.219)}$$

where we have used the fact that, at equilibrium, the temperature is uniform. The mass of the gas in the slab is $mn(z)A\,dz$. (For very small dz we can take the value of n at the bottom and ignore the variation of the density through the slab.) The z component of the gravitational force on the slab is

$$F_g = -mnAg\,dz \qquad \text{(S4.220)}$$

Setting $F_P + F_g$ equal to zero and dividing by dz gives the differential equation

$$\frac{\partial p}{\partial n}\frac{dn}{dz} + mgn = 0 \qquad \text{(S4.221)}$$

(b) For an ideal gas $p(n, T) = nkT$. Equation (S4.221) is then

$$\frac{dn}{dz} = -(mg/kT)n \qquad \text{(S4.222)}$$

which has the solution

$$n(z) = n(0)e^{-(mg/kT)z} \qquad \text{(S4.223)}$$

Supplement to Chapter 5

REVIEW QUESTIONS

5.1   Describe the content of Axioms 1, 2, and 3.

5.2   Give a precise statement of Axiom 4 (the Second Law).

5.3   Mathematically, what do we mean when we say that the entropy is extensive?

5.4   Show that the molar entropy function, $s(\varepsilon, v)$, is a convex function.

5.5   Describe the physical meaning of the three parameters $\alpha = \partial S/\partial N$, $\beta = \partial S/\partial E$, and $\gamma = \partial S/\partial V$.

5.6   Define an *empirical temperature*.

5.7   Prove that β^{-1} is an empirical temperature.

5.8   Prove that γ/β is the equilibrium mechanical pressure.

5.9   Mathematically, what do we mean when we say that α, β, and p are *intensive* variables?

5.10   Prove that β is an intensive variable.

5.11   If two phases of a simple substance are in equilibrium, how many independent variables are there in the set α, β, and p?

5.12   What is a quasistatic process?

5.13   Prove that $dS = 0$ if a substance is slowly compressed in an insulating cylinder.

5.14   How many arbitrary parameters are there in the entropy function of a simple substance?

5.15   What is an isothermal curve in the E–V plane?

5.16   Describe how one can determine an isothermal curve.

5.17   What is an adiabatic curve in the E–V plane?

5.18   Describe how one can determine an adiabatic curve.

5.19   Using only properties derived from the axioms of thermodynamics and mechanical measurements, describe how one can determine the entropy function of a simple substance.

5.20   How is S related to the energy spectrum of a system (in rational units and in SI units)?

5.21   What happens to the derivatives of the entropy function at the boundaries of the thermodynamic state space?

5.22 Explain how a system can be brought to a negative temperature state.

5.23 What is Nernst's law?

5.24 Show that $S = \alpha N + \beta E + \gamma V$.

5.25 Derive the Gibbs-Duhem relation.

5.26 From $pV = NkT$ and $E = \frac{3}{2}NkT$, derive the entropy function $S(N, E, V)$ for a monatomic ideal gas.

5.27 Define the canonical potential ϕ, and express its partial derivatives in terms of thermodynamic variables.

5.28 Do the same for the grand potential ψ.

5.29 For a system in contact with a reservoir at coldness β, what is the condition that determines the equilibrium state?

5.30 For a system in contact with an energy reservoir at coldness β, write the probability of finding the system with energy E in terms of the entropy function of the system.

5.31 Using the fact that, for a system that can exchange particles with a reservoir, the mean square fluctuation in N is given by $(\Delta N)^2 = -\partial N/\partial \alpha$, derive the affinity of the solute for a dilute solution.

EXERCISES

Exercise 5.1   Some imaginary substance satisfies the equations of state

$$E = N(kT)^2/\hbar\omega \quad \text{and} \quad pV = NkT \tag{S5.1}$$

What is the entropy function of the substance in rational units?

Solution   Dividing both equations by N, we obtain formulas relating $\beta = 1/kT$ and p to the energy per particle ε and the volume per particle v.

$$\beta = \frac{1}{\sqrt{\hbar\omega\varepsilon}} \tag{S5.2}$$

and

$$\beta p = \frac{1}{v} \tag{S5.3}$$

But if $s^o(\varepsilon, v)$ is the entropy per particle in rational units, then $\beta = \partial s^o/\partial\varepsilon$. Integrating this differential equation with respect to ε gives

$$s^o = 2\sqrt{\varepsilon/\hbar\omega} + f(v) \tag{S5.4}$$

where f is an unknown function. The second equation of state, combined with the fact that $\beta p = \partial s^o/\partial v$, can be integrated to yield the relation

$$s^o = \log v + g(\varepsilon) \tag{S5.5}$$

where g is another unknown function. Equations (S5.4) and (S5.5), taken together, imply that

$$s^o = \log v + 2\sqrt{\varepsilon/\hbar\omega} + C \tag{S5.6}$$

where C is an arbitrary constant. The full entropy function is then determined by the fact that the entropy is extensive.

$$S^o(N, E, V) = Ns^o(E/N, V/N) = N\left(\log(V/N) + 2\sqrt{E/N\hbar\omega} + C\right) \tag{S5.7}$$

Exercise 5.2   A function of three variables $f(x, y, z)$ is *convex* if, for every pair of points $\mathbf{r}_1 = (x_1, y_1, z_1)$ and $\mathbf{r}_2 = (x_2, y_2, z_2)$, the function satisfies the inequality

$$f\left(\frac{\mathbf{r}_1 + \mathbf{r}_2}{2}\right) \geq \frac{f(\mathbf{r}_1) + f(\mathbf{r}_2)}{2} \tag{S5.8}$$

If the relation $\geq$ can be replaced by the stronger relation $>$, then the function is *strictly convex*. For any simple substance, show that $S(N, E, V)$ is convex but not strictly convex.

Solution   Consider a uniform equilibrium state of N moles of the substance, with an energy E, in a rigid insulating container of volume V. Its entropy is $S(N, E, V)$. Because it is at equilibrium, its entropy must be larger than that of any other macroscopic state with the same total energy, particle number, and volume. Let us consider a different state, composed of two uniform phases occupying volumes V_1 and V_2, and having energies E_1 and E_2 and molar numbers N_1 and N_2, where $V_1 + V_2 = V$, $E_1 + E_2 = E$, and $N_1 + N_2 = N$. The entropy of that state would be $S(N_1, E_1, V_1) + S(N_2, E_2, V_2)$. At this point we must be careful—if $E_2/E_1 = N_2/N_1 = V_2/V_1$, then this "new" state is actually the original uniform macrostate, conceptually separated into two parts with volumes V_1 and V_2, and it must therefore have the same entropy. If any of those equalities is not satisfied, then the new state is a macroscopically distinct state and, by the second law, it must have a smaller entropy than the uniform equilibrium state. Thus

$$S(N_1 + N_2, E_1 + E_2, V_1 + V_2) \geq S(N_1, E_1, V_1) + S(N_2, E_2, V_2) \tag{S5.9}$$

where the equality is satisfied if and only if $E_2/E_1 = N_2/N_1 = V_2/V_1$. The fact that the entropy is extensive implies that $S(N, E, V) = 2S(N/2, E/2, V/2)$. Using this in the left-hand side of Eq. (S5.9), and dividing the equation by 2, gives

$$S\left(\frac{\mathbf{r}_1 + \mathbf{r}_2}{2}\right) \geq \frac{S(\mathbf{r}_1) + S(\mathbf{r}_2)}{2} \tag{S5.10}$$

where $\mathbf{r} \equiv (N, E, V)$. Since it is possible for the equality to hold, the function is not strictly convex.

Exercise 5.3   An equation of state proposed by Berthelot in 1906 as an improvement on the ideal gas equation of state is

$$\left(p + \frac{a}{kTv^2}\right)(v - v_o) = kT \tag{S5.11}$$

where a and v_o are constants (with the appropriate units) to be determined empirically for each gas and $v = V/N$. Suppose a diatomic gas satisfies Berthelot's equation and the added condition that, as $v \to \infty$, $\varepsilon \to \frac{5}{2}kT$, where v and ε are the volume and energy per particle. (a) Determine the canonical potential per particle up to an arbitrary constant. (b) Determine the constant-volume specific heat (per particle) of the gas, as a function of T and v.

Solution   (a) In terms of $\beta = 1/kT$, Berthelot's equation can be written as

$$\beta p = -\frac{a\beta^2}{v^2} + \frac{1}{v - v_o} \tag{S5.12}$$

But letting $\phi(\beta, v) \equiv N^{-1}\phi(N, \beta, vN)$, we know that $\beta p = \partial \phi/\partial v$. Thus Eq. (S5.12) is a differential equation for $\phi(\beta, v)$, which, when integrated with respect to v, gives

$$\phi(\beta, v) = \frac{a\beta^2}{v} + \log(v + v_o) + f(\beta) \tag{S5.13}$$

where $f(\beta)$ is an unknown function. The energy per particle ε is given in terms of ϕ by

$$\varepsilon = -\frac{\partial \phi}{\partial \beta} = -\frac{2a\beta}{v} - f'(\beta) \tag{S5.14}$$

The condition that $\varepsilon \to \frac{5}{2}kT$ as $v \to \infty$ implies that

$$f'(\beta) = -\frac{5}{2\beta} \tag{S5.15}$$

which has the solution

$$f(\beta) = -\tfrac{5}{2} \log \beta + \text{const.} \tag{S5.16}$$

Therefore, the canonical potential per particle is

$$\phi = \frac{a\beta^2}{v} + \log(v - v_o) - \tfrac{5}{2} \log \beta + \text{const.} \tag{S5.17}$$

(b) The energy per particle is

$$\varepsilon = -\frac{\partial \phi}{\partial \beta} = -\frac{2a\beta}{v} + \frac{5}{2\beta} \tag{S5.18}$$

Written in terms of T, this is

$$\varepsilon = -\frac{2a}{kTv} + \tfrac{5}{2}kT \tag{S5.19}$$

The specific heat per particle at constant volume is

$$C_v = \frac{\partial \varepsilon}{\partial T} = \frac{2a}{kT^2 v} + \tfrac{5}{2} \tag{S5.20}$$

Exercise 5.4   The almost comical limit of empirical equations of state for dense gases was reached in 1927 with the Beattie–Bridgeman equation, containing five constants, a, b, c, d, and e, whose values must be determined experimentally for each gas. The conjectured equation of state is

$$p = kT\left(1 - \frac{a}{v(kT)^3}\right)\left(\frac{1}{v} + \frac{b}{v^2} - \frac{c}{v^3}\right) - \frac{d}{v^2} + \frac{e}{v^3} \tag{S5.21}$$

where $v = V/N$ is the volume per particle. (a) Make the reasonable assumption that, as $v \to \infty$, the energy per particle approaches the translational kinetic energy, $3kT/2$, plus an internal energy $\varepsilon_{\text{int}}(\beta)$ that is a known function of the inverse temperature. Let $u(\beta) = \int \varepsilon_{\text{int}} \, d\beta$ be the indefinite integral of ε_{int}. Determine the canonical potential per particle for the Beattie–Bridgeman equation up to an arbitrary constant. (b) Determine the energy per particle as a function of β and v.

Solution   (a) Equation (S5.21) states that

$$\frac{\partial(\phi/N)}{\partial v} = \beta p = (1 - a\beta^3 v^{-1})(v^{-1} + bv^{-2} - cv^{-3}) - d\beta v^{-2} + e\beta v^{-3} \tag{S5.22}$$

which, when integrated with respect to v, gives

$$\frac{\phi}{N} = \log v + (a\beta^2 - b + d\beta)v^{-1} + \tfrac{1}{2}(c + ab\beta^3 - e\beta)v^{-2} - \tfrac{1}{3}ac\beta^3 v^{-3} + f(\beta) \tag{S5.23}$$

where $f(\beta)$ is an arbitrary function. The fact that, as $v \to \infty$,

$$\frac{\partial(\phi/N)}{\partial \beta} = -\frac{E}{N} \to -\tfrac{3}{2}\beta^{-1} - \varepsilon_{\text{int}}(\beta) \tag{S5.24}$$

implies that, for very large v,

$$\frac{\phi}{N} \sim -\frac{3}{2}\log \beta - u(\beta) + g(v) \tag{S5.25}$$

where g is another arbitrary function. But Eq. (S5.23) says that, as $v \to \infty$,

$$\frac{\phi}{N} - \log v \to f(\beta) \tag{S5.26}$$

These equations can be consistent only if

$$f(\beta) = -\frac{3}{2}\log \beta - u(\beta) + \text{const.} \tag{S5.27}$$

Thus

$$\begin{aligned}
\frac{\phi}{N} = {} & \log v + (a\beta^3 - b + d\beta)v^{-1} + \tfrac{1}{2}(c + ab\beta^3 - e\beta)v^{-2} \\
& - \tfrac{1}{3}ac\beta^3 v^{-3} - \tfrac{3}{2}\log \beta - u(\beta) + C
\end{aligned} \tag{S5.28}$$

(b) The energy per particle is given by the thermodynamic relation

$$\begin{aligned}
\frac{E}{N} = {} & -\frac{\partial(\phi/N)}{\partial \beta} \\
= {} & -(3a\beta^2 + d)v^{-1} - \left(\tfrac{3}{2}ab\beta^2 - e\right)v^{-2} + ac\beta^2 v^{-3} + \frac{3}{2\beta} + \varepsilon_{\text{int}}(\beta)
\end{aligned} \tag{S5.29}$$

Exercise 5.5   Some substance has the entropy function

$$S = \lambda V^{1/2}(NE)^{1/4} \tag{S5.30}$$

where N is in moles and λ is a constant with the appropriate units. A cylinder is separated by a fixed partition into two halves, each of volume $1\,\text{m}^3$. One mole of the substance with an energy of $200\,\text{J}$ is placed in the left half, while two moles of the substance with an energy of $400\,\text{J}$ is placed in the right half. (a) Assuming that the partition conducts heat, what will be the distribution of energy between left and right at equilibrium. (b) Assuming the the partition moves freely and also conducts heat, what will be the volumes and energies of the samples in both sides at equilibrium?

Solution   (a) Let S_L and S_R be the entropies of the left and right substances. The equilibrium condition for energy transfer is that $\beta_L = \beta_R$. This says that

$$\frac{\lambda V_L^{1/2} N_L^{1/4}}{4 E_L^{3/4}} = \frac{\lambda V_R^{1/2} N_R^{1/4}}{4 E_R^{3/4}} \tag{S5.31}$$

or

$$\left(\frac{E_R}{E_L}\right)^{3/4} = \left(\frac{V_R}{V_L}\right)^{1/2}\left(\frac{N_R}{N_L}\right)^{1/4} = 2^{1/4} \tag{S5.32}$$

Therefore

$$E_R = 2^{1/3}E_L \tag{S5.33}$$

By energy conservation, we know that $E_L + E_R = 600\,\text{J}$. These two equations can easily be solved to give

$$E_L = 265.5\,\text{J} \quad \text{and} \quad E_R = 334.5\,\text{J} \tag{S5.34}$$

(b) If the partition is free to move, we have the additional equilibrium condition that $\partial S_L/\partial V_L = \partial S_R/\partial V_R$, which gives

$$\frac{\lambda(N_L E_L)^{1/4}}{2V_L^{1/2}} = \frac{\lambda(N_R E_R)^{1/4}}{2V_R^{1/2}} \tag{S5.35}$$

The two equilibrium conditions are

$$\frac{N_L^{1/4}E_L^{1/4}}{V_L^{1/2}} = \frac{N_R^{1/4}E_R^{1/4}}{V_R^{1/2}} \tag{S5.36}$$

and

$$\frac{V_L^{1/2}N_L^{1/4}}{E_L^{3/4}} = \frac{V_R^{1/2}N_R^{1/4}}{E_R^{3/4}} \tag{S5.37}$$

Multiplying corresponding sides of the two equations gives an equation that does not involve V_L or V_R.

$$\frac{N_L^{1/2}}{E_L^{1/2}} = \frac{N_R^{1/2}}{E_R^{1/2}} \tag{S5.38}$$

This states that

$$\frac{E_R}{E_L} = \frac{N_R}{N_L} = 2 \tag{S5.39}$$

which implies that

$$E_R = 400\,\text{J} \quad \text{and} \quad E_L = 200\,\text{J}. \tag{S5.40}$$

Using this in Eq. (S5.37) shows that $V_R = 2V_L$, which, combined with the fact that $V_L = V_R = 2\,\text{m}^3$, determines the equilibrium volumes.

$$V_L = \tfrac{2}{3}\,\text{m}^3 \quad \text{and} \quad V_R = \tfrac{4}{3}\,\text{m}^3 \tag{S5.41}$$

This simply says that, at equilibrium, the gases on both sides have the same energy per particle and volume per particle—a fairly trivial result.

Exercise 5.6 A *negative definite* $N \times N$ real symmetric matrix can be defined in two equivalent ways. We say that the matrix A is negative definite if all of its eigenvalues are negative numbers *or* if the expectation value of A, namely $\langle v|A|v \rangle = \sum v_i A_{ij} v_j$, is negative for every nonzero vector $(v_1, v_2, \ldots, v_N)$. Show that the convexity property of the molar entropy function $s(\varepsilon, v)$ [see Eq. (5.5)] implies that the matrix of its second derivatives

$$M = \begin{bmatrix} s_{\varepsilon\varepsilon} & s_{\varepsilon v} \\ s_{\varepsilon v} & s_{vv} \end{bmatrix} \tag{S5.42}$$

is negative definite.

Solution   If we choose $\varepsilon_1 = \varepsilon + \delta\varepsilon$, $v_1 = v + \delta v$, $\varepsilon_2 = \varepsilon - \delta\varepsilon$, and $v_2 = v - \delta v$, where $\delta\varepsilon$ and δv are arbitrary but very small numbers, then Eq. (5.5) can be written as

$$s(\varepsilon + \delta\varepsilon, v + \delta v) + s(\varepsilon - \delta\varepsilon, v - \delta v) - 2s(\varepsilon, v) < 0 \tag{S5.43}$$

Expanding this to second order in $\delta\varepsilon$ and δv, we notice that the zero order and first order terms cancel exactly, leaving

$$s_{\varepsilon\varepsilon}\, \delta\varepsilon^2 + 2s_{\varepsilon v}\, \delta\varepsilon\, \delta v + s_{vv}\, \delta v^2 < 0 \tag{S5.44}$$

The quantity in this inequality can be expressed in matrix form as

$$\begin{bmatrix} \delta\varepsilon & \delta v \end{bmatrix} \begin{bmatrix} s_{\varepsilon\varepsilon} & s_{\varepsilon v} \\ s_{\varepsilon v} & s_{vv} \end{bmatrix} \begin{bmatrix} \delta\varepsilon \\ \delta v \end{bmatrix} < 0 \tag{S5.45}$$

Since the inequality must hold for all values of $(\delta\varepsilon, \delta v)$ except $(0,0)$, the matrix must be negative definite.

Exercise 5.7   Suppose the matrix

$$A = \begin{bmatrix} a & c \\ c & b \end{bmatrix} \tag{S5.46}$$

is negative definite. Then it is easy to prove that $a < 0$ and $b < 0$. But one can also get another inequality, involving the off-diagonal element c, by using the fact that the determinant of A is equal to the product of its two eigenvalues, which must be positive. That is,

$$ab - c^2 > 0 \tag{S5.47}$$

Use these inequalities to show that the convexity of $s(\varepsilon, v)$ is equivalent to the two fairly obvious conditions that

$$\left(\frac{\partial T}{\partial \varepsilon} \right)_v > 0 \quad \text{and} \quad \left(\frac{\partial p}{\partial v} \right)_T < 0 \tag{S5.48}$$

That is, the temperature increases if we add energy to a substance at fixed volume and the pressure decreases if we expand a substance at fixed temperature. These are both examples of a general principle, called *Le Châtelier's principle*, which states that, for a system in a stable equilibrium state, a change in any quantity must create a force that brings that quantity back to its equilibrium value.

Solution   First we notice that the combination of the two inequalities $a < 0$ and $ab > c^2$ implies the third inequality $b < 0$. Thus the third inequality can be ignored. Applied to the matrix of the second derivatives of $s(\varepsilon, v)$, the first inequality is

$$\left(\frac{\partial^2 s}{\partial \varepsilon^2} \right) < 0 \tag{S5.49}$$

But $\partial s/\partial \varepsilon = \beta = 1/T$. Thus this can be written as

$$\left(\frac{\partial (1/T)}{\partial \varepsilon}\right)_v = -\frac{1}{T^2}\left(\frac{\partial T}{\partial \varepsilon}\right)_v < 0 \tag{S5.50}$$

or $(\partial T/\partial \varepsilon)_v > 0$. The determinantal inequality is

$$s_{\varepsilon\varepsilon}s_{vv} - s_{\varepsilon v}^2 > 0 \tag{S5.51}$$

The difficulty in deriving the second inequality is that it involves a derivative with the temperature held constant. This is equivalent to a derivative with β held constant. To get the required identity we have to consider the free expansion coefficient as a function of β and v. Using the fact that

$$\gamma(\beta, v) = \gamma(\varepsilon(\beta, v), v) \tag{S5.52}$$

we can see that

$$\left(\frac{\partial \gamma}{\partial v}\right)_\beta = \left(\frac{\partial \gamma}{\partial \varepsilon}\right)_v \left(\frac{\partial \varepsilon}{\partial v}\right)_\beta + \left(\frac{\partial \gamma}{\partial v}\right)_\varepsilon \tag{S5.53}$$

First we note that

$$\left(\frac{\partial \gamma}{\partial \varepsilon}\right)_v = s_{\varepsilon v} \quad \text{and} \quad \left(\frac{\partial \gamma}{\partial v}\right)_\varepsilon = s_{vv} \tag{S5.54}$$

which gives us two of the terms in Eq. (S5.51). To handle $(\partial \varepsilon/\partial v)_\beta$ we need a partial derivative identity whose proof will be given in Chapter 6. For any function of two variables $z = z(x, y)$, it is always true that, when y is expressed as a function of x and z,

$$\left(\frac{\partial y}{\partial x}\right)_z = -\frac{\left(\partial z/\partial x\right)_y}{\left(\partial z/\partial y\right)_x} \tag{S5.55}$$

This says that

$$\left(\frac{\partial \varepsilon}{\partial v}\right)_\beta = -\frac{\left(\partial \beta/\partial v\right)_\varepsilon}{\left(\partial \beta/\partial \varepsilon\right)_v} = -\frac{s_{\varepsilon v}}{s_{\varepsilon\varepsilon}} \tag{S5.56}$$

Therefore, using Eqs. (S5.54) and (S5.56), we see that

$$s_{\varepsilon\varepsilon}\left(\frac{\partial \gamma}{\partial v}\right)_\beta = s_{\varepsilon\varepsilon}s_{vv} - s_{\varepsilon v}^2 > 0 \tag{S5.57}$$

But we know that $s_{\varepsilon\varepsilon} < 0$. Thus we can write this inequality as

$$\left(\frac{\partial \gamma}{\partial v}\right)_\beta = \beta\left(\frac{\partial p}{\partial v}\right)_\beta = \frac{1}{T}\left(\frac{\partial p}{\partial v}\right)_T < 0 \tag{S5.58}$$

Since the temperature is always a positive number this is equivalent to the relation $(\partial p/\partial v)_T < 0$.

Exercise 5.8   All of the equilibrium states that we have considered have been stable equilibrium states. As an example of an unstable equilibrium state, consider the system shown in Fig. S5.1, consisting of a soda straw with equal soap bubbles

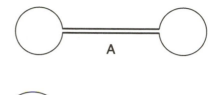

Fig. S5.1 (A) An unstable equilibrium
state of two soap bubbles. (B) One of the
two possible stable equilibrium states.

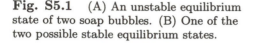

on the ends. (a) Using Laplace's law, which states that, for a bubble of radius R,
$p(\text{inside}) - p(\text{outside}) = 4\sigma/R$, where σ is the surface tension of the soap solution,
show that the situation shown in the figure will not persist indefinitely. (b) Describe
the final, stable, equilibrium state.

Solution (a) Suppose there is a small spontaneous fluctuation in which the radius
of the left bubble increases by an amount δR. The increase in the volume of the left
bubble would be $\delta V = 4\pi R^2\, \delta R$. But the extra gas must come from the right bubble,
which would then decrease in radius by an amount δR. The pressure difference in
the two ends would become (these pressures are inside pressures)

$$p_L - p_R = \frac{4\sigma}{R+\delta R} - \frac{4\sigma}{R-\delta R} \approx -\frac{8\sigma}{R^2}\delta R \tag{S5.59}$$

This pressure difference is in a direction that increases the size of the already larger
left bubble, which would then expand even more.
(b) Assuming that the initial fluctuation was in the direction described, the final
state would be one in which almost all of the gas was in the left bubble, which
would have a radius $R_o \approx 2^{1/3}R$. The right end of the soda straw would be capped
by a meniscus of radius R_o [see Fig. S5.1(B)]. It is now easy to see that this equilibrium state is stable. A small transfer of gas from right to left causes a small
drop in pressure in the left bubble but a much larger drop in pressure in the meniscus. Thus the relative pressure is such as to force gas from left to right, in accord
with Le Châtelier's principle. This is a simple mechanical example of spontaneous
symmetry breakdown. Although the system has obvious left–right symmetry, the
stable equilibrium states do not. As always, there is more than one stable equilibrium state, and the set of them does exhibit the system symmetry.

Exercise 5.9 One mole of a substance satisfies the equations $s^3/v = aT$ and
$\varepsilon = pv$, where a is a constant. Determine the molar entropy function.

Solution The relation $\partial s/\partial \varepsilon = 1/T$, combined with the first equation, gives the
differential equation

$$\frac{\partial s}{\partial \varepsilon} = \frac{av}{s^3} \tag{S5.60}$$

or

$$s^3 \frac{\partial s}{\partial \varepsilon} = av \tag{S5.61}$$

This can be integrated with respect to ε to give

$$s^4/4 = av\varepsilon + f(v) \tag{S5.62}$$

Multiplying the second equation of state by β and using the fact that $\beta p = \partial s / \partial v$, we get another differential equation for s.

$$\varepsilon \frac{\partial s}{\partial \varepsilon} = v \frac{\partial s}{\partial v} \tag{S5.63}$$

Multiplying this by s^3 gives

$$\varepsilon s^3 \frac{\partial s}{\partial \varepsilon} = v s^3 \frac{\partial s}{\partial v} \tag{S5.64}$$

But $s^3 \partial s / \partial \varepsilon$ can be evaluated using Eq. (S5.61).

$$a \varepsilon v = v s^3 \frac{\partial s}{\partial v} \tag{S5.65}$$

Cancelling the factors of v and integrating with respect to v gives

$$s^4/4 = a \varepsilon v + g(\varepsilon) \tag{S5.66}$$

This equation is consistent with Eq. (S5.62) only if

$$s^4/4 = a \varepsilon v + \text{const.} \tag{S5.67}$$

which gives

$$s = (4 a \varepsilon v + C)^{1/4} \tag{S5.68}$$

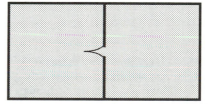

Fig. S5.2   A device, using a one-way valve, intended to defeat the Second Law.

Exercise 5.10   Shown in Fig. S5.2 is a device for defeating the Second Law. Two compartments, initially containing ideal gases at the same temperature and density, are connected by a passage that is closed by rubber flaps. No particles can pass from left to right because the flaps act as a one-way valve. Most particles cannot pass from right to left either, but every now and then a particle will have enough energy to push through the flaps. Since only high-energy particles get through and they only pass through in one direction, if we wait long enough, the gas on the left will become more dense and hotter than the gas on the right. But the uniform state is the state of maximum entropy, and thus the system will have gone from a higher to a lower entropy state. What's wrong with this?

Solution   We will do this analysis in two parts. First we will show, in a rough, pictorial way, that the statements made in the exercise are wrong and there are, at least approximately, as many particles that go from left to right as from right to left. But this rough argument cannot show that the valve has no effect at all on the particle densities in the right and left chambers. In order to show that, we will have to construct a much more precise argument that actually relies on some of the detailed properties of Hamiltonian dynamics.

Suppose that the amount of energy necessary to open the flaps wide enough to let a particle through is E and that the rate at which particles hit the space between the two flaps is R. If the system is at temperature T, then the probability that a particle coming from the right will have enough energy to open the flaps is roughly $e^{-E/kT}$. Thus a rough estimate of the rate at which particles pass from right to left is $Re^{-E/kT}$. Now, the valve is itself a mechanical system that can have both kinetic and potential energy. If the flaps are open then they have potential energy E. The probability of finding them in an open, rather than a closed state, is, again very roughly, $e^{-E/kT}$. But then the rate at which particles are let in from left to right by the spontaneous opening of the flaps is approximately $Re^{-E/kT}$. We can see that the argument made in the statement of this problem, that indicated that *no* particles moved from left to right, is seriously wrong. However, even this improved description of the operation of the system leaves one with the suspicion that *more* particles will go from right to left than vice versa. Let us now clearly eliminate that possibility.

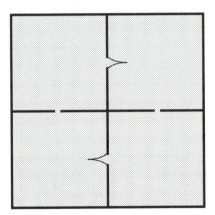

Fig. S5.3   If the one-way valve really works, then one should expect a steady clockwise circulation of particles through the valves.

We connect the system, via two holes, to an identical system that has a valve in the reverse direction. It is clear that, because of symmetry, in this new system no density differential will develop between the two sides. However, if in the original system the valve had any effect at all, then in this new system there would be a definite tendency for particles to circulate in a clockwise sense, going from right to left through the bottom valve and from left to right through the top one. We will now rigorously prove that there is no such tendency. We will describe the system by classical mechanics. This means that, for a detailed description, we would have to model the valve by some classical mechanism with a finite number of coordinates and corresponding canonical momenta. Our argument will not rely on any detailed model of the valve. We assume that the double system is completely isolated and has an energy E. Therefore all of its possible microstates are points on some huge energy surface $\omega(E)$ in the phase space of the system. We choose some fixed, long, time interval, $0 < t < T$. We choose some particular particle, which we call the marker particle. To every possible state on the energy surface we assign an integer, called the circulation number for that state. It is calculated by taking that state as an initial state for the system at time 0, letting the system progress according to Hamilton's equations during the time interval $(0, T)$, and adding $+1$ each time the marker particle passes through one of the valves in the normal direction and -1 each time it passes through one of the valves in the wrong direction. Now, if there is any tendency at all for particles to pass through the valves in the normal direction, then the average circulation number (averaged over all states on the energy surface) will be positive. Consider the set of states with circulation number 5. Call that set of states I_5, where the I indicates that, in computing the circulation number, we are considering these as initial states. Define as F_5 (F for final) the set of states that I_5 goes into at time T. Now, take any microstate in F_5 and reverse the velocities of every particle ($p_1 \rightarrow -p_1, p_2 \rightarrow -p_2$, etc.). The set of states we get this way we call F_5^*. Because Hamiltonian mechanics is exactly symmetrical with respect to time reversal (that is, a motion picture of a system moving according to Hamilton's equations is, if run backwards, still a system moving according to Hamilton's equations), each state in F_5^* will in the time interval $0 < t < T$ run exactly backwards and have a circulation number -5. In fact, it is easy to see that F_5^* is identical with I_{-5}. Let $P(I_5)$ be the probability that a state, picked at random on the energy surface (that is, with the microcanonical probability), will be in I_5. Then, by Liouville's theorem, $P(I_5) = P(F_5)$. But reversing the velocities does not change the probability of a state in the microcanonical ensemble. Therefore $P(F_5) = P(F_5^*) = P(I_{-5})$. This shows that averaging the circulation number over the microcanonical ensemble gives zero because the sets I_K and I_{-K} cancel in pairs for each K. There is no statistical tendency for particles to go through a valve in the "right" as opposed to the "wrong" direction.

Exercise 5.11   A container of volume V is partitioned into two subvolumes, V_A and V_B, that contain different ideal gases at the same temperature and pressure. A valve is opened in the wall separating the subvolumes so that the gases can mix. Obviously the system will now go to a new, uniformly mixed, equilibrium state, which must therefore have a larger entropy than the initial state. What is the change in entropy? This extra entropy is called the *entropy of mixing*.

Solution   The solution of this problem will illustrate a general method that is useful in many thermodynamics problems. The initial state of the system is composed of two substances at equilibrium for which we can calculate the entropy. The final state is a uniform equilibrium state whose entropy we would like to calculate. But the intermediate states, during the mixing process, will be complicated nonuniform states that cannot really be analyzed at all by equilibrium thermodynamics. What we will have to do is to find a different, quasistatic process that goes from the same initial to the same final state. It will have to be chosen so that we can calculate the entropy changes in the quasistatic process. The reason why this method is legitimate, even though the intermediate states are very different, is that the entropy of the system depends on its macrostate and not on how the system got to that macrostate.

The quasistatic process that we will use is one that utilizes *semipermeable membranes*. A semipermeable membrane is a membrane that allows one type of molecule to pass through it freely but not molecules of the other type. A semipermeable membrane must be clearly distinguished from a *one-way membrane* that allows particles to pass through in one direction but not in the other. By consider-

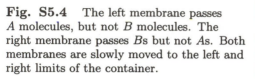

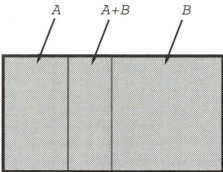

Fig. S5.4   The left membrane passes
A molecules, but not B molecules. The
right membrane passes Bs but not As. Both
membranes are slowly moved to the left and
right limits of the container.

ing the last exercise it is clear that a one-way membrane would violate the Second
Law of thermodynamics and is therefore impossible. Semipermeable membranes,
in contrast, work differently for different molecules, but they work the same way in
both directions. Semipermeable membranes that pass water molecules but not salt
are widely used in saltwater purification plants. Although it is often technically impossible to design a semipermeable membrane for two arbitrarily chosen molecules,
the difficulty does not arise because they involve any violation of thermodynamics.
It is purely an engineering problem. We begin with a pair of semipermeable membranes, positioned together between volumes V_A and V_B, as shown in Fig. S5.4.
The left membrane can pass A particles but not B particles. The right one does
the reverse. The two of them together act as a completely impermeable barrier.
We surround the cylinder by a large reservoir at temperature T. We first gradually
move the left membrane to the far left end of the cylinder. At any intermediate
stage the density of A molecules on both sides of the membrane will be equal since
the membrane freely passes A molecules and the density of B molecules to the right
of the membrane will be N_B/V_R, where V_R is the temporary volume to the right of
the left membrane. By the law of partial pressures, the excess force on the membrane, directed toward the left, will be $F_L = AN_BkT/V_R$. The amount of work
done by this excess force on the external agent that is keeping the membrane in
position is $dW = F\,dx = N_BkT\,dV_R/V_R$. The total work done on the outside agent
is

$$W_L = N_BkT \int_{V_B}^{V} \frac{dV_R}{V_R} = N_BkT \log(V/V_B)$$

In exactly the same way, the mechanical energy absorbed from the system in quasistatically moving the right membrane from its initial position to a position at the
far right is

$$W_R = N_AkT \log(V/V_A) \tag{S5.69}$$

The system within the cylinder began with N_A+N_B particles at temperature T and
therefore with an energy of $(N_A + N_B)kT$. This is also the energy of the gas within
the cylinder in its final state. Thus all the work done on the external agents (that
is, $W_L + W_R$) must have been compensated by thermal energy being conducted
to the gas in the cylinder by the reservoir. But, when a substance at temperature
T absorbs an amount of heat dQ, its entropy changes by $dS = dQ/T$. Since T is
constant throughout the procedure, the total entropy change of the gas within the
cylinder is

$$\Delta S = (W_L + W_R)/T = kN_A \log(V/V_A) + kN_B \log(V/V_B) \tag{S5.70}$$

Because the initial pressures were equal,

$$\frac{N_A}{V_A} = \frac{N_B}{V_B} = \frac{N_A + N_B}{V}$$

which allows us to write ΔS in terms of N_A and N_B alone.

$$\Delta S = kN_A \log\left(\frac{N_A + N_B}{N_A}\right) + kN_B \log\left(\frac{N_A + N_B}{N_B}\right) \qquad (S5.71)$$

The most revealing way to write this is to use Stirling's approximation and rational units to cast it into the form $(N_{AB} = N_A + N_B)$

$$\Delta S^o = \log(N_{AB}!) - \log(N_A!) - \log(N_B!) \qquad (S5.72)$$

Thus we see that, by purely thermodynamic methods, we can rediscover the N factorials that were found to be necessary in Eq. (3.83), based on the microscopic probabilistic interpretation of entropy.

Exercise 5.12   An air conditioner is a device that absorbs heat energy from a cooler reservoir (our room) and dumps it into a hotter reservoir (the outside). A refrigerator does the same thing between its cooler inside volume and our warmer room. Now everyone knows that air conditioners and refrigerators always come with electrical plugs—they never operate without electrical energy input. This may seem like a simple consequence of the law of conservation of energy. With regard to energy, one never gets something for nothing. But a more careful analysis shows that this is not true. If we absorb one joule of energy from inside our room and dump it into the hot air outside, we have completely satisfied the law of energy conservation. Why is it necessary to also take another joule of expensive electrical energy, convert it into useless heat, and dump that outside into the hot air? If it is only due to things like friction in the air conditioner machinery, then one might hope by careful engineering to make an air conditioner that will reduce our electrical bills to something much lower than the very high values that air conditioners currently cause. This exercise will show that it is the Second Law of thermodynamics, rather than energy conservation, that guarantees to the electric company a substantial summer income.

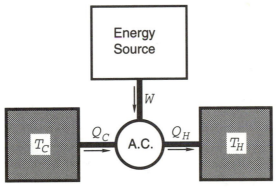

Fig. S5.5   An air conditioner is a device that uses energy in order to pass heat from a lower-temperature reservoir to a higher temperature one.

We model an air conditioner (or a refrigerator) as a device that absorbs heat energy from a cool reservoir at temperature T_C and also absorbs a certain amount of electrical or mechanical work and dumps the energy into a hot reservoir at temperature T_H. We will assume that the whole system, composed of the two reservoirs and the device, is isolated. The device absorbs an amount of heat Q_C from the cool reservoir and an amount of energy W in mechanical or electrical form and transfers an amount of heat Q_H to the hot reservoir. We define the *efficiency* of the device as the ratio

$$E = \frac{Q_C}{W} \tag{S5.73}$$

Clearly, it behooves us, if we do not own the electric company, to make the efficiency as large as possible. Show that, unfortunately, there is a fundamental limit, given by

$$E \leq \frac{T_C}{T_H - T_C} \tag{S5.74}$$

Solution   We will make the reasonable assumption that, after running for a certain time interval, the device itself is in the same state that it was initially. We will also assume that the reservoirs are large enough that their temperature changes are negligible. This assumption simplifies the analysis, but it is not really essential. The entropy change of the whole system is therefore

$$\Delta S = \frac{Q_H}{T_H} - \frac{Q_C}{T_C} \geq 0 \tag{S5.75}$$

where the final inequality reflects the fact that isolated systems only increase their total entropy. This inequality can be expressed as

$$\frac{Q_H}{Q_C} \geq \frac{T_H}{T_C} \tag{S5.76}$$

By energy conservation, $Q_H = Q_C + W$. Therefore

$$\begin{aligned}
E &= \frac{Q_C}{W} \\
&= \frac{Q_C}{Q_H - Q_C} \\
&= \frac{1}{Q_H/Q_C - 1} \\
&\leq \frac{1}{T_H/T_C - 1} = \frac{T_C}{T_H - T_C}
\end{aligned} \tag{S5.77}$$

Exercise 5.13   A *heat pump* is essentially an air conditioner turned around and used to pump heat into the house when the temperature outside is colder than that inside. A *radiant heater* is simply a device inside the house that converts electrical energy completely into heat energy. A typical situation is an inside temperature of 21° C ($\sim$ 70° F) and an outside temperature of 4° C ($\sim$ 39° F). Defining the efficiency of an electrical heater as the ratio of the heat added to the warm reservoir to the electrical power used, compare the best possible efficiencies of a heat pump and a radiant heater in the situation described.

Solution   For a heat pump we have defined the efficiency as

$$E = \frac{Q_H}{W} \tag{S5.78}$$

But $Q_H = W + Q_C$ and, according to the Second Law, $Q_H/T_H \geq Q_C/T_C$. Thus

$$\begin{aligned}
E &= \frac{Q_H}{Q_H - Q_C} \\
&= \frac{1}{1 - Q_C/Q_H} \\
&< \frac{1}{1 - T_C/T_H} = \frac{T_H}{T_H - T_C}
\end{aligned} \tag{S5.79}$$

In this case $T_H = 21 + 273 = 294\,\text{K}$ and $T_C = 4 + 273 = 277\,\text{K}$. Using Eq. (S5.79), we see that the maximum efficiency of the heat pump is 17.3. The efficiency of radiant heater is exactly 1, since $Q_H = W$. Thus, it is theoretically possible for a heat pump to be enormously more efficient than the most common style of electrical heater.

Exercise 5.14   The last two exercises involved devices in which mechanical (or electrical) energy is supplied to move heat between reservoirs in the direction opposite to that in which it would spontaneously flow. A *heat engine* does just the opposite. By allowing heat to move from a hotter to a cooler reservoir, it extracts a fraction of the heat energy and converts it into mechanical work, such as lifting a weight or compressing a spring. A nuclear power plant is an excellent example of a heat engine. The energy of nuclear fission reactions is used to create a high-temperature thermal reservoir within the core of the reactor. A water-driven steam turbine absorbs energy from the high-temperature reservoir, uses some of the energy to drive an electrical generator, and dumps the excess energy as heat into a cool reservoir, called a cooling pond. Again, the most desirable thing would be to convert 100% of the heat energy to electrical energy, but we will see that, on top of the necessary imperfections in steam turbines and electrical generators, there is a fundamental limitation, due to the Second Law, on the fraction of nuclear energy that can be converted to electrical power.

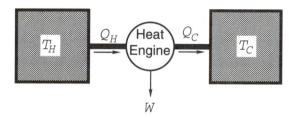

Fig. S5.6   A heat engine absorbs an amount of energy Q_H from a high-temperature reservoir, converts an amount W of it into mechanical energy, and dumps the rest into a low-temperature reservoir.

For a heat engine, the natural definition of efficiency is the ratio of mechanical or electrical work extracted to the thermal energy absorbed from the high-temperature source.

$$E = \frac{W}{Q_H} \tag{S5.80}$$

The aim of this exercise is to show that

$$E \le 1 - \frac{T_C}{T_H} \tag{S5.81}$$

Solution   In this case, after the device has run for a period of time, an amount of thermal energy Q_H has been extracted from the high-temperature reservoir, part of it W has been converted to mechanical energy, and the excess, $Q_C = Q_H - W$, has been transferred to the low-temperature reservoir. The change in entropy of the system, which must be nonnegative, is

$$\Delta S = \frac{Q_C}{T_C} - \frac{Q_H}{T_H} \ge 0 \tag{S5.82}$$

This means that

$$\frac{Q_C}{Q_H} \ge \frac{T_C}{T_H} \tag{S5.83}$$

and

$$E = \frac{Q_H - Q_C}{Q_H} = 1 - \frac{Q_C}{Q_H} \le 1 - \frac{T_C}{T_H} \tag{S5.84}$$

Exercise 5.15   Assuming that one could produce a cylinder with a frictionless piston, describe a heat engine that would actually achieve the theoretically maximum efficiency, that is, one that would satisfy Eq. (S5.84) as an equality. Such an engine is called a *Carnot* engine, after Sadi Carnot, the originator of thermodynamics.

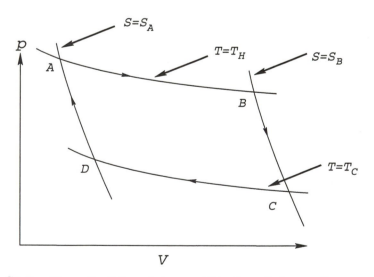

Fig. S5.7   The path of the substance within the cylinder, on the p-V plane, as it is taken through the cyclic process described in the exercise, is composed of two adiabats and two isotherms.

Solution   Our system consists of two large thermal reservoirs at temperature T_H and T_C plus a cylinder filled with a gas (not necessarily ideal), closed by a friction-less piston and with a heat-conducting upper end. We assume that, in its initial state, the gas has volume V_o and temperature T_H. We must assume that we have investigated the properties of the gas previously so that we can draw curves of constant temperature (isotherms) and curves of constant entropy (adiabats), for the sample of gas in the cylinder, on the p-V plane. We first draw the T_C and T_H isotherms (see Fig. S5.7). Since the intial state of the gas has temperature T_H, it is some point A on the high-temperature isotherm. We draw the adiabat through A and call the point at which it intersects the low-temperature isotherm D. We now put the cylinder in contact with the T_H reservoir and allow the gas to expand extremely slowly so that it will remain, during its expansion, at temperature T_H. Its thermodynamic state will therefore move down the T_H isotherm. We stop the expansion at some arbitrary point B. We now disconnect the cylinder from the reservoir, so that it is thermally isolated, and again allow it to expand very slowly. As we saw in Section 5.11, it will now move along an adiabat. We stop the expansion when the temperature reaches T_C. We place the cylinder in contact with the low-temperature reservoir and very gradually compress the gas, moving along the T_C isotherm. We stop when we have arrived at the previously determined point D. We again isolate the cylinder and gradually compress the gas, moving along the adiabat to our intial state A.

Let us now see how much work has been done on the external agent that is controlling the piston. Each time the piston moves a small amount, the work done on the external agent is $F\,dx = p\,dV$. Thus, in moving along the isotherm from A to B, the total work done is

$$W_{AB} = \int_A^B p(v)\,dV \tag{S5.85}$$

But this is exactly the area under the isothermal curve in the p-V plane. The same thing is true for the curve from B to C. In going from C to D and from D to A, it is the external agent that must do work in order to compress the gas. It is easy to see that the net work done on the external agent (that is, the net work that has been extracted from the system) is just the area in the p-V plane bounded by the four curves.

In going from A to B, the heat energy absorbed from the hot reservoir is

$$Q_{AB} = T_H(S_B - S_A) \tag{S5.86}$$

But, by energy conservation, the difference in the energy states of the gas at A and B must be

$$E_B - E_A = Q_{AB} - W_{AB} \tag{S5.87}$$

In going from B to C, the work done by the gas must equal its loss in internal energy.

$$E_B - E_C = W_{BC} \tag{S5.88}$$

Using similar arguments (W_{CD} is the work done *by* the external agent and Q_{CD} is the heat energy given *to* the cold reservoir),

$$Q_{CD} = T_C(S_B - S_A)$$
$$E_D - E_C = W_{CD} - Q_{CD} \tag{S5.89}$$
$$\text{and}\quad E_A - E_D = W_{DA}$$

The net work over the cycle is given by

$$
\begin{aligned}
W &= W_{AB} + W_{BC} - W_{CD} - W_{DA} \\
&= Q_{AB} - Q_{CD}
\end{aligned}
\tag{S5.90}
$$

The efficiency is defined as

$$
\frac{W}{Q_{AB}} = 1 - \frac{Q_{CD}}{Q_{AB}} = 1 - \frac{T_C}{T_H}
\tag{S5.91}
$$

Two things should be noted about this Carnot engine. First, because all the processes must be carried out extremely slowly, it is not, in spite of its maximal efficiency, a practical design for an industrial engine. The more important point is that every step in the cycle is completely reversible. It we traverse the path in the p-V plane in the opposite sense, namely $ADCBA$, then the Carnot engine acts as a refrigerator, absorbing an amount of work W and transferring an amount of heat Q_{AB} to the hotter reservoir. It is easy to verify that, as a refrigerator, it also has the maximum possible efficiency. That it is reversible is a direct consequence of the fact that the complete cycle does not change the sum of the entropies of the two reservoirs.

Exercise 5.16   The Berthelot equation of state, mentioned in Exercise 5.3, can be expressed in the form

$$
p = \frac{kT}{v - v_o} - \frac{a}{kTv^2}
\tag{S5.92}
$$

The fact that it contains two constants, v_o and a, that are different for different gases, is inconvenient for drawing general conclusions about systems that satisfy the equation. This contrasts with the ideal gas equation, which has the same form, free of arbitrary constants, for all substances that are adequately described by it. Show that, by measuring volumes, pressures, temperatures, and energies in appropriate units, the equations of state of all substances that satisfy the conditions described in Exercise 5.3 can be brought into the simplified universal forms

$$
\tilde{p} = \frac{1}{\tilde{v} - 1} - \frac{\tilde{\beta}^2}{\tilde{v}^2}
\tag{S5.93}
$$

and

$$
\tilde{E} = \frac{5}{2\tilde{\beta}} - \frac{2\tilde{\beta}}{\tilde{v}}
\tag{S5.94}
$$

Collections of substances whose equations of state can all be brought into a unified form by such rescaling procedures are said to satisfy a *law of corresponding states*.

Solution   It is best to make the necessary transformations on the canonical potential, from which both equations of state immediately follow. It has been shown that the canonical potential per particle is

$$
\phi = \frac{a\beta^2}{v} + \log(v - v_o) - \tfrac{5}{2}\log\beta + \text{const.}
\tag{S5.95}
$$

By extracting $-\log v_o$ from the constant term, which would have no effect on the equations of state, we can rewrite this as

$$\phi = \frac{a}{v_o}\frac{\beta^2}{\tilde{v}} + \log(\tilde{v} - 1) - \tfrac{5}{2}\log\beta + \text{const.} \qquad (S5.96)$$

where $\tilde{v} = v/v_o$. By extracting $-\tfrac{5}{2}\log\left(\sqrt{a/v_o}\right)$ from the constant term and defining a scaled inverse temperature $\tilde{\beta} = \sqrt{a/v_o}\beta$, we can eliminate all the arbitrary constants in ϕ.

$$\phi = \frac{\tilde{\beta}^2}{\tilde{v}} + \log(\tilde{v} - 1) - \tfrac{5}{2}\log\tilde{\beta} + \text{const.} \qquad (S5.97)$$

The equations of state are

$$E = -\frac{\partial\phi}{\partial\beta} = \left(\frac{v_o}{a}\right)^{1/2}\frac{\partial\phi}{\partial\tilde{\beta}} = \left(\frac{v_o}{a}\right)^{1/2}\left(\frac{5}{2\tilde{\beta}} - \frac{2\tilde{\beta}}{\tilde{v}}\right) \qquad (S5.98)$$

and

$$\beta p = \frac{\partial\phi}{\partial v} = v_o\frac{\partial\phi}{\partial\tilde{v}} = v_o\left(\frac{1}{v_o - 1} - \frac{\tilde{\beta}^2}{\tilde{v}^2}\right) \qquad (S5.99)$$

Further defining the dimensionless energies and pressures by

$$\tilde{E} = \frac{E}{\sqrt{v_o/a}} \quad \text{and} \quad \tilde{p} = \frac{p}{\sqrt{v_o a}} \qquad (S5.100)$$

we obtain dimensionless equations of state valid for all diatomic Berthelot gases.

$$\tilde{E} = \frac{5}{2\tilde{\beta}} - \frac{2\tilde{\beta}}{\tilde{v}} \qquad (S5.101)$$

and

$$\tilde{p} = \frac{1}{\tilde{\beta}(\tilde{v} - 1)} - \frac{\tilde{\beta}}{\tilde{v}^2} \qquad (S5.102)$$

Letting $\tilde{\beta} = 1/\tilde{\tau}$, where $\tilde{\tau}$ is related to the Kelvin temperature by $\tilde{\tau} = \sqrt{v_o/a}\,kT$, we can write these equations in terms of the dimensionless temperature variable $\tilde{\tau}$ as

$$\tilde{E} = \frac{5}{2}\tilde{\tau} - \frac{2}{\tilde{\tau}\tilde{v}} \qquad (S5.103)$$

and

$$\tilde{p} = \frac{\tilde{\tau}}{\tilde{v} - 1} - \frac{1}{\tilde{\tau}\tilde{v}^2} \qquad (S5.104)$$

Exercise 5.17 If, for notational simplicity, the tildes are dropped in the dimensionless Berthelot pressure equation, then it reads

$$p = \frac{\tau}{v - 1} - \frac{1}{\tau v^2} \qquad (S5.105)$$

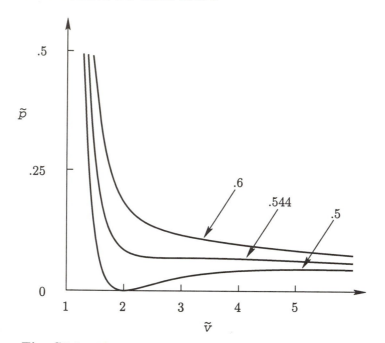

Fig. S5.8   Isotherms of the dimensionless Berthelot equation. The values of the dimensionless temperature τ are shown.

Shown in Fig. S5.8 are the isotherms corresponding to the dimensionless temperatures, $\tau = 0.5$, 0.544, and 0.6. Show that, for $\tau < 0.544$, there is a region in which the equation describes a substance that is unstable with respect to separation into two uniform phases of different densities.

Solution   In Exercise 5.7 it was shown that the conditions of stability with respect to phase separation were that $(\partial T/\partial \varepsilon)_v > 0$ and $(\partial p/\partial v)_T < 0$. Since the isotherms shown in the figure are plots of p at constant T, the fact that, below the *critical temperature* $\tau = 0.544$ (which will be calculated in a later exercise) they have sections of positive slope, indicates that the Berthelot equation of state cannot be taken as a valid description of any uniform phase for those values of T and v.

Exercise 5.18   In Fig. S5.9 are plotted the p-v isotherms for a typical simple substance such as neon or oxygen. The states within the region outlined by the dashed curve are two-phase states in which the substance is part liquid and part gas. A sample that begins in a liquid state at point A and is maintained at constant temperature by being kept in thermal contact with a reservoir while its volume is gradually increased will smoothly expand as a pure liquid until it reaches point B. Any further expansion will cause the liquid to evaporate (or boil, depending on the rate and details of the process), forming a state composed of two phases, liquid and gas, in equilibrium. As we saw in Section 5.10, a two-phase state of a simple substance has a pressure that is a unique function of its temperature. Therefore, the portion of the isotherm that connects the pure liquid state on the left segment of the dashed line with the pure gas phase on the right segment is perfectly horizontal, as it has been drawn. Once the substance has completely vaporized, reaching point C on the isotherm, any further expansion of the purely gaseous sample will cause a steady reduction in pressure. It often happens that one has reasonably accurate simple analytic expressions for the thermodynamic functions of the liquid phase and different analytic expressions for the thermodynamic functions of the gas phase. Choosing a temperature, let us say $T = T_o$, one could separately use these expressions to draw a liquid isotherm and a gas isotherm. One is then faced with the problem of where to draw the horizontal portion, connecting the two isotherms and representing the two-phase states. The question to be answered here is: What is the additional condition that determines the location of the two-phase instability and thus the locus of the dashed line in the figure?

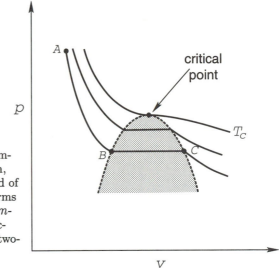

Fig. S5.9   The p-V isotherms of a simple substance. Within the shaded region, the states of the substance are composed of a mixture of liquid and gas. The isotherms for T greater or equal to the *critical temperature* T_c do not have a two-phase section. The critical isotherm touches the two-phase region at the *critical point*.

Note: Statistical mechanical arguments indicate that the *exact* thermodynamic functions have very subtle nonanalyticities at the location of the two-phase boundary that could, if one had an exact representation of the thermodynamic functions, be used to locate those boundaries. However, the deviations from analytic behavior as one approaches the two-phase boundary are so subtle that, to the knowledge of the author, they have never been experimentally detected. That is, when a pure gas sample in a very clean container, free of dust specks or other potential condensation sites, is slowly compressed at constant temperature beyond the dashed stability line, it does not immediately undergo a phase transition but instead moves along an apparently smooth extension of the gas-phase isotherm into a uniform undercooled state. It is undercooled in the sense that at that pressure the only stable gas phases are at higher temperatures. If the substance is maintained in that unstable uniform state, eventually something, such as a cosmic-ray ionization track or a spontaneous density fluctuation, will trigger the transition to the stable two-phase state. In careful experiments, however, "eventually" can be a very long time. The same phenomenon occurs on the other side of the two-phase region when a uniform liquid is carefully expanded beyond its point of stability.

Solution   As was pointed out in Section 5.10, the equilibrium between two simple phases requires that three conditions be fulfilled. The two phases must have equal pressures, temperatures, and affinities. Since we are discussing gas and liquid isotherms at the same temperature T_o, the temperature equality condition is satisfied trivially. The pressure equality condition simply means that the two-phase line must be drawn horizontally, which we have already noted. It does not tell us where it must be drawn. It is the condition that $\alpha_L = \alpha_G$ that determines the exact position of the phase transition. As we move down the liquid isotherm toward lower-density states, the affinity for particles steadily increases. Conversely, as we move up the gas isotherm toward higher-density states the affinity falls. At some pressure the affinities of the gas and liquid phases are equal. That is the location of the horizontal phase transition line.

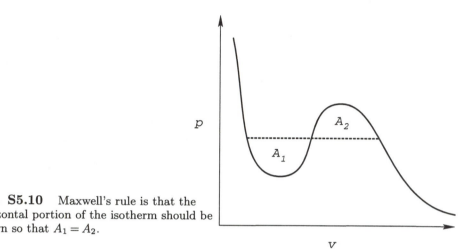

Fig. S5.10 Maxwell's rule is that the horizontal portion of the isotherm should be drawn so that $A_1 = A_2$.

Exercise 5.19 Because it exhibits isotherms with positive slope in the p-v plane, the Berthelot equation, discussed in Exercise 5.17, cannot possibly describe the equilibrium states of a substance for all values of p and v. Maxwell devised an argument that allows one to analyze liquid–gas phase transitions using empirical equations of state such as the Berthelot equation. It must be emphasized that Maxwell's scheme is an approximate method, not a true fundamental theory of phase transitions. Maxwell assumed that there is a two-phase region in which the isotherms are horizontal and that within that region the empirical equation of state is not valid. This assumption is certainly true. The essential approximation of Maxwell's method is the assumption that the single empirical equation of state *is* valid everywhere outside the two-phase region. Show that Maxwell's assumption implies that the horizontal portion of the isotherm satisfies the geometrical condition $A_1 = A_2$, where the areas A_1 and A_2 are shown in Fig. S5.10. Hint: Assume that the entropy function associated with the empirical equation of state, although it does not have the proper convexity properties, is at least an extensive function. (Certainly, no empirical equation that led to a nonextensive entropy function would ever be taken seriously.)

Solution In the last exercise we noted that the condition determining the location of the two-phase line is that $\alpha_L = \alpha_G$. But, by Eq. (5.46), along an isotherm

$$d\alpha = -\beta v \, dp \tag{S5.106}$$

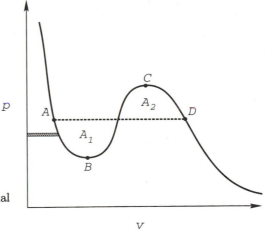

Fig. S5.11   The shaded area is equal to $-v\,dp$.

Also, this formula was derived using only the fact that the entropy was extensive. The derivation did not rely on any convexity properties of the entropy. Therefore we can expect this relation to be satisfied by the empirical equation, even in the region where that equation does not represent the true properties of the substance. Since β is a constant on the isotherm and $\alpha_L = \alpha_G$ at the limits of the two-phase region, Eq. (S5.106) implies that

$$\beta^{-1}\int_A^D d\alpha = -\int_A^D v\,dp = 0 \qquad (S5.107)$$

(See Fig. S5.11.) But, from A to B, $-v\,dp$ is a positive number equal to the area element shown in the figure. From B to C, $-v\,dp$ is negative, and, from C to D, $-v\,dp$ is again positive. It is easy to see that the full integral is given by

$$-\int_A^D v\,dp = A_2 - A_1 \qquad (S5.108)$$

which gives Maxwell's geometrical construction.

Exercise 5.20   The highest point of the two-phase region in the p-v plane is the liquid–gas critical point. The temperature of the isotherm passing through that point is the critical temperature T_c. The isotherms at higher temperature do not pass through the two-phase region. Therefore, there is no liquid–gas phase transition at temperatures higher than T_c. For a substance describable by the dimensionless Bertholet equation (with the Maxwell construction), determine τ_c.

Solution   At the critical point on the critical isotherm both $\partial p/\partial v$ and $\partial^2 p/\partial v^2$ are zero. This gives the following two simultaneous equations for v_c and τ_c.

$$\frac{\partial p}{\partial v} = -\frac{\tau_c}{(v_c - 1)^2} + \frac{2}{\tau_c v_c^3} = 0 \qquad (S5.109)$$

and

$$\frac{\partial^2 p}{\partial v^2} = \frac{2\tau_c}{(v_c - 1)^3} - \frac{6}{\tau_c v_c^4} = 0 \qquad (S5.110)$$

These can be rewritten as

$$\frac{\tau_c}{(v_c - 1)^2} = \frac{2}{\tau_c v_c^3} \tag{S5.111}$$

and

$$\frac{\tau_c}{(v_c - 1)^3} = \frac{3}{\tau_c v_c^4} \tag{S5.112}$$

Dividing the two sides of the first equation by the corresponding sides of the second gives a linear equation for v_c.

$$v_c - 1 = \frac{2}{3} v_c \tag{S5.113}$$

or $v_c = 3$. Putting this value into Eq. (S5.111) gives

$$\tau_c = \sqrt{8/27} = 0.544 \tag{S5.114}$$

Classical Thermodynamics

Our presentation of thermodynamics has deviated greatly from the traditional one. Its aim has been to bring together, under the title of thermodynamics, all those macroscopic features of systems that appear in the thermodynamic limit and that are not obvious consequences of the laws of mechanics or quantum mechanics. For the benefit of the reader who has not been exposed to a more standard treatment of thermodynamics, the next few exercises will be devoted to a very brief version of the concepts and methods of classical thermodynamics.

In the usual treatment of the subject, the first law of thermodynamics is the statement that heat is a form of energy, or, put differently, that the sum of the heat extracted from a body plus the mechanical work done by the body on any outside systems is equal to the decrease in the body's internal energy. During the period of time when thermodynamics was being developed, the proposition that the well-established but limited law of the conservation of mechanical energy, which was valid only in the absence of friction or other dissipative influences, could be extended into an absolute and universally applicable conservation principle was a daring and far from generally accepted idea. The fact that radiant energy could propagate through space devoid of matter meant that this universal "energy" could not include only the well-understood kinetic and potential energies of mechanics but must contain a radiant energy term. One must remember that, at the time, no electromagnetic theory encompassing radiant energy existed. In fact, the particulate nature of matter was still in serious dispute. However, in this day, when every student has seen photographs of individual atoms, when the general conservation of energy is an established theorem of mechanics, electromagnetics, and quantum theory, to include energy conservation as a physical postulate of thermodynamics, rather that treating it as a theorem of the logically prior subjects of mechanics and quantum mechanics, gives to the subject a peculiarly antique character. It seems, to most students, like a study of the astrolabe or Ptolemaic astronomy. Certainly the Pythagorean theorem is essential to almost any calculation in Newtonian mechanics. But, for good reasons, it is not considered to be a law of mechanics. That theorem, along with all of Euclidean geometry, is assumed as prior knowledge at the very start of an introduction to mechanics. In the same way, even in this "traditional" presentation of thermodynamics, which will be based on the postulates of Clausius and Kelvin, energy conservation will simply be taken as prior assumed knowledge.

We begin by postulating the existence of a large collection of substances, each separately isolated and in internal equilibrium. Our aim is to study what happens when these substances are brought into interaction with one another. The substances will be referred to as objects. It must be kept in mind that, at the start, no temperature scale of any kind has been defined.

Given two objects A and B, we define a *cyclic process* for transferring thermal energy from A to B as any means of doing so that, when it is completed, leaves everything else in the world (except A and B) in its original state. The phrase suggests some sort of cyclic machine that keeps coming back to the same state, and many particular examples are of that form. However, a much more trivial example of a cyclic process is the process of simply putting the objects into contact for a while and then separating them. Using the notion of a cyclic process, we now make an important definition. For any two objects, we say that "A is hotter than B" if any cyclic process exists that can transfer heat energy from A to B.

The central law of classical thermodynamics, in a form due to Rudolf Clausius, can now be given as the seemingly trivial statement: "If A is hotter than B, then B is not hotter than A." The appearance of triviality is entirely misleading. It might be compared with the apparent triviality of the axioms of Euclid that were carefully constructed so that, by stepwise logical deduction, an edifice of astonishing sophistication could be built upon them. First, let us convince ourselves that Clausius' statement of the second law is actually a strong statement with nonobvious consequences. It says, for instance, that if, when we place objects A and B in contact, heat spontaneously flows from A to B, then it is impossible to construct any device, no matter how ingenious and complicated, that will have the sole effect of transferring energy in the reverse direction (from B to A). The important restriction is expressed by the phrase "sole effect," which means that the device will leave everything else in the world, including itself, unchanged. That the second law is not trivial is well illustrated by a recent personal experience of the author. A few years ago I was shown a paper that a colleague had been asked to review for a prestigious physics journal. The paper had been submitted to the journal by a professor of physics at a respected American university. It described a device that purported to defeat the second law of thermodynamics by using magnetic fields. The reviewer revealed the flaws in the paper's reasoning and it was finally rejected for publication. But, although a century and a half has passed since the wide dissemination of the second law of thermodynamics, a steady stream of papers, mostly by clearly incompetent authors, that describe devices or processes in violation of the second law, are still submitted to scientific journals. Einstein expressed a widely held view of the second law of thermodynamics when he said that, of all the fundamental laws of physics, it is probably the one that has the most secure foundation. The reasoning behind this view is that, since it is really an expression of the law of large numbers, it has the character of a mathematical theorem rather than an empirically derived physical law that might have to be modified in the light of more careful experiments in the future.*

The basic arguments, beginning from the second law and leading to the defi-

---

* However, no law devised by man is ever completely secure. One of the predictions of Einstein's own theory of general relativity is the existence of black holes. When a chunk of matter falls into a black hole, the matter, along with all its degrees of freedom and possible quantum states, effectively disappears from the universe, taking some entropy with it. There is a corresponding increase in the radius of the black hole. In the presence of black holes, the law of increase of entropy must be modified so that it is the thermodynamic entropy plus some constant times the sum of the radii of all the black holes that must

nition of temperature and entropy, will be given as a series of exercises.

Exercise 5.21   Prove that, for any three objects, the statements "A is hotter than B" and "B is hotter than C" imply that A is hotter than C.

Solution   The proof is relatively trivial. If A is hotter than B, then a quantity of heat ΔQ can be transferred by a cyclic process from A to B. But, because B is hotter than C, that same quantity of heat could then be transferred from B to C. Since B is left in its initial state, the net effect is a cyclic process that has transferred heat from A to C.

Given two objects, A hotter than B, a *heat engine* operating between A and B is any device that absorbs an amount of heat energy ΔQ_A from A, uses some of the energy to do mechanical work ΔW, and dumps the remaining energy, $\Delta Q_B = \Delta Q_A - \Delta W$, into object B. It is convenient to picture the mechanical work as simply the lifting of a weight in a gravitational field. The fact that we have assumed that $\Delta Q_A = \Delta Q_B + \Delta W$ means that we have neglected friction or other dissipative processes within the heat engine. Therefore, it is somewhat of an idealized device. However, there is no fundamental limit to how close we can come to an ideal heat engine and it is such a useful concept that we will simply ignore all the technical complications that would occur if we restricted ourselves to truly realizable processes. A special category of heat engines are *reversible engines*. These are heat engines that can be operated in both directions. That is, by absorbing the heat energy ΔQ_B from the colder object and adding to it the mechanical work ΔW, it can transmit the heat energy $Q_A = Q_B + \Delta W$ to the hotter object. In this analysis we will just postulate the existence of these idealized reversible engines. In previous exercises we have shown that, ignoring internal friction, and using slow quasistatic processes, it is possible to construct such reversible engines.

Exercise 5.22   Given any heat engine operating between two objects, A hotter than B, we define the *efficiency ratio* of the heat engine as $R(A, B) = \Delta Q_A/Q_B$. (Note that this is not the definition of the *efficiency* used in Exercise 5.14.) A heat engine that could convert all the heat energy Q_A into work would have an efficiency ratio of infinity, something that we will show is impossible. Using the second law, prove that all reversible engines operating between the same two objects have the same efficiency ratio.

Solution   Suppose that two reversible heat engines, operating between objects A and B, have efficiency ratios $R(A, B)$ and $R'(A, B)$ with $R' < R$ (Fig. S5.12). We run the first one in the forward direction, extracting heat energy Q_A from A, storing mechanical energy ΔW, and dumping heat energy Q_B into B. We now use the stored energy ΔW to run the second heat engine in reverse, absorbing heat energy Q'_B from B and transmitting heat energy Q'_A into A. By conservation of energy,

$$\Delta W = Q_A - Q_B = Q'_A - Q'_B \tag{S5.115}$$

increase. Without a foundation in general relativity, we will have to ignore such, somewhat speculative, modern developments in thermodynamics.

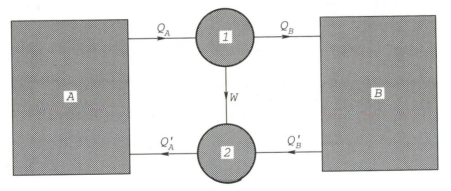

Fig. S5.12   If the efficiency ratios of the two reversible heat engines are not equal, then it is possible to use the pair of engines to transfer heat from a cooler to a hotter body.

But $Q_A = RQ_B$ and $Q'_A = R'Q'_B$, which says that

$$(R - 1)Q_B = (R' - 1)Q'_B \qquad (S5.116)$$

or

$$Q'_B = \frac{R - 1}{R' - 1}Q_B > Q_B \qquad (S5.117)$$

Thus the net effect has been to extract heat energy $Q'_B - Q_B$ from the colder object. Since no net work has been done, that heat energy must have been transferred to the hotter object, in violation of the second law. Therefore, all reversible heat engines operating between two given bodies have the same value of Q_A/Q_B.

Exercise 5.23   Given any three objects, with A hotter than B hotter than C, prove that

$$R(A, B)R(B, C) = R(A, C) \qquad (S5.118)$$

Solution   The argument needed is very similar to the one used in Exercise 5.20. We run a reversible engine between A and B in the forward direction, extracting heat Q_A from A and dumping heat Q_B into B. We then run a reversible engine between B and C, extracting the heat Q_B from B (which leaves B in its original state) and dumping heat Q_C into C. But the combination of the two processes constitutes a reversible heat engine between A and C with an efficiency ratio

$$\frac{Q_A}{Q_C} = \frac{Q_A}{Q_B}\frac{Q_B}{Q_C} \qquad (S5.119)$$

Exercise 5.24   The notion of "hotter" has been defined, but we still have no numerical scale of hotness. However, the result of the previous exercise provides a simple method of defining a convenient temperature scale that does not depend on the properties of any particular substance. A temperature scale is a set of numbers, assigned, one to each object, so that $T_A > T_B$ means that A is hotter than B. The thermodynamic temperature scale is defined in two steps. (a) We first declare that, given any two objects A and B, the ratio of their temperatures is just their efficiency ratio. That is

$$\frac{T_A}{T_B} = R(A, B) = \frac{Q_A}{Q_B} \tag{S5.120}$$

This allows the temperature ratio of any two objects to be measured by running any reversible engine between them. The result of the previous exercise guarantees that the ratios so obtained are mathematically consistent. That is, that

$$\frac{T_A}{T_B}\frac{T_B}{T_C} = \frac{T_A}{T_C} \tag{S5.121}$$

(b) The second step is completely trivial. We simply choose some convenient object, say O, and assign to it an arbitrary positive number T_o as its temperature.

Knowing one temperature and all ratios is obviously enough to assign temperatures to every object. Because a reversible engine always extracts more heat from the hotter reservoir than it delivers to the colder (or else the second law could easily be violated), we are assured that $T_A > T_B$ implies that A is hotter than B.

Exercise 5.25   There is another formulation of the second law, due to Kelvin. It postulates that it is impossible to construct any device whose sole effect would be to absorb heat energy from an object and convert it completely into mechanical work, such as lifting a weight. It is always possible to convert stored mechanical energy into the internal thermal energy of any chosen object. For example, one can use the mechanical energy to generate an electrical current through the object that will increase the object's thermal energy no matter how hot it was initially. Use this fact to show that the Kelvin and Clausius statements of the second law are logically equivalent.

Solution   Assume that Kelvin's law is false. Then we can construct an engine that will absorb an amount of heat Q from an object and use it to lift a weight, increasing its potential energy by an amount $\Delta W = Q$. But then we can take that mechanical energy and use it to heat up an object that is hotter than the original one. Thus the falsity of Kelvin's law implies the falsity of Clausius' law.

Now assume that Clausius' law is false. Then we can transfer heat Q from a colder to a hotter object. But, by running a reversible engine in the forward direction between the two objects, we could store up mechanical energy while removing from the hotter object the heat Q that had been transferred to it. The net result of the process would be to absorb heat from the colder body and convert it entirely into mechanical work. Thus the falsity of Clausius' law implies the falsity of Kelvin's law. Either both are true or both are false. That is the definition of logical equivalence.

The most remarkable discovery of Clausius was that *reversible* processes conserve a completely new physical property that he called the *entropy*. Since, at the time, there was not the slightest hint of the role played by probability and statistics in macroscopic phenomena, the discovery and precise definition of entropy by purely macroscopic, mechanical arguments was a work of tremendous genius. It was only many years later that Ludwig Boltzmann discovered the probabilistic meaning of the thermodynamic entropy.

Exercise 5.26   We now want to define the entropy and to show that it is conserved by reversible processes and increased by irreversible ones. Like the potential energy, the entropy of any object is only defined within an arbitrary additive constant. Therefore, beginning with all our objects in some particular equilibrium states, we assign to each an arbitrary entropy, S_{oA}, S_{oB}, ..., for that particular state. Now, if the object has a temperature T and it absorbs an amount of heat Q, then, by definition, its entropy changes by an amount $\Delta S = Q/T$. (a) Show that reversible processes conserve the total entropy of the system. (b) Show that the total entropy is not a conserved quantity by showing that it can be increased by cyclic but irreversible processes. (c) Show that no cyclic process can lower the total entropy.

Solution   (a) Let us run a reversible engine in the forward direction (the same result can be obtained by running it in the reverse direction) between two arbitrary objects, A and B, at temperatures T_A and T_B, with $T_A > T_B$. Then the heat extracted from A, namely Q_A, is related to the heat transmitted to B by

$$\frac{Q_A}{Q_B} = \frac{T_A}{T_B} \tag{S5.122}$$

or

$$\frac{Q_A}{T_A} = \frac{Q_B}{T_B} \tag{S5.123}$$

Since heat is extracted from A and added to B, $\Delta S_A = -\Delta Q_A/T_A$ and $\Delta S_B = Q_B/T_B$. Thus

$$\Delta S_A + \Delta S_B = 0 \tag{S5.124}$$

(b) Transferring heat by means of a reversible engine is not a cyclic process because the engine stores mechanical energy and therefore does not come back to its initial state. Suppose we transfer heat by a cyclic process from A to B (recall that the simplest cyclic process is just putting A and B in thermal contact for a while). In a cyclic process the heat lost by A must all be gained by B. Thus $Q_A = Q_B$. Then

$$\Delta S_A + \Delta S_B = Q_A\left(\frac{1}{T_B} - \frac{1}{T_A}\right) > 0 \tag{S5.125}$$

(c) We will consider a cyclic process involving three objects, A, B, and C, but it should be clear that the same analysis could be extended to any number of objects. Let Q_A be the heat energy added to or extracted from object A. If Q_A is positive, then heat has been added to A. If it is negative, then heat has been extracted. Since the process is cyclic,

$$Q_A + Q_B + Q_C = 0 \tag{S5.126}$$

Now we assume that the process has lowered the total entropy. Then

$$\frac{Q_A}{T_A} + \frac{Q_B}{T_B} + \frac{Q_C}{T_C} < 0 \tag{S5.127}$$

We now run a reversible engine between B and C that makes a change $-Q_C$ in the energy of C and a change Q'_B in B. Because it is reversible

$$\frac{Q'_B}{T_B} - \frac{Q_C}{T_C} = 0 \tag{S5.128}$$

This reversible process has either stored or used up some mechanical work. The net effect has been to leave C in its original state. Since the reversible process makes no change in the entropy, the net process gives an entropy change

$$\frac{Q_A}{T_A} + \frac{Q_B + Q'_B}{T_B} < 0 \tag{S5.129}$$

We now run a reversible engine between A and B that absorbs an amount of heat $Q_B + Q'_B$ from B and transmits an amount of heat $\Delta Q'_A$ to A. Again

$$\frac{Q'_A}{T_A} - \frac{Q_B + Q'_B}{T_B} = 0 \tag{S5.130}$$

Now B and C are in their original states. The total entropy change of the system is

$$\frac{Q_A + Q'_A}{T_A} < 0 \tag{S5.131}$$

But T_A is a positive number. Thus the total change in the thermal energy of A is

$$Q_A + Q'_A < 0 \tag{S5.132}$$

Thermal energy has been extracted from A and must therefore be stored as mechanical work in the reversible engines used in the last two stages. This violates Kelvin's law.

Certainly this set of exercises does not constitute a complete and adequate treatment of classical thermodynamics. They have been intended only to show the style of the subject and to demonstrate the ingenious arguments constructed by the early thermodynamicists in bridging the gap between the impossibility postulates, expressed in words, and a detailed mathematical structure that could be used to analyze and predict the properties of real substances and devices.

Supplement to Chapter 6

REVIEW QUESTIONS

6.1 For the chemical reaction $A_2 + 2B \leftrightarrow 2AB$ derive the equilibrium condition relating the affinities of A_2, B, and AB.

6.2 For the reaction shown above, derive the law of mass action.

6.3 For a reaction of the form $aA + bB \leftrightarrow rR$, derive Van't Hoff's law, $\partial \log K / \partial \beta = W$.

6.4 What does Van't Hoff's Law say about the temperature dependence of the equilibrium constant for an exothermic (energy-releasing) reaction?

6.5 For an ideal gas, how is the partition function of the gas Z related to the partition function of a single molecule?

6.6 In the formula for the partition function of a diatomic molecule, $z = z(\text{trans}) \times z(\text{int})$, what are $z(\text{trans})$ and $z(\text{int})$?

6.7 Derive $z(\text{int})$ for a diatomic molecule, using the usual approximations.

6.8 What is $z(\text{int})$ for an atom?

6.9 For the dissociation reaction $AB \leftrightarrow A + B$, derive the equilibrium constant K.

6.10 Derive the Saha equation for the ionization of hydrogen.

6.11 Write down the formula for dE in the energy representation.

6.12 Define the Helmholtz free energy $F(T, V)$, the enthalpy $H(S, p)$, and the Gibbs free energy $G(T, p)$.

6.13 Write equations for dF, dH, and dG.

6.14 Show that $(\partial V / \partial T)_p = -(\partial S / \partial p)_T$.

6.15 Define the constant pressure specific heat C_p, the coefficient of thermal expansion β_p, and the compressibility κ_T.

6.16 Show that $(\partial x / \partial y)_z = -(\partial z / \partial y)_x / (\partial z / \partial x)_y$.

6.17 How can we eliminate H in an expression of the form $(\partial H / \partial x)_y$?

6.18 How do we transform $(\partial S / \partial p)_V$ into an expression involving derivatives of the form $\left[\partial(S \text{ or } V)/\partial(p \text{ or } T)\right]_{(T \text{ or } p)}$?

6.19 Show that $C_p - C_v = TV\beta_p^2/\kappa_T$.

6.20 Explain the meaning of the terms in the Clausius–Clapeyron equation, $\partial p/\partial T = (S_G - S_L)/(V_G - V_L)$.

6.21 Derive the Clausius–Clapeyron equation.

6.22 What approximations are necessary in deriving the equation $\log p(T) = \text{const.} - L/RT$ for the two-phase equilibrium pressure?

6.23 Using a system composed of a dielectric capacitor and a vacuum capacitor, show that $E = \partial f/\partial D$, where $f(n, T, D)$ is the free energy of the dielectric.

6.24 What is the corresponding relation for magnetic fields?

6.25 Assuming that $E \sim D$, show that $f(n, T, D) = f_o(n, T) + D^2/2\epsilon$.

EXERCISES

Exercise 6.1 Both bromine and chlorine are highly soluble in water. In an aqueous solution of both gases the electron transfer reaction

$$Cl_2 + 2\,Br^- \leftrightarrow 2\,Cl^- + Br_2 \tag{S6.1}$$

has an equilibrium constant $K = 4 \times 10^4$. What is the ratio of the concentration of chlorine ions to that of bromine ions if the concentrations of chlorine and bromine molecules are equal?

Solution The mass action formula for the reaction is

$$\frac{n_{Cl^-}^2\, n_{Br_2}}{n_{Cl_2}\, n_{Br^-}^2} = 4 \times 10^4 \tag{S6.2}$$

If $n_{Br_2} = n_{Cl_2}$, then $n_{Cl^-}/n_{Br^-} = (4 \times 10^4)^{1/2} = 200$.

Exercise 6.2 A vessel contains a mixed gas of ammonia (NH_3), hydrogen (H_2), and nitrogen (N_2) in chemical equilibrium at 600 K and 10 atmospheres. The partial pressures, in atmospheres, of the three constituents are 0.22 (NH_3), 2.32 (N_2), and 7.46 (H_2). At fixed temperature, the pressure is increased to 20 atmospheres. What will be the partial pressures of the three gases at the higher total pressure? Treat all the gases as ideal.

Solution   The relevant chemical reaction is

$$N_2 + 3H_2 \leftrightarrow 2NH_3 \tag{S6.3}$$

According to the law of mass action [Eq. (6.9)],

$$\frac{n_A^2}{n_N n_H^3} = K \tag{S6.4}$$

where n_A, n_N, and n_H are the densities of ammonia, nitrogen, and hydrogen. For an ideal gas at fixed temperature the density is proportional to the partial pressure. Thus we could write Eq. (S6.4) in terms of the partial pressures.

$$\frac{p_A^2}{p_N p_H^3} = K_p \tag{S6.5}$$

where K_p is another constant. From the data given, we can calculate K_p.

$$K_p = \frac{(0.22)^2}{(2.32)(7.46)^3} \, \text{atm}^{-2} = 5.03 \times 10^{-5} \, \text{atm}^{-2} \tag{S6.6}$$

At a pressure of 20 atmospheres the partial pressures satisfy the two relations

$$p_A + p_N + p_H = 20 \, \text{atm} \tag{S6.7}$$

and

$$\frac{p_A^2}{p_N p_H^3} = K_p = 5.03 \times 10^{-5} \, \text{atm}^{-2} \tag{S6.8}$$

One more relationship is needed in order to fix the values of the three unknowns. The conservation of the number of nitrogen and hydrogen atoms is expressed by the equations

$$n_A V + 2n_N V = n_A^o V^o + 2n_N^o V^o \tag{S6.9}$$

and

$$3n_A V + 2n_H V = 3n_A^o V^o + 2n_H^o V^o \tag{S6.10}$$

where the superscripted quantities are the values at 10 atmospheres. In order to eliminate the new variables V^o and V, we equate the ratios of the corresponding sides of the two equations. Again using the fact that at fixed temperature the densities are proportional to the partial pressures, we obtain the relation

$$\frac{p_A + 2p_N}{3p_A + 2p_H} = \frac{p_A^o + 2p_N^o}{3p_A^o + 2p_H^o} = 0.312 \tag{S6.11}$$

where the final step uses the given data on the original partial pressures. This gives the linear equation

$$p_A + 2p_N = 0.312(3p_A + 2p_H) \tag{S6.12}$$

The two linear equations can be used to solve for p_N and p_H as linear functions of p_A. When these expressions are substituted for p_N and p_H in Eq. (S6.8), a fourth-order polynomial equation for p_A is obtained. There exist rather complicated exact formulas for solving fourth-order polynomial equations, or the equation may be solved numerically. However, modern symbolic mathematics programs, such as *Mathematica*, have made such calculations unnecessary. One need only enter the set of three simultaneous linear and nonlinear equations in order to obtain the four possible solutions. The only solution with positive values for all the pressures is $p_A = 0.84\,\text{atm}$, $p_N = 4.53\,\text{atm}$, and $p_H = 14.62\,\text{atm}$.

Exercise 6.3   Using the information in Table 6.1, calculate the internal partition functions for the atoms oxygen, magnesium, sulfur, and calcium in the temperature range $100 < T < 3000\,\text{K}$.

Solution   For oxygen, $g_0 = 5$, $g_1 = 3$, $\varepsilon_1 = 228\,\text{K}$, $g_2 = 1$, and $\varepsilon_2 = 326\,\text{K}$. Below $3000\,\text{K}$, one can ignore the higher excited states. Thus

$$z_O(\text{int}) = 5 + 3e^{-228/T} + e^{-326/T} \tag{S6.13}$$

Similar calculations for Mg, S, and Ca give

$$z_{Mg}(\text{int}) = 1 \tag{S6.14}$$

$$z_S(\text{int}) = 5 + 3e^{-571/T} + e^{-825/T} \tag{S6.15}$$

and

$$z_{Ca}(\text{int}) = 1 \tag{S6.16}$$

The most questionable approximation in this calculation is the neglect of the $^1D_2$ state in sulfur, which, at $3000\,\text{K}$, is equal to 0.7% of the sum of the terms given.

Exercise 6.4   For molecular oxygen,

$$\log z_{O_2}(\text{int}) = \varepsilon_B/kT + \log(kTI/\hbar^2) - \log(1 - e^{-\hbar\omega/kT}) \tag{S6.17}$$

By evaluating the three terms separately at a number of temperatures from $T = 100$ to $10,000\,\text{K}$, determine their relative importance in any calculation of the equilibrium constant for a reaction involving O_2.

Solution   From Table 6.1, we see that

$$\varepsilon_B/k = 60,000\,\text{K},$$
$$\hbar^2/kI = 4.14\,\text{K},$$
$$\text{and} \quad \hbar\omega/k = 2250\,\text{K}.$$

Therefore,

$$\log z_{O_2}(\text{int}) = 60,000/T + \log(T/4.14) - \log(1 - e^{-2250/T}) \equiv \phi_B + \phi_R + \phi_V \tag{S6.18}$$

T	ϕ_B	ϕ_R	ϕ_V
100	600	3.18	1.69×10^{-10}
500	120	4.79	1.12×10^{-2}
1000	60	5.49	0.11
3000	20	6.59	0.64
10000	6	7.79	1.60

Table S6.1   Values of terms in the internal canonical potential of O_2.

In terms of its absolute size and its rate of change with T, the term ε_B/kT clearly dominates the other two terms. Thus a reasonably good approximation for $z_{O_2}(\text{int})$ is simply

$$z_{O_2}(\text{int}) = e^{60000/T} \tag{S6.19}$$

Exercise 6.5   For the ideal gas dissociation reaction

$$A_2 \leftrightarrow 2A \tag{S6.20}$$

the *degree of dissociation* r is defined as the fraction of atoms in the dissociated state. Therefore, $r = n_A/(n_A + 2n_{A_2})$. Show that

$$r = (1 + 4\beta p/K)^{-1/2} \tag{S6.21}$$

where p is the pressure, $\beta = 1/kT$, and K is the equilibrium constant.

Solution   Letting $x = n_A$ and $y = n_{A_2}$, we can write the equilibrium equation and the ideal gas law as

$$x^2 = Ky \quad \text{and} \quad x + y = \beta p \tag{S6.22}$$

Using the second equation to eliminate y in the first gives the following quadratic equation for x.

$$x^2 + Kx - K\beta p = 0 \tag{S6.23}$$

The only positive solution of this equation is

$$x = \left[(K/2)^2 + K\beta p\right]^{1/2} - K/2 \tag{S6.24}$$

r is defined as $x/(x + 2y)$. Using the ideal gas law to eliminate y in this expression, we see that

$$
\begin{aligned}
r &= \frac{x}{2\beta p - x} \\
&= \frac{\left[(K/2)^2 + K\beta p\right]^{1/2} - K/2}{K/2 + 2\beta p - \left[(K/2)^2 + K\beta p\right]^{1/2}} \\
&= \frac{(1 + 4\beta p/K)^{1/2} - 1}{1 + 4\beta p/K - (1 + 4\beta p/K)^{1/2}}
\end{aligned} \tag{S6.25}
$$

Letting $\lambda = (1 + 4\beta p/K)^{1/2}$, we can write this as

$$r = \frac{\lambda - 1}{\lambda^2 - \lambda} = \lambda^{-1} \qquad \text{(S6.26)}$$

Exercise 6.6   Using Eq. (S6.13), determine the degree of dissociation of O_2 between $T = 1000$ and $3000\,\text{K}$ for a fixed pressure of 10^{-6} atm (Fig. S6.1).

Solution   From Eq. (S6.13) we get

$$z_O(\text{int}) = 5 + 3e^{-228/T} + e^{-326/T} \qquad \text{(S6.27)}$$

Using data given in Table 6.1, one can calculate

$$z_{O_2}(\text{int}) = \frac{(T/4.14)e^{60000/T}}{(1 - e^{-2250/T})} \qquad \text{(S6.28)}$$

For O_2 the reduced mass is half the mass of an oxygen atom.

$$\mu = 8\,u = 1.33 \times 10^{-26}\,\text{kg} \qquad \text{(S6.29)}$$

Therefore

$$\lambda_\mu = \frac{h}{\sqrt{2\pi \mu kT}} = \frac{6.17 \times 10^{-10}}{\sqrt{T}} \qquad \text{(S6.30)}$$

From Eq. (6.29) we see that

$$K = \frac{z_O^2}{z_{O_2}\lambda_\mu^3} \qquad \text{(S6.31)}$$

Using the formula

$$r = (1 + 4\beta p/K)^{-1/2} \qquad \text{(S6.32)}$$

with $p = 0.101\,\text{N/m}^2$ the function $r(T)$ has been plotted over the range $1000 < T < 3000\,\text{K}$. One can see that almost complete molecular dissociation takes place within the temperature interval $1500 < T < 2500\,\text{K}$. Notice that the actual dissociation temperature is much less that the naive estimate, $T \approx 60,000\,\text{K}$, based only on the binding energy of the diatomic molecule. The much greater amount of phase space available to two unbound atoms, in comparison to a single molecule, strongly tilts the equilibrium toward dissociation.

Exercise 6.7   Redo the previous exercise, using the approximation $z_{O_2}(\text{int}) = \exp(60000/T)$, mentioned in Exercise 6.4.

Solution   With this approximation

$$K = \frac{(5 + 3e^{-228/T} + e^{-326/T})^2}{(6.17 \times 10^{-10}/\sqrt{T})^3 e^{60000/T}} \qquad \text{(S6.33)}$$

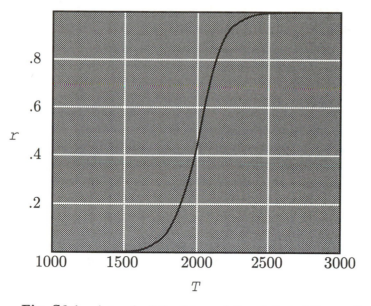

Fig. S6.1   A graph of the degree of dissociation of O_2 as a function of T for a fixed pressure of 10^{-6} atm.

The function

$$r = (1 + 4\beta p/K)^{-1/2} \tag{S6.34}$$

with $p = 0.101\,\text{N/m}^2$ is plotted below in Fig. S6.2. One can see that this approximation, which ignores the rotational and vibrational degrees of freedom of the molecule, exaggerates the phase-space effect that was mentioned in the last exercise, thus shifting the dissociation temperature to an even lower value.

Exercise 6.8   At a fixed temperature of 2000 K, plot the degree of dissociation of oxygen as a function of pressure in the range $0 < p < 1\,\text{N/m}^2 \approx 10^{-5}$ atm.

Solution   According to Exercise 6.6, the equilibrium constant for oxygen dissociation is

$$K = \frac{z_O}{z_{O_2}\lambda_\mu^3} \tag{S6.35}$$

where

$$z_O = 5 + 3e^{-228/T} + e^{-326/T} \tag{S6.36}$$

$$z_{O_2} = (T/4.14)e^{60000/T}/(1 - e^{-2250/T}) \tag{S6.37}$$

and

$$\lambda_\mu = (6.17 \times 10^{-10})/\sqrt{T} \tag{S6.38}$$

At a temperature of 2000 K,

$$\frac{4\beta}{K} = \frac{4z_{O_2}\lambda_\mu^3}{kTz_O^2} = 40.0\,\text{m}^2/\text{N} \tag{S6.39}$$

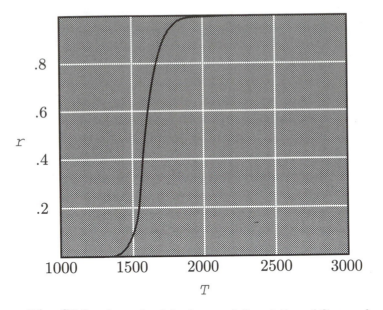

Fig. S6.2 A graph of the degree of dissociation of O_2 as a function of T, neglecting rotational and vibrational degreees of freedom.

The degree of dissociation as a function of pressure is therefore

$$r = (1 + 40p)^{-1/2} \tag{S6.40}$$

This is plotted, in the range $0 < p < 1\,N/m^2$ in Fig. S6.3.

Exercise 6.9 Plot $\log K$ for the ideal gas reaction

$$2CaO \leftrightarrow 2Ca + O_2 \tag{S6.41}$$

in the temperature range $1000 < T < 10,000\,K$.

Solution From Eq. (6.10), we see that

$$\log K = 2f^o_{Ca} + f^o_{O_2} - 2f^o_{CaO} \tag{S6.42}$$

Using Eq. (6.29), we can write this in the form

$$\log K = 2\log z_{Ca}(\text{int}) + \log z_{O_2}(\text{int}) - 2\log z_{CaO}(\text{int}) - 3\log(\lambda^2_{Ca}\lambda_{O_2}/\lambda^2_{CaO}) \tag{S6.43}$$

But

$$\frac{\lambda^2_{Ca}\lambda_{O_2}}{\lambda^2_{CaO}} = \frac{h}{\sqrt{2\pi kT}}\frac{m_{Ca} + m_O}{m_{Ca}\sqrt{2m_O}}$$
$$= 4.32 \times 10^{-10}/\sqrt{T} \tag{S6.44}$$

From the data in Table 6.2, we can calculate

$$z_{Ca}(\text{int}) = 1 + e^{-21808/T} + 3e^{-21883/T} + 5e^{-22036/T} \tag{S6.45}$$

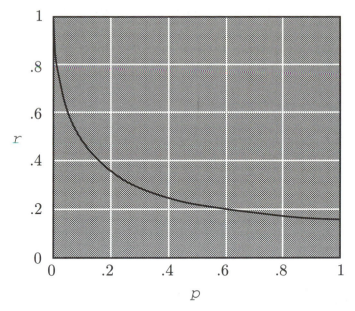

Fig. S6.3   The degree of dissociation of O_2 as a function of pressure (in N/m^2) for $T = 2000 \, K$.

From Table 6.1 and Eqs. (6.35) and (6.37), we get

$$z_{O_2}(\text{int}) = \frac{(T/4.14)e^{60000/T}}{(1 - e^{-2250/T})} \tag{S6.46}$$

and

$$z_{CaO}(\text{int}) = \frac{(T/0.5)e^{47000/T}}{(1 - e^{-1056/T})} \tag{S6.47}$$

Using any computer plotting program, it is now easy to plot $\log K$ from Eq. (S6.43), as shown in Fig. S6.4.

Exercise 6.10   An ideal gas is any substance that satisfies the equation $pV = NRT$, where N is the particle number in moles. Show that, for an ideal gas, the molar specific heats, C_p and C_V, differ by the gas constant. That is, that

$$C_p - C_V = R \tag{S6.48}$$

Solution   For one mole, the ideal gas equation is

$$pV = RT \tag{S6.49}$$

This equation allows us to evaluate the derivatives

$$\left(\frac{\partial V}{\partial T}\right)_p = \beta_p V = \frac{R}{p} \tag{S6.50}$$

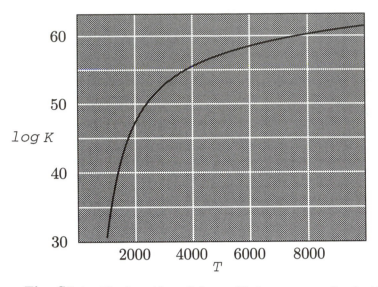

Fig. S6.4   The logarithm of the equilibrium constant for the ideal gas reaction $2CaO \leftrightarrow 2Ca + O_2$.

and

$$\left(\frac{\partial V}{\partial p}\right) = -\kappa_T V = -\frac{RT}{p^2} \tag{S6.51}$$

But, by Eq. (6.87),

$$C_p - C_V = \frac{TV\beta_p^2}{\kappa_T}$$

$$= TV\frac{(R/pV)^2}{(RT/p^2V)} = R \tag{S6.52}$$

Exercise 6.11   From the entropy function for a monatomic ideal gas, determine the affinity as a function of temperature and density and verify that it satisfies Eq. (5.78) for all densities. [The reason why there is no low-density restriction in applying Eq. (5.78) to an ideal gas is that, for an ideal gas, the basic assumption made in deriving that equation, namely the use of Poisson's formula, is valid at all densities because the particles are not statistically correlated.]

Solution   For a monatomic ideal gas

$$S^o = N(\log V + \tfrac{3}{2}\log E - \tfrac{5}{2}\log N + C) \tag{S6.53}$$

This gives

$$\alpha = \frac{\partial S^o}{\partial N} = \log V + \tfrac{3}{2}\log E - \tfrac{5}{2}\log N + C - \tfrac{5}{2} \tag{S6.54}$$

But $E = \tfrac{3}{2}NkT$. When this is used in Eq. (S6.54), one obtains

$$\alpha = -\log(N/V) + \tfrac{3}{2}\log T + C' \tag{S6.55}$$

Exercise 6.12 The affinity of a simple substance can be written as a function of the inverse temperature and the pressure. Starting from Eq. (5.45), which gives $\partial\alpha/\partial\beta$ along an isobar, derive a formula for the finite change, $\alpha(\beta_2, p) - \alpha(\beta_1, p)$, as an integral involving the constant pressure specific heat C_p and the enthalpy at β_1.

Solution Equation (5.45) states that

$$\left(\frac{\partial\alpha}{\partial\beta}\right)_p = -(\varepsilon + pv) \equiv -h \tag{S6.56}$$

where $h = H/N$ is the enthalpy per particle if everything is being expressed in rational units or the enthalpy per mole in practical units. Taking another derivative and using the fact that $C_p = (\partial h/\partial T)_p = -\beta^{-2}(\partial h/\partial\beta)_p$ gives

$$\left(\frac{\partial^2\alpha}{\partial\beta^2}\right)_p = \frac{C_p}{\beta^2} \tag{S6.57}$$

Using the formula, familiar from mechanics, that

$$x(t_2) = x(t_1) + v(t_1)(t_2 - t_1) + \int_{t_1}^{t_2} dt' \int_{t_1}^{t'} dt\, a(t) \tag{S6.58}$$

we can combine Eqs. (S6.56) and (S6.57) to get

$$\alpha(\beta_2, p) = \alpha(\beta_1, p) - h(\beta_1, p)(\beta_2 - \beta_1) + \int_{\beta_1}^{\beta_2} d\beta' \int_{\beta_1}^{\beta'} d\beta \frac{C_p(\beta, p)}{\beta^2} \tag{S6.59}$$

Exercise 6.13 Show that, according to Debye's theory, the entropy of a crystal is equal to one-third of its specific heat when $T \ll T_D$.

Solution For a crystal, expansion is such a small effect that there is no necessity to distinguish between C_p and C_V. The entropy is given by the integral

$$S = \int_0^T \frac{dQ}{T} = \int_0^T \frac{C\,dT}{T} \tag{S6.60}$$

But, at low T, the specific heat is proportional to T^3. By Eq. (4.33),

$$C = \frac{12\pi^4 NkT^3}{5T_D^3} \equiv C_o T^3 \tag{S6.61}$$

Thus

$$S = C_o \int_0^T T^2\,dT = \tfrac{1}{3}C_o T^3 \tag{S6.62}$$

Table S6.2   The Bridgeman transformation table*

$(\partial T)_p$	$-(\partial p)_T$	$\longrightarrow$	1
$(\partial V)_p$	$-(\partial p)_V$	$\longrightarrow$	$V\beta$
$(\partial S)_p$	$-(\partial p)_S$	$\longrightarrow$	C/T
$(\partial E)_p$	$-(\partial p)_E$	$\longrightarrow$	$C - pV\beta$
$(\partial H)_p$	$-(\partial p)_H$	$\longrightarrow$	C
$(\partial G)_p$	$-(\partial p)_G$	$\longrightarrow$	$-S$
$(\partial F)_p$	$-(\partial p)_F$	$\longrightarrow$	$-S - pV\beta$
$(\partial V)_T$	$-(\partial T)_V$	$\longrightarrow$	$V\kappa$
$(\partial S)_T$	$-(\partial T)_S$	$\longrightarrow$	$V\beta$
$(\partial E)_T$	$-(\partial T)_E$	$\longrightarrow$	$TV\beta - pV\kappa$
$(\partial H)_T$	$-(\partial T)_H$	$\longrightarrow$	$TV\beta - V$
$(\partial G)_T$	$-(\partial T)_G$	$\longrightarrow$	$-V$
$(\partial F)_T$	$-(\partial T)_F$	$\longrightarrow$	$-pV\kappa$
$(\partial S)_V$	$-(\partial V)_S$	$\longrightarrow$	$V^2\beta^2 - VC\kappa/T$
$(\partial E)_V$	$-(\partial V)_E$	$\longrightarrow$	$TV^2\beta^2 - VC\kappa$
$(\partial H)_V$	$-(\partial V)_H$	$\longrightarrow$	$TV^2\beta^2 - VC\kappa - V^2\beta$
$(\partial G)_V$	$-(\partial V)_G$	$\longrightarrow$	$SV\kappa - V^2\beta$
$(\partial F)_V$	$-(\partial V)_F$	$\longrightarrow$	$SV\kappa$
$(\partial E)_S$	$-(\partial S)_E$	$\longrightarrow$	$pV^2\beta^2 - pVC\kappa/T$
$(\partial H)_S$	$-(\partial S)_H$	$\longrightarrow$	$-VC/T$
$(\partial G)_S$	$-(\partial S)_G$	$\longrightarrow$	$SV\beta - VC/T$
$(\partial F)_S$	$-(\partial S)_F$	$\longrightarrow$	$pV^2\beta^2 + SV\beta - pVC\kappa/T$
$(\partial H)_E$	$-(\partial E)_H$	$\longrightarrow$	$pV^2\beta + pVC\kappa - VC - pTV^2\beta^2$
$(\partial G)_E$	$-(\partial E)_G$	$\longrightarrow$	$pV^2\beta + TSV\beta - VC - pSV\kappa$
$(\partial F)_E$	$-(\partial E)_F$	$\longrightarrow$	$pTV^2\beta^2 - pVC\kappa$
$(\partial G)_H$	$-(\partial H)_G$	$\longrightarrow$	$TSV\beta - VC - VS$
$(\partial F)_H$	$-(\partial H)_F$	$\longrightarrow$	$(TV\beta - V)(S + pV\beta) - pV\kappa$
$(\partial F)_G$	$-(\partial G)_F$	$\longrightarrow$	$pSV\kappa - SV - pV^2\beta$

*In this table $C \equiv C_p$, $\beta \equiv \beta_p$, and $\kappa \equiv \kappa_T$.

which shows that $S = C/3$.

Exercise 6.14   In Section 6.11 it was shown how an arbitrary partial derivative involving the variables p, T, S, V, E, F, H, G, and μ could be transformed into an expression involving p, T, and the five "handbook functions," S, V, C_p, β_p, and κ_T. In 1941, P. W. Bridgeman (*Physical Review*, **3**, p. 273) published a table that allows the results of such a calculation to be written down immediately. The table is reproduced as Table S6.2, and its use is illustrated by two examples.

$$\left(\frac{\partial G}{\partial p}\right)_V \to \frac{(\partial G)_V}{(\partial p)_V} \to \frac{SV\kappa_T - V^2\beta_p}{-V\beta_p} = V - \frac{S\kappa_T}{\beta_p} \tag{S6.63}$$

and

$$\left(\frac{\partial F}{\partial V}\right)_p \rightarrow \frac{(\partial F)_p}{(\partial V)_p} \rightarrow \frac{-S - pV\beta_p}{V\beta_p} = -\frac{S}{V\beta_p} - p \tag{S6.64}$$

That is, we first replace the partial derivative $(\partial x/\partial y)_a$ by a fraction made up of the mathematically meaningless symbols $(\partial x)_a$ and $(\partial y)_a$. We look up each symbol in the table (it will always be found either in the first column or the second, never in both), and replace it by the quantity shown in the last column. The resulting fraction gives the value of the partial derivative in the p-T representation.

Work out each of the following partial derivatives, first using Bridgeman's table, and then using the method of Section 6.11. (a) $(\partial F/\partial S)_p$; (b) $(\partial E/\partial T)_G$.

Solution   (a)

$$\left(\frac{\partial F}{\partial S}\right)_p \rightarrow \frac{(\partial F)_p}{(\partial S)_p} = -\frac{S + pV\beta_p}{C_p/T} \tag{S6.65}$$

To verify this by the method of Section 6.11, we first note that $dF = -S\,dT - p\,dV$, and thus

$$\left(\frac{\partial F}{\partial S}\right)_p = -S\left(\frac{\partial T}{\partial S}\right)_p - p\left(\frac{\partial V}{\partial S}\right)_p \tag{S6.66}$$

The first expression is eliminated using

$$\left(\frac{\partial T}{\partial S}\right) = \frac{1}{(\partial S/\partial T)_p} = \frac{T}{C_p} \tag{S6.67}$$

The second expression requires the third partial derivative identity.

$$\left(\frac{\partial V}{\partial S}\right)_p = \frac{(\partial V/\partial T)_p}{(\partial S/\partial T)_p} = \frac{V\beta_p}{C_p/T} \tag{S6.68}$$

The result agrees with Bridgeman's table. (b) By Bridgeman's table,

$$\left(\frac{\partial E}{\partial T}\right)_G \rightarrow \frac{(\partial E)_G}{(\partial T)_G} = \frac{-pV^2\beta_p - TSV\beta_p + VC_p + pSV\kappa_T}{V}$$
$$= C_p + pS\kappa_T - pV\beta_p - TS\beta_p \tag{S6.69}$$

By Section 6.11, we first use $dE = T\,dS - p\,dV$.

$$\left(\frac{\partial E}{\partial T}\right)_G = dT\left(\frac{\partial S}{\partial T}\right)_G - p\left(\frac{\partial V}{\partial T}\right)_G \tag{S6.70}$$

Both terms on the right-hand side require the second partial derivative identity.

$$\left(\frac{\partial S}{\partial T}\right)_G = -\frac{(\partial G/\partial T)_S}{(\partial G/\partial S)_T} \tag{S6.71}$$

and

$$\left(\frac{\partial V}{\partial T}\right)_G = -\frac{(\partial G/\partial T)_V}{(\partial G/\partial V)_T} \tag{S6.72}$$

These are followed by use of the differential relation $dG = -S\,dT + V\,dp$, giving

$$\left(\frac{\partial G}{\partial T}\right)_S = -S + V\left(\frac{\partial p}{\partial T}\right)_S \tag{S6.73}$$

$$\left(\frac{\partial G}{\partial S}\right)_T = -S \cdot 0 + V\left(\frac{\partial p}{\partial S}\right)_T = 1/\beta_p \tag{S6.74}$$

$$\left(\frac{\partial G}{\partial T}\right)_V = -S + V\left(\frac{\partial p}{\partial T}\right)_V \tag{S6.75}$$

$$\left(\frac{\partial G}{\partial V}\right)_T = -S \cdot 0 + V\left(\frac{\partial p}{\partial V}\right)_T = -1/\kappa_T \tag{S6.76}$$

To transform the remaining expressions, we need the second partial derivative identity.

$$\left(\frac{\partial p}{\partial T}\right)_S = -\frac{(\partial S/\partial T)_p}{(\partial S/\partial p)_T} = \frac{C_p/T}{V\beta_p} \tag{S6.77}$$

$$\left(\frac{\partial p}{\partial T}\right)_V = -\frac{(\partial V/\partial T)_p}{(\partial V/\partial p)_T} = \beta_p/\kappa_T \tag{S6.78}$$

Making all the back substitutions verifies Eq. (S6.68) (and gives one an appreciation for the efficiency of Bridgeman's table).

Supplement to Chapter 7

REVIEW QUESTIONS

7.1 Describe a system that is naturally analyzed using the grand canonical ensemble.

7.2 For a classical system, write down the probability density for the grand canonical ensemble and explain what it means? In other words, what is the physical interpretation of the function that you have written down?

7.3 Describe the model that was used for surface adsorption.

7.4 For the surface adsorption model, calculate the grand partition function.

7.5 Explain how a relationship can be obtained between the density of particles adsorbed on the surface and the density of particles in the gas.

7.6 Describe the possible energy eigenvalues of a many-particle ideal quantum system for the cases of bosons and fermions.

7.7 For an ideal quantum system, how are the single-particle energy values ε_k determined?

7.8 Derive the grand potential for an ideal many-boson system with a single-particle energy spectrum ε_k.

7.9 Do the same for an ideal many-fermion system.

7.10 Write formulas for N and E, as discrete sums involving the single-particle energy spectrum, for ideal Bose–Einstein and Fermi–Dirac systems.

7.11 For an ideal Fermi gas of spin-$\frac{1}{2}$ particles, derive integral formulas for N and E.

7.12 How is the chemical potential related to the affinity?

7.13 Prove that $p = \frac{2}{3}E/V$ for an ideal Fermi gas.

7.14 Derive a formula for the chemical potential of an ideal Fermi gas at zero temperature.

7.15 How is the Fermi energy defined?

7.16 Using the low temperature expansion for a Fermi–Dirac integral derived in the Mathematical Appendix, show that the specific heat of an ideal Fermi gas is proportional to T at low temperatures.

7.17 Explain why it is true that using the integral formula for N for the ideal Bose gas leads to difficulty and how one solves the problem.

7.18 For the ideal Bose gas, derive the formula for the critical density as a function of the temperature.

7.19 For the ideal Bose gas, derive the formula for the critical temperature as a function of the density.

7.20 Using variables n and $\tau = kT$, show the phase diagram for the ideal Bose gas.

7.21 What quantity is discontinuous across the Bose–Einstein transition line?

7.22 What is wrong with the ideal Bose gas as a model for any real physical system?

7.23 What is the rule that determines whether a particular atom satisfies Bose–Einstein or Fermi–Dirac statistics?

7.24 Draw a sketch of the phase diagram for ^{4}He.

7.25 In liquid helium, as $T \to 0$, what happens to the superfluid density?

7.26 In liquid helium, as $T \to 0$, what happens to the condensate density?

7.27 Draw a sketch of the excitation spectrum for liquid helium. Explain the physical meaning of what you have drawn.

7.28 Assuming that elementary excitations are the only possible excited states of the system (a false assumption), prove that liquid helium will be a superfluid below a critical velocity and relate that critical velocity to the excitation spectrum.

7.29 Why is ground glass (or some equivalent stuff) used in demonstrations of superfluid flow?

7.30 Describe quantized vortices in liquid helium. Are they the same as rotons?

7.31 For the quasiparticles in liquid helium, and for the photon gas, why is the affinity equal to zero?

7.32 Derive the phonon contribution to the specific heat of liquid helium at very low temperature.

7.33 Derive the energy density of radiation within a cavity at temperature T.

7.34 Derive Wien's displacement law for the location of the maximum in the frequency distribution function.

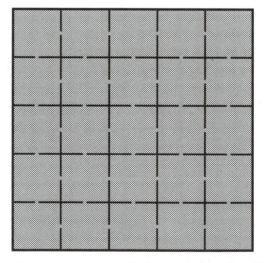

Fig. S7.1 A system composed of a large number of weakly interacting identical subsystems.

EXERCISES

Exercise 7.1   Consider a system composed of a large number M of identical boxes (Fig. S7.1). Each box has a pinhole that connects it with each of its nearest neighbors. The particles may have interactions, but we will neglect the interactions between particles that are in different boxes. The eigenstates of a set of N particles in one of the boxes are ψ_1^N, $\psi_2^N, \ldots$, with energies E_1^N, $E_2^N, \ldots$. Given that the complete closed system has N_T particles and a fixed energy E_T, use the method of the most probable distribution (from Chapter 2) to calculate the probability that a given system is in quantum state ψ_k^N. Show that it agrees with the grand canonical ensemble.

Solution   In order to make the analysis clearer, it is useful to assume that, once each minute, the pinholes simultaneously open for only 1 second and then remain closed for the other 59 seconds. We only describe the system while the pinholes are closed and each box is therefore a completely isolated subsystem. We number the boxes from 1 to M. A microstate is defined by giving the quantum state of the set of particles in each box. The macrostate is defined by saying how many boxes are in state ψ_k^N for every value of N and k. Let ν_k^N be the number of subsystems in state ψ_k^N. Certainly, the set of nonnegative integers ν_k^N must satisfy the three constraints

$$\sum_N \sum_k \nu_k^N = M \qquad \text{(the number of boxes)} \qquad (S7.1)$$

$$\sum_N \sum_k N\nu_k^N = N_T \qquad \text{(the number of particles)} \qquad (S7.2)$$

$$\sum_N \sum_k E_k^N \nu_k^N = E_T \qquad \text{(the total energy)} \qquad (S7.3)$$

The number of microstates that correspond to a given macrostate (defined by a particular set of ν_k^N's) is

$$K = \frac{M!}{\prod_N \prod_k \nu_k^N!} \qquad (S7.4)$$

Using Stirling's approximation and Lagrange's method, we maximize the function

$$F = \log M! - \sum_{N,k} \nu_k^N (\log \nu_k^N - 1) - \gamma \sum_{N,k} \nu_k^N - \alpha \sum_{N,K} N\nu_k^N - \beta \sum_{N,k} E_k^N \nu_k^N \qquad (S7.5)$$

Setting $\partial F / \partial \nu_k^N = 0$ gives

$$-\log \nu_k^N - \gamma - \alpha N - \beta E_k^N = 0 \qquad (S7.6)$$

or

$$\nu_k^N = Ce^{-\alpha N - \beta E_k^N} \qquad (S7.7)$$

The fraction of systems in state ψ_k^N is equal to the probability that a randomly chosen system will be in that state. It is

$$P_{Nk} = \frac{\nu_k^N}{\sum_k \nu_k^N} = \Lambda^{-1} e^{-\alpha N - \beta E_k^N} \tag{S7.8}$$

where

$$\Lambda = \sum_{N,k} e^{-\alpha N - \beta E_k^N} \tag{S7.9}$$

This result obviously agrees with the grand canonical ensemble prediction. We have simply used $M-1$ identical boxes as our reservoir for any given box.

Exercise 7.2   Derive the equations of state of a monatomic ideal gas, using classical mechanics and the grand canonical ensemble.

Solution   The Hamiltonian function for an N-particle monatomic ideal gas is

$$H_N = \sum_{n=1}^{N} p_n^2/2m \tag{S7.10}$$

To derive equations of state, one first calculates the grand partition function, defined as

$$\begin{aligned}
\Lambda(\alpha, \beta, V) &= \sum_{N=0}^{\infty} \frac{e^{-\alpha N}}{N! h^{3N}} \int_V d^{3N}r \int d^{3N}p \, e^{-\beta H_N} \\
&= \sum_{N=0}^{\infty} \frac{e^{-\alpha N}}{N! h^{3N}} V^N (2\pi m/\beta)^{3N/2}
\end{aligned} \tag{S7.11}$$

But

$$(2\pi m/\beta)^{1/2}/h = \frac{\sqrt{2\pi mkT}}{h} = 1/\lambda(\beta) \tag{S7.12}$$

where λ is the thermal de Broglie wavelength. Also $e^{-\alpha N} = (e^{-\alpha})^N$. Thus

$$\Lambda = \sum_{N=0}^{\infty} \frac{1}{N!} \left(\frac{e^{-\alpha}V}{\lambda^3} \right)^N = \exp\left(e^{-\alpha}V/\lambda^3\right) \tag{S7.13}$$

The grand potential is the logarithm of Λ.

$$\psi(\alpha, \beta, V) = e^{-\alpha}V/\lambda^3(\beta) \tag{S7.14}$$

N, E, and βp are given by the derivatives of ψ.

$$N = -\frac{\partial \psi}{\partial \alpha} = e^{-\alpha}V/\lambda^3 \tag{S7.15}$$

$$E = -\frac{\partial \psi}{\partial \beta} = \frac{3}{2\beta} e^{-\alpha}V/\lambda^3 \tag{S7.16}$$

and

$$\beta p = \frac{\partial \psi}{\partial V} = e^{-\alpha}/\lambda^3 \tag{S7.17}$$

Dividing the second and third equations by the first, one obtains the desired equations of state.

$$\frac{E}{N} = \frac{3}{2}kT \quad \text{and} \quad \frac{p}{NkT} = \frac{1}{V} \tag{S7.18}$$

Exercise 7.3   The inside surface of a $1\,\text{cm}^3$ box is covered with adsorption sites. Each site has an area of $4\,\text{Å}^2$ and a binding energy of $200\,\text{K}$ (the energy of a particle at the site is $-200\,\text{K}$). The box contains argon gas. At $300\,\text{K}$ the box is evacuated to a pressure of 10^{-7} atmospheres. It is completely sealed and the temperature is then reduced to $8\,\text{K}$. Without adsorption the pressure would drop to $(8/300) \times 10^{-7} = 2.67 \times 10^{-9}$ atmospheres. Taking adsorption into account, determine the actual pressure at $8\,\text{K}$.

Solution   Each of the six faces has an area of $1\,\text{cm}^2 = 10^{16}\,\text{Å}^2$. Thus the total number of adsorption sites is

$$N_s = 1.5 \times 10^{16} \tag{S7.19}$$

The mass of an argon atom is $39.95\,u = 6.64 \times 10^{-26}\,\text{kg}$. Therefore, the constant γ, appearing in Eq. (7.29), is

$$\gamma = k^{5/2}(2\pi m)^{3/2}/h^3 = 6.55 \times 10^5 \tag{S7.20}$$

The pressure at $300\,\text{K}$, in SI units, is $1.01 \times 10^{-2}\,\text{N/m}^2$. According to Eq. (7.30), the number of adsorbed particles at $300\,\text{K}$ is

$$N_a = \frac{N_s p}{p + \gamma T^{5/2} e^{-T_o/T}} = 290 \tag{S7.21}$$

which is completely negligible. Thus the number of particles in the system is

$$N = \frac{pV}{kT} = 2.44 \times 10^{12} \tag{S7.22}$$

At $8\,\text{K}$ we know that $p \leq NkT/V = 2.70 \times 10^{-4}\,\text{N/m}^2$. But

$$\gamma T^{5/2} e^{-T_o/T} = 1.65 \times 10^{-3} \tag{S7.23}$$

which indicates that the two terms in the denominator of Eq. (7.36) are of comparable size, so neither can be neglected. Using the fact that

$$N_a = N - \frac{pV}{kT} \tag{S7.24}$$

one obtains the following quadratic equation for p.

$$N - (V/kT)p = \frac{N_s p}{p + \gamma T^{5/2} e^{-T_o/T}} \tag{S7.25}$$

Putting in the known values of all constants and solving the quadratic equation, one finds a unique positive root.

$$p = 2.67 \times 10^{-7}\,\text{N/m}^2 = 2.64 \times 10^{-12}\,\text{atm} \tag{S7.26}$$

The disappearance of the gas onto the walls, which had no effect at $300\,\mathrm{K}$, reduces the pressure at $8\,\mathrm{K}$ by a factor of 1000.

Exercise 7.4   Consider an ideal gas composed of monatomic molecules of mass m and diatomic molecules of mass $2m$ and moment of inertia I. The molecules may undergo the chemical reaction $2A \leftrightarrow B$. A diatomic molecule has a ground-state energy of $E_g = -\varepsilon$ in comparison with two separated atoms. (a) Considering the monatomic and diatomic molecules as distinct species, calculate the grand potential $\psi(\alpha_1, \alpha_2, \beta, V)$ using classical mechanics and taking into account the rotational but not the vibrational degrees of freedom of the diatomic molecules. (b) Impose the condition for chemical reaction equilibrium, $2\alpha_1 = \alpha_2$, and obtain an equation for the pressure as a function of the temperature and the total density of atoms. (c) For Cl_2, $\varepsilon = 4.0 \times 10^{-19}\,\mathrm{J} = 29{,}000\,\mathrm{K}$, and $\hbar^2/2I = 4.95 \times 10^{-24}\,\mathrm{J} = 0.25\,\mathrm{K}$. Use these values to plot the pressure as a function of T for a density of $n(\mathrm{STP})$.

Solution   (a) The Hamiltonian for the system with K monatomic and L diatomic molecules is

$$H = \sum_{k=1}^{K} \frac{p_k^2}{2m} - \varepsilon L + \sum_{\ell=1}^{L} \left(\frac{P_\ell^2}{4m} + \frac{P_{\theta\ell}^2 + P_{\phi\ell}^2/\sin^2\theta_\ell}{2I} \right) \tag{S7.27}$$
$$= H_1 - \varepsilon L + H_2$$

The grand partition function is obtained by summing independently over all values of K and L.

$$\Lambda = \sum_{K=0}^{\infty} \sum_{L=0}^{\infty} \left(\frac{e^{-\alpha_1 K}}{K! h^{3K}} \int d^{3K}x \, d^{3K}p \, e^{-\beta H_1} \right)$$
$$\times \left(\frac{e^{-\alpha_2 L + \beta\varepsilon L}}{L! h^{5L}} \int d^{5L}x \, d^{5L}P \, e^{-\beta H_2} \right)$$
$$= \sum_{K} \left(\frac{e^{-\alpha_1 K}}{K! h^{3K}} \int d^{3K}x \, d^{3K}p \, e^{-\beta H_1} \right) \tag{S7.28}$$
$$\times \sum_{L} \left(\frac{e^{-(\alpha_2 - \beta\varepsilon)L}}{L! h^{5L}} \int d^{5L}x \, d^{5L}P \, e^{-\beta H_2} \right)$$

The sum over K is identical to the partition function that was calculated in the last exercise.

$$\sum_{K} \left(\frac{e^{-\alpha_1 K}}{K! h^{3K}} \int d^{3K}x \, d^{3K}p \, e^{-\beta H_1} \right) = \exp\left(e^{-\alpha_1} V/\lambda_1^3 \right) \tag{S7.29}$$

where $\lambda_1 = h/\sqrt{2\pi m k T}$. The integral in the second term is the Lth power of a five-dimensional integral of the form

$$I_5 = \frac{1}{h^5} \int_V d^3x \int d^3P \, e^{-\beta P^2/4m} \int d\theta \, d\phi$$
$$\times \int dP_\theta \, dP_\phi \, \exp\left[-\beta(P_\theta^2 + P_\phi^2/\sin^2\theta)/2I \right] \tag{S7.30}$$

This integral can be found in Eqs. (4.63) and (4.64).

$$I_5 = \frac{2IV}{\beta \hbar \lambda_2^3} \tag{S7.31}$$

The sum over L in Eq. (S7.28) can now be computed easily.

$$\sum_{L=0}^{\infty} \frac{1}{L!} \left(e^{-\alpha_2 + \beta\varepsilon} \frac{2IV}{\beta\hbar^2\lambda_2^3} \right)^L = \exp\left(e^{-\alpha_2 + \beta\varepsilon} \frac{2IV}{\beta\hbar^2\lambda_2^3} \right) \qquad (S7.32)$$

The grand potential, $\psi = \log\Lambda$, is

$$\psi(\alpha_1, \alpha_2, \beta, V) = \frac{e^{-\alpha_1}V}{\lambda_1^3(\beta)} + \frac{2e^{-\alpha_2 + \beta\varepsilon}IV}{\beta\hbar^2\lambda_2^3(\beta)} \qquad (S7.33)$$

(b) Imposing the condition for chemical reaction equilibrium, $\alpha_2 = 2\alpha_1$, and dropping the subscript on α_1, we obtain a grand potential with the single affinity variable α, which represents the affinity for A atoms.

$$\psi(\alpha, \beta, V) = e^{-\alpha}\frac{V}{\lambda_1^3} + e^{-2\alpha + \beta\varepsilon}\frac{2IV}{\beta\hbar^2\lambda_2^3} \qquad (S7.34)$$

The particle density is given in terms of ψ by the standard thermodynamic identity

$$n = -V^{-1}\frac{\partial\psi}{\partial\alpha} = \frac{e^{-\alpha}}{\lambda^3} + \frac{4\sqrt{2}I}{\hbar^2\beta}\frac{e^{-2\alpha + \beta\varepsilon}}{\lambda^3} \qquad (S7.35)$$

where we have used the fact that $\lambda_2 = \lambda_1/\sqrt{2}$ and have dropped the subscript on λ_1. In order to analyze the relationship between pressure and density, it is convenient to introduce a new variable, called the *activity*, that will play an important role when we study the nonideal gas in the next chapter. The activity is defined by

$$\zeta = \frac{e^{-\alpha}}{\lambda^3(\beta)} \qquad (S7.36)$$

If we define a rotational temperature, $T_R = \hbar^2/2kI$, we can write n in terms of ζ and T as

$$n = \zeta + 2\sqrt{2}\lambda^3(T/T_R)e^{\varepsilon/kT}\zeta^2 \equiv \zeta + 2C(T)\zeta^2 \qquad (S7.37)$$

The pressure can be expressed in terms of ζ and T by using the fact that $p/kT = \psi/V$.

$$p/kT = \zeta + \sqrt{2}\lambda^3(T/T_R)e^{\varepsilon/kT}\zeta^2 = \zeta + C(T)\zeta^2 \qquad (S7.38)$$

The pressure equation of state is obtained by solving for ζ as a function of n and then using that function in the second equation. Equation (S7.38) is a quadratic equation that can be easily solved to give (note that ζ must be positive)

$$\zeta = \frac{(1 + 8Cn)^{1/2} - 1}{4C} \qquad (S7.39)$$

Noting that $p/kT = n - C\zeta^2$, we get

$$\frac{p}{kT} = n - \frac{[(1 + 8Cn)^{1/2} - 1]^2}{16C} \qquad (S7.40)$$

For small n we can obtain the correction to the simple monatomic ideal gas law, $p = nkT$, by expanding the square root to first order in n. The result is

$$\frac{p}{kT} \approx n - C(T)n^2 \qquad (S7.41)$$

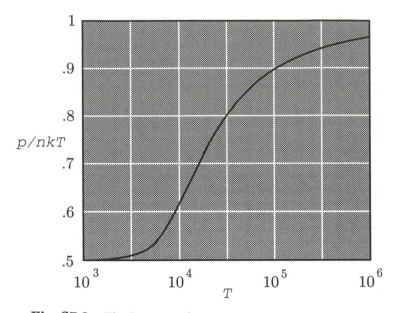

Fig. S7.2   The function p/nkT, where n is the density of *atoms*. At low temperature, the atoms form themselves into diatomic molecules and so $p = nkT/2$.

The pressure is reduced, in comparison with a monatomic gas, because the number of gas particles diminishes as atoms combine to form diatomic molecules. (c) For Cl_2 the binding energy and rotational temperature have been given. Using these values to calculate $C(T)$ and setting n equal to $n(STP)$ in Eq. (S7.32), we can plot $p(T)$ at constant n. At low T we expect $p \approx \frac{1}{2}nkT$ because most of the atoms will be bound into diatomic molecules. At high T we expect $p \approx nkT$. This is nicely illustrated in Fig. S7.2, where p/nkT is plotted on a logarithmic T scale.

Exercise 7.5 N_o molecules of hemoglobin are transported between the atmosphere, where the partial pressure of oxygen is 0.2 atm, to a region where the temperature is the same but the partial pressure of oxygen is 0.02 atm. Each molecule of hemoglobin has four sites that can hold an oxygen molecule. Assuming that Eq. (7.24) is applicable to the attachment of oxygen to hemoglobin (which is only a very rough approximation) and that the sites are 75% filled in the oxygen-rich atmosphere, calculate how many molecules of oxygen are released when the hemoglobin is transported to the oxygen-poor region.

Solution   Let the partial pressure of oxygen in the oxygen-rich atmosphere be p_H and that in the oxygen poor atmosphere be p_L. Then, by Eq. (7.24), in the oxygen-rich atmosphere

$$f_a = 0.75 = \frac{p_H}{p_H + g/z_a} \tag{S7.42}$$

This can be solved for the unknown quantity g/z_a.

$$g/z_a = p_H/3 \tag{S7.43}$$

According to the statement of the problem, $p_L = 0.1 \, p_H$. Thus, at p_L, the fraction of occupied sites is

$$f_a = \frac{0.1p_H}{0.1p_H + p_H/3} = \frac{3}{13} = 0.23 \tag{S7.44}$$

The number of oxygen molecules released is

$$4N_o(0.75 - 0.23) = 2.08N_o \tag{S7.45}$$

Exercise 7.6 In reality, the oxygen molecules attached to the four binding sites of a hemoglobin molecule have significant attractive interactions that were ignored in the last exercise. In order to determine qualitatively the effect of adsorbate interactions, consider the following extreme case. We assume that each hemoglobin molecule has two double binding sites. At any double site a single oxygen molecule will not stick but a pair of oxygen molecules will attach with an energy $\varepsilon = -kT_o$. (a) Derive the equivalent of Eq. (7.24) for this system. (b) For this system, repeat the last exercise.

Solution (a) The grand potential for the system is $2N_o$ times the grand potential for one double site. The grand partition function for a double site is (the possible values of N are 0 and 2)

$$\Lambda = 1 + e^{-2\alpha + T_o/T} \tag{S7.46}$$

Thus

$$\psi = 2N_o \log(1 + e^{-2\alpha + T_o/T}) \tag{S7.47}$$

The number of attached oxygen molecules is

$$N = -\frac{\partial \psi}{\partial \alpha} = 4N_o/(e^{2\alpha - T_o/T} + 1) \tag{S7.48}$$

Letting $f_a = N/4N_o$, we can solve this equation for $e^{2\alpha - T_o/T}$.

$$e^{2\alpha - T_o/T} = \frac{1 - f_a}{f_a} \tag{S7.49}$$

which gives, for α,

$$2\alpha - T_o/T = \log\left(\frac{1 - f_a}{f_a}\right) \tag{S7.50}$$

We assume that the hemoglobin is in equilibrium with an ideal gas, for which the affinity can be written in terms of the pressure and temperature as

$$\alpha = \log(g/p) \tag{S7.51}$$

Eliminating α between Eqs. (S7.50) and (S7.51) gives a relation between p and f_a.

$$2\log(g/p) - T_o/T = \log\left[(1 - f_a)/f_a\right] \tag{S7.52}$$

or

$$\frac{g^2}{p^2}e^{-T_o/T} = \frac{1 - f_a}{f_a} \tag{S7.53}$$

Solving this for f_a gives

$$f_a = \frac{p^2}{p^2 + g^2 e^{-T_o/T}} \tag{S7.54}$$

This formula is similar in structure to Eq. (7.24), but p has been replaced by p^2. (b) In the oxygen-rich atmosphere, $f_a = 0.75$. Therefore

$$0.75 = \frac{p_H^2}{p_H^2 + g^2 e^{-T_o/T}} \tag{S7.55}$$

which determines the value of $g^2 e^{-T_o/T}$.

$$g^2 e^{-T_o/T} = p_H^2/3 \tag{S7.56}$$

In the oxygen-poor environment, $p_L = 0.1\, p_H$, and

$$f_a = \frac{0.01\, p_H^2}{0.01\, p_H^2 + p_H^2/3} = \frac{3}{103} \approx 0.03 \tag{S7.57}$$

The number of oxygen molecules released is

$$4N_o(0.75 - 0.03) = 2.88\, N_o \tag{S7.58}$$

which is almost a 50% improvement in efficiency in comparison to the system without oxygen–oxygen interactions. Just such an interaction effect adds to the efficiency of the hemoglobin in our blood.

Exercise 7.7 A surface with K adsorption sites is in equilibrium with a mixed ideal gas containing molecules of types A and B. Each site can take either a single A molecule, with energy $\varepsilon_A = -kT_A$, or a single B molecule, with energy $\varepsilon_B = -kT_B$. The partial pressures of the two gases are p_A and p_B. f_A and f_B are the fractions of sites that are occupied by the two types of particles. (a) Derive equations for f_A and f_B in terms of p_A and p_B. (b) Assume that both gases are monatomic and that $T_A = T_B$. Determine the relation between p_A and p_B that will give equal numbers of adsorbed A and B atoms.

Solution   (a) The partition function for a single site is

$$\Lambda_1 = 1 + e^{-\alpha_A + T_A/T} + e^{-\alpha_B + T_B/T} \tag{S7.59}$$

and the grand potential per site is

$$\psi_1 = \log(1 + e^{-\alpha_A + T_A/T} + e^{-\alpha_B + T_B/T}) \tag{S7.60}$$

The probability that the site is occupied by an A molecule is f_A, which is the same as the average number of A molecules at the site. Thus

$$f_A = -\frac{\partial \psi_1}{\partial \alpha_A} = \frac{e^{-\alpha_A + T_A/T}}{\Lambda_1} \tag{S7.61}$$

Also

$$f_B = -\frac{\partial \psi_1}{\partial \alpha_B} = \frac{e^{-\alpha_B + T_B/T}}{\Lambda_1} \tag{S7.62}$$

The equation, $\alpha = \log(g/p)$, for the affinity of an ideal gas, implies that

$$e^{-\alpha_A} = p_A/g_A \quad \text{and} \quad e^{-\alpha_B} = p_B/g_B \tag{S7.63}$$

Using this in Eqs. (S7.61) and (S7.62) gives the desired equilibrium relations.

$$f_A = \frac{p_A e^{T_A/T}/g_A}{1 + p_A e^{T_A/T}/g_A + p_B e^{T_B/T}/g_B} \tag{S7.64}$$

and

$$f_B = \frac{p_B e^{T_B/T}/g_B}{1 + p_A e^{T_A/T}/g_A + p_B e^{T_B/T}/g_B} \tag{S7.65}$$

(b) If $f_A = f_B$ and $T_A = T_B$, then

$$\frac{p_A}{p_B} = \frac{g_A}{g_B} \tag{S7.66}$$

For a monatomic gas

$$g = (kT)^{5/2}(2\pi m)^{3/2}/h^3 \tag{S7.67}$$

Therefore, the condition for equal occupation of the adsorption sites is that

$$\frac{p_A}{p_B} = \frac{m_A^{3/2}}{m_B^{3/2}} \tag{S7.68}$$

Exercise 7.8 Let f_n be the average occupation number of the nth single-particle quantum state in an ideal Fermi Dirac system. Consider the expression (called the *variational entropy*)

$$S_{\text{var}} = -\sum_n [f_n \log f_n + (1 - f_n) \log(1 - f_n)] \tag{S7.69}$$

(a) Show that, if S_{var} is maximized under the conditions that

$$\sum_n f_n = N \quad \text{and} \quad \sum_n f_n E_n = E \tag{S7.70}$$

then the resulting distribution is the Fermi–Dirac distribution. (b) Show that, at the maximum, S_{var} is equal to the equilibrium entropy of the Fermi–Dirac system.

Solution   (a) Using Lagrange's method, we maximize the quantity

$$F = -\sum [f_n \log f_n + (1 - f_n) \log(1 - f_n)] - \alpha \sum f_n - \beta \sum f_n E_n \tag{S7.71}$$

without restrictions. $\partial F/\partial f_n = 0$ gives the relation

$$-\log f_n + \log(1 - f_n) - \alpha - \beta E_n = 0 \tag{S7.72}$$

or

$$\log\left(\frac{1 - f_n}{f_n}\right) = \alpha + \beta E_n \tag{S7.73}$$

Exponentiating both sides, this can easily be solved for f_n.

$$f_n = \frac{1}{e^{\alpha + \beta E_n} + 1} \tag{S7.74}$$

which is exactly the usual Fermi–Dirac distribution function. (b) In order to identify S^o with S_{var}, we note that, according to Eq. (7.39),

$$\psi = S^o_{\text{FD}} - \alpha N - \beta E = \sum_n \log(1 + e^{-\alpha - \beta E_n}) \tag{S7.75}$$

But $N = \sum f_n$ and $E = \sum f_n E_n$. Therefore,

$$S^o_{\text{FD}} = \sum_n \left[\log(1 + e^{-\alpha - \beta E_n}) + (\alpha + \beta E_n) f_n \right] \tag{S7.76}$$

Using the equilibrium relation $f_n = (e^{\alpha + \beta E_n} + 1)^{-1}$, we can derive the following identities, that will allow us to express S^o_{FD} entirely in terms of f_n.

$$e^{-\alpha - \beta E_n} = \frac{f_n}{1 - f_n} \tag{S7.77}$$

and

$$\alpha + \beta E_n = \log\left(\frac{1 - f_n}{f_n}\right) \tag{S7.78}$$

Thus

$$\begin{aligned}
S^o_{\text{FD}} &= \sum_n \left[\log\left(1 + \frac{f_n}{1 - f_n}\right) + f_n \log\left(\frac{1 - f_n}{f_n}\right) \right] \\
&= \sum_n \left[-\log(1 - f_n) + f_n \log(1 - f_n) - f_n \log f_n \right] \\
&= -\sum_n \left[f_n \log f_n + (1 - f_n) \log(1 - f_n) \right]
\end{aligned} \tag{S7.79}$$

which agrees with S_{var}.

Exercise 7.9   Repeat the last exercise for a boson system, using the following variational entropy function.

$$S_{\text{var}} = -\sum_n \left[f_n \log f_n - (1 + f_n) \log(1 + f_n) \right] \tag{S7.80}$$

Solution   (a) The quantity to be maximized is

$$F = -\sum_n \left[f_n \log f_n - (1 + f_n) \log(1 + f_n) \right] - \alpha \sum_n f_n - \beta \sum_n f_n E_n \tag{S7.81}$$

Setting $\partial F / \partial f_n = 0$ gives

$$-\log f_n + \log(1 + f_n) - \alpha - \beta E_n = 0 \tag{S7.82}$$

or

$$\log\left(\frac{1 + f_n}{f_n}\right) = \alpha + \beta E_n \tag{S7.83}$$

which implies that

$$f_n = \frac{1}{e^{\alpha + \beta E_n} - 1} \tag{S7.84}$$

(b) Equation (7.37) gives the following expression for the entropy of an ideal Bose–Einstein system.

$$S^o_{\text{BE}} = -\sum_n \log(1 - e^{-\alpha - \beta E_n}) + \sum_n (\alpha + \beta E_n) f_n \tag{S7.85}$$

and Eq. (7.44) yields the relation

$$e^{-\alpha - \beta E_n} = f_n/(1 + f_n)$$

These relations allow us to express S^o_{BE} entirely in terms of f_n.

$$
\begin{aligned}
S^o_{\text{BE}} &= -\sum_n \left[\log\left(1 - \frac{f_n}{1 + f_n}\right) - f_n \log\left(\frac{1 + f_n}{f_n}\right)\right] \\
&= -\sum_n \left[f_n \log f_n - (1 + f_n) \log(1 + f_n)\right]
\end{aligned}
\tag{S7.86}
$$

Exercise 7.10 If almost all single-particle quantum states have average occupation numbers much less than one, then both Fermi–Dirac and Bose–Einstein statistics can be approximated by Boltzmann statistics. Show that, in that case, it is convenient to work with a canonical ensemble and that the canonical potential is given by

$$\phi(N, \beta) = N\left[\log(Z_1/N) + 1\right] \tag{S7.87}$$

where the single-particle partition function, Z_1, is given by the following sum over single-particle quantum states.

$$Z_1(\beta) = \sum_k e^{-\beta \varepsilon_k} \tag{S7.88}$$

Solution If $e^{-\alpha - \beta \varepsilon_k} \ll 1$, then $\log(1 + \zeta e^{-\alpha - \beta \varepsilon_k}]) \approx \zeta e^{-\alpha - \beta \varepsilon_k}$, where $\zeta = \pm 1$, depending upon the type of statistics satisfied by the particles, and

$$
\begin{aligned}
\psi(\alpha, \beta) &= \zeta \sum_k \log\left(1 + \zeta e^{-\alpha - \beta \varepsilon_k}\right) \\
&\approx \sum_k e^{-\alpha - \beta \varepsilon_k} \\
&= e^{-\alpha} Z_1(\beta)
\end{aligned}
\tag{S7.89}
$$

Since

$$N = -\frac{\partial \psi}{\partial \alpha} = e^{-\alpha} Z_1 \tag{S7.90}$$

it is easy to solve for α as a function of N and β. (This cannot be done outside the Boltzmann approximation.)

$$\alpha = -\log(N/Z_1) \tag{S7.91}$$

The canonical potential is constructed by expressing $\psi + \alpha N$ as a function of N and β.

$$\begin{aligned}
\phi &= e^{-\alpha} Z_1(\beta) + \alpha N \\
&= N + N \log(Z_1/N) \\
&= N\big[\log(Z_1/N) + 1\big]
\end{aligned} \tag{S7.92}$$

Exercise 7.11   The density of states function, $D(\varepsilon)$, is defined by saying that the number of single-particle quantum states with energy in the interval $d\varepsilon$ is equal to $D(\varepsilon)\, d\varepsilon$. Calculate the density of states function for: (a) a three-dimensional gas of spin-0 particles satisfying periodic boundary conditions in a cube of volume L^3, and (b) the same system satisfying hard-wall boundary conditions $[\psi(x, y, z) = 0$ on the walls].

Solution   (a) For periodic boundary conditions in three dimensions, the allowed wave vectors form a three-dimensional cubic lattice in k space with a density $(2\pi/L)^3$. For a zero-spin particle there is only one eigenstate for each wave vector. The number of states with energy less than $\varepsilon = \hbar^2 k^2/2m$ is

$$\begin{aligned}
N(\varepsilon) &= (4\pi k^3/3)/(2\pi/L)^3 \\
&= \frac{L^3 k^3}{6\pi^2} \\
&= \frac{(2m)^{3/2} L^3}{6\pi^2 \hbar^3} \varepsilon^{3/2}
\end{aligned} \tag{S7.93}$$

The density of states function is the derivative of $N(\varepsilon)$.

$$D(\varepsilon) = N'(\varepsilon) = \frac{(2m)^{3/2} L^3}{4\pi^2 \hbar^2} \varepsilon^{1/2} \tag{S7.94}$$

(b) The eigenfunctions of the Schrödinger equation

$$-\frac{\hbar^2}{2m}\left(\frac{\partial^2 \psi}{\partial x^2} + \frac{\partial^2 \psi}{\partial y^2} + \frac{\partial^2 \psi}{\partial z^2}\right) = \varepsilon \psi \tag{S7.95}$$

that satisfy the boundary conditions

$$\begin{aligned}
0 &= \psi(0, y, z) = \psi(x, 0, z) = \psi(x, y, 0) \\
&= \psi(L, y, z) = \psi(x, L, z) = \psi(x, y, L)
\end{aligned} \tag{S7.96}$$

are of the form

$$\psi(x, y, z) = (2/L)^{3/2} \sin(n_1 \pi x/L) \sin(n_2 \pi y/L) \sin(n_3 \pi z/L) \tag{S7.97}$$

where n_1, n_2, and n_3 have the values $1, 2, 3, \ldots$. The corresponding energy eigenvalue is

$$\varepsilon_{n_1 n_2 n_3} = \frac{\hbar^2 \pi^2}{2mL^2}(n_1^2 + n_2^2 + n_3^2) \tag{S7.98}$$

Plotting the values of n_1, n_2, and n_3 in a three-dimensional n space, we see that there is one point for each unit cube in the positive octant. The condition that $\varepsilon_{n_1 n_2 n_3} < \varepsilon$

is the same as the condition that $n_1^2 + n_2^2 + n_3^2 < R^2$, where $R^2 = 2m\varepsilon/\pi^2\hbar^2$. The number of eigenstates that satisfy that condition is equal to $1/8$ of the volume of a sphere of radius R.

$$N(\varepsilon) = (1/8)(4\pi R^3/3)$$

$$= \frac{\pi}{6}\left(\frac{2mL^2\varepsilon}{\hbar^2\pi^2}\right)^{3/2} \tag{S7.99}$$

$$= \frac{(2m)^{3/2}L^3}{6\pi^2\hbar^3}\varepsilon^{3/2}$$

Since this is identical to Eq. (S7.93), the density of states function is unaffected by the change in the boundary conditions.

Exercise 7.12   For a system of fermions with a density of states function $D(\varepsilon)$, determine the dependence of the chemical potential on temperature for fixed N and $kT \ll \varepsilon_F$.

Solution   The average occupation of a quantum state of energy ε is $\{\exp[(\varepsilon - \mu)/kT] + 1\}^{-1}$. The number of such quantum states within the energy interval $d\varepsilon$ is $D(\varepsilon)\, d\varepsilon$. Thus N is given in terms of the density of states function by

$$N = \int_o^\infty \frac{D(\varepsilon)\, d\varepsilon}{e^{(\varepsilon - \mu)/kT} + 1} \tag{S7.100}$$

For $kT \ll \varepsilon_F$ the integral can be evaluated to order T^2 by using Eq. (7.55).

$$N = \int_o^\mu D(\varepsilon)\, d\varepsilon + \frac{\pi^2 k^2 T^2}{6}D'(\mu) \tag{S7.101}$$

At $T = 0$, $\mu = \varepsilon_F$. Therefore, at low temperature, one can write μ as $\varepsilon_F + \delta\mu$ and expand Eq. (S7.101) to first order in $\delta\mu$.

$$N = \int_o^{\varepsilon_F} D(\varepsilon)\, d\varepsilon + D(\varepsilon_F)\,\delta\mu + \frac{\pi^2 k^2 T^2}{6}\left[D'(\varepsilon_F) + D''(\varepsilon_F)\,\delta\mu\right] \tag{S7.102}$$

Using the fact that ε_F is defined by the relation

$$N = \int_o^{\varepsilon_F} D(\varepsilon)\, d\varepsilon \tag{S7.103}$$

and dropping the term proportional to $T^2\,\delta\mu$, which is of higher order than T^2, one obtains the following formula for the chemical potential shift.

$$\delta\mu = -\frac{\pi^2 k^2 T^2}{6}\frac{D'(\varepsilon_F)}{D(\varepsilon_F)} \tag{S7.104}$$

Therefore

$$\mu(T) = \varepsilon_F - \frac{\pi^2 k^2 T^2}{6}\frac{D'(\varepsilon_F)}{D(\varepsilon_F)} \tag{S7.105}$$

Exercise 7.13   Write the specific heat of a system of fermions as a one-dimensional integral involving the density of states function and show that at low temperature it is only the density of states at the Fermi energy, and not its derivative, that is important in determining C_V.

Solution   The average occupation of a quantum state of energy ε is $\{\exp[(\varepsilon - \mu)/kT] + 1\}^{-1}$. The number of such quantum states within the energy interval $d\varepsilon$ is $D(\varepsilon)\,d\varepsilon$. Thus the total energy of a fermion system at temperature T is

$$E = \int_o^\infty \frac{\varepsilon\, D(\varepsilon)\, d\varepsilon}{e^{(\varepsilon - \mu)/kT} + 1} \tag{S7.106}$$

Using the first two terms in the low-temperature expansion for a general Fermi–Dirac integral, given by Eq. (7.55), we can write E as

$$E = \int_o^\mu \varepsilon\, D(\varepsilon)\, d\varepsilon + \frac{\pi^2 k^2 T^2}{6}\big[D(\mu) + \mu D'(\mu)\big] \tag{S7.107}$$

In calculating $\partial E/\partial T$, to obtain the specific heat, we must remember that, at fixed N, the chemical potential μ is a function of T, given by Eq. (S7.105). Thus

$$C_V = \frac{\partial E}{\partial T} = \varepsilon_F D(\varepsilon_F)\frac{d\mu}{dT} + \frac{\pi^2 k^2 T}{3}\big[D(\varepsilon_F) + \varepsilon_F D'(\varepsilon_F)\big] \tag{S7.108}$$

where we have ignored terms that are proportional to $T^2\mu'(T)$, because they are of order T^3. Using the fact that, according to Eq. (S7.105),

$$\frac{d\mu}{dT} = -(\pi^2 k^2 T/3)\big[D'(\varepsilon_F)/D(\varepsilon_F)\big]$$

we see that the term involving $D'(\varepsilon_F)$ cancels in the formula for C_V, leaving

$$C_V = \frac{\pi^2 k^2 T}{3} D(\varepsilon_F) \tag{S7.109}$$

Exercise 7.14   At low temperature, the electronic specific heat of a metal is a linear function of temperature while the vibrational specific heat goes as T^3. For any metal, we can define a crossover temperature T_{cr} as the temperature at which $C_{el} = C_{vib}$. Far below T_{cr} the vibrational specific heat can be neglected. Far above T_{cr} the electronic specific heat is negligible. For sodium, plot C_{el}, C_{vib}, and C_{tot} in the range $0 < T < 2T_{cr}$.

Solution   We assume, correctly, that T_{cr} lies well below ε_F/k and T_D. Thus we can use the low-temperature approximations for both C_{el} and C_{vib}. Then the two contributions to the specific heat per atom are

$$C_{el} = \frac{\pi k^2 T}{2\varepsilon_F} \tag{S7.110}$$

and

$$C_{vib} = \frac{12\pi^4 k}{5}\frac{T^3}{T_D^3} \tag{S7.111}$$

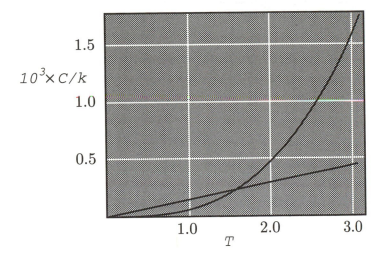

Fig. S7.3 The vibrational and electronic contributions to the specific heat of sodium at low temperatures.

Setting C_{el} equal to C_{vib}, we obtain, for the crossover temperature,

$$T_{cr}^2 = \frac{5kT_D^3}{24\pi^3\varepsilon_F} \tag{S7.112}$$

For sodium, $\varepsilon_F/k = 10,000\,\mathrm{K}$ and $T_D = 158\,\mathrm{K}$, which gives $T_{cr} = 1.63\,\mathrm{K}$. C_{el}/k and C_{vib}/k, over the range, $0 < T < 3.2\,\mathrm{K}$, are shown in Fig. S7.3.

Exercise 7.15 A nonuniform ideal gas of spin-$\frac{1}{2}$ fermions is in an external potential field $U(z)$ that depends only on the z coordinate. Assuming that the gas is restricted to the region $z > 0$, that the density of fermions at $z = 0$ is n_o, and that the density varies slowly enough that the equations for a uniform Fermi–Dirac gas can be used locally, derive an equation for the density function $n(z)$ that minimizes the total energy at $T = 0$.

Solution At $T = 0$ the Fermi energy of an ideal Fermi gas is related to the particle density by

$$n = \tfrac{4}{3}\gamma\varepsilon_F^{3/2} \tag{S7.113}$$

where $\gamma = 2\pi(2m/h^2)^{3/2}$. Because the density varies with height, the Fermi energy will be a function of z also. At height z, all the momentum states up to an energy $\varepsilon_F(z)$ will be filled and all others empty. This fact, that the momentum distribution of the particles is different at different locations in a nonuniform Fermi gas, contrasts with the situation in a classical gas, where the spatial density may vary from place to place but the momentum distribution is the same everywhere. Therefore, the problem boils down to that of finding the function $\varepsilon_F(z)$ that yields the lowest possible energy with a fixed number of particles. If $\varepsilon_F(z)$ is the desired optimal function, then it should be impossible to decrease the total energy by moving a particle from one height z_1 to another height z_2. If we wanted to lower the energy by moving a particle from z_1 to z_2, then we would take the most energetic particle at z_1, which would have a kinetic energy $\varepsilon_F(z_1)$ and a potential energy $U(z_1)$, and put it into the lowest-energy empty state at z_2, which would have a kinetic energy $\varepsilon_F(z_2)$ and a potential energy $U(z_2)$. For the minimum energy distribution, those two states should have the same energy. Thus

$$\varepsilon_F(z_1) + U(z_1) = \varepsilon_F(z_2) + U(z_2) \tag{S7.114}$$

for any z_1 and z_2. If we assume that $U(0) = 0$, then Eq. (S7.114) implies that

$$\varepsilon_F(z) + U(z) = \varepsilon_F(0) \tag{S7.115}$$

Using Eq. (S7.113) to relate $\varepsilon_F(z)$ to the density at z, we get the following equation for the density profile.

$$n(z) = \tfrac{4}{3}\gamma\big[\varepsilon_F(0) - U(z)\big]^{3/2} \tag{S7.116}$$

Note that, at any point where $U(z) > \varepsilon_F(0)$, the density is not given by Eq. (S7.116), which would yield a complex number for $n(z)$, but instead is given by the restriction that $n(z) \geq 0$, which implies that $n(z) = 0$ when $U(z) > \varepsilon_F(0)$.

Exercise 7.16   It is reasonable to treat the collection of electrons in a large atom as a nonuniform Fermi gas at zero temperature. The Fermi energy is then a function of the radial coordinate $\varepsilon_F(r)$. By combining Eq. (S7.116) with Poisson's equation for the electrostatic potential function, namely,

$$\nabla^2\phi = -\frac{\rho}{\epsilon_0} = \frac{e}{\epsilon_0}n(r) \tag{S7.117}$$

one obtains a differential equation for the electrostatic potential within any large atom. The equation so obtained is the *Thomas–Fermi equation.*

Solution   For a neutral atom, $\phi = 0$ at $r = \infty$. Equation (S7.116) should therefore be modified to relate the density at location r to the density at $r = \infty$, which is zero. Using the fact that $U(r) = -e\phi(r)$, the equation becomes

$$n(r) = \tfrac{4}{3}\gamma e^{3/2}\big[\phi(r)\big]^{3/2} \tag{S7.118}$$

But $n(r) = (\epsilon_0/e)\nabla^2\phi(r)$. When these two equations are combined, and the value of γ is recalled, one gets the Thomas–Fermi differential equation for $\phi(r)$.

$$\nabla^2\phi(r) = \sigma\phi^{3/2}(r) \tag{S7.119}$$

with

$$\sigma = \frac{(2m)^{3/2}e^{5/2}}{3\pi^2\epsilon_0\hbar^2} \tag{S7.120}$$

Exercise 7.17   The Thomas–Fermi equation does not contain the essential parameter Z, the atomic number of the atom. Certainly $\phi(r)$ cannot be the same for every atom. Determine where the atomic number Z should come into the Thomas–Fermi theory.

Solution   The atomic number Z tells one the amount of charge that resides in the nucleus at $r = 0$. As r goes to zero the electrostatic potential should approach a Coulomb potential due to a charge $Q = Ze$. That is, as $r \to 0$,

$$E_r(r) = -\frac{d\phi(r)}{dr} \to \frac{Ze}{4\pi\epsilon_0}\frac{1}{r^2} \tag{S7.121}$$

or

$$r^2\phi'(r) \to -\frac{Ze}{4\pi\epsilon_0} \tag{S7.122}$$

The atomic number appears only as a boundary condition at $r = 0$. The boundary condition at $r = \infty$ has already been used in the derivation of the equation. It is that $\phi(r) \to 0$ as $r \to \infty$.

Exercise 7.18   (a) Show that, by making a transformation of the form

$$\phi(r) = \frac{Ze}{4\pi\epsilon_0 r}u(r/a) \tag{S7.123}$$

one can eliminate all physical constants in both the Thomas–Fermi equation and the boundary condition at $r = 0$. Notice that the prefactor, $Ze/4\pi\epsilon_0 r$, is just the simple Coulomb potential of a point charge Ze. (b) Determine the size of the necessary scaling parameter a.

Solution   We first consider the boundary condition. From Eq. (S7.123), we see that

$$r^2\phi'(r) = -\frac{Ze}{4\pi\epsilon_0}u(r/a) + r\frac{Ze}{4\pi\epsilon_0 a}u'(r/a) \tag{S7.124}$$

The boundary condition $r^2\phi'(r) \to -Ze/4\pi\epsilon_0$ will obviously be satisfied if $u(0) = 1$ and $u'(0)$ is any finite number.

In order to transform the Thomas–Fermi equation, we introduce the variable $x = r/a$. Then

$$\frac{d}{dr} = \frac{1}{a}\frac{d}{dx} \tag{S7.125}$$

and

$$\nabla^2\phi(r) = \frac{1}{r^2}\frac{d}{dr}\left(r^2\frac{d\phi}{dr}\right) = \frac{1}{a^2 x^2}\frac{d}{dx}\left(x^2\frac{d\phi}{dx}\right) \tag{S7.126}$$

In terms of x, Eq. (S7.123) is

$$\phi = \frac{Ze}{4\pi\epsilon_0 a}\frac{u(x)}{x} \tag{S7.127}$$

Then

$$\frac{d}{dx}\left(x^2\frac{d\phi}{dx}\right) = \frac{Ze}{4\pi\epsilon_0 a}xu''(x) \tag{S7.128}$$

and the Thomas–Fermi equation becomes

$$\frac{Ze}{4\pi\epsilon_0 a^3}\frac{u''(x)}{x} = \sigma\left(\frac{Ze}{4\pi\epsilon_0 a}\right)^{3/2}\frac{u^{3/2}(x)}{x^{3/2}} \tag{S7.129}$$

The ugly constants on both sides cancel if $ze/4\pi\epsilon_0 a^3 = \sigma(Ze/4\pi\epsilon_0 a)^{3/2}$, which gives the required value of a.

$$a = \frac{1}{2}\left(\frac{3\pi}{4}\right)^{2/3}\frac{a_o}{Z^{1/3}} \tag{S7.130}$$

where the Bohr radius $a_o = 4\pi\epsilon_0\hbar^2/me^2$. With these substitutions, $u(x)$ satisfies the equation

$$\sqrt{x}\,u''(x) = u^{3/2}(x) \tag{S7.131}$$

and the boundary conditions

$$u(0) = 1 \quad \text{and} \quad u(\infty) = 0 \tag{S7.132}$$

Exercise 7.19   For an electron whose energy is much larger than $m_e c^2$, the relation between energy and momentum can be approximated by the extreme relativistic relation

$$\varepsilon = cp \tag{S7.133}$$

This approximation can be used for an electron gas in two situations: (a) if the electron density is so high that $\varepsilon_F \gg m_e c^2$, or (b) if the temperature is so high that $kT \gg m_e c^2$, regardless of the electron density.

For a highly relativistic electron gas, derive the density of states function $D(\varepsilon)$ and the relationship between the Fermi energy ε_F and the electron density.

Solution   Consider an electron in a periodic cube of side L. The relativistic single-particle states are still plane waves with wave vectors

$$\mathbf{k} = \left(\frac{2\pi}{L}K_x, \frac{2\pi}{L}K_y, \frac{2\pi}{L}K_z\right) \tag{S7.134}$$

where K_x, K_y, and K_z are integers and $\mathbf{p} = \hbar\mathbf{k}$. Therefore, the allowed momentum states will form a cubic lattice in momentum space with lattice spacing h/L, as shown in Fig. 2.5. For large ε, the number of eigenstates with energy less than ε, taking account of the two possible spin values and the fact that $V = L^3$, is

$$\begin{aligned} N(\varepsilon) &= 2\tfrac{4}{3}\pi\left(\frac{\varepsilon}{c}\right)^3 \Big/ (h/L)^3 \\ &= \frac{8\pi V}{3(ch)^3}\varepsilon^3 \end{aligned} \tag{S7.135}$$

The density of states function is

$$D(\varepsilon) = N'(\varepsilon) = \frac{8\pi V}{(ch)^3}\varepsilon^2 \tag{S7.136}$$

The Fermi energy is determined by the condition that the total number of single-particle states below ε_F is equal to the number of electrons in the gas.

$$N(\varepsilon_F) = \frac{8\pi V}{3(ch)^3}\varepsilon_F^3 = N \tag{S7.137}$$

or, in terms of the electron density $n = N/V$,

$$\varepsilon_F = (3/8\pi)^{1/3}chn^{1/3} \tag{S7.138}$$

Exercise 7.20   At what temperature is $m_e c^2 = kT$?

Solution

$$T = \frac{m_e c^2}{k} = 5.93 \times 10^9 \, \text{K} \tag{S7.139}$$

Such astronomical temperatures occur only in astronomy, in particular, during supernovas and other such explosive events.

Exercise 7.21   At what density is $\varepsilon_F = m_e c^2$?

Solution   In answering this question, one can use neither the nonrelativistic formula, which is valid only if $\varepsilon_F \ll m_e c^2$, nor the extreme relativistic formula, valid when $\varepsilon_F \gg m_e c^2$. Since the momentum spectrum is unchanged in going from the nonrelativistic to the relativistic case, it is best to define a Fermi momentum p_F that separates the filled from the unfilled momentum states. If n is the electron density, then

$$n = 2 \frac{4\pi}{3} p_F^3 / h^3 \tag{S7.140}$$

The relativistic relation between energy and momentum is

$$E = c\sqrt{p^2 + m^2 c^2} \tag{S7.141}$$

However, the energy given in this formula includes the rest mass energy, $m_e c^2$. The ε_F referred to in the question is the kinetic energy, $E - m_e c^2$. Thus, for a particle at the Fermi surface,

$$\varepsilon_F = c\sqrt{p_F^2 + m_e^2 c^2} - m_e c^2 \tag{S7.142}$$

Setting $\varepsilon_F = m_e c^2$ gives

$$p_F = \sqrt{3} m_e c \tag{S7.143}$$

and, using Eq. (S7.140), one can determine the required electron density.

$$n = 8\pi\sqrt{3}\left(\frac{m_e c}{h}\right)^3 = 3.04 \times 10^{36} \, \text{m}^{-3} \tag{S7.144}$$

Electron densities of this magnitude actually occur within white dwarf stars.

Exercise 7.22   Consider an ideal gas that is a mixture of spin-$\frac{1}{2}$ fermions, which we will call A particles, and spin-0 bosons, called B particles. The A particles have mass m, and the B particles have mass $M > 2m$. Assume that a particle transformation reaction

$$2A \leftrightarrow B \tag{S7.145}$$

is possible and that $M - 2m \ll m$. Determine the ratio of A particles to B particles as a function of the conserved particles density, $n_o = (2N_A + N_B)/V$, at very low temperature.

Solution   Ordinarily, nothing depends on what we take as the zero of our energy scale. However, when chemical reactions are possible (and we can treat this elementary particle reaction as a generalized chemical reaction), then it becomes important to measure the energies of both particles relative to the *same* zero-energy state. If we assume that a stationary A particle has zero energy, then a B particle of momentum $\mathbf{p}$ has an energy

$$E = \varepsilon_o + p^2/2M \tag{S7.146}$$

where $\varepsilon_o = (M - 2m)c^2$. This is what is needed in order for the reaction $B \to 2A$ to conserve energy. But this shift of the B particle energy scale has some effects on the formulas that were developed in the sections on the ideal Bose gas. For example, if we let ε denote the kinetic energy (that is, $\varepsilon = p^2/2M$), then the density of single-particle quantum states is $V\gamma\varepsilon^{1/2}$, and Eq. (7.65) for the particle density as a function of α becomes

$$n = \gamma \int_o^\infty \frac{\varepsilon^{1/2}\, d\varepsilon}{e^{\alpha+\beta(\varepsilon_o+\varepsilon)} - 1} \tag{S7.147}$$

The minimum value of the affinity is no longer $\alpha = 0$, but is given by $\alpha + \beta\varepsilon_o = 0$. That is,

$$\alpha = -\beta\varepsilon_o \tag{S7.148}$$

Equation (7.70) for the number of condensate particles in a phase II state is

$$N_{\mathbf{p}=0} = \frac{1}{e^{\alpha+\beta\varepsilon_o} - 1} \tag{S7.149}$$

At very low temperature it is safe to assume that the Bose gas is in a phase II state, which implies that its affinity is approximately $-\beta\varepsilon_o$.

$$\alpha_B \approx -\beta\varepsilon_o \tag{S7.150}$$

At very low temperature the chemical potential of the A particle gas is equal to its Fermi energy, and therefore its affinity is

$$\alpha_A = -\beta\varepsilon_F \tag{S7.151}$$

The reaction equilibrium equation is

$$2\alpha_A = \alpha_B \tag{S7.152}$$

which implies that

$$\varepsilon_F = \tfrac{1}{2}\varepsilon_o \tag{S7.153}$$

The condition $M - 2m \ll m$ allows us to use nonrelativistic formulas relating the Fermi energy to the density of A particles. That is,

$$n_A = \tfrac{4}{3}\gamma\varepsilon_F^{3/2} = \frac{8\pi m^{3/2}}{3h^3}\varepsilon_o^{3/2} \tag{S7.154}$$

As the temperature approaches zero, all other particles go into the zero-momentum boson state. Thus the density of B particles is given by the conservation equation

$$n_A + 2n_B = n_o \tag{S7.155}$$

That is

$$n_B = \frac{1}{2}\left(n_o - \frac{8\pi m^{3/2}}{3h^3}\varepsilon_o^{3/2}\right) \tag{S7.156}$$

If this equation gives a negative value for n_B, then actually $n_B = 0$ and the Fermi energy of the A particles is less that $\varepsilon_o/2$. It is given by setting $n_A = n_o$. That is

$$\varepsilon_F = (3n_o/4\gamma)^{2/3} \tag{S7.157}$$

Exercise 7.23 Redo the previous exercise for two-dimensional Fermi–Dirac and Bose–Einstein gases at finite temperature, assuming that $\varepsilon_o = 0$, so that $M = 2m$, and determine the density of fermions in the limits of high and low temperature at fixed n_o.

Solution   The most important difference between the two- and three-dimensional cases is that in two dimensions there is no Bose–Einstein condensation. Exercises 2.25 and 2.26 give the relationship between particle density and affinity for fermions and bosons in two dimensions. Equation (S2.114) is written for spinless fermions. For the spin-$\frac{1}{2}$ A particles the relation becomes

$$n_A = \frac{mkT}{\pi\hbar^2}\left[\log(1 + e^{\alpha_A}) - \alpha_A\right] \tag{S7.158}$$

For the B particles the relation between n_B and α_B is given by Eq. (S2.118), with m replaced by $2m$.

$$n_B = \frac{mkT}{\pi\hbar^2}\left[\alpha_B - \log(e^{\alpha_B} - 1)\right] \tag{S7.159}$$

n_A and n_B are to be determined by the two equations

$$n_A + 2n_B = n_o \quad \text{and} \quad 2\alpha_A = \alpha_B \tag{S7.160}$$

The first equation expresses the conservation of A-atoms and the second is the condition for reaction equilibrium. The fermion affinity can be any real number, but the boson affinity must be positive in order to keep the argument of the logarithm in Eq. (S7.159) positive. However, the condition for reaction equilibrium then demands that the fermion affinity must also be positive. Setting $\alpha_B = 2\alpha_A$, writing α_A simply as α, and defining a parameter $\nu_o = \pi\hbar^2 n_o/mkT$, the particle conservation equation, $n_A + 2n_B = n_o$, can be written in the form

$$\log(e^\alpha + 1) + 3\alpha - 2\log(e^{2\alpha} - 1) = \nu_o \tag{S7.161}$$

As $T \to 0$ at fixed n_o, the parameter $\nu_o \to \infty$. Equation (S7.161) then requires that $\alpha \to 0$. For very small values of α,

$$n_A \approx \frac{mkT}{\pi\hbar^2}\log 2 \tag{S7.162}$$

and

$$n_B = \tfrac{1}{2}(n_o - n_A) \approx \frac{n_o}{2} - \frac{mkT}{2\pi\hbar^2}\log 2 \tag{S7.163}$$

Thus, at low temperature, the reaction equilibrium shifts more and more towards the bosons.

At high temperatures $\nu_o \to 0$ and Eq. (S7.161) requires that $\alpha \to \infty$. This can be seen by assuming that α is very large and using the following expansions.

$$\log(e^\alpha + 1) = \log[e^\alpha(1 + e^{-\alpha})] \approx \alpha + e^{-\alpha} \tag{S7.164}$$

$$\log(e^{2\alpha} - 1) = \log[e^{2\alpha}(1 - e^{-2\alpha})] \approx 2\alpha - e^{-2\alpha} \qquad (S7.165)$$

But $e^{-2\alpha}$ is negligible in comparison with $e^{-\alpha}$ for large α. Thus Eq. (S7.161) becomes

$$e^{-\alpha} = \nu_o \qquad (S7.166)$$

or

$$\alpha = -\log \nu_o \qquad (S7.167)$$

In that limit the density of fermions is

$$
\begin{aligned}
n_A &= \frac{mkT}{\pi\hbar^2}\left[\log\left(1 + \frac{1}{\nu_o}\right) + \log \nu_o\right]\\
&= \frac{mkT}{\pi\hbar^2}\log(1 + \nu_o)\\
&\approx \frac{mkT}{\pi\hbar^2}\nu_o\\
&= n_o
\end{aligned}
\qquad (S7.168)
$$

Thus $n_B \approx 0$. This could have been predicted. At fixed n_o and high T the particles can be treated as classical particles. Since the binding energy is zero, the "molecules" all dissociate at high T.

Exercise 7.24   In a particular frame, a system of particles has an energy

$$E = \frac{1}{2}\sum m_i v_i^2 + U(\mathbf{r}_1, \ldots, \mathbf{r}_N) \qquad (S7.169)$$

and a momentum

$$\mathbf{P} = \sum m_i \mathbf{v}_i \qquad (S7.170)$$

The frame mentioned moves with velocity $\mathbf{V}$ through another Galilean inertial frame. Show that the energy and momentum of the system, as measured in the second frame, are

$$E' = E + \mathbf{V} \cdot \mathbf{P} + \tfrac{1}{2}MV^2 \qquad (S7.171)$$

and

$$\mathbf{P}' = \mathbf{P} + M\mathbf{V} \qquad (S7.172)$$

where $M = \sum m_i$ is the total mass of the system. This *Galilean transformation* of energy and momentum will be needed in order to calculate the normal fluid density in liquid helium at low temperature.

Solution   In the new frame, particle i will have velocity $\mathbf{v}_i + \mathbf{V}$. Since in a Galilean transformation relative distances of particles do not change, the potential energy is unchanged and therefore

$$
\begin{aligned}
E' - E &= \frac{1}{2}\sum m_i[(\mathbf{v}_i + \mathbf{V})^2 - v_i^2]\\
&= \mathbf{V} \cdot \sum m_i \mathbf{v}_i + \tfrac{1}{2}V^2 \sum m_i\\
&= \mathbf{V} \cdot \mathbf{P} + \tfrac{1}{2}MV^2
\end{aligned}
\qquad (S7.173)
$$

Also

$$\mathbf{P}' = \sum m_i(\mathbf{v}_i + \mathbf{V}) = \mathbf{P} + M\mathbf{V} \qquad (S7.174)$$

Exercise 7.25 Use the Galilean transformation equations to determine the energy and momentum of an excitation of wave vector **k** in a sample of liquid helium that is flowing with velocity **v**.

Solution Consider a bulk sample of the helium, at rest, containing a single excitation of wave vector **k**. Its energy and momentum are

$$E = E_g + \varepsilon_\mathbf{p} \quad \text{and} \quad \mathbf{p} = \hbar\mathbf{k} \tag{S7.175}$$

where E_g is the ground-state energy of the fluid at rest. Viewed in a frame in which the liquid is moving at velocity **v**, its energy and momentum are

$$\begin{aligned}
E' &= E + \mathbf{v} \cdot \mathbf{p} + \tfrac{1}{2}Mv^2 \\
&= E_g + \frac{1}{2}Mv^2 + \varepsilon_\mathbf{p} + \mathbf{v} \cdot \mathbf{p}
\end{aligned} \tag{S7.176}$$

and

$$\mathbf{P}' = M\mathbf{v} + \mathbf{p} \tag{S7.177}$$

Clearly, the first two terms in E', namely $E_g + MV^2/2$, represent the energy of the liquid without the excitation. Thus

$$\varepsilon'_\mathbf{p} = \varepsilon_\mathbf{p} + \mathbf{v} \cdot \mathbf{p} \tag{S7.178}$$

is the added energy, due to the existence of the excitation. But that is what we call the excitation energy. Similarly, since $M\mathbf{v}$ represents the momentum of the fluid without the excitation, the momentum of the excitation in the new frame is still $\mathbf{p} = \hbar\mathbf{k}$, just as it was in the old frame.

Exercise 7.26 At very low temperature, one can assume that the excitations in liquid helium constitute a low-density quasiparticle gas of simple phonons. With this assumption, calculate the normal fluid mass density ρ_n as a function of temperature.

Solution We consider a sample of helium, flowing steadily with a velocity **v** through a stationary pipe. The energy of an excitation of momentum **p** is $\varepsilon_\mathbf{p} = cp + \mathbf{v} \cdot \mathbf{p}$, where c is the velocity of sound in the liquid. Since the excitations are nonconserved bosons, $\alpha = 0$, and thus the average number of excitations per quantum state at temperature T is

$$(e^{\beta(cp+\mathbf{v}\cdot\mathbf{p})} - 1)^{-1} \equiv f(\beta cp + \beta\mathbf{v} \cdot \mathbf{p}) \tag{S7.179}$$

The number of quantum states in the momentum range $d^3\mathbf{p}$ is $(V/h^3)d^3\mathbf{p}$. Therefore, the momentum density due to the phonon gas is

$$\vec{\pi} = \frac{1}{h^3} \int \mathbf{p}f(\beta cp + \beta\mathbf{v} \cdot \mathbf{p})\, d^3\mathbf{p} \tag{S7.180}$$

If we choose the x axis in the direction of **v**, then, by symmetry, the only nonzero component of $\vec{\pi}$ must be π_x. We can then write Eq. (S7.180) as a scalar equation

$$\pi = \frac{1}{h^3} \int p_x f(\beta cp + \beta v p_x)\, d^3\mathbf{p} \tag{S7.181}$$

If $v \ll c$, which we may assume, since we are interested in calculating the ratio of π to v for small v, then we can expand the integrand to first order in v.

$$
\begin{aligned}
\pi &= \frac{\beta v}{h^3} \int p_x^2 f'(\beta c p) \, d^3\mathbf{p} \\
&= \frac{2\pi \beta v}{h^3} \int_{-1}^{1} \cos^2 \theta \, d(\cos\theta) \int_o^\infty p^2 f'(\beta c p) p^2 \, dp \\
&= \frac{4\pi \beta v}{3h^3} \int_o^\infty f'(\beta c p) p^4 \, dp \\
&= \frac{4\pi v}{3h^3 c^5 \beta^4} \int_o^\infty f'(x) x^4 \, dx \\
&= -\frac{16\pi (kT)^4 v}{3h^3 c^5} \int_o^\infty f(x) x^3 \, dx \\
&= -\frac{16\pi (kT)^4 v}{3h^3 c^5} \int_o^\infty \frac{x^3 \, dx}{e^x - 1} \\
&= -\frac{16\pi (kT)^4 v}{3h^3 c^5} 3! \zeta(4) \\
&= -\frac{16\pi^5 (kT)^4 v}{45 h^3 c^5} \\
&= -\frac{2\pi^2 (kT)^4 v}{45 \hbar^3 c^5}
\end{aligned}
\tag{S7.182}
$$

The normal fluid mass density is the ratio of $\vec{\pi}$ to the pipe velocity, calculated in the superfluid rest frame. Recall that it is the pipe, not the superfluid, that is dragging along the gas of excitations. In the superfluid rest frame the pipe velocity is $-v\hat{x}$. Thus the phonon contribution to the normal fluid mass density is

$$
\rho_n(\text{ph}) = \frac{2\pi^2 (kT)^4}{45 \hbar^3 c^5}
\tag{S7.183}
$$

It should be noted that it was necessary to compute the phonon distribution function in the rest frame of the pipe rather than the rest frame of the superfluid. In the superfluid frame the moving pipe interacts with the phonon gas, and therefore the phonon gas cannot be considered as an isolated system. In the pipe frame, on the other hand, the moving superfluid does not interact with the phonon gas as long as v is less than the critical velocity, because, as we have seen in Section 7.11, the two systems cannot exchange energy and momentum and still satisfy the fundamental conservation laws. Thus, in the pipe frame, the existence of the superfluid can be ignored except for its effect on changing the excitation spectrum of the phonons from $\varepsilon_\mathbf{p} = cp$ to $\varepsilon_\mathbf{p} = cp + \mathbf{v} \cdot \mathbf{p}$.

Exercise 7.27 Calculate the roton contribution to the normal fluid density in liquid helium.

Solution We consider the same situation as was described in Exercise 7.26. The roton spectrum, in the pipe frame, is

$$
\varepsilon_\mathbf{p} = \varepsilon_o + (p - p_o)^2 / 2\mu + v p_x
\tag{S7.184}
$$

Because, below the transition temperature, $kT \ll \varepsilon_o$, we can use the Boltzmann approximation to the Bose–Einstein distribution function. Then

$$\pi = \frac{1}{h^3} \int p_x e^{-\beta[\varepsilon_o + (p - p_o)^2/2\mu + v p_x]} \, d^3\mathbf{p} \tag{S7.185}$$

At $v = 0$ the integral vanishes. Expanding to first order in v, we get

$$\begin{aligned}
\pi &= -\frac{\beta v}{h^3} e^{-\beta \varepsilon_o} \int p_x^2 e^{-\beta(p - p_o)^2/2\mu} \, d^3\mathbf{p} \\
&= -\frac{2\pi \beta v}{h^3} e^{-\beta \varepsilon_o} \int_{-1}^{1} \cos^2\theta \, d(\cos\theta) \int_o^\infty p^2 e^{-\beta(p - p_o)^2/2\mu} p^2 \, dp \\
&= -\frac{4\pi \beta v}{3h^3} e^{-\beta \varepsilon_o} \int_o^\infty e^{-\beta(p - p_o)^2/2\mu} p^4 \, dp
\end{aligned} \tag{S7.186}$$

For the reason given after Eq. (7.113), we can extend the integral to $-\infty$, obtaining, with $q \equiv p - p_o$,

$$\begin{aligned}
\pi &= -\frac{4\pi \beta v}{3h^3} e^{-\beta \varepsilon_o} \int_{-\infty}^\infty e^{-\beta q^2/2\mu} (q^4 + 6 p_o^2 q^2 + p_o^4) \, dq \\
&= -\frac{4\pi^{3/2} \beta v}{3h^3} e^{-\beta \varepsilon_o} \left[\tfrac{3}{4}(2\mu kT)^{5/2} + 3 p_o^2 (2\mu kT)^{3/2} + p_o^4 (2\mu kT)^{1/2} \right]
\end{aligned} \tag{S7.187}$$

Thus

$$\rho_n(\text{rot}) = \frac{4\pi^{3/2}}{3h^3 kT} e^{-\varepsilon_o/kT} \left[\tfrac{3}{4}(2\mu kT)^{5/2} + 3 p_o^2 (2\mu kT)^{3/2} + p_o^4 (2\mu kT)^{1/2} \right] \tag{S7.188}$$

Exercise 7.28   (a) Using Eqs. (S7.183) and (S7.188), plot the theoretical prediction for the normal fluid density as a function of T between 1 and 3 K. (b) The transition temperature is the temperature at which the superfluid density becomes zero and thus ρ_n is equal to the mass density of helium, which is about 146 kg/m^3 at T_λ. (It does not vary strongly with temperature.) By finding the point on the curve at which $\rho_n = \rho$, estimate T_λ.

Solution   The graph of $\rho_n(T)$ in Fig. S7.4 was drawn using Eqs. (S7.183) and (S7.188) and the experimental values

$$\varepsilon_o/k = 8.6 \, \text{K}, \qquad p_o = 2.0 \times 10^{-24} \, \text{kg m/s}$$

$$\mu = 1.0 \times 10^{-27} \, \text{kg}, \qquad c = 244 \, \text{m/s}$$

As shown, $\rho_n = \rho$ at a temperature of about 2.8 K rather than at the actual transition temperature of 2.2 K. It is not surprising that this approximate calculation overestimates T_λ. The approximate excitation energies for both phonons and rotons all lie above the true excitation energy curve; thus the density of excitations at any temperature is underestimated by our approximations. This means that we must go to higher T in order to make $\rho_n = \rho$. Also, the whole theory, based upon noninteracting excitations, becomes less and less accurate as T approaches T_λ and the excitation density becomes very high.

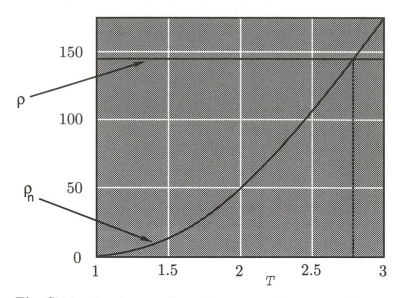

Fig. S7.4   The theoretically predicted value of the normal fluid density equals the experimental value of the total fluid density at about 2.8 K.

Exercise 7.29   Consider a tall evacuated pipe, standing up in the earth's gravitational field. Taking into account that there is a gravitational force on the radiant energy inside the pipe, determine the equilibrium energy density as a function of height.

Solution   Assume that the cross-sectional area of the pipe is A. The radiant energy within a slab of height dz is $E = \varepsilon(z)A\,dz$, where ε is the energy density. The gravitational force on that radiant energy is

$$F_g = \frac{E}{c^2}g = \frac{\varepsilon g}{c^2}A\,dz \tag{S7.189}$$

This force must be balanced by the pressure force, which is given by

$$F_p = Ap(z) - Ap(z + dz) = -\frac{dp}{dz}A\,dz \tag{S7.190}$$

Thus the condition of equilibrium is that

$$\frac{dp}{dz} = -\frac{g}{c^2}\varepsilon(z) \tag{S7.191}$$

Taking into account the fact that for a radiation gas $p = \varepsilon/3$, we obtain a single equation for the energy density.

$$\frac{d\varepsilon(z)}{dz} = -\frac{3g}{c^2}\varepsilon(z) \tag{S7.192}$$

This equation has the solution

$$\varepsilon(z) = \varepsilon(0)\,e^{-3gz/c^2} \tag{S7.193}$$

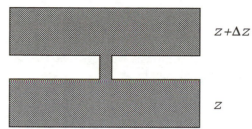

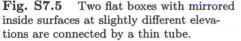

Fig. S7.5   Two flat boxes with mirrored inside surfaces at slightly different elevations are connected by a thin tube.

Exercise 7.30   Wait a minute! The energy density is proportional to the fourth power of the temperature. If the energy density varies with height as was suggested by the last exercise, then the temperature would have to vary with height, which is inconsistent with the condition for thermal equilibrium. We now have a flat contradiction between the condition for mechanical equilibrium and the condition for thermal equilibrium. Which one is right?

Solution   When faced with a paradox such as this, one must search for some fundamental principle that can be completely relied upon. It cannot be solved by manipulating formulas. The most reliable principle we have is that the total entropy of a system is a maximum at equilibrium. Because of the connection between entropy and probability, that principle simply states that a system will go from less probable to more probable macrostates.

Consider a system composed of two flat boxes, one at height z and the other at height $z + \Delta z$, connected by a thin tube (Fig. S7.5). The inside walls are perfect mirrors so that the radiation gas inside is perfectly isolated. When the system comes to equilibrium, the tube is closed, leaving only the two isolated boxes of radiation. What is the condition that they be in equilibrium? Any discussion that involves moving photons from one box to the other but does not use general relativity has to be viewed with suspicion. We must therefore avoid such an analysis. Instead, in the upper box, we picture a process in which an amount of radiant energy equal to $2m_e c^2$ is converted into an electron–positron pair at rest. The electron and positron are allowed to fall, in the weak gravitational field, to the lower box, where, by pair annihilation, their energy is converted back to radiation. Remembering that $\beta = \partial S/\partial E$ by definition, we see that the entropy of the upper box decreases by an amount $2\beta(z + \Delta z)m_e c^2$. At the lower box, the two particles have an energy $2m_e c^2 + 2m_e g \, \Delta z$, and so the entropy of the lower box is increased by an amount $2\beta(z)m_e(c^2 + g \, \Delta z)$. For equilibrium, the total entropy change due to this conceivable but unlikely process must be zero. This gives the equilibrium condition that

$$c^2 \beta(z + \Delta z) = (c^2 + g \, \Delta z)\beta(z) \tag{S7.194}$$

Expanding this to first order in Δz gives the following differential equation for the equilibrium temperature distribution.

$$\frac{d\beta(z)}{dz} = \frac{g}{c^2}\beta(z) \tag{S7.195}$$

When gravitational effects are taken into account, the equilibrium condition is not that the temperature be constant! Now we must check whether this agrees with our previous equation for $\varepsilon(z)$. According to Eq. (7.119),

$$\varepsilon = \frac{\pi^2}{15} \frac{\beta^{-4}}{(\hbar c)^3} \tag{S7.196}$$

Differentiating this with respect to z and using Eq. (S7.195) for $d\beta/dz$ gives the differential equation

$$\frac{d\varepsilon(z)}{dz} = -4\frac{g}{c^2}\varepsilon(z) \tag{S7.197}$$

We have oversolved the problem by a factor of 4/3! Where is our mistake?

In spite of the fact that we have done our best to avoid any situation in which general relativity would have to be invoked, a general relativistic effect has crept in to spoil our solution. Fortunately it is a simple thing that can be understood without the massive mathematical apparatus of general relativity. The condition for mechanical equilibrium was stated to be that the total force on a slab of radiation gas should vanish. It is better to say that the total momentum delivered to the slab of gas in any finite time interval should vanish. The pressure at heights z and $z+\Delta z$ give the rates at which momentum is delivered through the respective surfaces. But a *rate* unavoidably involves a local clock, and clocks do not run at the same rate at different locations in a gravitational field. If we consistently use a clock at z to define our rates, then, because the clock at $z+\Delta z$ runs faster by a factor of $(1+g\,\Delta z/c^2)$, the rate at which momentum is delivered to the slab through the upper surface is $p(z+\Delta z)(1+g\Delta z/c^2)$. The correct equation for mechanical equilibrium is

$$p(z) - (1 + g\,\Delta z/c^2)p(z + \Delta z) = \frac{g}{c^2}\varepsilon(z)\,\Delta z \tag{S7.198}$$

or

$$\frac{dp}{dz} = -\frac{g}{c^2}\left[p(z) + \varepsilon(z)\right] \tag{S7.199}$$

The pressure–energy relation, $p = \varepsilon/3$, then gives the correct differential equation

$$\frac{d\varepsilon}{dz} = -4\frac{g}{c^2}\varepsilon(z) \tag{S7.200}$$

in agreement with Eq. (S7.197). It is Eq. (S7.192) that is wrong.

Supplement to Chapter 8

REVIEW QUESTIONS

8.1 Derive the grand potential ψ for a monatomic ideal gas.

8.2 Derive the cluster expansion, up to order ζ^3, for the grand potential of a real gas.

8.3 In terms of the cluster expansion of ψ, write an expression for the potential energy.

8.4 What is the form of the Lenard–Jones potential?

8.5 From the cluster expansion for ψ, derive the virial expansion for p/kT up to terms of order n^2.

8.6 Describe the phase transition that occurs in a nematic liquid crystal.

8.7 Describe our simple two-dimensional model of a nematic liquid crystal.

8.8 In the following formula for the partition function of the model, what is the function $F(C_x, C_y)$?

$$Z(N_x, N_y, \beta, A) = \frac{1}{N_x! N_y! \lambda^{2(N_x+N_v)}} \int_A F(C_x, C_y) \, dC_x \, dC_y$$

8.9 How was the affinity for x-particles expressed as a ratio of partition functions?

8.10 What was the crucial approximation made in solving the model by the mean-field method?

8.11 Describe, qualitatively, the predictions of the mean-field solution.

8.12 For a system of adsorbed particles, with nearest-neighbor interactions, how was the canonical partition function calculated, using the mean-field approximation?

8.13 Using the mean-field approximation, calculate the partition function for a system of adsorbed particles with nearest-neighbor interactions.

8.14 For a system of interacting adsorbed particles, describe what is wrong with the expression obtained for $\alpha(n)$ by the mean-field method. (n is the particle density.)

8.15 What are the two conditions for equilibrium of the high-density and the low-density phases of the adsorbed particle system?

8.16 Describe the logic behind the Maxwell construction for the phase transition line in the α–n plane.

8.17 For a system with discrete configurations and a transition probability $T(C \to C')$, derive the master equation.

8.18 What is the "condition of detailed balance" in the Monte Carlo method?

8.19 What does the Monte Carlo method actually supply; that is, what is its fundamental output?

8.20 For a system of a single particle on a one-dimensional lattice with an energy function $E(n)$, describe a computer algorithm that implements the Monte Carlo method.

8.21 For a two-dimensional Ising lattice, describe a computer algorithm that implements the Monte Carlo method.

8.22 Describe how one can accurately calculate the specific heat and magnetic susceptibility of an Ising system, using the Monte Carlo method.

8.23 Qualitatively describe diamagnetism and paramagnetism.

8.24 For an atom with a z component of magnetic moment $m_z = \alpha M$, where $-J < M < J$, in a magnetic field B, calculate the magnetic partition function.

8.25 From the magnetic partition function, obtain a formula for the magnetic energy per atom.

8.26 What is Curie's law?

8.27 Describe the process of cooling by adiabatic demagnetization.

8.28 What is the source of the interaction between neighboring atoms that leads to ferromagnetism?

8.29 What is hysteresis?

8.30 Why are magnetic domains formed?

8.31 Describe the mean-field solution for the 3D Ising model.

8.32 Describe the phase diagram for the 3D Ising model.

8.33 Solve the equation $\tanh(T_c \bar{\sigma} / T) = \bar{\sigma}$ for the case where T is slightly below T_c.

8.34 Obtain a formula for the specific heat of a 3D Ising model in terms of the spontaneous magnetization $\bar{\sigma}$ according to the mean-field approximation.

8.35 How does the mean-field prediction of the specific heat $C(T)$ for the 3D Ising model compare with an accurate numerical solution for the same quantity?

EXERCISES

Exercise 8.1   In order to convert from the expansions for n and βp as power series in the activity to the virial equation of state that gives βp as a power series in n, one is faced with the following mathematical problem. Given two expansions

$$x = t + a_2 t^2 + a_3 t^3 + \cdots \tag{S8.1}$$

and

$$y = t + b_2 t^2 + b_3 t^3 + \cdots \tag{S8.2}$$

we wish to determine the coefficients A_2, $A_3, \ldots$ in the expansion

$$y = x + A_2 x^2 + A_3 x^3 + \cdots \tag{S8.3}$$

Illustrate the method of calculating any A_n in terms of the known coefficients a_n and b_n, by calculating A_2, A_3, and A_4.

Solution   In Eq. (S8.3), we substitute the power series expansions for x and y given by Eqs. (S8.1) and (S8.2). (We keep only terms up to order t^4.)

$$\begin{aligned}
t + b_2 t^2 + b_3 t^3 + b_4 t^4 &= t + a_2 t^2 + a_3 t^3 + a_4 t^4 + \cdots \\
&\quad + A_2 (t + a_2 t^2 + a_3 t^3 + \cdots)^2 \\
&\quad + A_3 (t + a_2 t^2 + \cdots)^3 \\
&\quad + A_4 (t + \cdots)^4 + \cdots
\end{aligned} \tag{S8.4}$$

Collecting terms of each order in t gives

$$\begin{aligned}
(b_2 - a_2 - A_2)t^2 &+ (b_3 - a_3 - 2a_2 A_2 - A_3)t^3 \\
&+ (b_4 - a_4 - a_2 A_2 - 2a_3 A_2 - 3a_2 A_3 - A_4)t^4 + \cdots = 0
\end{aligned} \tag{S8.5}$$

For this to be an identity in t, the coefficient of each power of t must vanish separately. Thus

$$A_2 = b_2 - a_2 \tag{S8.6}$$

$$A_3 = b_3 - a_3 - 2a_2 A_2 = b_3 - a_3 - 2a_2 b_2 + 2a_2^2 \tag{S8.7}$$

and

$$\begin{aligned}
A_4 &= b_4 - a_4 - a_2^2 A_2 - 2a_3 A_2 - 3a_2 A_3 \\
&= b_4 - a_4 - (a_2^2 + 2a_3)(b_2 - a_2) - 3a_2(b_3 - a_3 - 2a_2 b_2 + 2a_2^2)
\end{aligned} \tag{S8.8}$$

It is clear that one could calculate any desired A_n in this way, but it is also clear that the terms are becoming rapidly more complicated.

Exercise 8.2 Show that, for a system of particles with short-range forces, the cluster integral C_3 defined in Section 8.3 is independent of V for large V.

Solution $6VC_3$ is defined as the combination of integrals

$$I_3 - 3I_1I_2 + 2I_1^3 = \int d^3r_1 d^3r_2 d^3r_3 e^{-\beta(v_{12}+v_{13}+v_{23})}$$
$$- 3\int d^3r_1 d^3r_2 d^3r_3 e^{-\beta v_{23}} + 2\int d^3r_1 d^3r_2 d^3r_3 \tag{S8.9}$$

The second term is unsymmetrical under permutation of the variables $\mathbf{r}_1$, $\mathbf{r}_2$, and $\mathbf{r}_3$. This lack of symmetry can be eliminated by dropping the factor of 3 and writing 3 equivalent terms by renaming the variables. One then gets

$$6VC_3 = \int d^3r_1 \, d^3r_2 \, d^3r_3 \left(e^{-\beta(v_{12}+v_{13}+v_{23})} - e^{-\beta v_{12}} \right.$$
$$\left. - e^{-\beta v_{13}} - e^{-\beta v_{23}} + 2 \right) \tag{S8.10}$$

$v(r)$ is the interaction potential for two atoms separated by a distance r. Certainly $v(r) \approx 0$ unless r is less than some small distance R of the order of 10 Å. Therefore $e^{-\beta v(r)} \approx 1$ when $r \gg R$. We define a function, called the *Mayer f function* (after the physicist Joseph Mayer), by the formula

$$f(r) = e^{-\beta v(r)} - 1 \tag{S8.11}$$

Then $f(r) \approx 0$ for $r > R$. Writing Eq. (S8.10) in terms of the function f gives

$$6VC_3 = \int d^3r_1 \, d^3r_2 \, d^3r_3 \left(f_{12}f_{13}f_{23} + f_{12}f_{13} + f_{12}f_{23} + f_{13}f_{23} \right) \tag{S8.12}$$

We can do the integral of the first term by introducing variables $\mathbf{r}_1$, $\mathbf{x} = \mathbf{r}_2 - \mathbf{r}_1$, and $\mathbf{y} = \mathbf{r}_3 - \mathbf{r}_1$ and noting that $r_{23} = |\mathbf{x} - \mathbf{y}|$.

$$\int d^3r_1 \, d^3r_2 \, d^3r_3 \, f_{12}f_{13}f_{23} = \int d^3r_1 \int d^3x \, d^3y \, f(x)f(y)f(|\mathbf{x} - \mathbf{y}|) \tag{S8.13}$$

The integrand on the right vanishes if $x \gg R$ or $y \gg R$, and the integral may therefore be extended over all values of $\mathbf{x}$ and $\mathbf{y}$. $\mathbf{r}_1$ is integrated over the volume V. Therefore

$$\int d^3r_1 \, d^3r_2 \, d^3r_3 \, f_{12}f_{13}f_{23} = V \int d^3x \, d^3y \, f(x)f(y)f(|\mathbf{x} - \mathbf{y}|) \tag{S8.14}$$

In a similar way, the second term in Eq. (S8.12) gives

$$\int d^3r_1 \, d^3r_2 \, d^3r_3 \, f_{12}f_{13} = V \int d^3x \, d^3y \, f(x)f(y) = V \left(4\pi \int_0^\infty f(x)x^2 \, dx \right)^2 \tag{S8.15}$$

The integrals of the third and fourth terms in Eq. (S8.12) are obviously equal to that of the second term. Thus

$$6C_3 = \int d^3x \, d^3y \, f(x)f(y)f(|\mathbf{x} - \mathbf{y}|) + 3 \left(4\pi \int_0^\infty f(x)x^2 \, dx \right)^2 \tag{S8.16}$$

which is independent of V.

Exercise 8.3 For particles of radius a with hard-core interactions, the interaction potential is

$$v_{ij} = \begin{cases} \infty, & \text{if } r_{ij} < 2a \\ 0, & \text{otherwise} \end{cases} \tag{S8.17}$$

where $r_{ij} = |\mathbf{r}_i - \mathbf{r}_j|$. For such particles, the Mayer f function, defined in Exercise 8.2, can be written in terms of the step function $\theta(x)$.

$$f_{ij} = -\theta(2a - r_{ij}) \tag{S8.18}$$

Evaluate the cluster integrals, C_2 and C_3, for hard-core particles in two dimensions.

Solution The two-dimensional form of Eq. (8.15) is

$$\begin{aligned} C_2 &= \frac{1}{2} \int f(|\mathbf{r}|)\, d^2r \\ &= -\pi \int_0^\infty \theta(2a - r) r\, dr \\ &= -2\pi a^2 \end{aligned} \tag{S8.19}$$

Notice that the integral of $f(r)$ is equal to minus one times the area of a circle of radius $2a$, not one of radius a. This is a reflection of the fact that two hard-core particles must maintain a separation of at least twice their radius. The integral for C_3 is more difficult. According to Eq. (S8.16),

$$6C_3 = \int d^2x\, d^2y\, [f(x)f(y)f(|\mathbf{x} - \mathbf{y}|) + 3f(x)f(y)] \tag{S8.20}$$

The second term in the square brackets gives $12C_2^2$. [Compare with Eq. (S8.19).]

$$3 \int f(x)\, d^2x \int f(y)\, d^2y = 48\pi^2 a^4 \tag{S8.21}$$

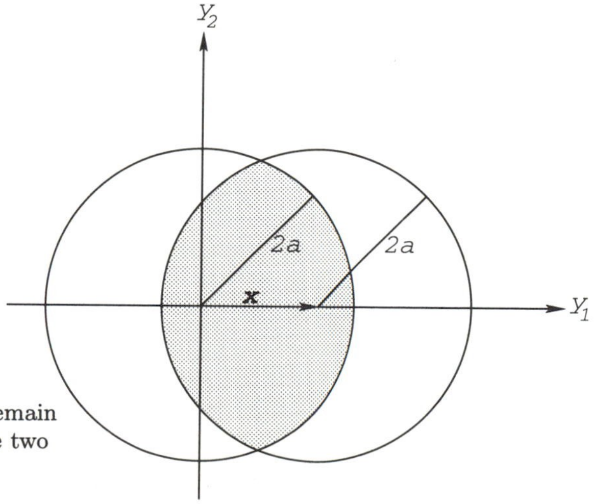

Fig. S8.1 The variable y must remain within the shaded intersection of the two circles.

As an aid in evaluating the integral of the first term in the brackets we have drawn the diagram in Fig. S8.1. We first draw a circle of radius $2a$ about the origin. Because of the factor $f(x)$, the variable $\mathbf{x}$ must remain within this circle, and we draw a second circle of radius $2a$, centered at $\mathbf{x}$. The variable $\mathbf{y}$ must remain within the shaded intersection of the two circles. Within that restricted region the value of the integrand is just the constant -1. The area of the shaded region depends only on the magnitude of $\mathbf{x}$. Let us call it $A(x)$. It is easy to see that

$$
\begin{aligned}
A(x) &= 2 \int_{x/2}^{2a} dy \, \sqrt{4a^2 - y^2} \\
&= 2 \left[ay\sqrt{1 - y^2/4a^2} + 2a^2 \sin^{-1}(y/2a) \right]_{x/2}^{2a} \qquad \text{(S8.22)} \\
&= 2\pi a^2 - \tfrac{1}{4}x\sqrt{16a^2 - x^2} - 2a^2 \sin^{-1}(x/4a)
\end{aligned}
$$

The needed integral is therefore

$$
\begin{aligned}
\int f(x)f(y)f(|\mathbf{x}-\mathbf{y}|) \, d^2x \, d^2y &= -2\pi \int_o^{2a} A(x)x \, dx \\
&= -2\pi^2 a^4 + \frac{\pi}{2} \int_0^{2a} x^2\sqrt{16a^2 - x^2}dx \qquad \text{(S8.23)} \\
&\quad + 4\pi a^2 \int_0^{2a} \sin^{-1}(x/4a)x dx \\
&= 2\pi\left(\sqrt{3} - \frac{\pi}{3}\right)a^4
\end{aligned}
$$

Thus

$$
C_3 = 8\pi^2 a^4 + \frac{\pi}{3}\left(\sqrt{3} - \frac{\pi}{3}\right)a^4 \qquad \text{(S8.24)}
$$

Exercise 8.4   Repeat the previous exercise for three-dimensional hard-core particles.

Solution   It is only necessary to indicate the changes in the equations for the two-dimensional case. Equation (S8.19) becomes

$$
\begin{aligned}
C_2 &= \frac{1}{2}\int f(r) \, d^3r \\
&= -2\pi \int_o^\infty \theta(2a - r)r^2 \, dr \qquad \text{(S8.25)} \\
&= -\frac{2\pi}{3}(2a)^3
\end{aligned}
$$

From this we see that

$$
3\int f(x) \, d^3x \int f(y) \, d^3y = 12C_2^2 = \tfrac{16}{3}\pi^2(2a)^6 \qquad \text{(S8.26)}
$$

In place of $A(x)$, we must calculate the volume of the intersection of two spheres of radius $2a$, separated by a distance $x < 2a$.

$$V(x) = 2 \int_{x/2}^{2a} dy \, \pi(4a^2 - y^2)$$
$$= \frac{32\pi}{3} a^3 - 4\pi a^2 x + \frac{\pi}{12} x^3 \tag{S8.27}$$

Finally

$$\int f(x)f(y)f(|\mathbf{x} - \mathbf{y}|) \, d^3x \, d^3y = -4\pi \int_0^{2a} V(x)x^2 \, dx$$
$$= -\tfrac{5}{6}\pi^2(2a)^6 \tag{S8.28}$$

and

$$C_3 = \tfrac{3}{4}\pi^2(2a)^6 \tag{S8.29}$$

Exercise 8.5   Calculate the virial coefficients, B_2 and B_3, for a gas that satisfies van der Waals' equation of state (see Problem 5.10).

Solution   Van der Waals' equation of state is

$$\frac{p}{RT} = \frac{1}{v - v_0} - \frac{a}{RTv^2} \tag{S8.30}$$

where v is the molar volume and a and v_o are empirical constants, different for different gases. The density, in moles/m^3, is given by $n = 1/v$. We can expand p/RT as a power series in n as follows.

$$\frac{p}{RT} = \frac{n}{1 - v_0 n} - \frac{a}{RT} n^2$$
$$= n + \left(v_0 - \frac{a}{RT}\right) n^2 + v_0^2 n^3 + \cdots \tag{S8.31}$$

Comparing this with the virial equation of state [Eq. (8.26)] and using the fact that n (in particles/m^3) $= N_A n$ (in moles/m^3), we see that

$$N_A B_2 = v_0 - \frac{a}{RT} \tag{S8.32}$$

and

$$N_A^2 B_3 = v_0^2 \tag{S8.33}$$

One sees that the van der Waals second virial coefficient, B_2, has very reasonable temperature dependence. It approaches a positive constant at large T and becomes negative at small T.

Exercise 8.6   (a) For neon,

$$N_A B_2(T) \rightarrow 1.3 \times 10^{-5} \text{m}^3/\text{mole}$$

as T becomes large and $B_2(125\,\mathrm{K}) = 0$. Use these facts to determine approximate van der Waals parameters, a and v_0, for neon. (b) Considering Fig. S5.9, one can see that, at the critical temperature and pressure, $(\partial p/\partial v)_T = 0$ and $(\partial^2 p/\partial v^2) = 0$. Use the van der Waals equation to estimate the critical temperature and pressure for neon. Do not expect to get an accurate result—neither the van der Waals equation nor the virial expansion is accurate near the critical point.

Solution   (a) According to Exercise 8.5,

$$N_A B_2 = v_0 - \frac{a}{RT} \tag{S8.34}$$

Therefore $v_0 = 1.3 \times 10^{-5}\ \mathrm{m}^3/\mathrm{mole}$. Also

$$a = (125\,\mathrm{K})Rv_0 = 1.35 \times 10^{-2}\mathrm{J\,m}^3/\mathrm{mole}^2 \tag{S8.35}$$

(b) Writing the molar volume and the temperature at the critical point as v_c and T_c, the conditions $\partial p/\partial v = 0$ and $\partial^2 p/\partial v^2$ give the two equations

$$\frac{1}{RT_c}\left(\frac{\partial p}{\partial v}\right) = -\frac{1}{(v_c - v_0)^2} + \frac{2a}{RT_c v_c^3} = 0 \tag{S8.36}$$

and

$$\frac{1}{RT_c}\left(\frac{\partial^2 p}{\partial v^2}\right) = \frac{2}{(v_c - v_0)^3} - \frac{6a}{RT_c v_c^4} = 0 \tag{S8.37}$$

Rewriting these equations as

$$\frac{1}{(v_c - v_0)^2} = \frac{2a}{RT_c v_c^3} \tag{S8.38}$$

and

$$\frac{1}{(v_c - v_0)^3} = \frac{3a}{RT_c v_c^4} \tag{S8.39}$$

and dividing the first equation by the second gives

$$v_c - v_0 = \tfrac{2}{3}v_c \tag{S8.40}$$

or

$$v_c = 3v_0 = 3.9 \times 10^{-5}\,\mathrm{m}^3 \tag{S8.41}$$

Using this in Eq. (S8.38), one finds that

$$T_c = \frac{8a}{27Rv_0} = 37\,\mathrm{K} \tag{S8.42}$$

and finally, using these values in the van der Waals equation gives the critical pressure.

$$p_c = 2.9 \times 10^6 \mathrm{N/m}^2 = 29\ \mathrm{atm} \tag{S8.43}$$

The experimental values of T_c and p_c for neon are 44.5 K and 26.9 atm, not too far from the values estimated using the second virial coefficient.

Exercise 8.7   Develop a cluster expansion for the grand potential ψ for a two-dimensional lattice gas with attractive nearest-neighbor interactions but only one particle allowed at each lattice site.

Solution   We choose the lattice constant as our unit of length. Then, in the lattice gas, the position of the nth particle, $\mathbf{x}_n$, is equal to one of the lattice points

$$\mathbf{x}_n = (k, l) \tag{S8.44}$$

where k and l range over the set of integers $1, 2, \ldots, L$. The interaction potential is

$$v_{ij} \equiv v(\mathbf{x}_i - \mathbf{x}_j) = \begin{cases} +\infty, & \text{if } \mathbf{x}_i = \mathbf{x}_j \\ -V, & \text{if } |\mathbf{x} - \mathbf{x}| = 1 \\ 0, & \text{otherwise} \end{cases} \tag{S8.45}$$

The infinite repulsive interaction at zero distance simply prevents double occupancy of a lattice site. The N-particle canonical partition function is

$$Z_N = \frac{1}{N!} \sum_{\mathbf{x}_1} \cdots \sum_{\mathbf{x}_N} \exp\left(-\beta \sum_{i<j} v_{ij}\right) \equiv \frac{S_N}{N!} \tag{S8.46}$$

where the $N!$ is necessary because we are now using numbered particles so that there are $N!$ of our present configurations for each physically distinct configuration. The grand partition function is

$$\Lambda = 1 + \sum_{N=1}^{\infty} \frac{e^{-\alpha N}}{N!} S_N \equiv 1 + \sum_{N=1}^{\infty} \frac{\zeta^N}{N!} S_N \tag{S8.47}$$

where the activity $\zeta = e^{-\alpha}$. Then, using the expansion of $\log(1 + x)$ given in Eq. (8.10), we get

$$\psi = \left(S_1\zeta + \frac{S_2}{2!}\zeta^2 + \frac{S_3}{3!}\zeta^3 + \cdots\right) - \frac{1}{2}\left(S_1\zeta + \frac{S_2}{2!}\zeta^2 + \cdots\right)^2 + \frac{1}{3}(S_1\zeta + \cdots)^3 - \cdots$$
$$= S_1\zeta + \frac{1}{2}(S_2 - S_1^2)\zeta^2 + \frac{1}{6}(S_3 - 3S_1 S_2 + 2S_1^3)\zeta^3 + \cdots \tag{S8.48}$$

Clearly, we are getting exactly the same expression as one gets for an ordinary gas, with continuous coordinates except for two things. (1) The definition of the activity in simply $e^{-\alpha}$, rather than $e^{-\alpha}/\lambda^3$. (2) The cluster integrals are replaced by cluster sums, taken over a discrete lattice.

According to Eq. (S8.48), $S_1 = L^2$. If we define cluster sums, $C_2, C_3, \ldots$ as in the continuous case, that is,

$$L^2 C_2 = \frac{1}{2}(S_2 - S_1^2) \tag{S8.49}$$

$$L^2 C_3 = \frac{1}{6}(S_3 - 3S_1 S_2 + 2S_1^3) \tag{S8.50}$$

and so forth, then

$$\psi = L^2(\zeta + C_2\zeta^2 + C_3\zeta^3 + \cdots) \tag{S8.51}$$

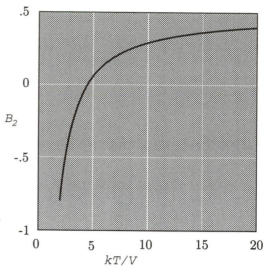

Fig. S8.2   The second virial coefficient B_2 as a function of the dimensionless temperature variable kT/V for a lattice gas with nearest-neighbor interactions.

Exercise 8.8   For the lattice gas considered in the previous exercise, calculate the cluster sum C_2 and the virial coefficient B_2.

Solution

$$L^2 C_2 = \tfrac{1}{2}(S_2 - S_1^2)$$
$$= \frac{1}{2}\sum_{\mathbf{x}_1}\sum_{\mathbf{x}_2}(e^{-\beta v_{12}} - 1) \tag{S8.52}$$

The quantity in the parentheses is the Mayer f function,

$$f(\mathbf{x}_{12}) = e^{-\beta v_{12}} - 1 = \begin{cases} -1, & \text{if } \mathbf{x}_{12} = 0 \\ e^{\beta V} - 1, & \text{if } |\mathbf{x}_{12}| = 1 \\ 0 & \text{otherwise} \end{cases} \tag{S8.53}$$

Introducing the relative variable $\mathbf{y} = \mathbf{x}_2 - \mathbf{x}_1$, and assuming that L is very large, we can write $L^2 C_2$ as

$$L^2 C_2 = \frac{1}{2}\sum_{\mathbf{x}_1}\sum_{\mathbf{y}} f(\mathbf{y})$$
$$= \tfrac{1}{2}L^2(4e^{\beta V} - 5) \tag{S8.54}$$

or

$$C_2 = \tfrac{1}{2}(4e^{\beta V} - 5) \tag{S8.55}$$

B_2 is related to C_2 by Eq. (8.27).

$$B_2 = -C_2 = \tfrac{5}{2} - 2e^{\beta V} \tag{S8.56}$$

B_2 is plotted in Fig. S8.2 as a function of kT/V. We see that it has the behavior that is expected of a second virial coefficient; it becomes large and negative at low temperatures and approaches a constant at high T.

Exercise 8.9   Consider a 3D gas of point particles with interparticle interactions given by a potential $v(r_{ij})$. We focus our attention on a particular particle (call it the zero particle) located at $\mathbf{r}_o$. We want to calculate the density of the other particles around the zero particle. If we take into account the interaction between the zero particle and every other particle in the system, but temporarily ignore the interactions among the other particles, then we would expect the other particles to be distributed in space with a density

$$n(\mathbf{r}) = n_o e^{-\beta v(\mathbf{r}-\mathbf{r}_o)} \tag{S8.57}$$

Therefore, within this approximation, the average of the interaction between the zero particle and all the other particles is

$$\langle v_o \rangle = n_o \int e^{-\beta v(\mathbf{r}-\mathbf{r}_o)} v(\mathbf{r}-\mathbf{r}_o) \, d^3\mathbf{r} = 4\pi n_o \int_o^\infty e^{-\beta v(R)} v(R) R^2 \, dR \tag{S8.58}$$

Multiplying this by N will double count the two-particle interaction terms. Thus our approximation for the total potential energy is

$$\left\langle \sum_{i<j} v_{ij} \right\rangle = 2\pi \frac{N^2}{V} \int_o^\infty e^{-\beta v(R)} v(R) R^2 \, dR \tag{S8.59}$$

Use this approximation to derive the pressure and energy equations of state and compare the result with the virial expansion.

Solution   For a classical gas the kinetic energy is always $\frac{3}{2}NkT$. Therefore, the total energy is

$$E(N,\beta,V) = \frac{3}{2}\frac{N}{\beta} + 2\pi \frac{N^2}{V} \int_o^\infty e^{-\beta v(R)} v(R) R^2 \, dR \tag{S8.60}$$

Integrating the identity $E = -\partial\phi(N,\beta,V)/\partial\beta$ with respect to β yields the following relation for the canonical potential

$$\phi = -\tfrac{3}{2}N \log\beta + 2\pi\frac{N^2}{V} \int_o^\infty (e^{-\beta v(R)} - 1) R^2 \, dR + g(N,V) \tag{S8.61}$$

where g is an arbitrary function and we have used the fact that

$$e^{-\beta v} v = -\frac{\partial(e^{-\beta v} - 1)}{\partial\beta} \tag{S8.62}$$

(The constant -1 must be added in order to prevent a divergence in the resulting integral.) When $v(r) = 0$, $\phi(N,\beta,V)$ should reduce to the canonical potential of an ideal gas. This determines the arbitrary function g and gives

$$\phi = \phi_{\text{ideal}} + 2\pi\frac{N^2}{V} \int_o^\infty (e^{-\beta v(R)} - 1) R^2 \, dR \tag{S8.63}$$

The pressure equation of state is then given by $\beta p = \partial\phi/\partial V$.

$$\beta p = n - 2\pi n^2 \int_o^\infty (e^{-\beta v(R)} - 1) R^2 \, dR \tag{S8.64}$$

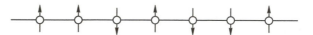

Fig. S8.3   The one-dimensional Ising model is a system of two-valued spins, equally spaced, on a line.

Equations (S8.60) and (S8.64) are identical with the corresponding equations derived from the two-term virial expansion.

The following few exercises concern the one-dimensional Ising model, with and without an external field. While the two-dimensional and three-dimensional Ising models are reasonable approximations to important physical systems, the one-dimensional version is not. However, it can be solved, both exactly and approximately, by a number of different techniques, and therefore is an excellent pedagogical exercise for learning calculational methods that may then be applied to more complex, realistic systems.

Exercise 8.10   Shown in Fig. S8.3 is a one-dimensional Ising model. It is a system of N quantized spins, located at the points of a one-dimensional lattice. The nth spin can have two possible states that we indicate with a variable $\sigma_n = \pm 1$. A pair of neighboring spins has an interaction energy equal to $-V$ if they are parallel and $+V$ if they are antiparallel. Therefore the total energy of any configuration, defined by specifying the values of $\sigma_1, \sigma_2, \ldots, \sigma_N$, is

$$E = -V(\sigma_1\sigma_2 + \sigma_2\sigma_3 + \cdots \sigma_{N-1}\sigma_N) \tag{S8.65}$$

By transforming to new variables, $\zeta_1 = \sigma_1$, $\zeta_2 = \sigma_2\sigma_1$, $\zeta_3 = \sigma_3\sigma_2, \ldots, \zeta_N = \sigma_N\sigma_{N-1}$, calculate the energy and specific heat of the system as functions of $x = \tau/V$.

Solution   In terms of the new variables

$$E = -V(\zeta_2 + \zeta_3 + \cdots + \zeta_N) \tag{S8.66}$$

It is not difficult to show that there is a one-to-one relationship between the set of variables $(\sigma_1, \sigma_2, \ldots, \sigma_N)$ and the set of variables $(\zeta_1, \zeta_2, \ldots, \zeta_N)$. The definition of the ζ variables gives a definite set of ζs for a given set of σs. We need only look at the inverse transformation. For a given set of ζs it is easy to see that

$$\zeta_1 = \sigma_1$$
$$\zeta_1\zeta_2 = \sigma_1^2\sigma_2 = \sigma_2$$
$$\zeta_1\zeta_2\zeta_3 = \sigma_1^2\sigma_2^2\sigma_3 = \sigma_3 \tag{S8.67}$$
$$\cdots$$
$$\zeta_1 \cdots \zeta_N = \sigma_1^2 \cdots \sigma_{N-1}^2\sigma_N = \sigma_N$$

Thus any set of σs gives a unique set of ζs and vice versa. By its definition, the possible values of ζ_n are ± 1. Thus the canonical partition function is

$$Z = \sum_{\zeta_1} \cdots \sum_{\zeta_N} \exp[-\beta V(\zeta_2 + \cdots + \zeta_N)]$$

$$= 2\left(\sum_\zeta e^{-\beta V\zeta}\right)^{N-1}$$

$$= 2^N (\cosh \beta V)^{N-1}$$

and

$$\phi(\beta) = N\big[\log(\cosh \beta V) + \log 2\big] \tag{S8.68}$$

where we have ignored the difference between $N-1$ and N. The average thermal energy is

$$E = -\frac{\partial \phi}{\partial \beta} = -NV \tanh(\beta V) \tag{S8.69}$$

and the specific heat is

$$C = \frac{\partial E}{\partial \tau} = N\frac{V^2}{\tau^2} \operatorname{sech}^2(V/\tau) \tag{S8.70}$$

The fact that E and C, as functions of temperature, are completely smooth shows that no phase transition occurs in this system at any temperature.

Exercise 8.11   In the Ising model considered in the previous exercise, the only source of energy was the interactions between neighboring spins. If the spins are associated with a magnetic moment μ and an external magnetic field is present, then, for each spin, there is another term in the energy, namely $\pm\mu B$, due to the interaction of the magnetic moment with the external field. Denoting the combination μB by the letter H, we can then write the energy for any configuration of a one-dimensional Ising model with an external field as

$$E = -H(\sigma_1 + \sigma_2 + \cdots + \sigma_N) - V(\sigma_1\sigma_2 + \cdots + \sigma_{N-1}\sigma_N) \tag{S8.71}$$

The reader can easily verify that the trick that was used in the last exercise to solve the problem when $H = 0$ will no longer work. However, the partition function and canonical potential of this model can also be calculated exactly, but the analysis is somewhat more complicated. It is best first to bend the lattice around to form a circle of length N. Then the Nth spin will interact with the first spin so that the expression for E will become

$$E = -H(\sigma_1 + \cdots + \sigma_N) - V(\sigma_1\sigma_2 + \cdots + \sigma_{N-1}\sigma_N + \sigma_N\sigma_1) \tag{S8.72}$$

This is referred to as using *periodic boundary conditions*.

Before attempting to calculate the partition function of this model, we will give a short preamble, reviewing some elementary linear algebra. A function of two integer variables $M(k, l)$, where k and l range from 1 to K, defines a K by K matrix in which the number in the ith row and the jth column is $M(i, j)$. In a similar way, a function of two spin variables $M(\sigma_1, \sigma_2)$ can be arranged as a 2×2 matrix, as shown.

$$M = \begin{bmatrix} M(+1, +1) & M(+1, -1) \\ M(-1, +1) & M(-1, -1) \end{bmatrix} \tag{S8.73}$$

$M(\sigma_1, \sigma_2)$ is *symmetric* if $M(+1, -1) = M(-1, +1)$. In the following, we will assume that M is symmetric.

The *trace* of M, written $\mathrm{Tr}(M)$, is the sum of its two diagonal elements. $\mathrm{Tr}(M)$ is also equal to the sum of the two eigenvalues of M.

$$\mathrm{Tr}(M) = \sum_{\sigma} M(\sigma, \sigma) = \lambda_+ + \lambda_- \tag{S8.74}$$

where we define λ_+ as the larger and λ_- as the smaller of the eigenvalues (which are always real for a symmetric matrix).

The determinant of M is equal to the product of its two eigenvalues.

$$\det(M) = M(+1, +1)M(-1, -1) - M(+1, -1)M(-1, +1) = \lambda_+ \lambda_- \tag{S8.75}$$

The matrix M^2 is defined by the usual rule for matrix multiplication.

$$M^2(\sigma_1, \sigma_3) = \sum_{\sigma_2} M(\sigma_1, \sigma_2)M(\sigma_2, \sigma_3) \tag{S8.76}$$

In this way any power of the matrix M can be calculated. For example,

$$M^3(\sigma_1, \sigma_4) = \sum_{\sigma_2} \sum_{\sigma_3} M(\sigma_1 \sigma_2)M(\sigma_2, \sigma_3)M(\sigma_3, \sigma_4) \tag{S8.77}$$

The eigenvalues of the matrix M^n are the nth power of the eigenvalues of M, that is, λ_+^n and λ_-^n.

We are now ready to describe the exercise. (a) Define a symmetric matrix $M(\sigma_1, \sigma_2)$ so that the partition function

$$Z = \sum_{\sigma_1} \cdots \sum_{\sigma_N} e^{-\beta E} = \mathrm{Tr}(M^N) \tag{S8.78}$$

(When such a matrix exists, it is called a *transfer matrix*.) (b) Assuming that $N \gg 1$, calculate the canonical potential for the one-dimensional Ising model with an external field. (c) For the special case $H = V$, plot the specific heat per spin as a function of the dimensionless temperature variable $t = \tau/V$ in the range $0 < t < 10$.

Solution (a) If we define parameters A and B by

$$A = \tfrac{1}{2}\beta H \quad \text{and} \quad B = \beta V \tag{S8.79}$$

then

$$-\beta E = 2A(\sigma_1 + \cdots + \sigma_N) + B(\sigma_1 \sigma_2 + \cdots + \sigma_N \sigma_1) \tag{S8.80}$$

and

$$e^{-\beta E} = (e^{A\sigma_1 + B\sigma_1\sigma_2 + A\sigma_2})(e^{A\sigma_2 + B\sigma_2\sigma_3 + A\sigma_3}) \cdots (e^{A\sigma_N + B\sigma_N\sigma_1 + A\sigma_1}) \tag{S8.81}$$

If we define $M(\sigma_1, \sigma_2)$ by

$$M(\sigma_1, \sigma_2) = e^{A\sigma_1 + B\sigma_1\sigma_2 + A\sigma_2} \tag{S8.82}$$

then

$$
\begin{aligned}
Z &= \sum_{\sigma_1} \cdots \sum_{\sigma_N} M(\sigma_1, \sigma_2) M(\sigma_2, \sigma_3) \cdots M(\sigma_N, \sigma_1) \\
&= \sum_{\sigma_1} M^N(\sigma_1, \sigma_1) \\
&= \mathrm{Tr}(M^N)
\end{aligned} \tag{S8.83}
$$

(b) We must now determine the eigenvalues of M. Written out as a 2×2 matrix [see Eq. (S8.73)],

$$M = \begin{bmatrix} e^{B+2A} & e^{-B} \\ e^{-B} & e^{B-2A} \end{bmatrix} \tag{S8.84}$$

From the matrix rules reviewed before, we know that

$$\lambda_+ + \lambda_- = \mathrm{Tr}(M) = e^B(e^{2A} + e^{-2A}) = 2e^B \cosh(2A) \tag{S8.85}$$

and

$$\lambda_+ \lambda_- = \det(M) = e^{2B} - e^{-2B} = 2\sinh(2B) \tag{S8.86}$$

Using the first equation to eliminate λ_- in the second one obtains a quadratic equation for λ_+ whose solution is

$$
\begin{aligned}
\lambda_+ &= e^B \left[\cosh(2A) + \sqrt{\sinh^2(2A) + e^{-4B}} \right] \\
&= e^{\beta V} \left[\cosh(\beta H) + \sqrt{\sinh^2(\beta H) + e^{-4\beta V}} \right]
\end{aligned} \tag{S8.87}
$$

λ_- is given by the same formula with the plus sign replaced by a minus sign. The trace of M^N is given by

$$
\begin{aligned}
\mathrm{Tr}(M^N) &= \lambda_+^N + \lambda_-^N \\
&= \lambda_+^N \left[1 + (\lambda_-/\lambda_+)^N \right] \\
&\approx \lambda_+^N
\end{aligned} \tag{S8.88}
$$

since the Nth power of a number less than one is extremely small for large N. Thus

$$
\begin{aligned}
\phi(\beta) &= \log Z \\
&= N \log(\lambda_+) \\
&= N \left\{ \beta V + \log \left[\cosh(\beta H) + \sqrt{\sinh^2(\beta H) + e^{-4\beta V}} \right] \right\}
\end{aligned} \tag{S8.89}
$$

(c) Having obtained the canonical potential, one can calculate the energy per spin and the specific heat by the usual formulas.

$$(E/N) = -\frac{\partial(\phi/N)}{\partial \beta} \tag{S8.90}$$

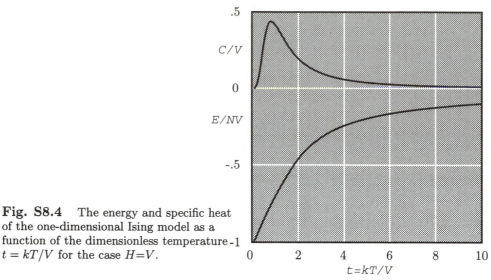

Fig. S8.4 The energy and specific heat of the one-dimensional Ising model as a function of the dimensionless temperature $t = kT/V$ for the case $H=V$.

and

$$C = \frac{\partial(E/N)}{\partial \tau} \qquad (S8.91)$$

However, because the second equation leads to an absolute morass of algebra, the use of a computerized algebra program is strongly recommended. For the case $H = V$, plots of (E/N) and C as functions of $t = \tau/V$ that were obtained using *Mathematica* are shown in Fig. S8.4.

Exercise 8.12   Show that no spontaneous magnetization takes place in the one-dimensional Ising model at any temperature.

Solution   For the 1D Ising model with an external field H, the canonical partition function is given by

$$Z = \sum_{\{\sigma\}} \exp\left(-\beta H \sum \sigma_i - \beta V \sum \sigma_i \sigma_{i+1}\right) \qquad (S8.92)$$

From this relation, and the fact that $\phi = \log Z$, it is easy to see that

$$\frac{\partial \phi}{\partial H} = \frac{\partial Z/\partial H}{Z} = -\beta \left\langle \sum \sigma_i \right\rangle = -\beta N \bar{\sigma} \qquad (S8.93)$$

From Eq. (S8.89), we see that

$$\frac{1}{\beta} \frac{\partial \phi}{\partial H} = N \sinh(\beta H) \left[1 + \cosh(\beta H)/\sqrt{\sinh^2(\beta H) + e^{-4\beta V}}\right] \bigg/ (\phi - N\beta V) \quad (S8.94)$$

But, because of the factor of $\sinh(\beta H)$, this function approaches zero as $H \to 0$. Regardless of the temperature, the system is never left with a finite magnetization when the external field is turned off.

For the one-dimensional Ising model, the coordination number c defined in Section 8.22 is two. Therefore, the mean-field approximation predicts that the system undergoes a ferromagnetic phase transition at a Curie temperature of $T_c=2V/k$. This is an entirely erroneous prediction—the exact solution shows that the system has no phase transition at any temperature.

Hans Bethe and Rudolf Peierls, many years ago, developed an improved version of mean-field theory that avoids that false prediction for the 1D Ising model and makes significantly improved (but still not very accurate) predictions for the 2D and 3D models. Applied to the 1D Ising model, the Bethe–Peierls approximation proceeds as follows.

Fig. S8.5 Replacing their neighbor spins by fixed variables of value $\bar{\sigma}$ isolates the triplet from the other spins in the lattice.

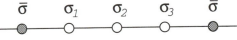

1. Consider any triplet of spins, σ_1, σ_2, and σ_3. Replace the spins neighboring the triplet by some as yet unknown average value $\bar{\sigma}$. This effectively replaces the very large lattice by a system composed of only three spins. (See Fig. S8.5.)
2. As functions of $\bar{\sigma}$, calculate the average values $\langle\sigma_1\rangle=\langle\sigma_3\rangle$ and $\langle\sigma_2\rangle$.
3. Determine the value of $\bar{\sigma}$ by demanding that $\langle\sigma_1\rangle=\langle\sigma_2\rangle$. (Note that this will not usually guarantee that $\bar{\sigma} = \langle\sigma_1\rangle$. It is impossible to get an approximate solution to satisfy all possible symmetry relations.)
4. With the value of $\bar{\sigma}$ given by 3, calculate $\langle\sigma_1\sigma_2\rangle$ and compute the thermal energy by using the formula

$$E = -NH\langle\sigma_1\rangle - NV\langle\sigma_1\sigma_2\rangle \qquad (S8.95)$$

Exercise 8.13 Show that the Bethe–Peierls approximation does not predict a phase transition for the 1D Ising model with no external field. (As a matter of fact, the Bethe–Peierls approximation gives the exact result for the energy of the 1D Ising model, with or without an external field.)

Solution Although we are interested only in the case $H=0$, it will be very convenient first to include an external field term H for spins 1 and 3 and a different external field term H' for the central spin 2. At the end, we will set H and H' equal to zero. With the definitions $h=\beta H$, $h'=\beta H'$, and $v=\beta V$, the value of βE for the three-spin system is

$$\beta E = -h(\sigma_1 + \sigma_3) - h'\sigma_2 - v(\bar{\sigma}\sigma_1 + \sigma_1\sigma_2 + \sigma_2\sigma_3 + \bar{\sigma}\sigma_3) \qquad (S8.96)$$

The partition function of the system is easily evaluated.

$$
\begin{aligned}
Z &= \sum_{\sigma_1}\sum_{\sigma_2}\sum_{\sigma_3} \exp\big[h(\sigma_1 + \sigma_3) + h'\sigma_2 + v(\bar\sigma\sigma_1 + \sigma_1\sigma_2 + \sigma_2\sigma_3 + \bar\sigma\sigma_3)\big] \\
&= \sum_{\sigma_2} \exp(h'\sigma_2) \sum_{\sigma_1} \exp[(h + v\bar\sigma + v\sigma_2)\sigma_1] \sum_{\sigma_3} \exp[(h + v\bar\sigma + v\sigma_2)\sigma_3] \\
&= 4\sum_{\sigma_2} \exp(h'\sigma_2)\cosh^2(h + v\bar\sigma + v\sigma_2) \\
&= 4\big[e^{h'}\cosh^2(h + v\bar\sigma + v) + e^{-h'}\cosh^2(h + v\bar\sigma - v)\big]
\end{aligned}
\tag{S8.97}
$$

Using the first line of Eq. (S9.97), it is easy to confirm that

$$
\langle\sigma_1 + \sigma_3\rangle = \frac{\partial Z/\partial h}{Z} \quad \text{and} \quad \langle\sigma_2\rangle = \frac{\partial Z/\partial h'}{Z}
\tag{S8.98}
$$

Thus

$$
\begin{aligned}
\langle\sigma_1 + \sigma_3\rangle =8\big[&e^{h'}\cosh(h + v\bar\sigma + v)\sinh(h + v\bar\sigma + v) \\
&+ e^{-h'}\cosh(h + v\bar\sigma - v)\sinh(h + v\bar\sigma - v)\big]/Z
\end{aligned}
\tag{S8.99}
$$

and

$$
\langle\sigma_2\rangle = 4\big[e^{h'}\cosh^2(h + v\bar\sigma + v) - e^{-h'}\cosh^2(h + v\bar\sigma - v)\big]/Z
\tag{S8.100}
$$

Setting $h = h' = 0$, and demanding that $\langle\sigma_1 + \sigma_3\rangle = 2\langle\sigma_2\rangle$, gives the following equation for $\bar\sigma$.

$$
\begin{aligned}
\cosh(v\bar\sigma + v)&\sinh(v\bar\sigma + v) + \cosh(v\bar\sigma - v)\sinh(v\bar\sigma - v) \\
&= \cosh^2(v\bar\sigma + v) - \cosh^2(v\bar\sigma - v)
\end{aligned}
\tag{S8.101}
$$

This equation can be converted to a solvable polynomial equation by the substitutions

$$
x = e^{v\bar\sigma} \quad \text{and} \quad y = e^{v}
\tag{S8.102}
$$

Expressing the hyperbolic functions in terms of exponentials, with a little algebra, one obtains the equation

$$
(x^2 - x^{-2})y^{-2} = 0
\tag{S8.103}
$$

Since y cannot be zero, this requires that $x=1$, which implies that $\bar\sigma=0$. Thus the Bethe–Peierls approximation does not predict spontaneous magnetization for any value of $v = V/kT$.

Exercise 8.14   Apply the Bethe–Peierls approximation to an Ising model, with no external field, on a 2D hexagonal lattice.

Solution   A portion of a hexagonal lattice is shown in Fig. S8.6. A central spin σ_1 and its three neighboring spins σ_2, σ_3, and σ_4, are isolated from the rest of the lattice by replacing all the shaded spins by some yet to be determined average value

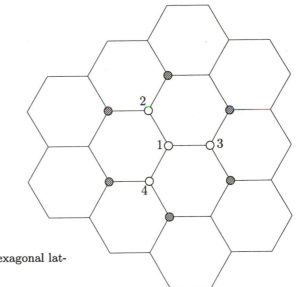

Fig. S8.6   A section of a hexagonal lattice.

$\bar{\sigma}$. Again, as in Exercise 8.13, it is best to include an external field term that has the value H for spins 2, 3, and 4, and a different value H' for spin 1. The value of βE for the four-spin system is then

$$\beta E = -h'\sigma_1 - h(\sigma_2 + \sigma_3 + \sigma_4) - v(2\bar{\sigma} + \sigma_1)(\sigma_2 + \sigma_3 + \sigma_4) \qquad (S8.104)$$

where, as before, $h = \beta H$, $h' = \beta H'$, and $v = \beta V$. The partition function is given by summing $e^{-\beta E}$ over all possible values of the four spin variables.

$$
\begin{aligned}
Z &= \sum_{\sigma_1}\sum_{\sigma_2}\sum_{\sigma_3}\sum_{\sigma_4} \exp\big[h'\sigma_1 + (h + 2v\bar{\sigma} + v\sigma_1)(\sigma_2 + \sigma_3 + \sigma_4)\big] \\
&= \sum_{\sigma_1} \exp(h'\sigma_1)\Big(\sum_{\sigma} \exp\big[(h + 2v\bar{\sigma} + v\sigma_1)\sigma\big]\Big)^3 \\
&= 8\sum_{\sigma_1} \exp(h'\sigma_1)\cosh^3(h + 2v\bar{\sigma} + v\sigma_1) \\
&= 8\big[e^{h'}\cosh^3(h + 2v\bar{\sigma} + v) + e^{-h'}\cosh^3(h + 2v\bar{\sigma} - v)\big]
\end{aligned}
\qquad (S8.105)
$$

Using the identities

$$\langle\sigma_1\rangle = \frac{\partial Z/\partial h'}{Z} \quad \text{and} \quad \langle\sigma_2 + \sigma_3 + \sigma_4\rangle = \frac{\partial Z/\partial h}{Z} \qquad (S8.106)$$

one obtains, after taking the partial derivatives and setting $h = h' = 0$,

$$\langle\sigma_1\rangle = 8\big[\cosh^3(2v\bar{\sigma} + v) - \cosh^3(2v\bar{\sigma} - v)\big]/Z \qquad (S8.107)$$

and

$$
\begin{aligned}
\langle\sigma_2 + \sigma_3 + \sigma_4\rangle = 24\big[&\cosh^2(2v\bar{\sigma} + v)\sinh(2v\bar{\sigma} + v) \\
&+ \cosh^2(2v\bar{\sigma} - v)\sinh(2v\bar{\sigma} - v)\big]/Z
\end{aligned}
\qquad (S8.108)
$$

Setting $\langle \sigma_2 + \sigma_3 + \sigma_4 \rangle = 3\langle \sigma_1 \rangle$ and making the substitutions,

$$x = e^{2v\bar{\sigma}} \quad \text{and} \quad y = e^v \tag{S8.109}$$

one obtains the formidable-looking polynomial equation,

$$\begin{aligned}
(xy + x^{-1}y^{-1})^2(xy - x^{-1}y^{-1}) + (xy^{-1} + x^{-1}y)^2(xy^{-1} - x^{-1}y) \\
= (xy + x^{-1}y^{-1})^3 - (xy^{-1} + x^{-1}y)^3
\end{aligned} \tag{S8.110}$$

Multiplying the equation through by x^3y^3 and collecting terms gives the equation

$$x^6 - (y^4 - 2y^2)x^4 + (y^4 - 2y^2)x^2 - 1 = 0 \tag{S8.111}$$

Letting $u = x^2$ and $A = y^4 - 2y^2$, we see that this is a cubic equation for u.

$$u^3 - Au^2 + Au - 1 = 0 \tag{S8.112}$$

Mean-field theories always have the trivial solution $\bar{\sigma} = 0$. This is reflected in the fact that $u = 1$ is a solution to the cubic equation. When we divide the equation by the factor $(u-1)$, we obtain the simple quadratic equation

$$u^2 - (A - 1)u + 1 = 0 \tag{S8.113}$$

Taking into account the fact that u must be positive, the only acceptable solution to this equation is

$$u = \tfrac{1}{2}(A - 1) + \tfrac{1}{2}\sqrt{(A - 1)^2 - 4} \tag{S8.114}$$

This solution is real if and only if $A \geq 3$, which implies that $y \geq \sqrt{3}$. Thus the Bethe–Peierls approximation predicts a ferromagnetic phase transition at a Curie temperature given by setting $e^v = \sqrt{3}$. That is, $\tau_c = 2V/\log 3 \approx 1.82V$. The exact relation, known from the Onsager solution, is $\tau_c/V = 1.518649\ldots$. Since the coordination number of the hexagonal lattice is three, simple mean-field theory would predict that the phase transition occurs as $\tau = 3V$. We see that the Bethe–Peierls method is a substantial improvement on the results of simple mean-field theory.

Exercise 8.15   A number of exercises in this chapter require that one numerically compute any solution, within the range $0 < \sigma < 1$, of an equation of the form

$$\sigma = f(\sigma) \tag{S8.115}$$

where $f(\sigma)$ is some relatively complicated function with the properties that $f(0) \geq 0$ and $f(1) < 1$. (a) Describe an algorithm for carrying out such a calculation. The algorithm should guarantee a certain fixed precision. (b) Write a Fortran program to implement the algorithm for the function $f(\sigma) = \tanh(A\sigma)$.

Solution   (a) The problem posed is equivalent to the problem of finding a solution of the equation $g(\sigma) = 0$ within the range $0 < \sigma < 1$, where $g(\sigma) = \sigma - f(\sigma)$. The conditions on $f(\sigma)$ guarantee that $g(0) \leq 0$ and $g(1) > 0$. The algorithm consists in finding two numbers x_{LO} and x_{HI} such that $g(x_{LO}) \leq 0$ and $g(x_{HI}) > 0$ but $x_{HI} - x_{LO} \to 1/2^N$, where N is an integer that determines the precision of the answer. We start with $x_{LO} = 0$ and $x_{HI} = 1$, and iterate the following steps N times.

1. Let $x = (x_{LO} + x_{HI})/2$.
2. If $g(x) > 0$, then $x_{HI} = x$, else $x_{LO} = x$. Go to 1.

Since $x_{HI} - x_{LO}$ is reduced by a factor of 2 by each iteration and the two numbers always bound the solution of $g(x) = 0$, we are guaranteed to obtain a root with an accuracy of $1/2^N$. In the case $g(0) = 0$ and $g(\sigma) > 0$ for $0 < \sigma \leq 1$, no positive root exists. In that case x_{LO} will stay at 0 and x_{HI} will end up at $1/2^N$, which means that we have obtained the zero root.

(b) The program, called Root.for, can be found in the Program Listings, and is included in the Program Disk.

Exercise 8.16   Two different systems have been considered that "live" on a two-dimensional square lattice. One is a lattice gas, with or without nearest-neighbor interactions. This was considered as a model for surface adsorption. The other is the Ising model, which is a model of a ferromagnetic two-dimensional solid. They are very similar in that in both models the configuration of the system is defined by specifying a variable at each lattice site that can have two possible values. For the lattice gas, the values are 0 and 1, while for the Ising model they are ± 1.

Show that the two systems are not only similar, but are completely equivalent, so that any way of calculating the thermodynamic functions for one system will immediately give the thermodynamic functions of the other.

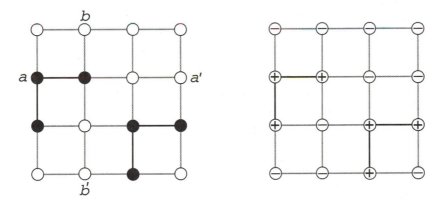

Fig. S8.7   A configuration of a lattice gas and the "corresponding" configuration of an Ising model.

Solution   Shown in Fig. S8.7 are lattice gas and Ising model configurations on 4×4 lattice. We are using periodic using boundary conditions, which means that the pair of points a and a' are neighbors, as are the pair of points b and b'. That is, the lattice "wraps around" in both the x and the y directions. In the lattice gas the occupied sites are black and, for each pair of neighboring particles, the bond between them has been darkened. In the Ising model the up spins are shown as + and the down spins as − and the bond has been darkened for each pair of neighboring up spins.

The grand potential of the lattice gas is the sum, over all configurations, of $\exp(-\alpha N - \beta E)$. We will assume that the binding energy of a particle to a site is zero. (A nonzero value, $-\varepsilon$, could be eliminated simply by redefining the parameter α to be $\alpha - \beta\varepsilon$.) If N_B is the number of darkened bonds in any configuration, and $-v$ is the bond energy, then

$$\Lambda = \sum \exp(-\alpha N + \beta v N_B) \tag{S8.116}$$

where the sum is over all possible configurations.

In any configuration of the Ising model, let N_+ be the number of spin-up sites, N_- the number of spin-down sites, N_{++} the number of bonds with up spins at both ends, N_{--} the number of bonds with down spins at both ends, and N_{+-} the number of bonds with different spins at the two ends. Then the energy of a configuration is

$$E = -H(N_+ - N_-) - V(N_{++} + N_{--} - N_{+-}) \tag{S8.117}$$

The canonical partition function for the Ising lattice is therefore

$$Z = \sum \exp\left[\beta H(N_+ - N_-) + \beta V(N_{++} + N_{--} - N_{+-})\right] \tag{S8.118}$$

where again the sum is over all configurations. As they stand, the expressions for Λ and Z look quite different, particularly since the expression for Z contains five parameters, while that for Λ contains only two. But we will see that the parameters in Z are not independent. They satisfy three linear relations, which can be used to eliminate three of them. The resulting expression will be equivalent to the expression for Λ.

The first linear relation is the trivial one that, for an $L \times L$ lattice,

$$N_+ + N_- = L^2 \tag{S8.119}$$

Now imagine that we place coins at both ends of each ++ bond and at the + end of each +− bond. The number of coins used will be $2N_{++} + N_{+-}$. But then each + vertex will be covered by four coins. Thus

$$2N_{++} + N_{+-} = 4N_+ \tag{S8.120}$$

By the obvious symmetry between + and −, we can get the third linear relation.

$$2N_{--} + N_{+-} = 4N_- \tag{S8.121}$$

It is not difficult to solve these equations for N_-, N_{--}, and N_{+-} in terms of the variables N_+ and N_{++}.

$$N_- = L^2 - N_+, \quad N_{--} = 2L^2 + N_{++} - 4N_+, \text{ and } N_{+-} = 4N_+ - 23N_{++} \tag{S8.122}$$

When these are substituted into the expression $H(N_+ - N_-) + V(N_{++} + N_{--} - N_{+-})$, one obtains

$$\begin{aligned} H(N_+ - N_-) &+ V(N_{++} + N_{--} - N_{+-}) \\ &= (2V - H)L^2 + (2H - 8V)N_+ + 4VN_{++} \end{aligned} \tag{S8.123}$$

and therefore

$$Z = e^{\beta(2V-H)L^2} \sum \exp\left[\beta(2H - 8V)N_+ + 4\beta V N_{++}\right]$$
$$= e^{\beta(2V-H)L^2} \Lambda(\alpha', v') \tag{S8.124}$$

where $\alpha' = \beta(8V - 2H)$ and $v' = 4V$. Thus a calculation of the grand partition function of the lattice gas is equivalent to a calculation of the canonical partition function of the Ising lattice.

Exercise 8.17 A square two-dimensional lattice has total of K sites. Located on the lattice are particles of two different types. There are N_A A particles and N_B B particles. Two A particles or two B particles at neighboring sites interact with an energy V, but an A particle and a B particle at neighboring sites interact with an energy $-V$. (a) Following the method of Section 8.9, use mean-field theory to calculate the canonical potential $\phi(N_A, N_B, \beta)$. (b) Assume that there is a chemical reaction $A \leftrightarrow B$. This would be true, for example, if A and B were really two different spin states of a single spin-$\frac{1}{2}$ particle. As a function of $n = (N_A + N_B)/K$ and β, calculate the equilibrium value of $(N_A - N_B)/(N_A + N_B)$.

Solution   (a) For N_A A particles and N_B B particles, to be distributed among K sites, there are

$$M = \frac{K!}{N_A!N_B!(K - N_A - N_B)!} \tag{S8.125}$$

possible configurations.

Consider any A particle. The mean-field approximations to the average number of A particles and B particles in each of its four nearest-neighbor sites are N_A/K and N_B/K. Thus the average interaction energy of the A particle is

$$4VN_A/K - 4VN_B/K$$

For a B particle, the equivalent expression is

$$4VN_B/K - 4VN_A/K$$

Taking into account the double counting problem, the average energy of a random distribution of N_A A particles and N_B B particles is

$$E = 2VN_A\frac{N_A - N_B}{K} + 2VN_B\frac{N_B - N_A}{K}$$
$$= 2V(N_A - N_B)^2/K \tag{S8.126}$$

Using this approximate energy for every configuration yields the partition function

$$Z = \frac{K!\exp\left[-2\beta V(N_A - N_B)^2/K\right]}{N_A!N_B!(K - N_A - N_B)!} \tag{S8.127}$$

and the canonical potential

$$\phi = K \log K - N_A \log N_A - N_B \log N_B$$
$$- (K - N_A - N_B) \log(K - N_A - N_B) - 2\beta V \frac{(N_A - N_B)^2}{K} \tag{S8.128}$$

With a little algebra, this can be written in terms of the densities $n_A = N_A/K$ and $n_B = N_B/K$, as

$$\phi = - K \big[n_A \log n_A + n_B \log n_B$$
$$+ (1 - n_A - n_B) \log(1 - n_A - n_B) + 2\beta V (n_A - n_B)^2 \big] \tag{S8.129}$$

(b) If we define new variables n and ν by $n = n_A + n_B$ and $\nu = n_A - n_B$, then $n_A = (n + \nu)/2$ and $n_B = (n - \nu)/2$. The canonical potential, expressed in terms of these variables, is

$$\phi = - K \Big[\tfrac{1}{2}(n + \nu) \log\Big(\frac{n + \nu}{2}\Big) + \tfrac{1}{2}(n - \nu) \log\Big(\frac{n - \nu}{2}\Big)$$
$$+ (1 - n) \log(1 - n) + 2\beta V \nu^2 \Big] \tag{S8.130}$$

In the presence of the chemical reaction, $A \leftrightarrow B$, n is a conserved quantity, but ν can adjust itself so as to maximize the value of ϕ. (Recall from Section 5.21 that the equilibrium condition for a system kept at fixed temperature is that ϕ, rather than S, is a maximum with respect to all nonconserved variables.) Setting $\partial\phi/\partial\nu = 0$ gives the equilibrium equation

$$\tfrac{1}{2} \log\Big(\frac{n + \nu}{2}\Big) - \tfrac{1}{2} \log\Big(\frac{n - \nu}{2}\Big) + 4\beta V \nu = 0 \tag{S8.131}$$

Letting $\gamma = (N_A - N_B)/(N_A + N_B) = \nu/n$, this equation can be written as

$$\log(1 + \gamma) - \log(1 - \gamma) = -8\beta V n \gamma \tag{S8.132}$$

But this is essentially identical to Eq. (8.48), developed for a model of nematic liquid crystals. Using the identity mentioned after that equation gives

$$\gamma = - \tanh(4\beta V n \gamma) \tag{S8.133}$$

This always has the solution $\gamma = 0$. But if V is negative and $4\beta|V|n > 1$, then it also has two solutions for nonzero values of γ that are equal in magnitude but opposite in sign. When these solutions exist, the solution $\gamma = 0$ does not represent the equilibrium state (see Problem 8.8). In Fig. S8.8, γ is plotted as a function of $\beta|V|n$ for negative V.

In the description of the computer programs at the end of this supplement, this model is called the AB model, and a Monte Carlo program for solving it is given. By investigating the model, using the display program, the reader will discover

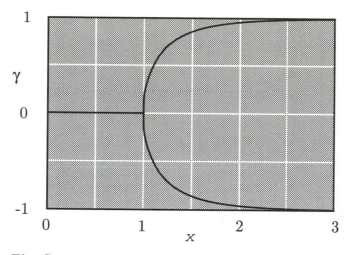

Fig. S8.8 $\gamma = (N_A - N_B)/(N_A + N_B)$ as a function of $x = 4\beta|V|n$. When $x > 1$ (that is, when $kT < 4|V|n$), there are two possible nonzero equilibrium values of γ.

that there is an antiferromagnetic phase that has been completely missed by this mean-field calculation.

Exercise 8.18   In Exercises 8.12 and 8.13, the Bethe–Peierls approximation, which is an improvement on simple mean-field theory, was presented. In this exercise we will illustrate another of the many possible "improvements" on the simple mean-field approximation. We consider a two-dimensional Ising model on a triangular lattice (see Fig. S8.9). The simple mean-field approximation would consist of looking at a single-spin variable and replacing all of its nearest-neighbor spins by their average values. In this new improved version of the method, we focus our attention on any triangle of three spins and replace all of the nearest neighbors of the triangle by fixed noninteger values $\bar{\sigma}$, to be determined later. This fudge effectively replaces the large lattice of spins by a system of only three spins, which has a total of $2^3 = 8$ possible configurations. Finally, the value of $\bar{\sigma}$ will be calculated by demanding that $\langle \sigma_1 + \sigma_2 + \sigma_3 \rangle = 3\bar{\sigma}$. (a) Assuming that there is an external field H and a nearest-neighbor interaction V, calculate $\bar{\sigma}$ as a function of H and β. (b) Assuming that the interaction energy within the triangle is typical of the interaction energy throughout the lattice, calculate the energy per spin as a function of β for the special case $H = 0$.

Solution   (a) For any configuration $(\sigma_1, \sigma_2, \sigma_3)$ of the three-spin system, its energy is

$$E = -(H - 4\bar{\sigma}V)(\sigma_1 + \sigma_2 + \sigma_3) - V(\sigma_1\sigma_2 + \sigma_1\sigma_3 + \sigma_2\sigma_3) \qquad \text{(S8.134)}$$

The probability of any configuration is

$$P(\sigma_1, \sigma_2, \sigma_3) = \frac{e^{-\beta E}}{Z} = Z^{-1} \exp[h(\sigma_1 + \sigma_2 + \sigma_3) + v(\sigma_1\sigma_2 + \sigma_1\sigma_3 + \sigma_2\sigma_3)] \quad \text{(S8.135)}$$

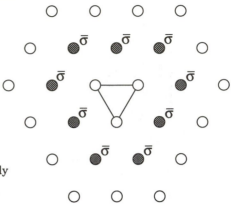

Fig. S8.9   The fixed $\bar{\sigma}$ spins effectively isolate the triangle of Ising spins.

where $h = \beta(H + 4\bar{\sigma}V)$ and $v = \beta V$. The partition function can be calculated by explicitly summing over the 8 possible states.

$$
\begin{aligned}
Z &= \sum_{\sigma_1\sigma_2\sigma_3} \exp\left[h(\sigma_1 + \sigma_2 + \sigma_3) + v(\sigma_1\sigma_2 + \sigma_1\sigma_3 + \sigma_2\sigma_3)\right] \\
&= e^{3h+3v} + e^{-3h+3v} + 3e^{h-v} + 3e^{-h-v} \\
&= 2e^{3v}\cosh(3h) + 6e^{-v}\cosh(h)
\end{aligned}
\tag{S8.136}
$$

The average value of $\sigma_1 + \sigma_2 + \sigma_3$ is given by

$$
\langle \sigma_1 + \sigma_2 + \sigma_3 \rangle = \sum_{\sigma_1}\sum_{\sigma_2}\sum_{\sigma_3}(\sigma_1 + \sigma_2 + \sigma_3)P(\sigma_1, \sigma_2, \sigma_3)
\tag{S8.137}
$$

Looking at Eq. (S8.135), we see that

$$
\langle \sigma_1 + \sigma_2 + \sigma_3 \rangle = \frac{\partial Z/\partial h}{Z} = 3\frac{e^{3v}\sinh(3h) + e^{-v}\sinh(h)}{e^{3v}\cosh(3h) + 3e^{-v}\cosh(h)}
\tag{S8.138}
$$

Demanding that $\bar{\sigma} = \langle\sigma_1 + \sigma_2 + \sigma_3\rangle/3$ gives the self-consistency condition

$$
\frac{\sinh[3\beta(H + 4V\bar{\sigma})] + e^{-4v}\sinh[\beta(H + 4V\bar{\sigma})]}{\cosh[3\beta(H + 4V\bar{\sigma})] + 3e^{-4v}\cosh[\beta(H + 4V\bar{\sigma})]} = \bar{\sigma}
\tag{S8.139}
$$

This equation for $\bar{\sigma}$ can be solved numerically. $\bar{\sigma}$ is shown as a function of the dimensionless temperature variable $t = kT/V$ for $H = 0$ in Fig. S8.10.

(b) If $H = 0$ then the energy in the real many-spin system is

$$
E = -V\left\langle\sum_{NN}\sigma(\mathbf{x})\sigma(\mathbf{x}')\right\rangle
\tag{S8.140}
$$

In the mean-field calculation we will approximate this sum over all bonds by one-half the number of triangles times the same sum over the three bonds in the chosen triangle. The one-half avoids double counting, since each bond lies between two triangles. Looking at Eq. (S8.135), one can see that

$$
\langle \sigma_1\sigma_2 + \sigma_1\sigma_3 + \sigma_2\sigma_3 \rangle = \frac{\partial Z/\partial v}{Z} = 3\frac{\cosh(3h) - e^{-4v}\cosh(h)}{\cosh(3h) + 3e^{-4v}\cosh(h)}
\tag{S8.141}
$$

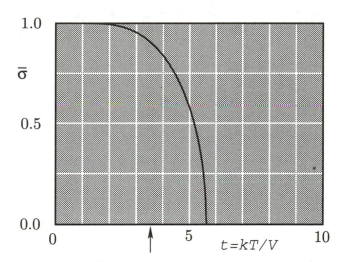

Fig. S8.10   The spontaneous magnetization $\bar{\sigma}$ as a function of the dimensionless temperature $t = kT/V$. In this approximation, the transition occurs at about $t = 5.65$. The exact transition temperature for a triangular lattice, indicated in the figure by an arrow, is $t \approx 3.64$. The simple mean-field approximation, which uses only a single spin, predicts a transition at $t = 6$. Thus this calculation does not give a large improvement on the simpler one.

In a triangular lattice of N spins, neglecting edge corrections, there are $2N$ triangles. Summing over all the triangles would count each bond twice, since it lies between two triangles. Therefore,

$$E = -3NV\langle\sigma_1\sigma_2 + \sigma_1\sigma_3 + \sigma_2\sigma_3\rangle \tag{S8.142}$$

and

$$\frac{E}{N} = -V\frac{\cosh(3h) - e^{-4v}\cosh(h)}{\cosh(3h) + 3e^{-4v}\cosh(h)} \tag{S8.143}$$

where $h = 4\beta V\bar{\sigma}$ and $v = \beta V$.

Exercise 8.19   In this exercise, the basic mechanisms of ferrimagnetism and antiferromagnetism will be explored. The two phenomena are closely related, and may be conveniently treated simultaneously. We will restrict ourselves to a two-dimensional model, but the mean-field analysis would be basically unchanged in treating the three-dimensional case.

Shown in Fig. S8.11 is a model of a two-dimensional ferrimagnetic substance. Two different types of atoms, which will be called A and B, are arranged in a checkerboard pattern. The magnetization direction is perpendicular to the plane of the lattice. A and B atoms have magnetic moments m_A and m_B, respectively, with $m_A > m_B$. There is an antiferromagnetic interaction between nearest neighbors (which are always A–B pairs). The magnetic moments also interact with a constant external magnetic field of strength B. Using the mean-field approximation,

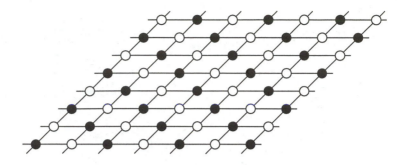

Fig. S8.11 · The checkerboard pattern of a ferrimagnetic crystal.

derive equations for the average magnetization of the substance as a function of the temperature and the external field strength.

Solution   We introduce spin variables $\sigma_A(\mathbf{x})$ and $\sigma_B(\mathbf{y})$, with the values ± 1, for the A and B atoms, respectively. The vector $\mathbf{x}$ ranges over the A lattice, while $\mathbf{y}$ ranges over the B lattice (that is, the location of the B atoms). The energy of a configuration in an external field B is

$$E = -B\left(m_A \sum_A \sigma_A(\mathbf{x}) + m_B \sum_B \sigma_B(\mathbf{y})\right) + V \sum_{\text{NN}} \sigma_A(\mathbf{x})\sigma_B(\mathbf{y}) \qquad \text{(S8.144)}$$

where $V > 0$ for an antiferromagnetic interaction. The average spin of an A atom can, in the mean-field approximation, be calculated by replacing its four nearest neighbors by their as yet unknown average values.

$$\bar{\sigma}_A = \sum_{\sigma_A}\left(\sigma_A e^{\beta(Bm_A - 4V\bar{\sigma}_B)\sigma_A}\right) \bigg/ \sum_{\sigma_A} e^{\beta(Bm_A - 4V\bar{\sigma}_B)\sigma_A} \qquad \text{(S8.145)}$$

Evaluating the sums, one obtains

$$\bar{\sigma}_A = \tanh\left[\beta(Bm_A - 4V\bar{\sigma}_B)\right] \qquad \text{(S8.146)}$$

A similar calculation for $\bar{\sigma}_B$ gives

$$\bar{\sigma}_B = \tanh\left[\beta(Bm_B - 4V\bar{\sigma}_A)\right] \qquad \text{(S8.147)}$$

Once these two simultaneous equations have been solved for $\bar{\sigma}_A$ and $\bar{\sigma}_B$, the magnetic moment per atom is given by

$$\bar{m} = (m_A\bar{\sigma}_A + m_B\bar{\sigma}_B)/2 \qquad \text{(S8.148)}$$

Exercise 8.20   The model that was introduced in the last exercise can describe both ferrimagnetism and antiferromagnetism. It exhibits antiferromagnetic behavior when $m_A = m_B$. This occurs when the crystal is, in fact, composed of only one type of magnetic atom with a nearest-neighbor interaction that favors antiparallel spin alignment. An important example is the set of copper oxide materials involved in high-temperature superconductivity. In these materials, the oxygen atoms play a role in creating an antiferromagnetic interaction between neighboring copper atoms but play no direct part in the magnetic properties of the crystal. For the antiferromagnetic case, $m_A = m_B = m$, take $B = 0$, and determine $\bar{\sigma}_A$ and $\bar{\sigma}_B$ as functions of $\tau = 1/\beta$.

Solution We introduce a parameter $\tau_c = 4V$, which will, of course, turn out to be the critical temperature for the antiferromagnetic transition. With $B = 0$, Eqs. (S8.146) and (S8.147) can be written as

$$\bar{\sigma}_A = -\tanh\left(\frac{\tau_c}{\tau}\bar{\sigma}_B\right) \tag{S8.149}$$

and

$$\bar{\sigma}_B = -\tanh\left(\frac{\tau_c}{\tau}\bar{\sigma}_A\right) \tag{S8.150}$$

It is clear that, if $\bar{\sigma}_A > 0$, then $\bar{\sigma}_B < 0$ and vice versa. In fact, the solutions of these equations for $\bar{\sigma}_A$ and $\bar{\sigma}_B$ all have the symmetry $\bar{\sigma}_A = -\bar{\sigma}_B$. Assuming that symmetry, one can write a single equation for $\bar{\sigma}_A$.

$$\bar{\sigma}_A = \tanh\left(\frac{\tau_c}{\tau}\bar{\sigma}_A\right) \tag{S8.151}$$

This equation is identical to Eq. (8.110), showing that, below the critical temperature τ_c, the A lattice is spontaneously magnetized to exactly the same degree as shown by a ferromagnetic model with the same value for the interaction strength V, but the opposite sign. However, because of the fact that the B lattice magnetization is exactly opposite to that of the A lattice, the total magnetization of the substance remains zero below the transition temperature. Although the antiferromagnetic phase transition does not exhibit spontaneous macroscopic magnetization, it can be clearly detected by monitoring the specific heat. In fact, the specific heat has precisely the same singularity as a two-dimensional ferromagnetic Ising model.

Exercise 8.21 Within the mean-field approximation, calculate the magnetic susceptibility in zero external field of a ferrimagnet for $\tau > \tau_c$.

Solution Above the critical temperature, if we assume that the external field B is very small, then $\bar{\sigma}_A$ and $\bar{\sigma}_B$ are also very small, and the right-hand sides of Eqs. (S8.146) and (S8.147) can be expanded to first order in B, $\bar{\sigma}_A$, and $\bar{\sigma}_B$, giving the linear equations

$$\bar{\sigma}_A = \frac{Bm_A - \tau_c\bar{\sigma}_B}{\tau} \tag{S8.152}$$

and

$$\bar{\sigma}_B = \frac{Bm_B - \tau_c\bar{\sigma}_A}{\tau} \tag{S8.153}$$

which can be rearranged as

$$\tau\bar{\sigma}_A + \tau_c\bar{\sigma}_B = Bm_A \quad \text{and} \quad \tau_c\bar{\sigma}_A + \tau\bar{\sigma}_B = m_B \tag{S8.154}$$

These simultaneous linear equations have the solution

$$\bar{\sigma}_A = \frac{\tau m_A - \tau_c m_B}{\tau^2 - \tau_c^2}B \quad \text{and} \quad \bar{\sigma}_B = \frac{\tau m_B - \tau_c m_A}{\tau^2 - \tau_c^2}B \tag{S8.155}$$

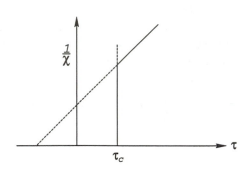

Fig. S8.12   One distinguishing feature of an antiferromagnet is that, above the transition temperature, the inverse of the magnetic susceptibility is a linear function of temperature with a negative intercept.

The susceptibility is given by

$$\chi = \frac{m_A \bar{\sigma}_A + m_B \bar{\sigma}_B}{B} = \frac{(m_A^2 + m_B^2)\tau - 2m_A m_B \tau_c}{\tau^2 - \tau_c^2} \tag{S8.156}$$

In the antiferromagnetic case, $m_A = m_B = m$, and

$$\chi = \frac{2m^2}{\tau + \tau_c} \tag{S8.157}$$

This relation implies that, above τ_c, the inverse of the susceptibility for an antiferromagnet is a linear function of τ with a negative intercept (Fig. S8.12). This is a prediction that is well satisfied by real antiferromagnetic materials (although the detailed formula for the value of the intercept is not reliable).

For a ferrimagnet, the susceptibility has a singularity at $\tau = \tau_c$. The mean-field calculation predicts that the singularity in χ has the form of a simple pole [that is, $\chi \approx C/(\tau - \tau_c)$], a prediction that does not agree with experiment.

The Programs

The remainder of this Supplement will be a description of the computer programs on the Program Disk that accompanies the text. Some of the programs come in two versions: a display version, which is a compiled program to be run on a Macintosh computer or an IBM PC with a graphics board, and a Fortran version, ready to be compiled and run on any computer with a Fortran compiler.

All of the programs described here are applications of the Monte Carlo method to various model systems. The display version of each program gives a visual representation of the sequence of physical states constructed by the Monte Carlo algorithm. It is intended to help the reader in developing physical intuition by actually seeing the sequence of states that the system passes through in approaching equilibrium. Also, by seeing the program in operation, the reader can more easily understand what the corresponding Fortran program, which operates without any display, is doing. The Fortran versions of the programs are necessary for two reasons. The first is that the Macintosh programs run quite slowly, particularly on a machine without a mathematical coprocessor, because they use a large part of their time in constructing screen displays. The second is that it would be unreasonable to expect the reader to be familiar with Macintosh programming and thus capable of modifying and augmenting the programs. But many of the problems associated with the Monte Carlo method do require that ability. Those problems will be done using the Fortran programs, which have been written with that consideration in mind.

Hot Spring 1 (Display)

This program is a Monte Carlo analysis of a particle on a one-dimensional lattice, under the influence of a harmonic oscillator potential. It follows the simple example presented in Section 8.12, with one minor revision. In order to keep the particle on the screen, it limits ℓ to the range $-L \leq \ell \leq L$, where the value of L is determined by the size of the Macintosh screen on the machine that is used to run the program. This is equivalent to setting $E(\ell) = \infty$ for $|\ell| > L$. Each time the particle, shown at the bottom of the screen, jumps to location ℓ, a white bar above that location grows by a fixed amount. When any white bar reaches the top of the screen, the program halts and waits for the mouse button to be pressed. The white bars therefore give a histogram of the time spent at each location. Theoretically, it should be a Gaussian curve, $f(\ell) = C \exp(-\varepsilon \ell^2 / kT)$, but in any particular run one will see statistical fluctuations about the ideal curve, since the sample is finite.

Hot Spring 2 (Display)

This program adds one additional feature to the previous program. Instead of always considering a move of one lattice unit, either left or right, it chooses a possible move from a uniform distribution over the possibilities $\Delta \ell = \pm 1, \pm 2, \ldots,$ $\pm MaxJump$, where the positive integer $MaxJump$ is set by the user. Once a possible move has been chosen, it is definitely made if it would lower the energy and it is made with a probability $P = \exp(-\Delta E / kT)$, if it would increase the energy. Although the value of $MaxJump$ has no effect on the equilibrium distribution, it can strongly affect the magnitude of the statistical error in a finite sample.

Hot Spring (Fortran)

This has three different Fortran versions. HotSpring_1.for is a Fortran version of the display program Hot Spring 1. The output of the program, in the file HotSpring.out, is the function $P(\ell)$, which gives the fraction of time that the particle spent at the lattice point ℓ. As the run time is made longer and longer, the function more and more closely approximates a Gaussian function. The parameter $s = \sqrt{kT/\varepsilon}$ is chosen by the user.

The second program, HotSpring_2.for, is a Fortran version of the display program Hot Spring 2. The user chooses values of the parameter $s = \sqrt{kT/\varepsilon}$, the maximum jump size, and the number of Monte Carlo moves. The output in the file HotSpring.out gives the value of $P(\ell)$, the fraction of time spent at location ℓ.

The third Fortran version, HotSpring_3.for, does not impose any artificial discreteness on the position variable x. It illustrates how one treats a system with continuous coordinates by means of the Monte Carlo method. This is a topic that was not covered in Chapter 8. The modifications to the discrete case are fairly minimal. One chooses a suggested jump Δx from a uniform distribution over the interval $-D < \Delta x < D$. If the jump $x \to x + \Delta x$ would lower the energy, then it is made. If it would increase the energy by an amount ΔE, then it is made with probability $P = \exp(-\beta \Delta E)$. The user chooses the two parameters, $s = \sqrt{kT/\varepsilon}$ and D, and the number of Monte Carlo moves. The output of the calculation is the set of three numbers $\langle x^2 \rangle$, $\langle x^4 \rangle$, and $\langle x^6 \rangle$, which can be compared with theoretical calculations of the same quantities using the canonical probability density.

Hot Ball (Display)

This is essentially the same as Hot Spring 2, but the coordinate is a vertical coordinate and the potential function is that for a uniform gravitational field; that is, $U(z) = mgz$. The parameter that defines the equilibrium distribution is the scale height, $s = kT/mg$. The coordinate z is restricted to positive values by setting $U(z) = \infty$ for negative z, and it is converted to an integer variable. The expected probability distribution is the exponential $P(z) = C \exp(-z/s)$.

Hot Ball (Fortran)

This is a Fortran version of the same program. The output of the program is the distribution $P(z)$, which is the fraction of time the particle spends at the height z and the theoretically exact probability, $P_o(z) = C \exp(-z/s)$, for comparison. The user must choose the values of the scale height s, the maximum jump size, and the number of Monte Carlo moves.

Ising 1 (Display)

This is an implementation of the algorithm described in Section 8.13. It is a Monte Carlo calculation on an L_x by L_y square lattice of two-valued spins. (The lattice structure is square but, if $L_x \neq L_y$, the whole sample is rectangular.) There is an external magnetic field variable H, whose value is chosen by the user. If H is positive, then up spins have lower energy than down spins. There is a nearest-neighbor interaction energy variable V, also chosen by the user. A positive value of V means that any pair of neighboring spins will have a lower energy if they are parallel than if they are antiparallel. If V is negative, antiparallel neighbors have lower energy than parallel ones. Thus positive V favors ferromagnetic ordering while negative V favors antiferromagnetic (checkerboard) ordering. An "up" spin is represented on the screen by a small white square. A "down" spin is drawn as a gray square. The program starts with a random configuration in which any spin is equally likely to be up or down. The lattice has periodic boundary conditions, which means that the rightmost spins are neighbors of the leftmost spins and the top spins are neighbors of the bottom spins.

The following step is repeated for a number of times N_{moves} that is chosen by the user. A random lattice point is chosen, and the program considers the possibility of changing the spin at that lattice point. If it would lower the energy, the change is made. If it would increase the energy by an amount ΔE, then the change is made with a probability $\exp(-\beta \Delta E)$.

Ising 1 (Fortran)

This is a Fortran version of the above program, written for an $L \times L$ lattice. The periodic boundary conditions are implemented by defining a function called $np1(n, L)$ that has the value

$$np1(n, L) = \begin{cases} n+1, & \text{for } n = 1, \ldots, L-1 \\ 1, & \text{for } n = L \end{cases} \tag{S8.158}$$

With this definition the nearest neighbor of spin $\sigma(x, y)$ in the $+x$ direction is always $\sigma(np1(x, L_x), y)$. The function

$$nm1(n, L) = \begin{cases} n-1, & \text{for } n = 2, \ldots, L \\ L, & \text{for } n = 1 \end{cases} \tag{S8.159}$$

gives the nearest neighbors in the $-x$ and $-y$ directions.

For given values of H and V, the program carries out calculations at a number of equally spaced temperatures in an interval $\tau_{min} \leq \tau \leq \tau_{max}$. At each temperature, the values of $\bar{M}$, $\bar{E}$, $(\Delta M)^2$, and $(\Delta E)^2$ are calculated, where $M = |\sum \sigma(x, y)|$. From the values of $(\Delta M)^2$ and $(\Delta E)^2$ one can determine the magnetic susceptibility and the specific heat by using Eqs. (8.82) and (8.83). M is defined as $|\sum \sigma|$, rather than as $\sum \sigma$, in order to avoid a problem that occurs when making calculations using a relatively small number of spins. For a real system of 10^{20} or more spins, below the ferromagnetic transition temperature the system will be found with its polarization in either the up or the down direction. It will never realistically make a spontaneous transition from one polarization state to the other. Therefore, the fluctuation in the polarization is negligibly small at temperatures below the transition temperature. However, for a system of only a few spins, occasional transitions between the two favored polarization states do occur. This causes the fluctuation in the polarization to have a large value. But the large fluctuation is only an uninteresting effect of small system size. It can be eliminated by looking at the observable $|\sum \sigma|$, rather than $\sum \sigma$.

Ising 2

This is identical to Ising 1, except for the following change. Instead of choosing the lattice point at which to consider a spin flip at random, the program methodically sweeps through the lattice in a raster pattern, going from top to bottom through one complete column and then moving to the next column. At each lattice point it uses the usual Monte Carlo algorithm to decide whether to flip the spin or leave it unchanged. It is left as a problem for the reader to show that this procedure is legitimate, in the sense that it produces a sampling sequence with the canonical distribution. Because this procedure requires the generation of much fewer random numbers, it is more efficient than the one used in Ising 1. There are both display and Fortran versions of this program.

Ising 3 (Fortran)

This program, which exists only in the Fortran version, is designed to study the effects of boundary conditions and to study correlation effects in the Ising system. In this program H is set to zero and V is set to one. The user specifies the inverse temperature β. Periodic boundary conditions are used in the y direction, but the first and last columns of spins (that is, those spins with $x = 0$ and $L_x + 1$) are set by the user. Each of the two fixed boundary columns can be set all up, all down, or zero. Setting the final column of spins, at $x = L_x + 1$, to zero has the same effect as having a "free boundary" at $x = L_x$. The user specifies how many sweeps through the lattice should be made. The output, given in the file Ising_3.out, consists of the average polarization of the spins in each column as a function of x, the column number. By looking at this output, the reader can see how far into the lattice the effects of the fixed boundaries extend. At high temperatures the effects are noticeable for only a few lattice distances away from the boundary, but below the ferromagnetic transition it is clear that the effects extend throughout the lattice. This is an example of the fact that, in the presence of spontaneous symmetry breakdown, ordinarily negligible aspects of the system, such as surface interactions, determine the observed equilibrium state.

Ising 4 (Fortran)

This Fortran program introduces random vacancies into a two-dimensional Ising model (with no external field). The user chooses a value for the vacancy rate, which must be less than one. The program first sweeps through the lattice, randomly setting spins equal to zero, with a probability equal to the chosen vacancy rate. A zero spin represents a vacancy in the lattice. During the subsequent Monte Carlo runs, the zero-spin values are never changed. The reader is encouraged to investigate, by means of this program, the effects of random impurities on phase transitions in crystalline materials.

Ising 3D (Fortran)

This is a three-dimensional version of Ising 2.

LatGas (Fortran)

This program makes a Monte Carlo calculation for the lattice gas model, analyzed with the mean-field approximation in Section 8.9. The user sets the lattice size, the range and number of values of the affinity, the value of βV, and the number of sweeps through the lattice in each run. The program gives, as output, the value of $\langle N \rangle$ in the file LatGas.out.

AB Model

This is the lattice gas model that was analyzed, with the mean-field approximation, in Exercise 8.17. Two types of particles, called A and B, can reside at the points of a square lattice. A particles are shown as white squares, B particles as gray squares, and empty lattice sites as black squares. Two A particles or two B particles interact with an energy V, but an A particle and a B particle that are nearest neighbors interact with an interaction energy $-V$. There is no energy binding the particles to the lattice sites. The number of particles is not fixed, but α_A is set equal to α_B. In the grand canonical ensemble, the probability of any configuration of the system is

$$P(C) = \Lambda^{-1} \exp[-\alpha(N_A + N_B) + \beta V(N_{AA} + N_{BB} - N_{AB})] \qquad (S8.160)$$

where N_A is the number of A particles, N_B is the number of B particles, and N_{AA}, N_{BB}, and N_{AB} are the number of nearest-neighbor AA, BB, and AB pairs, respectively.

The program comes in both display and Fortran versions. The user chooses the size of the lattice, the values of V/kT, the number of sweeps through the lattice, and the value (in the display program) or the range (in the Fortran program) of α. The configuration of the system is represented by a function $N(x, y)$ that has the value 0 for an empty site, $+1$ for an A particle, and -1 for a B particle. The program sweeps, raster style, through the points of the lattice. At each point, the program reads the value of $N(x, y)$ and then attempts a change of $N(x, y)$ to either of the other two possible values with equal probabilities. In "attempting a change" from, for instance, $N = +1$ to -1, the program evaluates ΔF, the change in the function

$$F(C) = \alpha(N_A + N_B) + \beta V(N_{AA} + N_{BB} - N_{AB}) \qquad (S8.161)$$

that would result. If ΔF is negative, then the change is made. If ΔF is positive, then the change is made with probability $\exp(-\Delta F)$.

The Fortran program, for each value of α, stores the values of $\langle N \rangle$, $\langle Q \rangle$, $(\Delta N)^2$, and $(\Delta Q)^2$ in the file AB.out, where $N = N_A + N_B$ and $Q = N_A - N_B$.

Supplement to Chapter 9

REVIEW QUESTIONS

9.1 What does it mean to say that $f(x, y)$ is analytic at the point (x, y)?

9.2 How is a phase transition point in the p-T plane defined?

9.3 Why are true phase transitions not possible in finite systems?

9.4 How do we distinguish between first-order and second-order phase transitions?

9.5 What is a critical point?

9.6 What is meant, with regard to critical phenomena, by universality?

9.7 How are the following critical exponents defined?
    1. The order parameter exponent β.
    2. The specific heat exponent α.
    3. The susceptibility exponent γ.
    4. The critical isotherm exponent δ.

9.8 What is the fundamental assumption of simple Landau theory?

9.9 In the Landau theory, what is the criterion for the critical temperature?

9.10 What is the behavior of $\phi(T, M)$ as a function of M, for T above and below T_c?

9.11 Calculate the equilibrium value of M as a function of t, near the critical point, using simple Landau theory.

9.12 What is the value of the order parameter exponent β in simple Landau theory?

9.13 Calculate $E(T)$, for T slightly less than T_c, in simple Landau theory, and use it to derive the specific heat exponent α.

9.14 Derive an equation for the magnetization M near the critical temperature in a weak magnetic field, and use it to derive the susceptibility exponent γ and the critical isotherm exponent δ.

9.15 How can one see that large-wavelength fluctuations are important at the critical point?

9.16 In the Landau–Ginsburg theory, what is the probability of a magnetization pattern $M(\mathbf{r})$?

9.17 How are the fluctuation amplitudes $M_{\mathbf{K}}$ related to the magnetization $M(\mathbf{r})$?

9.18 Why is it legitimate to take only a finite number of fluctuation amplitudes in calculating critical phenomena?

9.19 What is the partition function in the Landau–Ginsburg theory?

9.20 What is the Gaussian approximation to Z?

9.21   In the Gaussian approximation, calculate the specific heat.

9.22   What happens to the specific heat at $t = 0$ in the Gaussian approximation?

9.23   How is the critical dimension defined?

9.24   Explain how the critical dimension was shown to be $D = 4$.

9.25   What is the epsilon expansion?

9.26   In a general, pictorial, way, describe a renormalization transformation.

9.27   What is the energy exression for the chain of harmonic oscillators that was used to describe a renormalization transformation?

9.28   Describe how the renormalization transformation $(K_1, K_2) \to (\tilde{K}_1 \tilde{K}_2)$, was derived for the harmonic chain.

9.29   Derive the transformation $(K_1, K_2) \to (\tilde{K}_1, \tilde{K}_2)$ for the harmonic chain.

9.30   What is a fixed point of a renormalization flow diagram?

9.31   Why does any point (K_1, K_2), with $K_1 > 0$ flow to a trivial fixed point $(K_1^*, 0)$?

9.32   Describe the nontrivial fixed points of the renormalization flow for the harmonic chain.

9.33   What is a scale-invariant probability distribution?

9.34   What are the assumptions of Wilson's analysis of critical phenomena?

9.35   What are stable and unstable fixed points?

9.36   What kinds of fixed points are stable?

9.37   Given that, under the scaling transformation $\mathbf{r} \to s\mathbf{r}$ and $M \to \gamma M$, the correlation function satisfies the equation $\langle \gamma M(s\mathbf{r}) \gamma M(s\mathbf{r}') \rangle = \langle M(\mathbf{r}) M(\mathbf{r}') \rangle$, derive the relationship between γ, s, and the correlation function exponent λ.

9.38   Explain the meaning of all the terms $[\gamma, c(r), s, K, \text{and } \tilde{K}]$ in the renormalization relation $\gamma^2 c(sr|K) = c(r, \tilde{K})$.

9.39   Describe the four steps in a momentum-space renormalization transformation for the Landau–Ginsburg theory.

9.40   In terms of the parameters $x = A\Lambda^2/2\pi^2$, $y = 3b/2\pi^2$, and $\delta = s - 1$, the Wilson renormalization equations (in 3D) are

$$x \to x + (2x + y - xy)\delta \quad \text{and} \quad y \to y + (y - 3y^2)\delta$$

Derive the fixed point (x^*, y^*).

9.41   Derive the differential flow equations and give a rough sketch of the flowlines.

9.42   Describe what modifications are necessary in order to go beyond first order in ϵ for the Landau–Ginsburg renormalization transformation.

9.43   How are the fundamental exponents y_t and y_B defined?

9.44   What is the basic assumption that leads to the scaling formulas for critical exponents in terms of y_t and y_B?

9.45   Derive the relation $\beta = (3 - y_B)/y_t$.

9.46   Show that the susceptibility exponent $\gamma = (2y_B - 3)/y_t$.

9.47   In the real-space renormalization method, how is the renormalized Hamiltonian $\tilde{H}(\tilde{\sigma}_1, \tilde{\sigma}_2, \dots)$ defined?

9.48   What is meant by the proliferation of interactions?

9.49   What is meant by block spin ambiguity?

EXERCISES

Exercise 9.1 When we say that two atoms, let us say two oxygen atoms, are *hydrogen bonded*, we mean that there is a hydrogen atom located somewhere between the two atoms that is strongly bonded to one of them and weakly bonded to the other, the net effect being a weak bond between the two atoms. The hydrogen atom is not equally distant from the two bonded atoms, but is much closer to the atom to which it is strongly bonded. Ordinary water ice is a solid that is held together by hydrogen bonds. Each oxygen atom is strongly bonded to two hydrogens and weakly bonded to two others.

Fig. S9.1 A long chain of oxygen atoms, bonded by double hydrogen bonds, forms the imaginary substance, one-dimensional ice.

In Fig. S9.1 is a picture of a nonexistent substance, called *one-dimensional ice*. It is a very long chain of oxygen atoms, each connected to its nearest neighbors by double hydrogen bonds. As in real ice, there is a restriction that exactly two hydrogen atoms lie close to each oxygen atom. We can indicate the bonds by arrows, with the arrowhead pointing to the strongly bonded oxygen. Then each oxygen atom must have two arrows entering and two arrows leaving. The energy of the system can be written as the sum of the energies of the double bonds. We assume that the double bond patterns have energies as shown in Fig. S9.2.

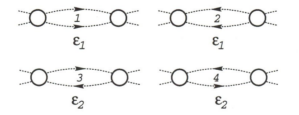

Fig. S9.2 The energies of the four possible double bond patterns.

The hydrogen bonds are geometrically quite localized and directional, so that it makes sense to distinguish between the two patterns shown with energy ε_2. (a) Assuming that the number of oxygens, N, is very large, calculate the partition function and show that the system undergoes a first order phase transition at a finite temperature. (b) Determine the latent heat of transition per oxygen atom.

Solution (a) If the first two oxygen atoms have either of the bonding patterns (1) or (2), then every pair of atoms will have the same bonding pattern as the first pair. Thus there are two states of the system with energies $-N\varepsilon_1$. If the first pair of oxygen atoms has bonding pattern (3) or (4), then every other pair has one of those two patterns, but not necessarily the same one. Thus there are 2^N states of the system with energies $-N\varepsilon_2$. The partition function is therefore

$$Z = 2e^{\beta N \varepsilon_1} + 2^N e^{\beta N \varepsilon_2}$$
$$= 2\left(e^{\beta \varepsilon_1}\right)^N + \left(2e^{\beta \varepsilon_2}\right)^N \tag{S9.1}$$

If $e^{\beta \varepsilon_1} > 2e^{\beta \varepsilon_2}$, then $2(e^{\beta \varepsilon_1})^N \gg (2e^{\beta \varepsilon_2})^N$, and the second term can be neglected. In the opposite case, in which $e^{\beta \varepsilon_1} < 2e^{\beta \varepsilon_2}$, the first term can be neglected. The transition temperature is determined by setting

$$e^{\beta \varepsilon_1} = 2e^{\beta \varepsilon_2} \tag{S9.2}$$

which gives

$$\tau_o = \frac{\varepsilon_1 - \varepsilon_2}{\log 2} \tag{S9.3}$$

Below τ_o,

$$\phi(\beta) = \log 2 + N\beta\varepsilon_1 \tag{S9.4}$$

Above τ_o,

$$\phi(\beta) = N(\log 2 + \beta\varepsilon_2) \tag{S9.5}$$

(b) The energy of the system, as a function of τ, is

$$E = -\frac{\partial \phi}{\partial \tau} = \begin{cases} -N\varepsilon_1, & \tau < \tau_o \\ -N\varepsilon_2, & \tau > \tau_o \end{cases} \tag{S9.6}$$

The latent heat of the transition per oxygen atom is therefore $\varepsilon_1 - \varepsilon_2$.

Exercise 9.2 In the one-dimensional ice system described in the last exercise assume that bonds (1) and (2) have a length (distance between neighboring oxygen atoms) of ℓ_1 and bonds (3) and (4) have a length of ℓ_2. The long molecule is held fixed at one end and a tension F is applied at the other. It may be helpful to imagine that the force is being created by a very long spring, attached at one end, that can store energy. One now has two variables to describe the equilibrium states of the system, namely τ and F (analogous to T and p). (a) In the τ–F plane, determine the phase transition line. (b) Show that the phase transition line satisfies the Clausius–Clapeyron equation.

Solution (a) We assume that $\ell_2 > \ell_1$. The system now has two macrostates—a *long state*, of length $N\ell_2$, and a *short state*, of length $N\ell_1$. The long state has 2^N corresponding microstates, while the short state has only 2. If we take the energy of the spring as zero when the ice molecule is in a long state, then the system in the short state has an extra energy of $N(\ell_2 - \ell_1)F = N\,\Delta\ell\,F$. The system now has two states of energy $-N(\varepsilon_1 - \Delta\ell\,F)$ and 2^N states of energy $-N\varepsilon_2$. Everything proceeds as before, except for the change of replacing ε_1 by $\varepsilon_1 - \Delta\ell\,F$. Letting $\Delta\varepsilon = \varepsilon_1 - \varepsilon_2$, the transition temperature is given by

$$\tau_o = \frac{\Delta\varepsilon - F\,\Delta\ell}{\log 2} \tag{S9.7}$$

The transition line in the τ–F plane is thus the straight line

$$\tau \log 2 + F \Delta \ell = \Delta \varepsilon \qquad \text{(S9.8)}$$

(b) The entropy change in going from the short to the long state is

$$S_{\text{long}} - S_{\text{short}} = \log 2^N - \log 2 \approx N \log 2 \qquad \text{(S9.9)}$$

The length change is

$$L_{\text{long}} - L_{\text{short}} = N \Delta \ell \qquad \text{(S9.10)}$$

Solving Eq. (S9.8) for $F(\tau)$ gives

$$F = \frac{\Delta \varepsilon - \tau \log 2}{\Delta \ell} \qquad \text{(S9.11)}$$

Thus

$$\frac{dF}{d\tau} = -\frac{\log 2}{\Delta \ell} = -\frac{S_{\text{long}} - S_{\text{short}}}{L_{\text{long}} - L_{\text{short}}} \qquad \text{(S9.12)}$$

Comparing this with Eq. (6.90), one sees that they disagree in sign. This shows that the tension force F corresponds to the negative of the compression force p.

Exercise 9.3   According to the Landau–Ginsburg theory, the conditional canonical potential associated with a magnetization pattern $M(\mathbf{r})$ is

$$\phi(M) = \phi_o - \int_V \left(at M^2 + b M^4 + c |\nabla m|^2 \right) d^3 r \qquad \text{(S9.13)}$$

For temperatures not too close to the critical temperature, the observed magnetization pattern would be one that maximizes the value of $\phi(M)$. In Problem 9.2, the reader was asked to show that the maximizing function, with any given boundary conditions, is a solution of the differential equation

$$-c \nabla^2 M + at M + 2b M^3 = 0 \qquad \text{(S9.14)}$$

In this exercise we consider the case of temperatures below the critical temperature; that is, of negative t. If $t < 0$, then the absolute minimum of $\phi(M)$ is given by the trivial solution of Eq. (S9.14) in which $M(\mathbf{r}) = M_o$ is constant. The value of M_o is given by Eq. (S9.14) without the gradient term.

$$-a|t| M_o + 2b M_o^3 = 0 \qquad \text{(S9.15)}$$

or

$$M_o = \pm \sqrt{\frac{a|t|}{2b}} \qquad \text{(S9.16)}$$

It is easy to see that the other solution of Eq. (S9.15), namely $M_o = 0$, gives a local minimum, rather than a maximum of $\phi(M)$.

A more interesting solution of Eq. (S9.14) is one that describes a two-phase state, such as that shown in Fig. S9.3. To describe such a state, one can take a function $M(x)$ that depends only on x and has the form shown in the figure. (a) Assuming that $M(-x) = -M(x)$ and that $M(x) \to M_o$ as $x \to \infty$, determine an

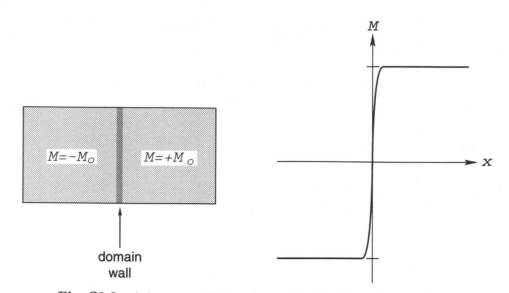

Fig. S9.3 A ferromagnetic two-phase state. The down phase on the left is separated from the up phase on the right by a domain wall. The magnetization function is shown to the right. One can calculate the domain wall energy, using the Landau–Ginsburg theory.

exact solution of Eq. (S9.14). (b) Using the solution obtained in (a), and the fact that the free energy is related to the canonical potential by $F = -\tau\phi$, determine the free energy per unit area associated with an interface between two magnetic domains of opposite magnetization.

Solution   (a) For a function of x alone, and for negative t, Eq. (S9.14) becomes

$$-c\frac{d^2M}{dx^2} - a|t|M + 2bM^3 = 0 \tag{S9.17}$$

The most obvious elementary function with the general characteristics shown in Fig. S9.3 is

$$M(x) = A\tanh(\alpha x) \tag{S9.18}$$

where A and α are as yet undetermined. We will see that A and α may be chosen so as to produce an exact solution of the nonlinear differential equation. Taking two derivatives, one obtains

$$\frac{d^2M}{dx^2} = -2A\alpha^2\frac{\tanh(\alpha x)}{\cosh^2(\alpha x)} \tag{S9.19}$$

Substituting this into Eq. (S9.17) gives the relation

$$2Ac\alpha^2\frac{\tanh(\alpha x)}{\cosh^2(\alpha x)} - a|t|\tanh(\alpha x) + 2bA^3\tanh^3(\alpha x) = 0 \tag{S9.20}$$

or

$$2c\alpha^2\mathrm{sech}^2(\alpha x) + 2bA^2\tanh^2(\alpha x) = a|t| \tag{S9.21}$$

But $\tanh^2(\alpha x) = 1 - \mathrm{sech}^2(\alpha x)$. Thus, if the constants α and A are chosen so that

$$2c\alpha^2 = a|t| \quad \text{and} \quad 2bA^2 = a|t| \tag{S9.22}$$

then Eq. (S9.21) is satisfied. The second equation states that the asymptotic values of $M(x)$ are $\pm M_o$, as one could have guessed.

The thickness of the interface is approximately $2/\alpha$. But, by Eq. (S9.22),

$$\frac{2}{\alpha} = 2\sqrt{\frac{2c}{a|t|}} = 2\sqrt{2}\ell \tag{S9.23}$$

where $\ell = \sqrt{c/a|t|}$ is the correlation length, defined in Eq. (9.34). One should notice that, as the critical temperature is approached, the domain wall thickness increases in proportion to the correlation length.

(b) We consider a sample in the form of a long rectangular cylinder, with $L_y = L_z = 1\,\text{m}$ and $L_x = 2L$, where L is some very large number. The canonical potential of the uniformly magnetized state would be given by

$$\phi(M) = \phi_o - \int_{-L}^{L} (-a|t|M_o^2 + bM_o^4)\, dx \tag{S9.24}$$

The canonical potential per unit area of interface is the canonical potential of the nonuniform state minus the canonical potential of the uniform state. After we have made the subtraction (but not before), it is possible to let $L \to \infty$. Then

$$\Delta\phi(M) = -\int_{-\infty}^{\infty} \left[-a|t|[M^2(x) - M_o^2] + b[M^4(x) - M_o^4] + c\left(\frac{dM}{dx}\right)^2 \right] dx \tag{S9.25}$$

Using the facts that $M(x) = M_o \tanh(\alpha x)$ and $M_o^2 = a|t|/2b$, we see that

$$-a|t|(M^2 - M_o^2) + b(M^4 - M_o^4) = \frac{a^2 t^2}{2b}\left[1 - \tanh^2(\alpha x)\right] - \frac{a^2 t^2}{4b}\left[1 - \tanh^4(\alpha x)\right]$$

$$= \frac{a^2 t^2}{4b}\left[1 - \tanh^2(\alpha x)\right]^2$$

$$= \frac{a^2 t^2}{4b}\operatorname{sech}^4(\alpha x) \tag{S9.26}$$

and

$$c\left(\frac{dM}{dx}\right)^2 = \frac{ca|t|}{2b}\alpha^2 \operatorname{sech}^4(\alpha x) = \frac{a^2 t^2}{4b}\operatorname{sech}^4(\alpha x) \tag{S9.27}$$

Thus

$$\Delta\phi(M) = -\frac{a^2 t^2}{2b}\int_{-\infty}^{\infty} \operatorname{sech}^4(\alpha x)\, dx$$

$$= -\frac{2a^2 t^2}{3b}\alpha \tag{S9.28}$$

$$= -\frac{\sqrt{2}(a|t|)^{5/2}}{3\sqrt{c}}$$

The interface free energy per unit area is therefore

$$\Delta F = \frac{\sqrt{2}\tau(a|t|)^{5/2}}{3\sqrt{c}} \tag{S9.29}$$

Notice that the interface free energy needed to create a smaller domain of reversed polarization within a larger domain of a given polarization goes to zero at the critical temperature. This is the magnetic equivalent of the fact that the surface tension of a liquid drop vanishes at the critical point. Both effects show why fluctuations become so important near the critical point.

Exercise 9.4 In Section 9.19, it was suggested that renormalization theory is, in some sense, a generalization of the Central Limit Theorem. If that is so, then the Gaussian probability density that appears in the CLT should be the solution of some appropriately constructed fixed-point problem. Given a sequence of random variables x_1, x_2, ... that satisfy the conditions of the CLT, construct a reasonable renormalization procedure whose fixed point gives the unique Gaussian probability density predicted by that theorem.

Solution The statistically independent random variables x_1, x_2, ... have probability distributions $P_1(x_1)$, $P_2(x_2)$, ... that satisfy the conditions that $\bar{x}_n = 0$ and

$$\lim_{N \to \infty} \left(\frac{\Delta x_1^2 + \cdots + \Delta x_N^2}{N} \right) = a^2 \tag{S9.30}$$

We construct a single renormalization step by defining new random variables that are weighted sums of pairs of neighboring variables. That is,

$$y_1 = \lambda(x_1 + x_2), \quad y_2 = \lambda(x_3 + x_4), \quad y_3 = \lambda(x_5 + x_6), \ldots \tag{S9.31}$$

A second renormalization would produce the sequence of random variables

$$\begin{aligned} z_1 &= \lambda(y_1 + y_2) = \lambda^2(x_1 + x_2 + x_3 + x_4) \\ z_2 &= \lambda(y_3 + y_4) = \lambda^2(x_5 + x_6 + x_7 + x_8) \end{aligned} \tag{S9.32}$$

After n steps, each variable would represent the sum of 2^n of the original variables, multiplied by a scaling factor λ^n. If a fixed point exists, then it must be possible to choose the scaling factor λ so that the probability distributions for the variables are unaffected by the renormalization transformation when n is sufficiently large. It is clear from Eq S9.31 that, because the original variables were statistically independent, the renormalized variables y_1, y_2, ... are also statistically independent. Thus each renormalization retains the property of independence.

If we write the nth renormalization transformation as $u_1, u_2, \ldots \to \tilde{u}_1, \tilde{u}_2, \ldots$, where $\tilde{u}_1 = \lambda(u_1 + u_2)$, etc., then Eq. (1.51) says that the probability distribution for $\tilde{u}_1$ is related to those for u_1 and u_2 by

$$\tilde{P}_1(\tilde{u}_1) = \int \delta\big(\tilde{u}_1 - \lambda(u_1 + u_2)\big) P_1(u_1) P_2(u_2) \, du_1 \, du_2 \tag{S9.33}$$

The fixed-point property is that, as $n \to \infty$, all probability distributions approach some limit function $P(u)$. By Eq. (S9.33), $P(u)$ must satisfy the relation

$$P(u) = \int \delta\big(u - \lambda(u_1 + u_2)\big) P(u_1) P(u_2) \, du_1 \, du_2 \tag{S9.34}$$

The Fourier transform of $P(u)$, which we write as $Q(k)$, is defined by

$$
\begin{aligned}
Q(k) &= \int e^{iku} P(u)\, du \\
&= \int e^{iku} \delta(u - \lambda(u_1 + u_2)) P(u_1) P(u_2)\, du\, du_1\, du_2 \\
&= \int e^{ik\lambda u_1} P(u_1)\, du_1 \int e^{ik\lambda u_2} P(u_2)\, du_2 \\
&= Q(\lambda k) Q(\lambda k) \\
&= Q^2(\lambda k)
\end{aligned}
\tag{S9.35}
$$

Before we try to determine the probability density function $P(u)$, let us try to determine the scaling factor λ. After n renormalization steps,

$$
u_1 = \lambda^n (x_1 + \cdots + x_N)
\tag{S9.36}
$$

where $N = 2^n$. But then, by the Law of Large Numbers, the uncertainty in u_1 is

$$
\begin{aligned}
\Delta u_1 &= \lambda^n \Delta(x_1 + \cdots + x_N) \\
&= \lambda^n \sqrt{N} a \\
&= (\sqrt{2}\lambda)^n a
\end{aligned}
\tag{S9.37}
$$

In order for this to be independent of n, we must choose $\lambda = 1/\sqrt{2}$. Thus Eq. (S9.35) can be written as

$$
Q(k) = Q^2(k/\sqrt{2})
\tag{S9.38}
$$

If we let $F(k) = \log Q(k)$, then

$$
F(k) = 2F(k/\sqrt{2})
\tag{S9.39}
$$

An obvious solution to this functional equation is $F(k) = Ak^2$. From the results of Exercise 1.31, we know that $F(0) = 0$, $F'(0) = 0$, and $F''(0) = -a^2$. To show that $F(k) = -\frac{1}{2}ak^2$ is the only solution, we let $F(k) = -\frac{1}{2}ak^2 G(k)$, where $G(0) = 1$. Equation (S9.39) then says that

$$
-\tfrac{1}{2}ak^2 G(k) = -\tfrac{1}{2}ak^2 G(k/\sqrt{2})
\tag{S9.40}
$$

or

$$
G(k) = G(k/\sqrt{2})
\tag{S9.41}
$$

Beginning with any value of k, we can use this repeatedly to get

$$
G(k) = G(k/2^K) \to G(0) = 1
\tag{S9.42}
$$

This tells us that $Q(k) = \exp(-\frac{1}{2}ak^2)$ and, taking the Fourier transform,

$$
\begin{aligned}
P(u) &= \frac{1}{2\pi} \int_{-\infty}^{\infty} e^{-iku} e^{-\frac{1}{2}ak^2}\, dk \\
&= e^{-u^2/2a} \int e^{-\frac{1}{2}a(k+iu/a)^2}\, dk \\
&= \frac{e^{-u^2/2a}}{\sqrt{2\pi a}}
\end{aligned}
\tag{S9.43}
$$

which is exactly the result predicted by the Central Limit Theorem.

Exercise 9.5 The Ising model and the Landau–Ginsburg theory describe systems with a real scalar order parameter, namely, the magnetization M. Such a system is said to have a one-component order parameter. In contrast, a ferromagnetic system in which the magnetization is a vector $\mathbf{M}$ has a three-component order parameter. By a long, complicated analysis, using quantum field theory, it is possible to express the fundamental critical exponents y_t and y_B [see Eq. (9.120)] for a system with an n-component order parameter as an expansion in the parameter $\epsilon = 4 - D$, where D is the dimensionality of the system. Through second order, the formulas are

$$y_t = 2 - \frac{n+2}{n+8}\epsilon - \frac{(n+2)(13n+44)}{2(n+8)^3}\epsilon^2 \tag{S9.44}$$

and

$$y_B = 3 - \frac{1}{2}\epsilon - \frac{n+2}{4(n+8)}\epsilon^2 \tag{S9.45}$$

(a) For a one-component order parameter, in 3D, calculate y_t and y_B to second order. (b) Use the result obtained in (a) to compute the magnetization exponent β.

Solution   (a) Setting n and ϵ equal to one in Eqs. (S9.44) and (S9.45) gives

$$y_t = 2 - \frac{1}{3} - \frac{19}{162} = \frac{251}{162} \approx 1.55 \tag{S9.46}$$

and

$$y_B = 3 - \frac{1}{2} - \frac{1}{12} = \frac{29}{12} \approx 2.42 \tag{S9.47}$$

(b) When these values are used in Eq. (9.126), one obtains

$$\beta = \frac{3 - y_B}{y_t} = 0.374 \tag{S9.48}$$

Exercise 9.6 For a one-dimensional Ising lattice, with periodic boundary conditions, and $\beta V = K$, the probability of a configuration is

$$P(\sigma_1, \sigma_2, \ldots, \sigma_N) = Ce^{K(\sigma_1\sigma_2 + \sigma_2\sigma_3 + \cdots + \sigma_N\sigma_1)} \tag{S9.49}$$

Assume that N is an even number and, by summing over the even spins $\sigma_2, \sigma_4, \ldots$, determine the probability distribution for the odd spins $\sigma_1, \sigma_3, \ldots$.

Solution   Following the analysis of Exercise 8.11, we can write the probability in the form

$$P(\sigma_1, \sigma_2, \ldots) = CM(\sigma_1, \sigma_2)M(\sigma_2, \sigma_3) \cdots M(\sigma_N, \sigma_1) \tag{S9.50}$$

where the 2×2 matrix M is

$$M = \begin{bmatrix} e^K & e^{-K} \\ e^{-K} & e^K \end{bmatrix} \tag{S9.51}$$

Summing over the even spins will simply replace the product of N matrices M in Eq. (S9.50) by a product of $N/2$ matrices M^2. That is,

$$P(\sigma_1, \sigma_3, \sigma_5, \ldots) = CM^2(\sigma_1, \sigma_3)M^2(\sigma_3, \sigma_5)\cdots M^2(\sigma_{N-1}, \sigma_1) \qquad \text{(S9.52)}$$

where

$$M^2 = \begin{bmatrix} e^K & e^{-K} \\ e^{-K} & e^K \end{bmatrix} \begin{bmatrix} e^K & e^{-K} \\ e^{-K} & e^K \end{bmatrix} = 2 \begin{bmatrix} \cosh(2K) & 1 \\ 1 & \cosh(2K) \end{bmatrix} \qquad \text{(S9.53)}$$

Exercise 9.7   Determine the value of $\tilde{K}$ as a function of K, for which we can write the matrix M^2 in the form

$$M^2 = C \begin{bmatrix} e^{\tilde{K}} & e^{-\tilde{K}} \\ e^{-\tilde{K}} & e^{\tilde{K}} \end{bmatrix} \qquad \text{(S9.54)}$$

Solution   Comparing Eqs. (S9.53) and (S9.54) gives two equations for C and $\tilde{K}$.

$$Ce^{\tilde{K}} = 2\cosh(2K) \qquad \text{(S9.55)}$$

and

$$Ce^{-\tilde{K}} = 2 \qquad \text{(S9.56)}$$

Since we are not interested in determining the constant C, we divide the first equation by the second to obtain

$$e^{2\tilde{K}} = \cosh(2K) \qquad \text{(S9.57)}$$

or

$$\tilde{K} = \tfrac{1}{2}\log\cosh(2K) \qquad \text{(S9.58)}$$

For positive K, the flow diagram for this transformation is shown in Fig. S9.4. The flow diagram for negative values of K is left as a problem for the reader (Problem 9.9). This procedure, in which the spin variables in the renormalized probability distribution are not, in any sense, averages over a block of the original spin variables, but are simply a subset of the original spin variables, is called a *decimation transformation*.

Fig. S9.4   The curve shown is $\tilde{K}(K)$. The renormalization flow of any point on the K axis can be calculated by first projecting the point upward onto the curve, then horizontally onto the $\tilde{K}$ axis, and finally at 45 degrees back onto the K axis. Due to the fact that the curve is always below a 45 degree line through the origin, any point on the K axis moves toward the origin under repeated renormalization.

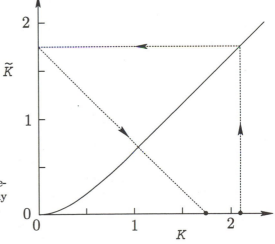

Exercise 9.8   The previous exercise shows that the probability distribution for the variables $(\sigma_1, \sigma_3, \sigma_5, \ldots)$ is of the form

$$P(\sigma_1, \sigma_3, \sigma_5, \ldots) = Ce^{\tilde{K}(\sigma_1\sigma_3 + \cdots)} \tag{S9.59}$$

Is it now correct to search for critical points of the 1D Ising model by looking for solutions of the equation

$$K = \tfrac{1}{2} \log \cosh(2K) \tag{S9.60}$$

which is just Eq. (S9.58) with $\tilde{K}$ set equal to K?

Solution   No. That simple procedure is guaranteed to miss any critical points that exist in the model. Actually, since the 1D Ising model has no critical points, the false procedure would not really miss anything, but it is nevertheless important to understand why it is wrong.

The transformation from K to $\tilde{K}$ allows the probability distribution for the renormalized variables

$$(\tilde{\sigma}_1, \tilde{\sigma}_2, \tilde{\sigma}_3, \ldots) \equiv (\sigma_1, \sigma_3, \sigma_5, \ldots) \tag{S9.61}$$

to be written in the same form as the probability distribution for the original variables $(\sigma_1, \sigma_2, \sigma_3, \ldots)$, except that K is replaced by $\tilde{K}$. This immediately implies that the correlation functions satisfy the renormalization relation

$$\langle \tilde{\sigma}_1 \tilde{\sigma}_2 \rangle_{\tilde{K}} = \langle \sigma_1 \sigma_2 \rangle_K \tag{S9.62}$$

and, for any n,

$$\langle \tilde{\sigma}_1 \tilde{\sigma}_{n+1} \rangle_{\tilde{K}} = \langle \sigma_1 \sigma_{n+1} \rangle_K \tag{S9.63}$$

where

$$\langle \sigma_i \sigma_j \rangle_K \equiv \sum_{\{\sigma\}} \sigma_i \sigma_j e^{K(\sigma_1\sigma_2 + \sigma_2\sigma_3 + \cdots)} / Z(K) \tag{S9.64}$$

[In Eq. (S9.64), we have not indicated the dependence of $\langle \sigma_i \sigma_j \rangle_K$ on the lattice length N, because, for large N, the correlation function becomes independent of N, and, in renormalization theory, we are interested only in the properties of the system in the large-N limit.] If a solution of the equation $\tilde{K}(K^*) = K^*$ existed, then it would imply that the renormalized variables $(\tilde{\sigma}_1, \tilde{\sigma}_2, \tilde{\sigma}_3, \ldots)$ had the same probability distribution as the original variables, and therefore that

$$\langle \tilde{\sigma}_1 \tilde{\sigma}_{n+1} \rangle_{K^*} = \langle \sigma_1 \sigma_{n+1} \rangle_{K^*} \tag{S9.65}$$

or that, in terms of the original variables,

$$\langle \sigma_1 \sigma_{2n+1} \rangle_{K^*} = \langle \sigma_1 \sigma_{n+1} \rangle_{K^*} \tag{S9.66}$$

By this simple procedure, we are looking for states in which the correlation function does not change with distance *at all*. We are thus guaranteed to come up with only the following two trivial cases.

1. $K = 0$ (equivalent to $T = \infty$), where $\langle \sigma_1 \sigma_{n+1} \rangle = 0$ for all n.
2. $K = \infty$ (equivalent to $T = 0$), where $\langle \sigma_1 \sigma_{n+1} \rangle = 1$ for all n.

To search for critical points we should be looking for cases in which the correlation function falls off as some inverse power of n. That is, for solutions of [see Eq. (9.74)]

$$\langle \sigma_1 \sigma_{n+1} \rangle \sim \frac{C}{n^\lambda} \tag{S9.67}$$

where $\lambda > 0$. Equation (S9.66) would then be replaced by the less demanding equation,

$$\langle \sigma_1 \sigma_{2n+1} \rangle = \frac{1}{2^\lambda} \langle \sigma_1 \sigma_{n+1} \rangle \tag{S9.68}$$

In using a decimation transformation with discrete variables, one cannot expect nontrivial critical points to be revealed by simple fixed points of the renormalization flow.

The next few exercises present some typical, but very small, real-space renormalization calculations, of gradually increasing complexity, on two-dimensional Ising lattices.

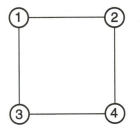

Fig. S9.5   A 2×2 square lattice.

Exercise 9.9   We consider a square Ising lattice with a Hamiltonian that contains nearest-neighbor (NN), next-nearest-neighbor (NNN), and four-spin interaction terms. That is,

$$-\beta E = H = K_1 \sum_{\text{NN}} \sigma_i \sigma_j + K_2 \sum_{\text{NNN}} \sigma_i \sigma_j + K_3 \sum_{\text{sq}} \sigma_i \sigma_j \sigma_k \sigma_l \tag{S9.69}$$

The first sum is taken over all bonds, the second sum over the diagonals of all squares, and the third sum over the corner spins of all squares. (a) For a 2×2 lattice, number the spins as shown in Fig. S9.5, and write the value of H as an explicit function of σ_1, σ_2, σ_3, and σ_4. (b) Let $z(\sigma_1, \sigma_2, \sigma_3, \sigma_4) = e^H$, and evaluate $z(+, +, +, +)$, $z(+, +, +, -)$, $z(+, +, -, -)$, and $z(+, -, -, +)$ in terms of K_1, K_2, and K_3. Using rotational and spin-flip symmetries, one could calculate the value of $z(\sigma_1, \sigma_2, \sigma_3, \sigma_4)$ for any other configuration in terms of the four values listed here. (c) Calculate the partition function and the probability for any configuration of the 2×2 lattice in terms of K_1, K_2, and K_3. (d) Express the Hamiltonian parameters K_1, K_2, and K_3 in terms of the probabilities $P(+, +, +, +)$, $P(+, +, +, -)$, $P(+, +, -, -)$, and $P(+, -, -, +)$.

Solution   (a)

$$\sum_{\text{NN}} \sigma_i \sigma_j = \sigma_1 \sigma_2 + \sigma_3 \sigma_4 + \sigma_1 \sigma_3 + \sigma_2 \sigma_4 \tag{S9.70}$$

$$\sum_{\text{NNN}} \sigma_i \sigma_j = \sigma_1 \sigma_4 + \sigma_2 \sigma_3 \tag{S9.71}$$

and

$$\sum_{sq} \sigma_i \sigma_j \sigma_k \sigma_l = \sigma_1 \sigma_2 \sigma_3 \sigma_4 \qquad (S9.72)$$

and, therefore,

$$H = K_1(\sigma_1\sigma_2 + \sigma_3\sigma_4 + \sigma_1\sigma_3 + \sigma_2\sigma_4) + K_2(\sigma_1\sigma_4 + \sigma_2\sigma_3) + K_3\sigma_1\sigma_2\sigma_3\sigma_4 \quad (S9.73)$$

(b) From Eq S9.73, $z(+,+,+,+) = \exp(4K_1 + 2K_2 + K_3)$, $z(+,+,+,-) = \exp(-K_3)$, $z(+,+,-,-) = \exp(-2K_2 + K_3)$, and $z(+,-,-,+) = \exp(-4K_1 + 2K_2 + K_3)$.

(c) There are two configurations equivalent to $(+,+,+,+)$, there are eight equivalent to $(+,+,+,-)$, there are four equivalent to $(+,+,-,-)$, and there are two equivalent to $(+,-,-,+)$. (Notice that $2 + 8 + 4 + 2 = 2^4$, the total number of configurations.) Therefore,

$$Z = 2e^{4K_1+2K_2+K_3} + 8e^{-K_3} + 4e^{-2K_2+K_3} + 2e^{-4K_1+2K_2+K_3} \qquad (S9.74)$$

The probability function, $P(\sigma_1, \sigma_2, \sigma_3, \sigma_4) = Z^{-1} z(\sigma_1, \sigma_2, \sigma_3, \sigma_4)$.

(d) Taking the logarithms of the equations given in the answer to (B), and letting $\phi = \log Z$, we can write the following linear equations for the Hamiltonian parameters.

$$\begin{aligned}
4K_1 + 2K_2 + K_3 &= \log P(+,+,+,+) - \phi \\
-K_3 &= \log P(+,+,+,-) - \phi \\
-2K_2 + K_3 &= \log P(+,+,-,-) - \phi \\
-4K_1 + 2K_2 + K_3 &= \log P(+,-,-,+) - \phi
\end{aligned} \qquad (S9.75)$$

These equations can easily be solved to give

$$\begin{aligned}
K_1 &= \log\left[P(+,+,+,+)/P(+,-,-,+)\right]/8 \\
K_2 &= \log\left[P(+,+,+,+)/P(+,+,-,-)\right]/4 + K_1 \\
K_3 &= \log\left[P(+,+,-,-)/P(+,+,+,-)\right]/2 + K_2
\end{aligned} \qquad (S9.76)$$

This exercise allows one to convert from the probability distribution on the 2×2 lattice to the Hamiltonian parameters that will produce that probability distribution. The next exercise shows how to go from the Hamiltonian parameters on a 4×4 lattice to the probability distribution on the corresponding 2×2 lattice of block spins. Combining the two transformations, we can then go from the Hamiltonian parameters on a 4×4 lattice to the Hamiltonian parameters on the 2×2 lattice of block spins. The two taken together give a complete renormalization transformation on a very small system.

Exercise 9.10 Consider the 4×4 lattice shown in Fig. S9.6. For notational simplicity, the spin variables are named $a, b, \ldots, p$, rather than σ_a, σ_b, etc. The lattice is broken up into four 2×2 blocks, as shown. For the block having the spins a, b, c, and d, the block spin variable $B(a, b, c, d)$ is defined as

$$B(a,b,c,d) = \begin{cases} \text{sgn}(a+b+c+d), & \text{if } a+b+c+d \neq 0 \\ a, & \text{if } a+b+c+d = 0 \end{cases} \qquad (S9.77)$$

This assigns the block spin $+1$ to the patterns 1, 2, and 3 in Fig. 9.16 and the value -1 to the patterns 4, 5, and 6. (a) Assuming that the Hamiltonian on the 4×4

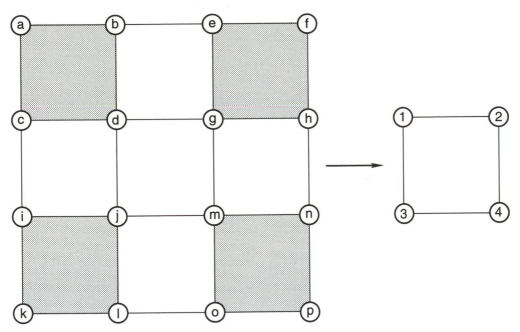

Fig. S9.6   The single spins on the 2×2 lattice on the right represent the block spins for the shaded squares on the 4×4 lattice on the left.

lattice is of the form given in Eq. (S9.69), write a formula for the probability distribution on the 2×2 lattice of block spins. (b) Write a Fortran computer program to carry out the complete renormalization transformation from the parameters K_1, K_2, and K_3 on the 4×4 lattice to the parameters $\hat{K}_1$, $\hat{K}_2$, and $\hat{K}_3$ on the 2×2 lattice.

Solution   (a) The canonical probability density on the 4×4 lattice is

$$P(a, b, \ldots, p) = Z^{-1} \exp\left[H(a, b, \ldots, p)\right] \tag{S9.78}$$

The Kronecker delta function, $\delta\left(B(a, b, c, d) - s_1\right)$, is equal to one if the block spin of the upper left block has the value s_1, otherwise it is equal to zero. Therefore, if we let $P(s_1, s_2, s_3, s_4)$ be the probability that the four block spins have the values s_1, s_2, s_3, and s_4, then, by Eq. (1.51),

$$\begin{aligned}
ZP(s_1, s_2, s_3, s_4) = \sum_a \cdots \sum_p &\, \delta\left(B(a, b, c, d) - s_1\right)\delta\left(B(e, f, g, h) - s_2\right) \\
&\times \delta\left(B(i, j, k, l) - s_3\right)\delta\left(B(m, n, o, p) - s_4\right) \\
&\times \exp\left[H(a, \ldots, p)\right]
\end{aligned} \tag{S9.79}$$

Since the formulas that were given in the last exercise require only the ratios of probabilities, there will be no need to calculate the partition function Z.

(b) The Fortran program, called Renorm_1 in the Program Listing, and included on the program disk, carries out a single renormalization transformation. The user gives the values of K_1, K_2, and K_3, which are called X_1, X_2, and X_3 so that, by Fortran conventions, they will be real numbers. The array $z(s_1, s_2, s_3, s_4)$ will contain the unnormalized block spin probabilities. Their initial values are first set to zero. The unnormalized probability of a configuration of the 4×4 lattice, called the *Boltzmann weight* of the configuration, is

$$W = e^{X_1 \cdot nn + X_2 \cdot nnn + X_3 \cdot nsq} \tag{S9.80}$$

where nn is the sum over all products of nearest-neighbor spins, nnn is the sum over all products of next-nearest-neighbor spins, and nsq is the sum over all products of the corner spins for each square.

For a given configuration, the Boltzmann weight and the values of the four block spins are calculated, and then, in accordance with Eq. (S9.80), the value of the Boltzmann weight is added to the appropriate array element in $z(s_1, s_2, s_3, s_4)$.

When $z(s_1, s_2, s_3, s_4)$ is calculated, the renormalized interaction parameters, called Y_1, Y_2, and Y_3, are computed using Eq. (S9.76).

We now make the somewhat wild assumption that the renormalization transformation,

$$(K_1, K_2, K_3) \rightarrow (\tilde{K}_1, \tilde{K}_2, \tilde{K}_3) \tag{S9.81}$$

is so insensitive to the size of the lattice that one can use the result obtained from a 4×4 lattice to predict the properties of very large lattices. It would then make sense to iterate the transformation, using the output values of one run as input values for the next, and to search for fixed points.

We begin with a simple Ising model; that is, one with only nearest-neighbor interactions. Such a model has parameters $(K_1, 0, 0)$, where $K_1 = \beta V$. If the value of K_1 were exactly equal to the critical value $(\beta V)_c = 0.4407 \ldots$ (known from Onsager's exact solution) and if the renormalization transformation were also exact, then, under repeated renormalizations,

$$(K_1, 0, 0) \rightarrow (K_1', K_2', K_3') \rightarrow (K_1'', K_2'', K_3'') \rightarrow \cdots \tag{S9.82}$$

the Hamiltonian parameters would approach a fixed point (K_1^*, K_2^*, K_3^*).

Exercise 9.11 Starting at points $(K_1, 0, 0)$ for various values of K_1, iterate the renormalization transformation derived in the previous exercises and plot the projection of the renormalization flow diagram onto the (K_1, K_2) plane. Table S9.1 gives the result of iterating the renormalization transformation, starting with the values $(K_1, 0, 0)$, where $K_1 = 0.3$, 0.4, 0.4216, 0.5, and 0.6. (Naturally, the middle value was arrived at by trial and error.) For all values of K_1 less than the value $K_1^o = 0.4216$ the points flow to the trivial noninteracting $(T = \infty)$ fixed point at $(0, 0, 0)$. For values of K_1 larger than K_1^o, the points flow to infinity, which is just the other $(T = 0)$ trivial fixed point. The point $(K_1^o, 0, 0)$ flows into a nontrivial, unstable fixed point at $(K_1^*, K_2^*, K_3^*) \approx (0.292, 0.086, -0.048)$. Thus this calculation predicts that the value of $K_1 = V/\tau$ at the critical point is $K_1 = 0.4216$. This is much closer to the true critcal value $K_1 \approx 0.4407$ than is the mean-field prediction of $K_1 = 0.25$. The flow diagram is given in Fig. S9.7. Although this is a very crude calculation, a couple of things should be noted about the results, because they also hold for more accurate (and more complicated) calculations of this type.

1. As we proceed from nearest-neighbor to next-nearest-neighbor and to four-spin and still more complicated interaction terms, the Hamiltonian parameters at the fixed point steadily diminish. That is, $K_1^* > K_2^* > K_3^*$. It is only this

K_1	K_2	K_3
0.30000	0.00000	0.00000
0.18117	0.02484	0.01156
0.09911	0.01569	0.00878
0.04202	0.00505	0.00305
0.01386	0.00102	0.00063
0.00393	0.00016	0.00010
0.00104	0.00002	0.00001
0.00027	0.00000	0.00000
0.00007	0.00000	0.00000
0.40000	0.00000	0.00000
0.29986	0.05380	0.02588
0.25176	0.06172	0.03313
0.20543	0.05221	0.02950
0.14567	0.03378	0.01969
0.08039	0.01466	0.00878
0.03265	0.00390	0.00238
0.01039	0.00072	0.00044
0.00289	0.00011	0.00006
0.42160	0.00000	0.00000
0.32937	0.06159	0.03007
0.30275	0.07898	0.04224
0.29500	0.08392	0.04653
0.29276	0.08534	0.04796
0.29217	0.08576	0.04844
0.29212	0.08593	0.04861
0.29235	0.08606	0.04871
0.29288	0.08628	0.04884
0.50000	0.00000	0.00000
0.44621	0.09342	0.04872
0.54526	0.16326	0.08818
0.87410	0.29197	0.15899
1.73588	0.60966	0.34877
0.60000	0.00000	0.00000
0.61224	0.13954	0.08008
0.94781	0.29900	0.17416
1.88057	0.65228	0.40050

Table S9.1 The result of iterating the renormalization transformation described in Exercise 9.10.

fortunate accident that allows one to simplify the calculation by cutting off the proliferating set of Hamiltonian parameters at some fairly small number of terms.

2. Because the nontrivial fixed point is unstable, the sequence of points obtained by any numerical renormalization scheme eventually will diverge from it. Trying to find a sequence that truly converges to the fixed point is like trying to balance a needle on its point. This effect creates serious difficulties when

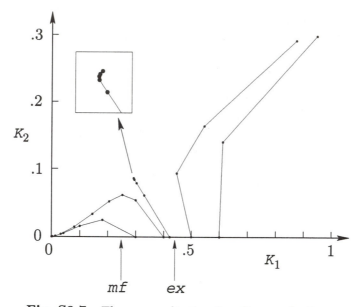

Fig. S9.7   The renormalization flow diagram for the parameters K_1 and K_2. The predicted critical value for $K_1 = V/\tau$ is the point that flows into the fixed point. The exact critical value and the mean-field predictions are both indicated by arrows under the K_1 axis. In the box is shown a magnification of the flow near the fixed point.

one attemps to do renormalization calculations on larger systems (where one cannot carry out a complete sum over all configurations) by using the Monte Carlo method to evaluate the probability distribution for the block spins. The Monte Carlo method introduces unavoidable statistical fluctuations into each renormalization step. One is then trying to balance a needle on its point with a trembling hand.

The most unsatisfactory aspect of this calculation, aside from the very small size of the system, is the rather arbitrary treatment of the block spin ambiguity problem. A better method of calculating the block spin probability function $z(s_1, s_2, s_3, s_4)$ is the following.

1. We temporarily allow the block spins to have the values, ± 1 *and* 0. There is then no block spin ambiguity at all.

2. Suppose, in the sum over states on the 4×4 lattice, we generate a state with a Boltzmann weight W and a block spin configuration $(0, s_2, s_3, s_4)$, where s_2, s_3, and s_4 are nonzero. In the previous method, we would have assigned the weight W either to $z(-1, s_2, s_3, s_4)$ or to $z(+1, s_2, s_3, s_4)$, depending upon the details of the four site spins that make up s_1. Clearly, it would be more reasonable to assign half the weight W to each of the two block spin configurations. In a similar way, the Boltzmann weight for a block spin state of the form $(0, 0, s_3, s_4)$ should be evenly redistributed among the four states $(\pm 1, \pm 1, s_3, s_4)$ and so forth for all block spin states involving any number of zeros.

Exercise 9.12   (a) Modify the program used in the previous exercise to handle the zero block spin values in the way suggested. (b) Iterate the renormalization transformation so obtained and plot the renormalization flow diagram in the (K_1, K_2) plane.

Solution (a) In the program Renorm_1, there are storage locations in $z(s_1, s_2, s_3, s_4)$ for the variables s_1, s_2, s_3, and s_4 to take the three values, -1, 0, and $+1$, although the storage locations associated with the zero values are never actually used. In the modified program, Renorm_2 (see the Program Listings), the unnormalized block spin probability function $z(s_1, s_2, s_3, s_4)$ is first calculated by allowing the block spins to have the values ± 1 and 0. This has been accomplished by a small modification of the function Iblock(K). When that calculation has been completed, the unnormalized probabilities associated with any zero block spin values are redistributed, according to the method described, by the subroutine Redistribute.

(b) With this new method of handling block spin ambiguity, the renormalization tranformation was iterated to produce the flow diagram shown in Fig. S9.8. The predicted critical value of K_1 is now $K_1^o = 0.438$ in comparison with the exact value of 0.4407. The error, which is about 0.6%, is not readily discernible on the scale of the diagram.

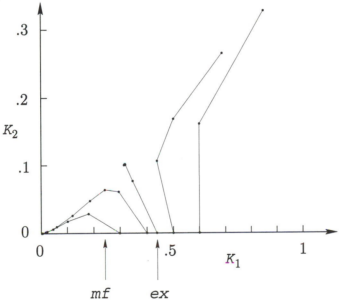

Fig. S9.8   The renormalization flow diagram produced by the program Renorm_2.

Exercise 9.13   Consider the set of seven spins shown in Fig. S9.9. For simplicity, we call them a, b, ..., g. We define S_1 as the sum of the products of all pairs of nearest-neighbor spins, S_2 as the sum of the products of all pairs of next-nearest-neighbor spins, and S_3 as the sum, over all four-spin parallelograms, of the product of the four vertex spins.* That is,

$$S_1 = (a + b + c + d + e + f)g + ab + bc + cd + de + ef + fa \qquad (S9.83)$$

* It might seem more natural that the term following the next-nearest-neighbor spins would be the products of the three vertex spins over all triangles. In the absence of an

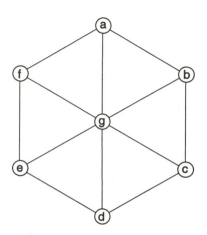

Fig. S9.9   A set of seven spins, arranged in a hexagon. The underlying lattice is triangular.

$$S_2 = ac + bd + ce + df + ea + fb \tag{S9.84}$$

and

$$S_3 = (abc + bcd + cde + def + efa + fab)g \tag{S9.85}$$

Assume that the Hamiltonian of the system has the form

$$-\beta E = H = K_1 S_1 + K_2 S_2 + K_3 S_3 \tag{S9.86}$$

We define the three correlation functions

$$C_n(K_1, K_2, K_3) \equiv \frac{\sum S_n e^{K_1 S_1 + K_2 S_2 + K_3 S_3}}{\sum e^{K_1 S_1 + K_2 S_2 + K_3 S_3}} \tag{S9.87}$$

where the sum is over the 2^7 configurations of the system. Show how one may calculate the value of the Hamiltonian parameters that will give some specified set of correlation functions, C_1^o, C_2^o, and C_3^o. That is, give a practical method of solving the equations

$$C_n(K_1^o, K_2^o, K_3^o) = C_n^o \tag{S9.88}$$

for K_1^o, K_2^o, and K_3^o, with some preassigned accuracy, given C_1^o, C_2^o, and C_3^o.

Solution   The calculational method is based on the observation that, according to Eq. (S9.87),

$$
\begin{aligned}
\frac{\partial C_m}{\partial K_n} &= \frac{\sum S_m S_n e^H}{\sum e^H} - \frac{(\sum S_m e^H)(\sum S_n e^H)}{(\sum e^H)^2} \\
&= \langle S_m S_n \rangle - \langle S_m \rangle \langle S_n \rangle \\
&\equiv M_{mn}(K_1, K_2, K_3)
\end{aligned}
\tag{S9.89}
$$

Using this formula, it is relatively easy to calculate M_{mn} for given values of K_1, K_2, and K_3. The calulational procedure for solving Eq. (S9.88) is as follows.
1. Begin with some starting values of K_1, K_2, and K_3. Define δK_n by

$$\delta K_n = K_n^o - K_n \tag{S9.90}$$

external magnetic field, because the initial Hamiltonian has up–down spin symmetry, the expectation value of any product of three spins is exactly zero. The three spin terms become important when an external magnetic field is introduced.

where (K_1^o, K_2^o, K_3^o) is the, as yet unknown, solution of Eq. (S9.88). Then

$$C_m^o = C_m(K_1 + \delta K_1, K_2 + \delta K_2, K_3 + \delta K_3)$$

$$\approx C_m(K_1, K_2, K_3) + \sum_n \frac{\partial C_m}{\partial K_n} \delta K_n \qquad (S9.91)$$

$$= \langle S_m \rangle + \sum_n M_{mn} \delta K_n$$

The equations

$$\sum_n M_{mn} \delta K_n = C_m^o - \langle S_m \rangle \qquad (S9.92)$$

are three linear equations for the unknowns, δK_1, δK_2, and δK_3.

2. Replace the original values of K_1, K_2, and K_3 with $K_1 + \delta K_1$, $K_2 + \delta K_2$, and $K_3 + \delta K_3$. Calculate $|\delta K| = (\sum \delta K_n^2)^{1/2}$. If $|\delta K|$ is less then the desired accuracy we are finished; if not, go back to (1).

This algorithm is implemented in the subroutine FindHamPars, which is part of the program Renorm_3 in the Program Listings (and the Program Disk). The Subroutine Solve(A,B,X), called by FindHamPars, is simply a linear equation solver for three variables.

This calculation should be compared with part (d) of Exercise 9.9, where the Hamiltonian parameters were computed from the probability function $P(\sigma_1, \sigma_2, \sigma_3, \sigma_4)$ for a 2×2 lattice. In that case there were just enough independent probabilities to determine uniquely the values of the three Hamiltonian parameters. For the system of seven spins being considered here, the number of independent probability values would be much larger than the number of Hamiltonian parameters to be determined unless one allowed the number of parameters to increase beyond three. One cannot exactly match an arbitrary probability distribution on a lattice of seven spins with only three Hamiltonian parameters. That is exactly why the proliferation of interactions takes place in an exact calculation. If one demands that the system have only three parameters, then they must be determined from only three relations, as they were in this exercise. One cannot expect to reproduce exactly the detailed probability distribution on a lattice of more than four spins.

Exercise 9.14 In Fig. S9.10, the shaded triangles are associated with the block spins A, B, ..., G. The site spins associated with block spin A are called a_1, a_2, and a_3, counting clockwise from the top of the triangle. There is no block spin ambiguity for a triangular lattice. For example,

$$A = \text{sgn}(a_1 + a_2 + a_3) \qquad (S9.93)$$

always gives a unique value for A of plus or minus one. The Hamiltonian of the system of 21 spins is of the same form as that in the system considered in the previous exercise.

$$-\beta E = H = K_1 \underbrace{\sum \sigma_i \sigma_j}_{\text{NN}} + K_2 \underbrace{\sum \sigma_i \sigma_j}_{\text{NNN}} + K_3 \underbrace{\sum \sigma_i \sigma_j \sigma_k \sigma_\ell}_{\text{four.}} \qquad (S9.94)$$

We define block spin correlation functions, C_1, C_2, and C_3, as the expectation value in the 21-spin system of the observables [compare with Eqs. (S9.83–85)],

$$S_1 = (A + B + C + D + E + F)G + AB + BC + CD + DE + EF + FA \qquad (S9.95)$$

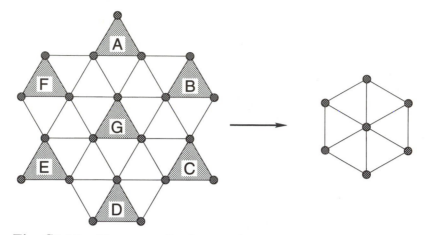

Fig. S9.10 The renormalization transformation of a small system on a triangular lattice.

$$S_2 = AC + BD + CE + DF + EA + FB \tag{S9.96}$$

and

$$S_3 = (ABC + BCD + CDE + DEF + EFA + FAB)G \tag{S9.97}$$

Write a Fortran program to compute C_1, C_2, and C_3, given K_1, K_2, and K_3.

Solution The program is part of Renorm_3, in the form of a subroutine (called FindCorrFtns) whose input is the set of three Hamiltonian parameters, here called X_1, X_2, and X_3, and whose output is the set of correlation functions C_1, C_2, and C_3 (called Y_1, Y_2, and Y_3 in the program). The peculiar business in the middle of the subroutine that involves the function Ibits is a trick that allows the needed 21 levels of Do Loop nesting in spite of the fact that most Fortran compilers do not permit such deep nesting.

Exercise 9.15 (a) Using the subroutines FindCorrFtns and FindHamPars write a program to carry out a renormalization transformation $(K_1, K_2, K_3) \rightarrow (\tilde{K}_1, \tilde{K}_2, \tilde{K}_3)$, on the system shown in Fig. S9.10. (b) Iterate the transformation, plot the flow diagram for the parameters K_1 and K_2, and compare the value of $(K_1^o, 0, 0)$ that flows to a fixed point with the exact critical value for the triangular lattice, namely $K_c = 0.27465\ldots$.

Solution (a) The program, called Renorm_3, is in the Program Listings. Each iteration involves converting the Hamiltonian parameters into correlation functions and then converting the correlation functions back into Hamiltonian parameters. The progrm has not been written to run interactively because it is much too slow. Each iteration requires a sum over the $2^{21} \approx 2 \times 10^6$ states of the 21-spin system. The values of K_1, K_2, and K_3 for each iteration are printed in the file Renorm_3.out.
(b) The flow diagram produced by Renorm_3 is shown in Fig. S9.11. The value of $(K_1^o, 0, 0)$ that flows to the fixed point is about 0.288, which is reasonably close to the exact value of $V/\tau_c \approx 0.275$. The error is larger than that obtained for the square lattice by Renorm_2, but much smaller than the error that would be given by mean-field theory, which predicts a critical value of $K_c = 1/6$.

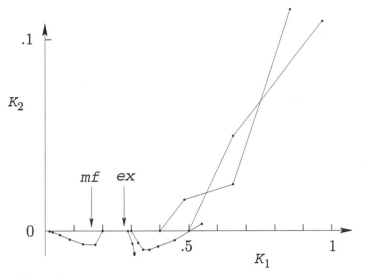

Fig. S9.11   The renormalization flow diagram produced by Renorm_3. The exact critical value of K_1 and the mean-field prediction are shown by arrows.

Exercise 9.16   Write a Fortran program to carry out the renormalization transformation indicated in Fig. S9.12.

Solution   The system considered in the previous exercise was clearly close to the maximum size that could be analyzed by summing over all of its configurations. For the system shown in Fig. S9.12, the number of configurations is $2^{36} \approx 7 \times 10^{10}$—too large to be completely summed over. Some approximation method is necessary. The large sum would occur only in calculating the block spin correlation functions on the 6×6 system. Since they are canonical expectation values of observables, the obvious thing to do is to calculate them by the Monte Carlo method. In the second step,

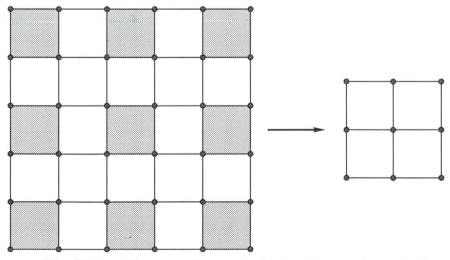

Fig. S9.12   The renormalization of a 6×6 lattice to produce a 3×3 lattice.

namely finding the Hamiltonian parameters from the given correlation functions, we are working on the 3×3 lattice, which is small enough to be treated exactly. Therefore, it is only the subroutine FindCorrFtns that must be significantly revised. The subroutine FindHamPars must be changed in order to make it applicable to a square, rather than a triangular lattice, but that is completely straightforward. The revised subroutine FindCorrFtns is part of the program Renorm_4 in the Program Listings. A number of comments are necessary to explain certain details of the program.

1. I_s and I_b are the arrays of site spins and block spins, respectively. The free boundary conditions on the site spin system are taken care of by extending the 6×6 lattice to an 8×8 lattice and setting all of the outer edge spins equal to zero. Those spins are never changed. When making Monte Carlo moves on the spins of the internal 6×6 lattice, it is not necessary to check whether a spin is an edge or a corner spin. All spins now have neighbors in all directions. (The outer spins are actually set to zero in the subroutine Initialize, which is not listed here.)

2. The block spin ambiguity problem is treated here in still another new way. If the sum of the site spins is zero, then the block spin is randomly set to plus or minus one. In a long enough Monte Carlo run, this will have the same effect as the redistribution program that is used in Renorm_2. Half of the zero block spin contributions will be assigned to $+1$ and half to -1.

When the full program, Renorm_4, is used to construct a flow diagram, the result is not visually distinguishable from the diagram produced by Renorm_2. This seems to support the assumption, made in that calculation, that the renormalization flow pattern is quite insensitive to the size of the system, which has been increased by a factor of 1.5 in going from Renorm_2 to Renorm_4.

Supplement to Chapter 10

REVIEW QUESTIONS

10.1 Derive the probability distribution function for the charge on the capacitor in a simple RC circuit at temperature T.

10.2 Derive the probability distribution function for the current through the inductor in an LCR circuit at temperature T.

10.3 Define the time-averaged correlation function, $\langle Q(0)Q(t)\rangle$.

10.4 Given n observables $A_1(z),\ldots, A_n(z)$, define the conditional entropy $S(\mathbf{a}, E)$.

10.5 How is $S(\mathbf{a}, E)$ related to $S^o(E)$?

10.6 For any function of the n observables $A_1,\ldots,A_n$, write $\langle F\rangle$ as an integral in the macrostate space.

10.7 How is the time evolution operator T_t defined?

10.8 How is the macroscopic equation of motion $\mathbf{a}(t,\mathbf{a}^o)$ defined?

10.9 For a given set of macroscopic observables, $A_1(z),\ldots, A_n(z)$, what is the property of macroscopic determinism?

10.10 From the property of macroscopic determinism, show that the value of the entropy $S(\mathbf{a}, E)$ increases in time.

10.11 How does one define the thermodynamic force β_i associated with an observable $A_i(z)$?

10.12 For n classical observables $A_1(z),\ldots, A_n(z)$ in a microcanonical ensemble, how is the time-dependent correlation matrix $C_{ij}(t)$ defined?

10.13 Write $C_{ij}(t)$ as an integral over the macrostate space.

10.14 What is the "standard form" of the linear equation of motion $\mathbf{a}(t,\mathbf{a}^o)$ close to equilibrium?

10.15 Derive the Fluctuation-Dissipation Theorem.

10.16 What is the relationship between the circuit equations and the macroscopic equation of motion $\mathbf{a}(t,\mathbf{a}^o)$ for an LCR circuit?

10.17 For an LCR circuit, derive the functions $q(t,q^o,i^o)$ and $i(t,q^o,i^o)$.

10.18 Write the solutions for $q(t)$ and $i(t)$ in the "standard form."

10.19 What is the underdamped oscillator condition?

10.20 Determine the correlation function $C_{qi}(t)$ for an LCR circuit.

10.21 Describe the property of time-reversal symmetry of classical mechanics.

10.22 What is the momentum reversal operator R?

10.23 Explain why $RT_t = T_{-t}R$.

10.24 What is meant by the momentum-reversal parity ϵ_i of the observable A_i?

10.25 Prove Onsager's Theorem, that $F_{ij}(t) = \epsilon_i \epsilon_j F_{ji}(t)$.

10.26 Describe the system composed of a capacitor, two reservoirs, and a thin wire that was used in analyzing the thermoelectric effect.

10.27 Write the flow equations, in the standard form, for the device described in the previous question.

10.28 What is the prediction of the Reciprocity Theorem for the device?

10.29 How is the thermoelectric power of a substance defined?

10.30 How is the quantum mechanical response function $R_{BA}(t)$ defined?

10.31 Explain the meaning of the symbols in the Fluctuation-Dissipation Relation, $i\hbar \tilde{R}_{BA}(\omega) = (1 - e^{-\beta\hbar\omega})\tilde{C}_{BA}(\omega)$.

10.32 What are the characteristics of the Boltzmann entropy function?

10.33 What is the definition of the collision rate $P(\mathbf{p}_1, \mathbf{p}_2 | \mathbf{p}_1', \mathbf{p}_2')$?

10.34 What are the symmetry relations satisfied by $P(\mathbf{p}_1, \mathbf{p}_2 | \mathbf{p}_1', \mathbf{p}_2')$?

10.35 Derive the Boltzmann equation for $\partial F(\mathbf{p}, t)/\partial t$.

10.36 What is the physical meaning of the Boltzmann entropy function S_B?

10.37 What is the formula for s_B in terms of $F(\mathbf{p})$?

10.38 Derive the Boltzmann entropy theorem, $ds_B/dt \geq 0$.

10.39 What are the basic approximations made in the derivation of the Boltzmann equation?

10.40 Why does there seem to be a discepancy between the time-reversal properties of mechanics and those of the Boltzmann equation?

10.41 Explain how the "Boltzmann theory" can be time symmetrical although the Boltzmann equation is not.

10.42 Describe Ritz's electromagnetic explanation of time asymmetry.

10.43 Describe the cosmological explanation of time asymmetry.

EXERCISES

Exercise 10.1   Using the time translational invariance of a system with a time independent Hamiltonian, show that, for any two classical observables, the correlation function satisfies the time-inversion symmetry

$$C_{BA}(t) = \epsilon_A \epsilon_B C_{BA}(-t) \tag{S10.1}$$

where ϵ_A and ϵ_B are the time-reversal parities of the observables [see Eq. (10.57)].

Solution

$$
\begin{aligned}
C_{BA}(t) &= \langle B(\tau + t)A(\tau)\rangle, &&\text{for any } \tau \\
&= \langle B(0)A(-t)\rangle, &&\text{if } \tau = -t \\
&= \langle A(-t)B(0)\rangle \\
&= C_{AB}(-t) \\
&= \epsilon_A \epsilon_B C_{BA}(-t), &&\text{by the Reciprocity Theorem}
\end{aligned}
\tag{S10.2}
$$

Notice that:

1. Any *autocorrelation function*, such as $C_{qq}(t)$ or $C_{ii}(t)$, is an even function, and thus is a function of $|t|$ only.
2. Because, for quantum mechanical operators, $B(0)A(-t) \neq A(-t)B(0)$, the theorem, as stated, is true only for classical observables.

The next two exercises will present enough basic circuit theory to allow the reader to appreciate what is the mother of all fluctuation-dissipation theorems, called Nyquist's Theorem. That theorem, which involves voltage fluctuations in an impedance, was derived in 1928, by H. Nyquist, to explain the results of a series of beautiful experiments done by J. B. Johnson on the thermally produced electrical noise in passive impedances (since called Johnson noise). It was not until 1951 that Nyquist's Theorem was extended to general systems.

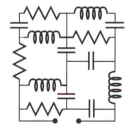

Fig. S10.1   A passive linear device.

Exercise 10.2   A circuit with two input points, composed of an arbitrary combination of ideal resistors, capacitors, and inductors, is the classic example of a passive linear device (Fig. S10.1). Passive means that the device contains no internal source of energy. Linear means that, if the voltage $V_1(t)$ applied to the input terminals produces the current $I_1(t)$ and the voltage $V_2(t)$ produces the current $I_2(t)$, then the voltage $V_1 + V_2$ will produce the current $I_1 + I_2$. In general, the current that flows through the terminals at time t is not a function only of the voltage at time t. Because the inductors and capacitors can store energy in the system, $I(t)$ depends upon $V(t')$ for all $t' < t$. The linearity property guarantees that $I(t)$ may be written in the form

$$I(t) = \int_{-\infty}^{t} A(t - t')V(t') \, dt' \tag{S10.3}$$

That $I(t)$ depends on $V(t')$ only for $t' < t$ shows that no current flows before an external voltage is impressed—an obvious consequence of the fact that the device contains no internal energy source. The function $A(t)$ is called the *admittance function* of the device. The instantaneous rate at which energy is being supplied to the device is $P(t) = V(t)I(t)$. Suppose an arbitrary voltage signal $V(t)$ [with $V(t) \to 0$ as $t \to \pm \infty$] is applied to the input terminals. Using Fourier transforms, derive a compact formula for the total energy absorbed by the device.

Solution   The total energy is equal to the integral

$$
\begin{aligned}
E &= \int_{-\infty}^{\infty} dt \, V(t)I(t) \\
&= \int_{-\infty}^{\infty} dt \, V(t) \int_{-\infty}^{t} dt' \, A(t - t')V(t') \tag{S10.4} \\
&= \int_{-\infty}^{\infty} dt \int_{-\infty}^{\infty} dt' \, V(t)V(t')A(t - t')\theta(t - t')
\end{aligned}
$$

Let

$$V(t) = \frac{1}{2\pi} \int_{-\infty}^{\infty} V(\nu)e^{i\nu t} \, d\nu \tag{S10.5}$$

and

$$A(t)\theta(t) = \frac{1}{2\pi} \int_{-\infty}^{\infty} A(\omega)e^{i\omega t} \, d\omega \tag{S10.6}$$

Then

$$
\begin{aligned}
E &= \frac{1}{(2\pi)^3} \int_{-\infty}^{\infty} V(\nu)V(\nu')A(\omega)e^{i(\nu+\omega)t}e^{i(\nu'-\omega)t'} \, dt \, dt' \, d\nu \, d\nu' \, d\omega \\
&= \frac{1}{2\pi} \int_{-\infty}^{\infty} V(\nu)V(\nu')A(\omega) \, \delta(\nu + \omega) \, \delta(\nu' - \omega) \, d\nu \, d\nu' \, d\omega \tag{S10.7} \\
&= \frac{1}{2\pi} \int_{-\infty}^{\infty} V(-\omega)V(\omega)A(\omega) \, d\omega
\end{aligned}
$$

Both $V(t)$ and $A(t)\theta(t)$ are real functions. Using the relationship $F(t) - F^*(t) = 0$, it is easy to see that, for any real function, the Fourier transform must have the symmetry relation $F(-\omega) = F^*(\omega)$. Thus we can write E in the more compact form

$$E = \frac{1}{2\pi} \int_{-\infty}^{\infty} |V(\omega)|^2 A(\omega) \, d\omega$$

$$= \frac{1}{2\pi} \int_{0}^{\infty} |V(\omega)|^2 \left[A(\omega) + A^*(\omega) \right] d\omega \qquad (S10.8)$$

$$= \frac{1}{\pi} \int_{0}^{\infty} |V(\omega)|^2 G(\omega) \, d\omega$$

where $G(\omega)$, called the conductance, is the real part of the complex admittance $A(\omega)$.

Exercise 10.3 Suppose that the voltage signal applied to the input terminals of a passive linear device is a fluctuating voltage $V(t)$ that does not approach zero as $t \to \infty$. Assume that the correlation function of the voltage

$$C(t) = \lim_{T \to \infty} \left(\frac{1}{T} \int_{-T/2}^{T/2} V(t')V(t' + t) \, dt' \right) \qquad (S10.9)$$

exists for all values of t. Obtain a formula for the average rate at which power is transferred to the device in terms of the Fourier transform of the correlation function $C(\omega)$. The result is called the Wiener–Khinchine Theorem.

Solution We define the envelope function $H_T(t)$ by

$$H_T(t) = \begin{cases} 1, & -T/2 < t < T/2 \\ 0, & \text{otherwise} \end{cases} \qquad (S10.10)$$

and we temporarily replace $V(t)$ by the finite voltage signal

$$V_T(t) = V(t)H_T(t) \qquad (S10.11)$$

The total energy delivered to the device by the signal $V_T(t)$ would be

$$E_T = \frac{1}{\pi} \int_{0}^{\infty} |V_T(\omega)|^2 G(\omega) \, d\omega \qquad (S10.12)$$

Clearly, the average power delivered to the device by the fluctuating signal is the limit of E_T/T as $T \to \infty$. Also

$$V_T(\omega) = \int_{-T/2}^{T/2} V(t)e^{-i\omega t} \, dt \qquad (S10.13)$$

and, therefore,

$$\frac{1}{T}|V_T(\omega)|^2 = \frac{1}{T} \int_{-T/2}^{T/2} dt \int_{-T/2}^{T/2} dt' \, V(t)V(t')e^{-i\omega(t-t')}$$

$$= \frac{1}{T} \int_{-T/2}^{T/2} dt' \int_{t'-T/2}^{t'+T/2} ds \, V(t')V(t' + s) \, e^{-i\omega s} \qquad (S10.14)$$

For large s, $V(t'+s)$ and $V(t')$ become statistically independent, which allows one to extend the integral over s to infinity in both directions. Then, by interchanging the order of integration, it follows that

$$\frac{1}{T}|V_T(\omega)|^2 \to C(\omega) = \int_{-\infty}^{\infty} C(t)e^{-i\omega t}\,dt \qquad (S10.15)$$

which says that the average power delivered to the device is

$$P = \frac{1}{\pi}\int_o^{\infty} C(\omega)G(\omega)\,d\omega \qquad (S10.16)$$

The quantity $P(\omega) = 2C(\omega)$ is called the *power spectral density* of the fluctuating voltage signal. In terms of $P(\omega)$, the average power is

$$P = \frac{1}{2\pi}\int_o^{\infty} P(\omega)G(\omega)\,d\omega \qquad (S10.17)$$

Exercise 10.4 Calculate the charge–charge correlation function $\langle Q(0)Q(t)\rangle$ in the simple RC circuit shown in Fig. 10.1.

Solution One might be tempted to solve this problem simply by setting $L=0$ in the formulas for the LCR circuit, but those formulas were derived using the assumption that the LCR circuit was underdamped. That is, that $R<2\sqrt{L/C}$, which is not satisfied if $L=0$. The circuit equation for a simple RC circuit is

$$R\frac{dq}{dt} + \frac{1}{C}q = 0 \qquad (S10.18)$$

which has the solution

$$q(t) = q^o e^{-\alpha t} \qquad (S10.19)$$

where $\alpha = 1/RC$. Using Eq. (10.4) to evaluate the thermodynamic force, one obtains $\beta_q = -q/CkT$. Therefore, Eq. (S10.19) may be written in the standard form

$$q(t) = -CkTe^{-\alpha t}\beta_q^o \qquad (S10.20)$$

The Fluctuation-Dissipation Theorem and the result of Exercise 10.1 then imply that

$$\langle Q(0)Q(t)\rangle = CkTe^{-\alpha|t|} \qquad (S10.21)$$

Exercise 10.5 Calculate the correlation function for the voltage fluctuations across the resistor in a simple RC circuit and, by taking the limit $C \to 0$, obtain the correlation function for the voltage fluctuations across an isolated resistor at temperature T.

Solution In an RC circuit, the voltage across the resistor at time t is equal to $Q(t)/C$. Therefore, the voltage correlation function is

$$\langle V(0)V(t)\rangle = \frac{\langle Q(0)Q(t)\rangle}{C^2} = kT\frac{e^{-|t|/RC}}{C} \qquad (S10.22)$$

The limit $C \to 0$ can be taken by using the following representation for the Dirac delta function.

$$\lim_{\epsilon \to 0} \left(\frac{e^{-|t|/\epsilon}}{2\epsilon} \right) = \delta(t) \tag{S10.23}$$

Thus one finds that the voltage–voltage correlation function for a bare resistor is the singular function

$$C(t - t') = \langle V(t)V(t') \rangle = 2RkT \, \delta(t - t') \tag{S10.24}$$

This formula says that the voltages across the resistor at any two different times are statistically independent. Of course, this is only an approximation of reality. It is based on the assumptions that the resistor has zero inductance and capacitance—a condition that never holds for a real device. Taking the Fourier transform of $C(t)$, one sees that the power spectral density of a resistor at temperature T is

$$P(\omega) = 4RkT \tag{S10.25}$$

This result is Nyquist's Theorem (see Fig. S10.2).

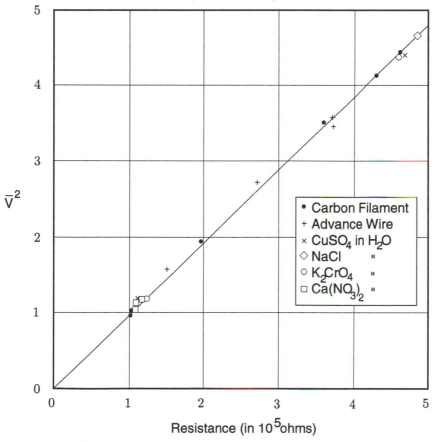

Fig. S10.2   A figure from the Physical Review article by J. B. Johnson, showing the temperature dependence of $\langle V^2 \rangle$ for a variety of different resistor types, including ordinary resistance wires and columns of electrolytic solutions.

The next few exercises lead up to the equation of motion version of the quantum mechanical Fluctuation-Dissipation Theorem.

Exercise 10.6   For an operator A with a continuous spectrum, define the conditional canonical potential $\phi(a)$.

Solution   For a classical observable $A(z)$ the conditional canonical potential is defined by the relation

$$e^{\phi(a)} = \int \delta(a - A(z)) e^{-\beta H(z)} \, d^{2K} z / N! h^K \tag{S10.26}$$

For a quantum system, the integral will obviously be replaced by a sum over the states in a quantum canonical ensemble. The nontrivial part of this exercise is to define the peculiar quantity $\delta(a - A)$, when A is a Hermitian operator. An operator has been well defined if one gives its effect on any complete set of states. The operator A has some complete set of eigenstates $|n, \alpha\rangle$ that satisfy the eigenvalue equation

$$A|n, \alpha\rangle = \alpha|n, \alpha\rangle \tag{S10.27}$$

The parameter n numbers the possibly degenerate states with a given value of α. Any function of the operator A is then defined by the equation

$$F(A)|n, \alpha\rangle \equiv F(\alpha)|n, \alpha\rangle \tag{S10.28}$$

A special case is the function $\delta(a - A)$ for which

$$\delta(a - A)|n, \alpha\rangle = \delta(a - \alpha)|n, \alpha\rangle \tag{S10.29}$$

With this definition of $\delta(a - A)$, $\phi(a)$ can be defined by

$$e^{\phi(a)} = \sum_n e^{-\beta E_n} \langle E_n|\delta(a - A)|E_n\rangle \tag{S10.30}$$

Exercise 10.7   Given n commuting operators, $\mathbf{A} = (A_1, A_2, \ldots, A_n)$, define an equation of motion $\mathbf{a}(t, \mathbf{a}^o)$, and show that $\mathbf{a}(0, \mathbf{a}^o) = \mathbf{a}^o$. (We must restrict ourselves to commuting operators because it makes no quantum mechanical sense to set the simultaneous initial values of noncommuting operators.)

Solution   Following the lead of Eq. (10.19), $\mathbf{a}(t, \mathbf{a}^o)$ is defined as

$$\mathbf{a}(t, \mathbf{a}^o) = e^{-\phi(\mathbf{a}^o)} \sum_n e^{-\beta E_n} \langle E_n|\mathbf{A}(t) \, \delta(\mathbf{a}^o - \mathbf{A})|E_n\rangle \tag{S10.31}$$

where the n-dimensional delta function $\delta(\mathbf{a}^o - \mathbf{A})$ is the product $\delta(a_1^o - A_1) \cdots \delta(a_n^o - A_n)$. The following identity is an obvious consequence of Eq. (S10.29), which defines $\delta(\mathbf{a}^o - \mathbf{A})$.

$$A_i \, \delta(a - A_i) = a \, \delta(a - A_i) \tag{S10.32}$$

Using this identity and the definition of $\phi(\mathbf{a})$ given in Eq. (S10.30), it is a simple matter to show that $a_i(0, \mathbf{a}^o) = a_i^o$. A useful identity, which follows from Eqs. (S10.29) and (S10.32), is the operator identity

$$A_i = \int a_i \, \delta(\mathbf{a} - \mathbf{A}) \, d^n\mathbf{a} \qquad (S10.33)$$

Exercise 10.8   Assume that, for sufficiently small $\mathbf{a}^o$, the macroscopic equation of motion can be linearized in the standard form,

$$a_i(t, \mathbf{a}^o) = \sum_k F_{ik}(t)\beta_k(\mathbf{a}^o) \qquad (S10.34)$$

where $\beta_k(\mathbf{a}) = \partial\phi(\mathbf{a})/\partial a_k$, and define the time-dependent correlation matrix $C_{ij}(t) = \langle A_i(t)A_j \rangle$, where the average is over a quantum canonical ensemble. Prove the equation of motion form of the Fluctuation-Dissipation Theorem, namely, that

$$C_{ij}(t) = -F_{ij}(t) \qquad (S10.35)$$

Solution

$$C_{ij}(t) = e^{-\phi} \sum_n e^{-\beta E_n} \langle E_n|A_i(t)A_j|E_n \rangle \qquad (S10.36)$$

Using the identity given in Eq. (10.33), we see that

$$\begin{aligned} C_{ij}(t) &= e^{-\phi} \int a_j^o \sum_n e^{-\beta E_n} \langle E_n|A_i(t)\, \delta(\mathbf{a}^o - \mathbf{A})|E_n \rangle \, d^n\mathbf{a}^o \\ &= \int e^{\phi(\mathbf{a}^o)-\phi} a_i(t, \mathbf{a}^o)a_j^o \, d^n\mathbf{a}^o \end{aligned} \qquad (S10.37)$$

But this is identical with Eq. (10.28). Therefore, the same analysis that was used for the classical case [see Eq. (10.31)] is applicable without change to the quantum mechanical version of the theorem. It should be pointed out that, although the form of the theorem is the same for a classical and a quantum system, the actual functions $F_{ij}(t)$ would be different for a classical system and the corresponding quantum mechanical one.

The theorem can be extended to pairs of operators, A and B, that do not necessarily commute. We define a relative equation of motion $b(t, a)$ and a correlation function $C_{BA}(t)$ by

$$b(t, a) = e^{-\phi(a)} \sum_n e^{-\beta E_n} \langle E_n|B(t)\, \delta(a - A)|E_n \rangle \qquad (S10.38)$$

and

$$C_{BA}(t) = e^{-\phi} \sum_n e^{-\beta E_n} \langle E_n|B(t)A|E_n \rangle \qquad (S10.39)$$

The linearized equation of motion has the form

$$b(t, a) = F_{BA}(t)\beta(a) \qquad (S10.40)$$

The same manipulations used above will then show that

$$C_{BA}(t) = -F_{BA}(t) \tag{S10.41}$$

Exercise 10.9   In Section 10.8, the correlation functions for thermal fluctuations of charge and current in an LCR circuit were calculated for the case of an underdamped oscillator, that is, for $R < 2\sqrt{L/C}$. Find the correlation functions in the overdamped case and carefully analyze their behavior in the limit $L \to 0$.

Solution   The circuit equation $L\ddot{q} + R\dot{q} + q/C = 0$ is unchanged. If we try a simple exponential solution, $q(t) = \exp(-\lambda t)$, then we obtain the quadratic equation $L\lambda^2 + R\lambda + 1/C = 0$, which has the two solutions $\lambda = \gamma + \alpha$ and $\gamma - \alpha$, where $\gamma = R/2L$ and $\alpha = (\gamma^2 - \omega_0^2)^{1/2}$ with $\omega_o = 1/\sqrt{LC}$. Thus we can write the general solution as

$$q(t) = e^{-\gamma t}(Ae^{\alpha t} + Be^{-\alpha t}) \tag{S10.42}$$

We must now evaluate the constants A and B in terms of q^o and i^o, the initial values of the charge and current. $q(0) = q^o$ and $\dot{q}(0) = i^o$ give the linear equations $A + B = q^o$ and $\alpha(A - B) - \gamma(A + B) = i^o$, which have the solution

$$A = \tfrac{1}{2}(1 + \gamma/\alpha)q^o + \frac{1}{2\alpha}i^o \quad \text{and} \quad B = \tfrac{1}{2}(1 - \gamma/\alpha)q^o - \frac{1}{2\alpha}i^o \tag{S10.43}$$

Thus the complete solution of the circuit equation is

$$
\begin{aligned}
q(t) &= e^{-\gamma t}\left(\frac{1 + \gamma/\alpha}{2}e^{\alpha t}q^o + \frac{1}{2\alpha}e^{\alpha t}i^o + \frac{1 - \gamma/\alpha}{2}e^{-\alpha t}q^o - \frac{1}{2\alpha}e^{-\alpha t}i^o\right) \\
&= e^{-\gamma t}\left(\cosh \alpha t + \frac{\gamma}{\alpha}\sinh \alpha t\right)q^o + e^{-\gamma t}\frac{\sinh \alpha t}{\alpha}i^o
\end{aligned}
\tag{S10.44}
$$

and similarly

$$i(t) = -e^{-\gamma t}\frac{\omega_0^2 \sinh \alpha t}{\alpha}q^o + e^{-\gamma t}\left(\cosh \alpha t - \frac{\gamma}{\alpha}\sinh \alpha t\right)i^o \tag{S10.45}$$

Of course, this solution could have been obtained much more quickly simply by replacing ω by $i\alpha$ in Eqs. (10.38) and (10.39).

Using the facts that $q^o = -CkT\beta_q^o$ and $i^o = -(kT/L)\beta_i^o$, we can follow the analysis of Section 10.8 to obtain the correlation function for positive t and use the symmetry relations of Exercise 10.1 to extend them to all values of t. The result is

$$C_{qq} = kTCe^{-\gamma|t|}\left(\cosh \alpha t + \frac{\gamma}{\alpha}\sinh \alpha|t|\right) \tag{S10.46}$$

$$C_{ii} = \frac{kT}{L}e^{-\gamma|t|}\left(\cosh \alpha t - \frac{\gamma}{\alpha}\sinh \alpha|t|\right) \tag{S10.47}$$

and

$$C_{qi} = -C_{iq} = \text{sgn}(t)\frac{kTC}{\alpha}e^{-\gamma|t|}\sinh \alpha|t| \tag{S10.48}$$

In the limit $L \to \infty$, α and γ both go to infinity, but

$$\gamma - \alpha = \gamma\left(1 - \sqrt{1 - \omega_0^2}\right) \approx \tfrac{1}{2}\omega_o^2/\gamma = \frac{1}{RC} \tag{S10.49}$$

and

$$\frac{\gamma}{\alpha} = \left(1 - \frac{\omega_o^2}{\gamma^2}\right)^{-1/2} \approx 1 + \frac{2L}{CR^2} \tag{S10.50}$$

Also, $\cosh \alpha t + \sinh \alpha t = e^{\alpha t}$, $\cosh \alpha t - \sinh \alpha t = e^{-\alpha t}$, and $\sinh \alpha |t| \approx \frac{1}{2} e^{\alpha |t|}$. Using these substitutions in Eqs. (S10.46), (S10.47), and (S10.48) gives

$$\begin{aligned} C_{qq}(t) &\approx kTCe^{-\gamma|t|}\left(e^{\alpha|t|} + \frac{L}{CR^2}e^{\alpha|t|}\right) \\ &\approx kTCe^{-|t|/RC} \end{aligned} \tag{S10.51}$$

$$\begin{aligned} C_{ii}(t) &\approx \frac{kT}{L}e^{-\gamma|t|}\left(e^{-\alpha|t|} - \frac{L}{CR^2}e^{\alpha|t|}\right) \\ &\approx \frac{kT}{L}e^{-(R/L)|t|} - \frac{kT}{CR^2}e^{-|t|/RC} \end{aligned} \tag{S10.52}$$

and

$$C_{qi}(t) = -C_{iq}(t) \approx \text{sgn}(t)\frac{kTLC}{R}e^{-|t|/RC} \approx 0 \tag{S10.53}$$

Using the representation of the delta function given in Eq. (S10.23), one can write $C_{ii}(t)$ as

$$C_{ii}(t) \approx 2RkT\,\delta(t) - \frac{C_{qq}(t)}{(RC)^2} \tag{S10.54}$$

As $L \to 0$, the direct current–current correlation, caused by the inertial effect of the inductor, goes to zero for any finite time interval, leaving only a weaker corelation caused by the fact that a positive current at one time will tend to charge the capacitor, leading at any later time to a negative discharging current.

Exercise 10.10   In Section 8.8, we considered a two-dimensional fluid composed of two types of particles, called x particles and y particles. In that analysis, it was assumed that it was possible for an x particle to be transformed into a y particle and vice versa. In this exercise we will look at the same system, but we will assume that no such transformation is possible, so that the two types of particles are separately conserved. If it is also assumed that the total numbers of x- and y particles are equal, then the system is an example of a *symmetric binary fluid*.
Assume that the flows of x- and y particles are driven by gradients in α_x and α_y and that they satisfy the diffusion equations

$$\mathbf{J}_x = D\nabla\alpha_x \quad \text{and} \quad \mathbf{J}_y = D\nabla\alpha_y \tag{S10.55}$$

and that α_x and α_y are given by their mean-field formulas

$$\alpha_x = -\log n_x - \ell^2 n_y - 2\log\lambda \quad \text{and} \quad \alpha_y = -\log n_y - \ell^2 n_x - 2\log\lambda \tag{S10.56}$$

(a) Taking $n = n_x + n_y$ as constant, derive a linear differential equation for the function $\psi(\mathbf{r}, t) = n_x(\mathbf{r}, t) - n_y(\mathbf{r}, t)$.   (b) By expanding the function $\psi(\mathbf{r}, t)$ as a Fourier series,

$$\psi(\mathbf{r}, t) = \sum_{\mathbf{k}} \psi_{\mathbf{k}}(t)e^{i\mathbf{k}\cdot\mathbf{r}} \tag{S10.57}$$

show that $\psi_{\mathbf{k}}(t) = \psi_{\mathbf{k}}(0) \exp(-t/\tau_k)$ and show that the relaxation time τ_k for a mode of wave vector $\mathbf{k}$ goes as $\tau_k \sim k^{-2}|n_c - n|^{-1}$ near the critical density $n_c = 2/\ell^2$. The fact that the relaxation time $\tau_k \to \infty$ at the critical point is approached is a phenomenon called *critical slowing down*. Critical slowing down is a serious problem for the experimentalist who is trying to measure the properties of states close to the critical point. It means that, as the critical point is approached, the system takes longer and longer to come to equilibrium. It is also a great nuisance to theoretical physicists trying to simulate systems close to the critical point by the Monte Carlo method because the same phenomenon appears in Monte Carlo calculations—that is, the system takes an extremely large number of Monte Carlo steps to reach equilibrium near the critical point. Note: Both the use of the diffusion equation and the use of the mean-field formulas for α_x and α_y are actually invalid very close to the critical point. Therefore, although the qualitative prediction of the existence of critical slowing down is correct, the quantitative predictions regarding the degree of slowing down are not.

Solution   (a) The conservation equations for x particles and y particles are

$$\dot{n}_x = -\nabla \cdot \mathbf{J}_x = -D \nabla^2 \alpha_x \tag{S10.58}$$

and

$$\dot{n}_y = -\nabla \cdot \mathbf{J}_y = -D \nabla^2 \alpha_y \tag{S10.59}$$

Therefore,

$$\dot{\psi} = \dot{n}_x - \dot{n}_y = -D \nabla^2 (\alpha_x - \alpha_y) \tag{S10.60}$$

Since $n_x = (n + \psi)/2$ and $n_y = (n - \psi)/2$,

$$\alpha_x - \alpha_y = -\log\left(\frac{n + \psi}{n - \psi}\right) + \ell^2 \psi$$
$$\approx -\left(\frac{2}{n} - \ell^2\right)\psi \tag{S10.61}$$

where, in the last line, it is assumed that $\psi \ll n$. Using this in Eq. (S10.60) gives a linear differential equation for $\psi(\mathbf{r}, t)$.

$$\frac{\partial \psi}{\partial t} = \left(\frac{2}{n} - \ell^s\right) D \nabla^2 \psi \equiv \tilde{D} \nabla^2 \psi \tag{S10.62}$$

(b) Letting

$$\psi(\mathbf{r}, t) = \sum_{\mathbf{k}} \psi_{\mathbf{k}}(t) e^{i\mathbf{k}\cdot\mathbf{r}} \tag{S10.63}$$

and using the fact that $\nabla^2 e^{i\mathbf{k}\cdot\mathbf{r}} = -k^2 e^{i\mathbf{k}\cdot\mathbf{r}}$, the linear equation can be written as

$$\sum_{\mathbf{k}} (\dot{\psi}_{\mathbf{k}} + \tilde{D} k^2 \psi_{\mathbf{k}}) e^{i\mathbf{k}\cdot\mathbf{r}} = 0 \tag{S10.64}$$

which implies that, for each $\mathbf{k}$,

$$\dot{\psi}_{\mathbf{k}} + \tilde{D} k^2 \psi_{\mathbf{k}} = 0 \tag{S10.65}$$

This equation has the solution

$$\psi_{\mathbf{k}}(t) = \psi_{\mathbf{k}}(0)e^{-\tilde{D}k^2 t} = \psi_{\mathbf{k}}(0)e^{-t/\tau_k} \tag{S10.66}$$

with $\tau_k^{-1} = D(2/n - \ell^2)k^2$. From the fact that $\ell^2 = 2/n_c$, one can see that

$$\frac{2}{n} - \ell^2 = 2\left(\frac{1}{n} - \frac{1}{n_c}\right) = \frac{2}{nn_c}(n_c - n) \tag{S10.67}$$

Clearly, for n close to but less than n_c,

$$\tau_k \sim k^{-2}(n_c - n)^{-1} \tag{S10.68}$$

Exercise 10.11   Certainly the present state of the earth, with trees and flowers and people, seems to be much further from an equilibrium state than the original state of molten rocks and hot gases. Is not the evolution of life a violation of the Second Law of thermodynamics? Explain why it is not.

Solution   The Second Law implies that the entropy of an *isolated* system can only increase. Life has evolved on the surface of the earth, which is not at all an isolated system. It has a source of high-temperature energy—the sun. It has a low-temperature reservoir—outer space, at 2.7 K.

   Let us consider what would happen to life if we made the earth an isolated system. We could do that by putting a big shiny steel sphere around the earth; let us say, at a distance half way to the moon. It would immediately become black night. All plants would soon die, after which all herbivores would die, after which all carnivores would die.

   Of course, one might object that this disaster was caused only by leaving out the sun. Suppose, instead, we construct a bigger shiny sphere that surrounds the sun and all the inner planets through Mars. (It is necessary that the sphere be shiny in order to truly isolate the system.) Then, within a very short time, every direction in the sky would look like the surface of the sun. The surface of the earth would soon come into equilibrium with the surface of the sun, which is about 6,000 K. Obviously, this vaporization death is even surer than the freezing death that was described before. It is clear that our lucky position, situated between a hot reservoir and a cold reservoir, has been essential for our biological prosperity.

Exercise 10.12   In Exercise 2.3, the rate at which particles and energy are transferred through a small hole separating two very dilute gases slightly out of equilibrium was calculated. Since this process is a correlated exchange of two different quantities, the Reciprocity Theorem is relevant. Show that it is satisfied by the formulas derived there [Eqs. (S2.6) and (S2.8)].

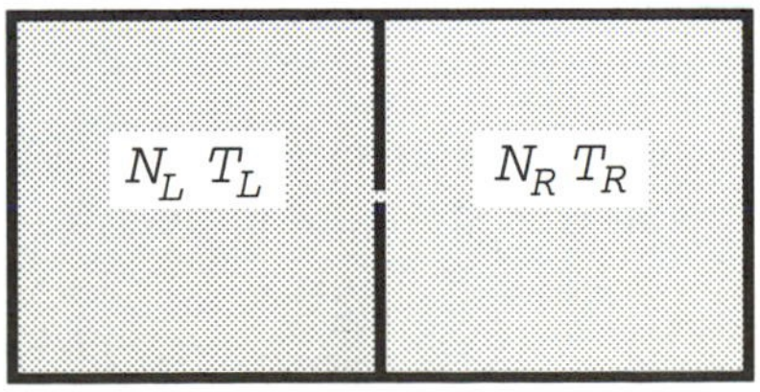

Fig. S10.3

Solution   For simplicity, we will assume that the two volumes shown in Fig. S10.3 are unit volumes. Let N_L, E_L, N_R, and E_R be the particle numbers and energies of the gases on the left and the right. If we make a transformation to the four variables $N = (N_L + N_R)/2$, $\nu = (N_L - N_R)/2$, $E = (E_L + E_R)/2$, and $\varepsilon = (E_L - E_R)/2$, then N and E are fixed and equilibrium occurs at $\nu = \varepsilon = 0$. According to Eqs. (S2.6) and (S2.8),

$$\dot{\nu} = (\dot{N}_L - \dot{N}_R)/2 = -\frac{A}{\sqrt{2\pi m/k}}\left(2T^{1/2}\nu + \frac{N\Delta T}{2T^{1/2}}\right) \tag{S10.69}$$

and

$$\dot{\varepsilon} = (\dot{E}_L - \dot{E}_R)/2 = -\frac{2kA}{\sqrt{2\pi m/k}}\left(2T^{3/2}\nu + \tfrac{3}{2}NT^{1/2}\Delta T\right) \tag{S10.70}$$

These equations are not in the standard form. We must transform the right-hand sides from the variables ν and ΔT to the thermodynamic forces $\beta_\nu = \partial S/\partial \nu$ and $\beta_\varepsilon = \partial S/\partial \varepsilon$. The entropy of the system is

$$S^o = S_L^o(N_L, E_L) + S_R^o(N_R, E_R) = S_L^o(N + \nu, E + \varepsilon) + S_R^o(N - \nu, E - \varepsilon) \tag{S10.71}$$

which implies that

$$\beta_\nu = \alpha_L - \alpha_R \quad\text{and}\quad \beta_\varepsilon = \frac{1}{kT_L} - \frac{1}{kT_R} \tag{S10.72}$$

We must now solve for ν and ΔT in terms of β_ν and β_ε. Letting $T = (T_L + T_R)/2$, we can write β_ε as

$$\beta_\varepsilon = \frac{1}{k(T + \Delta T/2)} - \frac{1}{k(T - \Delta T/2)} \approx -\frac{\Delta T}{kT^2} \tag{S10.73}$$

For an ideal gas,
$$\alpha = -\log N + \tfrac{3}{2}\log T + C \tag{S10.74}$$

and therefore,

$$\begin{aligned}
\beta_\nu &= \alpha_L - \alpha_R \\
&= -\log(N + \nu) + \tfrac{3}{2}\log(T + \Delta T/2) + \log(N - \nu) - \tfrac{3}{2}\log(T - \Delta T/2) \\
&\approx -\frac{2\nu}{N} + \frac{3}{2}\frac{\Delta T}{T} \\
&= -\frac{2\nu}{N} - \tfrac{3}{2}kT\beta_\varepsilon
\end{aligned}$$
$$\tag{S10.75}$$

which gives
$$2\nu = -N\beta_\nu - \tfrac{3}{2}NkT\beta_\varepsilon \tag{S10.76}$$

Making these replacements in Eqs. (S10.69) and (S10.70), we get the equations for $\dot{\nu}$ and $\dot{\varepsilon}$ in the desired standard form.

$$\dot{\nu} = \frac{A}{\sqrt{2\pi m/k}}(NT^{1/2}\beta_\nu + 2NkT^{3/2}\beta_\varepsilon) \tag{S10.77}$$

and

$$\dot{\varepsilon} = \frac{2kA}{\sqrt{2\pi m/k}}(NT^{3/2}\beta_\nu + 3NkT^{5/2}\beta_\varepsilon) \tag{S10.78}$$

It is now obvious that the coefficient of β_ε in the first equation is equal to the coefficient of β_ν in the second, which is Onsager's relation.

Exercise 10.13 In Section 10.16 it is stated that, for negative times, one should not use the Boltzmann equation directly to go from time 0 to time t, but rather one should use the time-reversal symmetry property, $F(\mathbf{p}, -t) = F(\mathbf{p}, t)$, to relate the momentum distribution at negative t to the same quantity at positive t, which can be calculated with the Boltzmann equation. This may seem like a very ad hoc prescription. To show that it is actually a reasonable and natural one, it is useful to look at a similar but simpler situation, involving the system analyzed in the previous exercise. For a system of two very dilute ideal gases in adjoining compartments, connected by a small hole, show that: (a) the values of $\nu(t)$ and $\varepsilon(t)$ for negative t should not be calculated by using Eqs. (S10.77) and (S10.78), but, rather, should be computed by using the symmetry relations $\nu(-t) = \nu(t)$ and $\varepsilon(-t) = \varepsilon(t)$ with positive t; and (b) the entropy of the system increases in both directions in time.

Solution (a) For any microstate of the N-particle system, one can determine what will happen if the system were run backwards in time simply by reversing the velocity of every particle and running the system forward in time. If all we know about the system are the values of N_L, E_L, N_R, and E_R at $t = 0$, then the most probable distributions on left and right, at that instant, are separate uniform Maxwell distributions with the given densities and energies. For a Maxwell distribution, reversing the velocities of all particles makes no significant change because the number of particles with any velocity $\mathbf{v}$ is equal to the number with the velocity $-\mathbf{v}$. Therefore, it is clear that the time evolution of the particle and energy densities in the velocity-reversed state will be the same as the time evolution of the same quantities in the original state. This shows that $\nu(-t) = \nu(t)$ and $\varepsilon(-t) = \varepsilon(t)$. It also shows that, at $t = 0$, one can use the Maxwell distributions on left and right to calculate the particle and energy fluxes through the hole. It does *not* justify the use of the Maxwell distribution to calculate the fluxes at any time after $t = 0$. That calculation is only an approximate one, based on the assumption that the mean free path of the particles is significantly larger than the size of the hole. (Recall that that is the criterion for a *very dilute* ideal gas.) Just inside the hole, to the left, the probability distribution of leftgoing particles is that of the right-hand gas, because that is where those particles came from. Therefore, close to the hole, the velocity distribution is not a Maxwell distribution. Our use of the Maxwell distribution in calculating the flux through the hole is based on the assumption that *the particles that are headed toward the hole* came from deep within the gas where it is reasonable to assume that the distribution is not significantly affected by the existence of the small hole far away. This is equivalent to the assumption used in deriving the Boltzmann equation that *particles that are headed toward one another* are uncorrelated.

Thus the recipe for calculating $\nu(t)$ and $\varepsilon(t)$ for negative t by using the symmetry relations $\nu(-t) = \nu(t)$ and $\varepsilon(-t) = \varepsilon(t)$ is exact, but the method used to calculate $\nu(t)$ and $\varepsilon(t)$ for positive t is only approximate. Exactly the same thing can be said of Boltzmann's prescription for calculating the momentum distribution at positive and negative t, given the same distribution at $t = 0$.

It should be pointed out that the approximation used in calculating the flux through the hole is based on the ratio of the mean free path to the size of the hole, but the approximation used in deriving Boltzmann's equation is based on the ratio of the mean free path to the two-particle interaction distance. Since the interaction distance is very much smaller than any realistic hole in a partition, Boltzmann's approximation is very much more accurate.

(b) Because of the time-reversal symmetry, we need only show that the entropy increases for positive t. The entropy of the system is equal to $S_L^o(N_L, E_L) + S_R^o(N_R, E_R)$, but it will be more convenient to express the total entropy in terms of the variables, N, E, ν, and ε. Then, because N and E are constant,

$$\frac{d}{dt} S(N, E, \nu, \varepsilon) = \frac{\partial S}{\partial \nu} \dot{\nu} + \frac{\partial S}{\partial \varepsilon} \dot{\varepsilon} \tag{S10.79}$$
$$= \beta_\nu \dot{\nu} + \beta_\varepsilon \dot{\varepsilon}$$

Using Eqs. (S10.77) and (S10.78) for $\dot{\nu}$ and $\dot{\varepsilon}$ gives

$$\dot{S} = \frac{A}{\sqrt{2\pi m/k}} (N T^{1/2} \beta_\nu^2 + 4 N k T^{3/2} \beta_\nu \beta_\varepsilon +^k N T^{5/2} \beta_\varepsilon^2)$$
$$= \frac{A N T^{1/2}}{\sqrt{2\pi m/k}} \left[(\beta_\nu + 2kT\beta_\varepsilon)^2 + 2kT^2 \beta_\varepsilon^2 \right] \tag{S10.80}$$

which is a manifestly nonnegative expression.

Root

```
Program Root
Write(*,*)'Give a positive value for A.'
Read(*,*)A
xlo=0.0
xhi=1.0
Do 1 i=1,10
x=(xlo+xhi)/2
If (g(x,A).GT.0) then
xhi=x
Else
xlo=x
End If
1     Continue
Write(*,*)'Root = ',x
End
```

```
Function g(x,A)
g=x-Tanh(A*x)
Return
End
```

Hot Spring 1

```
Program HotSpring_1
Real P(-100:100)
Integer dL
Write(*,*)
Write(*,*)'Give a value for s=Sqrt(kT/epsilon).'
Read(*,*)s
Write(*,*)
Write(*,*)'Give a value for the number of moves.'
Read(*,*)Nmoves
L=0
BetaE=0
Do 1 i=-100,100
1 P(i)=0
Iran=99991
Do 2 n=1, Nmoves
dL=Move(Iran)
dBetaE=(L+dL)**2/s**2-BetaE
If ((dBetaE.LT.0).OR.(Ran(Iran).LT.Exp(-dBetaE))) then
L=L+dL
BetaE=BetaE+dBetaE
End If
P(L)=P(L)+1
2 Continue
Do 3 i=-100, 100
3 P(i)=P(i)/Nmoves
Open(Unit=1,File='HotSpring.out')
Do 4 i=-100, 100
4 If(P(i).GT.0) Write(1,100)i,P(i)
100 Format(I4,F12.6)
Close(1)
End
```

```
Function Move(I)
If (Ran(I).GT.0.5) then
```

```
Move=1
Else
Move=-1
End If
Return
End
```

Hot Spring 2

```
Program HotSpring_2
Real P(-100:100)
Integer dL
Write(*,*)
Write(*,*)'Give a value for s=Sqrt(kT/epsilon).'
Read(*,*)s
Write(*,*)
Write(*,*)'Give a value for the number of moves.'
Read(*,*)Nmoves
Write(*,*)
Write(*,*)'Give a value for the maximum jump size (an integer).'
Read(*,*)MaxJump
L=0
BetaE=0
Do 1 i=-100,100
1 P(i)=0
Iran=99991
Do 2 n=1, Nmoves
dL=Move(MaxJump,Iran)
dBetaE=(L+dL)**2/s**2-BetaE
If ((dBetaE.LT.0).OR.(Ran(Iran).LT.Exp(-dBetaE))) then
L=L+dL
BetaE=BetaE+dBetaE
End If
P(L)=P(L)+1
2 Continue
Do 3 i=-100, 100
3 P(i)=P(i)/Nmoves
Open(Unit=1,File='HotSpring.out')
Do 4 i=-100, 100
4 If(P(i).GT.0) Write(1,100)i,P(i)
100 Format(I2,F12.6)
Close(1)
End
```

This Function gives a nonzero random integer from -Max to Max

---

```
Function Move(Max,I)
K=2*Max*Ran(I)
If (K.GE.Max) then
Move=K-Max+1
Else
Move=K-Max
End If
Return
End
```

Hot Spring 3

```
Program HotSpring_3
Write(*,*)
```

```
Write(*,*)'Give a value for s=Sqrt(kT/epsilon).'
Read(*,*)s
Write(*,*)
Write(*,*)'Give a value for the number of moves.'
Read(*,*)Nmoves
Write(*,*)
Write(*,*)
#'Give a value for the maximum jump size (a positive number).'
Read(*,*)DelPosMax
Position=0
BetaE=0
Sum2=0
Sum4=0
Sum6=0
Iran=99991
Do 2 n=1, Nmoves
DelPos=2*(Ran(Iran)-0.5)*DelPosMax
dBetaE=(Position+DelPos)**2/s**2-BetaE
If ((dBetaE.LT.0).OR.(Ran(Iran).LT.Exp(-dBetaE))) then
Position=Position+DelPos
BetaE=BetaE+dBetaE
End If
Sum2=Sum2+Position**2
Sum4=Sum4+Position**4
Sum6=Sum6+Position**6
2 Continue
AvePos2=Sum2/Nmoves
AvePos4=Sum4/Nmoves
AvePos6=Sum6/Nmoves
Write(*,*)
Write(*,*)'<X**2> = ',AvePos2
Write(*,*)
Write(*,*)'<X**4> = ',AvePos4
Write(*,*)
Write(*,*)'<X**6> = ',AvePos6
End
```

Hot Ball

```
Program HotBall
Real P(0:1000)
Integer dL
Write(*,*)
Write(*,*)'Give a value for s=Sqrt(kT/mg).'
Read(*,*)s
Write(*,*)
Write(*,*)'Give a value for the number of moves.'
Read(*,*)Nmoves
Write(*,*)
Write(*,*)'Give a value for the maximum jump size (an integer).'
Read(*,*)MaxJump
L=0
Do 1 i=0, 1000
1 P(i)=0
Iran=99991
Do 2 n=1, Nmoves
dL=Move(MaxJump,Iran)
dBetaE=dL/s
NewL=L+dL
If((dBetaE.LT.0).OR.(Ran(Iran).LT.Exp(-dBetaE))) then
```

```
If(NewL.GE.0) L=NewL
End If
P(L)=P(L)+1
2 Continue
Do 4 i=0, 1000
4 P(i)=P(i)/Nmoves
Open(Unit=1,File='HotBall.out')
Cutoff=0.000001
Constant=(1-Exp(-1/s))
Do 5 i=0, 1000
Exact=Constant*Exp(-i/s)
5 If((P(i).GT.0).OR.(Exact.GT.Cutoff)) Write(1,100)i,P(i),Exact
100 Format(I4,2F12.6)
Close(1)
End
```

```
Function Move(Max,I)   (See HotSpring_2 for listing)
```

Ising 1

```
Program Ising_1
Integer*2 Lat(10,10)
Common/Params/H,V,Tau,M,E,Iran
Real*8 SumM, SumE, SumM2, SumE2
L=10
H=0.1
V=0.5
TauMin=1.0
TauMax=10.0
Nruns=10
Nmoves=4000
Call Initialize(Lat,L)
DelTau=(TauMax-TauMin)/(Nruns-1)
Do 4 Tau=TauMin, TauMax, DelTau
Do 2 i=1,100*L**2
2 Call MakeMove(Lat,L)
SumM=0
SumM2=0
SumE=0
SumE2=0
Do 3 i=1,Nmoves
Call MakeMove(Lat,L)
SumM=SumM+Abs(M)
SumM2=SumM2+M**2
SumE=SumE+E
SumE2=SumE2+E**2
3 Continue
AveM=SumM/Nmoves
AveM2=SumM2/Nmoves
AveE=SumE/Nmoves
AveE2=SumE2/Nmoves
DelSqM=AveM2-(AveM)**2
DelSqE=AveE2-(AveE)**2
AveM=AveM/L**2
AveE=AveE/L**2
Chi=DelSqM/(Tau**2*L**2)
SpecHeat=DelSqE/(Tau**2*L**2)
Write(1,100)Tau, AveM, AveE, Chi, SpecHeat
4 Continue
Close(1)
```

```
100    Format(F6.2,4F18.6)
End
```

```
Subroutine MakeMove(Lat,L)
Integer*2 Lat(L,L)
Common/Params/H,V,Tau,M,E,Iran
i=1+L*Ran(Iran)
j=1+L*Ran(Iran)
DelE=2*Lat(i,j)*(H+V*(Lat(i,nm1(j,L))+Lat(i,np1(j,L))
#+Lat(nm1(i,L),j)+Lat(np1(i,L),j)))
If ((DelE.LT.0).OR.(Ran(Iran).LT.Exp(-DelE/Tau))) then
Lat(i,j)=-Lat(i,j)
M=M+2*Lat(i,j)
E=E+DelE
End If
End
```

```
Function np1(n,L)
If(n.EQ.L) then
np1=1
Else
np1=n+1
End If
Return
End
```

```
Function nm1(n,L)
If(n.EQ.1) then
nm1=L
Else
nm1=n-1
End If
Return
End
```

```
Subroutine Initialize(Lat,L)
Integer*2 Lat(L,L)
Common/Params/H,V,Tau,M,E,Iran
Iran=99991
Open(Unit=1,File='Ising_1.out')
Write(1,*)'L, H, V =',L,H,V
Write(1,50)
50 Format(//'   Tau',14X,'AveM',14X,'AveE',15X,'Chi',10X,'SpecHeat')
Write(1,*)
Do 1 i=1,L
Do 1 j=1,L
Lat(i,j)=1
1    Continue
M=L**2
E=-(H+2*V)*L**2
End
```

Ising 2

```
Program Ising_2
Integer*2 Lat(10,10)
Common/Params/H,V,Tau,M,E,Iran
L=10
H=0.0
V=1.0
```

```
TauMin=1.0
TauMax=10.0
Nruns=10
Nsweeps=4000
Call Initialize(Lat,L)
DelTau=(TauMax-TauMin)/(Nruns-1)
Do 4 Tau=TauMin, TauMax, DelTau
SumM=0
SumM2=0
SumE=0
SumE2=0
Do 2 i=1,100
2 Call Sweep(Lat,L)
Do 3 i=1,Nsweeps
Call Sweep(Lat,L)
SumM=SumM+Abs(M)
SumM2=SumM2+M**2
SumE=SumE+E
SumE2=SumE2+E**2
3 Continue
AveM=SumM/Nsweeps
AveM2=SumM2/Nsweeps
AveE=SumE/Nsweeps
AveE2=SumE2/Nsweeps
DelSqM=AveM2-(AveM)**2
DelSqE=AveE2-(AveE)**2
AveM=AveM/L**2
AveE=AveE/L**2
Chi=DelSqM/(Tau**2*L**2)
SpecHeat=DelSqE/(Tau**2*L**2)
Write(1,100)Tau, AveM, AveE, Chi, SpecHeat
4 Continue
Close(1)
100     Format(F6.2,4F18.6)
End
```

```
Subroutine Sweep(Lat,L)
Integer*2 Lat(L,L)
Common/Params/H,V,Tau,M,E,Iran
Do 1 i=1,L
Do 1 j=1,L
DelE=2*Lat(i,j)*(H+V*(Lat(i,nm1(j,L))+Lat(i,np1(j,L))
#+Lat(nm1(i,L),j)+Lat(np1(i,L),j)))
If ((DelE.LT.0).OR.(Ran(Iran).LT.Exp(-DelE/Tau))) then
Lat(i,j)=-Lat(i,j)
M=M+2*Lat(i,j)
E=E+DelE
End If
1       Continue
End
```

```
Function np1(n,L)   (See Ising_1 for listing)
```

```
Function nm1(n,L)   (See Ising_1 for listing)
```

```
Subroutine Initialize(Lat,L)   (See Ising_1 for listing)
```

Ising 3

```
Program Ising_3
```

```
Integer*2 Lat(0:21,10)
Real*4 SumM(0:21),Pol(0:21)
Common/Params/BetaV,Ileft,Iright,Iran
Lx=20
Ly=10
LP1=Lx+1
BetaV=0.4
Ileft=1
Iright=0
Nsweeps=4000
Call Initialize(Lat,Lx,Ly,LP1)
Do 1 i=0,LP1
SumM(i)=0
1 Pol(i)=0
Do 2 i=1,1000
2 Call Sweep(Lat,Lx,Ly,LP1)
Do 3 n=1,Nsweeps
Call Sweep(Lat,Lx,Ly,LP1)
Do 3 i=0,LP1
Do 3 j=1,Ly
3 SumM(i)=SumM(i)+Lat(i,j)
Do 4 i=0,LP1
4 Pol(i)=SumM(i)/(Ly*Nsweeps)
Do 5 i=0,LP1
5 Write(1,100)i,Pol(i)
100    Format(I4,F9.5)
Close(1)
End
```

```
Subroutine Sweep(Lat,Lx,Ly,LP1)
Integer*2 Lat(0:LP1,Ly)
Common/Params/BetaV,Ileft,Iright,Iran
Do 1 i=1,Lx
Do 1 j=1,Ly
DelBetaE=2*BetaV*Lat(i,j)*
#(Lat(i,nm1(j,Ly))+Lat(i,np1(j,Ly))+Lat(i-1,j)+Lat(i+1,j))
If ((DelBetaE.LT.0).OR.(Ran(Iran).LT.Exp(-DelBetaE))) then
Lat(i,j)=-Lat(i,j)
End If
1      Continue
End
```

```
Function np1(n,L)   (See Ising_1 for listing)
```

```
Function nm1(n,L)   (See Ising_1 for listing)
```

```
Subroutine Initialize(Lat,Lx,Ly,LP1)
Integer*2 Lat(0:LP1,Ly)
Common/Params/BetaV,Ileft,Iright,Iran
Open(Unit=1,File='Ising_3.out')
Write(1,*)' Lx, Ly, BetaV =',Lx,Ly,BetaV
Write(1,*)
Write(1,*)' Ileft, Iright =',Ileft,Iright
Write(1,*)
Write(1,*)'  I     Pol(I)'
Write(1,*)
Iran=99991
Do 1 j=1,Ly
Lat(0,j)=Ileft
Lat(LP1,j)=Iright
```

```
Do 1 i=1,Lx
If (Ran(Iran).LT.0.5) then
Lat(i,j)=1
Else
Lat(i,j)=-1
End If
1 Continue
End
```

Ising 3D

```
Program Ising_3D
Integer*2 Lat(10,10,10)
Common/Params/H,V,Tau,M,E,Iran
Real*8 SumM, SumE, SumM2, SumE2
L=10
H=0.0
V=1.0
TauMin=4.0
TauMax=5.0
Nruns=6
Nsweeps=1000
Call Initialize(Lat,L)
DelTau=(TauMax-TauMin)/(Nruns-1)
Do 4 Tau=TauMin, TauMax, DelTau
SumM=0
SumM2=0
SumE=0
SumE2=0
Do 2 i=1,100
2 Call Sweep(Lat,L)
Do 3 i=1,Nsweeps
Call Sweep(Lat,L)
SumM=SumM+Abs(M)
SumM2=SumM2+M**2
SumE=SumE+E
SumE2=SumE2+E**2
3 Continue
AveM=SumM/Nsweeps
AveM2=SumM2/Nsweeps
AveE=SumE/Nsweeps
AveE2=SumE2/Nsweeps
DelSqM=AveM2-(AveM)**2
DelSqE=AveE2-(AveE)**2
AveM=AveM/L**3
AveE=AveE/L**3
Chi=DelSqM/(Tau**2*L**3)
SpecHeat=DelSqE/(Tau**2*L**3)
Write(1,100)Tau, AveM,  AveE, Chi, SpecHeat
4 Continue
Close(1)
100   Format(F6.2,4F18.6)
End
```

```
Subroutine Sweep(Lat,L)
Integer*2 Lat(L,L,L)
Common/Params/H,V,Tau,M,E,Iran
Do 1 i=1,L
Do 1 j=1,L
Do 1 k=1,L
```

```
DelE=2*Lat(i,j,k)*(H+V*(Lat(i,j,nm1(k,L))+Lat(i,j,np1(k,L))+
#Lat(i,nm1(j,L),k)+Lat(i,np1(j,L),k)+
#Lat(nm1(i,L),j,k)+Lat(np1(i,L),j,k)))
If ((DelE.LT.0).OR.(Ran(Iran).LT.Exp(-DelE/Tau))) then
Lat(i,j,k)=-Lat(i,j,k)
M=M+2*Lat(i,j,k)
E=E+DelE
End If
1       Continue
End
```

```
Function np1(n,L)   (See Ising_1 for listing)
```

```
Function nm1(n,L)   (See Ising_1 for listing)
```

```
Subroutine Initialize(Lat,L)
Integer*2 Lat(L,L,L)
Common/Params/H,V,Tau,M,E,Iran
Iran=99991
Open(Unit=1,File='Ising_3D.out')
Write(1,*)' L, H, V =',L,H,V
Write(1,*)
Write(1,50)
50 Format('   Tau',14X,'AveM',14X,'AveE',15X,'Chi',10X,'SpecHeat')
Write(1,*)
Do 1 i=1,L
Do 1 j=1,L
Do 1 k=1,L
Lat(i,j,k)=1
1       Continue
M=L**3
E=-(H+3*V)*L**3
End
```

Ising 4

```
Program Ising_4
Integer*2 Lat(10,10)
Common/Params/BetaV,M,Iran,VacancyRate
Real*8 SumM, SumM2
L=10
VacancyRate=0.1
BetaVmin=0.1
BetaVmax=1.0
Nruns=10
Nsweeps=1000
Call Initialize(Lat,L)
DelBetaV=(BetaVmax-BetaVmin)/(Nruns-1)
Do 4 BetaV=BetaVmin, BetaVmax, DelBetaV
SumM=0
SumM2=0
Do 2 i=1,1000
2 Call Sweep(Lat,L)
Do 3 i=1,Nsweeps
Call Sweep(Lat,L)
SumM=SumM+Abs(M)
SumM2=SumM2+M**2
3 Continue
AveM=SumM/Nsweeps
AveM2=SumM2/Nsweeps
```

```
DelSqM=AveM2-(AveM)**2
Write(1,100)BetaV, AveM, DelSqM
4 Continue
Close(1)
100    Format(F6.2,2F18.6)
End
```

```
Subroutine Sweep(Lat,L)
Integer*2 Lat(L,L)
Common/Params/BetaV,M,Iran,VacancyRate
Do 1 i=1,L
Do 1 j=1,L
If (Lat(i,j).NE.0) then
DelBetaE=2*BetaV*Lat(i,j)*(Lat(i,nm1(j,L))+Lat(i,np1(j,L))
#  +Lat(nm1(i,L),j)+Lat(np1(i,L),j))
If ((DelBetaE.LT.0).OR.(Ran(Iran).LT.Exp(-DelBetaE))) then
Lat(i,j)=-Lat(i,j)
M=M+2*Lat(i,j)
End If
End If
1      Continue
End
```

```
Function np1(n,L)   (See Ising_1 for listing)
```

```
Function nm1(n,L)   (See Ising_1 for listing)
```

```
Subroutine Initialize(Lat,L)
Integer*2 Lat(L,L)
Common/Params/BetaV,M,Iran,VacancyRate
Iran=99991
Open(Unit=1,File='Ising_4.out')
Write(1,*)' L, VacancyRate =',L,VacancyRate
Write(1,*)
Write(1,50)
Write(1,*)
50 Format(' BetaV',14X,'Mave',12X,'DelSqM')
Do 1 i=1,L
Do 1 j=1,L
If (Ran(Iran).LT.VacancyRate) then
Lat(i,j)=0
Else If (Ran(Iran).LT.0.5) then
Lat(i,j)=1
Else
Lat(i,j)=-1
End If
1      Continue
M=0
Do 2 i=1,L
Do 2 j=1,L
M=M+Lat(i,j)
2      Continue
End
```

Lat Gas

```
Program LatGas
Integer*2 Lat(10,10)
Common/Params/Alpha,BetaV,F,N,Iran
L=10
```

```
BetaV=-2
AlphaMin=0.0
AlphaMax=5.0
Nruns=11
Nsweeps=1000
Call Initialize(Lat,L)
DelAlpha=(AlphaMax-AlphaMin)/(Nruns-1)
Do 4 Alpha=AlphaMin, AlphaMax, DelAlpha
Do 2 i=1,1000
2 Call Sweep(Lat,L)
Nsum=0
Do 3 i=1,Nsweeps
Call Sweep(Lat,L)
Nsum=Nsum+N
3 Continue
AveN=Float(Nsum)/Float(Nsweeps*L**2)
Write(1,100)Alpha,AveN
4 Continue
Close(1)
100    Format(F7.2,F18.6)
End
```

```
Subroutine Sweep(Lat,L)
Integer*2 Lat(L,L)
Common/Params/Alpha,BetaV,F,N,Iran
Do 1 i=1,L
Do 1 j=1,L
DelF=(1-2*Lat(i,j))*(Alpha+BetaV*(Lat(i,nm1(j,L))+
#Lat(i,np1(j,L))+Lat(nm1(i,L),j)+Lat(np1(i,L),j)))
If ((DelF.LT.0).OR.(Ran(Iran).LT.Exp(-DelF))) then
Lat(i,j)=1-Lat(i,j)
N=N-1+2*Lat(i,j)
End If
1      Continue
End
```

```
Function np1(n,L)   (See Ising_1 for listing)
```

```
Function nm1(n,L)   (See Ising_1 for listing)
```

```
Subroutine Initialize(Lat,L)
Integer*2 Lat(L,L)
Common/Params/Alpha,BetaV,F,N,Iran
Open(Unit=1,File='LatGas.out')
Write(1,*)' BetaV =',BetaV
Write(1,*)
Write(1,50)
Write(1,*)
50 Format('  Alpha,              AveN')
Iran=99991
Do 1 i=1,L
Do 1 j=1,L
Lat(i,j)=0
1      Continue
N=0
Return
End
```

AB Model

```
Program AB_Model
Integer*2 Lat(10,10),NNsum(10,10)
Common/Params/Alpha,BetaV,Nocc(-1:1),Iran
L=10
BetaV=1
AlphaMin=-10
AlphaMax=10
NumberOfAlphas=11
Nsweeps=4000
Call Initialize(Lat,NNsum,L)
Write(1,*)' L=',L,' BetaV=',BetaV,'  Nsweeps=',Nsweeps
Write(1,*)
Write(1,*)'·    Alpha            AveN            AveQ',
#'         DelSqN        DelSqQ'
Write(1,*)
DeltaAlpha=(AlphaMax-AlphaMin)/(NumberOfAlphas-1)
Do 4 Alpha=AlphaMin, AlphaMax, DeltaAlpha
AveN=0
AveQ=0
DelSqN=0
DelSqQ=0
Do 2 i=1,1000
2 Call Sweep(Lat,NNsum,L)
Do 3 i=1,Nsweeps
Call Sweep(Lat,NNsum,L)
NA=Nocc(1)
NB=Nocc(-1)
AveN=AveN+NA+NB
AveQ=AveQ+NA-NB
DelSqN=DelSqN+(NA+NB)**2
DelSqQ=DelSqQ+(NA-NB)**2
3 Continue
AveN=AveN/Nsweeps
AveQ=AveQ/Nsweeps
DelSqN=DelSqN/Nsweeps-AveN**2
DelSqQ=DelSqQ/Nsweeps-AveDif**2
Write(1,100)Alpha,AveN,AveQ,DelSqN,DelSqQ
100 Format(F10.4,4(1PE15.6))
4 Continue
Close(1)
End
```

```
Subroutine Sweep(Lat,NNsum,L)
Integer*2 Lat(L,L),NNsum(L,L)
Common/Params/Alpha,BetaV,Nocc(-1:1),Iran
Do 1 i=1,L
Do 1 j=1,L
Nold=Lat(i,j)
Nnew=NewOccNumber(Nold)
DeltaF=Alpha*(Iabs(Nnew)-Iabs(Nold))+BetaV*NNsum(i,j)*(Nnew-Nold)
If ((DeltaF.LT.0).OR.(Ran(Iran).LT.Exp(-DeltaF))) then
Lat(i,j)=Nnew
ip1=np1(i,L)
im1=nm1(i,L)
jp1=np1(j,L)
jm1=nm1(j,L)
Nchange=Nnew-Nold
```

```
NNsum(ip1,j)=NNsum(ip1,j)+Nchange
NNsum(im1,j)=NNsum(im1,j)+Nchange
NNsum(i,jp1)=NNsum(i,jp1)+Nchange
NNsum(i,jm1)=NNsum(i,jm1)+Nchange
Nocc(Nold)=Nocc(Nold)-1
Nocc(Nnew)=Nocc(Nnew)+1
End If
1       Continue
End
```

```
Function np1(n,L)   (See Ising_1 for listing)
```

```
Function nm1(n,L)   (See Ising_1 for listing)
```

```
Function NewOccNumber(Nold)
Common/Params/Alpha,BetaV,Nocc(-1:1),Iran
If(Nold.EQ.0) then
If(Ran(Iran).LT.0.5) then
NewOccNumber=1
Else
NewOccNumber=-1
End If
Else
If(Ran(Iran).LT.0.5) then
NewOccNumber=-Nold
Else
NewOccNumber=0
End If
End If
Return
End
```

```
Subroutine Initialize(Lat,NNsum,L)
Integer*2 Lat(L,L),NNsum(L,L)
Common/Params/Alpha,BetaV,Nocc(-1:1),Iran
Open(Unit=1,File='AB.out')
Iran=99991
Do 1 i=1,L
Do 1 j=1,L
Lat(i,j)=0
NNsum(i,j)=0
1       Continue
Nocc(-1)=0
Nocc(1)=0
Nocc(0)=L**2
Return
End
```

Renorm 1

```
Program Renorm_1
Real z(-1:1,-1:1,-1:1,-1:1)
Integer a,b,c,d,e,f,g,h,i,j,k,l,m,n,o,p
Write(*,*)'Give values for X1, X2, and X3.'
Read(*,*)X1,X2,X3
W1=Exp(X1)
W2=Exp(X2)
W3=Exp(X3)
Do 1 I1 = -1, 1, 2
Do 1 I2 = -1, 1, 2
```

```
Do 1 I3 = -1, 1, 2
Do 1 I4 = -1, 1, 2
1 z(I1,I2,I3,I4)=0
Do 2 a = -1, 1, 2
Do 2 b = -1, 1, 2
Do 2 c = -1, 1, 2
Do 2 d = -1, 1, 2
Do 2 e = -1, 1, 2
Do 2 f = -1, 1, 2
Do 2 g = -1, 1, 2
Do 2 h = -1, 1, 2
Do 2 i = -1, 1, 2
Do 2 j = -1, 1, 2
Do 2 k = -1, 1, 2
Do 2 l = -1, 1, 2
Do 2 m = -1, 1, 2
Do 2 n = -1, 1, 2
Do 2 o = -1, 1, 2
Do 2 p = -1, 1, 2
nn=a*b+b*e+e*f+c*d+d*g+g*h+i*j+j*m+m*n+k*l+l*o+o*p+a*c+c*i+i*k+
#    b*d+d*j+j*l+e*g+g*m+m*o+f*h+h*n+n*p
nnn=a*d+b*c+c*j+d*i+i*l+j*k+b*g+e*d+d*m+g*j+j*o+m*l+e*h+f*g+
#    g*n+h*m+m*p+n*o
nsq=a*b*d*c+c*d*j*i+i*j*l*k+b*e*g*d+d*g*m*j+j*m*o*l+e*f*h*g+
#    g*h*n*m+m*n*p*o
Weight=W1**nn*W2**nnn*W3**nsq
I1=Iblock(a,b,c,d)
I2=Iblock(e,f,g,h)
I3=Iblock(i,j,k,l)
I4=Iblock(m,n,o,p)
z(I1,I2,I3,I4)=z(I1,I2,I3,I4)+Weight
2 Continue
Y1=Alog(z(1,1,1,1)/z(1,-1,-1,1))/8
Y2=Alog(z(1,1,1,1)/z(1,1,-1,-1))/4-Y1
Y3=Alog(z(1,1,-1,-1)/z(1,1,1,-1))/2+Y2
Write(*,*)' Y1 =',Y1,' Y2 =',Y2,' Y3 =',Y3
End
```

```
Function Iblock(Ia,Ib,Ic,Id)
Iabcd=Ia+Ib+Ic+Id
If(Iabcd.EQ.0)then
Iblock=Ia/Iabs(Ia)
Else
Iblock=Iabcd/Iabs(Iabcd)
End If
End
```

Renorm 2

```
Program Renorm_2
Real z(-1:1,-1:1,-1:1,-1:1)
Integer a,b,c,d,e,f,g,h,i,j,k,l,m,n,o,p
Write(*,*)'Give values for X1, X2, and X3.'
Read(*,*)X1,X2,X3
W1=Exp(X1)
W2=Exp(X2)
W3=Exp(X3)
Do 1 I1=-1, 1
Do 1 I2=-1, 1
Do 1 I3=-1, 1
```

```
Do 1 I4=-1, 1
1 z(I1,I2,I3,I4)=0
Do 2 a=-1, 1, 2
Do 2 b=-1, 1, 2
Do 2 c=-1, 1, 2
Do 2 d=-1, 1, 2
Do 2 e=-1, 1, 2
Do 2 f=-1, 1, 2
Do 2 g=-1, 1, 2
Do 2 h=-1, 1, 2
Do 2 i=-1, 1, 2
Do 2 j=-1, 1, 2
Do 2 k=-1, 1, 2
Do 2 l=-1, 1, 2
Do 2 m=-1, 1, 2
Do 2 n=-1, 1, 2
Do 2 o=-1, 1, 2
Do 2 p=-1, 1, 2
nn=a*b+b*e+e*f+c*d+d*g+g*h+i*j+j*m+m*n+k*l+l*o+o*p+
#    a*c+c*i+i*k+b*d+d*j+j*l+e*g+g*m+m*o+f*h+h*n+n*p
nnn=a*d+b*c+c*j+d*i+i*l+j*k+b*g+e*d+d*m+
#    g*j+j*o+m*l+e*h+f*g+g*n+h*m+m*p+n*o
nsq=a*b*d*c+c*d*j*i+i*j*l+k+b*e*g*d+d*g*
#    m*j+j*m*o*l+e*f*h*g+g*h*n*m+m*n*p*o
Weight=W1**nn*W2**nnn*W3**nsq
I1=Iblock(a+b+c+d)
I2=Iblock(e+f+g+h)
I3=Iblock(i+j+k+l)
I4=Iblock(m+n+o+p)
z(I1,I2,I3,I4)=z(I1,I2,I3,I4)+Weight
2 Continue
Call Redistribute(z)
Y1=Alog(z(1,1,1,1)/z(1,-1,-1,1))/8
Y2=Alog(z(1,1,1,1)/z(1,1,-1,-1))/4-Y1
Y3=Alog(z(1,1,-1,-1)/z(1,1,1,-1))/2+Y2
Write(*,*)' Y1 =',Y1,' Y2 =',Y2,' Y3 =',Y3
End
```

```
Function Iblock(Iabcd)
If(Iabcd.EQ.0)then
Iblock=0
Else
Iblock=Iabcd/Iabs(Iabcd)
End If
Return
End
```

```
Subroutine Redistribute(z)
Real z(-1:1,-1:1,-1:1,-1:1)
Do 1 I2 = -1, 1
Do 1 I3 = -1, 1
Do 1 I4 = -1, 1
z(1,I2,I3,I4)=z(1,I2,I3,I4)+z(0,I2,I3,I4)/2
z(-1,I2,I3,I4)=z(-1,I2,I3,I4)+z(0,I2,I3,I4)/2
z(0,I2,I3,I4)=0
1 Continue
Do 2 I1 = -1, 1, 2
Do 2 I3 = -1, 1
Do 2 I4 = -1, 1
z(I1,1,I3,I4)=z(I1,1,I3,I4)+z(I1,0,I3,I4)/2
```

```fortran
z(I1,-1,I3,I4)=z(I1,-1,I3,I4)+z(I1,0,I3,I4)/2
z(I1,0,I3,I4)=0
2 Continue
Do 3 I1 = -1, 1, 2
Do 3 I2 = -1, 1, 2
Do 3 I4 = -1, 1
z(I1,I2,1,I4)=z(I1,I2,1,I4)+z(I1,I2,0,I4)/2
z(I1,I2,-1,I4)=z(I1,I2,-1,I4)+z(I1,I2,0,I4)/2
z(I1,I2,0,I4)=0
3 Continue
Do 4 I1 = -1, 1, 2
Do 4 I2 = -1, 1, 2
Do 4 I3 = -1, 1, 2
z(I1,I2,I3,1)=z(I1,I2,I3,1)+z(I1,I2,I3,0)/2
z(I1,I2,I3,-1)=z(I1,I2,I3,-1)+z(I1,I2,I3,0)/2
z(I1,I2,I3,0)=0
4 Continue
Return
End
```

Renorm 3

```fortran
Program Renorm_3
Open(Unit=1,File='Renorm_3.out')
Write(1,*)'      X1          X2          X3'
Xmin = .2
Xmax = .5
Nruns = 4
Niters = 7
DelX = (Xmax-Xmin)/(Nruns-1)
Do 2 Xstart = Xmin, Xmax, DelX
X1 = Xstart
X2 = 0
X3 = 0
Write(1,100)X1,X2,X3
Do 1 I = 1, Niters
Call FindCorrFtns(X1,X2,X3,C1,C2,C3)
Call FindHamPars(C1,C2,C3,X1,X2,X3)
1 Write(1,100)X1,X2,X3
2 Write(1,*)
100 Format(3F12.6)
Close(1)
End
```

```fortran
Subroutine FindCorrFtns(X1,X2,X3,Y1,Y2,Y3)
Implicit Integer*2 (A-G)
Real*4 Zsum,Sum1,Sum2,Sum3
Integer*4 KK
Iblock(a)=a/Iabs(a)
W1=Exp(X1)
W2=Exp(X2)
W3=Exp(X3)
Zsum=0
Sum1=0
Sum2=0
Sum3=0
Do 1 KK=0, 2**21-1
a1=2*Ibits(KK,0,1)-1
a2=2*Ibits(KK,1,1)-1
a3=2*Ibits(KK,2,1)-1
```

```
b1=2*Ibits(KK,3,1)-1
b2=2*Ibits(KK,4,1)-1
b3=2*Ibits(KK,5,1)-1
c1=2*Ibits(KK,6,1)-1
c2=2*Ibits(KK,7,1)-1
c3=2*Ibits(KK,8,1)-1
d1=2*Ibits(KK,9,1)-1
d2=2*Ibits(KK,10,1)-1
d3=2*Ibits(KK,11,1)-1
e1=2*Ibits(KK,12,1)-1
e2=2*Ibits(KK,13,1)-1
e3=2*Ibits(KK,14,1)-1
f1=2*Ibits(KK,15,1)-1
f2=2*Ibits(KK,16,1)-1
f3=2*Ibits(KK,17,1)-1
g1=2*Ibits(KK,18,1)-1
g2=2*Ibits(KK,19,1)-1
g3=2*Ibits(KK,20,1)-1
nn=
#f1*a3+a3*a2+a2*b1+f3*f2+f2*g1+g1*b3+b3*b2+e1*g3+g3*g2+
#g2*c1+e3*e2+e2*d1+d1*c3+c3*c2+d3*d2+f3*f1+e3*e1+e1*f2+
#f2*a3+a3*a1+e2*g3+g3*g1+g1*a2+d3*d1+d1*g2+g2*b3+b3*b1+
#d2*c3+c3*c1+c1*b2+b1*b2+a1*a2+a2*b3+b3*c1+c1*c2+a3*g1+
#g1*g2+g2*c3+f1*f2+f2*g3+g3*d1+d1*d2+f3*e1+e1*e2+e2*d3
nnn=
#a1*g1+f1*e1+a3*g3+a2*g2+b1*c1+f3*e3+f2*e2+g1*d1+b3*c3+
#b2*c2+g3*d3+g2*d2+f3*g3+f1*g1+e3*d3+e1*d1+f2*g2+a3*b3+
#a1*b1+e2*d2+g3*c3+g1*c1+a2*b2+g2*c2+b1*g1+b2*g2+a1*f1+
#a2*f2+b3*g3+c1*d1+c2*d2+a3*f3+g1*e1+g2*e2+c3*d3+g3*e3
nfour=
#a1*g1*a2*a3+f1*e1*f2*f3+a3*g3*g1*f2+a2*g2*g1*b3+b1*c1*b2*b3+
#f2*e2*e1*g3+g1*d1*g2*g3+b3*c3*c1*g2+g3*d3*d1*e2+g2*d2*d1*c3+
#f3*g3*e1*f2+f1*g1*a3*f2+e1*d1*e2*g3+f2*g2*g1*g3+a3*b3*a2*g1+
#e2*d2*d1*d3+g3*c3*g2*d1+g1*c1*b3*g2+a2*b2*b1*b3+g2*c2*c1*c3+
#b1*g1*a2*b3+b2*g2*c1*b3+a2*f2*g1*a3+b3*g3*g1*g2+c1*d1*c3*g2+
#a3*f3*f1*f2+g1*e1*f2*g3+g2*e2*d1*g3+c3*d3*d1*d2+g3*e3*e1*e2
Weight=W1**nn*W2**nnn*W3**nfour
A=Iblock(a1+a2+a3)
B=Iblock(b1+b2+b3)
C=Iblock(c1+c2+c3)
D=Iblock(d1+d2+d3)
E=Iblock(e1+e2+e3)
F=Iblock(f1+f2+f3)
G=Iblock(g1+g2+g3)
Zsum=Zsum+Weight
Sum1=Sum1+Weight*((A+B+C+D+E+F)*G+A*B+B*C+C*D+D*E+E*F+F*A)
Sum2=Sum2+Weight*(A*C+B*D+C*E+D*F+E*A+F*B)
Sum3=Sum3+Weight*(A*B*C+B*C*D+C*D*E+D*E*F+E*F*A+F*A*B)*G
1 Continue
Y1=Sum1/Zsum
Y2=Sum2/Zsum
Y3=Sum3/Zsum
Return
End
```

```
Subroutine FindHamPars(C10,C20,C30,X1,X2,X3)
Real*4 CoefMat(3,3),DelC(3),DelX(3)
Integer*2 A,B,C,D,E,F,G
Data Tolerance /0.00001/
Do 2 Iter = 1, 100
```

```
W1=Exp(X1)
W2=Exp(X2)
W3=Exp(X3)
Z=0
S1=0
S2=0
S3=0
S11=0
S12=0
S13=0
S22=0
S23=0
S33=0
Do 1 A = -1, 1, 2
Do 1 B = -1, 1, 2
Do 1 C = -1, 1, 2
Do 1 D = -1, 1, 2
Do 1 E = -1, 1, 2
Do 1 F = -1, 1, 2
Do 1 G = -1, 1, 2
nn=(A+B+C+D+E+F)*G+A*B+B*C+C*D+D*E+E*F+F*A
nnn=A*C+B*D+C*E+D*F+E*A+F*B
nfour=(A*B*C+B*C*D+C*D*E+D*E*F+E*F*A+F*A*B)*G
Weight=W1**nn*W2**nnn*W3**nfour
S1=S1+Weight*nn
S2=S2+Weight*nnn
S3=S3+Weight*nfour
S11=S11+Weight*nn**2
S12=S12+Weight*nn*nnn
S13=S13+Weight*nn*nfour
S22=S22+Weight*nnn**2
S23=S23+Weight*nnn*nfour
S33=S33+Weight*nfour**2
Z=Z+Weight
1 Continue
C1=S1/Z
C2=S2/Z
C3=S3/Z
DelC(1)=C10-C1
DelC(2)=C20-C2
DelC(3)=C30-C3
CoefMat(1,1)=S11/Z-C1**2
CoefMat(1,2)=S12/Z-C1*C2
CoefMat(1,3)=S13/Z-C1*C3
CoefMat(2,1)=CoefMat(1,2)
CoefMat(2,2)=S22/Z-C2**2
CoefMat(2,3)=S23/Z-C2*C3
CoefMat(3,1)=CoefMat(1,3)
CoefMat(3,2)=CoefMat(2,3)
CoefMat(3,3)=S33/Z-C3**2
Call Solve(CoefMat,DelC,DelX)
X1=X1+DelX(1)
X2=X2+DelX(2)
X3=X3+DelX(3)
If(Sqrt(DelX(1)**2+DelX(2)**2+DelX(3)**2).LT.Tolerance)GoTo 3
2 Continue
3 Continue
End
```

```
Subroutine Solve(A,B,X)
```

```
Real*4 A(3,3),B(3),X(3)
Do 2 i=1,3
C=Sqrt(A(i,1)**2+A(i,2)**2+A(i,3)**2)
Do 1 j=1,3
1 A(i,j)=A(i,j)/C
2 B(i)=B(i)/C
C=A(1,1)*A(2,1)+A(1,2)*A(2,2)+A(1,3)*A(2,3)
D=Sqrt(1-C**2)
Do 3 i=1,3
3 A(2,i)=(A(2,i)-C*A(1,i))/D
B(2)=(B(2)-C*B(1))/D
C1=A(1,1)*A(3,1)+A(1,2)*A(3,2)+A(1,3)*A(3,3)
C2=A(2,1)*A(3,1)+A(2,2)*A(3,2)+A(2,3)*A(3,3)
D=Sqrt(1-C1**2-C2**2)
Do 4 i=1,3
4 A(3,i)=(A(3,i)-C1*A(1,i)-C2*A(2,i))/D
B(3)=(B(3)-C1*B(1)-C2*B(2))/D
Do 5 i=1,3
5 X(i)=B(1)*A(1,i)+B(2)*A(2,i)+B(3)*A(3,i)
Return
End
```

Renorm 4

```
Program Renorm_4
Common Is(0:7,0:7),Ib(3,3),Nsweeps,Iran
Call Initialize
Nsweeps=100000
Xmin=0.437
Xmax=0.438
NumberOfRuns=6
NumberOfIters=7
DelX=(Xmax-Xmin)/(NumberOfRuns-1)
Do 2 Xstart=Xmin, Xmax, DelX
X1=Xstart
X2=0
X3=0
Write(1,100)X1,X2,X3
Do 1 I = 1, NumberOfIters
Call FindCorrFtns(X1,X2,X3,C1,C2,C3)
Call FindHamPars(C1,C2,C3,X1,X2,X3)
1 Write(1,100)X1,X2,X3
2 Write(1,*)
Close(1)
100 Format(3F12.6)
End
```

```
Subroutine FindCorrFtns(X1,X2,X3,C1,C2,C3)
Common Is(0:7,0:7),Ib(3,3),Nsweeps,Iran
Do 1 n=1,1000
1 Call Sweep(X1,X2,X3)
Sum1=0
Sum2=0
Sum3=0
Do 2 n=1,Nsweeps
Call Sweep(X1,X2,X3)
Call MakeBlockSpins
Sum1=Sum1+
#Ib(1,1)*Ib(1,2)+Ib(1,2)*Ib(1,3)+Ib(2,1)*Ib(2,2)+Ib(2,2)*Ib(2,3)+
#Ib(3,1)*Ib(3,2)+Ib(3,2)*Ib(3,3)+Ib(1,1)*Ib(2,1)+Ib(2,1)*Ib(3,1)+
```

```
#Ib(1,2)*Ib(2,2)+Ib(2,2)*Ib(3,2)+Ib(1,3)*Ib(2,3)+Ib(2,3)*Ib(3,3)
Sum2=Sum2+
#Ib(1,1)*Ib(2,2)+Ib(1,2)*Ib(2,1)+Ib(1,2)*Ib(2,3)+Ib(1,3)*Ib(2,2)+
#Ib(2,1)*Ib(3,2)+Ib(2,2)*Ib(3,1)+Ib(2,2)*Ib(3,3)+Ib(2,3)*Ib(3,2)
Sum3=Sum3+
#Ib(1,1)*Ib(1,2)*Ib(2,2)*Ib(2,1)+Ib(1,2)*Ib(1,3)*Ib(2,3)*Ib(2,2)+
#Ib(2,1)*Ib(2,2)*Ib(3,2)*Ib(3,1)+Ib(2,2)*Ib(2,3)*Ib(3,3)*Ib(3,2)
2 Continue
C1=Sum1/Nsweeps
C2=Sum2/Nsweeps
C3=Sum3/Nsweeps
Return
End
```

```
Subroutine MakeBlockSpins
Common Is(0:7,0:7),Ib(3,3),Nsweeps,Iran
Ib(1,1)=Iblock(Is(1,1)+Is(1,2)+Is(2,1)+Is(2,2))
Ib(1,2)=Iblock(Is(1,3)+Is(1,4)+Is(2,3)+Is(2,4))
Ib(1,3)=Iblock(Is(1,5)+Is(1,6)+Is(2,5)+Is(2,6))
Ib(2,1)=Iblock(Is(3,1)+Is(3,2)+Is(4,1)+Is(4,2))
Ib(2,2)=Iblock(Is(3,3)+Is(3,4)+Is(4,3)+Is(4,4))
Ib(2,3)=Iblock(Is(3,5)+Is(3,6)+Is(4,5)+Is(4,6))
Ib(3,1)=Iblock(Is(5,1)+Is(5,2)+Is(6,1)+Is(6,2))
Ib(3,2)=Iblock(Is(5,3)+Is(5,4)+Is(6,3)+Is(6,4))
Ib(3,3)=Iblock(Is(5,5)+Is(5,6)+Is(6,5)+Is(6,6))
Return
End
```

```
Function Iblock(K)
Common Is(0:7,0:7),Ib(3,3),Nsweeps,Iran
If(K.NE.0)then
Iblock=K/Iabs(K)
Return
End If
If(Ran(Iran).GT.0.5)then
Iblock=1
Else
Iblock=-1
End If
Return
End
```

```
Subroutine Sweep(X1,X2,X3)
Common Is(0:7,0:7),Ib(3,3),Nsweeps,Iran
Do 1 i=1,6
Do 1 j=1,6
DelE=2*Is(i,j)*(X1*(Is(i-1,j)+Is(i,j+1)+Is(i+1,j)+Is(i,j-1))+
#X2*(Is(i-1,j-1)+Is(i-1,j+1)+Is(i+1,j+1)+Is(i+1,j-1))+
#X3*(Is(i,j-1)*Is(i-1,j-1)*Is(i-1,j)+Is(i-1,j)*Is(i-1,j+1)*
#Is(i,j+1)+Is(i,j+1)*Is(i+1,j+1)*Is(i+1,j)+Is(i+1,j)*
#Is(i+1,j-1)*Is(i,j-1)))
If((DelE.LT.0).OR.(Ran(Iran).LT.Exp(-DelE)))Is(i,j)=-Is(i,j)
1 Continue
Return
End
```

```
Subroutine FindHamPars(C10,C20,C30,X1,X2,X3)
Real*4 CoefMat(3,3),DelC(3),DelX(3)
Integer*2 A,B,C,D,E,F,G,H,I
Data Tolerance /0.00001/
```

```
Do 2 Iter=1,100
W1=Exp(X1)
W2=Exp(X2)
W3=Exp(X3)
Z=0
S1=0
S2=0
S3=0
S11=0
S12=0
S13=0
S22=0
S23=0
S33=0
Do 1 A = -1, 1, 2
Do 1 B = -1, 1, 2
Do 1 C = -1, 1, 2
Do 1 D = -1, 1, 2
Do 1 E = -1, 1, 2
Do 1 F = -1, 1, 2
Do 1 G = -1, 1, 2
Do 1 H = -1, 1, 2
Do 1 I = -1, 1, 2
nn=A*B+B*C+D*E+E*F+G*H+H*I+A*D+D*G+B*E+E*H+C*F+F*I
nnn=A*E+B*D+B*F+C*E+D*H+E*G+E*I+F*H
nsq=A*B*E*D+B*C*F*E+D*E*H*G+E*F*I*H
Weight=W1**nn*W2**nnn*W3**nsq
S1=S1+Weight*nn
S2=S2+Weight*nnn
S3=S3+Weight*nsq
S11=S11+Weight*nn**2
S12=S12+Weight*nn*nnn
S13=S13+Weight*nn*nsq
S22=S22+Weight*nnn**2
S23=S23+Weight*nnn*nsq
S33=S33+Weight*nsq**2
Z=Z+Weight
1 Continue
C1=S1/Z
C2=S2/Z
C3=S3/Z
DelC(1)=C10-C1
DelC(2)=C20-C2
DelC(3)=C30-C3
CoefMat(1,1)=S11/Z-C1**2
CoefMat(1,2)=S12/Z-C1*C2
CoefMat(1,3)=S13/Z-C1*C3
CoefMat(2,1)=CoefMat(1,2)
CoefMat(2,2)=S22/Z-C2**2
CoefMat(2,3)=S23/Z-C2*C3
CoefMat(3,1)=CoefMat(1,3)
CoefMat(3,2)=CoefMat(2,3)
CoefMat(3,3)=S33/Z-C3**2
Call Solve(CoefMat,DelC,DelX)
X1=X1+DelX(1)
X2=X2+DelX(2)
X3=X3+DelX(3)
If(Sqrt(DelX(1)**2+DelX(2)**2+DelX(3)**2).LT.Tolerance)GoTo 3
2 Continue
3 Continue
```

```
Return
End
```

```
Subroutine Solve(A,B,X)
Real*4 A(3,3),B(3),X(3)
Do 2 i=1,3
C=Sqrt(A(i,1)**2+A(i,2)**2+A(i,3)**2)
Do 1 j=1,3
1 A(i,j)=A(i,j)/C
2 B(i)=B(i)/C
C=A(1,1)*A(2,1)+A(1,2)*A(2,2)+A(1,3)*A(2,3)
D=Sqrt(1-C**2)
Do 3 i=1,3
3 A(2,i)=(A(2,i)-C*A(1,i))/D
B(2)=(B(2)-C*B(1))/D
C1=A(1,1)*A(3,1)+A(1,2)*A(3,2)+A(1,3)*A(3,3)
C2=A(2,1)*A(3,1)+A(2,2)*A(3,2)+A(2,3)*A(3,3)
D=Sqrt(1-C1**2-C2**2)
Do 4 i=1,3
4 A(3,i)=(A(3,i)-C1*A(1,i)-C2*A(2,i))/D
B(3)=(B(3)-C1*B(1)-C2*B(2))/D
Do 5 i=1,3
5 X(i)=B(1)*A(1,i)+B(2)*A(2,i)+B(3)*A(3,i)
Return
End
```

```
Subroutine Initialize
Common Is(0:7,0:7),Ib(3,3),Nsweeps,Iran
Iran=99991
Open(Unit=1,File='Renorm_4.out')
Write(1,*)'     X1          X2          X3'
Write(1,*)
Do 1 i=1,6
Do 1 j=1,6
If(Ran(Iran).LT.0.5)then
Is(i,j)=1
Else
Is(i,j)=-1
End If
1 Continue
Do 2 i=0,7
Is(0,i)=0
Is(7,i)=0
Is(i,0)=0
2 Is(i,7)=0
End
```

Mathematical Appendix

A.1 PROOF OF THE CENTRAL LIMIT THEOREM

We first need an important formula involving the Fourier transform of the probability density associated with the sum of any number of independent random variables. Let $S = X + Y + Z$, where X, Y, and Z are independent. From Eq. (1.51), we see that the probability distribution associated with S is

$$P_S(s) = \int \delta(s - x - y - z) P_X(x) P_Y(y) P_Z(z) \, dx \, dy \, dz \qquad (A.1)$$

The Fourier transform of $P_S(s)$ is given by

$$
\begin{aligned}
\tilde{P}_S(k) &= \int e^{iks} P_S(s) \, ds \\
&= \int e^{iks} \delta(s - x - y - z) P_X(x) P_Y(y) P_Z(z) \, dx \, dy \, dz \, ds \\
&= \int e^{ik(x+y+z)} P_X(x) P_Y(y) P_Z(z) \, dx \, dy \, dz \\
&= \tilde{P}_X(k) \tilde{P}_Y(k) \tilde{P}_Z(k)
\end{aligned}
\qquad (A.2)
$$

The Fourier transform of the probability density is called the *characteristic function* associated with a random variable. Thus, for independent random variables, the characteristic function for their sum is the product of their characteristic functions. In the proof of the central limit theorem, the following easy generalization of this theorem will be used. The proof is left as an exercise for the reader.

Given N independent random variables $x_1, x_2, \ldots, x_N$, with probability densities, $P_1, P_2, \ldots, P_N$, if we define a random variable $T = \lambda(x_1 + \cdots + x_N)$ where $\lambda > 0$, then

$$\tilde{P}_T(k) = \tilde{P}_1(\lambda k) \cdots \tilde{P}_N(\lambda k) \qquad (A.3)$$

We are now ready to prove the theorem. We consider an infinite sequence of independent random variables $x_1, x_2, \ldots$ with probability densities $P_1(x_1), P_2(x_2), \ldots$. It is no serious restriction to assume that the average value of each of the separate variables vanishes.

$$\int P_n(x_n) x_n \, dx_n = 0 \qquad (A.4)$$

For some positive integer N we define the random variable $T = (x_1 + \cdots + x_N)/\sqrt{N}$. According to Eq. (A.3), the characteristic function associated with T is given by

$$\tilde{P}_T(k) = \tilde{P}_1(k/\sqrt{N}) \cdots \tilde{P}_N(k/\sqrt{N}) \tag{A.5}$$

We must now derive a certain inequality involving $\tilde{P}_n(k)$. It is a theorem of calculus that there exists a positive constant C such that

$$|e^{i\theta} - (1 + i\theta - \tfrac{1}{2}\theta^2)| \equiv |R(\theta)| < C|\theta|^3 \tag{A.6}$$

for any real number θ. Thus

$$e^{ikx} = 1 + ikx - \tfrac{1}{2}k^2x^2 + R(kx) \tag{A.7}$$

where $|R(kx)| < C|k^3||x^3|$. Using this and Eq. (A.4) we see that

$$\begin{aligned}
\tilde{P}_n(k) &= \int e^{ikx_n} P_n(x_n)\, dx_n \\
&= \int [1 + ikx_n - \tfrac{1}{2}k^2x_n^2 + R(kx_n)] P_n(x_n)\, dx_n \\
&= 1 - \tfrac{1}{2}k^2(\Delta x_n)^2 + r_n(k)
\end{aligned} \tag{A.8}$$

where $|r_n(k)| < C|k^3|\langle|x_n^3|\rangle$ and Δx_n is the uncertainty in the random variable x_n. From Eqs. (A.5) and (A.8) we find that

$$\begin{aligned}
\log \tilde{P}_T(k) &= \sum_{n=1}^{N} \log \tilde{P}_n(k/\sqrt{N}) \\
&= \sum_{n=1}^{N} \log\left(1 - \frac{1}{2N}k^2(\Delta x_n)^2 + r_n(k/\sqrt{N})\right)
\end{aligned} \tag{A.9}$$

As $N \to \infty$ we can use the expansion $\log(1 + \epsilon) \approx \epsilon$.

$$\log \tilde{P}_T(k) \approx -\frac{k^2}{2}\left(\frac{1}{N}\sum_{n=1}^{N}(\Delta x_n)^2\right) + \sum_{n=1}^{N} r_n(k/\sqrt{N}) \tag{A.10}$$

We now make two assumptions regarding the sequence of random variables. They are that

$$\lim_{N\to\infty}\left(\frac{1}{N}\sum_{n=1}^{N}(\Delta x_n)^2\right) \equiv a^2 < \infty \tag{A.11}$$

and

$$\lim_{N\to\infty}\left(\frac{1}{N^{3/2}}\sum_{n=1}^{N}\langle|x_n^3|\rangle\right) = 0 \tag{A.12}$$

Equation (A.12) guarantees that the remainder term can be ignored in the limit, giving, as $N \to \infty$

$$\log \tilde{P}_T(k) \to -\tfrac{1}{2}a^2k^2 \tag{A.13}$$

or

$$P_T(t) = \frac{1}{2\pi} \int \tilde{P}_T(k)e^{-ikt} \, dk$$

$$\rightarrow \frac{1}{2\pi} \int e^{-\frac{1}{2}a^2k^2 - ikt} \, dk \qquad (A.14)$$

$$= \frac{e^{-t^2/2a^2}}{\sqrt{2\pi}\, a}$$

The average of the first N random variables [that is $X = (x_1 + \cdots + x_N)/N$] is related to T by $X = T/\sqrt{N}$. The probability density associated with X is related to that associated with T by

$$P_X(x) \, dx = P_T(t) \, dt \qquad (A.15)$$

where $dx = dt/\sqrt{N}$. Thus, if N is large enough to allow the use of the asymptotic formula [Eq. (A.14)], then

$$P_X(x) = \sqrt{N} P_T(\sqrt{N}x)$$

$$= \sqrt{N/2\pi a^2}\, e^{-Nx^2/2a^2} \qquad (A.16)$$

which is the result we set out to prove.

A.2   STIRLING'S APPROXIMATION

$$\log N! = \log(1 \cdot 2 \cdots N) = \sum_{n=1}^{N} \log n \qquad (A.17)$$

In Fig. A.1, two curves are superimposed. The discontinuous curve is a graph of $\log n$, with n ranging from 1 to 40. The smooth curve is the function $\log(x)$. $\log N!$ is given by the area under the discontinuous curve. Stirling's approximation to $\log N!$ is defined as the area under the smooth approximating curve.

$$\log N! \approx \int_1^N \log(x) \, dx = N \log N - N + 1 \qquad (A.18)$$

Clearly, the ratio of the area under the smooth curve to the area under the discontinuous curve approaches one as $N \rightarrow \infty$. For large N, the 1 can be dropped. For more mathematically minded readers, the next section supplies a detailed evaluation of the error in Stirling's approximation and a generalization of the analysis to noninteger N.

A.3   THE GAMMA AND FACTORIAL FUNCTIONS

For any number N greater than -1, the factorial function $N!$ and the gamma function $\Gamma(N+1)$ are defined by

$$N! = \Gamma(N+1) = \int_o^\infty e^{-x}x^N \, dx \qquad (A.19)$$

By a partial integration, it can be seen that $N! = N(N-1)!$ or, equivalently, $\Gamma(N+1) = N\,\Gamma(N)$. For $N=0$, the integral is easily done, giving $0! = 1$. Therefore, for any positive integer

$$N! = \Gamma(N+1) = N(N-1)\cdots 2\times 1 \qquad (A.20)$$

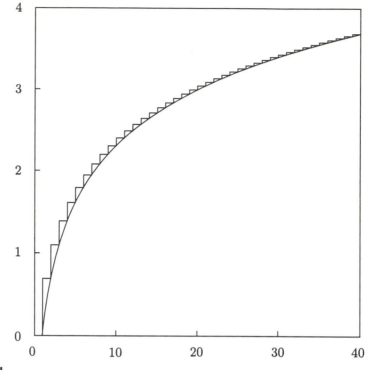

FIGURE A.1

Another important case is $\Gamma(\frac{1}{2}) = \sqrt{\pi}$. From Eq. (A.19), we see that

$$N! = \int_o^\infty e^{g(x)} \, dx \qquad (A.21)$$

where $g(x) = N \log x - x$. $g(x)$ is a function with a single maximum at $x = N$. The function approaches $-\infty$ at both limits of integration. The expansion to second order of $g(x)$ about its maximum yields

$$g(N + \xi) \approx N \log N - N - \xi^2/2N \qquad (A.22)$$

which gives the following approximation for $N!$, valid for large N.

$$N! \approx e^{N \log N - N} \int_{-N}^\infty e^{-\xi^2/2N} \, d\xi$$
$$\approx e^{N \log N - N}(2\pi N)^{1/2} \qquad (A.23)$$

From this we can easily obtain the following four-term approximation to $\log N!$.

$$\log N! \approx N \log N - N + \tfrac{1}{2} \log N + \tfrac{1}{2} \log 2\pi \qquad (A.24)$$

The use of the first two terms in this expansion is Stirling's approximation.

A.4   THE ERROR FUNCTION
The complementary error function is defined by the integral

$$\mathrm{erfc}(x) = \frac{2}{\sqrt{\pi}} \int_x^\infty e^{-t^2} \, dt \qquad (A.25)$$

For large values of x, the following expansion is useful.

$$\text{erfc}(x) = \frac{e^{-x^2}}{\sqrt{\pi}\, x} \sum_{m=0}^{\infty} \frac{(-1)^m (2m)!}{m!(2x)^{2m}}$$

$$= \frac{e^{-x^2}}{\sqrt{\pi}\, x} \left(1 - \frac{1}{2x^2} + \frac{3}{4x^4} - \cdots \right) \tag{A.26}$$

A.5 LAGRANGE PARAMETERS

The following is Lagrange's method of calculating the stationary points of a function $f(\mathbf{x})$ when the N-dimensional variable $\mathbf{x}$ is subjected to K constraints of the form $u_n(\mathbf{x}) = C_n$ ($n = 1, \ldots, K$), where the u_n are K known functions, the C_n are K given constants, and $K < N$. What we seek is a point $\mathbf{x}^o$ with the property that it satisfies the constraints and that for any infinitesimal $d\mathbf{x}$ satisfying $u_n(\mathbf{x}^o + d\mathbf{x}) = u_n(\mathbf{x}^o)$ ($n = 1, \ldots, K$) it is also true that $f(\mathbf{x}^o + d\mathbf{x}) = f(\mathbf{x}^o)$. That is, the function f is stationary within the set of points defined by the constraint equations.

We define a function $g(\mathbf{x}, \lambda)$ of $\mathbf{x}$ and the K *Lagrange parameters* $\lambda_1, \ldots, \lambda_K$ by

$$g(\mathbf{x}, \lambda) = f(\mathbf{x}) + \sum_{n=1}^{K} \lambda_n u_n(\mathbf{x}) \tag{A.27}$$

We next find the solution (or solutions) of the following system of $N + K$ equations in $N + K$ unknowns. We call the solution $\mathbf{x}^o$ and λ^o.

$$\frac{\partial g(\mathbf{x}^o, \lambda^o)}{\partial x_k^o} = 0, \qquad k = 1, \ldots, N \tag{A.28}$$

and

$$u_n(\mathbf{x}^o) = C_n, \qquad n = 1, \ldots, K \tag{A.29}$$

By Eq. (A.28), we know that $g(\mathbf{x}^o + d\mathbf{x}, \lambda) = g(\mathbf{x}^o, \lambda)$ for *any* $d\mathbf{x}$. But, by the definition of $g(\mathbf{x}, \lambda)$, if we choose a $d\mathbf{x}$ that does not change the value of any of the constraint functions, then $f(\mathbf{x}^o + d\mathbf{x}) = f(\mathbf{x}^o)$. Thus $\mathbf{x}^o$ is a solution of the constrained variational problem.

A.6 THE N-DIMENSIONAL SPHERE

The volume of an N-dimensional sphere of radius R is given by the integral

$$V_N(R) = \int \theta(R^2 - x^2)\, d^N x \tag{A.30}$$

where $x^2 = \sum x_n^2$ and θ is the unit step function. Using the transformation of variables $y_n = x_n/R$ in Eq. (A.30) and making use of the fact that $\theta(\lambda x) = \theta(x)$ for any positive value of λ, one obtains

$$\int \theta(R^2 - x^2)\, d^N x = R^N \int \theta(1 - y^2)\, d^N y = R^N V_N(1) \tag{A.31}$$

Differentiating Eq. (A.31) and using the fact that $d\theta(R^2 - x^2)/dR = 2R\,\delta(R^2 - x^2)$, where $\delta(x)$ is the Dirac delta function, gives

$$2R \int \delta(R^2 - x^2)\, d^N x = N V_N(1) R^{N-1} \tag{A.32}$$

We now multiply both sides by $e^{-R^2} dR$ and integrate from 0 to ∞ to get

$$\int d^N x \int_0^\infty e^{-R^2} \delta(R^2 - x^2) \, dR^2 = \int e^{-\sum x_n^2} \, d^N x$$
$$= N V_N(1) \int_0^\infty e^{-R^2} R^{N-1} dR \qquad (A.33)$$

The integral on the second line is easily calculated in terms of the factorial function. The integral on the right-hand side of the first line factors into a product of N simple Gaussian integrals, each having the value $\pi^{1/2}$. Thus

$$\pi^{N/2} = V_N(1)(N/2)! \qquad (A.34)$$

where, for odd N, the factorial function $(N/2)!$ is defined in terms of the gamma function. This formula for $V_N(1)$ implies that

$$V_N(R) = \frac{\pi^{N/2}}{(N/2)!} R^N \qquad (A.35)$$

The surface area of an N-dimensional sphere is the derivative with respect to R of its volume. Thus

$$S_N(R) = \frac{dV_N(R)}{dR} = \frac{N}{R} V_N(R) \qquad (A.36)$$

A.7 NORMAL MODES AND SMALL VIBRATIONS

We consider a classical system of N identical particles, each of mass m. The complete configuration of the system can be defined by $K = 3N$ scalar coordinates, $(x_1, x_2, \ldots, x_K) \equiv (x)$. The potential energy of the system is some function of the coordinates, $V(x_1, x_2, \ldots, x_K)$. Let $(x^o) = (x_1^o, x_2^o, \ldots, x_K^o)$ be the configuration of minimum potential energy (the mechanical equilibrium configuration). Then

$$\frac{\partial V(x^o)}{\partial x_i} = 0, \qquad i = 1, \ldots, K \qquad (A.37)$$

We introduce K new coordinates, $\xi_i = x_i - x_i^o$ $(i = 1, \ldots, K)$ that describe the displacement of the system particles from their equilibrium positions. If we restrict our analysis to small vibrations about the equilibrium configuration, then we can expand $V(x)$ in a Taylor expansion about the point (x^o). Noting that the first derivatives all vanish at (x^o), we get

$$V(x^o + \xi) = V(x^o) + \frac{1}{2} \sum_{i,j} V_{ij} \xi_i \xi_j \qquad (A.38)$$

where $V_{ij} = \partial^2 V(x^o)/\partial x_i \partial x_j$. We are free to add any constant to the potential energy function without changing the equations of motion. We can thus assume that $V(x^o) = 0$. The Lagrangian function is defined as $T - V$. Therefore, for small vibrations,

$$L = \tfrac{1}{2} m \sum_{i=1}^K \dot{\xi}_i^2 - \frac{1}{2} \sum_{i,j} V_{ij} \xi_i \xi_j \qquad (A.39)$$

The crucial part of this analysis is to realize that V_{ij} is a symmetric $K \times K$ matrix. It therefore has K orthogonal and normalized eigenvectors, with corresponding eigenvalues. That is, there are K nonzero solutions of the set of linear equations

$$\sum_{j=1}^{K} V_{ij} u_j^k = \lambda_k u_i^k, \qquad i = 1, \ldots, K \tag{A.40}$$

and those eigenvectors satisfy the orthogonality conditions

$$\sum_{i=1}^{K} u_i^k u_i^l = \delta_{kl} = \begin{cases} 0, & k \neq l \\ 1, & k = l \end{cases} \tag{A.41}$$

We now make another coordinate transformation, to K *normal-mode coordinates* $Q_1, Q_2, \ldots, Q_K$, defined by

$$\xi_i = \sum_{k=1}^{K} u_i^k Q_k \tag{A.42}$$

and note that $\dot{\xi}_i = \sum u_i^k \dot{Q}_k$. Writing the Lagrangian in normal-mode coordinates, we see that

$$\begin{aligned} L &= \tfrac{1}{2} m \sum_{k,l,i} u_i^k u_i^l \dot{Q}_k \dot{Q}_l - \frac{1}{2} \sum_{k,l,i,j} V_{ij} u_j^k u_i^l Q_k Q_l \\ &= \tfrac{1}{2} m \sum_{k,l} \delta_{kl} \dot{Q}_k \dot{Q}_l - \frac{1}{2} \sum_{k,l,i} \lambda_k u_i^k u_i^l Q_k Q_l \\ &= \frac{1}{2} \sum_{k} \dot{Q}_k^2 - \frac{1}{2} \sum_{k} \lambda_k Q_k^2 \end{aligned} \tag{A.43}$$

The momentum canonical to the normal-mode coordinate Q_k is

$$P_k = \frac{\partial L}{\partial \dot{Q}_k} = m \dot{Q}_k \tag{A.44}$$

Writing the Hamiltonian function in terms of the normal-mode coordinates and momenta gives

$$H = T + V = \sum_{k=1}^{K} \left(\frac{P_k^2}{2m} + \tfrac{1}{2} \lambda_k Q_k^2 \right) \tag{A.45}$$

This is exactly the Hamiltonian for a system composed of K noninteracting harmonic oscillators with angular frequencies $\omega_1, \omega_2, \ldots, \omega_K$ given by

$$\omega_k = \sqrt{\frac{\lambda_k}{m}} \tag{A.46}$$

The assumption that the configuration (x^o) gives a minimum for the potential energy $V(x)$ guarantees that all the eigenvalues λ_k, are nonnegative and, therefore, that all the angular frequencies are real.

A.8   THE TRANSFORMATION OF MULTIPLE INTEGRALS

Consider the two-dimensional integral of a function $f(x_1, x_2)$ over some region R in the x_1–x_2 plane.

$$I = \int_R f(x_1, x_2) \, dx_1 \, dx_2 \tag{A.47}$$

In this section we will describe how to express the integral I in terms of two new variables, y_1 and y_2, that are related to the original variables by some known transformation equations, $y_1(x_1, x_2)$ and $y_2(x_1, x_2)$.

1. First we must express the function $f(x_1, x_2)$ in terms of the new variables. We do that by defining a new function $F(y_1, y_2)$ by

$$F(y_1, y_2) = f(x_1(y_1, y_2), x_2(y_1, y_2)) \qquad (A.48)$$

2. Next we must transform the region of integration. The region of integration in the y_1–y_2 plane can be defined by

$$(y_1, y_2) \in Q \quad \text{iff} \quad (x_1(y_1, y_2), x_2(y_1, y_2)) \in R \qquad (A.49)$$

3. Finally the volume element $dx_1 \, dx_2$ must be transformed into the new volume element $dy_1 \, dy_2$. That is done by the replacement

$$dx_1 \, dx_2 \rightarrow |J(y_1, y_2)| \, dy_1 \, dy_2 \qquad (A.50)$$

where the *Jacobian determinant* J is the determinant of the partial derivatives.

$$J(y_1, y_2) = \begin{vmatrix} \partial x_1/\partial y_1 & \partial x_2/\partial y_1 \\ \partial x_1/\partial y_2 & \partial x_2/\partial y_2 \end{vmatrix} \qquad (A.51)$$

The final result is

$$I = \int_Q F(y_1, y_2)|J(y_1, y_2)| \, dy_1 \, dy_2 \qquad (A.52)$$

The generalization to three or more variables is very simple. Steps (1) and (2) are unchanged, except for the fact that they involve three or more variables. The only change in step (3) is that, for a transformation involving N variables $(x_1, x_2, \ldots, x_N) \rightarrow (y_1, y_2, \ldots, y_N)$, the Jacobian determinant becomes an $N \times N$ determinant.

$$J = \begin{vmatrix} \partial x_1/\partial y_1 & \cdots & \partial x_N/\partial y_N \\ \cdots & \cdots & \cdots \\ \partial x_1/\partial y_N & \cdots & \partial x_N/\partial y_N \end{vmatrix} \qquad (A.53)$$

A simple but important example is the transformation of an integral involving the coordinates of two equal-mass particles

$$I = \int_o^L dx_1 \int_o^L dx_2 \, f(x_1, x_2) \qquad (A.54)$$

to center-of-mass and relative coordinates, defined by

$$y_1 = (x_1 + x_2)/2 \quad \text{and} \quad y_2 = x_1 - x_2 \qquad (A.55)$$

The inverse transformation is

$$x_1 = y_1 + y_2/2 \quad \text{and} \quad x_2 = y_1 - y_2/2 \qquad (A.56)$$

1. The new function is

$$F(y_1, y_2) = f(y_1 + y_2/2, y_1 - y_2/2) \qquad (A.57)$$

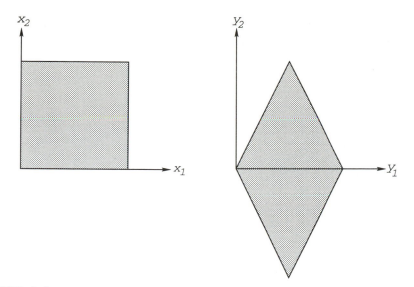

FIGURE A.2

2. The vertices of the rectangle in the x_1–x_2 plane are $(x_1, x_2) = (0,0)$, $(L,0)$, $(0, L)$, and (L, L). These are transformed into $(y_1, y_2) = (0, 0)$, $(L/2, L)$, $(L/2, -L)$, and $(L, 0)$. (See Fig. A.2.)

3. The Jacobian determinant is

$$\begin{vmatrix} \partial x_1/\partial y_1 & \partial x_2/\partial y_1 \\ \partial x_1/\partial y_2 & \partial x_2/\partial y_2 \end{vmatrix} = \begin{vmatrix} 1 & 1 \\ 1/2 & -1/2 \end{vmatrix} = -1 \tag{A.58}$$

and therefore

$$dx_1 \, dx_2 \rightarrow |-1| \, dy_1 \, dy_2 = dy_1 \, dy_2 \tag{A.59}$$

which completes the transformation

$$\int_o^L dx_1 \int_o^L dx_2 \, f(x_1, x_2) = \int \int_{R'} F(y_1, y_2) \, dy_1 \, dy_2 \tag{A.60}$$

A.9   INTEGRALS WITH LARGE EXPONENTS

We want to develop an approximation to the integral

$$I = \int_a^b e^{\lambda f(x)} \, dx \tag{A.61}$$

that is valid for large values of λ. We assume that the function $f(x)$ has a single maximum at $x = X$ and that $a < X < b$. For very large λ, the ratio $e^{\lambda f(x)}/e^{\lambda f(X)}$ is extremely small if $x \neq X$. Thus, the integral is dominated by the values of the integrand close to $x = X$. We may therefore extend the range of integration from (a, b) to $(-\infty, \infty)$ and expand $f(x)$ about its maximum, retaining only the first two nonvanishing terms. [The linear term vanishes because $f(x)$ is stationary at X.]

$$f(x) \approx f(X) - \tfrac{1}{2} a \xi^2 \tag{A.62}$$

where $\xi = x - X$ and $a = -\partial^2 f(X)/\partial X^2$. a must be positive, since $f(x)$ has a *maximum* at $x = X$. Therefore, for large λ,

$$
\begin{aligned}
I &\approx e^{\lambda f(X)} \int_{-\infty}^{\infty} e^{-\frac{1}{2}\lambda a \xi^2} \, d\xi \\
&= \left(\frac{2\pi}{a\lambda}\right)^{1/2} e^{\lambda f(X)}
\end{aligned}
\tag{A.63}
$$

where the Gaussian integral can be found in the Table of Integrals.

A.10   THE CLASSICAL LIMIT

We want to show that, in the limit $\hbar \to 0$, the number of N-particle quantum states with energy less than E approaches the phase-space volume enclosed by the classical energy surface $\omega(E)$ divided by $h^K N!$. This is a difficult theorem, particularly for any reader who is not secure in the mathematics of quantum mechanics. Therefore, to simplify the analysis slightly, we will treat a system of N one-dimensional particles. Nothing really depends in any important way on the dimensionality of the system; treating a one-dimensional system simply allows us to avoid a number of shifts, back and forth, between scalar coordinates $(x_1, \ldots, x_K)$ and vector coordinates $(\mathbf{r}_1, \ldots, \mathbf{r}_N)$. In fact, in this section we will use boldface letters to distinguish operators, rather than vectors, from ordinary scalars.

The Hamiltonian operator for the system is

$$
\mathbf{H} = \mathbf{T} + \mathbf{V} = \sum_{1}^{N} \frac{\mathbf{p}_i^2}{2m} + V(x_1, \ldots, x_N)
\tag{A.64}
$$

where the coordinates and momenta satisfy the Heisenberg commutation relations

$$
x_i \mathbf{p}_j - \mathbf{p}_j x_i = i\hbar \delta_{ij}
\tag{A.65}
$$

We are interested in the classical limit of the function $\sum_n \theta(E - E_n)$. However, it will help if we first determine the classical limit of a simpler function, namely $F(\alpha) = \sum_n \exp(\alpha E_n)$, where α is any (possibly complex) constant. First we recall a few facts from quantum mechanics. 1. The exponential of any operator is the operator defined by the power series expansion,

$$
e^{\mathbf{A}} = 1 + \mathbf{A} + \tfrac{1}{2}\mathbf{A}^2 + \cdots
\tag{A.66}
$$

2. If $|\psi\rangle$ is an eigenfunction of $\mathbf{A}$ with eigenvalue a, then

$$
e^{\mathbf{A}}|\psi\rangle = e^a|\psi\rangle
\tag{A.67}
$$

This is easy to verify, using the expansion of $e^{\mathbf{A}}$. 3. The *trace* of an operator, written $\mathrm{tr}(\mathbf{A})$, is defined as the sum of its diagonal elements with respect to any complete set of functions.

$$
\mathrm{tr}(\mathbf{A}) = \sum_n \langle \phi_n | \mathbf{A} | \phi_n \rangle
\tag{A.68}
$$

The value of $\mathrm{tr}(\mathbf{A})$ does not depend on the set of functions used to evaluate it.

Using these facts, we can see that

$$
\begin{aligned}
\mathrm{tr}(e^{\alpha \mathbf{H}}) &= \sum_n \langle E_n | e^{\alpha \mathbf{H}} | E_n \rangle \\
&= \sum_n \langle E_n | E_n \rangle e^{\alpha E_n} \\
&= \sum_n e^{\alpha E_n} \\
&= F(\alpha)
\end{aligned}
\tag{A.69}
$$

In the limit $\hbar \to 0$, the right hand side of Eq. (A.65) can be neglected. Then $\mathbf{T}$ and $\mathbf{V}$ can be treated as commuting operators. Thus in that limit

$$
e^{\alpha \mathbf{H}} = e^{\alpha \mathbf{T} + \alpha \mathbf{V}} \approx e^{\alpha \mathbf{V}} e^{\alpha \mathbf{T}}
\tag{A.70}
$$

Because the trace of an operator is independent of the set of functions that is used to evaluate it, we can, in evaluating the trace of $e^{\alpha \mathbf{H}}$, use the set of eigenfunctions of $\mathbf{T}$ rather than $\mathbf{H}$. The reason why we do this is that the eigenfunctions of the kinetic energy are known, simple functions, while those of the complete Hamiltonian are usually not known for a system of interacting particles. In writing the kinetic energy eigenfunctions we will temporarily ignore any symmetry requirements on the wave functions (Bose–Einstein or Fermi–Dirac). Then, if we assume that the system is in a periodic box of length L, a correctly normalized kinetic energy eigenfunction can be constructed by choosing any N allowed momentum values, $p_1, \ldots, p_N$, and writing the N-particle plane wave function

$$
|p_1, \ldots, p_N \rangle = L^{-N/2} \exp\left(\frac{i}{\hbar} (p_1 x_1 + \cdots + p_N x_N) \right)
\tag{A.71}
$$

With this function

$$
\mathbf{T} | p_1, p_2, \ldots, p_N \rangle = \sum_1^N \frac{\mathbf{p}_i^2}{2m} | p_1, p_2, \ldots, p_N \rangle = \left(\sum_1^N p_i^2 / 2m \right) | p_1, p_2, \ldots, p_N \rangle
\tag{A.72}
$$

and

$$
e^{\alpha \mathbf{V}} e^{\alpha \mathbf{T}} | p_1, p_2, \ldots, p_N \rangle = \exp\left[\alpha \sum p_i^2 / 2m + \alpha V \right] | p_1, p_2, \ldots, p_N \rangle
\tag{A.73}
$$

Using these functions to evaluate the trace, we get

$$
\begin{aligned}
F(\alpha) &= \sum_{\{p\}} \langle p_1, p_2, \ldots, p_N | e^{\alpha \mathbf{H}} | p_1, p_2, \ldots, p_N \rangle \\
&= L^{-N} \sum_{\{p\}} \int d^N x \, e^{-(i/\hbar) \sum p_i x_i} \left(\exp\left[\alpha \sum p_i^2 / 2m + \alpha V \right] \right) e^{(i/\hbar) \sum p_i x_i} \\
&= L^{-N} \sum_{\{p\}} \int d^N x \, e^{\alpha \sum p_i^2 / 2m + \alpha V}
\end{aligned}
\tag{A.74}
$$

The sum, indicated by $\{p\}$, is over all allowed values of $p_1, p_2, \ldots, p_N$. Each sum over momentum eigenvalues can be converted to an integral by using the fact that, in a periodic box of length L, the density of momentum eigenvalues is L/h (see Fig. 2.5). $F(\alpha)$ may then be written as

$$F(\alpha) = \frac{1}{h^N} \int d^N p \int d^N x \, e^{\alpha H(x,p)} \qquad (A.75)$$

where $H(x,p)$ is the classical Hamiltonian function $\sum p_i^2/2m + V(x)$. We must now correct for the fact that we have ignored the wave function symmetry. First we notice that, in the classical limit, the density of momentum eigenvalues L/h becomes greater and greater, while the number of particles N remains fixed. Therefore, the cases in which two or more momenta are equal become a smaller and smaller fraction of the sum over all N momentum values and can be ignored. If the N momenta $p_1, p_2, \ldots, p_N$ are different, then the correction we need to make is simple. We have overcounted the number of momentum eigenstates by a factor of $N!$ and so we must divide the result in Eq. (A.75) by $N!$ in order to obtain the trace only over properly symmetrized states. This gives

$$F(\alpha) = \frac{1}{N!h^N} \int d^N p \, d^N x \, e^{\alpha H(x,p)} \qquad (A.76)$$

To get the classical limit for $\sum \theta(E - E_n)$ we use the fact that any function $f(x)$ can be written in terms of its Fourier transform $\tilde{f}(k)$.

$$f(x) = \int dk \, e^{ikx} \tilde{f}(k) \qquad (A.77)$$

Then

$$\sum_n f(E_n) = \int dk \sum_n e^{ikE_n} \tilde{f}(k)$$
$$\xrightarrow[\hbar \to 0]{} \frac{1}{N!h^N} \int d^N x \, d^N p \int dk \, e^{ikH(x,p)} \tilde{f}(k) \qquad (A.78)$$
$$= \frac{1}{N!h^N} \int d^N x \, d^N p \, f\big(H(x,p)\big)$$

Specializing this result to the function $f(x) = \theta(E - x)$ gives the result we need.

$$\sum_n \theta(E - E_n) = \frac{1}{N!h^N} \int \theta\big(E - H(x,p)\big) \, d^N x \, d^N p \qquad (A.79)$$

A.11   A THEOREM IN ELECTROSTATICS

The relationship between the electric field $\mathbf{E}$ and the polarization field $\mathbf{P}$ within a dielectric is

$$\nabla \cdot \left(\mathbf{E} + \frac{1}{\epsilon_o} \mathbf{P} \right) = \rho_{\text{free}} \qquad (A.80)$$

where ρ_{free} is the density of free charge. Thus, if ρ_{free} is kept fixed but $\mathbf{E}(\mathbf{r})$ and $\mathbf{P}(\mathbf{r})$ are changed to $\mathbf{E}(\mathbf{r}) + \Delta\mathbf{E}(\mathbf{r})$ and $\mathbf{P}(\mathbf{r}) + \Delta\mathbf{P}(\mathbf{r})$, then

$$\nabla \cdot \left(\Delta\mathbf{E} + \frac{1}{\epsilon_o} \Delta\mathbf{P} \right) = 0 \qquad (A.81)$$

Since $\nabla \times \mathbf{E} = 0$, we can express $\mathbf{E}$ in terms of a scalar potential by $\mathbf{E} = -\nabla\phi(\mathbf{r})$. The needed steps in transforming the integral $\epsilon_o \int \mathbf{E} \cdot \Delta\mathbf{E}\, d^3\mathbf{r}$ are explained following the equation.

$$
\begin{aligned}
\epsilon_o \int_{E_3} \mathbf{E} \cdot \Delta\mathbf{E}\, d^3\mathbf{r} &= -\epsilon_o \int_{E_3} \nabla\phi \cdot \Delta\mathbf{E}\, d^3\mathbf{r} \\
&= -\epsilon_o \int_{E_3} \nabla \cdot (\phi\Delta\mathbf{E})\, d^3\mathbf{r} + \epsilon_o \int_{E_3} \phi\nabla \cdot (\Delta\mathbf{E})\, d^3\mathbf{r} \\
&= \epsilon_o \int_{E_3} \phi\nabla \cdot (\Delta\mathbf{E})\, d^3\mathbf{r} \\
&= -\int_{E_3} \phi\nabla \cdot (\Delta\mathbf{P})\, d^{\mathbf{r}} \\
&= -\int_{E_3} \nabla \cdot (\phi\Delta\mathbf{P})\, d^3\mathbf{r} + \int_{E_3} \nabla\phi \cdot \Delta\mathbf{P}\, d^3\mathbf{r} \\
&= -\int_{V_S} \mathbf{E} \cdot \Delta\mathbf{P}\, d^3\mathbf{r}
\end{aligned}
\tag{A.82}
$$

In going from line 1 to line 2, the vector identity $\nabla \cdot (\phi\mathbf{A}) = \nabla\phi \cdot \mathbf{A} + \phi\nabla \cdot \mathbf{A}$ was used. The first term in line 2 is zero by Gauss's theorem. Line 3 $\rightarrow$ line 4 uses Eq. (A.81). Line 4 $\rightarrow$ line 5 uses the vector identity again, and line 6 follows from Gauss's theorem and the fact that $\mathbf{P} = 0$ in free space.

A.12   A THEOREM IN MAGNETOSTATICS

The relationship between the magnetic field $\mathbf{B}$ and the magnetization field $\mathbf{M}$ within any substance is

$$
\nabla \times (\mathbf{B} - \mu_o\mathbf{M}) = \mu_o(\mathbf{J} + \nabla \times \mathbf{M}_{\text{ext}})
\tag{A.83}
$$

where $\mathbf{J}$ is the electric current density and $\mathbf{M}_{\text{ext}}$ is the magnetization of any fixed external magnets. If $\mathbf{J}$ and $\mathbf{M}_{\text{ext}}$ are kept fixed but $\mathbf{B}$ and $\mathbf{M}$ are changed by the amounts $\Delta\mathbf{B}(\mathbf{r})$ and $\Delta\mathbf{M}(\mathbf{r})$, then

$$
\nabla \times (\Delta\mathbf{B} - \mu_o\Delta\mathbf{M}) = 0
\tag{A.84}
$$

Since $\nabla \cdot \mathbf{B} = 0$, we can express $\mathbf{B}$ in terms of a vector potential by $\mathbf{B} = \nabla \times \mathbf{A}(\mathbf{r})$. The needed steps in transforming the integral $\mu_o^{-1} \int \mathbf{B} \cdot \Delta\mathbf{B}\, d^3\mathbf{r}$ are explained following the equation.

$$
\begin{aligned}
\frac{1}{\mu_o} \int_{E_3} \mathbf{B} \cdot \Delta\mathbf{B}\, d^3\mathbf{r} &= \frac{1}{\mu_o} \int_{E_3} (\nabla \times \mathbf{A}) \cdot \Delta\mathbf{B}\, d^3\mathbf{r} \\
&= \frac{1}{\mu_o} \int_{E_3} \nabla \cdot (\mathbf{A} \times \Delta\mathbf{B})\, d^3\mathbf{r} + \frac{1}{\mu_o} \int_{E_3} \mathbf{A} \cdot \nabla \times (\Delta\mathbf{B})\, d^3\mathbf{r} \\
&= \int_{E_3} \mathbf{A} \cdot \nabla \times (\Delta\mathbf{M})\, d^3\mathbf{r} \\
&= \int_{E_3} (\nabla \times \mathbf{A}) \cdot \Delta\mathbf{M}\, d^3\mathbf{r} - \int_{E_3} \nabla \cdot (\mathbf{A} \times \Delta\mathbf{M})\, d^3\mathbf{r} \\
&= \int_{V_S} \mathbf{B} \cdot \Delta\mathbf{M}\, d^3\mathbf{r}
\end{aligned}
\tag{A.85}
$$

Line 1 $\rightarrow$ line 2 uses the vector identity $\nabla \cdot (\mathbf{A} \times \mathbf{B}) = (\nabla \times \mathbf{A}) \cdot \mathbf{B} - \mathbf{A} \cdot (\nabla \times \mathbf{B})$. Line 2 $\rightarrow$ line 3 uses Gauss's theorem and Eq. (A.84). Line 3 $\rightarrow$ line 4 uses vector identity again and line 4 $\rightarrow$ line 5 uses Gauss's theorem and the fact that $\Delta \mathbf{M} = 0$ outside the sample.

A.13   EXPANSIONS OF FERMI–DIRAC INTEGRALS

We want to obtain a power series expansion in the temperature $\tau = kT$ of an integral of the form

$$I = \int_o^\infty \frac{g(\varepsilon)}{e^{(\varepsilon-\mu)/\tau} + 1} d\varepsilon \qquad (A.86)$$

At $\tau = 0$ the integral has the value

$$I_o = \int_o^\mu g(\varepsilon) \, d\varepsilon \qquad (A.87)$$

The difference between I and I_o, which we call I_1, may be written as

$$I_1 = I - I_o = \int_o^\infty \left[\frac{1}{e^{(\varepsilon-\mu)/\tau} + 1} - \theta\left(\frac{\mu - \varepsilon}{kT}\right) \right] g(\varepsilon) \, d\varepsilon \qquad (A.88)$$

We now make a transformation from the variable ε to a variable $x = (\varepsilon - \mu)/\tau$, noting that $d\varepsilon = \tau \, dx$.

$$I_1 = kT \int_{-\mu/kT}^\infty \left(\frac{1}{e^x + 1} - \theta(-x) \right) g(\mu + kTx) \, dx \qquad (A.89)$$

The function $[(e^x + 1)^{-1} - \theta(-x)]$ becomes exponentially small if $|x|$ is much larger than one. Therefore the lower limit of integration, for small T, may be extended to $-\infty$. The function, $g(\mu + kTx)$ can be expanded in a power series in the small quantity kTx, giving

$$I_1 = \sum_{n=0}^\infty (kT)^{n+1} \frac{d^n g(\mu)}{d\mu^n} J_n \qquad (A.90)$$

where

$$J_n = \int_{-\infty}^\infty \left[\frac{1}{e^x + 1} - \theta(-x) \right] x^n dx \qquad (A.91)$$

If we let $f(x) = (e^x + 1)^{-1} - \theta(-x)$, then it is easy to see that, for $x > 0$,

$$f(-x) = \frac{1}{e^{-x} + 1} - 1 = -\frac{1}{e^x + 1} = -f(x) \qquad (A.92)$$

Thus the factor in large parentheses is an odd function of x. This implies that $J_n = 0$ for even n and, for odd n,

$$J_n = 2 \int_o^\infty \frac{x^n}{e^x + 1} dx = 2(1 - 2^{-n})n!\zeta(n + 1) \qquad (A.93)$$

where $\zeta(x)$ is the Riemann zeta function. Using the special values of ζ given in the Table of Integrals, we get the following expansion.

$$I = \int_o^\mu g(\varepsilon) \, d\varepsilon + \frac{\pi^2}{6} g'(\mu)(kT)^2 + \frac{7\pi^4}{360} g'''(\mu)(kT)^4 + \cdots \qquad (A.94)$$

A.14   TABLE OF INTEGRALS

$$\int_0^\infty e^{-ax^2} x^{2n}\, dx = \frac{1 \times 3 \cdots (2n-1)}{2^{n+1} a^{(2n+1)/2}} \sqrt{\pi} \qquad \int_0^\infty e^{-ax^2} x^{2n+1}\, dx = \frac{n!}{2a^{n+1}}$$

$$\int_0^\infty \frac{x^s}{e^x - 1}\, dx = s!\zeta(s+1) \qquad s > 0 \qquad \int_0^\infty \frac{x^s}{e^x + 1}\, dx = (1 - 2^{-s})s!\zeta(s+1)$$

$$\int_0^\infty \frac{dx}{e^x + 1} = \log 2 \qquad\qquad \int_0^\infty e^{-x} x^u\, dx = \Gamma(u+1)$$

Special Values of the Riemann Zeta Function and the Gamma Function

$\zeta(1) = \infty$	$\zeta(2) = \dfrac{\pi^2}{6}$	$\zeta(3) = 1.202$	$\zeta(4) = \dfrac{\pi^4}{90}$
$\zeta(5) = 1.037$	$\zeta(6) = \dfrac{\pi^6}{945}$	$\zeta(3/2) = 2.612$	$\zeta(5/2) = 1.341$
$\Gamma(u+1) = u\Gamma(u)$	$\Gamma(N) = (N-1)!$	$\Gamma(1/2) = \sqrt{\pi}$	

A.15   PHYSICAL CONSTANTS AND CONVERSIONS

c	2.998×10^8	m/s	Speed of light
e	1.602×10^{-19}	C	Electronic charge
$\hbar$	1.055×10^{-34}	J s	Planck's constant
h	6.626×10^{-34}	J s	Planck's constant
k	1.381×10^{-23}	J/K	Boltzmann's constant
u	1.661×10^{-27}	kg	Atomic mass unit
m_e	9.1095×10^{-31}	kg	Electron mass
N_A	6.022×10^{23}		Avogadro's number
R	8.314	J/mol K	Molar gas constant
ϵ_o	8.854×10^{-12}	F/m	Permittivity of vacuum
μ_o	$4\pi \times 10^{-7}$	H/m	Permeability of vacuum
Å	10^{-10}	m	Ångstrom

Energy   1 cal $= 4.186$ J

 1 eV $= 1.602 \times 10^{-19}$ J

Pressure   1 atm $= 1.013 \times 10^5$ N/m^2

 1 bar $= 10^5$ N/m^2

 1 mm Hg $= 1.333 \times 10^5$ N/m^2

Temperature $0°$ C $= 273.15$ K

STP $= 0°$ C and 1 atm n(STP) $= 2.69 \times 10^{25}$ part./m^3

GENERAL REFERENCES

D. Chandler, *Introduction to Modern Statistical Mechanics* (Oxford Univ. Press, 1987).

C. Kittel, *Thermal Physics*, (John Wiley and Sons, 1969)

F. Reif, *Fundamentals of Statistical and Thermal Physics* (McGraw–Hill Book Co., 1965).

O. Penrose, *Foundations of Statistical Mechanics* (Pergamon Press, 1970).

F. Mandl, *Statistical Physics* (John Wiley and Sons, 1971).

G. H. Wannier, *Statistical Physics* (Dover, 1966).

H. B. Callen, *Thermodynamics and an Introduction to Thermostatistics* (John Wiley and Sons, 1985).

E. Fermi, *Thermodynamics* (Dover, 1956).

PROBABILITY THEORY

Y. A. Rozanov, *Probability Theory, A Concise Course* (Dover, 1969).

W. Feller, *An Introduction to Probability Theory and Its Applications* (John Wiley and Sons, 1961).

KINETIC THEORY OF IDEAL GASES

R. D. Present, *Kinetic Theory of Gases* (McGraw–Hill Book Co., 1958).

S. G. Brush, *The Kind of Motion We Call Heat* (North Holland, 1976).

LIQUID HELIUM

J. Wilks and D. S. Betts, *An Introduction to Liquid Helium* (Oxford Univ. Press, 1987).

MONTE CARLO CALCULATIONS

K. Binder and D. W. Heerman, *Monte Carlo Simulations in Statistical Physics: An Introduction* (Springer Verlag, 1992).

M. H. Kalos and P. A. Whitlock, *Monte Carlo Methods* (John Wiley and Sons, 1986).

CRITICAL PHENOMENA AND RENORMALIZATION THEORY

J. J. Binney, N. J. Dowrick, A. J. Fisher, and M. E. J. Newman, *The Theory of Critical Phenomena* (Oxford Univ. Press, 1992).

R. J. Creswick, H. A. Farach, and C. P. Poole, *An Introduction to Renormalization Group Methods in Physics* (John Wiley and Sons, 1992).

THE FLUCTUATION-DISSIPATION THEOREM

H. Nyquist, Physical Review **32**, 110 (1928).

H. B. Callen and T. A. Welton, Physical Review **83**, 34 (1951).

R. Kubo, Reports on Progress in Physics **29**, 255 (1966).

THE BOLTZMANN EQUATION AND TIME ASYMMETRY

L. Boltzmann, *Lectures on Gas Theory* (Univ. of California Press, 1964).

S. G. Brush, *Kinetic Theory* (Pergamon Press, 1972).

P. Ehrenfest and T. Ehrenfest, *The Conceptual Foundations of the Statistical Approach in Mechanics* (Cornell Univ. Press, 1959).

J. L. Lebowitz, *Boltzmann's Entropy and Time's Arrow*, in Physics Today (Sept. 1993).

P. C. W. Davies, *The Physics of Time Asymmetry* (Univ. of California Press, 1974).

H. D. Zeh, *The Physical Basis of the Direction of Time* (Springer Verlag, 1991).